HISTORY OF URBAN FORM

To MY MOTHER; PAT, SARAH, JOANNA and JONATHAN;

FREDERIC, JOSHUA and SOPHIE

HISTORY OF URBAN FORM

Before the Industrial Revolutions

THIRD EDITION

A E J MORRIS

DiplArch (UCL), DiplTP (Lond), ARIBA, FRGS

Prentice
Hall

An imprint of **Pearson Education**

London • New York • Toronto • Sydney • Tokyo • Singapore • Hong Kong
Cape Town • New Delhi • Madrid • Paris • Amsterdam • Munich • Milan • Stockholm

Pearson Education Limited
Edinburgh Gate
Harlow
Essex CM20 2JE,
England

and Associated Companies throughout the world

Visit us on the World Wide Web at:
http://www.pearsoned.co.uk

ISBN-13: 978-0-582-30154-2

First published 1972 by George Godwin Limited under the
subtitle 'Prehistory to the Renaissance'

Second edition published 1979
Third edition published 1994

ARP Impression 98

British Library Cataloguing in Publication Data
A CIP record for this book is available from the British
Library.

Library of Congress Cataloging-in-Publication Data
A catalogue entry for this title is available from the Library
of Congress.

Set by 3 in 10/11pt Baskerville

Printed in Great Britain by Clays Ltd, St Ives plc

PPLC/10

Contents

Foreword

In 1974 I came across *History of Urban Form* in a London bookshop. Its effective combination of text and illustrations and its coherent organization led me to seek out the author. We met several times in London and later at Cornell. When the publishers decided to make his work available in the United States, it was a pleasure to accept Morris's invitation to write the foreword to the American printing. I have found no reason to change the following three paragraphs from that statement.

This book is the best single-volume general history of urban planning and development that has yet appeared, and I think it unlikely that a better one will be written. Skilfully drawing on more specialized studies by others of individual countries or periods, Mr Morris has incorporated the results of his own research and observations. He has accomplished the seemingly impossible in being both complete and concise.

Carefully chosen quotations from other authorities placed alongside the admirably clear and straightforward text provide additional or contrasting commentary on the topics selected for discussion. Equally helpful to the reader are the numerous illustrations – many of them drawn especially for this book – which facilitate understanding of the text.

History of Urban Form is the ideal book for those seeking an introductory treatment of the history of urban planning and physical growth. The more advanced student can also read it with profit because of the author's fresh interpretations of familiar themes and his insights gained through first-hand explorations of the cities about which he writes.

Revised throughout and substantially expanded, this book now includes a valuable examination and interpretation of town and city planning in Spain and its colonial possessions and an exploration of urban development in the Islamic countries of the Middle East. Additional illustrations enhance a valuable feature of earlier editions, and material in the text bringing the reader into the early years of the modern era usefully extends the scope of this comprehensive survey.

John W Reps, Professor Emeritus
Department of City and Regional Planning
College of Architecture, Art and Planning
Cornell University

Introduction

This expanded third edition of *A History of Urban Form* has been brought out twenty years after the book first appeared in 1972 as a consolidated development of my lecture series and articles. It is also now more than forty years since, as an undergraduate architectural student at University College London, I first became fascinated by dimly perceived reasons for the shapes of historic towns – which gradually crystallized into my personal system of historic 'urban form determinants'. Meanwhile, the second edition of 1979 had been expanded with considerable broadening of the spread of city examples in Europe and the USA; with existing first edition text revision restricted to correction of errors.

In setting out, twenty years on, to further expand and to fully revise an established book of this kind it was difficult to know how much to change. There was no need to depart from the structural arrangement of sequential historical periods, even had economy in the use of my time not determined retention of as much as possible of the standing text. Other than adding the entire new chapters and main parts to existing chapters, revision of the second edition main text has been to introduce only vital new material.

The main updating and expansion changes to the book have taken the form of that which comprises, in effect, a third edition overlay of explanatory notes, additional quotations, and up to date references. Happily this process (which was unimaginable twenty years ago) has been greatly facilitated by my original page layout with its parallel major and secondary text columns.

The third edition

The major expansion of the third edition has taken the form of adding two entirely new chapters and new parts in six of the nine original chapters. The two new chapters are: Nine – Spain and her Empire, which is derived from the chapter originally written for the Spanish language edition published in 1984 by Editorial Gustavo Gili, Barcelona (now in its fourth impression); and Eleven – Islamic Cities of the Middle East, an entirely new chapter which I present as the new book's major contribution to international urban history, not only because it fully establishes the direct unbroken link between the first Sumerian

INTRODUCTION

cities of Mesopotamia and those of Islamic tradition, but also because during the period of its completion I have been able to benefit from availability of the most recent English language publications by Muslim scholars. For a Westerner working out of his culture, these invaluable books and articles clarified for me the nature and effects of the uniquely Islamic control of historic urban form.

Individual chapter expansion has involved:

One – Ur has been consolidated as the archetypal ancient Mesopotamian city and linked forward to the origins of Islamic urban culture in the Arabian Peninsular; in addition the twin systems of urban form, and locational determinants are fully introduced.

Two – the city of Athens is brought through its often overlooked centuries of Islamic occupation, to the redevelopment plan of the 1850s; and new examples of ancient Greek urbanism have been included.

Three – the coverage of Roman provincial urbanism has been extended to include Spain and North Africa; and there is my personal explanation of the determinants of the morphology of the City of London, as it was re-established after Romano-British desertion of the site.

Four – the European medieval period includes two new sections dealing with the Spanish Christian and Islamic cities; and Dutch urbanism – which has been derived from my introductory historical chapter contributed to *Public Planning in the Netherlands*, Oxford University Press, 1985.

Seven – the chapter has been generally re-organized and the European city coverage increased with the addition of new end-of-period, mid-nineteenth century urban cartography, and aerial photographs.

Ten – a new introductory section deals with the Spanish settlement of parts of the area of the future USA; and the conclusion to the pre-industrial historical period in the making of American cities has been rewritten.

Urban form determinants

Although in no way pretending to the status of a 'theory of urban form', as sought after by academic geographers of various persuasions and specialities; nevertheless my consolidated system of 'urban form determinants' is presented as a structured basis for analysis of underlying reasons for historic, as also present-day urban morphology. (As a third edition change I have become reconciled to use of the term 'urban morphology' as a synonym for my 'urban form'. Hitherto and irrationally, I had always shied off the word morphology because of one eminent urban historian's main subheading – 'the Morphology of a Wurt' – when writing of Eastern European village settlement.)

My list of determinants is not inclusive and readers will be able to add from their own backgrounds and experience. There is a widespread and growing interest in local historical matters for which this book can assist in establishing a general background, and as I have now stressed in three editions, *History of Urban Form* has been written for students of urban history in the widest sense, and not just for those concerned with gaining particular professional qualifications. In which connection, over the years it is pleasing to have learnt of readers who have found the book useful in planning special-interest vacations.

Although present in the background, as it were, the earlier editions did not include formal introductory establishment of the urban form

If you can face the prospect of no more public games
Purchase a freehold house in the country. What it will cost you
Is no more than you pay in annual rent for some shabby
And ill-lit garret here. A garden plot's thrown in
With the house itself, and a well with a shallow basin –
No rope-and-bucket work when your seedlings need some water!...
Insomnia causes more deaths amongst Roman invalids
Than any other factor (the most common complaints of course,
Are heartburn and ulcers, brought on by over-eating.)
How much sleep, I ask you, can one get in lodgings here?
Unbroken nights – and this is the root of the trouble –
Are a rich man's privilege. The wagons thundering past
Through those narrow twisting streets, the oaths of draymen
Caught in a traffic jam – these alone would suffice
To jolt the doziest sea-cow of an Emperor into
Permanent wakefulness. If a business appointment
Summons the tycoon, he gets there fast, by litter.
Tacking above the crowd. There's plenty of room inside:
He can read, or take notes, or snooze as he jogs along –
Those blinds drawn down are most soperific. Even so
He outstrips us; however fast we pedestrians hurry
We're blocked by the crowds ahead, while those behind us
Tread on our heels. Sharp elbows buffet my ribs,
Poles poke into me; one lout swings a crossbeam
Down on my skull, another with a barrel.
My legs are mud-encrusted, big feet kick me, a hobnailed,
Soldier's boot lands squarely on my toes...
(Juvenal, The Sixteen Satires)

What is to be the future development of the great city?... On this question the division is clear and sharp, especially in the United States, where mechanization is so much more advanced than in Europe. One opinion is that the metropolis cannot be saved and must be broken up; the other that, instead of being destroyed, the city must be transformed in accordance with the structure and genius of our time...

(This point of view) holds likewise that men cannot be separated from nature, and consequently that the city cannot continue to exist in its present form. But it immediately points out that the city is more than a contemporary and passing phenomenon. It is a product of many differentiated cultures, in many different periods. Thus the question of its life or death cannot be settled simply on the basis of present-day experience or conditions. The city cannot be damned to extinction merely because it has been misused since industrialization or because its whole structure has been rendered impotent by the intrusion of a technical innovation, the motorcar. The question has to be considered from a broader view and extended into other queries. Are cities connected with every sort of society and civilization?... Or are they eternal phenomenon based on the contact of man with man despite the interference of any mechanization? For myself, I believe that the institution of the city is one native to every cultural life and every period.
(Sigfried Giedion, Space, Time and Architecture)

determinants; an omission which has now been rectified by the inclusion of a major new part in chapter one. Listing other than salient determinants for each of the 312 urban and village settlements now illustrated in the book would have rapidly become repetitively self defeating. Readers will find my indexing of particular use in this respect.

While I believe that my coining of the term urban form determinants is original, the underlying idea, of course, is not. Yet I do not know of any comparably extensive general basis for the analysis of historic urban morphology.

The earlier editions were seen as the first volume of a two-part international history of urban form, bringing the subject through to more or less recent times. I now know that more than the one volume and several authors are necessary for that purpose. The international scope is too wide and it broadens all the time. While recognizing that I can no longer contemplate my own comprehensively international volume two, there are those favoured parts that have been extensively researched and which may yet be taken through into print.

This book in its three editions has been written in a rooftop studio in a mostly old enlarged cottage in a North Hampshire village. The preceding lecture series, articles and early drafts took shape in 1930s North London suburbia; a domestic base which could not be improved on locally, however, or in one of the historic inner parts, or old 'villages' of London. Hence this urban historian's emigration to a country village. In my previous two Introductions I sought to justify this move by explaining: 'my own kind of "ideal city" no longer exists – if indeed it ever did.' Twenty years on – and approaching thirty in the village, I know that still to be true – especially in terms of family enjoyment. And yet, on the other hand, many of life's most memorable moments have been in cities: none more so than Paris – which could also have been included in the dedication.

A E J Morris
Lower Froyle
Hampshire, England
April 1993

Is the city a natural triumph of the herd instinct over humanity, and therefore a temporal necessity as a hangover from the infancy of the race, to be outgrown as humanity grows?

Or is the city only a persistent form of social disease eventuating in the fate all cities have met?

Civilization always seemed to need the city. The city expressed, contained, and tried to conserve what the flower of the civilization that built it most cherished, although it was always infested with the worst elements of society as a wharf is infested with rats. So the city may be said to have served civilization. But the civilizations that built the city invariably died with it. Did the civilizations themselves die of it?

Acceleration invariably preceded such decay.

Acceleration in some form usually occurs just before decline and while this acceleration may not be the cause of death it is a dangerous symptom. A temperature of 104 in the veins and arteries of any human being would be regarded as acceleration dangerous to life...

I believe the city, as we know it today, is to die.

We are witnessing the acceleration that precedes dissolution.

(Frank Lloyd Wright, The Future of Architecture)

Acknowledgements

This book, in the earlier and now its third expanded editions could not have been possible without the assistance in many ways of numerous people in many countries. My sincere apologies, as well as my grateful thanks are due to those whose names are excluded from the necessarily brief list which follows.

Above all – and they can but be nameless – from my lecturing days at South Bank Polytechnic, London, and at other places, I acknowledge the response on the parts of innumerable students who, if they but knew it, were on the receiving end of a lifelong learner's compulsive communicating. Departmental colleagues at South Bank Polytechnic to whom I owe particular debts include Shean McConnell; Peter Inch, a fellow urban historian who gave early criticism when needed and whose draft improvements I can still recognize; Falkner Carson; and latterly Reg Pearce. The Polytechnic authorities allowed research time which assisted completion of the first edition.

Real-world practitioners who have encouraged and assisted include Jim Antoniou, architect, planner and sometime journalistic colleague; John R Harris, architect, in particular for help in the Arabian Gulf; and George Duncan, architect, who set me really thinking about the historic Islamic city.

Of numerous American friendships, professional and personal, my thanks are due to Professor John W Reps of Cornell University – the academia where I would have most liked to have been based – for his hospitality and advice and, not least, for consolidating my interests in the fascinating world of urban cartography. Also from academia, my gratitude to Professors Bill Hendon, Jim Richardson, and Ashok Dhutt of the University of Akron, Ohio; and, of many other practitioners, to Edmund Bacon, properly 'of Philadelphia'.

Editors to whom I owe grateful thanks for commissioning articles thereby assisting my international urban-world travels, include A R Davis, *Official Architecture and Planning* (latterly *Built Environment*) and *Middle East Construction*; Derek Weber, *The Geographical Magazine*; and Ian Murray Leslie, *The Builder*.

Bibliographical references

A wide-ranging general international history of a complex subject must inevitably draw from specialist works of others. In updating the books listed in my Select bibliography as recommended further reading, I make no apologies for retaining several that are now quite elderly. I grew up with those texts and age has not dimmed their perception. Of old friends, metaphorically speaking, special debt is due to R E Wycherley, *How the Greeks Built Cities*; Jerome Carcopino, *Daily Life in Ancient Rome*; John Summerson, *Georgian London*; and J W Reps, *The Making of Urban America*. Each of those books was invaluable in laying the basis of individual chapters in the first and second editions. Indeed without the availability of Rep's masterly work Chapter Ten would not have been possible in its present form. His is the specialist urban history that I would most like to have written. For this edition's new chapters, comparably weighty acknowledgements are made to J H Parry, *The Spanish Seaborne Empire* – Chapter Nine; and Besim Selim Hakim, *Arabic-Islamic Cities* – Chapter Eleven.

General urban histories in the 'elderly' category that I have found of greatest value include: Lewis Mumford, *The City in History*, a generally admirable, enduring work lacking only the breadth of illustrations for it to have made this book unnecessary; Patrick Abercrombie, *Town and Country Planning*, a miniscule heavyweight book with an excellent historical summary, based on his prolific article writing which, had it been properly compiled in book form, might also have made subsequent efforts superfluous; Paul Zucker, *Town and Square*; and S E Rasmussen, *Towns and Buildings*.

Of more recent general studies there are: Edmund Bacon, *Design of Cities* – wide-ranging by way of background to his city planning work in Philadelphia; Leonardo Benevolo, *The History of the City* – a single-volume international history from urban origins through to the 1970s; Michael Webb, *The City Square* – on this special topic; Spiro Kostoff, *The City Shaped: urban patterns and meanings through history*; and James E Vance Jr, *The Continuing City* – with a wide coverage of the French bastides.

Acknowledgement is due to the following for permission to reproduce illustrations: Aerofilms (Figures 1.10, 4.9, 4.46, 4.53, 5.19, 6.19, 8.21, 8.26, 8.30, 11.28); the Athlone Press, publishers of *A History of Architecture on the Comparative Method* by Banister Fletcher (Figure 3.11); Karl Baedeker, publishers of *Russia*, 1914 edition (Figures 7.31, 7.44); the British Library (Figures 4.52, 4.54); the British Museum (Figure 11.11); Cambridge University Collection – copyright reserved (Figure 4.19); Czech Historic Buildings Ministry (Figures 4.90 to 4.101); Denver Public Library, Western History Department – photograph by W H Jackson (Figure 10.60); Greek Tourist Office (Figure 2.19); Institut Géographique, Paris (Figures 4.39, 6.29); Italian State Tourist Office (Figure 5.35); Kingston-upon-Hull City Library (Figure 4.48); KLM Aerocarto (Figure 5.8); Museum of the City of New York (Figures 10.29, 10.31); the National Library of Wales (Figure 4.59); Office du Tourisme du Havre et de la Région (Figure 6.32); the State Historical Society of Colorado (Figure 10.59); Swiss National Tourist Office/Swiss Federal Railways (Figures 4.65, 4.67); Topham/*Geographical Magazine* (Figure 4.49). Whilst every effort has been made to trace the owners of copyright material, in a few cases this has proved impossible

ACKNOWLEDGEMENTS

and we take this opportunity to offer our apologies to any copyright holders whose rights we may have unwittingly infringed.

The maps of Lima (Figure 9.34), St Augustine (10.2), Santa Fe (10.3), Tucson (10.10), Boston Harbour (10.16), San Francisco (10.48) and Los Angeles (10.50) are taken from the extremely wide range of international urban cartographic reproductions published by Historic Urban Plans, Box 276, Ithaca, NY, USA.

Many of the early and mid-nineteenth century city maps included in this edition have been reproduced by Historical Town Maps, Stobs Castle, Hawick, Scotland; using original cartography as published in the four atlases listed in the adjoining column.

The redrawing of maps and plans from originals which could not be reproduced at inevitable small size is by the author, who also compiled the page layouts on the basis of his original book design.

My thanks for their sympathetic treatment of a sometimes unavoidably distracted author are due to Robert McKown, Julia Burden, John Brooks and George Mockridge of my first publishers, George Godwin Limited; and now to the staff of Longman. I am also grateful to Warren Sullivan of Halsted Press, New York for their promotion of the book in North America; and to Gustavo Gili, Barcelona for publication of the Spanish language edition, now in its fourth impression.

Lastly but not least my gratitude to my wife Pat (a history graduate of University College, London) who has lived through the preparation of three editions, and who has shared with me – variously accompanied by Sarah, Joanna and Jonathan – the fascination of visiting and revisiting places old and new.

The following original maps, made available by Whitehall Press Limited, have been reproduced in whole or in part in this book:

Durham and Newcastle-upon-Tyne from *The British Atlas.*

Amsterdam, Antwerp, Berlin, Copenhagen, Dresden, Frankfurt am Main, Lisbon, Madrid, Oporto, Rome ('Modern' in 1830), St Petersburg (Leningrad), Stockholm, Toulon and Vienna from *The Society for the Diffusion of Useful Knowledge Atlas.*

Brighton, Bristol, Plymouth, York, Edinburgh, Brussels and New York City from *The Tallis Illustrated Atlas.*

Oxford and Cambridge from *The Weekly Dispatch Atlas.*

1 – The Early Cities

In the historical evolution of the first urban civilizations and their cities it is possible to discern three main phases. Each of these involved 'radical and indeed revolutionary innovations in the economic sphere in the methods whereby the most progressive societies secure a livelihood, each followed by such increases in population that, were reliable statistics available, each would be reflected by a conspicuous link in the population graph'.[1]

The first of these phases covers the whole of the Palaeolithic Age, from its origins, at least half a million years ago, until around 10000 BC, followed by the proto-Neolithic and Neolithic Ages. These in turn lead to the fourth phase, the Bronze Age, starting between 3500 and 3000 BC and lasting for some 2,000 years. During this last period the first urban civilizations were firmly established.

In his most valuable book *The First Civilisations: The Archaeology of their Origins*, Dr Glyn Daniel states that 'we now believe that we know from archaeology the whereabouts and the whenabouts of the first civilisations of man – in southern Mesopotamia, in Egypt, in the Indus Valley, in the Yellow River in China, in the Valley of Mexico, in the jungles of Guatemala and Honduras, and the coastlands and highlands of Peru. We will not call them primary civilisations because this makes it difficult to refer to Crete, Mycenae, the Hittites, and Greece and Rome as other than secondary civilisations and this term secondary seems to have a pejorative meaning. We shall talk rather of the first, the earliest civilisations, and of later civilisations'. Figure 1.2 gives the locations of these seven original urban civilizations and relates them to the earliest known, or assumed, agricultural regions.[2]

As shown by the time chart (Figure 1.1), the seven civilizations occurred at markedly different times. The first three – the Mesopotamian, the Egyptian, and the Indus Valley – are the so-called 'dead' cultures, out of which there evolved, in direct line of descent, the Greek, Roman and Western European Christian civilizations. Mesopotamia is also of fundamental importance for its formative influence on the evolution of urban settlement in the Arabian Peninsular, where Islamic culture originated in the seventh century AD.[3]

Although occurring much more recently than Chinese civilization, the fourth oldest, the three American cultures – Mexican, Central

1. Childe, *What Happened in History*, 1964. Childe is still a valuable general reference, but see also C Redman, *The Rise of Civilisation: From Early Farmers to Urban Society in the Middle East*, 1978; A B Knapp, *The History and Culture of Ancient Western Asia and Egypt*, 1988; C Renfrew, *Before Civilisation: The Radio-Carbon Revolution in Prehistoric Europe*, 1976; M Brawer, *Atlas of the Middle East*, 1988.

2. It must be kept in mind from the outset that archaeologists are continuing to rewrite the pages of humankind's early history. The two earlier editions of *History of Urban Form* were written in the knowledge of archaeological evidence that removed the title 'Father of Town Planning' from the fifth-century BC Greek architect-planner, Hippodamus of Miletus, and passed it back in history to the unknown Harappan priestly planner who, around 2150 BC in the Indus Valley of western India, laid out the first of that civilization's planned new cities. However, see the concluding section of this chapter for the intriguing archaeological questions posed by the Harappans having arrived alongside the Indus with a fully formed urban planning system. Where did they come from? Because that is where there are the most likely remains of still earlier planned cities.

In this edition the major archaeological reassessment of the 1970s and 1980s concerns the nature of the city in Ancient Egypt, in particular not only the elevation of Kahun to urban status from that of a workmen's village encampment, but also its claim to have been the first use of the gridiron system of town planning. Otherwise gaps continue to be filled in the record, notably that of the desertion of Roman London and its re-establishment by the Anglo-Saxons (see Chapter 3).

3. For a new analytical description of the traditional Islamic City, see in this chapter for the pre-Islamic Arabian introduction, and Chapter 11.

American and Peruvian – are also dead: brutally destroyed by Spanish *conquistadores* during the decade or so after 1519. 'There, in the sixteenth century,' writes Daniel, 'Europe met, if not its own past, at least a form of its past',[4] where, for example, metal technology was either extremely limited or yet to be discovered.[5]

China is the fascinating exception, from its origins in the Yellow River basin during the late third millennium BC, its culture has lasted to the twentieth century without permanent interruption. Furthermore, during the eighth century AD – one of its peaks of power and influence – Chinese urban civilization was introduced into Japan, where until then only essentially agricultural settlements had existed.

This chapter will deal with the origins, and describe key examples of the urban settlements of three of the earliest civilizations: first, the Sumerian civilization of the Tigris/Euphrates floodplains of Mesopotamia (present-day Iraq) – with post-scripted extension through into the adjoining Arabian Peninsular; second, the Egyptian civilization of the Nile Valley and Delta; and third, the Harappan civilization of the Indus Valley.

Urban origins in Mexico (the Aztecs), Central America (the Maya) and Peru (the Incas) are included in Chapter 9 (Spain and her Empire). The continuing Chinese civilization is summarized in Appendix A; Appendix B describes the related origins of urban settlement in Japan, through to that country's industrial revolution commencing in the second half of the nineteenth century. Urban beginnings in Europe generally, and Britain in particular, are dealt with in Chapter 4 as part of the background to the European medieval period.

In some parts of the world, notably North America, Australasia and southern Africa, European urban culture was either introduced into uninhabited territories, or imposed on indigenous peoples. There are still a few remotely isolated societies which, left to their own devices, are no further advanced than the Palaeolithic phase.

This account of the origins of cities is based on the ordinarily accepted belief that the development of settled agriculture was the essential prerequisite for the evolution of urban settlements. In the late 1960s this doctrine was challenged by Jane Jacobs, who pronounced 'the dogma of agricultural primacy is as quaint as the theory of spontaneous combustion ... in reality agriculture and animal husbandry arose in cities', on which basis, it was claimed that 'cities must have preceded agriculture'.[6] Jane Jacobs had devised that theory in support of her short-lived personal interpretation of economic failings of later twentieth-century American cities, which sought to gain historical support from the recently published archaeological findings of James Mellaart at Çatal Hüyük, in Anatolia.[7] This extraordinary site was seemingly qualified for 'urban' status by the seventh millennium BC – perhaps even earlier, thus anticipating the beginnings of Mesopotamian civilization by 3,000 or more years. Jericho, of comparably ancient origin, has also occasioned controversy. The significance of these two exceptional settlements is assessed later in the chapter, with the summary argument against the theory of urban primacy.

Early settlements

Human-like creatures first appear on the earth perhaps as long as 1 million years ago, and become 'dispersed from England to China, and

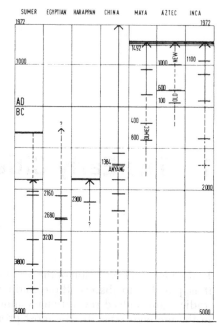

Figure 1.1 – Time chart showing the comparative dates of the seven first civilizations.

It is impossible to make an exact determination of the world's population in remote ages because firm data cannot be established. Nevertheless, scientists have done their best. Here is a recent estimate, rough as it necessarily is (E S Deevey, Human Population, Scientific American, September 1960, pp. 195–6):

Prehistoric world population
Lower Paleolithic (1,000,000 years ago)
125,000
Middle Paleolithic (300,000 years ago)
1,000,000
Upper Paleolithic (25,000 years ago)
3,340,000
Mesolithic (before 10,000 years ago)
5,320,000

If these figures are even fairly correct, there were little more than five million human beings when the hunting and food-gathering phase of mankind's existence reached its full development. The long, slow increase in population was brought about by improved weapons, better hunting techniques, and more capable methods of dealing with cold weather, predatory beasts, and other natural threats to existence. Getting more food made it possible for more people to stay alive and breed even more people.
(P van Doren Stem, Prehistoric Europe)

4. G Daniel, *The First Civilisations: The Archaeology of their Origins*, 1968.
5. See A Street and W Alexander, *Metals in the Service of Man*, rev 9th edn, 1990.
6. J Jacobs, *The Economy of Cities*, 1969,
7. J Mellaart, *Çatal Hüyük*, 1967.

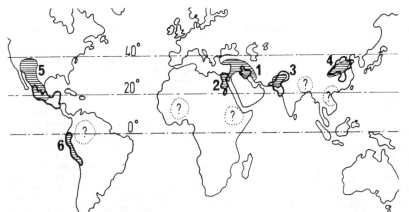

Figure 1.2 – The location of the first civilizations (in heavy outline) related to the location of the earliest known agricultural communities (the hatched areas) and possible other early agricultural centres. Key: 1, Southern Mesopotamia (Sumerian civilization); 2, Nile valley (Egyptian); 3, Indus valley (Harappan); 4, Yellow River (Shang); 5, Mesoamerica (Aztec and Maya); 6, Peru (Inca). The lines of latitude 20° and 40° north of the Equator contain five of the first civilizations.

Most of the major technological innovations of antiquity were made within the limited area of the Near East and the eastern end of the Mediterranean, and little could be more fatal than imagining that those regions were in antiquity as we know them today. Even in the past ten thousand years enormous changes have taken place which owe nothing to population changes (either migrations or explosions), nor to the recent development of cities, roads and railways. Far more fundamental is that fact that the entire ecology of the region has undergone drastic changes. What we know today as open, dusty plains or rich farmlands were, ten thousand years ago, more or less thickly forested, and within the forest lived a wide variety of wild animals. This is not to say that deserts did not exist, but rather that many hills that we know of today as barren ranges of rock were then at least lightly covered with trees, while the river valleys probably carried very dense forest cover.

(H Hodges, Technology in the Ancient World)

from Germany to the Transvaal'.[8] By about 25000 BC the physical and organic evolution of Homo sapiens is considered to have come to an end and the modern processes of cultural evolution start.

From their first appearance, down to the beginning of the Neolithic Age, humans existed on much the same basis as any of the other animals, by gathering naturally occurring foodstuffs in the form of berries, fruits, roots and nuts and, somewhat later, by preying on other animals and by fishing. The social unit was the family, but the society was of necessity a mobile one, always having to move to fresh sources of food, carrying its few possessions from one crudely fashioned temporary shelter to another. There was no permanent physical unit about 14000 BC when 'as the last great ice age was approaching men were sufficiently well equipped to evict other denizens and themselves to find shelter in caves. There we find true homes'.[9] Permanence of residence was, however, determined by the continuing availability of food within reach of the 'home'.

Professor Childe notes that this gathering economy corresponds to what Morgan[10] calls savagery and that it 'provided the sole source of livelihood open to human society during nearly 98 per cent of humanity's sojourn on this planet'.[11] Such an economy imposed a limit on population with a direct relationship to the prevailing climatic and geological conditions. The entire population of the British Isles around 2000 BC has been put by Childe at no more than 20,000, with an increase to a maximum of 40,000 during the Bronze Age. In France the Magdalenian culture between 15000 to 8000 BC, with at first exceptionally favourable food resources, had a maximum population density of one person per square mile, with the general figure around 0.1 to 0.2.[12] Other examples given by Childe are that 'in the whole continent of Australia the aboriginal population is believed never to have exceeded 200,000 – a density of only 0.03 per square mile',[13] while on the prairies of North America he quotes Kroeber's estimate that 'the hunting population would not have exceeded 0.11 per square mile'.[14]

Somewhere around 8,000 to 10,000 years ago humankind started to exercise some measure of control over the supply of food by systematic cultivation of certain forms of plants, notably the edible wild grass seeds, ancestors of barley and wheat, and by the domestication of animals. 'The escape from the impasse of savagery was an economic and scientific revolution that made the participants active partners with nature instead of parasites on nature.'[15] This Neolithic agricultural revolution transformed the economy into one with an increasing

8. L Mumford, *The City in History*, 1991; it is pleasing that this major work has been republished in a new paperback imprint. For the evolution of *Homo sapiens* see also J Waechter, *Man Before History*, 1976.

9. Childe, *What Happened in History*.

10. L H Morgan, *Ancient Society; or Researches in the Lines of Human Progress from Savagery through Barbarism to Civilisation*, 1877 (republished as *Ancient Societies*, Harvard University Press, 1964). Morgan defined these terms more exactly according to the enlargements of human sources of subsistence. He distinguished seven periods – seven ethnic periods as he called them. The first were: Lower Savagery, from the emergence of humans to the discovery of fire; Middle Savagery, from the discovery of fire to the discovery of the bow and arrow; Upper Savagery, from the discovery of the bow and arrow to the discovery of pottery; Lower Barbarism, which began with the discovery of pottery (which, to Morgan, was the dividing line between Savagery and Barbarism) and ended with the domestication of animals; Middle Barbarism, from the domestication of animals to the smelting of iron ore; and Upper Barbarism, from the discovery of iron to the invention of the phonetic alphabet. Finally, the seventh period was civilization with writing and the alphabet. (Quoted Daniel, *The First Civilisations*.)

11. Childe, *What Happened in History*.

12. G Childe, *The Dawn of European Civilisation*, 1957.

13. Childe, *What Happened in History*.

14. A L Kroeber, *A Roster of Civilisations and Culture*, 1962.

15. Childe, *What Happened in History*; see also D Zohary and M Hapf, *Domestication of Plants in the Old World*, 1988; I L Mason, *Evolution of Domesticated Animals*, 1984; C A Reed, *The Origins of Agriculture*, 1977; I Hodder, 'An Interpretation of Çatal Hüyük and a Discussion of the Origins of Agriculture', *Bulletin no 24*, 1987, Institute of Archaeology, University College London.

food producing basis, enabling the social unit to expand, if only marginally so, to that of the clan.

Permanence of residence in one place was now far more possible, with the physical unit becoming that of the village, although the earliest settlements would have amounted to no more than a cluster of rudimentary huts. Morgan terms this stage in the development of civilization, barbarism.

Neolithic humans did not bring about controlled food production solely by their own efforts. On the contrary, evidence perhaps points to the fact that left to their own devices 'Homo sapiens would have remained a rare animal – as the savage in fact is'.[16] The decisive step forward towards eventual urban civilization had to await the external stimulus provided by the climatic changes resulting from the ending of the last of the ice ages around 7000 BC. This melting of the vast northern ice sheets 'not only converted the steppes and tundras of Europe into temperate forest, but also initiated the transformation of the prairies south of the Mediterranean, and in Hither Asia, into deserts interrupted by oases'.[17]

On these prairies 'when Northern Europe was tundra or ice sheet ... grew the wild grasses that under cultivation became our wheats and barleys; sheep and cattle fit for domestication roamed wild. In such an environment human societies could successfully adopt an aggressive attitude to surrounding nature and proceed to the active exploitation of the organic world. Stock breeding and the cultivation of plants were the first revolutionary step in man's emancipation from dependence on the external environment'.[18]

The Neolithic, although often referred to as a period for the sake of convenience, was not confined to a particular period of time, but its duration varied in different areas. In some cases people continued to depend on hunting, fishing and gathering while their more progressive neighbours practised a Neolithic economy. Similarly, Neolithic peoples in certain areas continued to use stone tools long after others were making tools and weapons of bronze or iron. The word Neolithic, in fact, simply implies food production based on crops and domesticated stock, without metals.

Although there is no doubt that the Neolithic was a 'revolution' in man's way of life, it has been suggested that the word 'evolution' is more appropriate since the transformation was so gradual. Recent research has shown that there were semi-settled communities, from about 8900 BC onwards, among peoples formerly known as Mesolithic but now generally referred to as proto-Neolithic. The development of full food production was an evolution rather than a sudden revolution; yet there is no doubt that the consequences of this change were revolutionary in the fullest sense of the word.

(S Cole, The Neolithic Revolution)

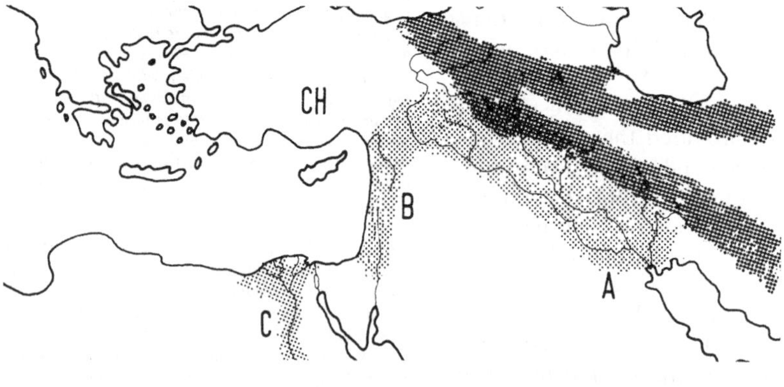

Figure 1.3 – Map of the Near East showing the 'Fertile Crescent' in light tint, and ancient sources of copper in dark tint. Key: A, Southern Mesopotamia, the valleys of the Tigris and the Euphrates; B, Palestine; C, Egypt, the valley and delta of the Nile. CH locates Çatal Hüyük.

It is generally accepted that favourable conditions for the agricultural revolution first occurred south and east of the Mediterranean, around what is known as the 'Fertile Crescent' – a term introduced by Professor Breasted[19] and synonymous with the phrase 'the Cradle of Civilization'. This fertile zone, related to which were all the village and later urban civilizations of the Near and Middle East, is shown in the light tint on Figure 1.3. The zone is shaped, appropriately enough, in the form of a sickle, starting at the head of the Persian Gulf and extending northwards towards the mountain sources of the Tigris before turning westwards across to the Euphrates. From here the zone curves out through Syria and the valleys and plains of Palestine. It is interrupted by the Sinai desert but the broad delta and narrow valley of the Nile form a substantial continuation further south into Egypt.

In Mesopotamia the record of Neolithic settlement 'begins in small

16. Childe, *What Happened in History*.
17. Childe, *What Happened in History*; see also T M L Wrigley, M J Ingram and G Farmer, *Climate and History: Studies in Past Climates and Their Impact on Man*, 1985.
18. Mumford, *The City in History*; see also Zohary and Hopf, *Domestication of Plants in the Old World*; Mason, *Evolution of Domesticated Animals*.

oases on steppes and plateaux. Despite the threat of drought the difficulties of taming the soil were less formidable there than on the floodplains of the major rivers'.[20] By 5500 BC, following at least 3,000 years of slow development, farming communities were firmly established on the higher ground and were gradually moving down the valleys of the Tigris and Euphrates as the alluvial deposits dried out, and as techniques, especially irrigation, were improved.

In Egypt at Merimde, in the north-west delta area, Professor Fairman records that 'perhaps as early as 4000 BC the settlement occupied an area of at least 600 by 400 yards, and in one part some of the huts are arranged in two definite rows with a lane between'.[21] Other Egyptian Neolithic village sites are recorded at Fayum, beside a lake west of the Nile Valley, as being firmly established during the first half of the fifth millennium.

Bronze Age

Before describing the transformation between 3500 and 3000 BC of Neolithic folk society villages into the first cities – Professor Childe's 'urban revolution' – a definition of the term 'city' is needed. This has been concisely provided by Gideon Sjoberg as 'a community of substantial size and population density that shelters a variety of non-agricultural specialists, including a literate élite'.[22]

Two requirements for the urban revolution are implicit in this definition: first, the production of a surplus of storable food, and other primary materials, by a section of the society, in order to support the activities of the specialists; second, the existence of a form of writing, without which permanent records cannot be kept and the development of mathematics, astronomy and other sciences is not possible.

There are other requirements which must be considered, the most important of which are social organization to ensure continuity of supplies to the urban specialists and to control labour forces for large-scale communal work, and technological expertise, providing the means for transporting materials in bulk, and significant improvements in the nature and quality of tools.

As Childe has said, 'the possibility of producing the requisite surplus was inherent in the very nature of the Neolithic economy; its realisation, however, required additions to the stock of applied science at the disposal of all barbarians, as well as modification in social and economic relations'.[23]

Throughout the fourth millennium BC sufficient technological requirements for the urban revolution were met, either by invention or discovery. To quote Mumford again, 'as far as the present record stands, grain cultivation, the plough, the potter's wheel, the sail boat, the draw-loom, metallurgy, abstract mathematics, exact astronomical observations, the calendar, writing and other modes of intelligible discourse in permanent form, all came into existence around 3000 BC, give or take a few centuries'.[24]

The critical requirement for the urban revolution is the production of a food surplus. So far as is known this first became a possibility on the alluvial plains of the Tigris/Euphrates.[25] Between 4000 and 3000 BC – perhaps earlier – some of the village communities in the lower Tigris/Euphrates region not only increased greatly in size but also changed in structure. These processes culminated in the Sumerian city states of

Figure 1.4 – Urban centres in Mesopotamia, mountain foothills shown tinted. Key 1, Eridu; 2, Ur; 3, Erech (all Sumerian cities); 4, Babylon; 5, Assur; 6, Arbela (Erbil); 7, Nineveh; 8, Baghdad; E, River Euphrates; T, River Tigris. The dotted coastline at the head of the Arabian Gulf is that of the period around 2000 BC.

19, 20. J Breasted, Ancient Times, 1935. Although Breasted is now one of the more venerable of the book's references, his concept of 'the fertile crescent' retains validity, as confirmed by A B Knapp, The History and Culture of Ancient Western Asia and Egypt, 1988.

21. H W Fairman, 'Town Planning in Pharaonic Egypt', Town Planning Review, April 1949.

22. G Sjoberg, 'The Origin and Evolution of Cities', Scientific American, September 1965 (also in Cities, a Scientific American book, 1967). Other definitions of civilization include:

'A society to be called civilised must have two of the following: towns of upwards of 5,000 people; a written language; and monumental ceremonial centres' (Professor Clyde Kluckhohn).

'Writing is of such importance that civilisation cannot exist without it, and conversely, that writing cannot exist except in a civilisation' (I J Gelb, A Study of Writing: The Foundations of Grammatology, 1952).

'A civilisation was a society with a functionally interrelated set of social institutions as: (i) class stratification marked by highly different degrees of ownership of control of the main productivity resources; (ii) political and religious hierarchies complementing each other in the administration of territorially organised states; and (iii) complex division of labour with full-time craftsmen, servants, soldiers and officials alongside the great mass of primary peasant producers.' (Professor Robert Adam). These quotations are given by Daniel, as set out in C H Kraeling and R C Adams (eds) City Invincible: A Symposium on Urbanisation and Cultural Development in the Ancient Near East, 1960.

23. Childe, What Happened in History; see also T I Williams, A Short History of Technology: From Earliest Times to AD 1900, 1960/1979; H Hodges, Technology in the Ancient World, 1970.

24. Mumford, The City in History; see also O H W Dilke, Mathematics and Measurements, 1987.

25. See H E W Crawford, 'Stimuli towards Urbanization in South Mesopotamia', in P J Ucko, R Tringham and G W Dimbleby (eds) Man, Settlement and Urbanism, 1972; see also p. 18 for the critical assessment of the Jacobs theory of urban pre-eminence.

3000 BC onwards, with their tens of thousands of inhabitants, elaborate religions, political and military class structure, advanced technology and extensive trading contacts.

Agriculture on the alluvium depended on irrigation, at first only in localized, rudimentary forms but later utilizing large-scale canal and embankment works associated with the advent of fully established cities. 'The land that was to become Sumer lacked building-stone or even timber (apart from palm-stems), let alone minerals; its climate was arid and its rivers did not give rise to annual inundations like those provided by the Nile. Yet it was a land of opportunity.'[26]

It is not certain when the first settlements on the alluvium were founded. Grahame Clarke records that 'the first inhabitants well known to us are those named after al 'Ubaid, a humble village set on a low mound or island of river silt in the Euphrates valley. These people first appear in the archaeological record in the latter part of the fifth millennium'.[27] Through to about 2750 BC, when Sargon founded the

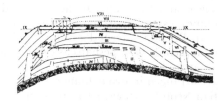

Figure 1.5 – Cross-section through the *tell* at Troy showing the various stages whereby 'ground' level within the successive defensive walls was gradually raised above bedrock. A vertical slice through Ur (Figure 1.7) or Erbil (Figure 1.10) would reveal a similar succession of levels.

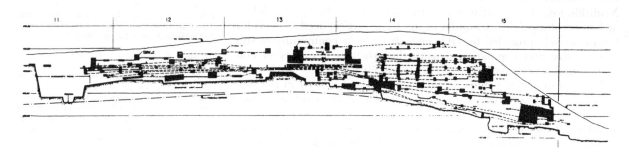

Figure 1.6 – Megiddo, in Palestine; cross-section through the *tell* looking north.

city of Agade near Babylon as the capital of a united Sumerian state, the major urban settlements were effectively autonomous city-states with 'at least eleven of these, including Ur, Erech, Larsa, Kish and Nippur at one time supporting independent and sometimes warring dynasties'.[28] In turn the Akkadian dynasty was overthrown and the city of Ur assumed control of the Sumerian empire during the period of the Third Dynasty – about 2110 to 2015 BC. Ur is the most revealing example of a Sumerian city both on account of its importance as the capital of one of the dynasties and the greater extent of excavations there. Its location is about one-third of the way between the present-day head of the Persian Gulf and Baghdad. During the Third Dynasty it was alongside the Euphrates (which now flows 10 miles away to the west) only a few miles from the sea.

Before describing the city of Ur, a brief explanation is needed of the formation of *tells*, both in early Mesopotamia and in subsequent urban history. The word *tell* is of pre-Islamic origin and refers to those clearly defined man-made settlement mounds which are such an archaeological feature of Iran, Iraq, Palestine, Turkey, southern Russia and a few places in Europe. In recent times these mounds have generally been uninhabited; nevertheless they are the result of site occupation over several millennia. Indeed Erbil (ancient Arbela – Figure 1.10) – and Kirkuk are still lived in, or following Glyn Daniel 'perhaps one should say lived on; they have been more or less continuously occupied from very early times to the present day – perhaps for six to eight thousand years'.[29]

26. G Clark, *World Prehistory – An Outline*, 1961; see also H Crawford, *Sumer and the Sumerians*, 1991.
27. Clark, *World Prehistory – An Outline*; see also R McC Adams, 'Patterns of Urbanization in Early Southern Mesopotamia', in P J Ucko, R Tringham and G W Dimbleby (eds) *Man, Settlement and Urbanism*, 1972; and S A Kubba, *Mesopotamian Architecture and Town Planning 10,000–3,500 BC*, 1987.
28. Sir Leonard Woolley, *Ur of the Chaldees*, rev. updated edn, P R S Moorey (ed.), 1982; see below for related references to the work of this most important of Mesopotamian archaeologists.
29. G Daniel, *The First Civilisations*, 1968; the determining reason for continued site occupancy was usually that of the need to minimize capital investment in the fortifications, given that a move to another location would have entailed construction of a completely new defensive system. Repair and strengthening of an existing system required comparatively much less resource investment. See below in this chapter for introductory description of the role of fortification as an urban form determinant, and Chapter 5 for its later manifestations.

In ancient history, however, it is stressed that other determinants would have contributed to any decision to stay put; notably (but *inter alia*) mystical/religious considerations in respect of city temples and shrines, and commercial vested interests deriving from local agricultural and proto-industrial trading resources.

A *tell* was created from a city's new buildings being constructed on the ruins of old ones. In Mesopotamia, and other river valley locations, most buildings were made of sun-dried mud-brick; kiln-fired bricks were used only for facing city walls or for palaces and temples. The life of a mud-brick house in Mesopotamia was probably limited to about 75 years, by which time weathering brought about collapse. The rubble was levelled to provide foundations for the new house, thereby raising the effective ground level. This process was normally continuous, the city regenerating itself cell by cell. Complete rebuilding, perhaps after destruction and an unoccupied period, sometimes also occurred.

Analogous processes have raised present-day ground levels in other cities to considerable heights above their original levels; London and Rome, among many other old-established cities, are characterized by historic buildings with ground floors well below adjoining street levels. Sir Leonard Woolley records that 'the mosaic pavements of Roman Londinium lie 25 to 30 feet below the streets of the modern City'.[30] The hilly topography of Rome itself, as described by Professor Lanciani, was totally changed even before the end of the ancient period; the Palatine Hill, for example, became covered 'with a layer of rubbish from 6 to 67 feet thick'.[31] Where occupation has been continuous, streets have risen because new surfaces have been laid on old levels, often necessitating the incorporation of steps. Where cities have been deserted for lengthy periods, dust accumulates naturally. Lanciani notes that 'if the Forum of Trajan, excavated by Pius VII (1800–23), was not cleaned or swept once a week, at the end of each year it would be covered by an inch of dust, by one hundred inches at the end of a century'.[32]

Mesopotamian civilization: the Sumerian cities

UR OF THE CHALDEES

The most consistently preserved level of ruins is that of the Isin-Larsa period of around 1700 BC, the excavation of which is described in Sir Leonard Woolley's fascinating book *Ur of the Chaldees*. In this later period the layout retained the basic form of the Third Dynasty city 'and work upon other sites makes it clear that Ur was, in all essentials, typical of the Sumerian state capitals from the Persian Gulf right up to Mari on the middle Euphrates'.[33]

There are three basic parts of Third Dynasty Ur: the old walled city, the *temenos* or religious precinct, and the outer town. The walled city was an irregular oval shape, about three-quarters of a mile long by half a mile wide. It stood on the mound formed by the ruins of the preceding buildings with the Euphrates flowing along the western side and a wide navigable canal to the north and east. Two harbours to the north and west provided protected anchorages and it is possible that a minor canal ran through the city area.

The defensive wall was essentially that constructed during the eighteen-year reign of Ur-Nammu – the founder of the Third Dynasty. Sir Leonard Woolley describes it as 'rising 26 feet or more above the plain and acting as retaining-wall to the platform on which the town buildings were raised. There was a rampart constructed of unbaked brick throughout, which at its base was no less than 77 feet thick. The wall proper, built of baked brick, which ran along the top of the rampart, has disappeared at the point where the trial excavations were

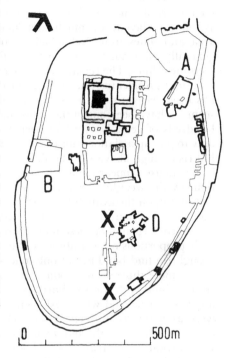

Figure 1.7 – Ur; general layout plan of the period 2100–1900 BC (as excavated by Sir Leonard Woolley). The city area within the wall was 89 hectares and its possible maximum population was 35,000. A figure of 250,000 has been estimated for the total city-state population. Key: A, North harbour; B, West Harbour; C, the *temenos* (as Figure 1.8); D, housing area of the period around 1900 BC (as Figure 1.9). The main stream of the Euphrates flowed alongside the western side. While the outline of the city and the layout of the organic growth housing district D are both of the period 2100–1900 BC, it is important to note that the regular orthogonal plan of the *temenos* is that of about 600 BC. (See also Figure 1.28 as a conjectural plan of 'Greater Ur'.) Woolley believed that there may have been a straight processional avenue of the same late period leading up to the *temenos*, perhaps similar to that for which there is evidence at Nineveh (see Note 66). The avenue at Ur may have been on the line 'X–X', to the west of the housing area D.

30. Sir Leonard Woolley, *Digging up the Past*, 1952.
31, 32. R Lanciani, *The Ruins and Excavation of Ancient Rome*, 1897/1968.
33. P R S Moorey (ed.) *Ur of the Chaldees*, 1982 (updated rev. edn of Woolley, *Excavations at Ur*, 1954, which was based on Woolley, *Ur of the Chaldees*, 1929); H Crawford, *Sumer and the Sumerians*, 1991; M Roaf, *Cultural Atlas of Mesopotamia and the Ancient Near East*, 1990.

made but, judging from the unusually large size of the bricks employed for it, it must have been a very solid structure'.[34]

The *temenos* occupied most of the north-western quarter of the city. With the exception of the harbours, it contained the only significant open spaces in the city, even though their use was essentially reserved for the priests and members of the royal household.[35] As recorded by Woolley, its layout was much later than that of the Isin-Larsa housing area and dates from Nebuchadnezzar's reign (*c.* 600 BC) when the generally unplanned arrangement of the area was reorganized along rectilinear lines. The remainder of the city within the walls was densely built up as residential quarters. A considerable part of one such district has been excavated to the south-east of the *temenos*. This housing area seems to have been one of the oldest parts of the city, 'where for many hundreds of years houses had been built, and had fallen into decay, only to pile up a platform for fresh building, so that by 1900 BC it was a hill rising high above the plain'.[36]

The houses appear to have been occupied by the middle class rather than by the wealthy. Their size varied, as did their ground plan, depending upon the availability of space and the owner's means. But on the whole the houses were built according to a general arrangement.

The construction of these houses proved far more sophisticated and their proportions more ambitious than Woolley had expected. He had thought to find buildings of only one storey, built in mud brick, and with a mere three or four rooms. Instead he discovered houses of two storeys, built with walls of burnt brick below and rising in mud brick above – plaster and whitewash hiding the change in material. There were as many as thirteen or fourteen rooms round a central paved court, which gave light and air to the house. In Woolley's words, this was obviously a great city, whose sophisticated living conditions proved that it had inherited the traditions of an ancient and highly organized civilization.[37]

Organic growth and planned towns

At Ur, it was clear to Woolley that the residential district south-east of the *temenos* was the result of a long evolutionary process. Yet, as he emphasized, the houses are grouped together in layouts that had 'grown out of the conditions of the primitive village and are not laid out on any *system of town planning*' [added italics].[38] As one of the basic tenets of this book, the natural, unplanned process whereby an urban settlement evolves from a village origin is termed organic growth, and it represents by far the broadest of two directly contrasted, continuing streams of activity whereby mankind, through to the present day, has created our urban settlements.

The organic growth of Sumerian cities is exemplified by Ur, as revealed by excavation, and, intriguingly, by arguably the most important (for the author the most fascinating) of numerous aerial photographs included throughout the book. This photograph (Figure 1.10) which shows Erbil in north-east Iraq, some 200 miles north of Baghdad, was taken during the 1930s; yet, if we mentally erase the adjoining suburbs and scale down the extent of the historic Islamic city, we could well be looking at an aerial picture of ancient Arbela – as it was then named – during the Sumerian period.[39] The site is known to have been more or less continuously occupied for as long as 6,000–8,000 years,

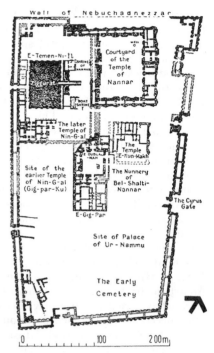

Figure 1.8 – Ur: plan of the *temenos*, the city's religious citadel, enclosed by massive walls and dominated by the multi-stage Ziggurat in the western corner. The arrangement of Ziggurat, temples, palaces and associated government buildings is as organized along *planned* lines under Nebuchadnezzar. Woolley believes that the form of the *temenos* at the beginning of the second millennium BC (i.e. contemporary with the housing area shown below) would also have been the result of *organic growth* processes, although individual *temenos* buildings of that date would have had rectilinear plans. See the *Antiquaries Journal*, vol X, no 4, October 1930, for a sequence of plans of the *temenos* at periods in its history.

34. Woolley, *Ur of the Chaldees*, 1982.

35. The absence of other than incidental urban open spaces available to the general public in the Sumerian city was subsequently to become a characteristic of the Arabian and then – in direct line of descent – the Islamic cities.

36. Woolley, *Ur of the Chaldees*, 1982.

37. Woolley, *Ur of the Chaldees*; see also Woolley's series of published papers describing the excavations at Ur in *Antiquaries Journal*, most importantly that of October 1931 (vol XI, no 4) which describes the housing area in great detail. See also H U F Winstone, *Woolley of Ur – The Life of Sir Leonard Woolley*, 1990.

38. Woolley, *Ur of the Chaldees*.

39. See S Lloyd: 'the houses themselves were so exactly like those of a small town in Iraq today – say, for instance, Najaf – that if one could see air photographs of both they would be almost indistinguishable, except of course for the character of the religious buildings in the centre' (*Foundations in the Dust*, 1947). The morphological continuity between the cities of ancient Mesopotamia and those of Islamic culture is a basis of Chapter 11's analysis of the Islamic city.

Figure 1.9 – Ur: detail plan of part of the housing area of the period 1900–1674 BC, excavated by Sir Leonard Woolley south-east of the *temenos* (as Figure 1.7, D). Key: A, Baker's Square, a small market space; B, Bazaar Alley leading to it from the main street; C, small local shrines. Streets are in random tint; house courtyards in dotted tint. See Figure 1.30 for the original plan of the entire organic growth housing layout as recorded by Woolley.

Figure 1.10 – Erbil (ancient Arbela) in north-east Iraq about 200 miles north of Baghdad at the foot of the Kurdistan mountains. The ancient *tell*, which has never been systematically excavated, is believed to have been continuously occupied for as many as 6,000 years. This key aerial photograph, personally the most intriguing and arguably the most important in the book, illustrates a close-knit cellular urban morphology, evocative of biological sections, which not only epitomizes the organic growth form of ancient history but also provides the direct link between the Sumerian cities of Mesopotamia and those of Islamic urban culture. See Figure 1.30 for the detail plan of the housing area excavated by Woolley at Ur, and Chapter 11 generally.

It is stressed, however, that the organic growth urban form of Ur, and Erbil both ancient and modern, is based on the climate-response Middle Eastern courtyard house type. For the contrasted medieval morphology of more northerly climes see the English village aerial photographs as Figures 4.18, 19 and 20, and Chapter 4 generally.

making it perhaps the oldest present-day city anywhere in the world. The close-knit cellular urban grain of Erbil (Arbela) and also Ur is analysed below as exemplifying Sumerian organic growth.

The second stream, which has produced only a comparatively small number of towns, had a more recent, yet still ancient origin. It comprises the planned towns of history and today, which have been created at urban status, at a given moment in time. The planned town is introduced later in this chapter in the sections dealing with the Egyptian and Harappan civilizations. (In modern times, the phrase 'planned town' has become synonymous with 'new town', the British series since 1947 being the best known international examples.)[40]

Organic growth and planned urban form

As one of several introductory explanations, the term 'organic growth' is also used to describe the kind of urban form which has evolved

40. See F J Osborn and A Whittick, *The New Towns*, 3rd edn, 1977.

without preconceived planned intervention; similarly, in direct contrast, 'planned' urban form, which is the result of predetermined intention.[41] Later in the book, as will be seen, it is an historical characteristic to find on the one hand that planned extensions, or renewals, were added to organic growth settlements; with organic growth changes affecting towns of planned origin on the other. Cities combining both organic growth and planned forms include Edinburgh (8) and Vienna (7) as two personal European favourites, with Boston, Mass. (10) and Kyoto (B.4) from further afield. (Chapter references for examples are given in parentheses.) (As an aside, bringing our subject right up to date, late-twentieth-century city planners throughout the world are preoccupied with making 'planned' urban sense out of organic growth inheritance.)

Urban form determinants

In history the forms of settlements at both rural/village and urban/city status have been determined by factors and influences which, as the book's second basic concept, we shall call urban form determinants.[42] These determinants are of two different origins. First, there are three which derive from geographical 'natural-world' attributes of the location of a settlement; most importantly its climate, the topography, and the available construction materials. Each of these determinants has played a part in the shaping of all historic urban form, both organic growth and planned. Second, there are the numerous determinants which have their origin in humankind's intervention in the natural settlement processes, which we shall term 'man-made' determinants. As will be seen, some of this latter group apply to both organic growth and planned towns (most importantly fortification); whereas there are others (notably the gridiron system of land subdivision) whose relevance is restricted to planned towns only.

Characteristically the form of an urban settlement at any given period is the result of a number of locally effective determinants. In the uncomplicated, distant past only a few held sway; compared with circumstances prevailing in late twentieth-century urban/metropolitan agglomerations, where numerous determinants can be identified, many (if not most) of which are seemingly at odds with one another. (Not that we should regard present-day metropolitan predicaments as without historic parallel: circumstances in ancient Rome of the second century AD provide a powerful example of urban history anticipating itself: see Chapter 3.)

Analysis of past and present urban forms against a check-list of determinants can be one way to use the book. Readers, whether students and professionals, or 'just-interesteds' of whatever persuasion, may well discover others of local significance to add to those of general relevance. An understanding of the 'why' and the 'how' of present-day urban circumstances, it is argued, can but precede any attempt to instigate change, which usually means to strive after improvement. It is not that history necessarily has answers to offer: rather it is a question of needing to gain an understanding of the relevant local past in order to identify the nature of its present-day problems. It is stressed that neither sequence nor comparative written length accorded the determinants is of significance. That which is of fundamental importance is relative effect as the cause of the form of individual settlements.

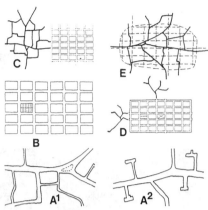

Figure 1.11 – Organic growth and planned urban form diagrams. Key: A, two characteristic kinds of Organic Growth: Western European, providing for street frontage plot development, and Mesopotamian/Islamic with housing access culs-de-sac; B, the gridiron as the usual basis of Planned Urban Form (see also Figures 1.18 and 19); C, an organic growth nucleus with planned gridiron extension, loosely based on Edinburgh (8); D, a planned gridiron nucleus with organic growth extension, loosely based on Timgad (3); E, the special three-dimensional Western European circumstances whereby an early medieval organic growth pattern was superimposed on the abandoned gridiron of a temporarily deserted Roman city – based on Cirencester, England (3).

41. 'Organic growth' and 'planned' are my personal terms for the two types of urban origin, dating back to my first lectures and articles and consolidated in this book's first edition (1972). In my opinion, synonyms resorted to by other urban historians are cumbersome if not misleading. 'Unplanned' does not convey the essential natural-world organic cellular accretion processes; the *Villes Spontanées and Villes Créés en Islam* of E Pauty (1951) imply other than long-term evolution of the former; J E Vance's 'organic layout' and 'modular preconceived design' complicate the issues (*The Continuing City*, 1990); while B S Hakim's 'incrementally grown' and 'spontaneously created' may lose in the translation from his native Arabic (*Arabic–Islamic Cities: Building and Planning Principles*, 1986).

42. 'Urban form determinants' is not only a second instance of personal terminology but also there is no other urban historian who consistently follows this approach to the analysis of historic urban morphology. B G Trigger in his 'Determinants of Growth in Pre-Industrial Societies' (in P J Ucko, R Tringham and G W Dimbleby (eds) *Man, Settlement and Urbanism*, 1972) examines 'the range of factors which have been noted as promoting the increases in the size of urban populations', but he is not concerned with 'those which influence either the specific location of cities or their internal layout'. (See also B G Trigger, 'The Determinants of Settlement Patterns', in K L Chang (ed.) *Settlement Archaeology*, 1968.)

Hakim's approach is along similar lines, without being taken as far. His stated first objective in *Arabic–Islamic Cities* is 'to identify and record the building and planning principles which shaped the traditional Arabic–Islamic city'.

Natural world determinants

TOPOGRAPHY

The terrain on which a settlement became established, or over which it expanded,[43] could have an underlying effect not only on geographical extent, ranging from the finite hilltop area of a medieval European burg (4) to the limitless prairie of Oklahoma City (10), but also on direction of growth, as exemplifed by seaside or major riverside cities such as Cologne (Figure 5.6) and Cairo and Baghdad (11); and by ridge-top cities such as Edinburgh (8). Shaping within river valleys, or on peninsulars, is another typical topographical response, for which see Durham, England (4) and Prague and Cesky Krumlov in Czechoslovakia (7 and 4), Miletus (2), Pittsburg (10), and Alexandria, Egypt (11).

Only exceptionally in history could a society achieve engineering feats of radical topographical adjustment, of which the excavation that made possible the Imperial Fora in ancient Rome is an extraordinary example (3). On the other hand, exceptional climatic, or natural disaster adjustment has facilitated or dictated change, as exemplified by the drying out of Lake Texcoco and the expansion thereover of Mexico City (9); and the thirteenth-century need to replace Winchelsea and Ravenser, in England (4), when the original harbour towns were engulfed by the sea.

In history, as also today, topography has played a main part in the creation of the urban third dimension; its visual effects ranging from the virtually non-existent of an open prairie settlement, to the dramatic skyline of San Francisco (10); and with Athens, Rome and Edinburgh as other outstanding European examples. Of Muslim cities, the characters of Muscat, Ghardaia and Sa'na (11) owe much to their respective topographical settings.

CLIMATE

Shelter has been a fundamental human need, of varying significance and taking different forms depending on local climatic circumstances.[44] At Ur, where Woolley unearthed the houses of around 1700 BC, daytime temperatures could be dangerously high and those at night unpleasantly low. In these hot-dry conditions, natural climate-response shelter – as we shall term it – takes the form of a grouping of rooms on one, sometimes two storeys, seldom higher, around a courtyard where there is always some shade and, consequently, a degree of convected air-movement (Figure 1.13). The courtyard house (described below in the section dealing with Sumerian urban form) is also of fundamental importance in respect of its formative influence on the arrangement of houses in Islamic culture. In addition to the courtyard houses at Ur, other examples include Olynthus (2); Pompeii and Timgad as two of many Roman towns (3) and Cordoba from the Islamic period in Spain (4).

However, there is a crucially important historic alternative in cities of the Arabian Peninsular whereby inland in the south-west and in hot-humid climates, houses constructed on several storeys are provided with screened bay windows (*mashrabiya*) opening on to public streets.[45] This most attractive architectural element enabled viewers, in particular the women of the family, to look out in privacy, and also took advantage of such slight breezes as may occur. Inland examples are Makka, pre-eminently, and San'a, with Jeddah on the Red Sea coast.

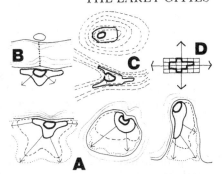

Figure 1.12 – Characteristic topographical circumstances for urban settlement and subsequent growth. Key: A, seafront, island and peninsular origins – respectively Brighton (8), Manhattan Island (10), Miletus (2); B, riverbank origin, initially with ferry or ford limitation on cross-river settlement and awaiting later (modern) construction technology for significant bridgehead development. Both locations usually resulted in growth directions away from the nucleus – London (8), Kiev (7); C, hill and ridge-top origins – Edinburgh (8) – and the valley base converse – Makka (11); D, flat, open 'prairie' location, with no major topographical growth constraints – Oklahoma City (10).

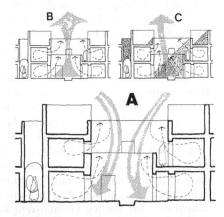

Figure 1.13 – Climate: the courtyard house and its performance in high sun-angle (Mesopotamian/Arabian) locations. Key: A (night), the courtyard and rooms fill with cool air; B (noon), the courtyard is heated by the sun, and hot air rises creating cooling convection currents in the rooms; C (afternoon), courtyard and rooms at their hottest but convection currents increasingly caused by shadow cooling. (Solid traditional construction walls with small openings, and thick flat roofs, minimize internal heat gain.)

43. The topographical determinant is dealt with in urban and historical geography references, see C T Smith, *An Historical Geography of Western Europe Before 1800*, 1967.

44. For effects of climate see H Fathy, *Natural Energy and Vernacular Architecture: Principles and Examples with Reference to Hot Arid Climates*, 1986; K Talib, *Shelter in Saudi Arabia*, 1984; M N Bahadori, 'Passive Cooling Systems in Iranian Architecture', *Scientific American*, no 238, 1978.

45. For *mashrabiya* (or *shanashil* as locally termed) see J Warren and I Fethi, *Traditional Houses in Baghdad*, 1982.

In northerly damp-cool climes, the response was markedly different (see Chapter 4) and when introduced by Roman conquerors the courtyard house never found favour. The usual northern arrangement is that of outward looking, extroverted houses either joined together in street-frontage terrace form, or free-standing in their own 'garden' space. A main consideration, arguably a determinant in its own right, was human need to cluster together around the fire in the centre of the dwelling.

CONSTRUCTION MATERIALS AND TECHNOLOGY

The performance characteristics of locally available construction materials have imposed constraints on builders and latterly architects, that were overcome only by the revolutionary new construction technology made available by the industrial revolutions at the end of our period.[46] In history, for all but the comparatively few religious, royal, civic and other monumental buildings, for which materials could be transported in bulk, there was no alternative but to use local materials. These are mainly brick (either mud or burnt clay); stone, depending on geological circumstance; or timber, and the process is known as 'vernacular building', or the 'local vernacular', with dialect in speech as the explanatory analogy.[47]

Traditional materials and the techniques for their use, have limited the height of walls, the width of openings, and the clear span of floors and roofs, thereby determining the fundamentally human-scale, third-dimensional aspects of everyday buildings in historic cities (Figure 1.14). This in contrast to the structural opportunities inherent in modern steel frame (in succession to iron) and concrete construction, whereby the sky is the limit. Not only has the urban third dimension been altered out of recognition, but also bridge and tunnel technology has enabled suburban expansion beyond previous natural boundaries.[48] New York City exemplifies both these changes (10).

Examples of the use of local materials include Mesopotamian mudbrick (as above) and stone in other favoured Islamic locations such as Makkah and San'a (11); marble, supremely so, in the hands of Greek architects and sculptors (2); masonry stonework for the European cathedrals; and timber for non-religious counterparts. The spans of openings, floors and roofs in Islamic construction were generally constrained by lack of suitable masonry material, or timber.

Man-made determinants

By comparison with their natural-world counterparts, the man-made determinants are considerably more numerous and have continued to increase in number as urban societies, and technologies have evolved from earliest times.[49] A considerable proportion are also much more complex in nature and effect, the most important and wide-ranging of which are the three 'primary motivating forces in the generation of urban forms' (added italics): trade, political and social power, and religion.[50] Not only have these forces had major determining effects on historic urban morphology, but also singly, or in combination, they have been mainly responsible for urban formation and growth. Before turning to their roles as urban form determinants, our primary concern, it should be noted that urban historians and geographers of the various persuasions are strongly divided as to their relative importance.

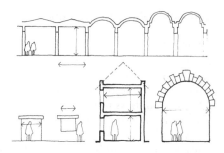

Figure 1.14 – Construction materials: characteristic room sizes, height of building and size of openings in walls, as constrained by traditional 'local vernacular' materials and technology, related to the scale of the human figure.

In a period of three centuries, from 1050 to 1350, several million tons of stone were quarried in France to build eighty cathedrals, five hundred large churches, and tens of thousands of parish churches. More stone was quarried in France during these three centuries than in ancient Egypt during its whole history – and the Great Pyramid alone has a volume of 40,500,000 cubic feet. Foundations of some of the great cathedrals were imbedded as deep as thirty feet – the average depth of a Paris subway station – and in certain cases formed a stone mass as large as that above ground.

In the Middle Ages there was one church or chapel for approximately every two hundred people. These cult edifices consequently covered a considerable area in ratio to the small cities of the time. For example, in Norwich, Lincoln, and York – cities of five to ten thousand inhabitants – there were, respectively, fifty, forty-nine, and forty-one churches. The ambitious who wanted to reconstruct their church on a larger plot were always faced with serious problems: it was often necessary to destroy one or two neighboring churches and build new lodgings for the expropriated. The area of Amiens Cathedral, covering about 208,000 square feet, permitted the entire population of nearly ten thousand

46. On construction materials and technology, see J E Gordon, *Structures: or Why Things Don't Fall Down*; J E Gordon, *The New Science of Strong Materials: or Why You Don't Fall through the Floor*, 1991; R Cotterill, *The Cambridge Guide to the Material World*, 1985; H W M Hodges, 'Domestic Building Materials and Ancient Settlements', in P J Ucko, R Tringham and G W Dimbleby (eds) *Man, Settlement and Urbanism*, 1972; J Gimpel, trans. C F Barnes jr, *The Cathedral Builders*, 1961.

47. On vernacular architecture in the Middle East see J-L Bourgeois with C Pelos, *Spectacular Vernacular: A New Appreciation of Traditional Desert Architecture*, 1983; and for the British Isles see R W Brunskill, *Illustrated Handbook of Vernacular Architecture*, 3rd exp. edn, 1987.

48. For the origins of modern high-rise, wide-span architectural engineering see C W Condit, *American Building: Materials and Techniques from the beginning of the Colonial Settlements to the Present*, 1968.

49. J A Agnew, J Mercer and D Sopher (eds) Introduction, *The City in Cultural Context*, 1984.

50. See Agnew et al, *The City in Cultural Context*, for discussion of this topic.

Childe accepts that 'the economic base in the form of the emergence of a sufficient surplus is the fundamental prerequisite'.[51] Boulding believes that 'cities arise when political means are employed to convey food surpluses into authoritarian hands',[52] and Jones also subscribes to power.[53] Wheatley, however, makes a powerful case for the pre-eminence of 'the symbolic integrative functions of the early cities'; explaining that 'it is doubtful if a single autonomous, causative factor will ever be identified in the nexus of social, economic, and political transformations which resulted in the emergence of urban forms, but one activity does seem in a sense to command a sort of priority. . . . This does not mean that religion . . . was a primary causative factor, but rather that it permeated all activities, all institutional change; and afforded a consensual focus for social life'.[54] Although much the same can be said of trade and power, my own view identifies with that of Wheatley in accepting religion as the characteristic consensual focus for social life; while recognizing that the three forces were ordinarily ever-present with locally effective emphases of time and place. I also believe that within their context, throughout urban history generally the most influential everyday causative factor has been that of 'practical expediency'. In other words, while it is urban history's exceptional results of trade, power, religion and other such determining influences that are conventionally highlighted, the great background mass of ordinary urban development has been shaped by mundane everyday requirements.

Effects of the three primary motivating forces are summarized first, before introducing the most significant man-made urban form determinants.

ECONOMIC

The role of the city as a market-place has required buildings and spaces for the making and sale of goods. During our period 'industry' was typically on a small-scale domestic-adjunct basis with products being made or finished to order by individual craftsmen employing few assistants. Although in European towns there was an urban space, or several, that accommodated communal trading activity, in medieval times, as stressed in Chapter 4, 'the entire city was a market'.[55] Circumstances in the Islamic city were different, with a clear distinction between home and workshop; the latter accommodated in the *Suq* (11). A main purpose of the market square, or the wide market street alternative, was to provide for the temporary stalls of travelling salesmen and merchants during their regular visits. In prosperous later medieval market towns, covered market halls were provided, in which the largest European cities gradually consolidated their separate major markets, exemplified by the *horrea* of ancient Rome (3) and the Covent Garden fruit and vegetable market in London (8).

POLITICAL

The city as a military and latterday ballot-box power base embodied the citadels, castles and palaces of past ruling elites, and now boasts the city halls of modern democracy. (Aggrandizement for political and other purposes is introduced below as a separate determinant.) In Western Europe, Haussmann's Parisian boulevards were in main part a response to the political need for mob-control (6), while internal security was a hidden yet ever-present political consideration in the

people to attend the same service. For comparison, imagine in a modern city of one million people a stadium built in the middle of town large enough to accommodate the whole population, remembering that the largest stadium in the world seats only 180,000.

Nave, tower, and spire heights are astonishing. An architect could build fourteen-story building in the choir of Beauvais Cathedral before reaching the vault, 157 feet 6 inches above the floor. To equal the masters of Chartres, who pushed their cathedral's spire up to 345 feet 6 inches, the present municipality would have to build a thirty-story skyscraper. To equal the Strasbourgers who raised their spire to 466 feet would require a forty-story skyscraper.

(J Gimpel, The Cathedral Builders, 1961)

Whenever, in any of the seven regions of primary urban generation we trace back the characteristic urban form to its beginnings we arrive not at a settlement that is dominated by commercial relations, a primordial market, or at one that is focused on a citadel, an archetypal fortress, but rather at a ceremonial complex. Of course, the modes of religious expression are often more stylized and repetitive than are those of, say, petty commerce or political organization, and are consequently likely to be more readily discernible in archeological assemblages. It is also true that the material manifestations of cult, and ritual are likely to be cast in a durable form capable of surviving the vicissitudes of time. Indeed some of the most ancient are of striking impressiveness even today. Moreover, writing and respresentational art, on which we are dependent for a large part of our knowledge of these centers, but which were both intimately associated with, and were perhaps born of, ritual, may induce us to exaggerate the role of religious ceremonial. But even allowing for the biases in interpretation thus induced by the nature of the evidence, and discounting the number and visual preponderance of religiously prescribed elements in the morphology of these complexes, the predominantly religious focus to the schedule of social activities associated with them leaves no room to doubt that we are dealing primarily with centers of ritual and ceremonial.

(P Wheatley, The Pivot of the Four Quarters, 1971)

51. Childe, *What Happened in History*.
52. K Boulding, 'The Death of the City', in O Handlin and J Burchard (eds) *The Historian and the City*, 1963.
53. E Jones, *Metropolis: The World's Great Cities*, 1990.
54. P Wheatley, *The Pivot of the Four Quarters: A Preliminary Enquiry into the origins and Character of the Ancient Chinese City*, 1971; see also J N Postgate, 'The Role of the Temple in the Mesopotamian Secular Community', in P J Ucko, R Tringham and G W Dimbleby (eds) *Man, Settlement and Urbanism*, 1972.
55. H Saalman, *Mediaeval Cities*, 1988.

traditional Muslim city, where the political and the religious forces are combined within Islam (11).

RELIGIOUS

The city as a devotional centre was given direct expression through if not always its largest buildings, then ordinarily its tallest and visually most assertive spire, tower and dome skyline elements. In addition to cathedrals, churches, temples and shrines for devotional ceremony, the church in its widest sense was a major urban landowner, as it remains today in some countries.[56] Its activities could require extensive building programmes ranging from the pyramids and temples of Ancient Egypt (1) to the diverse urban activities of the Spanish Catholic Church, exemplified by Santiago de Compostela (9). Exceptionally, in direct contrast, Philadelphia, Pennsylvania, was a Quaker city initially planned without churches for communal worship (10).

THE PRE-URBAN CADASTRE

This most useful term[57] refers to the pattern of pre-existing man-made rural property boundaries, regional routes, drainage ditches, and so on, over which an organic growth settlement expanded, or which had to be recognized in the planning of new urban form. From earliest times, as soon as land came into individual ownership, the boundaries were 'legally' protected, and could be changed only by autocratic action or democratic intervention.

Although out of our present period, it is the pre-urban cadastre that explains otherwise baffling forms of later nineteenth- and twentieth-century suburbs, epitomized by London's expansion during the 1920s and 1930s. The secret is in the excellent large-scale UK Ordnance Survey maps, available in sequential editions from the mid-nineteenth century, and which reveal, when overlaid, the rural pattern of farms, individual fields, lanes and regional routes, that determined the morphology of the suburbs (Figure 1.15).[58]

Historical examples of a pre-urban cadastre include the English village field patterns of Figures 4.18, 4.19 and 4.20, with further description in the captions, the comparable Dutch *sloten* or drainage ditches (4) and the Great American Grid, on which the western lands were settled (10).

DEFENCE

From the moment when material and human resources of early settlements first became envied by other societies, need for defence against external attack has been an ever-present preoccupation. 'Star Wars' wizardry of the 1990s is but its latest manifestation, removed now to outer space as the technologically most advanced stage in geographical distancing of the attackers and the defended.[59] Commencing with simple two-dimensional length and height palisades and walls, of which Jericho is the melodious favourite, it gradually became vital to keep assault artillery at ever-increasing distance from the urban soft-centre. In so doing, geometric design of three-dimensional defensive systems became ever more complex, reaching its peak in the works of the French military engineer, Vauban (6), and his Dutch contemporaries (4). Notable European defensive systems include Constantinople (Istanbul) (3), Naarden (5), Antwerp and Vienna (7), Berwick-upon-Tweed, England (8), and most importantly, Paris (6). New World

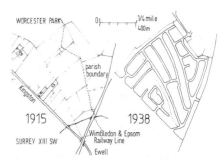

Figure 1.15 – Pre-urban cadastre: a traced extract of a United Kingdom Ordnance Survey London suburban map of 1938 (at the scale of 6 inches to 1 mile) with the pre-existing pattern of open fields and roads as shown on the comparable 1912 map.

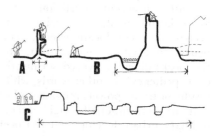

Figure 1.16 – Defence: three diagrammatic stages in the evolution of a typical defensive system. Key: A, wall or palisade of minimal horizontal dimension; B, pre-artillery early medieval wall with ditch or moat; C, the mature artillery defensive zone.

56. The most powerful Church in history was probably the Catholic Church in Spain and Latin America; see J E Hardoy and C Aranovitch, *The Scale and Functions of Spanish American Cities around 1600*: 'the Church exercised an unquestionable political pressure through the possession of valuable properties and the granting of loans'.

57. I encountered the valuable term 'pre-urban cadastre' in D Ward's article, 'The Pre-Urban Cadastre and the Urban Pattern of Leeds', *Annals of the Association of American Geographers*, vol 52, 1962. Previously I had first recognized the role of the underlying (suburban) landownership patterns as a primary determinant of urban morphology from reading Woolley: 'anyone looking at the plans can see at once that there is no such thing as town-planning at Ur … the town in general preserved the form, or lack of form, of the primitive village, and there are no straight streets or broad thoroughfares, only winding lanes whose course had been dictated by the accidents of land ownership'. See also P Beaumont, M Bonine and K McLachlan (eds) *Qanat, Kariz and Khattara: Traditional Water Systems in the Middle East and North Africa*, 1987, who note that rectangular fields on the Iranian plateau were determined by the water supply system; and the determining role of *sloten*, water drainage ditches, on the form of Dutch urban expansion in Chapter 4.

58. On the British suburbs, see H J Dyos, *Victorian Suburb: A Study of the Growth of Camberwell*, 1968.

59. On fortification see M Bateman and R Riley (eds) *The Geography of Defence*, 1987; M J Rowlands, 'Defence: A Factor in the Organisation of Settlements', in P J Ucko, R Tringham and G W Dimbleby (eds) *Man, Settlement and Urbanism*, 1972; C L Woolley, 'Excavations

counterparts, comparatively few in number, include San Juan in Puerto Rico, Cartagena, Havanna and Manila (9); and Louisberg, constructed to a French plan in Canada.

AGGRANDIZEMENT

From earliest times, religious, monarchical, political and other vested interests have been glorified in cities, if not actually raised over and above. The ruling elite's ziggurat dominated Ur, just as Norman castles lorded it over Saxon England (4). Throughout Europe countless medieval spires point the way to church, while papal policy was concerned to re-create Rome as a fitting capital of the Catholic world (5). Autocratic aggrandizement reached its apogee at Versailles (6), from where it could but decline, with Washington DC (10) its democratic political counterpart.

THE GRIDIRON

Although in history there were exceptional uses of other systems of town planning, it was a rectilinear gridiron pattern of land subdivision that was the rule in providing the geometrical basis of planned settlements. As shown diagrammatically by Figure 1.18, a gridiron plan ordinarily comprises an orthogonal network of streets dividing the planned area into primary building blocks (or *insulae*) which are further divided into rectilinear individual plots as the basis of urban land distribution. Notable uses of the gridiron in history were made by the Greeks for their planned cities (2), the Romans in their imperial urban planning policy (3), numerous medieval monarchs and landowners (4), European city planners in the USA for Philadelphia, Manhattan Island, Oklahoma City, San Francisco and countless others, and Latin America, where the Laws of the Indies made it the mandatory basis of Spanish colonial city planning (9 and 10).

However, a clear distinction must be maintained between a gridiron plan for an entire city, as defined above, and the rectilinear planning of only a primary street system forming comparatively extensive 'super-blocks', the subdivision of which was left to the occupiers to agree between themselves (see Figure 1.19).[60] As one of this edition's major reassessments, Mohenjo-daro and the other Harappan civilization cities of the Indus Valley are now best regarded as being of this latter type, rather than, as believed hitherto, the earliest known instances of gridiron planning. This revised opinion does not affect their standing as the earliest known planned cities: it is the nature and extent of their planning which is now open to doubt. Furthermore, there are related changes of opinion concerning the city in Ancient Egypt, most importantly the evidence supporting Kahun's claim as the earliest known instance of gridiron planning. Neither does this reassessment support the claim of Hippodamus of Miletus to have been the 'father of town planning': if it was not an Harappan priest who took over that mantle, as presently recognized, then it was the Egyptian planner of Kahun. (See the Harappan and Egyptian sections below, and Chapter 2 for further discussion.)

URBAN MOBILITY

Need for urban street systems to provide for the movements of pedestrians, pack-animals and wheeled transport has had two essentially different effects: first, there are the measures taken to increase the

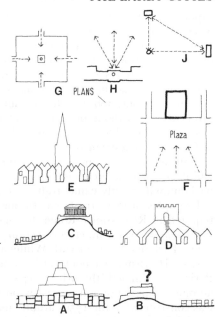

Figure 1.17 – Aggrandizement: diagrammatic cross-sections (the vertical dimension is exaggerated) through. Key: A, a typical Sumerian city, with its ziggurat – as Ur (Figure 1.7); B, an Harappan city, with its western citadel – as Mohenjo-daro (Figure 1.41); C, an ancient Greek city, with the Temple on its acropolis – as Athens (2); D, a Norman castle in eleventh-century England, lording it over a conquered Saxon town – as Wallingford (4); E, the church in a medieval European village (4); F, the church in a Latin American city – as Mexico City (9); G, a royal statue-square in Paris (6); H, royal aggrandizement at Versailles, France (6); J, democratic aggrandizement at Washington, DC (10).

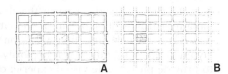

Figure 1.18 – The Gridiron as the basis of the plan of a complete urban entity. Key: A, shown as a closed, finite plan – as Monpazier, France (4); B, shown as an open-ended urban development, or land distribution framework – as Oklahoma City, USA (10).

at Ur 1928–29', *Antiquaries Journal*, vol IX, no 4, October 1929 (for the walls at Ur); J R Kenyon, *Mediaeval Fortification*, 1990; see also the general references on this topic in Chapter 5.

60. For the available ancient mathematical basis of the gridiron see S A Kubba, *Mesopotamian Architecture and Town Planning*, 1987. 'Expediency supergrid' is my personal term for the large-scale, main routes planning of modern new cities such as Chandigarh, Brasilia and Milton Keynes, and the comparable expansion of existing cities such as Kuwait City, Jeddah and Islamabad. In principle, Milton Keynes has been planned along the same lines as the ancient cities of el-'Amarna, Egypt, and Mohenjo-daro, Pakistan (India) (see below in this chapter).

capacity of an existing system, without affecting the urban form; whereas the second – the planning of extensions or alterations to the system – can have radical effect. In recent automobile-dominated decades the former response has become known as 'traffic management', with one-way streets, gyratory intersections, computer-controlled traffic lights and the like, in order to squeeze ever more vehicles through the system. It is only too easy to imagine that present-day urban traffic problems are without precedent, yet the opposite holds true. Julius Caesar was forced to ban carts from Rome's streets during the daytime in favour of pedestrian traffic (3), and the City of London included a seventeenth-century limit on horse-power (8). Islamic urban culture had its own requirements in respect of street and lane widths based on pack-animal, mainly camel traffic (11).

Examples of new streets cut through existing cities, include those required in Renaissance Rome, to cope with weight of pilgrim traffic (5), London's early nineteenth-century Regent Street, linking the Regent's Park development with Whitehall (8), and, although out of our period, Haussmann's boulevards system of the 1870s and 1880s in Paris (6). Although the urban bypass is essentially a response of the automobile age, London was early on this scene when the New Road was constructed in the 1750s around its northern perimeter (8). Before availability of urban mass transit, similarly outside our period, a journey-to-work on foot was a constraint on the horizontal spread of cities.

AESTHETIC

The organization of a part of a city according to 'a set of principles of good taste and appreciation of beauty',[61] is linked with aggrandizement in its ancient origins. Records show that the Assyrian king Sennacherib took pride in having had the streets of Nineveh widened and straightened as a royal *via triumphalis*[62] (see the description below of Mesopotamian urban form); although we may never know if truly aesthetic considerations played a part in the shaping of the ziggurat at Ur, the hanging gardens of Babylon, or the citadel at Mohenjo-daro. Neither do we know which came first when the Greeks so carefully planned the layout of the Acropolis: was it to glorify the Athenian goddess, or was it to provide pleasurable visual experience? This question is further considered in Chapter 2.

Leaving aside considerations of autocratic or religious aggrandizement, the unifying colonnade constructed in front of the several disparate Forum buildings at Pompeii is agreed as one of the earliest 'urban beautification' projects (3), while the architectural harmony that contributes so greatly to the beauty of Siena's Piazza del Campo was a fourteenth-century municipal requirement. Subsequently, during the Renaissance and its aftermath, aesthetic considerations became of paramount importance in those parts of cities of autocratic, political or social significance – never forgetting, though, that through until recent times it was only an exceedingly small fraction of urban development that ever saw the hand of an architect or urban designer. Much still does not. Symmetry, balance and uniformity are primary Renaissance principles, exemplified by the Capitol Piazza, Rome – Michelangelo's supreme achievement (5); the eighteenth-century urbanism at Nancy, France (6); the squares and parades of Bath in England (8); and Washington DC (9) at the scale of the entire city.

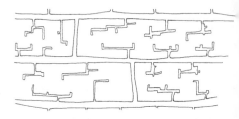

Figure 1.19 – The rectilinear planned main routes 'supergrid', with *laissez-faire* infilling of the grid blocks – as el-'Amarna (Figure 1.33) and Mohenjo-daro (Figure 1.41).

The city of Berlin, had Hitler's Nazi Germany won the Second World War: Let us suppose we are settlers from Gotenburg in Reichskommissariat Ukraine, travelling to the capital for a week's sightseeing. We would catch one of the special wide-gauge double-decker railway cars for the overnight journey from the Black Sea. Our destination would be Berlin's Central railway station – an astonishing feat of engineering, larger than New York's Grand Central station, with traffic arriving and departing on four levels. We would descend from the platform on a vast outside escalator. 'The idea,' according to Speer, 'was that as soon as they stepped out of the station, visitors would be overwhelmed, or rather stunned, by the urban scene and thus the power of the Reich.'
(R Harris, Fatherland, 1992)

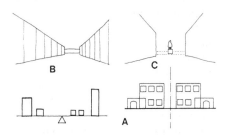

Figure 1.20 – Example aesthetic considerations in urban design. Key: A, balance and symmetry about an axis; B, skyline uniformity in perspective; C, focal interest terminating a vista. (See also Chapter 5 generally.)

61. *Concise Oxford Dictionary*, 1991; for ancient aesthetic considerations see S A A Kubba, *Mesopotamian Architecture and Town Planning*, 1987, and historical period references below.
62. A L Oppenheim, *Ancient Mesopotamia: Portrait of a Dead Civilisation*, rev. edn, 1977.

LEGISLATION

The introduction of mandatory legal measures intended to control the form of cities in general, or, and the more usual, its residential parts, were essentially a result of social conditions in nineteenth-century industrial cities, but there are many examples of municipal legislation within our period, including the city of Amsterdam (4; 7) pre-eminent among other Dutch examples, which worked in practice; whereas seventeenth-century attempts to control the extent of London (8) were doomed to failure in the absence of an administering bureaucracy. The only instance of nationally applicable bureaucratically enforced city planning legislation of our period is the Spanish colonial 'Laws of the Indies' (9). There are also the comparable effects of the Islamic building 'guidelines', although they were not formally codified legislation.

URBAN INFRASTRUCTURE

Urban improvements of previous generations characteristically retain validity only to serve as constraints on their successors, unable to pay for replacement, or otherwise act for removal of those urban necessities. The sewers of Ancient Rome, with the potable water supply system, were unavoidable constraints on comprehensive replanning ambition (3); just as the nineteenth-century railway viaducts in south-east London; or the city's Charing Cross railway bridge across the Thames, proposals to remove which date from but a few years after its opening in the 1880s. With its water supply system Constantinople acquired history's most extensive infrastructure system, including open air cisterns of enormous capacity, one of which accommodates a present-day football stadium (3).

SOCIAL, RELIGIOUS AND ETHNIC GROUPING

Other than castles, citadels and enclaves of a ruling elite, it is only from Renaissance times that social class segregation has had an evident effect on the overall form of cities and the visual appearance of their residential districts. Distinction between the wealthy and the poor was a question of size of house, with small one-room dwellings mixed in with large residences. Pompeii is a clear example (3). There was separation on religious and ethnic grounds, notably in the case of Jewish communities, but other than their possible walled enclosure, their presence also had little effect on urban form. 'In the historic Islamic city,' Hakim writes, 'at the macro scale influential groups and individuals tended to select "better" sites with good water or defensive potential; but at the micro scale of neighbourhoods or housing clusters, the practice was egalitarian with small humble dwellings tucked in between large ones.'[63]

Social class distinction in European cities began to take effect on urban morphology from the seventeenth century when the Place Royale in Paris was developed by the French monarchy as an aristocratic upper-class residential quarter (6). The London counterpart was Covent Garden, planned in 1630 as the first of numerous squares culminating in the early nineteenth-century Regent's Park development (8).

LEISURE

Whereas Greek and Roman cities characteristically contained buildings and open spaces used for either participatory sporting activities, or

Figure 1.21 – Legislation: as control of height of buildings, and distances in between. Outside our period, the British 'by-law' legislation of the later nineteenth century served to enforce the gridiron as the basis of mass housing development.

Among the relevant legal texts, the most prominent is Papinian, on the Care of Cities, concerned with the maintenance and repair of the streets. The city overseers (astunomikoi – translated into Latin as curatores urbium) are to take care of the streets of the city so that they are kept level, so that the houses are not damaged by overflows (fluminia – rheumata), and so that there are bridges where they are needed. 1. And they are to take care that private walls and enclosure walls of houses facing the streets are not in bad repair, so that the owners should clean and repair them as necessary. If they do not clean or repair them, [the overseers] are to fine them until they make them safe. 2 They are to take care that nobody digs holes in the streets, encumbers them, or builds anything on them. In the case of contravention, a slave may be beaten by anyone who detects him, a free man must be denounced to the overseers and the overseers are to fine him according to law and make good the damage. 3. Each person is to keep the public street outside his own house in repair and clean out the open gutters and ensure that no vehicle is prevented from access. Occupiers of rented accommodation must carry out these repairs themselves, if the owner fails to do so, and deduct their expenses from the rent.
(D F Robinson, Ancient Rome: City Planning and Administration, 1992)

The importance of television in providing cheap – and sometimes nourishing – entertainment, particularly for the poor, should stop us moralising against 'circuses', even if we find the particular form some of the Roman entertainments took revolting. In the first place, it is clear that, even had they wanted to, the Roman urban masses could not have spent their days in idleness at the races, the theatre or at other shows. Just after the Social War, at the start of the period considered in this book, there were 57 days of games, taking up approximately three separate weeks in April, one in July, a fortnight in September and another in November. These regular, public games were part of the religion of the state commemorating triumphs and the averting of disasters. On the same grounds, further sessions were added by Sulla and then by Julius Caesar; the tendency to establish regular festivals to celebrate imperial occasions had led by AD 354 to there being a total of 177 days on which public games took place. (The modern office worker in Britain has over 130 free days in the year.)
(Robinson, Ancient Rome: City Planning and Administration)

63. B S Hakim, Arabic–Islamic Cities: Building and Planning Principles, 1986.

passive spectator enjoyment, these provisions are not usually part of either their Mesopotamian predecessors or their European medieval and Renaissance successors. Neither are they present in the historic Islamic city. A main reason for these differences was availability of 'spare time', which at Rome in the third century AD amounted to about 200 days in the year (3). Climate was also a determining consideration, with public gatherings of crowd proportions requiring a degree of shelter unavailable in climatic extremes until recent times.[64]

Locational determinants

Introductory consideration is also required of underlying reasons for the existence on their sites of settlements of both organic growth and planned types. These we shall call locational determinants. Throughout the book, descriptions of major towns and cities ordinarily establish the original *raison d'être*, and those of significant subsequent change.[65] Chapter 4, which covers the formative medieval period in European urban history, includes the main explanation of locational determinants, which are summarized here under the separate headings 'organic growth' (from village origins) and 'planned' (at urban status). Chapter 11 includes the Islamic counterparts.

ORGANIC GROWTH SETTLEMENTS

Availability of a permanent potable water supply and one or more reliable continuing sources of food were the essential prerequisites for village settlement. These are the primary locational determinants. Without drinking water, settlement was not possible other than under totally exceptional circumstances, as exemplified by pre-industrial (for which read 'oil') Kuwait City, which relied on water brought in by boat (11). Food could be 'imported', as has been suggested for Çatal Hüyük (see note adjoining); but in history such artificial settlements could only have been exceedingly few and far between.

A settlement's defensive potential, against either natural forces or military aspirations of rivals, ordinarily resulted in the choice of a higher-ground location, which could be either as little as a topographically insignificant – but absolutely vital – mere metre or two above the highest remembered tide or river flood level, or as much as a hundred or more metres, with a few so favoured fastness settlements, of which Ronda, in southern Spain, is a pre-eminent example (9). Valley locations, or the lee side of high ground or forest, could provide welcome shelter from the climate.

Trade (the 'market-place' factor) is the economic function that distinguishes between rural (village) settlements and the proportion that were elevated to urban (town) status. Usually this resulted from location at a crossing of land routes, or intersection of land and water (river) routes, convenient of access to traders and their customers. As stressed in Chapter 4, the resultant market towns, serving their hinterland of agricultural villages, jealously guarded their economic advantage. Market-places also grew up to serve castle towns (burgs) or religious establishments (monasteries).

PLANNED URBAN SETTLEMENTS

These locational determinants are further divided into two types: first, those immediate locational criteria used to select the site for a planned

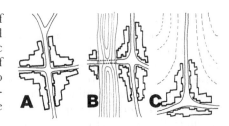

Figure 1.22 – Locational determinants: the 'market-place' factor exemplified at: A, land route crossings; B, intersections of land and water routes; C, at entrances to mountain passes.

Catal Huyuk and agricultural primacy

At the beginning of the chapter Mellaart's archaeological findings at Catal Huyuk were introduced as having been used by Jane Jacobs, in her book *Economy of Cities*, as the basis of the claim that 'agriculture and animal husbandry arose in cities;' contrary to the accepted belief that the development of settled agriculture was the essential pre-requisite for the evolution of urban settlements. In turn, the theory of urban primacy was central to her interpretation of social and economic failings of late twentieth-century American cities.

For this purpose Mrs Jacobs invented the imaginary city of New Obsidian which was 'the centre of a large trade in obsidian, the tough black natural glass produced by some volcanoes. The city is located on the Anatolian plateau of Turkey.' In 8,500 BC New Obsidian's population numbered about 2,000 persons, a large proportion of whose food is imported 'from foreign hunting territories. This food, which consists overwhelmingly of live animals and hard seeds, is traded at the barter square for obsidian and for other exports of the city.'

However, the claim for New Obsidian, and with it that of urban primacy, can be dismissed on two archaeological grounds:

first, as established earlier, a hunting-gathering catchment area of at least 20,000 square miles would have been needed for an urban population of 2,000 persons, giving a circle of radius of about 80 miles. How could the food supply have been taken into the city in quantity, over such considerable distances, when wheeled transport was still 5,000 years in the future? Or stored, without pottery technology?

Second, it is reasonable to expect that the trading activity and meting out of stores required some kind of permanent recording system? Yet, as far as is known, the earliest form of writing originated several millenia later.

---

64. See Chapter 11 for differences between the traditional Islamic city and its Western-world counterparts.
65. Locational determinants are considered in varying detail in urban geography books, for which see H Carter, *The Study of Urban Geography*, 2nd edn, 1975; H Carter, *An Introduction to Urban Historical Geography*, 1983/1989 – conceived as a subject in its own right; C T Smith, *An Historical Geography of Western Europe Before 1800*, 1967.

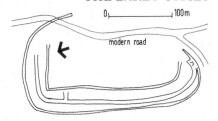

Figure 1.23 – Jericho: outline of the walls and excavated areas.

new town, which are in effect the same as those of organic growth, and second, the social, political and economic reasons for their regional or national siting. This latter group comprises the following criteria.

First, strategic (military) control of conquered or disputed territory, which is characterized by the way the Romans held on to their important provinces (3); a very similar method was adopted in the thirteenth century by the English King Edward I in troublesome parts of Wales and south-west France (4). In addition, numerous individual European lords founded fortified towns (burgs) to control key centres of local communication; Muslim army commanders also established a number of military bases that evolved into urban settlements.

Second, commercial and agricultural opportunity (often together with politico-military purpose) has provided an incentive for new towns throughout history; introductory mention is made of the Greeks with their colonial settlements (2) and, beyond our present period, the post-1917 new towns programme in the Soviet Union.

Third, mineral wealth exploitation, typically following conquest, which brought the Romans to Britain and Spain, where they founded new towns. As one of history's supreme ironies, it took Spanish *conquistadores* to their Latin American 'New World' (9).

Fourth, resettlement of excess population where agricultural land was available (or where employment and improved urban circumstances could be provided) was a second purpose of the Greek colonies. It was a primary concern to the Roman authorities when demobilizing their legions (3); a driving force in the eastern expansion of medieval Europe (4); one reason for European emigration to the Americas and other new worlds (9; 10); and a primary purpose of the British New Towns programme after the Second World War.

In addition to the above locational determinants, there were settlements that were founded for mystical or religious reasons. If their number was minimal compared with those serving everyday functions, their significance cannot be underestimated. As centres of pilgrimage these sanctuaries often acquired considerable economic strength, as exemplified by Delphi, Olympia and Epidaurus in ancient Greece (2) and Makkah in pre-Islamic Arabia (1; 11). In more recent times, a small number of existing cities have become centres of pilgrimage, notably Medina, the second city, with others in Islam, and Lourdes within the Roman Catholic faith.

JERICHO AND ÇATAL HÜYÜK

Ancient Jericho and Çatal Hüyük represent the two most powerful challenges to the accepted belief that urban civilization first emerged in Mesopotamia. Jericho is known to have been a densely developed settlement with formidable defences and an evolved administration as early as 8000 BC. There is radio-carbon dating of approximately 9000 BC for what is assumed to have been a sanctuary established by Mesolithic hunters beside a spring, which was later used for irrigated cultivation in the Jordan Valley, which at Jericho is some 900 ft below sea level. 'The descendants of these hunters,' notes Kathleen Kenyon, 'must have made remarkable progress to have achieved full transition from a wandering to a settled existence in what must have been a community of considerable complexity, in about one thousand years,' adding 'after the settlement had expanded to its full size, it was surrounded by massive defences, and assumes an urban character'.[66]

Figure 1.24 – Çatal Hüyük; reconstruction of a sector of the town showing the arrangement of houses, characteristically entered by ladder from the roof, and grouped around an open space.

Radio-carbon dating has established an estimate of *c.* 5600 BC for levels 1 to 0. After 5600 BC the old mound of Çatal Hüyük was abandoned and a new settlement established to the west. In his book *Çatal Hüyük*, Mellaart observes that 'the need for defense may be the original reason for the peculiar way in which the people of Çatal Hüyük constructed dwellings without doorways, and with sole entry through the roof. Villages of this type are still found in central and eastern Anatolia ... a solid outer wall built of stone is the alternative, but stone was not available on the plain ... moreover the city of Çatal Hüyük was extensive and would need considerable manpower to man the entire circuit against enemy attack. Once the wall was breached an enemy would have been able to break into the city. The solution adopted was a different one: the builders did not construct a solid wall but surrounded the site with an unbroken row of houses and storerooms, accessible only from the roof.'

However, Mellaart was able to excavate only an area of about one acre (about one-thirtieth of the evident *tell*) and he seems mistaken in assuming that renewal of a part of a city comprising a number of more or less rectangular cells, or the addition of a similar unit, can be seen as representing deliberate town planning. Rather, it is a form of controlled organic growth.

66. K Kenyon, *Archaeology in the Holy Land*, 3rd edn, 1979.

Sir Mortimer Wheeler, however, remained unconvinced by a case for urban status as early as the eighth millennium, arguing 'in common usage, civilisation is held to imply certain qualities in excess of the attainment ascribable to Jericho'.[67] Nevertheless, he suggested that 'an approach to this condition is represented by a substantial town discovered at Çatal Hüyük', located 50 km south-east of Konya in southern Turkey, on the Anatolian plateau some 3000 ft above sea level. The most important of two mounds rising above the level plain is oval-shaped, approximately 450 m long by 275 m wide. It covers an area of about 12 ha rising to a height of 17.5 m above the plain and was found to contain 19 m depth of Neolithic deposits, compared with the 13.7 m at Jericho. In the three seasons 1961–3 James Mellaart excavated an area of about 1 acre which yielded twelve successive building levels, representing twelve different cities, not phases or repairs of existing buildings.[68] Radio-carbon dating established c. 6500 BC for the oldest level and c. 5600 BC for levels 1 to 0. After c. 5600 BC the site was abandoned and a new settlement established to the west, across the river. The caption to Figure 1.24 describes the unusual form of Çatal Hüyük.

Daniel asserts 'neither Jericho nor Çatal Hüyük were civilizations: they were large settlements that could be called towns, or proto-towns. They did not have the (other) requirements of the Kluckhohn formula'.[69] They may have been unsuccessful experiments towards civilization, a synoecism that did not succeed, or we might label them just as very overgrown peasant villages.

Jerusalem

Jerusalem's long urban history has been traced back for nearly 4,000 years but fortunately for the archaeologist the area of the modern city does not include the site of the earliest settlements which lies to the south-east. Kathleen Kenyon in *Jerusalem: Excavating 3000 Years of History*, describes how the importance of the city from probably the third millennium onwards lies in the fact that its location enabled it to control the vital north-south route through the central highland of Palestine.

The first settlement occupied the southern end of a ridge bounded to the east by the valley called Siloam (ancient Kedron) and on the west by the valley called the Tyropoeon. The city's written history predates the extensive Biblical records by several centuries in that it is mentioned in letters of c. 1390 to 1360 BC sent by local governors to Akhenaten's officials in Egypt. The present record shows that the earliest settlement occupied an area of some 10.87 acres and that the first defensive wall dated from around 1800 BC. The line of this wall was followed by that of Jebusite Jerusalem, captured by David in around 996 BC. David, and his son and successor Solomon, established Jerusalem as the religious centre uniting the tribes of Judah and Israel. Solomon built the first temple on an extensive artificial terrace north of the old city area, most probably combined with his palace complex. However, nothing is known of these buildings: whatever remained by the time of Herod the Great (37–4 BC) was buried within the vast platform constructed for a new temple.

Herod's temple has also completely disappeared but the great platform, bounded by massive retaining walls, has survived as the most striking feature of the modern city.

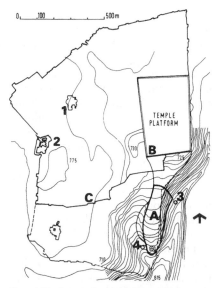

Figure 1.25 – Jerusalem: a general plan relating the site of the first settlement to the area of medieval Jerusalem within the line of Suleman the Magnificent's walls of AD 1538–41. Key: A, King David's city on original ridge top site; B, the Temple Platform with the Dome of The Rock; C, the first wall of the upper city; 1, Holy Sepulchre on Golgotha Hill; 2, the Citadel with David's Tower and David's Gate; 3, Gihon Spring; 4, Siloam Pool. See Moshe Safdie, *The Harvard Jerusalem Studio*, 1986; and Arieh Sharon, *Planning Jerusalem: the Old City and Its Environs*, 1973. See K. Kenyon, *Jerusalem: Excavating 3000 Years of History*, 1967; K J Asali (ed.) *Jerusalem in History*; D Bahot with C T Rubinstein, *The Illustrated Atlas of Jerusalem*, 1989, with excellent historical maps.

Figure 1.26 – Babylon: general plan of Nebuchadnezzar's city. The total area of about 90 acres was enclosed

67. Sir Mortimer Wheeler, *Civilisations of the Indus Valley*, 1966.
68. J Mellaart, *Çatal Hüyük*, 1967; see also I Hodder, 'An Interpretation of Çatal Hüyük and a Discussion of the Origins of Agriculture', *Bulletin No. 24*, 1987, Institute of Archaeology, University College London.
69. G Daniel, *The First Civilisations: The Archaeology of their Origins*, 1968.

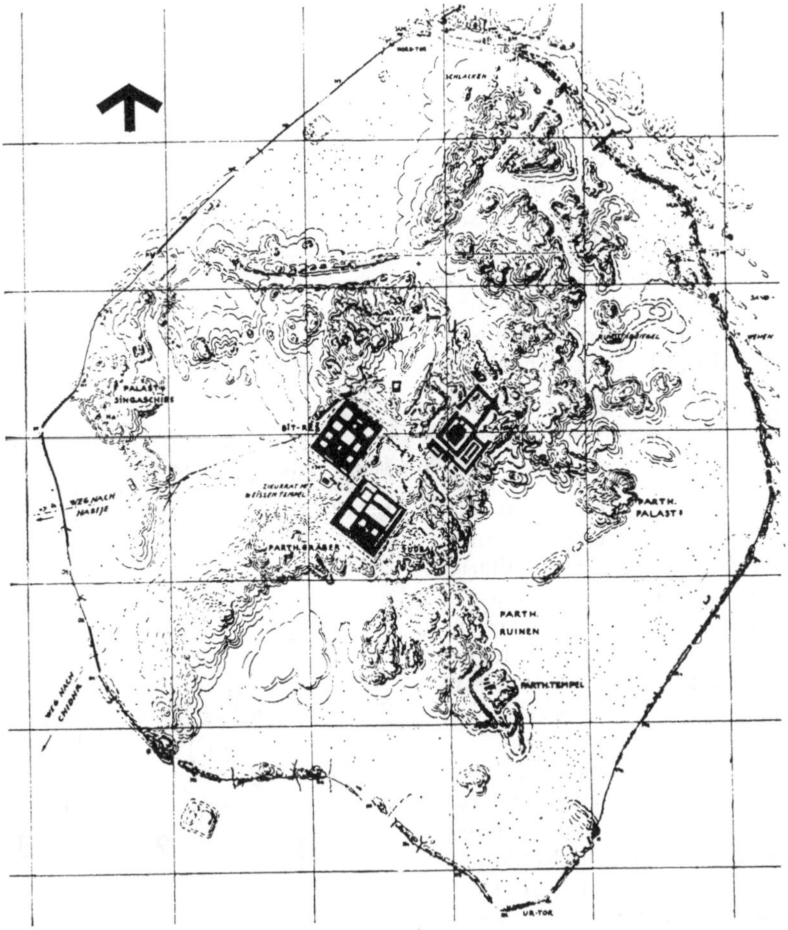

within a double wall. Greater Babylon was surrounded by an outer wall some 11 miles in length; estimates of the total population thus enclosed range as high as 500,000. Babylon was originally located on the left bank of the central arm of the old Euphrates at a junction of trading routes between the Persian Gulf and the Mediterranean. The city had an ancient history and was fought over on numerous occasions before being re-built for the last time under Essarhaddon from 680 BC along rectilinear supergrid lines. The archaeological record is unclear concerning whether or not the gridiron was used for the minor residential streets. The old city was on the east side of the stone-embanked Euphrates, connected across to the 'New Town' by a permanent bridge. As excavated the plan is essentially that of the city of Nebuchadnezzar, who reigned from 605 to 561 BC, succeeding shortly after the Babylonians had destroyed the Assyrian Empire. Following Nebuchad-nezzar's capture of Jerusalem in 587 BC, it was the city to which Johoakim, King of Judah, and thousands of his tribesmen were taken in exile.

Figure 1.27 – Uruk: general plan of the city showing the line of the third millennium BC wall and the location of the centre which was occupied by the Eanna temple complex. During the Uruk period (c. 3500 to 3000 BC) this consisted of the usual group of temples, palaces and administrative and storage buildings. The impress-ive Ziggurat of Ur Nammu dated from around 2100 BC. Uruk, known also as Warka, and the Erech of the Old Testament, was situated near the Euphrates some 60 miles upstream from Ur. It was the largest of the known Sumerian cities with an area of 1,235 acres within the third millennium BC ramparts. This defensive perimeter has been traced in its entirety and consisted of a double wall some 6 miles in length strengthened by nearly a thousand semi-circular bastions. Uruk flourished from about 3500 to 2300 BC.

Mesopotamian urban form

Illustrated by the city of Ur and the aerial photograph of Erbil (Arbela), this analysis of the form of Mesopotamian cities serves four interrelated purposes: first, it demonstrates, as an introductory model example, the roles played by determinants in the evolution of the urban form of a particular period in history; second, there is further more detailed description of the effect of climate on the creation of the courtyard house and the related urban route system; third, it serves to establish the roots of Islamic urban culture, in that 'the urban form and organ-isational system prevalent in most traditional cities within the Islamic world originated in pre-Islamic models: in particular the Mesopota-mian model'; and fourth, it introduces the author of that statement, Professor Besim S. Hakim, without whose invaluable reference work, *Arabic–Islamic Cities: Building and Planning Principles*, the Islamic content of this new edition could not have been possible in its present form.[70]

With the introduction to the Arabian Peninsular, which follows this section, a link forward is provided to Chapter 11's detailed considera-tion of Islamic urban form, thereby establishing a context for the inclu-sion in Chapter 4 of the Islamic cities of medieval southern Spain. (After this Islamic prelude, readers may prefer to proceed directly to Chapter 11, thereby maintaining continuity within Middle Eastern Islamic culture, before returning to Chapter 2 and Greek beginnings of the Western European Christian counterpart.)

The city stands on a broad plain and is an exact square, a hundred and twenty furlongs each way ... It is sur-rounded in the first place by a deep moat, full of water, behind which rises a wall fifty royal cubits in width and two hundred feet in height ... In the circuit of the wall there are a hundred gates, all of brass, with brazen lin-tels and sideposts ... The city is divided into two por-tions by the river [Euphrates] which runs through the middle of it ... The houses are mostly three and four stories high, the streets all run in straight lines, not only those parallel to the river, but also the cross streets which lead down to the waterside ... The centre of each division of the town is occupied by a fortress ... In the one stood the palace of the kings, surrounded by a wall of great strength and size; in the other was the sacred precinct of Jupiter Belus, a square enclosure, two fur-longs each way, with gates of solid brass. (In the middle of this precinct there was a ziggurat of eight platforms.) At the topmost tower (of the ziggurat) there is a spa-cious temple.

(Herodotus, Histories, I, 178–181. Trans. G. Rawlinson, Everyman's Library, London, 1910)

70. Hakim, *Arabic–Islamic Cities*.

HISTORY OF URBAN FORM

THE SUMERIAN CITY

Oppenheim informs us that 'the typical Sumerian city, and probably most of the later cities, consisted of three parts. First, the city proper – the walled area which contains the temple or temples, the palace with the residences of the royal officials, and the houses of the citizens. ... next came the "suburb", in which we find agglomerations of houses, farms, cattle-folds, fields and gardens, all of which provide the city with food and raw materials; and third, there was the harbour section which functioned beyond its actual use as a harbour, as the centre of commercial activity'. When the Old Testament speaks of the three days it took to cross the city of Nineveh (Jonah 3:3), he adds, 'the reference may be to the green reaches of the outer city'.[71]

THE CITY OF UR

In the absence of indications to the contrary, the city of Ur can be assumed to have originated either as one or a close grouping of agricultural settlements, or, as Oppenheim hypothesizes, it was when rural landed owners 'began to maintain "town houses" at nearby sanctuaries and eventually moved their main residence into the agglomeration of dwellings that grew up around the temple complex'.[72] Whatever the reason, the location would have been determined on marginally higher ground, secure from flooding yet near enough to the Euphrates to benefit from riverside trade as it developed, and the opportunity of intensive, irrigated agriculture.

Analysis of the city's subsequent form starts with topographical considerations which suggest that the 'off-centre' location of the *temenos*, assumed originally to have been in the middle, could have resulted from need for the residential city to expand away from the Euphrates, which flowed along the northern and western sides (Figure 1.29). There is a possible alternative cause in the characteristic Mesopotamian location of a temple or palace complex, 'in such a way that it straddles the circumvallation of the settlement';[73] thereby placing it off-centre from the beginning. The roles of fortification, mud-brick construction technology, and aggrandizement have already been described in the preceding general description of determinants.

COURTYARD HOUSING AND THE URBAN STREET SYSTEM

Several determinants played their parts in the formation of the residential areas of the Mesopotamian city, most importantly climate, the pre-urban cadastre, and urban mobility (Figure 1.30). 'The concept of a house planned around an open space or courtyard, appeared in the Middle East with the earliest cities there; [and] a prime example is the city of Ur,' confirms Hakim,[74] although the present writer prefers the concept of house rooms 'grouped' (or naturally arranged) around a courtyard, rather than planned as such, there being no evidence that those Sumerian houses were other than organic growth descendants of primitive village dwellings.[75] (However, when assimilated into Islamic culture, a consciously organized, if not formally 'planned' layout of rooms around the courtyard was usual.)

If water, as a fountain or pool, could be provided in the courtyard, then significant domestic micro-climatic improvements could be effected. Latterly and notably in parts of Arabia and Persia (Iran), natural 'air-conditioning' systems were evolved that not only achieved very worthwhile interior environmental improvements,[76] but also were

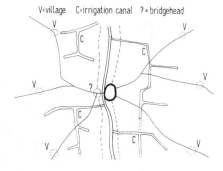

Figure 1.28 – Ur: conjectural regional 'plan' with the urban nucleus surrounded by the irrigated agricultural area with its 'village' settlements. Key: C, canals; V, villages. (For a comparable detailed study, see R McC Adams and H J Nissen, *The Uruk Countryside: The Natural Setting of Urban Societies*, 1972.)

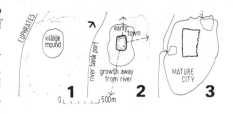

Figure 1.29 – Ur: analytical diagram illustrating possible growth stages of an original settlement on the east bank of the Euphrates.

71. Oppenheim, *Ancient Mesopotamia*; see also T C Young jr, 'Population Densities and Early Mesopotamian Urbanism', in P J Ucko, R Tringham and G W Dimbleby (eds) *Man, Settlement and Urbanism*, 1972.
72. Oppenheim, *Ancient Mesopotamia*; see also S A Kubba, *Mesopotamian Architecture and Town Planning 10,000–3,500 BC*, 1987, whose period is extended to include the Sumerian cities.
73. Oppenheim, *Ancient Mesopotamia*; see also H J Nissen, 'The City Wall of Uruk', in P J Ucko, R Tringham and G W Dimbleby (eds) *Man, Settlement and Urbanism*, 1972.
74. Hakim, *Arabic–Islamic Cities*; see also Woolley, *Antiquaries Journal*, vol. xi, no 4, October 1931; S Lloyd, *Foundations in the Dust*, 1947.
75. Woolley, *Ur of the Chaldees*.
76. On natural air-conditioning see E Beazley and M Harverson, *Living with the Desert: Working Buildings of the Iranian Plateau*, 1982; H Fathy, *Natural Energy and Vernacular Architecture: Principles and Examples with Reference to Hot Arid Climates*, 1986.

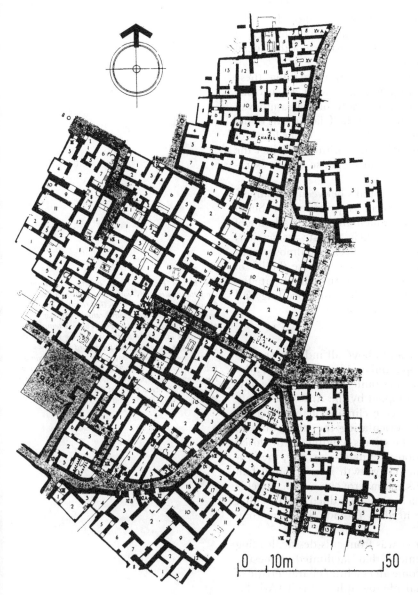

A ▣ courtyard

B chapel

Figure 1.30 – Ur: the detail plan of the housing area excavated by Leonard Woolley, located as D in Figure 1.7; and a diagram showing A, the organic growth pattern of the access lanes and the house courtyards, and B, the 'neighbourhood' chapels.

0   10m   50

responsible for some of history's most distinctive architectural forms, as illustrated in Chapter 11.

Room windows, such as they were, faced into the protected domestic-domain courtyard, with only the one doorway into the public domain. In addition to mitigating effects of solar excess, the courtyard also provided a measure of protection against windblown dust and sand. Early versions of white-washing provided a degree of protection for the mud-brick construction and reflected some of the heat of the sun. Although evolved in terms of defence against climatic assault, the courtyard house also enabled the men of a family, or clan, to defend their domain – in particular its vulnerable women and children – in face of human attack, should need arise. It is not known when protection of women became a dominant Mesopotamian social requirement:

Among the buildings we excavated were a number which could, with tolerable certainty, be identified as shops. These always had a narrow frontage on the street but ran back to a considerable depth; in front was a very small room, behind it a long store-room which might be simple or might be divided up into compartments by cross-walls. In No. 14 Paternoster Row, when the private house which had occupied the site was turned into a cook-shop, a proper window was cut through the front wall with its sill about a metre above street level. The actual cooking was done in the front room, as in the modern Near Eastern cook-shop, where the kitchen is always open to view. The standard shop is a small booth whose whole front is taken up by a 'window' opening and a door, and behind it a magazine in which goods could be stored. The front of the booth would be closed by a wooden shutter – of which of course no trace would survive, and most of its interior would be taken up by a raised wooden bench serving at once as a counter and as a seat for the shopkeeper; a few shelves or pegs against the wall would complete the furniture. The owner of No. 9 Church Street had a whole range of magazines attached to his house, and it is difficult to explain these otherwise than as store-rooms for merchandise, though there is nothing here in the way of a shop.

(Sir Leonard Woolley, Ur of the Chaldees, rev. ed. 1982)

arguably it was always so. Neither can it be known if (and when) the related emphasis on domestic privacy for the women and children of the family became established. Nevertheless the existence of this social determinant can be reasonably assumed.

Subsequently, as embodied in the pre-Islamic cities of Arabia, these interrelated defensive and privacy requirements were to become assimilated as an essential determinant of the traditional Islamic way of life, urban and rural. However, at the outset of the Islamic part of the book, we must not lose sight of the fact that there were other forms of multi-storey climate response housing in Islamic culture that met the privacy requirements, notably that evolved in response to the extra-special circumstances of the city of Makka.

Further confirmation of the direct cultural links between the Sumerian city and traditional Islamic urban form is provided by Seton Lloyd, who writes 'there is no need to say much about the daily life of the Sumerians at home, as illustrated by their private possessions which remained in the ruined houses, again because almost every domestic detail exactly resembled those which characterise the life of the Iraqi peasants today. They had the same domestic animals, ate the same food cooked in the same way, slept on the same bedding and bought their provisions in the same sort of market'.[77]

THE URBAN STREET SYSTEM

Writing of determinants of the narrow and above all indirect organic growth street patterns of the Mesopotamian city, Hassan Fathy observes that 'by simple analysis it becomes quite understandable how such a pattern came to be universally adopted by the Arabs. It is only natural for anybody experiencing the severe climate of the desert to seek shade by narrowing and properly orientating the streets and to avoid the hot desert winds by making the streets winding, with closed vistas'.[78] Hakim stresses that in addition to climate-response, the processes of land allocation and incremental growth over a pre-urban cadastre of irregularly shaped gardens, palm groves and small fields were also contributory factors, to which must be added the fact that hardly anywhere in Islam was there need to provide for other than minimal-width pack-animal traffic.[79]

Immediate access to individual houses was from culs-de-sac, leading from the thoroughfares; an arrangement that facilitated domestic-domain security and which was to become an essential characteristic of Islamic urban morphology. The aerial photograph of Erbil (Arbela) exemplifies the essential timelessness of the climatic and related determinants (see also Chapter 11).

By later Mesopotamian times there were exceptions, however, and there is evidence at Nineveh of a straight, open vista street, laid out in the late eighth century BC by the king, Sennacherib, as a *via triumphalis* from the main gate up to his palace. (It is recorded that 'to encroach upon the new avenue by building new houses was to be punished with death by impalement', the ultimate penalty for failing to comply with city planning regulations.)[80]

The Arabian Peninsular: an Islamic prelude

Extension, both geographical and historical, from Mesopotamia into the adjoining Arabian Peninsular completes this introductory link to

The Middle East has been baking in near-record temperatures recently with many places in Iraq, Iran, Saudi Arabia and the Persian Gulf reaching 45c. In the north, Basra recorded a temperature of almost 49c on Monday last, close to its record of 50c – a temperature that Kirkuk in Iraq climbed to later in the week. In the Gulf itself, Kuwait equalled the record max of 48c for July on two successive days – the average maximum being a relatively cool 39c. Unfortunately the high temperatures were preceded by blowing dust and sand, reducing visibility to 500 m at times.
(The Guardian, London, mid-July 1989)

77. Lloyd, *Foundations in the Dust.*
78. H Fathy, *Constancy, Transportation and Change in the Arab City*, 1973.
79. Besim Salim Hakim in correspondence with the author; see also R W Bulliet, *The Camel and the Wheel*, 1975, who observed that 'the street plan of the traditional Islamic city was a natural result of lack of wheeled vehicles'; and S Piggott, *The Earliest Wheeled Transport*, 1983.
80. Oppenheim, *Ancient Mesopotamia.*

Chapter 11's consideration of Islamic urban form; introducing, most importantly, the cities of Makkah and Medina – respectively the birthplace and refuge of the Prophet Muhammad – where Islam originated in the seventh century AD.[81]

THE CLIMATIC CONTEXT

The Arabian Peninsular is joined to Mesopotamia (Iraq) on its northern side across waterless sand-dune desert, and it is bounded on its three other sides by the sea (Figure 1.31). It slopes from west to east, with the watershed close to the Red Sea at distances varying from only 40 km (25 miles) in the north to 128 km (80 miles) down south near the Gulf of Aden. Even in the south, where there is annual monsoon rainfall of 30–50 cm (12–20 inches), there are no permanent rivers or bodies of water. Elsewhere there is less than 10 cm (4 inches) annual rainfall and over much of the central and northern desert there can be none at all for several years.[82] In hot–dry desert regions, oases supplied by underground water sources are few and far between. Daytime temperatures on the Central Plateau can reach 120 °F (49 °C), and on higher ground night-time temperatures well below freezing are experienced. Windblown dust and sand, often of extreme severity, is another climatic characteristic. In hot–humid coastal regions summer temperatures can reach 42 °C with all-year average relative humidity of 75–80 per cent.

THE ARABS

'In prehistoric and pagan times,' writes Mansfield, 'two races inhabited Arabia: one was largely nomadic and wandered with their flocks over the great [northern] deserts which lie between the river Euphrates and the centre of the peninsular. The others were the inhabitants of the rain-fed uplands in the south – the Yemen. It was the former who were the "Arabs", with the first recorded use of their name in 853 BC, on the inscription of the Assyrian king Shalmaneser III.'[83] Subsequently, by the time of Herodotus and later Greek and Roman writers, the terms 'Arabia' and 'Arab' had come to refer to the entire peninsular.

Mansfield provides an evocative description of the northern Arab way of life, emphasizing 'the great horizons of the desert provided a sense of liberty [yet] the harsh environment made its own iron laws which moulded the structure of tribal life. Survival depended on the solidarity and self-protection of the tribe'.[84] The Arabs relied greatly on the date palm: 'the mother and the Aunt of the Arabs', as traditionally revered. The date palm was one of the earliest domesticated plants, indeed no wild growing species has been discovered, according to Oppenheim, probably because 'it requires the services of an horticulturalist in pollination if a substantial crop of dates is to be harvested. Its fruits can be easily preserved and represents an essential source of the calories needed in the diet of a working population'.[85] Not the least of its attributes is that it has a high tolerance to brackish water. Otherwise the Arabs lived 'as parasites' on the camel, which they had domesticated between 1500 and 1200 BC, 'drinking its milk and very occasionally eating its meat'. The camel was their primary means of transport and their tent fabric was woven from its hair, with that of the goat.[86] (The traditional Bedouin tent and the way of life it housed is described in Chapter 11.)

Figure 1.31 – The Arabian Peninsular: showing the main geographical features and locating key cities.

The area where the date-palm was of prime importance, and where irrigation could have been used earliest to 'bathe its feet in water', was the whole region between Euphrates and Nile. I am inclined to believe that the first cultivation was nearer to the Euphrates than to the Nile. The fact that King David (c. 1000 BC) did not include the palm in the list of trees to be planted in his garden is not significant: only at Jericho did the date-palm flourish in Palestine. The earliest paintings on Egyptian monuments include representations of the date-palm, but Mesopotamian plantation agriculture was at least as early as, and probably earlier than, the Egyptian. Whether the first cultivators of the palm were Egyptian or Asian, it cannot be much less than 8,000 years since that cultivation began, and the date-palm is doubtless as ancient a partner as the grape-vine in man's business of making himself comfortable on this planet.
(E Hyams, Plants in the Service of Man, 1971)

81. For Makkah and Medina see P Crone, *Meccan Trade and the Rise of Islam*, 1987; N Groom, *Frankincense and Myrrh: A Study of the Arabian Incense Trade*, 1981; M Bloch, 'The Social Influence of Salt', *Scientific American*, vol 209, 1963; E Essin, *Mecca the Blessed: Medina the Radiant*, 1963; M S Makki, *Medina, Saudi Arabia*, 1982; see also the main Chapter 11 references.
82. For the climate of the Arabian Peninsular see K Talib, *Shelter in Saudi Arabia*, 1984; *The Times Atlas of the World*, 1990.
83. P Mansfield, *The Arabs*, 1976/1980; see also A Hourani, *A History of the Arab Peoples*, 1991.
84. Mansfield, *The Arabs*.
85. Oppenheim, *Ancient Mesopotamia*.
86. Mansfield, *The Arabs*; see also I L Mason, 'Camels', in Mason (ed.) *Evolution of Domesticated Animals*, 1984; T Faegre, *Tents: Architecture of the Nomads*, 1979.

HISTORY OF URBAN FORM

MAKKAH AND MEDINA

Contrary to possible preconception, not all of the northern Arabs were nomads. There were tribes that came to prefer a settled agricultural existence in one of the village oases, or that of merchants in one of the exceedingly few and far between cities. The nomads needed market-place opportunity to trade their animals for basic commodities, notably salt, and there was thriving, comparatively safe overland caravan trade, through from ports on the Indian Ocean to ready markets in Egypt and Syria. Frankincense and myrrh were particularly prized items.

The city of Makkah, in the eastern mountain foothills about 60 miles from the Red Sea, was one of the oldest and probably the most prosperous of the trading centres when, in AD 571, the Prophet of Islam, Muhammad ibn (son of) Addullah, was born into the Hashim clan of the leading Quraysh tribe.[87] Makkah's potable water supply came only from wells and was not adequate to support oasis agriculture. Medina, then known as Yathrib, some 300 miles north-east, was by contrast a sizeable agricultural oasis city, or (according to some authorities) more like a close grouping of proto-urban village settlements.

Before Islam, the religion of the Arabs was a polytheism based on belief in a variety of spirits, often embodied in a particular rock or tree shrine. The most famous of these was the shrine of the *Ka'aba* at Makkah, where the great Black Stone was an established centre of pilgrimage long before it became central to Islam (see Chapter 11).[88]

THE ISLAMIC PRELUDE: AN INTERIM CONCLUSION

The main description of Islamic urban form is presented in Chapter 11, which brings this pre-industrial international history of urban form through to the mid-twentieth century. Islam, however, has not been confined to the broader 'Middle East', as defined for our purpose in Figure 11.1. In Europe its hold over the southern two-thirds of Spain endured for upwards of 800 years, while Istanbul (Constantinople) remains to this day the exceptional Muslim city of European Turkey.

Departing from a chronological sequence of chapters by including the Spanish Islamic cities in Chapter 4 (before Chapter 11's analysis of Islamic urban form) enables them to be described side by side with their Western European Christian contemporaries. As Hakim explains, 'Islam emerged from the heartland of Arabia, which had strong pre-Islamic Arab traditions in various spheres of life, including building practice.' He adds 'while many of the traditions that were not compatible with Islamic values were prohibited ... others, with modifications, became part of Islamic civilisation. Some of these influences could be traced back to earlier Semitic and Arab civilisations, particularly in the region of Mesopotamia'.[89] With the exception of the introduction of the mosque, other religious function buildings, and the main *Suq* (covered or semi-covered market) around the primary mosque of the city, Islam in effect served to perpetuate the essential morphology of the ancient Mesopotamian city, through a system of decision-making governed by semi-legislative urban guidelines.

Egypt

Although superficially comparable with Mesopotamia, in that both countries contained great rivers flowing through immensely fertile valleys and plains which offered parallel opportunities to early humans,

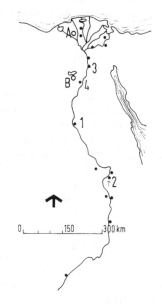

Figure 1.32 – Urban centres in Egypt. Key: 1, Akhenaten (Tel-el-Amarna); 2, Thebes (Luxor); 3, Memphis (southern Cairo); 4, Kahun. (A, Neolithic villages at Merimde; B, at Fayum.) See also D O'Connor, *The Geography of Settlement in Ancient Egypt*; P J Ucko, R Tringham and G W Dimbleby (eds) *Man, Settlement and Urbanism*, 1972.

87. For the Prophet Muhammad see Chapter 11 references.
88. See G R D King, *The Historical Mosques of Saudi Arabia*, 1986, for detailed description, and photographs of the *Ka'aba* in 1884–5.
89. Hakim, *Arabic–Islamic Cities*.

the evolution of urban settlements in Egypt took place along markedly different lines.[90] Writing in the early 1960s, Jacquetta Hawkes and Sir Leonard Woolley stated: 'nothing could be more unlike the mosaic of city states that divided between them the valleys of the Euphrates and the Tigris, than the unified kingdom of Egypt, in which the city was really non-existent'.[91] Leaving aside, for the moment, questions raised by what is meant by the Egyptian 'city', the conventional wisdom expressed in that influential statement is emphasized here because subsequent archaeological findings and reinterpretation of the ancient Egyptian record, indicate not only that there were 'cities' at various sites along the Nile Valley, but also that there was early-date 'town planning' of a previously unacknowledged nature.[92] Together, these related reassessments represent the first major change of opinion in this third edition.

Egypt, in effect the Nile Valley and Delta, was a unified state from about 3100 BC, with the river its centralizing factor. Although in some respects a later civilization than the Mesopotamian, nevertheless it is clear that only a comparably advanced society could have organized and carried through the immensely demanding early third millennium monumental construction programme – of which the Great Pyramid of Cheops at Gizeh, dated from about 2600 BC, is arguably the greatest wonder of all time.[93] Archaeological evidence confirms that the Egyptian cities of this period included Memphis, known to have been founded in about 3100 BC by Egypt's first historic king, Menes.[94] However, reasons why they took markedly different forms also account for the scarcity of urban remains.

The principal, arguably determining reason for this is the internal peace which existed in Egypt from earliest times; there was no economic necessity, as in Mesopotamia, continually to occupy the same site in order to take advantage of the enormous capital investment represented by the defensive walls. A second reason directly relates to the first – given urban mobility each successive Pharaoh was free to spend his reigning life on earth preparing his tomb for the life after death (the basis of Egyptian religion) in a different location from that of his predecessor.

A further related reason for the paucity of urban remains, as compared with the many surviving religious buildings, is that almost all of the resources of the building industry, together with all the durable materials, were made available for the process of tomb and temple construction. The Egyptian urban areas were built of mud-brick, as in Mesopotamia, but failing the creation of a recognizable *tell*, resulting from long-term site occupation, there is no way of locating the ancient cities, even if worthwhile remains could have survived unprotected by later layers of buildings. As Henri Frankfort aptly explains, 'each Pharaoh took up residence near the site chosen for his tomb, where, during the best part of his lifetime, the work on the pyramid and temple was carried out whilst government was based on the nearest town. After the death of the Pharaoh the place was abandoned to the priests, who maintained his cult and managed his mortuary estate unless the successor also decided to build his tomb in the area'.[95]

So as not to delay the mortuary work, city building under the Pharaohs was generally a quick one-stage process. This is illustrated by the still only partially excavated ancient Egyptian city of Tel-el-Amarna. This settlement, situated about halfway between Cairo and Luxor, was

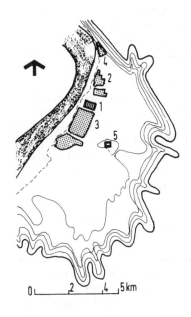

Figure 1.33 – Layout of Akhenaten (Tel-el-Amarna). Key: 1, central area; 2, north suburb; 3, south city; 4, customs house; 5, workers village (Figure 1.35).

90. For Egypt generally see W J Murnane, *The Penguin Guide to Ancient Egypt*, 1983; J-L de Cenival, *Egyptian Architecture*, 1964; M Brawer, *Atlas of the Middle East*, 1988; P A Clayton, *The Rediscovery of Ancient Egypt: Artists and Travellers in the Nineteenth Century*, 1982; M A Murray, *The Splendour that was Egypt*, 1977; P Johnson, *The Civilisation of Ancient Egypt*, 1978.
91. J Hawkes and L Woolley, *Prehistory and the Beginnings of Civilisation*, 1963. These two eminent British archaeologists were not alone in misrepresenting Ancient Egypt in this respect; see also an American counterpart, J Wilson, 'Egypt through the New Kingdom: Civilization without Cities', in C Kraeling and R McC Adams (eds) *City Invincible*, 1960.
92. See B J Kemp, 'The Early Development of Towns in Egypt', *Antiquity* 51, 1977; and D O'Connor, 'The Geography of Settlement in Ancient Egypt', in P J Ucko, R Tringham and G W Dimbleby (eds) *Man, Settlement and Urbanism*, 1972, who confirms: 'the definition once given of Egypt as a "civilisation without cities" can only be accepted if "city" is understood in a most narrow and specialised sense'.
93. For the Great Pyramid of Cheops, the only survivor of the seven great wonders of the ancient world, see I E S Edwards, *The Pyramids of Egypt*, 1947/1972; R David, *The Pyramid Builders of Ancient Egypt*, 1986: Ch 4, 'The Towns of the Royal Workmen').
94. B J Kemp, 'Temple and Town in Ancient Egypt', in P J Ucko, R Tringham and G W Dimbleby (eds) *Man, Settlement and Urbanism*, 1972.
95. H Frankfort, *The Birth of Civilization in the Near East*, 1951.

occupied for the space of only forty years. The city was built on the eastern bank of the Nile 'at a spot where the cliffs recede to form a huge semi-circle some seven miles long by two and a half to three miles wide'.[96] The reason for starting the new city was that the Pharaoh Akhenaten found it difficult to institute religious reforms in the existing capital of Thebes and moved down river to the new site. Two years after his death, in 1356 BC, his successor returned to Thebes and the old faith. Amarna was abandoned and never occupied again.

EL-'AMARNA

Reassessment of the ancient Egyptian city of el-'Amarna is based on Kemp's paper 'Temple and Town in Ancient Egypt'.[97] The plan of the city shows a form of linear development alongside the east bank of the Nile, with three main routes parallel to the river connecting the various parts. The overall length is about 7 km, with a width in from the bank varying from about 800 metres to 1,500 metres. The general form (shown by Figure 1.33) is that of a central area with north and south Suburbs and, of significance disproportionate to its size, a planned necropolis workmen's village located about 1,000 m to the east. Although the city was unfortified, there were walls, probably symbolical, enclosing the smaller central temple and a major northern palace. Akhenaten's new sun temples and their administrative and storage buildings were in the centre, with nearby palaces, government offices, and barracks (Figure 1.34).

'Since many buildings had a rectangular plan of their own,' writes Kemp, 'a certain regularity of overall layout was perhaps inevitable, though the degree of real planning seems to have been small. Palaces and temples were built to a common street frontage and intervening buildings took their alignment from them, but further back alignment seems to have been dictated mostly by the incoming streets from the South Suburb, which arrived at an angle.'[98] Within the housing suburbs Professor Fairman states: 'there were no defined blocks of *insulae*, no standardised sizes of estates. What appears to have happened is that the wealthiest people selected their own house sites, and built along the main streets, to whose line in general they adhered. Less wealthy people then built in vacant spaces behind the houses of the rich and finally the houses of the poor were squeezed in, with little attempt at order, wherever space could be found. The houses of all types were found in a single quarter, and though there are slum areas it is evident that there was no zoning'.[99]

Kemp concluded that 'the combination of these organic neighbourhood units with a system of wide thoroughfares leading to the city centre represents a probably very effective and acceptable type of urban layout'.[100] On which basis it is probable – because we cannot be certain – that el-'Amarna was an early example of that kind of partial urban planning whereby the 'authorities' laid out a main route structure and left the infilling to the occupants, thereby establishing an intriguing link between el-'Amarna and the Harappan cities.[101]

On the other hand there can be no doubting the planned intention underlying the form of the workmen's village located in the desert fringe about 1,000 metres to the east (Figures 1.33 and 1.35). This rigidly rectilinear plan is variously misrepresented in urban histories as an early example of the gridiron. By our definition, however, because it is neither a separate town, nor even a contiguous part of el-'Amarna, it

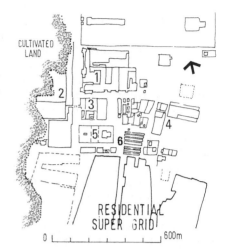

Figure 1.34 – El-'Amarna: the central area. Key: 1, larger Aten sun temple; 2, 'The Great Palace'; 3, king's residence in the central area; 4, police and military barracks; 5, smaller Aten sun temple; 6, central employees' village; 7, the south suburb planned, probably, as a main routes 'supergrid'.

Figure 1.35 – Detail arrangement of the workers' village at Tel-el Amarna. Sir Leonard Woolley, who excavated this city, wrote, 'We dug out a model village put up for the labourers who excavated the rock-cut-out tombs made in the desert hills. A square walled enclosure was entirely filled with rows of small houses divided by narrow streets; except for the foreman's quarters near the gate they were all monotonously alike, each with its kitchen-parlour in front, its bedrooms and cupboard behind, the very pattern of mechanically devised industrial dwellings' (*Digging up the Past*). As defined above,

96. H W Fairman, 'Town Building in Pharaonic Egypt', *Town Planning Review*, April 1949; for el-'Amarna see also B J Kemp, 'Temple and Town in Ancient Egypt', in Ucko et al. (eds) *Man, Settlement and Urbanism*.
97, 98. Kemp, 'Temple and Town in Ancient Egypt'.
99. Fairman, 'Town Building in Pharaonic Egypt'.
100. Kemp, 'Temple and Town in Ancient Egypt'.
101. And, in direct comparability, the modern 'supergrid' cities exemplified by Milton Keynes.

is only an example, albeit most important, of the rectilinear layout of a contractors encampment.[102]

KAHUN

Whereas third edition reassessment of el-'Amarna is that of degree only, the significance of Kahun in urban history has had to be radically revised in the light of Kemp's work.[103] Instead of being merely another version of an orthogonally planned construction camp, laid out for workers engaged in building the pyramid of Amenemhat III – as previously believed[104] – Kahun must now be regarded not only as deserving of urban status, but also as the earliest known example of the gridiron system of town planning, albeit applied for the specifically Egyptian type of temple-town. An historical context for Kahun is provided by Kemp who distinguishes between two different situations whereby the Ancient Egyptian temple was either 'a new foundation on a hitherto non-urban site and where it was to remain the principal or even sole reason for a settlement's existence'; or where the temple was 'one part of a community which had other independent bases for its economic life'.[105] Kahun exemplifies the first category and el-'Amarna is of the second.

the rectilinear village housing layout is no longer to be accorded status as an early instance of urban gridiron planning. In previous editions I wrote 'Use of the gridiron for only a small and relatively insignificant part of Tel-el-Amarna would seem to be a clear example of fourteenth century BC political expediency – alternatively expressed as town planning being the art of the practical. In this way it is possible to resolve the apparent anomaly whereby the main urban area of the city was allowed to develop along *laissez-faire* organic growth lines – even though the value of the gridiron in laying out a new town was understood. The implementation of any town plan implies political control, either autocratic or democratic, to ensure that the inhabitants conform to its requirements. Clearly it was possible to impose a plan on the workmen; unfortunately, we may never know whether or not Akhenaten would have preferred to impose a similar planning control on his wealthy powerful relatives and his political and religious officials.'

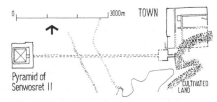

Figure 1.36 – Kahun: the general plan relating the town to the Pyramid of Senwosret II situated 4,500 m to the west (after Kemp, 'Temple and Town in Ancient Egypt').

Figure 1.37 – Kahun: the gridiron plan of the town as partially excavated, at the same orientation as Figure 1.36. Key: A, the 'acropolis'; B, the town temple (presumed). At about 1,300 m square, the town of Kahun is larger than might be imagined from the plan, and it is size, compared with the smaller el-'Amarna village, combined with the presence of religious and administrative central functions, that justifies urban status. 'The town was quite separate from the large lower mortuary temple ... and the preserved area seems to be almost exclusively occupied by houses. Of this a good half is taken up with the small often back-to-back houses of the ordinary inhabitants. The remainder consists largely of nine much larger house units on either side of a single street. The history of Kahun,' Kemp concludes, 'illustrates the ultimate total dependence of the town on the mortuary temple' (Kemp, 'Temple and Town in Ancient Egypt').

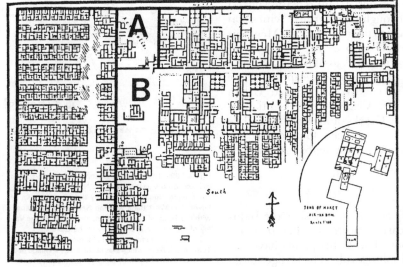

Kahun was constructed about 1835 BC in three parts (shown by Figure 1.36) comprising the large mortuary temple and pyramid of King Senwosret II, and the planned gridiron town which initially accommodated the construction workers, before being taken over by the priests and officials responsible for their administration. 'Given the need to build a settlement paid for from a single source in one stage,' observes Kemp, 'it is hard to see what alternative to a gridiron could have existed.'[106] Although the mortuary temple was its primary economic *raison d'être*, there is evidence that a proportion of the inhabitants of Kahun were otherwise employed in agriculture, and, on one occasion, on building work for King Amenemhat III, possibly, suggests Kemp, for a part of his pyramid complex.[107] Kahun existed for about 150 years until it was abandoned when disestablishment of the mortuary temple removed its economic basis. As a special-function town Kahun was essentially no different from the numerous medieval European

102. The Egyptian archaeological record includes numerous other such short-lived encampment villages, including several near the main pyramid group at Giza, but their 'planning' is not at urban status.
103. Kemp, 'Temple and Town in Ancient Egypt', and 'The Early Development of Towns in Egypt'; see also R David, *The Pyramid Builders of Ancient Egypt*, 1986: ch. 4, 'The Towns of the Royal Workmen'.
104. A E J Morris, *History of Urban Form*, 1st and 2nd editions: 'Neither el-'Amarna nor Kahun amounts to more than contractors' hutments.'
105. Kemp, 'Temple and Town in Ancient Egypt'.
106. Kemp, 'Temple and Town in Ancient Egypt'. My comparable terminology used in this book is 'expediency gridiron' planning.
107. Kemp, 'Temple and Town in Ancient Egypt'; it is this alternative construction industry employment that misled earlier historians.

counterparts that were founded, or which grew up to serve monasteries and other religious establishments, or Delphi, foremost among Greek shrines.

Only about half of the town had survived when first excavated around the beginning of this century, mainly by Sir Flinders Petrie.[108] It was approximately square, of walled sides about 1,300 metres, as shown in Figure 1.37. Petrie wrote of the interior: 'each street was of a uniform type of house; there were no gardens but each house no matter how small, had its own open court just as the presentday Egyptian houses have. The ordinary workman has at least three rooms, plus the courtyard, and the other houses – depending on the status of the occupants had four, five and six rooms, with some of the larger houses being on two floors'.[109] The mortuary temple itself was not a place of popular worship and there is documentary evidence showing that the town had at least one temple of its own, with a part-time priesthood.

MEMPHIS AND THEBES

'The most important towns of Egypt,' writes O'Connor, 'were the national capitals or administrative centres, of which only the aberrantly located Akhenaten (Tell el-'Amarna) can be studied in detail.'[110] Memphis and Thebes were the respective capitals of the northern and southern parts of Egypt, and 'their histories and extant remains suggest that at times both were probably large enough to be termed "cities", at least when compared with the other chief towns'.[111] Both Memphis and Thebes are included in a paper by Kemp of 1977;[112] they are described in the respective captions to Figures 1.38 and 1.39. Next in importance at intervals along the Nile, there were the capitals of the provinces (or *nomes*) into which the unified country was divided. Urban settlement in the Delta region came later. Alexandria is essentially a Phoenician colony consolidated by the Greeks and Romans, while Cairo is a comparatively recent city having been founded by the Muslim army at the beginning of its drive west across Africa (see Chapter 11).

India: Harappan cities

The Indian subcontinent's two major river systems, those of the Indus and Ganges basins, in the west and east respectively (Figure 1.40), have markedly different climatic regions. The Indus plains have low rainfall but, when irrigated, the alluvial soil is fertile. The Ganges basin is much wetter, with over 80 inches of rain per annum in Bengal. Archaeological records indicate that urban settlement began in the western Indus basin, where conditions would have been much the same as in the Tigris/Euphrates region, and later spread across to the Ganges.[113] Annual natural inundation, probably with simple irrigation control, allowed relatively large communities to become established on the Indus plains by the third millennium BC, but, in contrast, the rich damp soils of the Ganges basin 'must originally have consisted of forests and marshes, which would have needed a considerable labour force, equipped with effective tools, to bring them under cultivation'.[114] Bridget and Raymond Allchin give the sixth century BC, at the earliest, for the origins of urban settlement along the Ganges, observing that the situation is probably analogous to that found in many parts of Europe where heavier and richer soils could not be, or were not, utilized until well into the Iron Age.[115]

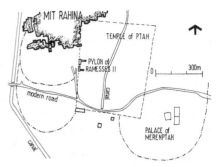

Figure 1.38 – Memphis; located about 21 km south of the medieval Islamic nucleus of modern Cairo (see Figure 11.45), the extensive ancient civil and administrative royal city, with its numerous temples and palaces, was in the fertile zone on the west bank of the Nile, with the pyramid complex of Sakkara, which includes the step pyramid of Djoser, on the edge of the desert 4 km to the northwest. Little is known of Memphis, part of which lies beneath the modern village of Mit Rahina. (After Kemp, *The Early Development of Towns in Egypt.*)

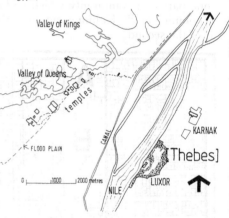

Figure 1.39 – Thebes; the civil and administrative royal city was on the east bank of the Nile, where it has become increasingly built over by the modern town of Luxor. The West Bank was the Necropolis of Thebes, with the Tomb of Tutankhamun, and others, in the Valley of the Kings and the Valley of the Queens.

The two outstanding east bank monuments are the Temple of Luxor – which was occupied by a village until the end of the nineteenth century, the mosque of which still exists; and the temple complex at Karnak about 2.5 km north of modern Luxor, with the superb Great Temple of Amun (c. 1991–c. 785 BC).

108, 109. W M Flinders Petrie, *Some Sources of Human History*, 1922; see also M S Drower, *Flinders Petrie: A Life in Archaeology*, 1985.

110, 111. O'Connor, 'The Geography of Settlement in Ancient Egypt', in Ucko et al (eds) *Man, Settlement and Urbanism.*

112. Kemp, 'The Early Development of Towns in Egypt'.

113. For the climate of the Indus Valley see T M L Wrigley, M J Ingram and G Farmer, *Climate and History: Studies in Past Climates and their Impact on Man*, 1985.

114. See G L Possehl (ed), *Ancient Cities of the Indus*, 1979.

115. B Allchin and R Allchin, *Birth of Indian Civilization: India and Pakistan before 500 BC*, 1968.

Development of settlement in the Indus basin seems to have paralleled that in Mesopotamia, with Neolithic farming communities establishing villages on the higher plains away from the actual river courses during the fifth millennium BC, before becoming sufficiently well organized socially and technically to take on the challenge of farming the flood-plains. The civilization which produced the principal known urban centres is the Harappan, named after Harappa, one of its two most important examples. Harappa itself existed between 2150 and 1750 BC, allowing a tolerance of 100 years at the beginning and the Allchins quote radio-carbon dates obtained by G F Dales[116] from Mohenjo-daro (the other most important Harappan city) which give the mature Harappan civilization a span of 2154–1864 BC. Although the origins of this culture are only vaguely known, it is clear that around 1750 BC it came to a sudden end. Archaeologists are divided as to whether the cause was natural – possibly extreme long-term flooding, or desiccation – or of human origin, resulting from the Aryan invasion of India.

Figure 1.40 – Outline map of the Indus valley locating the (assumed) three main cities of the Harappan civilization and minor urban settlements. Running parallel to the Indus, to the south-east, is the dried-up bed of the river Sutlej (Ghaggar). The small insert map locates this area within the Indian sub-continent.

THE ORIGINS OF TOWN PLANNING

Before concentrating on Mohenjo-daro as the most important Harappan city example – Harappa, Kalibangan and Lothal are discussed more briefly later in the section – the remarkable consistenty of Harappan urban form and the intriguing questions this poses must first be considered. There is a basic form to these three examples and others of lesser significance. Each has an imposing citadel sited to the west of, and completely separate from, the main 'lower city' urban area. These citadels are raised on high mud-brick platforms and are surrounded by massive walls, probably as much for protection against river flooding as against military assault. Although the lower cities were similarly protected, the higher stronger citadels provided refuge when, as excavations reveal, they suffered periodic inundation.

The isolation of citadel from lower city, and the assumptions which have to be made concerning its function, present problems which archaeologists have yet to answer. If Harappan cities had developed along the same organic growth lines as in Mesopotamia we should expect to find the citadel *within* the urban area, most probably at its highest levels. For the citadel to be located in one instance *outside* the main urban area could be regarded as an exception to this rule – the result of some chance determinant. However, this cannot be the case with the Harappan cities, separated as they are by hundreds of miles and with the major examples all following the same basic plan form. We must look for reasons other than chance.

THE FIRST PLANNED TOWNS

The momentus answer to that question brings us to the earliest known planned towns in history. All the Harappan cities must have been laid out according to the same 'system of town planning', a conclusion reinforced by similarities in the layout of their lower cities on a rectilinear basis of main east–west routes directed to the citadels, and north–south cross routes.

Hitherto, in support of the belief that the Harappan cities were the earliest known instances of gridiron town planning, I had suggested 'irregularities in the plan of the Mohenjo-daro lower city (Figure 1.43)

More than one archaeologist writing about these early cities in the Indus valley has commented upon the dreariness, the repetitiveness and the dead-level nature of the various objects produced by the Indus craftsmen. Looking at the way in which the cities were planned, with their rows of mean huts for the workers, grouped close to furnaces for metal production or to pottery kilns, one has the uncomfortable feeling that here were city states in which production was ruthlessly organised but in which the techniques employed were probably not very efficient. One gets, in fact, the feeling that here again the dead hand of the civil servant was in operation, such as was surmised during the declining years of Rome.
(Hodges, Technology in the Ancient World)

On the other hand the general configuration of the Indus civilization was quite distinct. The leading characteristics betrayed by the excavated remains include a high degree of civic discipline and organization uniformity over an extensive area and stability over long periods of time. The rectangular grid of the street plan differing so notably from the cramped irregularity prevalent in ancient Mesopotamia, the elaborate system of sewers and rubbish shoots, the carefully maintained walls, the great communal granaries and the standardized systems of weights and measures all reflect an ordered and settled society. What was the basis of the discipline maintained over so long a period of time? There is no indication that military authority was at all prominent: weapons were relatively inconspicuous and defences were apparently confined to the citadels, a reflection no doubt of the comparative isolation of the Indus people from neighbours at a comparable level of technology. Nor has any obvious sign of royal authority been found, whether in the form of palaces, outstanding tombs or regalia. A likely alternative is that the sanctions behind this ordered way of life were religious and that the Indus Valley civilization, like that of modern Tibet, was essentially theocratic.
(G Clark, World Prehistory – An Outline)

116. Allchin and Allchin, *Birth of Indian Civilization*; see also Sir Mortimer Wheeler, *Civilisation of the Indus and Beyond*, 1966.

are the results of continual rebuilding, over several centuries, gradually modifying the street lines, with periodic flooding necessitating more or less complete new starts'.[117] In the more recent light of archaeologist interpretations of el-'Amarna, and Islamic new city 'planning', I now think it more likely that the 'irregularities' resulted from only those main cross-axes having been determined on the ground, leaving the arrangement of the 'super-blocks' to their occupants. Inevitably their rectilinear construction layouts conformed more or less with the axial grid. As established in the preceding section introducing the gridiron, this reassessment confirms Mohenjo-daro and its contemporaries as the earliest known planned cities, but not to the extent of also having the first gridiron plans.[118]

Caution, however, is still necessary in awarding the title of 'first planned towns in history'. Until the full import of Harappan excavations was appreciated, urban historians believed that it was a Greek architect-planner, Hippodamus of Miletus, who was the first to organize formally a complete new urban entity, when he drew up his famous plan for the rebuilding of Miletus from 479 BC. On the basis of the available evidence, it now seems that it was not a Greek who first consciously put together the component parts of a city in a predetermined relationship. If it was any one person, it was more than likely to have been an anonymous Harappan priest of as yet unknown date.

As has now happened to claims that the Greeks originated the art (or science) of town planning, such supposition could well be invalidated by archaeologists proving theories that urban civilization was introduced into the Indus basin by an already advanced people. The Allchins suggest that if the Harappan civilization came into existence only around 2150 BC, it is necessary to admit that not only the end of the cities, but also their initial impetus, may have been due to the Indo-European speaking peoples.[119] If this is indeed the case then we have the answer to the remaining problem of the consistency of Harappan urban form. The cities are variants of a standard plan, already developed by newcomers to the Indus basin, who established their urban culture in much the same way as the Romans did throughout their Empire and as Europeans later did in their 'new world' colonies. Where, when, and by whom were the earlier versions of the Harappan cities built?[120]

MOHENJO-DARO

The best documented centre of the Harappan civilization is Mohenjo-daro, located on the right bank of the Indus some 3 miles from the present river course. Little is known of the early history of the city. Continuing alluvial deposition has, according to the Allchins, 'raised the whole land surface in this area more than 30 feet since Harappan times, and as the water table has risen correspondingly archaeologists have so far been unable to plumb the lower levels of this vast site'.

The citadel mound at Mohenjo-daro was raised above the level of the flood-plain and surrounded by a burnt brick embankment some 43 feet in height. It would seem to have included neither the palace of an absolute ruler nor a dominating religious symbol – like the ziggurat at Ur – but rather a number of buildings for various civic-religious purposes.[121] These include granaries for the protected storage of the food surplus; what are taken to be administrative offices, with possibly a large assembly hall; and, the most intriguing find of all, the Great

Figure 1.41 – Mohenjo-daro: general layout relating the citadel to the west of the lower city, with contours at 10 metre intervals on an otherwise flat plain location. The present course of the Indus is some 5 km to the west. The plan of the lower city is to be regarded as that of a masterplan 'supergrid', rather than that of a gridiron *per se*.

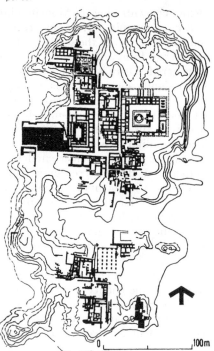

Figure 1.42 – Mohenjo-daro: detail plan of the citadel mound, as excavated. The precise rectilinear planning of this impressive architectural ensemble is closely comparable with that of the temenos precinct at Ur of the Chaldees (Figure 1.8).

117. A E J Morris, *History of Urban Form*, 1st and 2nd editions.
118. See Chapter 2 for the Greek gridiron city plans.
119. Allchin and Allchin, *Birth of Indian Civilization*.
120. At time of writing (February 1992) archaeologists have not yet found an answer.
121. Allchin and Allchin, *Birth of Indian Civilization*.

Bath – 8 feet deep and 39 by 23 feet on plan. The bath was surrounded by a portico and other rooms on more than one floor; bitumen was used as a waterproofing membrane between the outer and foundation brick layers, with water supplied from an adjoining well and overflowing through a corbelled drain. 'The significance of the extraordinary structure can only be guessed. It has been generally agreed that it must be linked with some sort of ritual bathing such as has played so important a part in later Indian life.'[122] Consideration of possible purposes for the citadels must not lose sight of their west of city locations, whereby the setting sun would have dramatically silhouetted any rooftop civic or religious ritual in the eyes of the masses. Aggrandizement of interests unknown would seem to be the best answer to questions posed by the citadels.

The street pattern of the lower city has the primary gridiron already discussed with entrances to the houses from minor lanes more or less at right angles to the main routes. There is evidence of a wide range of house types, from small single-room 'tenements' to large houses, with several dozen rooms and several courtyards. The larger houses were all inward-looking, with no openings on to the main streets. In many cases brick stairways led up to upper floors, or usable flat roofs. Most houses had bathrooms, connected by drainage channels to main drains with access manholes running under the streets. Some of the bathrooms may have been on upper floors. Sir Mortimer Wheeler includes two intriguing photographs of elaborate sanitary installations and observes that 'the high quality of the sanitary arrangements at Mohenjo-daro could well be envied in many parts of the world today. They reflect decent standards of living coupled with an obviously zealous municipal supervision. Houses sometimes had a privy on the ground or upper floor connected with the attendant drains and water-chutes which in their turn gave onto the main sewers'.[123] It is possible that a branch of the main stream of the Indus may have been taken through the lower town to flush the sewers and serve as a 'sanitary' canal. Water supply was on the basis of both private and public wells and climate and the high water-table in the porous alluvial soil must have required some form of downstream sewage disposal system.

Shops have been identified along main streets in Mohenjo-daro; one such building, which could perhaps have been a restaurant, measured 87 by 64½ feet on plan and had separate living quarters arranged around a courtyard. Wheeler notes that 'temples have not been clearly identified, but further examination would probably reveal two or three in the areas already excavated'. The Allchins endorse an estimate of 35,000 as a probable population figure for Mohenjo-daro and believe that this figure would also apply to Harappa.[124]

HARAPPA

Harappa was situated some 4,000 miles away to the north-east, in the Punjab alongside the Ravi, a tributary of the Indus. Its ancient remains were plundered for brick rubble by railway constructors around the middle of the nineteenth century but the general outline of the citadel has been identified and enough disclosed of the lower-town layout to confirm its essential similarity to Mohenjo-daro. The citadel was enclosed by a mud-brick rampart or embankment, constructed on a 40 ft wide base and faced with burnt brick. Within this wall a mud-filled platform carried the citadel buildings, the remains of which were

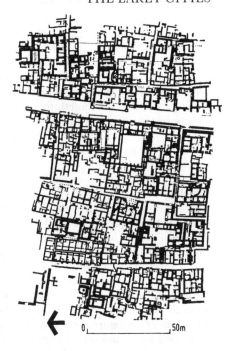

Figure 1.43 – Mohenjo-daro: detail plan of the excavated housing area in the south-west corner of the lower city, with one of the main north–south city streets at the top. 'The lower town must also have contained a wide range of shops and craft workshops: among these potters' kilns, dyers' vats, metal-workers', shell-ornament-makers' and bead-makers' shops have been recognized, and it is probable that had the earlier excavators approached their task more thoughtfully much more information would have been obtained about the way in which these specialists' shops were distributed through the settlement. Another class of building to be expected in the lower city is the temple' (*Birth of Indian Civilisation*). The differing rectilinear alignments of the housing groups comprising this residential district, and by association the remainder of the lower city, are to be regarded as a result of *laissez-faire* development within the main routes 'supergrid'.

122. See G Urban and M Jansen (eds) *The Architecture of Mohenjo-daro*, 1984.
123. Wheeler, *Civilisation of the Indus and Beyond*; see also J Hawkes, *Mortimer Wheeler: Adventurer in Archaeology*, 1982.
124. Allchin and Allchin, *Birth of Indian Civilization*.

unfortunately too badly damaged for the interior layout to be established. Outside the citadel, in the 300 yard wide space between it and the river, Wheeler records the existence of 'barrack-like blocks of workmen's quarters, serried lines of circular brick floors, formerly with central wooden mortars for pounding grain, and two rows of ventilated granaries, twelve in all, marshalled on a podium. The total floorspace of the granaries was something over 9,000 square feet, approximating closely to that of the Mohenjo-daro granary before enlargement. The whole layout, in the shadow of the citadel, suggests close administrative control of the municipal food-stocks within convenient proximity to the river-highway.'

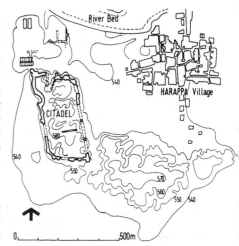

Figure 1.44 – Harappa: general layout. Less extensively excavated than Mohenjo-daro and more disturbed by subsequent site occupation, the city of Harappa closely resembles Mohenjo-daro in most important respects.

LOTHAL

This smaller Harappan town was an important trading centre on the coast south-east from the Indus delta and 450 miles from Mohenjo-daro. Lothal was approximately rectilinear in outline, with its long axis running north–south. It was surrounded by a massive embankment and a level platform some 12 feet in height, which formed the south-eastern quarter of the town, is believed to have served similar functions to the separate citadels of other Harappan cities. Alongside this platform and running north for almost the entire length of the eastern wall, a rectilinear enclosure some 239 by 40 yards in area has been identified as a dock for shipping: its high baked-brick revetments, 15 feet high, are still perfectly preserved. The Allchins note that at one end of this dock 'a spillway and locking device were installed to control the inflow of tidal water and permit the automatic desilting of the channels'.[125]

HARAPPAN CIVILIZATION: A POSTSCRIPT

After the fall of the Harappan civilization at the hands of savage, nomadic, light-skinned Aryans who did not know what to do with the urban centres they found on the Indus plain, Wheeler writes of a 'long phase of cultural fragmentation, not altogether unlike that from which it sprang, but including, perhaps, remoter exotic elements'.[126] The newcomers gradually became settled agriculturalists, and as Andreas Volwahsen notes: 'gradually the villages of their tribal chiefs developed into towns, the centres of small principalities and republics. The ancestors of these new city-builders had completely destroyed the urban civilisation of the Indus valley and their otherwise very detailed legends contain hardly any mention of them ... for this reason the transformation of their simple village culture into an urban civilisation of far greater complexity took place without any connection with, and even without any recollection of, the skilful town planning of their predecessors'.[127] In this respect there is a close parallel with medieval Western Europe, when there were numerous instances of organic growth, 'born-again' settlements becoming established on top of a long-since forgotten imperial Roman gridiron. The City of London (discussed in Chapter 3) is a classic of this kind.

One highly significant aspect of this renewed Indian civilization is the evolution of a theoretical and practical basis of urban planning, according to strictly religious principles, which involved the selection and application of a suitable predetermined plan-form, the term for which is *mandala*. A brief description of the role of the *mandala* in Indian town planning, based on an excellent section of Volwahsen's book, is given as Appendix C.[128]

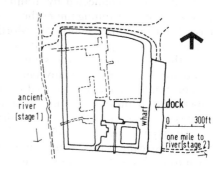

Figure 1.45 – Lothal, general layout. The 'citadel' occupies the south-eastern quarter, with an impressive wharf between it and the dock basin. The Allchins describe how 'an important part of the raised platform contained further brick platforms intersected by ventilating channels, representing no doubt the foundations of warehouses or granaries comparable to those of the other sites. The overall dimensions of this block were 48.5 by 42.5 m.

Evidently there were no other buildings on the platform, for a row of twelve bathrooms and drains were discovered there. The remaining three-quarters of the town seem to have been the principal living area, divided by streets of 4 to 6 m in width and narrower lanes to 2 to 3 m. The main street ran from north to south. In this area numerous traces of specialists' workshops were found, including copper and goldsmiths' shops, a bead factory, etc.' (*Birth of Indian Civilisation*). (See also S R Rao, *Lothal 1955–62: An Harappan Port Town*, vol I, 1979.)

125. Allchin and Allchin, *Birth of Indian Civilization*; see also S R Rao, *Lothal: A Harappan Port Town – Volume One*, 1979.

126. Wheeler, *Civilisation of the Indus and Beyond*.

127. A Volwahsen, *Living Architecture Indian*, 1970.

128. Volwahsen, *Living Architecture Indian*.

2 – Greek City States

Townscape, writes Anthony Kriesis, is the true reflection of the way of living and the attitude to life of its inhabitants. Although this observation holds true throughout urban history, the twentieth century no less, at no time is it more clearly exemplified than by the Greek cities of the sixth to third centuries BC. In addition, seldom in history has this attitude been so clearly determined by factors inherent in its geographical context.[1]

Regional topography

The first of these influences was topography. This determined Greek territorial organization on the basis of clearly defined, separate city states, rather than through a single unified nation. The two main foci of the city states, Greece itself and the Ionian coastline of Asia Minor, are mountainous, with only limited fertile areas in the form of isolated valleys, plains and plateaux (Figure 2.1). Such conditions favoured the existence of small and independent states, each of which generally came to consist of an urban nucleus, surrounded by countryside and subordinate agricultural village communities.[2]

Two more or less synonymous terms are given to this wholly typical urban/rural entity – 'city state' and 'polis'. Kitto explains these terms by saying: 'polis is the Greek word which we translate as city state; it is a bad translation, because the normal polis was not much like a city and was very much more than a state ... since we have not got the thing the Greeks called the polis, we do not possess an equivalent word'.[3] For the purposes of his general history, *The Greeks*, Kitto elects to avoid 'the misleading term city state, and use the Greek word instead'.

In urban history, however, there were other comparable distinct urban/rural entities, both in theory and practice, during subsequent periods. For Kitto's reason that polis is specifically a Greek word, this history of urban form will use the more generally applicable term city state. On occasions the Greek city states joined together to face a common enemy, notably the Persians, but they were also intermittently in conflict with each other.

The Greek city (the urban nucleus of the city state) with its clearly defined limits, compact urban form and – superficially at least – integrated social life, often represents unparalleled achievements to modern planners. Caught up in the intricacies of today's predicaments it is only too easy to look back nostalgically across the centuries to what is

There are a few plains – not large ones, but extremely important in the economy and the history of the country. Of these, some are coastal, like the narrow and fertile plain of Achaea that runs along the southern coast of the Gulf; others lie inland, like Lacedaemon (Sparta), perhaps almost entirely barred from the sea by mountains, like the plains of Thessaly and Boeotia. The Boeotian plain is particularly lush, and with a heavy atmosphere; 'Boeotian pig' the more nimble witted Athenians used to call their neighbours.
(H D F Kitto, The Greeks)

1. The main chapter references are R E Wycherley, *How the Greeks Built Cities*, 2nd edn, 1962 (still the most useful English language single-volume work); A Kriesis, *Greek Town Building*, 1965 (part German); R Martin, *L'Urbanisme dans la Grèce antique*; von Gerkan, *Griechische Stadteanlagen*, 1924 (with his most important plan of Miletus). (The Greek period is in need of a fully illustrated general urban history.)

For a general history of Greek architecture, see W B Dinsmoor, *The Architecture of Ancient Greece*, 1950/1973; for the general historical background, H D F Kitto, *The Greeks*, 1951, remains my preferred reference.

2. Glyn Daniel in his *The First Civilisations: the Archaeology of their Origins*, writes that 'the trouble with the word Urban, and with referring as Childe did to the Urban Revolution, is that this word is to most people overlaid with ideas of conurbations, skyscrapers, factories, underground railways and double-decker buses, with commuters and big business. I would prefer to use the English version of the Greek word "synoecismus" which was used by Thucydides and meant the union of several towns and villages under one capital city. Garner in 1902 spoke of the time "when the town was first formed by the synoecism of the neighbouring villages"'

3. Kitto, *The Greeks*.

Figure 2.1 – General map of the central and eastern Mediterranean showing the Greek sphere of influence. (Miletus, Priene, Pergamon and Troy are on the Ionian coastline of Asia Minor.) Key: 1, Olynthus; 2, Olympia; 3, Delphi; 4, Thebes; 5, Corinth.

believed to have been a veritable golden age of cities.[4] Greek culture, however, was far from being exclusively urban. Professor Kitto stresses this: 'Town and country were closely knit – except in those remoter parts, like Arcadia and western Greece, which had no towns at all. City life, where it developed, was always conscious of its background of country, mountains and sea, and country life knew the usages of the city. This encouraged a sane and balanced outlook; classical Greece did not know at all the resigned immobility of the steppe-mind, and very little the short-sighted follies of the urban mob.'[5] Similarly, Wycherley notes that 'the life of the Greek city state was founded upon agriculture and remained dependent on it; city state and city were not necessarily the same even though the former was most visibly embodied in the latter'.[6]

Climate

Climate, as a second determinant, had a pleasantly beneficial effect on the basis of everyday life in ancient Greece – throughout the year it was generally both agreeable and reliable. As Kitto puts it, 'Greece is one of those countries which have a climate and not merely weather. Winter is severe in the mountains; elsewhere moderate and sunny. Summer sets in early and is hot, but, except in the land-locked plains, the heat is not enervating for the atmosphere is dry, and the heat is tempered with the daily alternation of land and sea breezes. Rain in summer is almost unknown; late winter and autumn are the rainy seasons.'[7] This attractive situation encouraged an open-air, communally orientated attitude to life, which assisted the development of Greek democracy. In direct contrast, however, the domestic Greek world was that of privacy within the ubiquitous courtyard house.

In theory at least, all citizens had a voice in the affairs of their city state. Here numbers were never large: only three cases of more than 20,000 citizens are known – Athens (the city state occupying the plain of Attica and in most respects a non-typical example), Syracuse, and Acragas (Girgenti) both in Sicily. Many never exceeded 5,000 citizens; those which did invariably developed from humble origins. The possibility for all citizens to gather throughout the year in one place, at one time, made feasible the Greek self-governing innovation.[8] Meetings had to take place in the open air; only late in Greek history were construction techniques sufficiently advanced to enable the, by then, representative assemblies to take place indoors, in the bouleuterion. Similarly, large-scale open-air theatrical ceremonies were performed

... when the reader has calculated how much of his working time is consumed in helping him to pay for things which the Greeks simply did without – things like settees, collars and ties, bedclothes, laid-on water, tobacco, tea and the Civil Service – let him reflect on the time-using occupations that we follow and he did not – reading books and newspapers, travelling daily to work, pottering about the house, mowing the lawn – grass being, in our climate, one of the bitterest enemies of social and intellectual life.

Again, the daily round was ordered not by the clock but by the sun, since there was no effective artificial light. Activity began at dawn. In Plato's 'Protagoras' an eager young man wants to see Socrates in a hurry, and calls on him so early that Socrates is still in bed (or rather, 'on' bed, wrapped presumably in his cloak) and the young man has to feel his way to the bed because it is not yet light. Plato obviously thinks that this call was indeed made on the early side, but it was nothing outrageous. We envy, perhaps, the ordinary Athenians who seem to be able to spend a couple of hours in the afternoon at the baths or a gymnasium (a spacious athletic and cultural centre provided by the public for itself).

We cannot afford to take time off in the middle of the day like this. No: but we get up at seven, and what with shaving, having breakfast, and putting on the complicated panoply which we wear, we are not ready for anything until 8.30. The Greek got up as soon as it was light, shook out the blanket in which he had slept, draped it elegantly around himself as a suit, had a beard and no breakfast, and was ready to face the world in five minutes. The afternoon, in fact, was not the middle of his day, but very near the end of it.
(H D F Kitto, The Greeks)

4. Lewis Mumford, in his influential *The City in History*, 1966/1991, contributed to twentieth-century city planning escapism in writing, for example, 'the recovery of the essential values that first were incorporated in the ancient cities, above all those of Greece, is a primary condition for the further development of the city in our time'. (Wycherley concluded his second edition preface with those words.)
5. Kitto, *The Greeks*.
6. Wycherley, *How the Greeks Built Cities*.
7. Kitto, *The Greeks*; see also Gustav Renier who cautioned, in his *Dutch Nation*, 1944 'far too much is said about climate by those who study comparative national character' (see the Netherlands section in Chapter 4). Suffice to agree that climate plays a part in determining national character in addition to urban form.
8. The small numbers of citizens in ancient Greek cities, meant (in theory at least) that decisions of a city planning nature could be arrived at through democratic participation. There is a strong appeal for modern city planners schooled in this way.

initially at the foot of conveniently sloping natural auditoria. Later these were frequently laid out as beautifully conceived architectural and landscape entities.

Climate also gave Greek citizens the leisure to enjoy these and other civic privileges. Greek living standards were low, certainly as compared with Roman and more recent ones. Few of the city states had particularly fertile agricultural situations, even if relatively little effort was needed to produce the basic essentials of life. The availability of slave labour must be taken into account, but its role should not be exaggerated. While the Greeks were slave-owners 'like all civilized peoples in antiquity, and many since',[9] it is certainly false to believe that Athenian culture was dependent on slavery. The average small farm produced little more than the proprietor's domestic requirements and although larger farms could support a few slaves there is no comparison with the rural depopulation situation in Italy which produced the Roman *latifundia* – large estates worked by slaves. Professor A W Gomme estimates that before the Peloponnesian War Attica alone had about 125,000 slaves – approximately 65,000 as domestic servants, 50,000 in industry and 10,000, by far the worst off, in the mines. At the same time there were about 45,000 male adult citizens, giving a total population in excess of 100,000.[10] Industrial slavery was on a small scale; few concerns are likely to have employed more then twenty slaves. Kitto describes how the buildings on the Acropolis were built through thousands of separate contracts: 'one citizen with one slave contracts to bring ten cartloads of marble ... [another] employing two Athenians and owning three slaves contracts for the fluting of one column'.[11] Slaves could hold responsible positions as, for example, 'policemen', without either the obligation or the honour of serving in the Athenian army or navy. Slave labour did not force wages down to a subsistence level. On the contrary, as Childe notes, working at the minimum wage an Athenian day labourer in the fifth century would earn in 150 days enough to provide the subsistence minimum of food and clothing for the whole year.[12] General acceptance of this minimum is a basic reason why the Greeks had so much 'spare time' to spend on their civic activities.

Construction materials

Another factor, with a more immediate impact on the character of Greek cities, was the ready availability of high quality marble. Worked to fine details, marble was the medium by which Greek architecture attained standards of perfection seldom reached in later history. The important civic buildings were conceived as three-dimensional, free-standing sculptural *objets d'art*, in whose construction neither expense nor effort were spared. Some care was given to the organization of spatial relationships between these buildings, notably the group on the Acropolis in Athens (Figure 2.19). If rivalled by a few isolated Roman and Renaissance examples, the attention that the Greeks gave to detail when they fashioned civic spaces was neither understood nor relevant during the Middle Ages and although vitally needed in the twentieth century is neither appreciated nor, seemingly, attainable today.[13]

Yet, and this is entirely in keeping with Greek values, there was only minimal concern for domestic comforts. Homes, in direct contrast to civic buildings, were but rudimentary structures, either grouped

Democracy in the age of Pericles produced that inherent dignity of the individual born of free speech, a sense of unity with one's fellow men, and a full opportunity for participation in affairs of the community. The Athenian citizen experienced the exhilaration of freedom and accepted the challenge of responsibility it thrust upon him with honour and with pride. The discovery of freedom gave impetus to the search for truth as honest men desire it. Philosophy was nurtured, and there were no depths which the wise and intelligent were afraid to plumb. Reason was encouraged, logic invited and science investigated. There was no truth which might be covered and remain undisclosed. Inspired by this atmosphere it was no wonder that great philosophy was born; only in freedom can such greatness be cultivated, not freedom from care but freedom of the spirit. This was the environment of culture which produced Socrates, Plato and Aristotle.
(R D Gallion, The Urban Pattern)

The year 500 BC represents in antiquity almost the end of one aspect of technological development in the Near East and indeed in the whole of the western world. It is true to say that virtually no new raw material was to be exploited for the next thousand years and that no really novel method of production was to be introduced. What new advances were made were to be almost entirely in the field of engineering, and most of the principles involved had themselves already been discovered and applied, although usually on a smaller scale.

Historians reviewing this state of affairs ... have tended to attribute it to a number of causes, the first of which was the widespread use of slave labour. The dirty end of production, it is argued, was put entirely in the hands of slaves, and increased output could be achieved only by one of two means: either by acquiring more slaves or making those that one had work the harder. Since it is not in the nature of slaves either to invent new means of production or to exploit new materials, the possibility of any further technological development came to an abrupt end.
(H Hodges, Technology in the Ancient World)

9. Kitto, *The Greeks.*
10. A W Gomme, *History of Greece.*
11. Kitto, *The Greeks.*
12. Childe, *The Dawn of European Civilisation,* 1957.
13. 'Today' was written in respect of the 1960s when my views on the contemporary international city were crystallizing. Now, in the 1990s, while recognizing internationally that there are numerous examples of quality modern urban-placemaking – most consistently in what were West German cities – it still saddens to have to record that the British remain mostly uninterested in their everyday surroundings, with only a small minority able to identify with that which has been exceptionally achieved.

together by chance, in organic growth districts, or rigidly organized along basic gridiron lines. Such marked contrast between the splendour of civic areas and the squalor of housing is entirely typical of ancient Greek cities.

Emergence of Greek civilization

Greek civilization emerged by way of direct antecedents in the Mycenaean and Minoan cultures which were established respectively on the Greek mainland and on the island of Crete. In this way Greece is known to have had direct links with the Sumerian and Egyptian civilizations. The Minoan civilization 'emerges into literacy about 2000 BC',[14] following more than 1,000 years of interaction between Cretan Neolithic villagers, functioning in a mixed farming and fishing subsistence economy, and immigrants from the Nile delta and the mainland of Asia Minor. Initially Minoan culture developed in eastern Crete, under the continuing influence of immigrants from Asia Minor, before spreading to the Messara Plain where it was cross-fertilized by Egyptian contacts and 'the richer culture of Early Minoan II–III developed'.[15] From around 2000 BC, during the Middle Minoan period (which lasted until 1580 BC), a combination of wealth gained by trade and of inspiration derived by contact with the civilized peoples to the south and east made possible the first distinctively European civilization.

Civilization in Crete was at its peak during Late Minoan I–II times. The beginning of this period is marked by the reconstruction of the Palace of Knossos after earthquake devastation; its end comes probably at the hands of the Mycenaeans. The typical Minoan city was concentrated around a centre formed by the palace and a kind of agora – an open space for festal and possibly political gatherings.[16] The most important example of such a palace-town is Knossos, situated some 3½ miles inland from the northern coasts of the island (Figure 2.2).

In mainland Greece the Early Helladic people were conquered, about 1800 BC, by more war-like farmers, probably Greek-speaking Indo-Euroeans. Their civilization – the Late Helladic, better known as Mycenaean – grew to strength in the sixteenth century BC. Although there were obvious Minoan influences, Clark stresses that the Mycenaeans and their culture were firmly rooted in mainland Greece.[17] Culminating with the destruction of Knossos around 1400 BC the Mycenaeans conquered Crete and established their dominion throughout the Aegean world. Childe writes disparagingly about their civilization, calling it semi-barbarous, barely literate and highly militarist.[18] Their 'cities', notably Mycenae itself (Figure 2.4) and Tiryns (Figure 2.5), were like their Cretan precursors, little more than fortified castle-towns. Mycenae covered only 11 acres; Tiryns, as it stood within its walls 26 feet thick and 60 feet high, stretched over a mere 4.94 acres, 2 of which were taken by the palace. Troy (illustrated in cross-section on page 6) did not exceed 4 acres.

During the thirteenth century Mycenaean power declined. 'Other conquerors, the Dorians, came down from north-central Greece, making a sudden end of a long civilization and begining a Dark Age, three centuries of chaos, after which Classical Greece began to emerge.'[19] It is now thought most likely that European Greece, rather than Ionia, recovered first from this set-back and, following the re-establishment of

Figure 2.2 – Knossos was one of a number of palace-towns built in central Crete by Minoan rulers. It was notable for the waterborne sewage disposal system serving part of the domestic quarter of the palace.

Figure 2.3 – Gournia in eastern Crete, near the large Bay of Hierapetra, consisted of about sixty houses, mostly two-storeyed, crowded together on a limestone ridge. The site area was about 6½ acres. On the summit, the palace faced onto a large public space which may have been used as a market.

14. G Childe, *What Happened in History*, 1964.
15. G Clark, *World Prehistory*, 1963.
16. R W Hutchinson, 'Prehistoric Town Planning in and around the Aegean', *Town Planning Review*, vol XXIII, no 4, 1953.
17. Clark, *World Prehistory*.
18. Childe, *What Happened in History*.

urban culture, took the lead in the colonizing movement after 750 BC. Between 900 and 600 BC city states were evolving both in Greece and Ionia, with Sparta 'asserting her primacy in the Peloponnese and becoming the acknowledged leader of the Hellenic race'.[20] Athens, although at this time only a second- or third-class power, succeeded in carrying through the unification of Attica. Her power was slow to develop and not until the twenty years of beneficent administration under Pisistratus (546–527 BC) did she become a city of international significance. By the sixth century BC Greek cities generally had attained high levels of civilization.

The contribution made by fifth-century Athens to Greek and European culture is held by Kitto to be quite astonishing; he says that, unless our standards of civilization are comfort and contraptions, Athens from (say) 480 to 380 BC was clearly the most civilized society that has yet existed. Under the leadership initially of Sparta, and latterly of Athens, the Greeks defeated the Persians between 499 and 479 BC although many cities were destroyed by the invaders, including Miletus in 494 BC and Athens itself in 480 BC. (The totally different ways with which these two opportunities to reconstruct were followed up are described on pp. 43–6.)

Victory inspired the Athenians. During the fifty years between the Persian and Peloponnesian war they 'aimed at, and for a short time held, an empire which comprised or controlled not only the whole Aegean, but also the Corinthian Gulf and Boetia: and there were those who dreamed, and continued to dream, of conquering distant Sicily'.[21] This supreme period in Athenian history is known as the Periclean Age, after their most famous leader, Pericles, who dominated the Assembly from 461 until his death in 429. His policy made Athens the undisputed artistic centre of Greece: architects created the incomparable new buildings of the Acropolis; her sculptors, painters and potters were unequalled; and, 'the most Athenian art of all, tragic drama, was growing more assured every year'.[22]

However, as Professor W B Dinsmoor observes, 'the supremacy of Athens in the Aegean portion of the Greek world was but short-lived; for a succession of long wars, the Peloponnesian (431–404 BC) and the Corinthian (395–387 BC), drained all her energies and deprived her of political leadership. Thus the fall of Athens in 404 BC may justly be taken as the beginning of a new epoch; humiliated and impoverished she was in no condition to maintain the high artistic excellence which she had reached under Pericles'.[23] Subsequently, first Sparta (404–473 BC), then Thebes (371–362 BC), became dominant powers, until, as a result of the battle of Chaeronea in 338 BC, the Greeks were forced to surrender their independence to the Macedonians under King Philip II. His celebrated son – Alexander the Great (336–323 BC) – consolidated their victory before turning his attentions eastward to conquer the Persians. Alexander sought to maintain power throughout his huge empire by founding new Greek cities – notably Alexandra (331 BC). As such he anticipated the policy adopted by imperial Rome.

The term 'Hellenic' is usually applied to Greek civilization prior to the Macedonian conquest. Afterwards, 'the Greek cities lost something vital, though not without certain gains, and some of the finer qualities of Greek art and architecture evaporated; and the modified form of Greek culture which is found in the succeeding centuries can be conveniently distinguished as Hellenistic, though the lower limit of the age

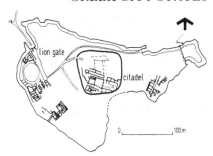

Figure 2.4 – The citadel at Mycenae was first occupied around 3000 BC and the 'city' which had grown up around it reached a peak of prosperity from 2200 to 1600 BC. During this time it is believed that the citadel was occupied by the ruling clan and possibly special craftsmen. Old Mycenae was destroyed around 1100 BC but was re-established for a period, sending troops to fight against the Persians in 480 and 479 BC before eventual final destruction in 468 BC.

Figure 2.5 – Tiryns is regarded as a second, smaller 'capital' of the Mycenaean rulers. The area within the citadel was only 4.94 acres, but outside the walls the settlement spread for some distance over the plain to the east of the hills. The citadel, as illustrated, was in two parts: the palace, at the southern higher end and an area within the wall which served as a refuge for the neighbouring population. Tiryns (in the Peloponnese) is a disappointing site to visit. Just off the main road from Argos (the legendary 'oldest' town in Greece) across the plain to Nauplion on the coast, Tiryns is an abruptly encountered, not very impressive jumble of stones, with which it is difficult to relate Homer's description of 'wall-girt Tiryns'. Mycenae, however, is extremely impressive and is an essential visit.

The Athenians occupied a territory, Attica, which is slightly smaller than Gloucestershire, and in their greatest period were about as numerous as the inhabitants of Bristol – perhaps rather less. Such was the size of the state which, within two centuries and a half, gave birth to Solon, Pisistratus, Themistocles, Aristeides and Pericles among statesmen, to Aeschylus, Sophocles, Euripides, Aristophanes and Menander among dramatists, to Thucydides, the most impressive of all historians, and to Demosthenes, the most impressive of orators, to Mnesicles and Ictinus, architects of the Acropolis, and to Phidias and Praxiteles the sculptors, to Phormio, one of the most brilliant of naval commanders, to Socrates and to Plato – and this list takes no account of mere men of talent.

(H D F Kitto, The Greeks)

19, 20. Kitto, *The Greeks*.
21, 22. Kitto, *The Greeks*.
23. Dinsmoor, *The Architecture of Ancient Greece*.

to which the name is applicable is not clear; from the first century BC onwards, when Roman power had supplanted that of the Hellenistic monarchs in the eastern Mediterranean, one usually speaks of Roman times in Greece'.[24] This chapter limits itself to the Hellenic city states of the sixth and fifth centuries BC.

The Greek contribution

The Greeks made several immensely significant contributions to urban history. In this chapter these are described on the basis of historical sequence rather than comparative importance. First came the colonizing movement, whereby urban growth pressures were contained by sending out emigrant expeditionary parties to found new cities in other parts of the Mediterranean. More or less contemporary with this movement was the evolution of the twin foci of Greek cities – the acropolis as the religious centre and the agora as the multi-purpose everyday heart; these are described with other Greek urban form components. Lastly there was the use made of the gridiron by Greek town planners from the early part of the fifth century BC as the basis of a systematic approach to the organization of cities. Chapter 1 has shown how, contrary to previously accepted opinion, the Greeks were not the first to plan cities, the honour probably going to Harappan priests if not to their antecedents of as yet undetected origin outside the Indus basin (see page 31).

The Greek period is also notable for the clear contrasts revealed between the two streams of urban development: planned urban form, either as new towns or redeveloped city districts, and organic growth pattern, of which the city of Athens is by far the most important example.

COLONIZING MOVEMENT

Beginning around 750 BC and lasting for something over 200 years, Greek city states were involved in a process of urban growth control the value of which was appreciated only intermittently over the succeeding centuries until taken by Ebenezer Howard as the basis of his revolutionary early twentieth-century garden-city movement.[25] This process involved the Greeks in the creation of new city states, as colonies, one major reason for which was to take excess population from the parent city.

It was a process forced on the Greeks in that, as Wycherley says, 'they have usually been a fertile race and the nature of the country imposes a very definite limit on the population',[26] rather than as the result of any intellectual reasoning concerning the ideal size of cities. In addition, but probably secondary to growth pressures, the Greeks were quick to develop trading contacts which their colonies established for them throughout virtually the length and breadth of the Mediterranean. In this they were following the examples of the Phoenicians, who also had substantial trade in the area.

By 734 BC Corinth founded Syracuse in Sicily. Marseilles also has early Greek colonial origins dating from the seventh century BC, with 'the site of the ancient Phoenician-Greek agora remaining the site of the later Roman forum and even of the mediaeval market square'.[27] Naples and Pompeii are other distant examples of early Greek foundations. Miletus, later to become a planning byword coupled with the name of

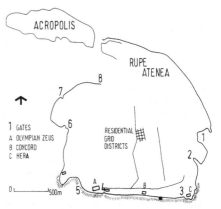

Figure 2.6 – Agrigento (ancient Akragas), on the south coast of Sicily, was founded in 581 BC by a group originating partly from the existing colony of Gela (located some 75 km to the east) and partly from Rhodes, which had itself founded Gela in 688 BC. The site is that of a shallow bowl of land contained between two linked ridges, both running south-east–north-west approximately parallel to the sea. The much higher northern ridge is now occupied by the surviving medieval town of Agrigento, while the other is the deserted site of the ancient city's superb fifth-century BC temples, including the extremely well-preserved Temple of Concord. (See further note with 'The Theseion' in Athens, Figure 2.17.) Sheltered between the ridges there were the ancient gridiron-planned housing areas, only a small part of which has been excavated. In 480 BC Akragas was powerful enough to defeat the Carthaginians, only to be devastated by them in 406 BC. Revival followed under the Romans (who named it Agrigentum) and then the medieval town which, fortunately, occupied the high ground.

Ever since Neolithic times the coasts and islands of the Aegean had been the subject of exploration by mariners eager for trade or new areas for colonisation; at the height of their power the Mycenaeans had posts as far west as South Italy, Sicily and the Lipari Islands and trade connections over broad tracts of Europe from Iberia and southern Britain in the west to the Ukraine and even Transcaucasia in the east.

Already during the eighth century Ionian Greeks had begun to explore the northern shore of Asia Minor and in due course trading stations were established at Trebizond and Sinope, the first for loading iron, copper and gold from Transcaucasia and the latter perhaps for transhipment into larger craft.
(G Clark, World Prehistory – An Outline)

24. Wycherley, How the Greeks Built Cities.
  For Greek cities generally during Roman times, see T Cornell and J Matthews, Atlas of the Roman World, 1982.
25. Ebenezer Howard was truly the 'father' of the modern British New Towns Movement. Few others in urban history, ancient or modern, have played such a pivotal role in altering the course of events. It was his 1898 book, Tomorrow: A Peaceful Path to Real Reform (republished four years later as Garden Cities of Tomorrow) that gradually focused attention on the new town concept, leading eventually to the British New Towns Act 1948. See A Sutcliffe, Towards the Planned City 1780–1914, 1981.
26. Wycherley, How the Greeks Built Cities.
27. P Zucker, Town and Square, 1959.

Hippodamus, was the starting-point of a gigantic colonization, and at least sixty colonies originated from there.[28]

Each of the colonies was a city state organized along the social and economic lines of its parents, but in contrast to the generally unplanned, uncontrolled organic growth patterns of the parents the majority of the offspring were developed along planned lines. The chapter is concluded with summary descriptions of important Greek colonies.

GREEK URBAN FORM COMPONENTS

The basic elements of the typical Greek city plan comprise the acropolis, the enclosing city wall, the agora, residential districts, one or more leisure and cultural areas, a religious precinct (if separate from the acropolis), the harbour and port, and possibly an industrial district. The organization of these parts – with the exception of the last two – into a city is most clearly illustrated by the example of Priene.

The acropolis is the general term for the original defensive hill-top nucleus of the older Greek cities and the fortified citadel of many of the colonial foundations. From being the site of the total urban area, the acropolis either gradually evolved into the religious sanctuary of the city, as with the most famous example at Athens, or became deserted and left outside the city limits, as at Miletus.[29] As long as the city remained of limited size, centred directly on the acropolis, there was no need for a perimeter defensive wall. When attacked the citizens withdrew on to the acropolis until either they capitulated or their enemies gave up. With all the important buildings located on the acropolis, only a proportion of the relatively expendable housing would be lost.

From about the sixth and fifth centuries, however, starting in Ionia, the real and sentimental value of investment outside the acropolis was great enough for it to require protection. Democratic Greek society of this time also required security for the whole community; separate fortification of the acropolis was felt to be antidemocratic and a symbol of tyranny.[30] It was Aristotle, on the subject of town walls, who said that an acropolis was suitable for oligarchy and monarchy and level ground for democracy.[31]

Not all Greek cities were fortified. The typical arrangement is at Athens, Miletus and Priene, where the walls are loosely spread around both unplanned and planned urban areas taking maximum advantage of the terrain. The wall is more of an afterthought, in contrast to the initial, rigidly rectilinear perimeter of the average Roman Empire foundation (see page 57). Two reasons why Greek city walls are more flexible than the urban form constrictors of later ages are a balance of population between urban and rural parts of city states, and the policy of limiting population by founding new cities.

'The word agora,' as Wycherly observes, 'is quite untranslatable, since it stands for something as peculiarly Hellenic as polis, or sophrosyne. One may doubt whether the public places of any other city have ever seen such an intense and sustained concentration of varied activities. The agora was in fact no mere public place, but the central zone of the city – its living heart. In spite of an inevitable diffusion and specialization of functions, it retained a real share of its own miscellaneous functions. It remained essentially a single whole, or at least strongly resisted division. It was the constant resort of all citizens, and it did not spring to life on occasions but was the daily scene of social life, business and politics.'[32] As a focal point of a planned city, the

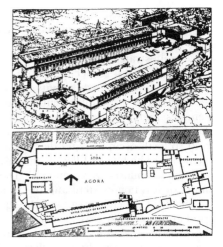

Figure 2.7 – Assos; restored plan of the Agora, and a conjectural perspective view. The trapezoidal shape, wide at the western end, was determined by the shape of the narrow terrace on which the Agora was located. This city is the most northerly of the sites on the mainland of Asia Minor (see R E Wycherley, *How the Greeks Built Cities*, 1962).

Only after 500 BC did genuine squares develop in Greece. City planning as such, conscious collective and integrated action beyond the mere construction of individual houses, existed already in India and Egypt in the third millennium BC, but never the impulse to shape a void within the town into a three-dimensional area which we call a 'square'. This may be explained sociologically; only within a civilisation where the anonymous human being had become a 'citizen', where democracy had unfolded to some extent could the gathering place become important enough to take on a specific shape. This sociological development was paralleled by an aesthetic phenomenon: only when a full consciousness of space evolved and at least a certain sensitive perception of spatial expansion began to spread – one may compare the essentially frontal sculpture of Egypt and Mesopotamia with the roundness of Greek classical sculpture – only then could the void before, around, and within a structure become more than a mere counterpart to articulated volume.
(P Zucker, Town and Square)

28. S E Rasmussen, *Towns and Buildings*, 1969. For Marseilles, which is not included in this book, see P Lavedan, *Les Villes françaises*; G Duby, *Histoire de la France urbaine*, Volumes 1, 2, 3.

29. Wycherley writes that for those ancient Greek cities that originated on and around an 'acropolis', 'at an early stage there might be no distinction in meaning between "polis" and "acropolis"' (*How the Greeks Built Cities*). With the growth of Athens and Miletus the effect of the topographical determinant was markedly different. The Athenian acropolis was essentially an isolated hilltop in a plain, and growth could take place in all directions; whereas Miletus could expand only downhill towards the sea, gradually moving the centre away from its acropolis.

30. Wycherley, *How the Greeks Built Cities*.

31. Aristotle, *Politics* vii 10.4.

32. Wycherley, *How the Greeks Built Cities*.

agora was as near the middle as possible, or in the harbour cities alongside the port. With unplanned cities a natural place for the agora was between the main gate and the entrance to the acropolis. Athens clearly illustrates this.

To the Greeks, preoccupied with intellectual matters, home life was secondary to communal activity. As a result – to quote Wycherly again – 'the Greeks of the fifth century put their best, architecturally, into temples and public buildings; in the scheme of the Greek city the houses were subordinate. The agora, shrines, the theatre, gymnasia and so forth occupied sites determined by traditional sanctity or convenience. The houses filled in the rest'.[33]

Home comforts were minimal. Drainage and refuse disposal were more or less non-existent and the resulting contrast between the magnificence of the civic areas and the squalor of the housing districts was probably as marked as at any time in urban history. The courtyard house was the rule, but there were no standard or typical room arrangements; individual dwellings within the same grid block were of different sizes and plans, as most clearly exemplified by the north hill housing district at Olynthus (Figure 2.8).[34]

The economy of the Greek city states, being based to a considerable extent on slave labour, gave the citizens ample leisure time to be employed on either intellectual discussion or collective activities. For these latter communal requirements, specialized building types were developed, including the theatre, gymnasium and stadium, each of which was regarded as essential in every city. The theatre required a suitable natural auditorium slope and in many cities it is to be found on the southern side of the acropolis. Often, these leisure and cultural functions are grouped together.

Two of these Greek components represent revolutionary additions to those found in the Mesopotamian city; neither the agora, as the multipurpose area for democratic assembly, nor the leisure centre (to use a modern term) had a place in the Sumerian schema, whereby the urban mass was, most probably, denied opportunity for public gathering and such leisure time, as was available, was spent resting at home. 'Revolutionary addition' is fully justified because these two Greek innovations constitute one of the main divides between Christian and Islamic urban culture. Whereas the former evolved from Greek and Roman precedents; the historic Islamic city, in direct line of descent from Mesopotamia, likewise had no need for either democratic political assembly, or communal leisure activity in public. This fundamental difference is further described in Chapter 11.

Systematic city planning

By coincidence the two most important planned Greek cities – Miletus and Priene – are located within a short distance of each other on the Ionian coastline of Asia Minor. Priene is the clearer example, described by Wycherley as 'containing everything which makes a polis, all very neatly and ingeniously arranged and subordinated to the Hippodamian plan'.[35] Compared to Miletus, which in Roman times during the first century AD acquired 80,000 to 100,000 inhabitants, Priene contained only some 400 houses, with at most 4,000 population. Miletus will be described first because its rebuilding predates that of its neighbour by over a century; it moreover introduces here the systematic Hippodamian method of planning cities.

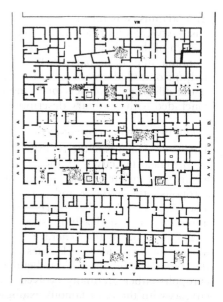

Figure 2.8 – Detail of north hill housing at Olynthus, built after 432 BC to a gridiron layout based on main streets running the length of the ridge (as Avenues A and B) with numerous cross-streets alternating with back alleys which divide the housing blocks into two parts. The tinted areas within the houses mark the internal courtyard around which the rooms are planned. As in the preceding historical periods, courtyard housing was the general pattern of Greek urban housing. This layout could well be part of a 1960s housing scheme, perhaps designed in accordance with ideas put forward by Alexander and Chermayeff in their *Community and Privacy*, but note the variety of room arrangements within the standard house areas, indicative perhaps, of do-it-yourself building within a tightly controlled planned framework. However, it has long since been clear that the modern introverted courtyard house (or 'patio-housing' as it is known in Britain) is of negligible interest to private housing developers, or their clientele with whose house-in-garden expectations they can but conform. It has had only occasional use in public housing projects and is generally disliked and derided by tenants. (See Chapter 11 for Islamic courtyard housing types.)

33. Wycherley, *How the Greeks Built Cities*.
34. See the comparable courtyard housing at Pompeii and Timgad (Chapter 3) and Islamic counterparts (Chapter 11, with their Mesopotamian predecessors described in Chapter 1).
35. Wycherley, *How the Greeks Built Cities*.

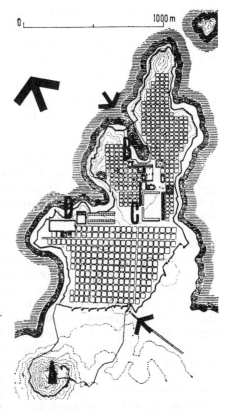

MILETUS AND HIPPODAMUS

Miletus had played a major role in the gradual establishment of Greek commercial and military power between the tenth and sixth centuries BC, founding, as described earlier, a large number of colonies and becoming as a result the head of a powerful confederacy of city states. But early in the fifth century Ionia was overrun by the Persians and in 494 BC Miletus was captured, sacked and destroyed. It can safely be assumed that the old city was the product of centuries of haphazard, organic growth, in contrast to the planned forms of at least some of its more recent colonies. In rebuilding Miletus from 479 BC onwards the opportunity was taken to plan 'an entirely new and modern city unlike the Athenians, who also returned to find their city destroyed, but gradually restored the *status quo ante* with the addition of more magnificent temples'.[36]

The master plan for the reconstruction of the city was prepared by a Milesan architect, Hippodamus of Miletus, whose planning work, at Miletus and elsewhere had earned for him the title 'father of town planning', but only until the archaeological record was taken back further from the Greek to the Harappan period. A previous generation of urban historians unwittingly misrepresented his achievements, rather in the same way that Sir Christopher Wren's admirers have claimed unwarranted planning expertise on his behalf and critics of Baron Haussmann have distorted his role in renewing mid-nineteenth-century Paris.[37]

Clearly Hippodamus was not the inventor of the gridiron: its application to the layout of planned *parts* of cities as early as 2670 BC, and its probable general role as the urban form regulator of the early third millennium BC Harappan cities have been described in the preceding chapter. The claim that he was the 'father of town planning' may, however, still retain a measure of truth. If the Harappan cities are recognized as the first known planned urban settlements then Hippodamus has been anticipated by at least 2,000 years. (Caution is necessary because of the continuing archaeological work in the Indus basin, with always the added possibility of significant finds elsewhere.) But if Mohenjo-daro, Harappa and other contemporary Indus basin cities are not accorded planned urban status, then Hippodamus might still conceivably have set a significant precedent with his plan for Miletus.[38]

The key consideration is whether or not it was Hippodamus who first, at a given moment in time, organized all of the component parts of a new town – the central area, housing districts, commerce, cultural and leisure facilities and a defensive wall – to make an integrated planned urban entity. It may seem improbable on the basis of scanty knowledge of what actually happened at Miletus, but we do have extensively documented modern examples of individuals forcing a change of direction in accepted urban growth trends – Ebenezer Howard as the recognized instigator of the garden-city/new-town movement, and Constantine Doxiadis as the *éminence grise* behind many mid-twentieth-century developments, are but two examples.

For all his significance in the history of town planning, Hippodamus remains a mysterious personality concerning whom there is little information of value. Aristotle, no less, has much to answer for in creating a myth with his observation that Hippodamus was 'the son of Euryphon, a native of Miletus, the same who invented the art of planning cities, and who also laid out the Piraeus – a strange man, whose

Figure 2.9 – Miletus: the general plan as excavated by von Gerkan. The original peninsular situation on the southern side of the estuary of the River Meander, facing across to Priene, has long since disappeared as a result of silting-up of the bay. (A similar fate to that experienced by Winchelsea in the fourteenth and fifteenth centuries – see page 124). Key: A, early fortified hilltop settlement, a form of acropolis; B, the main harbour; C, the Agora complex; D, theatre and other cultural/leisure activity facilities. See E Bacon, *Design of Cities*, 1974, for a sequence of historical model photographs of the agora area at Miletus. Looking at the pristine, rectangular grid-block shapes of the second-century AD Roman model (which are the same as those of the fourth-century BC model) has revived my misgivings that the von Gerkan 'plan' of Miletus is too precise to be a true record of some 500 years of urban history. Because we now know of more than the one instance of archaeological misrepresentation, and with von Gerkan's plan enshrined in all the urban histories, it is essential that Miletus, Hippodamus and von Gerkan are reappraised.

36. Wycherley, *How the Greeks Built Cities*.
37. See Chapter 8 for critical assessment of the 1666 Wren Plan for London, and L Pinkney, *Baron Haussmann and the Rebuilding of Paris*, 1958.
38. At this date of revision (late 1991), the Harappan cities of the Indus Valley are still the earliest known examples of urban planning.

fondness of distinction led him into a general eccentricity of life, which made some think him affected ... besides aspiring to be an adept in the knowledge of nature, he was the first person not a statesman who made inquiries about the best form of government'.[39]

Various city planning schemes have been attributed to him. He can certainly be regarded as responsible for Miletus, his first work; he then moved on to Athens with a commission from Pericles to lay out the new harbour town of Piraeus about 450 BC. Piraeus was followed by the colonization of Thurii, in southern Italy, from 443 BC onwards: this was described by Kriesis as a progressive community which believed in planning and employed Hippodamus to build them a model town.[40] Claims that he was responsible for planning the new city of Rhodes in 408 BC are now discounted. Zucker eliminates these as highly improbable, since by then he would have reached an age unusual even for city planners.[41]

MILETUS: THE PLAN

Of even greater significance than the detailed form of the plan of Miletus is the far-sighted attitude of the Milesians, who seem to have had visions of their city regaining much of its former greatness, and to have planned accordingly.[42] Although there must have been every reason for the returning survivors to restart on a small scale along reinstated organic lines (as with the rebuilding of the City of London after 1666), the citizens decided instead to adopt a plan which not only served for the initial rebuilding phases but also was later able to serve as a basis for the vastly increased area of the first century AD Roman city. There was no loss of form resulting from this unanticipated expansion, nor was demolition of houses needed to accommodate the greatly enlarged agora area with its eventual complex of spaces and buildings.

As rebuilt from 479 BC, Miletus occupied the whole of a rocky indented peninsula to the north of the original acropolis (Figure 2.9). At first the new walls included this hill, but in the course of time a new wall was built excluding it. The agora area is centrally placed in the form of a rectangle with the long side leading from the defended harbour inlet. Of the three distinct residential groups, the southernmost with its considerably larger house blocks dates from Roman times. West of the agora and grouped around a second inlet are the theatre, gymnasium and stadium building. The area within the walls is about 220 acres and the maximum dimensions along and across the peninsular are 1,960 yards and 1,200 yards respectively. Rasmussen records that Miletus prospered even more under the Romans and grew to be a city of 80,000 to 100,000 inhabitants.[43] Its final decay dates from the second century AD.

PRIENE

Across the valley of the aptly named river Meander from Miletus, on a south-facing spur of Mount Mycale, Priene was constructed from about 350 BC onwards to replace an abandoned nearby settlement. The city is built on four broad terraces which descend some 320 feet from the acropolis to the stadium and gymnasium on its southern edge (Figure 2.11). The Temple of Athena Polias, the theatre, a second gymnasium, and the agora are located on the two intermediate terraces. The basis of the plan is formed by seven east–west streets following the contours, and a total of fifteen north–south stepped paths, giving access between

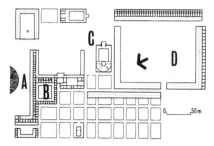

Figure 2.10 – Miletus: detail plan of the Agora area in the second century BC. (Note: harbour to the left.) Key: A, harbour and port facilities; B, large colonnaded courtyard, surrounded by shops and offices; C, Council House (175–164 BC) with a colonnaded court in front; D, South Agora (third century BC). Later additions to the Agora area were made in the first and second centuries AD under Roman rule. (See plan of second century AD in *How the Greeks Built Cities*, p. 83).

R E Wycherley notes that 'the harbour building was the first important architectural scheme of the new agora, and besides giving the town a fine water-front, provided facilities for the merchants as Miletus' mercantile prosperity returned. The harbour zone was, naturally, developed first ... The architects who succeeded one another in carrying out the great work maintained its unity and also its subordination to the general street-plan'.

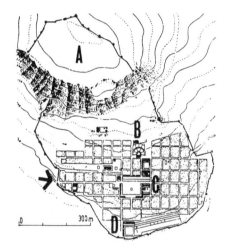

Figure 2.11 – Priene: general plan (north at the top). The contour lines, both solid and dotted, are at 25-metre intervals; the 50-metre contour passes through the main western gate into the city. The ancient course of the River Meander provided water access a short distance further down the slope to the south. Key: A, Acropolis, rising to more than 375 m above sea level (some 300 m above the level of the Agora); B, Theatre; C, Agora complex; D, gymnasium and stadium.

39. Aristotle, quoted Kriesis, *Greek Town Building*.
40. Kriesis, *Greek Town Building*.
41. Zucker, *Town and Square*.
42. Wycherley, *How the Greeks Built Cities*.
43. Rasmussen, *Towns and Buildings*.

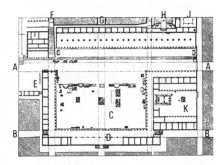

them, up and down the hillside. The main streets are 23 feet wide and the remainder 13 feet wide. Streets and paths are orientated north–south and east–west: housing blocks so formed are of a regular 51.4 by 38.6 yards' size and contain four dwellings on average.[44] It has been estimated that there were about 400 houses in Priene, thus giving a total population not exceeding 4,000.

The agora is in the centre of the city occupying two complete housing blocks and parts of others, on both sides of the main street running from the western gate (Figure 2.13). This street is widened to 10 yards across the agora. Along its northern side a continuous flight of six steps lead up 1.6 yards to the Sacred Stoa (127 yards in length) built in about 150 BC to replace an earlier building. At the rear of this stoa is a line of magistrates' offices. The Bouleuterion and Prytaneion are at the eastern end of the stoa. At the western end an access stairway lets on to the western portico of the main open southern section of the agora, a space some 82 by 50 yards in area originally containing only one altar in the middle but acquiring in the course of time a number of other monuments and statues. Directly to the west of this portico is the main food market. The portico continues around the southern and eastern sides of the space, and then eastward to face the Sacred Stoa across what is in effect a 46-yard-long colonnaded street. The Temple of Zeus is in its own religious precinct immediately to the east. The main civic buildings are mostly on the north side of the main street.

Figure 2.12 – An artist's impression of Priene: a clear example of the form of a small Greek city of the period. The close-knit but ordered urban grain can be contrasted with that of Erbil – an organic growth counterpart – as shown by the aerial photograph reproduced in Figure 1.10. Key: A, Agora; B, Temple of Zeus; C, Gymnasium; E, Temple of Athena; F, Stadium; G, on far left main entrance into city. (Drawing after A Zippelius)

Figure 2.13 – Priene: detail plan of the Agora (north at the top). Key: A–A, main east–west street across city; B–B, grid street continued through the southern stoa with access up by way of steps; C, main Agora space; D, colonnaded hall; E, fish and meat-market; F, stepped grid footpath on hillside; G, the north stoa; H, Bouleuterion; J, Prytaneion; K, Temple of Zeus.

Athens: organic growth

In direct contrast to Miletus and other systematically planned Greek cities of the fifth century BC, Athens was never planned as a whole.[45] As with Miletus, an opportunity for comprehensive reconstruction was offered, following devastation during the Persian wars but, perhaps because of its greater size or the need for more immediate rebuilding,

44. Kriesis, *Greek Town Building*.
45. For Athens see L Benevolo, *The History of the City*, 1980; for a sequence of six maps from early times to the present day, see J Travlos, *Pictorial Dictionary of Ancient Athens*, 1971; J Travlos, *Athens au fil du temps*, 1975; and the excellent descriptive *Historical Map of Athens*, Ministry of Culture, 1989.

the Athenians preferred to reinstate their old city. The two main groups of civic buildings – the Acropolis and the Agora – were rebuilt with great care and considerable attention to spatial relationships, but in each instance their layouts were determined by inherited constraints. Similarly, later in history, the city of Rome did not lose its organic growth structure, although it also contained a number of methodically organized building groups.

The Athenian Acropolis, site of the Neolithic village nucleus of the city, must have been one of the best natural fortresses of the ancient world. At its highest point, north-east of the Parthenon, it rises some 300 feet above the general level of the plain with sheer rock faces on all sides except the west, where there is an accessible slope. It is of irregular shape, roughly 350 yards by 140 yards, with the long dimension oriented east–west. The Acropolis is situated some 4 miles from the Aegean sea, on the plain of Attica (Figure 2.14). Including the Acropolis there were five hills within the ancient walls and the city, which now occupies the greater part of the Attic plain, is surrounded by an amphitheatre of mountains, nowhere far distant and to the east in Hymettos barely 5 miles from the centre.[46]

From earliest times humans had been attracted to the area by the presence of natural springs: it is known to have been occupied as early as 2800 BC, although the traditionally accepted date of the foundation of Athens is 1581 BC, when the worship of Athena was established on the Acropolis.[47] At first the city was confined to its hilltop site, with the main approach path winding its way up the western slope.

Following the unification of Attica under the leadership of Athens during the eighth century BC, the city grew steadily in authority and extent. New housing districts were added by haphazard growth on the plain, around the lower slopes of the Acropolis, which gradually assumed the religious precinct function which it was to keep throughout the city's ancient history. The agora area developed from a market and meeting place which had been established where the Panathenaic Way started its ascent up the western Acropolis slope (Figure 2.16).

At first the Athenian fleet was based on the Bay of Phaleron, which by its gently shelving sands was admirably adapted for beaching ships in accordance with the customs of early times.[48] When the city became a major naval power, with a fleet of 200 ships, a more permanent anchorage was required and in 493 BC the Piraeus peninsula was chosen as a new fortified naval base. To overcome the problem of ensuring access between Athens and its harbour in time of war Themistocles proposed moving the city to the Piraeus. Pericles in effect did this, in about 456 BC, when he built the 'Long Walls' linking the two. The northern and southern walls were respectively about 4½ and 4 miles in length.[49] It is known that Hippodamus planned Piraeus, towards the middle of the fifth century, giving it his gridiron structure and, apparently, an unusual arrangement of one agora near the sea and another inland.

THE AGORA

As it approached the Acropolis, the Panathenaic Way was deflected towards the north-east by a spur of higher ground running north from the Areopagus hill.[50] Along the eastern slope of this spur were situated the first civic buildings, thereby establishing the basis of a north–south, east–west orientated agora space, traversed diagonally by the ceremo-

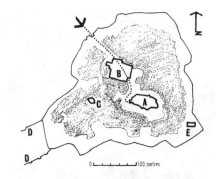

Figure 2.14 – Athens, general plan of the fifth century BC showing the relationship of the main parts of the city. Key: A, the Acropolis; B, the Agora; C, the Pnyx; D, the Long Walls to Piraeus; E, the Olympieion Temple. The dotted line through the north-western part of the city, leading to the Acropolis, shows the route of the Panathenaic Way.

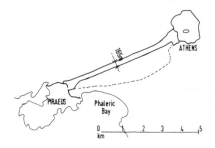

Figure 2.15 – Diagrammatic plan of Athens and Piraeus showing the line of the Long Walls and (dotted) that of a possible earlier wall. The Long Walls enclosed a protected corridor some 6 km in length, averaging 165 m in width.

46. L R Muirhead, *Athens and Environs*, 1962.
47. The date 1581 BC is that of Greek legend: it has as little real meaning as 753 BC, when Romulus and Remus reputedly founded Rome or that of 25 March AD 421 at midday exactly, when Venice is supposed to have been established.
48. E A Gardner, *Greece and the Aegean*, 1938.
49. See Chapter 10 for description of the way the emergent nineteenth-century Los Angeles acquired the corridor of land linking the historic nucleus to a port facility on the Pacific.
50. See H A Thompson and R E Wycherley, *The Agora at Athens*, 1972.

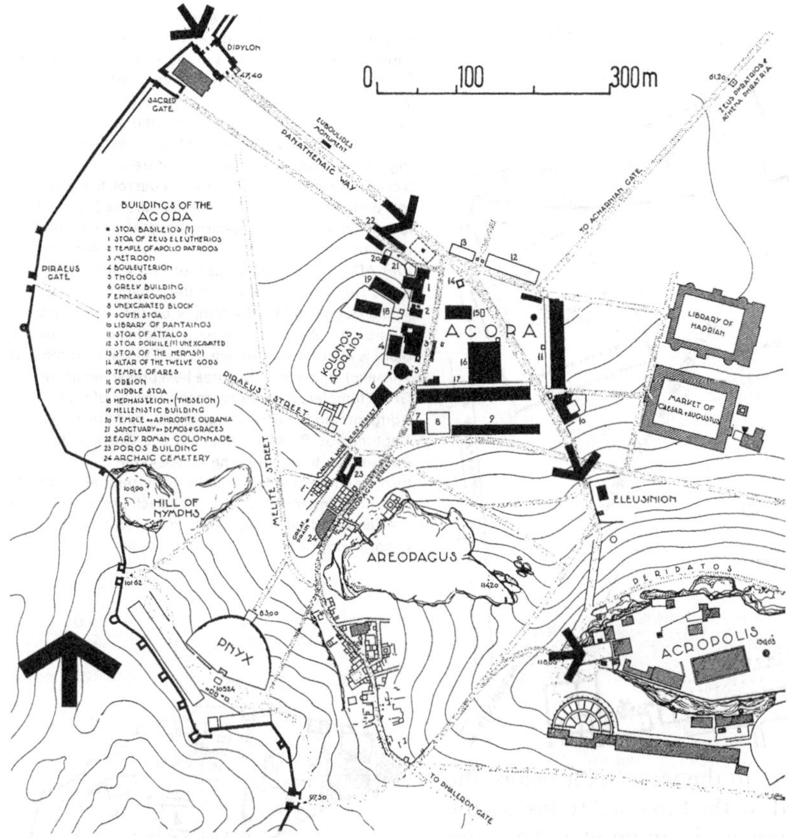

Figure 2.16 – Detail plan of the north-western part of Athens in the second century AD. The Agora is shown in its final form with the 'huge, clumsy structure of the Odeion' (Bacon) and that of the Middle Stoa, inserted into the central space. The line of the Panathenaic Way is shown by a sequence of arrows. East of the Agora, two important Roman buildings are shown: the library of Hadrian and the market of Caesar Augustus. A typical linear group of organic growth housing is shown south of Areopagus hill. This was the general pattern of housing in Athens.

nial route. The earliest known remains are possibly of the seventh century BC, but most monuments of the archaic agora date from the sixth. The later history of the agora can be divided into three separate phases: Classical fifth century; Hellenistic second century BC; and Roman second century AD.

The first agora buildings were largely destroyed by the Persians; subsequent rebuilding was slow but by the end of the fifth century the classical Athenian agora was completed (Figure 2.17). Most of the buildings were on previously occupied sites. The western side comprised the Tholos (a circular committee-room in the south-western corner); the old Bouleuterion, with the new one behind, further up the slope; and the site of the Metroön (a special temple) not built on until later and forming the greater part of the open area in front of the Hephaisteion. This imposing Doric temple (c. 428 BC) still survives, almost intact, on top of the spur, opposite and dominating the centre of the agora.[51] The Temple of Apollo, north of the Hephaisteion axis, was not rebuilt until the fourth century. The last building on this side was the Stoa of Zeus. In front of it, alongside the Panathenaic Way, a late sixth-century altar was dedicated as a central milestone of the Attic road system showing that the spot was thought of as the centre not only of Athens but of all Atica.[52]

Little is known of the buildings along the northern and eastern sides. There is evidence of a square peristyle at their intersection – possibly a law court – but it was not completed. To the south was a sequence of buildings: the central south stoa of the late fifth century, flanked by a

This closeness to rural ways no doubt partly accounts for the primitive housing accommodations and sanitary facilities that characterised the Greek cities right into the fourth century and even later. The houses were lightly built of wood and sun-dried clay; so flimsy were the walls that the quickest way for a burglar to enter a house was by digging through the wall. Residentially speaking the biggest cities were little better at first than overgrown villages; indeed, precisely because of their overgrowth and density of site occupation, they were certainly much worse, because they lacked the open spaces of the farmyard and neighbouring field.

Thus the highest culture of the ancient world, that of Athens, reached its apex in what was, from the standpoint of town planning and hygiene, a deplorably backward municipality. The varied sanitary facilities that Ur and Harappa had boasted two thousand years before hardly existed even in vestigial form in fifth-century Athens. The streets of any Greek city, down to Hellenistic times, were little more than alleys, and many of these alleys were only passageways, a few feet wide. Refuse and ordure accumulated at the city's outskirts, inviting disease and multiplying victims of the plague. The stereotyped, largely false image of the 'medieval town', which many people who should know better still retain, would in fact be a true image for the growing cities of sixth- and fifth-century Greece, particularly in Attica and the Peloponnese. Certainly it applies with far more justice to these cities than to many towns in Western Europe in the thirteenth century AD.

(L Mumford, The City in History)

51. The Hephaisteion in Athens is one of a trio of marvellously surviving Greek Doric temples – architecturally, by comparison the Parthenon is a much less complete fourth. The other two are the Temple of Concord at Agrigento (Akragas) in Sicily, and the Temple of Poseidon at Paestum in Italy, my personal favourite.
52. Wycherley, How the Greeks Built Cities.

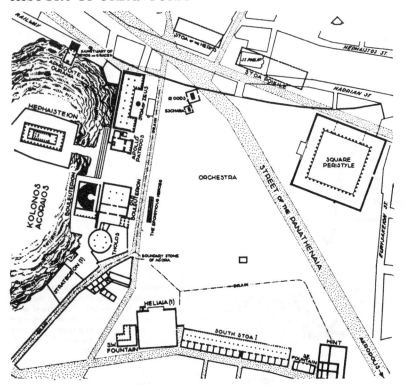

Figure 2.17 – Athens: the Agora area in the third century BC (compare with Figure 2.16 which gives the final form of the ancient Agora in the second century AD). The route of the Panathenaic Way, leading up to the western end of the Acropolis, was deflected north and east by a spur of high ground (to the left of the plan) along the eastern side of which the earliest Agora buildings were located. The precise nature of the buildings along the northern and eastern sides of the Agora of this period have not been determined. The Stoa of Attalus which formed the eastern side of the final arrangement of the Agora was not built until the second century AD. (See Bacon, *Design of Cities*, for a series of plans showing the growth of the Athenian Agora.)

The Hephaisteion, the temple on the higher ground outside the Agora to the west (left-hand edge of the plan) is alternatively known as the Theseion (Temple of Theseus). It is one of the three best preserved of ancient Greek temples; the others are the Temple of Concord at Agrigento, in Sicily (see Figure 2.6), and the Temple of Neptune, at Paestum (the ancient Greek colony of Poseidonia, founded in the mid-seventh century BC). I have seen all three and collectively they are marvels beyond comparison. If I had to select one, it would be the Temple of Neptune, for which the location holds the richer memories.

mint to the east and possibly a law court to the west. 'Not before the second century BC was the agora framed by the Stoa of Attalus (to the east), one more proof that nothing was less in the mind of archaic and classical Athens than the creation of a regularised, closed or half-closed square.'[53] It seems that buildings were not permitted in the classical agora. Part of the area was used for theatrical performances, up to the time when a specialist theatre was built on the southern Acropolis slopes, with spectators perhaps sitting on the broad flight of steps leading up to the Hephaisteion.

During the Hellenistic period a number of improvements were made to the agora. A new south stoa was added in front of the old one, and its colonnade, together with that of the Stoa of Attalus, must have provided a highly effective and simple repetitive foil to the important individual buildings along the western side, where the old Bouleuterion had been replaced by a new, larger Metroön. This latter building also had a colonnaded front, complementing the Stoa of Zeus on each side of the Hephaisteion axis. Subsequent developments were of retrograde quality: 'symptomatic of the architectural disaster to come is the huge, clumsy structure of the Odeion, an indoor meeting hall ... its ungainly mass throws the sensitive and delicate buildings of the earlier periods out of scale'.[54] A second building – the Temple of Ares – was also allowed in the agora, which became increasingly cluttered by statues, fountains and shrines. The destruction of the agora occurred near the end of the third century AD.

THE ACROPOLIS

In the course of time the original rocky outcrop was terraced up with massive retaining walls in order to form a number of interconnected

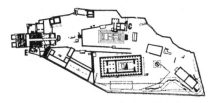

Figure 2.18 – Athens: plan of the Acropolis, north at the top. (See Figure 2.16 for relationship of the Acropolis with the Agora, to the north-west.) Key: A, Athena Temple (Parthenon); B, Archaic Athena Temple site; C, Erechtheum; D, Propylaea; E, Nike Temple.

At Athens, the single-handed victory over the Persian hosts at Marathon on 11 October, 490 BC, supplied the motive for an architectural renaissance. Another factor was the opening, at about the same time, of the Pentelic marble quarries just outside Athens; hitherto very little material had been extracted from this source, and for architectural and sculptural purposes it had been necessary to import marble at considerable expense from the islands, especially Paros, so that it had been employed very sparingly. The conjunction of these events led to a scheme, presumably sponsored by Aristeides, for the rebuilding of the Acropolis in a material unrivalled in the Greek world.
(W B Dinsmoor, *The Architecture of Ancient Greece*)

53. Zucker, *Town and Square*.
54. E Bacon, *Design of Cities*, 1974.

Figure 2.19 – Athens: aerial view of the Acropolis from the south-west, looking across the partially restored ruins of the Roman Theatre of Herodes.

platforms. The most important phase in its history was the second half of the fifth century BC when, as the culmination of the city's reconstruction after the Persian sack,[55] four world-famous buildings were completed: the incomparable Parthenon (447–432), the Propylaea (437–432), the beautiful little Temple of Nike Apteros (426) and the three-part Erechtheum (420–393).

The Parthenon – officially the temple of Athena Polias – was built on a magnificent limestone platform 237 feet long by 110 wide, centrally located on the southern side of the Acropolis. This platform is some 35 feet above bedrock in one corner. It was previously intended for an earlier temple – itself the second on the site – started after the defeat of the Persians in 490, only to be destroyed at an early construction stage when the city had to be abandoned ten years later. The Erechtheum

55. E A Gutkind, *Urban Development in Southern Europe: Italy and Greece*, 1969, writes that 'the devastation of Athens by the Persians in 480 BC was so great that, as the story goes, Themistocles considered the transfer of the city to Piraeus. Yet in 479 BC Athens was surrounded by walls. This was the first fortification of a city on mainland Greece [and] finally ended the role of the acropolis as the citadel of Athens.

was built to the north of the Parthenon, separated from it by the site of the Peisistratid Temple of Athena also destroyed by the Persians. The main central space on the Acropolis between these buildings and the Propylaea was dominated by the colossal bronze statue of Athena Promachos.[56] The Propylaea, which stood on the site of earlier gateways to the Acropolis, remained incomplete at the outbreak of the Peloponnesian War. The Temple of Nike Apteros stands on a separate bastion to the right of the Propylaea, as viewed from the outside. The southern slope of the Acropolis was well on the way to becoming the city's cultural precinct, with the impressive Theatre of Dionysos providing seats for some 12,000 spectators.

Greek urban planning is assessed below, when further reference is made to the Athenian Acropolis. Here it remains only to stress that it was topography and respect for traditional siting which effectively determined the location of these four buildings, even though the axis of the Propylaea is parallel to that of the Parthenon and also points approximately toward the statue of Athena Promachos, thus constituting one of the rare instances of formal relations between buildings antedating Hellenistic times.[57]

Athens: a Roman and later postscript

Urban historians have usually lost interest in Athens by the close of the fifth-century BC 'Age of Pericles', with, characteristically, little more than concluding dismissal of the Roman contribution.[58] Yet in major respects the city's history was only beginning, with an extraordinary story of down and up during the following two or so millennia, culminating in the re-establishment of Athens as the Greek capital in the 1830s, after the War of Independence against the Turks. Not least in this formative experience were almost four centuries of Islamic occupation, which left typically indelible marks on a city, as indeed a country, still poised – in limbo as it were – between east and west in numerous fascinating respects.

FROM PERICLES TO ISLAM

After the Golden Age of Pericles, the city's political and economic fortune fluctuated and cultural investment was curtailed. The Hellenistic period saw the Stoa of Eumenes (197–159 BC) added to the Theatre of Dionysus as a superbly situated promenade, but it awaited the Romans for renewed major public building works. At first they allowed the city considerable autonomy but in 86 BC the Athenians were punished by Roman legions under their general, Sulla, and the city walls were razed. However, as an 'open city' within the pax Romana, Athens regained her cultural leadership of the Mediterranean world. The city continued to expand and with funding by the emperors Julius Caesar and Augustus a new commercial forum was added in the first century BC on a 'redevelopment site' east of the old agora, to which it was directly linked (Figure 2.16). The Emperor Hadrian favoured Athens, building an extensive new Library in AD 131–132, north of the Forum, and completing the enormous, long-delayed Temple of Olympian Zeus in his new Roman district east of the historic nucleus. Later Roman buildings include the Odeion of Herodes Atticus, originally a roofed theatre with seating for 5,000 to 6,000 spectators constructed in 160–174 at the south-western corner of the Acropolis, completing the

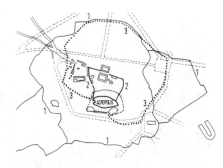

Figure 2.20 – Athens: the sequence of defensive walls after J Travlos, *Athens au fil du temps*, 1975. Key: 1, the ancient city wall of Themistocles, extended by the Roman Emperor Hadrian; 2, the period 1204–1456; 3, the Turkish wall of Ali Hasekis, during the War of Independence, 1778–1833.

... somewhere between the second and the first century BC, Dicaerchus could observe: 'The road to Athens is a pleasant one, running between cultivated fields the whole way. The city is dry and ill-supplied with water. The streets are nothing but miserable old lanes, the houses mean, with a few better ones among them. On his first arrival a stranger would hardly believe that this is the Athens of which he has heard so much.' The best one can say of the housing situation in Athens is that the quarters of the rich and the poor were side by side, and that except perhaps in size and inner furnishings, were scarcely distinguishable: in the fifth century, noble poverty was more esteemed than ignoble riches, and public honours and family repute counted for more than private wealth.
(L Mumford, *The City in History*)

Figure 2.21 – Athens: the street layout when Greek independence was achieved in 1834, and Leo von Klenze produced his master plan for the renewal and expansion of the city. (See Chapter 11 for comparative Islamic examples.)

56. For contrasted descriptions of the Acropolis see Dinsmoor, *The Architecture of Ancient Greece*, and Le Corbusier, *Towards a New Architecture*, 1946/1970.
57. Dinsmoor, *The Architecture of Ancient Greece*.
58. For the position of Athens under Rome, see J C Stobart, *The Grandeur that was Rome*, 4th rev. edn 1964.

cultural precinct; and the Stadium, originally that of the fourth-century BC Panathenaic Games, was rebuilt by Herodes Atticus in AD 144 to provide some 60,000 white marble terraced seats. These were 'quarried' for use elsewhere during the city's centuries of oblivion, but they were replaced along the original lines when the Stadium was reconstructed for the modern revival of the Olympic Games in 1896.[59]

In the third century in face of threats by the Goths and Heruli, the Emperor Valerian repaired the old wall of Themistocles, with extension to the east to include the new Hadrianic town. This was in vain, however, because in 267 the lower city beneath the Acropolis was devastated by their attack. Afterwards, Athens was rebuilt on a greatly reduced scale contained to 40 acres within a new wall immediately north of the Acropolis. 'This district, from the Acropolis to Hadrian's Library and the Roman Agora was to form the centre of the town from the end of antiquity until the beginning of the 19th century.'[60]

In 1461 Athens fell to the Turks and the city entered its Islamic period. This endured for nearly four centuries, until Greek independence was regained in 1830. Under the Turks the city was divided into two parts: the Acropolis-Kastro, the bastion of the occupying power, and the lower city, divided into Greek and Turkish quarters. Up on the Acropolis, the Parthenon was converted into a mosque, with a minaret at the south-west corner, the Erechtheion housed a harem, the Propylaea became the military headquarters and an arsenal, and latterly, in 1684, the marvellous little temple of Nike Apteros was demolished to provide stone for a gun emplacement. The space between buildings was tightly packed with Turkish dwellings. Then in 1687, during the siege of Athens by the Venetians, the still more or less complete Parthenon was used as a powder magazine, with disastrous effect when it exploded, demolishing the roof, the naos (inner walls) and twenty-eight of the enclosing columns.[61] Down below, covering the area of present-day Plaka and Monastiraki districts, there were the densely packed courtyard houses and narrow winding access lanes of a typical Islamic city.

Reversion to citadel of the Acropolis was to occur on two further nineteenth-century occasions: first, when the Athenians rebelled on 25 April 1821, the Turks holding out therein until 10 June; the second, in 1827 when the Greeks in turn were forced to surrender, after an eleven-month siege that devastated the lower city and its environs. Only in 1834, four years after the War of Independence, did the Turks finally relinquish the Acropolis: in that year Otho of Bavaria, the newly appointed King of Greece, made his ceremonial entry into a devastated city of perhaps, 4,000 inhabitants, bringing with him the Bavarian architect Leo von Klenze with plans to recreate Athens as a fitting new capital city.[62] It was all there to be done: Athens, in ruins, extended not much further than the present-day Plaka and Monastiraki districts, while the Agora was wasteland and south of the Acropolis was open country.

ATHENS RESURGENT

The von Klenze plan of 1834 is shown as Figure 2.22,[63] and there was a broadly similar proposal by S. Kelanthis and E Schaubert (1833). Their common basis was the creation of an archaeological zone, clear of new buildings on the northern slopes of the Acropolis and to the west, with a main new axial street leading away from the Acropolis to the north.[64] Symmetrical diagonal avenues skirt the edge of the old city,

Figure 2.22 – Athens: the von Klentze Master Plan. The combination of symmetry and orthogonal 'drawing board' lines are characteristically those of mid-nineteenth-century Austrian urban design (for which see also the Ringstrasse in Vienna, in Chapter 7).

The Hephaisteion in Athens is one of a trio of marvellously surviving Greek Doric temples – architecturally, by comparison the Parthenon is a much less complete fourth. The other two are the Temple of Concord at Agrigento (Akragas) in Sicily, and the Temple of Poseidon at Paestum in Italy, my personal favourite.

59. It had been proposed that the Stadium would also form a central part of the facilities for the 1996 Olympic Games, which would have been an appropriate venue for the centenary Games, but the bid by Athens to be the host city was not accepted. Among reasons given was the unacceptable levels of atmospheric pollution created by Athenian vehicle traffic.

60. Bacon, *Design of Cities*; see Travlos, *Athens au fil du temps*, for the sequence of maps showing the varying extent of Athens.

61. For one of the most extraordinary might-have-beens of urban and architectural history, see R Carter, 'Karl Friedrich Schinkel's Project for a Royal Palace on the Acropolis', *Journal of the Society of Architectural Historians*, vol XXXVIII, 1979.

62. See M Webb, *The City Square*, 1990, for illustration of the work of Leo von Klenze in his native Munich, where he was court-architect for the Bavarian king, Max I Josef. Von Klenze was an admirer of ancient Greek architecture and his 'temple' buildings at Königsplatz are interesting, of their kind.

63. For the von Klenze Plan for Athens see L Benevolo, *The History of the City* (an urban historian who brings historic Athens through to the nineteenth century, but who omits mention of the city's formative centuries within Islam).

64. The common basis for the plans was the creation of an archaeological zone clear of new building on the northern slopes of the Acropolis, and to the west; with a main axial street leading up to it through the new districts. Symmetrical diagonal avenues skirted the edge of the old city. By 1861 the renewing city had a population of about 40,000.

with that to the south-east setting the angle in the direction of the historic stadium. The intersection is present-day Omonia Square, and the main avenues were used as the alignments of the new gridiron districts. By 1861 the city's population had reached 41,000.

Stadiou Street, however, never made it as far as the Stadium, ending at Sintagma Square;[65] likewise the archaeological precinct was to remain a planner's pipedream, submerged in main part beneath expedient house rebuilding by the Plaka landowners. As the oldest district of Athens, the Plaka is dismissed in some circles because its location and comparative age of small-scale buildings and spaces has made it the centre of the city's burgeoning tourist industry. Sheer pressure of in-season numbers may well militate against serious enjoyment of the Plaka's urban character; yet one need only look beyond the tourist trinkets, over and above, perhaps, with glimpses of the Acropolis, in order to be part of a truly exceptional human experience. And even the crowds, in their way, are recreating the contrast between everyday ancient city bustle and the impressive skyline dignity of the Acropolis.

Greek urbanism: minor examples

DURA-EUROPOS

Dura-Europos is believed to have been founded about 280 BC as one of several fortress colonies safeguarding the Euphrates river-crossings and trade route.[66] It was laid out on the right bank of the river with a straightforward gridiron plan and, anticipating Roman imperial policy, was populated with Greek soldiers. Around 100 BC the town became part of the Parthian Empire and enjoyed a period of great prosperity as its northernmost garrison town, well situated on the main caravan route west to Palmyra.

OLYNTHUS

Olynthus was located on the northern Aegean coastline, alongside the Toronaic Gulf.[67] The original village settlement had occupied the southern spur of two hills from about 1000 BC before being destroyed during the Persian invasion of 479 BC, following which the site was redeveloped as one of several Chalcidian towns. Olynthus soon became the dominant one of these centres and from around 430 BC the Chalci-

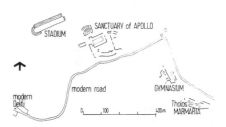

Figure 2.23 – Delphi: outline plan of one of history's most improbably located cities, whose *raison d'être* was the sanctuary of Apollo where hosts of pilgrims travelled to consult the oracle. The comparatively isolated, steeply mountainous area can appear mysterious to this day, an effect compounded for the ancients by frequent thunderstorms and occasional earthquakes. And, as a particular attraction, there was a deep rocky chasm from whence emanated unearthly noise and magical vapours. There it was, from the second millennium BC, that the earth goddess Ge (or Gaia) hid herself, guarded by her son, the snake Python, and with her divine pronouncements interpreted by the oracle. Apollo killed the snake (according to Homer) and took over as the god Python who gave oracles through a local woman known first as the Pythia and later as the Delphic Sybil.

There was a charge made for consultations, a proportion of which was used to improve the amenities and to attract yet more pilgrims. Most of the investment went towards the magnificent buildings of the Sanctuary of Apollo, for the detail plan of which see W B Dinsmoor, *Ancient Greek Architecture*. Known to have been still prosperously in business during the second century AD, the Sanctuary of Apollo was closed down in 381 by the Christian Byzantine Emperor Theodosius the Great.

The plan of Delphi is also one of the more misleading in urban history. From the modern road it is a steep climb up the mountainside through the Sanctuary and theatre to the stadium. Views back down and across the valley are stupendous.

Figure 2.24 – Dura Europos: general plan (north at the top); the Euphrates ran north-east of the town. See C Hopkins, *The Discovery of Dura-Europos*, 1979; A Perkins, *The Art of Dura-Europa*, 1973.

65. Sintagma Square (Platia Sintagmatos), overlooked by the Vouli, the Greek National Parliament building, is the popular tourist centre of Athens, which otherwise – in common with many large cities over-imbued with history – lacks a central focus (see the *Historical Map of Athens*, Ministry of Culture, 1989).
66. For Dura-Europos see R Stillwell, W L MacDonald and M A MacAllister, *The Princeton Encyclopedia of Classical Sites*, 1976.
67. For Olynthus see Wycherley, *How the Greeks Built Cities*.

Figure 2.25 – Olynthus: general plan of the city as built from 430 BC (north at the top), locating the gridiron housing area shown in detail in Figure 2.8. New Olynthus was laid out on the gentle slopes of the plateau north of the original site and covered an area of approximately 3,600 feet from north to south, by 600 feet wide. The gridiron formed the basis of the plan: four main north–south streets between 21 and 15 feet wide were intersected by minor cross-streets 15 feet wide. (Figure 2.8 shows in detail only the central three of the five housing grid blocks.) The main entrance was in the north-west corner; the agora area was probably located between the western grid section of New Olynthus and the original site, which remained in more or less unplanned occupation.

dians abandoned their other settlements to concentrate around Olynthus. From 379 BC the city grew rapidly until 348 BC when, at the time of its final destruction by Philip of Macedonia, its population including many slaves may have exceeded 15,000.

SELINUS

Selinus in western Sicily was founded around 630 BC by local tribesmen. It was situated on two flattish hills, the southern of which, alongside the sea, formed the original site.[68] Early expansion took place on to the second, northern hill and adjoining slopes. The original hilltop gradually lost its predominantly residential character and became from 580 BC onwards an acropolis with about 7½ acres of its total area of nearly 22 acres forming a religious precinct containing several extremely important examples of Greek temple architecture. Unplanned housing groups, however, continued to occupy the easier slopes of the hill. It has been estimated that the total area of the city in the late fifth century BC amounted to some 70 acres, with a population of around

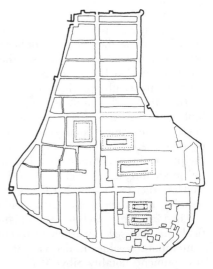

Figure 2.26 – Selinus (north at the top), as laid out from 409 BC with the temple group in the south-eastern corner. The main cross-axes are a departure from usual Greek practice.

68. For Selinus see M I Finley, *Ancient Sicily*, 1968.

30,000, excluding slaves. In 409 BC the city was completely destroyed by the Carthaginians. It was then rebuilt on the original southern hilltop, the layout being based on the gridiron, incorporating the temple in its south-eastern quarter.

Greek urbanism, theory and practice

The planned Greek city, for all its regularity and formal building relationships, was never the result of academic urban planning rules. Indeed Wycherley comments that 'as far as we can tell there was no recognised body of theory'.[69] This is illustrated by the various views on 'ideal city' populations. Aristotle perceptively argues that 'ten citizens do not make a polis, whereas with ten thousand it is a polis no longer'. Plato, basing his estimate on mathematical theory, writes of 5,040 shares in the land, while Hippodamus divides his preferred total of 10,000 into three parts: artisans, husbandmen and armed defenders of the state. (He also divided the land into three zones: sacred, private and public.) More generally, Aristotle wanted city-state populations to be 'self-sufficient for the purpose of living the good life after the manner of a political community', and argued that 'to decide questions of justice and in order to distribute the offices according to merit it is necessary for the citizens to know each other's personal characters'. Aristotle also made several deceptively simple general pronouncements including 'a town should be built so as to give its inhabitants security and happiness' and, 'the difficulty with such things is not so much in the matter of theory but in that of practice'.

Greek urban form of the Hellenic period was therefore essentially the result of applying uncomplicated planning principles to the site in question and of accepting, seemingly without question, that town planning is indeed the art of the practical. Paul Zucker regards this overriding concern with practicalities as 'all the more astonishing since after all Aristotle was the first philosopher to deal with aesthetic problems in general ... yet he never discusses the city planning from an aesthetic point of view'.[70] Urban space as such had no aesthetic meaning: it existed only as a by-product of placing two or more buildings together on the same site, even if, as happened frequently in Hellenistic times, buildings around a space are 'no longer spatially isolated but are anchored in some system of mutual reference, as are the stoas and porticos'. Greek architects and artists were preoccupied with volume – the physical mass of individual buildings or sculpture. Interest in spatial modelling, as it developed slowly from the fifth century, was centred on the agora and hardly ever applied to the arrangements of temples, monuments and statues on an acropolis.

The Greek approach to urban planning was then essentially a practical one. Yet several urban historians make the mistake of applying contemporary values in decrying the assumed quality of Greek urban environment, notably Sibyl Moholy-Nagy who writes of Miletus as 'a nightmare to anyone who would think three-dimensionally. It would have been a rat's maze of blank walls'[71] (see the section on Miletus on page 43). In the absence of contemporary Greek observations on their cities it is pointless to pursue the argument further except to note that if the Greeks had not found their surroundings acceptable then they, of all people, surely had the ability to change them.

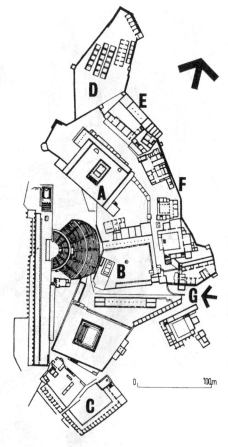

Figure 2.27 – Pergamon, detail plan of the acropolis complex. Key: A, Temple of Trajan; B, precinct with Temple of Athena; C, the upper agora with the Great Altar of Zeus on the terrace to the north; D, storehouses; E, barracks; F, palace complex; G, main gateway.

Pergamon had poor communications inland but combined a strong acropolis site with a fertile coastal plain giving easy sea access. Little is known of the plan and buildings of the lower town; archaeological interest has been centred on the acropolis and its remarkable group of buildings and civic spaces. The extravagance of their layout on extremely difficult, steeply sloping ground is typically Hellenistic in character, compared with the simple approach adopted by Hellenic planners at, for example, Priene. Nevertheless the skill with which the great crescent of buildings was arranged around the central natural theatre (each on its own terraced platform) has seldom been equalled.

69. Wycherley, *How the Greeks Built Cities*.
70. Zucker, *Town and Square*.
71. S Moholy-Nagy, *Matrix of Man*, 1968.

3 – Rome and the Empire

The legendary date for the foundation of the city of Rome is 753 BC.[1] Subsequent Roman history is usually divided into three phases: the Kings, 753–510 BC; the Republic 509–27 BC; and the Empire (imperial Rome) 27 BC–AD 330. Before about 270 BC the Romans were fully committed, establishing their mastery of the Italian peninsular, after which the Punic Wars against the Carthaginians (264–146 BC) further decided that Rome should become a world power, and that the lands of the west should be ruled by an Aryan, not a Semitic race.[2] For the next 300 years the imperial boundaries were extended ever further from Rome, reaching their maximum extent under the Emperor Trajan (AD 98–117); in conquering Mesopotamia and Assyria he made the Tigris instead of the Euphrates the eastern boundary (see Figure 3.2). His successor, Hadrian (AD 117–138), abandoned these territories, and Armenia, but successfully consolidated the frontier defences for an uneasy but peaceful period. However, under Marcus Aurelius (161–180) the tide began to turn. From then on Rome was always on the defensive; the period of the decline and fall of the Empire begins.

Dacia was lost to the Goths in AD 270. They also invaded east Germany, Transylvania, Illyricum, and Greece as far as Athens and Sparta. The Juthungi reached North Italy; the Alemanni, who first appear about AD 210, thrust into Gaul and Italy, and for a moment appeared before Rome.[3] Gradually the centre of gravity of the Empire moved to the east, Rome becoming increasingly unsuited to the strategic requirements of military government. Diocletian (284–305) ruled mainly from Nicomedia, on the eastern shore of the Sea of Marmora, and his successor, Constantine (306–337), finally deserted Rome for the new eastern capital of Constantinople, founded in 330 on the site of Byzantium (for which see the Epilogue to this chapter). The decline and fall of the city of Rome is outlined in the introduction to Chapter 5.

It has been fashionable in certain historical circles to dismiss the artistic, architectural and urban design work of the Romans as at best mediocre copies of Greek originals. While the Greeks were recognized as 'artists' in the fullest sense, the Romans were discounted as 'practical engineers' without any significant aesthetic ideas of their own. Lewis Mumford is strongly anti-Roman; his book *The City in History* will be quoted later in this chapter for its critique of the city of Rome. In general he has this to say: 'The special Roman contribution to planning was chiefly a matter of sturdy engineering and flatulent exhibitionism: the taste of *nouveaux riches*, proud of their pillaged bric-à-brac, their numerous statues and obelisks, stolen or meticulously copied, their

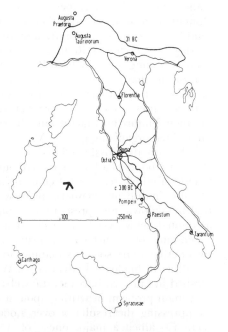

Figure 3.1 – Italy: showing the extent of Roman territory around 300 BC and latterly the Alpine frontier as existing before the reign of Augustus (29 BC–14 AD). The main roads are also shown. (Augusta Praetoria is modern Aosta; Augusta Taurinorum is Turin.)

---

1. Rome, with her Romulus and Remus legend – twin baby brothers set adrift in a basket on the Tiber, washed ashore beneath the Palatine Hill, suckled by a she-wolf, adopted by shepherds and founders of Rome – is one of several cities with mystical, and mythical, birthdays. See also the legends concerning Athens and Venice.

2, 3. R. H. Barrow, *The Romans*, 1970. For general background, see also H H Scullard, *A History of the Roman World 753 to 146 BC*, 4th rev. edn, 1978; E T Salmon, *A History of the Roman World 30 BC to AD 138*, 1949; M Cary and H H Scullard, *A History of Rome*, 1975; J E Stambaugh, *The Ancient Roman City*, 1988; T Cornell and J Matthews, *Atlas of the Roman World*, 1982; E T Salmon, *Roman Colonization under the Republic*, 1969; J Carcopino, *Daily Life in Ancient Rome*, 1941/1991.

imitative acquisitions, their expensive newly commissioned decorations.' Such criticism is unduly harsh. There were of course pronounced negative aspects to the Roman way of life, particularly in Rome itself, but for a more balanced assessment we need only turn to the contemporaneous conclusions of a fellow-American author Paul Zucker in his book *Town and Square*; he observes that 'in architecture as well as in sculpture the Romans created entirely new and original artistic values, although taking over the artistic vocabulary of the Greeks'.[4]

Artistic attainment apart, the main achievement of the Romans was surely the creation and administration of their vast Empire during the course of which they introduced urban civilization into all of Europe east of the Rhine and Danube. With more recent comparable circumstances in mind it may perhaps be argued that many recipients of Roman civilization would rather have been left alone in their furs and mud huts. Certainly exploitation of mineral and agricultural resources to supply the city of Rome's population of 1.25 million played an ever-increasing role; a side to Roman activities which is directly comparable to that of the British in many parts of their Empire, and if native societies were dragged, protesting, into the nineteenth and twentieth centuries, are the long-term benefits really to be doubted? Accepting their Empire as a fact it must be recognized that the Romans had undoubted organizing genius. Merely to keep their capital functioning at all, with its grossly inflated population, is proof of the devoted, efficient labours of a host of anonymous bureaucrats able to deploy the talents of gifted, resourceful civil engineers. Mumford, it seems, finds pleasure in drawing comparisons between Roman society and its eventual breakdown, and present-day situations, particularly in the USA, as quoted below, yet he neglects to stress that Rome's unimaginative acceptance of its urban circumstances is in no way different from twentieth-century society's inability to break out of its urban impasse, for all its technological advantages. We are, as the Romans were, enmeshed in our own bureaucratic webs.

A main problem in writing about ancient Rome is that of selection. Compressing the results of over 1,000 years of rapid change into a section – albeit a major one – of this general study has inevitably resulted in some aspects of the subject being omitted, and others only too briefly mentioned. Rome has been described in great detail in a multitude of archaeological and related studies. The present work relies in the main on three outstanding books: two from the later nineteenth century – Professor Lanciani's *The Ruins and Excavations of Ancient Rome* and J H Parker's *Archaeology of Rome* – are not ordinarily available;[5] the third – Jerome Carcopino's *Daily Life in Ancient Rome*, first published in 1941, has been made available in paperback by Penguin Books.[6] The method adopted here for presenting an ordered description of such a disorganized subject is to commence with a brief introductory general history of Rome, and to follow this by a more detailed description of the most important elements of urban form both in Rome itself and throughout the Empire.

ROMAN URBAN PLANNING

The total contrast between the chaotic organic growth of the city of Rome and the regulated formality of the great majority of Roman provincial towns is even more marked than that between Athens and

Despite the fact that the Romans were quite capable of indulging in gigantic undertakings, their technologies remained at the small-equipment level. Thus, for example, if it was required to increase the output of iron the number of furnaces was multiplied, but the furnaces themselves remained the same size. Whatever the cause, the idea of building a larger furnace and devising machinery to work it seems to have been beyond the Roman mind. As a result, the last few centuries of Roman domination produced very little that was technologically new. No new raw materials were discovered, no new processes invented, and one can indeed say that long before Rome fell all technological innovation had ceased.
(H Hodges, Technology in the Ancient World)

A great deal of nonsense has been talked about the luxury of the Romans as one of the causes of their decline. Even Mommsen relates with shocked emotion that they imported anchovies from the Black Sea and wine from Greece. Two hot meals a day they had and 'frivolous articles' including bronze-mounted couches. There were professional cooks, and actually baker's shops began to appear about 171 BC. It is true that all this luxury would pale into insignificance before the modern worker's breakfast-table with bacon from Canada, tea from Ceylon or coffee from Brazil, sugar from Jamaica, eggs from Denmark, and marmalade from South Africa. Cato would have swooned at the sight of our picture frames coated with real gold, for he taxed table-ware worth more than 1,000 'denarii'. The truth is that Rome, having grown rich was just beginning to grow civilised. It is the everlasting misfortune of Rome that events occurred in that order.
(R M Haywood, Ancient Rome)

4. The reader is referred to G Picard, *Living Architecture: Roman*, 1966, for a general illustrated description of Roman architecture, and to F Baumgart, *A History of Architectural Styles*, 1970, for the relationship of Roman architecture to the preceding and succeeding styles.

5. *The Ruins and Excavations of Ancient Rome* was published by Macmillan, London, in 1897; *The Archaeology of Rome* was published in 1874, from which an abridged version entitled *The Architectural History of Rome* was published by Parker, London, in 1881.

6. Jerome Carcopino was Director of the Ecole Française de Rome; his *Daily Life in Ancient Rome* was first published in 1941 and by Penguin Books in 1956. *Daily Life in Ancient Rome* (on the daily round of business appointments, social calls, work, and leisure amid the tumult and power of Rome in the second century AD) is unquestionably one of the most informative and easy to read specialist works on urban history. It is pleasing to note that Carcopino's extraordinarily fascinating account was republished in paperback in 1991.

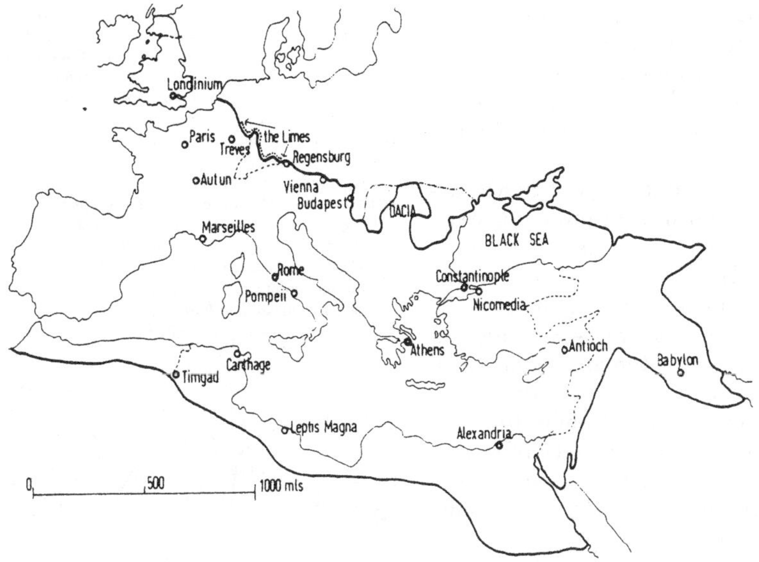

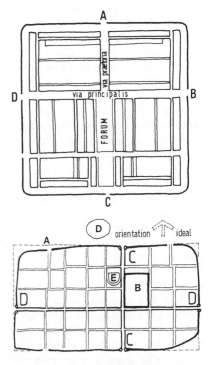

Figure 3.2 – The Roman Empire at its greatest extent, locating the examples of imperial urban planning illustrated in this chapter. (See Figure 3.1 for the Italian peninsular and Figure 3.40 for the map of Roman Britain.)

Polybius, a Greek statesman and historian of the second century BC, who spent many years among the Romans, left a careful account of the camps in his discussion of the Roman army. Every camp was constructed according to the same master plan; although natural features were sometimes made a part of it, ordinarily it was pitched on reasonably flat land and constituted a fort without rivers or cliffs to aid in its defense. It was square; each side was 2,150 feet long and had a gate.

(R M Haywood, Ancient Rome)

Figure 3.3 – Top, the typical Roman army castra layout. Below, the component parts of a typical imperial urban plan, loosely based on Camulodunum in Roman Britain (see Figure 3.41). Key: A, the defensive wall, rectangular in theory but characteristically of curving outline when constructed around an existing city; B, the Forum at the intersection of the Cardo (C) and Decumanus (D); E, the theatre; and D, the arena. The housing forming the main part of the town was usually of the courtyard type, but see Calleva Atrebatum (Silchester) in Britain for an exception (Figure 3.46).

the numerous systematically planned Greek cities of the post-Hippodamian period. This is due partly to the extreme size and population of the city of Rome – estimated to have been at least 1.2 million in the second century AD – and its complexity of building relationships and inherited constraints and partly to the direct simplicity of layout adopted by the Roman engineers for smaller, planned, provincial urban settlements.

Before describing significant aspects of the city, the general principles and practice of Roman town planning must be established. In imposing and maintaining their authority throughout their vast empire, the Romans built thousands of fortified legionary camps known as *castra*, which characteristically existed only as temporary centres for local military activities. Such camps had to be operational in the shortest time and, following strictly applied rules of castremetation,[7] they were invariably laid out according to the gridiron, within predetermined rectilinear defensive perimeters (Figure 3.3). Though *castra* were essentially temporary, a large number did form the basis of permanent towns. In addition, many other towns were founded for economic and political reasons.

Permanent urban settlements, whether developed from the *castra* or of special origin, were given equally simple standardized plans. Venta Silurum (Caerwent), a Romano-British town (Figure 3.45), and Timgad in North Africa (Figure 3.28) are good examples although, as described later, they differ considerably in the way the housing blocks are developed. The perimeter is usually square or rectangular; within this two main cross streets form the basis of the street structure – the *decumanus*, through the centre of the town, and the *cardo*, usually bisecting the *decumanus* at right angles, towards one end. Secondary streets complete the grid layout, and form the building blocks, known as *insulae*. The forum area – the Roman equivalent of the Greek agora – is typically located in one of the angles formed by the intersection of the *decumanus* and the *cardo*, and normally comprises a colonnaded courtyard with a meeting hall built across one end. The main temple, the theatre, and the public baths – the latter made the Roman occupation of foggy,

7. *Castrametation* – the practice of laying out Roman military camps, and an important part of legionary standing orders. (Nineteenth-century British army engineers were similarly involved, mainly in India but also elsewhere.) The equivalent term for the rectilinear surveying basis of Roman land distribution is *centuriatio* (see Note 56).

damp Britain tolerable in part – were also located near the forum in the centre of the town. The amphitheatre, a large spatial unit characteristically requiring sloping ground for seating, was normally located outside the town. Fortifications were sometimes omitted at first because of the strong imperial frontier defences, only to prove necessary at later, insecure stages in the history of these towns.

The Romans fully appreciated that to attempt to hold newly acquired territories by military force could result in continuing guerrilla warfare which would distract the legions from their task of extending and maintaining the imperial frontiers. It could also prejudice the development of commerce. Native tribesmen therefore had to be brought into the Empire on advantageous terms; this was achieved, explains R G Collingwood, by equating Romanization with urbanization.[8] Tribal centres, which usually amounted to little more than crudely assembled villages, were redeveloped as Roman towns of varying status and leading tribesmen, with varying degrees of enthusiasm, were invited to share the advantages of Roman urban culture and trading prospects. Other towns were also founded for economic and political reasons, their populations provided by time-expired legionnaires, or emigrant settlers from Rome and other older towns. The most important imperial towns were directly connected to each other and back to Rome by the magnificent system of main roads which facilitated strategic military and trading communications. Less important towns were linked to it by minor roads.

The three main classes of imperial towns were *coloniae* – either newly founded settlements or native towns, allied to Rome with full Roman status and privileges; *municipia* – usually important tribal centres, taken over with formal chartered status but only partial Roman citizenship for their inhabitants; and *civitates* – market and administrative centres for tribal districts which were retained in a Romanized form.[9] Throughout the Empire status was not necessarily reflected by size. London with a walled area of 326 acres was more than half as big again as York, which, with a total area of some 200 acres, was the largest of the four British *coloniae* while St Albans, the only known *municipium*, was rebuilt on 203 acres in AD 150. But London's exact status remains unknown, except that it was not in the category of *colonia*, *municipium* or *civitas* (see page 85). Cirencester, at 240 acres the second largest Romano-British city, was apparently an important and prosperous *civitas*. Neither did status have any significant effect on urban form. Roman imperial urbanization invariably implied gridiron structures for new and rebuilt towns alike although, as explained later, local topographical features usually determined the details of the layout and perimeter of the individual towns.[10]

The function of the Roman camps, explains C T Smith, was more often offensive than defensive; they were the supply bases and troop headquarters of armies that depended greatly on movement.[11] Accessibility was therefore a main siting requirement and in place of the comparatively isolated but more readily defended hilltop sites preferred by their Celtic predecessors and German successors, the Romans opted for river crossings and route intersections. Towns developed from such favourably located *castra* most frequently survived the Dark Ages, or, if they were abandoned, stood the best chance of resuscitation, stimulated by the early medieval commercial revival. Similarly many Celtic *oppida* were moved from their hilltops on being taken over as Roman

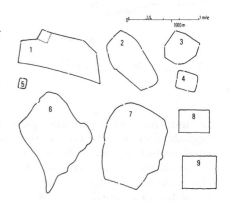

Figure 3.4 – Imperial towns: some comparative sizes. Key: 1, Londinium (London); 2, Corinium Dobunnorum (Cirencester); 3, Calleva Atrebatum (Silchester); 4, Venta Silurum (Caerwent); 5, Viroconium (Wroxeter); 6, Augustodunum (Autun); 7, Augusta Treverorum (Trier); 8, Augusta Praetoria (Aosta); 9, Thamugadi (Timgad). (See also Appendix E relating Londinium and Timgad to the sizes of other cities in history.)

8. R G Collingwood, 'Britain', in *Cambridge Ancient History*, Volume 12. (See later references in the Romano-British section of this chapter.)

9. See E T Salmon, *Roman Colonization under the Republic*, 1969, for the legal status of Roman imperial towns; see also Cornell and Matthews, *Atlas of the Roman World*.

10. The caption to Figure 3.3 has described the component parts of Roman imperial urban planning. The primary determinants were topography and the gridiron, always allowing for local exceptions. The courtyard house was the usual response in Mediterranean, African and other hot climate circumstances, but the gridiron, *per se*, denies that determinant.

In northern cool/cold, damp climes the courtyard house was less appropriate and it was frequently replaced by houses-in-gardens, for which see Calleva Atrebatum (Silchester) in England (Figure 3.46). Civic architecture excepted, other than in Rome and economically favoured provincial capitals, construction materials and technology imposed a general two-storey height limit on residential buildings. Only latterly in the history of the Empire did the fortification determinant become a major consideration, for which see J Maloney and B Hobley (eds) *Roman Urban Defences in the West*, 1983.

civitates. Autun, laid out below Bibracte in Gaul, and Dorchester by the Frome in Britain, replacing Maiden Castle are but two of many examples.

The importance of *colonia* in the imperial planning schema is returned to in this chapter's section illustrating examples of Roman imperial urbanism (pages 69–70).

Rome

THE FIRST SETTLEMENT

Rome, City of the Seven Hills, had its origins in the several villages built by Latin tribes when they moved down to the Tiber plain from mountains to the south-east. Six of the seven hills – in reality only low but steep-sided mounds rising above the flood plain on the Tiber's left bank – are the Palatine, Capitoline, Caelian, Esquiline, Viminal and Quirinal. They are ranged around a small depression or valley which in earliest times must have been a swamp,[12] and this land subsequently

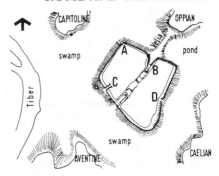

Figure 3.5 – Rome: outline plan of the Kingly Palatine Hill related to the adjoining hills and the Tiber and showing the lower ground conditions prior to the construction of the surface water drainage system. Key: A, Porta Romanula; B, Porta Mugonia; C, Steps of Cacius; D, possible gateway (after Lanciani).

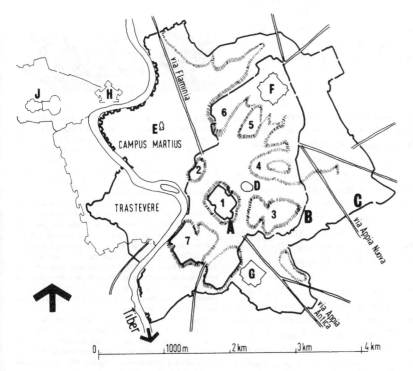

Figure 3.6 – Rome: diagrammatic plan of the city at its greatest extent. The figures indicate the seven hills – 1, Palatine; 2, Capitoline; 3, Caelian; 4, Esquiline; 5, Viminal; 6, Quirinal; 7, Aventine.

The three defensive perimeters are indicated by letters A to C: A, the Palatine, site of the dominant of the original village settlements (see Figure 3.5); B, the Republican Wall of 367–352 BC; C, the Aurelian Wall of 272–280 AD, hurriedly constructed in response to the Barbarian invasion of 271. This wall did not enclose the entire area of the city but followed the line of an existing customs barrier (see description, page 62). It has been estimated that the total area of 1,386 hectares (3,465 acres) within the Aurelian Wall was used as follows: streets, 138 ha; public buildings, 84 ha; warehouses, 24 ha; gardens, 98 ha; Campus Martius, 60 ha; River Tiber, 22 ha; residential (all classes), 970 ha. (J C Russell, *Late Ancient and Mediaeval Population*, 1948.)

West of the Tiber, later defences including those for the Vatican, are shown in light line.

Letters from D onward represent major buildings: D, the Colosseum, at the south-eastern end of the Forum area; E, the Pantheon; F, Baths of Diocletian; G, Baths of Caracalla; H, Mausoleum of Hadrian; J, St Peter's, built 1506–1626 on the site of the ancient basilican church of St Peter.

became the centre of Rome. The seventh hill, the Aventine, is to the south of this group.

It is believed that each of the hills had its village settlement, with origins dating back well before 753 BC. The two most important of these were the Palatine, traditionally the home of the early Romans, and the Capitoline, that of the Sabine tribe. The Palatine was approximately square, with an area of about 65 acres. Each side was about 490 yards in length and for this reason Roman writers refer to the hilltop village as Roma Quadrata (Figure 3.5).[13] Lanciani argues against the traditional view that this first settlement had a regular plan based on *cardo* and *decumanus* cross-axes, believing that the shepherds who

11. C T Smith, *An Historical Geography of Western Europe before 1800*, 1967.
12. F R Cowell, *Everyday Life in Ancient Rome.*
13. There is no general agreement on the form taken by the ancient layout of the Palatine Hill; various authorities give differing assumed plans. It seems certain, however, that whatever shape Roma Quadrata did take, it was not that of a regular rectangle. (Figure 3.5 shows the plan assumed by Lanciani in *The Ruins and Excavations of Ancient Rome*.) There is still no reason to believe that the first village settlement on the Palatine Hill was other than organic growth in form.

occupied the hill in 753 BC had no idea whatever of grammatical or astronomical rules of their own.[14] The Palatine, in common with the other hills, was admirably suited for early defensive settlement, with its steeply sloping sides rising above lower ground which was marshy at all times and under water whenever the Tiber was in flood. The location had a further strategic advantage in its central position on the Italian peninsular, where routes converged to cross the River Tiber.

The basic requirements of early hill-village communities are, however, very different from those of a great metropolis. From the time when the first three settlements (those on the Palatine, Capitoline and Quirinal) expanded down their hill slopes on to the lower ground to coalesce into one continuous urban area, through to that of the present-day, Roman planners, architects and engineers have been engaged in a continuing struggle to overcome the intrinsic deficiencies of the site. Flooding, disease – in particular malaria – river pollution and the related drinking water problem, poor bearing capacity, and the hilly topography have been ever-present hazards. Add to these natural problems the planning constraints that resulted from preceding generations' attempts to overcome them – in particular the large-scale sewer and aqueduct systems – and it is by no means surprising that ancient Rome, like so many large modern urban centres, was incapable of being comprehensively restructured. At best there could be only piecemeal 'town-patching' measures.

SEWERS AND WATER SUPPLY

A description of the component parts of ancient Rome is given later in this chapter. The role of the sewer and aqueduct systems as urban form determinants warrants, however, their separate consideration now.[15] The numerous shallow-depth culverted streams and surface water sewers constructed to drain the marshes, and Rome's magnificent system of elevated aqueducts and associated fresh-water reservoirs, created impediments to subsequent redevelopment similar to those presented by railway viaducts and other such barriers in modern cities – notably in South London. In both ancient and modern examples the resource investment has been such that, although planning considerations would imply re-routing, economic constraints all too frequently dictate their retention.

The centre of Rome enclosed by the 5½-mile long Servian Wall was the Forum Romanum. Here stood the traditional market and public gathering area in the valley between the Palatine, Capitol and Quirinal Hills, and overlooked by the Temple of Jupiter on the Capitol. The Cloaca Maxima, the first of the great sewers of Rome, was constructed in about 578 BC as an open drain (Figure 3.7). In 184 BC it was roofed with a magnificent 11-feet diameter semi-circular stone vault, which is still in use as part of central Rome's present-day surface water drainage system. For most of its length it ran at or near surface level, as did many of the less famous sewers.

In addition to draining the lower-lying parts of the city and taking surface water from the streets, the system served to collect the sewage of the *rez-de-chausée* and of the public latrines which stood directly along the route, but no effort was made to connect the *cloacae* with the private latrines of the separate *cenacula* (upper-storey flats).[16] The great majority of ordinary people in Rome had to rely on the public latrines. This network of underground services, although not as great a planning

Sanitary reform was accomplished first, by the draining of marshes and ponds; secondly, by an elaborate system of sewers; thirdly, by the substitution of spring water for that of polluted wells; fourthly, by the paving and multiplication of roads; fifthly, by the cultivation of land; sixthly, by sanitary engineering, applied to human dwellings; seventhly, by substituting cremation for burial; eighthly, by the drainage of the Campagna; and lastly, by the organisation of medical help. The results were truly wonderful. Pliny says that his 'ville-giatura' at Laurentum was equally delightful in winter and summer, while the place is now a hotbed of malaria. Antoninus Pius and M. Aurelius preferred their villa at Lorium (Castel di Guido) to all other imperial residences and the correspondence of Fronto proves their presence there in midsummer. No one would try the experiment now. The same can be said of Hadrian's villa below Tivoli, of the Villa Quinctiorum on the Appian Way, of that of Lucius Versus at Aqua Traversa etc. The Campagna must have looked in those happy days like a great park, studded with villages, farms, lordly residences, temples, fountains and tombs.

The cutting of the aqueducts by the barbarians, the consequent abandonment of suburban villas, the

Figure 3.7 – Rome: the line of the Cloaca Maxima across the north-western end of the Forum area to the Tiber. Lanciani describes how the hills of the left bank made three valleys, each drained originally by its own river, and that 'the first step towards the regulation of these three rivers was taken even before the advent of the Tarquins. Their banks were then lined with great square blocks of stone, leaving a channel about five feet wide, so as to prevent the spreading and the wandering of flood-water, and provide the swampy valleys with a permanent drainage; but, strange to say, the course of the streams was not straightened nor shortened. If the reader looks at the map above, representing the course of the Cloaca Maxima through the Argiletum and the Velabrum, he will find it so twisted and irregular as to resemble an Alpine torrent more than a drain built by skilful Etruscan engineers. The same thing may be repeated for the other main lines of drainage in the valleys Sallustiana, Murcia etc.

When the increase of the population and the extension of the city beyond the boundaries of the Palatine made it necessary to cover those channels and make them run underground, it was too late to think of straightening their course, because their banks were already fixed and built over.'

14. R Lanciani, *The Ruins and Excavations of Ancient Rome*, 1897, reprinted 1968. The most useful, divertingly readable single-volume reference source.
15. The determining effect on urban futures of past improvements to the infrastructure is exemplified by the ancient Roman sewer and potable-water aqueduct systems. (Railway viaducts in south-east London are nineteenth-century equivalents.)
16. Carcopino, *Daily Life in Ancient Rome*.

constraint as the aqueduct system, certainly inhibited large-scale redevelopment.

Rome's water supply, in common with all riverside settlements, was originally taken mainly from its river. This source met demands until the end of the fourth century BC, by which time the volume of sewage discharged into the Tiber had resulted in an unacceptable pollution level. Drinking water was having to be laboriously brought in from outside the city area. On a grand civil engineering scale, typical of the Romans, this problem was solved by constructing the system of aqueducts and reservoirs which eventually attained a total length of some 316 miles. The first aqueduct was the Aqua Appia of 312 BC; the longest was the Aqua Marcia at just over 56 miles – built between 144 and 140 BC when increase of the population had diminished distribution of water from 116 gallons to 94 gallons per head daily.[17] The largest volume of water was supplied by the Anio Novus; this was supported on arches 105 feet in height for 8 miles of its length. Most aqueducts reached Rome at considerable heights above the level of the valley, in order to supply the prestigious hilltop districts. (The wealthy had their own private reservoirs and paid the State for the water supplied; about one-third of supplies was expended in this way.) Where the aqueducts, the Anio Novus and the Claudia, entered Rome massive structures were necessary, and these formed barriers to subsequent redevelopment proposals.

The total volume of water carried by the aqueducts was 60 million cubic feet per day, but, as Carcopino notes, very little of this immense supply found its way to private houses,[18] and that which did had to be carried from distribution fountains to the upper-level dwellings. During the latter part of the third century AD there were in Rome 11 great *thermae*, 926 public baths, 1,212 public fountains, 247 reservoirs and a 'stagnum Agrippae'.[19] During the Empire, 'planning consent' for the construction of new baths for public use was granted only if the applicant could prove that he had arranged for a special supply of water.

DEVELOPMENT UNDER THE CAESARS

After the first century BC, civil wars inside Italy and the collapse of the Republic, attributable in part to extremes of wealth and squalor and poverty in the city, Augustus reconstituted the State, and reorganized the city between 27 BC and AD 14. He boasted that he had found Rome a city of brick and left it one of marble – a claim which could have been only partially true, but which has inspired numerous subsequent urban planning ambitions. Augustus in 7 BC completed the reorganization of Rome into the fourteen regions which lasted as long as the Empire; five were contained within the ancient circuit of the city, five others were partly within, and four were completely outside. The regions were divided into *vici* – quarters separated from each other by the streets which bounded them; Augustus had granted each *vicus* a special administration presided over by its own *magister*.[20] In AD 73 the elder Pliny records that Rome was divided into 265 such *vici*.

Several years before this census, Nero's famous fire of AD 64 had left only four of the regions untouched: 'three had been completely obliterated and seven others hopelessly damaged'.[21] Premeditated or not, this fire was needed to remove the worst excesses of high density, shoddy buildings and grossly inadequate streets, in order to give an opportun-

permanent insecurity, the migration of the few survivors under cover of the city walls, and the choking up of drains, caused a revival of malaria. Mediaeval Romans found themselves in a condition worse than that of the first builders of the city.
(R Lanciani, The Ruins and Excavations of Ancient Rome)

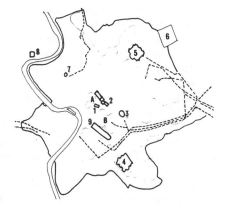

Figure 3.8 – Rome: the aqueduct system. Key: A, Capitoline Hill; B, Palatine Hill; 1, Forum Romanum Magnum; 2, sequence of imperial fora; 3, Colosseum; 4, Baths of Caracalla; 5, Baths of Diocletian; 6, Praetorium barracks; 7, Pantheon; 8, Mausoleum of Hadrian; 9, Circus Maximus.

The majority of the aqueducts entered the city at its eastern corner. Identified from north to south these were: Aqua Appia, which runs through to the Aventine Hill; Aqua Marcia and, on the same aqueduct into the city, the combined Aqua Tepula and Aqua Julia; Anio Vetus; Aqua Claudia and Anio Novus, both carried on the same arches for the last 7 miles into the city. The Aqua Virgo entered Rome from the north and served the low-lying Campus Martius district; Aqua Traiana and Aqua Aurelia served the Trastevere district from the west and water was also carried to this part of the city by service aqueducts crossing the Tiber on the bridges. The Aqua Marcia Antoniniana served the Baths of Caracalla from the south-east (not shown on map).

17. Lanciani, *The Ruins and Excavations of Ancient Rome*.
18. Carcopino, *Daily Life in Ancient Rome*.
19. Lanciani, *The Ruins and Excavations of Ancient Rome*.
20. Carcopino, *Daily Life in Ancient Rome*.
21. Cowell, *Everyday Life in Ancient Rome*.

ity for comprehensive rebuilding, which the citizens would not otherwise have accepted.

In succession to Nero, three emperors – Vespasian, Titus and Domitian (AD 69–96) – established an era of imperial peace and prosperity which made possible the second-century AD 'Golden Age' of Rome under the Emperors Nerva, Trajan, Hadrian, Antonius Pius and Marcus Aurelius (AD 96–180). During this period, according to F R Cowell, 'grand and glorious edifices were then added to the city, so that buildings, markets, baths, temples, statues and other monuments accumulated in Rome, until it had attained a splendour and a magnificence which made it truly a wonder of the world. Perhaps no city before or since has ever so captured the admiration and the imagination of mankind'.[22] This quotation shows, however, only one side of the picture: that of the magnificence of Rome's civic areas. As a contrasting view, Lewis Mumford has argued that, 'from the standpoint of both politics and urbanism, Rome remains a significant lesson of what to avoid: its history presents a series of classic danger signals to warn one when life is moving in the wrong direction. Wherever crowds gather in suffocating numbers, wherever rents rise steeply and housing conditions deteriorate, wherever a one-sided exploitation of distant territories removes the pressures to achieve balance and harmony nearer at hand, there the precedents of Roman building almost automatically revive, as they have come back today: the arena, the tall tenement, the mass contests and exhibitions, the football matches, the international beauty contests, the strip-tease made ubiquitous by advertisement, the constant titillation of the senses by sex, liquor and violence – all in true Roman style'.[23]

Urban form components

FORTIFICATION

Lanciani notes that Rome has been fortified seven times, within seven lines of walls: by the first king, by Servius Tullius, by Aurelian, by Honorius, by Leo IV, by Urban VIII, and by the Italian Government (Figure 3.6).[24] The first fortifications were those around the Palatine Hill. The second wall, attributed to Servius Tullius of 550 BC, is, according to Carcopino,[25] that of the Republic of two centuries later, constructed between 378 and 352 BC. As already described, need for defences gradually diminished and it is not until 650 years later that a new system was required.

The Aurelian Wall was constructed between AD 272 and c. 280, in response to the barbarian invasion of 271. It was 11½ miles in length, enclosed an area of 5⅓ square miles, and required a strip of land some 62 feet wide: 16 feet for the inner perimeter route, 13 feet for the wall itself, and 33 feet for the external cleared defensive zone.[26] The wall had projecting towers at intervals of 100 Roman feet (32 yards) which after the restoration of the defences by Honorius in 402, totalled 381 in number. The Aurelian Wall did not enclose the entire city but, as recorded by Lanciani,[27] followed the line of the existing octroi – the customs barrier encircling the city proper within the surrounding suburban areas. One-sixth of this barrier was formed by natural features and substantial engineering works, including retaining walls to hillside developments, the wall of the Praetorian camp and lengths of the Marcian and Claudian aqueducts. For those sections of the octroi which

Whether it was accidental or caused by the emperor's criminal act is uncertain – both versions have supporters. Now started the most terrible and destructive fire which Rome had ever experienced. It began in the Circus, where it adjoins the hills. Breaking out in shops selling inflammable goods, and fanned by the wind, the conflagration instantly grew and swept the whole length of the Circus. There were no walled mansions or temples, or any other obstructions which could arrest it. Nero was in Antium. He only returned to the city when the fire was approaching the mansion he had built to link the Gardens of Maecenas to the Palatine. The flames could not be prevented from overwhelming the whole of the Palatine, including his palace. Nevertheless, for the relief of the homeless, fugitive masses he threw open the Field of Mars, including Agrippa's public buildings, and even his own Gardens. Nero also constructed emergency accommodation for the destitute multitude. Food was brought from Ostia and neighbouring towns, and the price of corn was cut. Yet these measures, for all their popular character, earned no gratitude. For a rumour had spread that, while the city was burning, Nero had gone to his private stage and, comparing modern calamities with ancient, had sung of the destruction of Troy ... Nero profited by his country's ruin to build a new palace. Its wonders were not so much customary and commonplace luxuries like gold and jewels, but lawns and lakes and faked rusticity – woods here, open spaces and views there. With their cunning, impudent artificialities, Nero's architects and contractors outbid Nature ... In parts of Rome unfilled by Nero's palace construction was not – as after the burning by the Gauls – without plan or demarcation. Street-fronts were of regulated dimensions and alignment, streets were broad and houses spacious. Their height was restricted, and their frontages protected by colonnades. ... A fixed proportion of every building had to be massive, untimbered stone from Gabii or Alba (these stones being fireproof). Furthermore, guards were to ensure a more abundant and extensive public water-supply, hitherto diminished by irregular private enterprise.

Householders were obliged to keep fire-fighting apparatus in an accessible place; and semi-detached houses were forbidden – they must have their own walls. These measures were welcomed for their practicality, and they beautified the new city. Some, however, believed that the old town's configuration had been healthier, since its narrow streets and high houses had provided protection against the burning sun, whereas now the shadowless open spaces radiated a fiercer heat. (Tacitus, The Annals of Imperial Rome, translated by Michael Grant)

22. Cowell, Everyday Life in Ancient Rome.
23. L Mumford, The City in History, 1961. Mumford cited Rome under the later emperors as an example of what could become of Western society ... unless it changed its ways. On re-reading, it would appear that Mumford was not averse to dramatic emphasis when it suited his purpose.
24. Lanciani, The Ruins and Excavations of Ancient Rome.
25. Carcopino cites G Saeflund's masterly work on the Servian Wall, Le Mura di Roma Republicana (Lund, 1932), as superseding all previous discussions and establishing the date 378–352 BC for this wall. Lanciani would seem to be one of the authorities accepting the earlier 550 BC date; he was, however, writing of archaeological work of the second part of the nineteenth century.
26, 27. Lanciani, The Ruins and Excavations of Ancient Rome.

required strengthening, Aurelian expropriated an area of about 80 acres of private property[28] involving the demolition, or incorporation into the defences, of innumerable houses, garden walls and tombs. The succeeding fortifications of Rome, down to those of the nineteenth century, all follow in the main the lines of the Aurelian wall.

STREET SYSTEM

The functional zones were linked together by a route system of streets, which always smacked of their ancient origin and maintained the old distinctions which had prevailed at the time of their rustic development: the *itinera*, which were tracks only for men on foot; the *actus*, which permitted the passage of only one cart at a time; and finally the *viae* proper, which permitted two carts to pass each other, or to drive abreast.[29] Within the Republican Wall only the Sacra Via, which ran through the original forum area between the Colosseum and the Capitoline Hill, and the Via Nova, parallel to it against the Palatine Hill to the south-west, are considered by Carcopino as of *via* status. Leading out from the limits of Republican Rome, through the subsequent built-up areas, 'not more than a score of others deserved the title: the roads which led out of Rome to Italy, the Via Appia, the Via Latina ... etc. They varied in width from 15 feet to 21 feet, a proof that they had not been greatly enlarged since the day when the Twelve Tables had prescribed a maximum width of 15 feet'.[30] Few of the *vici* which formed the general route network were of this width. They were not only inconveniently narrow but also extremely tortuous. By the time of Julius Caesar the street system was grossly overloaded, with continual conflict between pedestrian and vehicle traffic. As a result Caesar was forced to ban transport carts from the city during hours of daylight, with the exceptions of builders' carts and a few categories of official chariots. During the daytime Rome was therefore largely free from wheeled traffic, but at night, as observed by Juvenal, the crossing of wagons in the narrow winding streets and the swearing of the drivers brought to a standstill would snatch sleep from a sea-calf or the Emperor Claudius himself.[31] Subsequent emperors extended these regulations first to other Italian municipalities and then generally throughout the Empire. Hadrian further controlled the traffic by limiting the teams and load of the carts allowed to enter the city.[32]

HOUSING

There were two basic types of housing found in the city: the *domus* for privileged single-family occupation and the *insula* (building block) divided up into a number of flats or *cenacula*. In the middle of the fourth century AD 1,797 *domus* were recorded; this compared with 46,602 *insulae* each, according to Carcopino, with an average of five flats whose average occupancy was at least five or six persons. (Taking these figures Carcopino deduces a population of 1.2 million in the early second century AD, the period of his *Daily Life in Ancient Rome*.) Rome was therefore predominantly a city of flat-dwellers, living in buildings which as early as the third century BC had reached up three floors. As the population increased still further so did the height of buildings: Julius Caesar in his *Lex Julia de Mode Aedificorum* had to impose a limit of 70 feet in order to minimize the ever-present dangers of structural collapse. Augustus reaffirmed this limit and Trajan later reduced it to 60 feet. Carcopino remarks that necessity knows no laws and in the fourth century the sights of the city included that giant apartment

The great dictator had realised that in alleyways so steep, so narrow and so traffic-ridden as the 'vici' of Rome the circulation by day of vehicles serving the needs of the population of so many hundreds of thousands caused an immediate congestion and constituted a permanent danger. He therefore took the radical and decisive step which his law proclaimed.

From sunrise until nearly dusk no transport cart was henceforward to be allowed within the precincts of the 'urbs'. Those which had entered during the night and had been over-taken by the dawn must halt and stay empty. To this inflexible rule four exceptions alone were permitted: on days of solemn ceremony, the chariots of the Vestals, of the Rex Sacrorum and of the Flamines; on days of triumph, the chariot necessary to the triumphal procession; on days of public games those which the official celebration required. Lastly one perpetual exception was made for every day of the year in favour of the carts of the contractors who were engaged in wrecking a building to reconstruct it on better and hygienic lines.

(J Carcopino, Daily Life in Ancient Rome)

---

28. Athens, when refortified after devastation in the late third century AD, could afford only a comparably expedient use of existing structures; until its removal in 1990 and 1991, there had been a modern parallel with the construction of the Berlin Wall, dividing the city into its western and eastern sectors, which also incorporated numerous buildings (see Chapter 7).

29, 30. Carcopino, *Daily Life in Ancient Rome*.

31. Juvenal, *The Sixteen Satires, Satire III*, translated by P Green, Penguin, 1967.

32. See Chapter 8 for a description of the closely similar traffic situation in London of the century and a half before the Fire of 1666.

house, the Insula of Felicula.[33] Only the very rich could afford a *domus*, which in plan was generally a sequence of rooms facing into courtyards (as at Pompeii, illustrated on page 72). By turning blind unbroken walls to the street a *domus* afforded a degree of privacy to its fortunate occupants, in direct contrast to the *insula* which always opened to the outside and, when it formed a quadrilateral around a central courtyard, had doors, windows and staircases opening both to the outside and to the inside.[34] Either way there was no escaping noise and dust. Fire was obviously an ever-present hazard; construction relied to a great extent on timber and lighting was by naked flames. Such heating as there was took the form of movable, mainly charcoal-fuelled stoves. Running water was not supplied to the upper floors.

Julius Caesar's legislation required the use of tiles as an incombustible roof material and stipulated that between buildings an open space – the *ambitus* – 28.75 inches wide, had to be maintained. This 'fire-break' was lost when the legislation was amended to permit party-wall construction. Augustus created a corps of fire-fighting night watchmen, or *vigiles*, but dread of fire was such an obsession among rich and poor alike that Juvenal was prepared to quit Rome to escape it saying, 'No, No, I must live where there is no fire and the night is free from alarms.'[35]

Topographically Rome was essentially an egalitarian city. With the exception of the emperors' palaces built on the Palatine Hill and possibly separate working-class districts on the downstream banks of the Tiber and the slopes of the Aventine, 'high and low, patricians and plebeian, everywhere rubbed shoulders without coming into conflict'. On the subject of workers' housing Carcopino states that 'they did not live congregated in dense, compact, exclusive masses; their living quarters were scattered about in almost every corner of the city but nowhere did they form a town within a town'.[36]

MARKETS

Economically Rome was sustained only by imports on a vast scale. She had three ports: Ostia (Figures 3.9 and 3.21), Portus and, within the city, the pool of the Tiber. The latter began with the last of the city bridges (Pons Sublicius) and extended as far as the reach of the Vicus Alexandri, 1½ miles below the Porta Ostiensis. Here larger sea-going vessels were obliged to take to moorings to avoid sandbanks and the exceedingly sharp turns of the upper channel. Between these limits the left bank of the river was divided into wharves, called *porti* or harbours. Each was used for a particular kind of trade: marble, wine, oil, lead, pottery, etc. Associated with the *porti* and occupying the left-bank area between the Aventine and the city walls, as well as a considerable right-bank area, were the warehouses, or *horrea*. Some of these, notably the grain stores, which were government property, specialized in one commodity, most, however, comprised a sort of general store where all kinds of wares lay cheek by jowl.[37] Lanciani records that there were 290 of these public warehouses, named after either their contents or their owner or builder. He notes that the Horrea Galbana alone occupied a space of 218 yards by 169.[38] Carcopino, in the light of more recent excavations, describes how their foundations went back to the end of the second century BC, and states that they were enlarged under the Empire and possessed rows of *tabernae* ranged around three large intermediate courtyards which covered more than 8 acres.[39] In addition

The via Portuensis became important on the construction of Portus Augusti, the new harbour and city, now Porto, at the Tiber's mouth, and on the northern or right bank. These works were projected by Augustus, and carried out by Claudius, for the double purpose of remedying the silted-up condition of the natural port of Ostia, and of providing a straighter and deeper course for the waters of the river, which, being held back by the bar at its mouth, aggravated the frequent floods at Rome. The process of silting, however, proceeds so rapidly in the Tiber, owing to the quantity of sand washed down from the hills, that by the time of Trajan it was necessary to form both a new harbour and a new channel. The site of the Claudian works has become

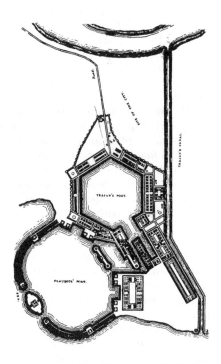

Figure 3.9 – Ostia (Port of Rome): the artificial harbours constructed under Claudius (AD 42–54) and Trajan (100–106) to augment the capacity of the original facilities alongside the Tiber.

33. Carcopino, *Daily Life in Ancient Rome*. This was not a uniformly designed apartment building, constructed at one time, but rather an entire multi-storey city block (*insula*) comprising individual disparate dwellings.
34. Carcopino, *Daily Life in Ancient Rome*.
35. Juvenal, *The Sixteen Satires, Satire XIV*.
36. Carcopino, *Daily Life in Ancient Rome*. Ancient Rome exemplifies the absence of social grouping as a determinant of urban form, but see Chapters 6 and 8 for discussion of seventeenth- and eighteenth-century privileged class developments in Paris and London.
37, 38. Lanciani, *The Ruins and Excavations of Ancient Rome*.
39. Carcopino, *Daily Life in Ancient Rome*.

to the *horrea* – the 'hyper-markets' of their day – Rome was a city of small shopkeepers, the great majority trading from ground-floor *insulae* premises but with major concentrations in and around the fora area, notably the Forum of Trajan.

Supplies needed for daily consumption were brought into specialist wholesale trade markets, e.g. the *holitorium* for vegetables, *boarium* for horned cattle, *suarium* for pigs, *vinarium* for wine-merchants, and *piscarium* for fishmongers. Other trades gradually established themselves in their own districts and streets. The *holitorium* was centrally located between the Capitoline Hill and the Tiber, with the *boarium* some distance downstream between the Palatine and the river. Near the latter, again centrally located, was the *suarium*. A third meat market, the Campus Pecuarius for sheep, has not been located. Lanciani describes how these three markets were all used for actual trade, on the spot, and suggest that to avoid yet further congestion in the street the oxen came to the *boarium* by the river, by barge-loads.[40] However, cattle once bought were driven down city streets to the individual butcher's shops for slaughter.

CITY CENTRE

The original centre of the city, on the valley floor between the Palatine and Capitoline Hills and the end of the Quirinal ridge, responded to population increase in two contrasting ways. First it was extended towards the south-east by 'controlled' organic growth which was ultimately to be halted in AD 82 against the massive walls of the Colosseum, some 600 yards distant from its origins under the Capitoline (Figure 3.12). The second response was the construction of a carefully planned sequence of linked imperial fora between 50 BC and AD 114, with a main axis at an angle to the north, between the Capitoline and the Quirinal. The original centre formed part of Regio VIII and is properly termed the Forum Romanum Magnum. The title Forum Romanum is generally applied to the sequence of buildings and spaces extending to the south-east.

The linearity of the Forum Romanum was determined mainly by topography. The valley between the Palatine and the Oppian was not only the logical direction for growth; it already contained in addition the immensely important Sacra Via ('Queen of Streets') as a further reason for locating there the city's new civic buildings, both religious and secular. The Forum Romanum Magnum had commercial origins: Lanciani records that at the time of the foundation of Rome the bartering trade between the various tribes settled on the heights of the left bank of the Tiber was concentrated in the hollow ground between the Palatine, Capitoline and Quirinal.[41] This was the marshy area subsequently drained, from around 509 BC, by the Cloaca Maxima.

During the 'Kingly' period in the city's history, i.e. to 509 BC, the embryonic forum gradually took on the regular shape of a parallelogram, which it retained down to the end of the Empire. At first it would have had a multi-purpose function, combining market and civic activities with political business. This simple situation could not last for long, however: the beaten earth and rudimentary huts were the nucleus of history's most complex city centre. From serving a population numbered in hundreds it grew to meet the demands of considerably over 1 million.[42] Exemplifying Rome's general organic growth character the centre was never planned in its entirety: new facilities were added as

completely solid land, while the Port of Trajan, which is hexagonal in form and about 2,400 yards in circuit still exists, with a depth of about ten feet of water. The canal communicating with this port is still open and is the only navigable channel into the river. Porto itself is now two miles from the sea.
(J H Parker, The Architectural History of the City of Rome)

The student wishing to survey the ground formerly occupied by these great establishments connected with the harbour of Rome, must make the ascent of the Monte Testaccio, which rises to the height of 115 feet in the very heart of the region of Horrea. The hill itself may be called a monument of the greatness and activity of the harbour of Rome. The investigations of Reiffersheid and Bruzza completed in 1878 by Heinrich Dressel, prove that the mound is exclusively formed of fragments of earthen jars (amphorae, diotae), used in ancient times for conveying to the capital the agricultural products of the province, especially of Baetica and Mauretania.

Baetica supplied not only Rome but many parts of the western Empire with oil, wine, wax, pitch, linseed, salt, honey, sauce and olives prepared in a manner greatly praised by Pliny. Potters' stamps and painted or scratched inscriptions of Spanish origin, identical with those of Monte Testaccio, have been discovered in France, Germany and the British Islands. It appears that the harbour regulations obliged the owners of vessels or the keepers of warehouses to dump in a space marked by the Commissioners the earthen jars which happened to be broken in the act of unloading or while on their way to the sheds. The space was at first very limited; in progress of time part of a public cemetery, containing among others the tomb of Rusticelli, was added to it. At the beginning of the fourth century the rubbish heap had gained a circumference of half a mile, and a height of over a hundred feet.
(R Lanciani, The Ruins and Excavations of Ancient Rome)

40, 41. Lanciani, The Ruins and Excavations of Ancient Rome.

42. It is reasonable to assume that whatever the population of the city proper, the administrative and commercial centre must have served well over 1 million people.

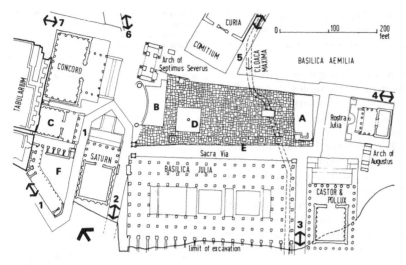

Figure 3.10 – The Forum Romanum Magnum, detail plan. (For the view as reconstructed, from the direction shown top right, see Figure 3.11.) The forum contains too many buildings for individual descriptions; the two most important basilicae – Aemilia and Julia – and the Temple of Caesar have been described in the main text. Brief notes on some of the other major buildings are as follows:

Right-hand side (looking from the Rostra. B)
Temple of Saturn (rebuilt in 42 BC) on a raised platform, reached by an impressive flight of stairs up from the Clivus Capitolinus. Between this temple and the Basilica Julia the Vicus Jugarius led from the forum, against the Capitoline Hill to the Forum Olitorium.

Temple of Castor and Pollux (AD 6), across the Vicus Tuscus from the Basilica Julia. This street rivalled the Sacra Via in importance and linked the forum with the Circus Maximus.

Within the forum at its eastern end, in front of the Temple of Caesar, nineteenth-century archaeologists mistakenly demolished a multi-cell building (keyed A) possibly of the late fourth century AD. Opinions have varied as to whether this was a wine shop, or, more likely, a public office building.

Left-hand side
The Comitium, across the Argiletum – the main street leading from the forum to the Subura – from the Basilica Aemilia, was the centre of civil and political business during Rome's early days, while the forum functioned solely as a market-place. The Comitium itself served as a semi-private forecourt to the Curia, the Senate House, 'politically speaking,' observes Lanciani, 'the most important building in the Roman world'.

The Arch of Septimus Severus (AD 204) was erected on the edge of the Rostra platform, some six or seven feet above the level of the forum and the Comitium, and could not therefore have been across the main street bordering the forum, as incorrectly shown on some plans.

Behind the Rostra, against the Capitoline Hill, there were two important temples – the Temple of Concord, reconstructed for a second time in AD 10, and the Temple of Vespasian (AD 94) – and the Tabularium, the Roman public records office, rebuilt in 78 BC after the fire of 83 BC, on the steep slope of the Capitoline facing the forum. The upper part of this building coincides almost exactly with the area of the medieval Palazzo del Senatore which formed the south-eastern side of the Renaissance Capitoline Piazza (page 183).

Key: A, unknown multi-cell building; B, Rostra; C, Temple of Vespasian; D, Column of Phocas; E, the eight monumental columns; 1, Clivus Capitolinus; 2, Vicus Jugarius; 3, Vicus Tuscus; 4, Sacra Via; 5, the Argiletum; 6, link to the imperial fora; 7, slope up to the Capitoline Hill.

This archaeological record is difficult, if not impossible, to relate to the reality when visited. Guidebooks are not particularly helpful and it is best not to expect to be able to conjure up images of the ancient Roman city centre.

need arose. Development was controlled to a considerable extent by inherited constraints: regions, sites and routes, sacrosanct for various reasons; topographical limitations beyond even Roman engineering capabilities; and perhaps the major reason, the need to maintain adequate open space for movement by foot and wheel, and for civic assemblies. In describing the Forum Romanum Magnum and its enclosing buildings, where 'for so many centuries the destinies of the ancient world were swayed',[43] it is most convenient to take the Rostra at its north-western end (under the Capitoline) as a viewpoint and looking south-east refer to right and left-hand sides. (The detailed plan as Figure 3.10.)

By the first century BC the gross overcrowding in the forum area, and its multifarious activities, forced the start of an extension and redevelopment programme which was to last for some 150 years. As a first measure, fishmongers were removed from the steps of the basilicae to their own specialist market, the *forum piscatorium*. New civic buildings were required and in 54 BC work started on the Basilica Aemilia, to the south-east of the Curia, on a site acquired at great expense from a number of owners. It was completed in 34 BC, during the construction period of the new Forum Julium on the other side of the Curia. This latter consisted of a sacred enclosure around the Temple of Venus Genetrix, consecrated by its founder, Julius Caesar, in 45 BC. Pliny, obviously impressed, wrote 'we wonder at the Egyptian pyramids, when Caesar as dictator spent one hundred millions of sesterces merely for the land on which to build his forum'.[44] (Lanciani writing at the end of the nineteenth century calculated that this was equivalent to 44.5 dollars per square foot.) The Forum Julium was especially intended for legal business.

On the far right-hand side of the Forum Romanum Magnum and occupying almost its entire length, Caesar, in about 54 BC, also laid out his magnificent Basilica Julia. Dedicated while still in an unfinished state in 46 BC, it was completed, after fire damage, by Augustus in 12 BC. About 350 feet by 185 feet this basilica was one of Rome's largest buildings. It was used for the court of the *centumviri*. The younger Pliny has described an important trial day: eighty judges sat on their benches, while on either side of them stood the eminent lawyers

43. Lanciani, *The Ruins and Excavations of Ancient Rome*.
44. Quoted by J H Middleton in *The Remains of Ancient Rome*.

TEMPLE OF
JUPITER
CAPITOLINUS

TEMPLE OF
JUNO MONETA

Ⓐ

Figure 3.11 – The Forum Romanum, an artist's impression of the buildings and spaces of the later Empire period. The direction of this view is given in Figure 3.10. In the distance can be seen the temples of the twin eminences of the Capitoline Hill. The hollow on top of the Capitoline Hill was to become the location of Michelangelo's incomparable sixteenth-century Capitoline Piazza in front of the medieval Palazzo del Senatore which, in turn, was built in part on ruins of the ancient Tabularium. (From B Fletcher, *History of Architecture on the Comparative Method*, 1987.)

who had to conduct the prosecution and to defend the accused. The great hall could hardly contain the mass of spectators; the upper galleries were occupied by men on one side, by women on the other, all anxious to hear, which was very difficult, and to see, which was easier.[45] At ground level on the right-hand side of the forum there were the shops of bankers and money-lenders. At the far south-eastern end, on the spot where he was murdered on 15 March 44 BC, the Temple of Julius Caesar was erected (33–29 BC) to his memory. The site of this building was taken out of the Forum Romanum, thereby reducing its length to 110 yards and necessitating a change in the line of the Sacra Via.

The Forum Julium was the first of the imperial fora, constructed to the left (north-east) of the Forum Romanum. It was not on the new south–north axis but was to constitute in effect a link between that and the Forum Romanum. The second of the new fora was built by Augustus in 42 BC, beyond the Forum Julium, between it and the Quirinal. The layout of the Forum Augusti is basically that of a forecourt to the magnificent Temple of Mars Ultor, with flanking hemicycles to north and south. The most remarkable feature of the place, noted Carcopino, was a wall of blocks of peperino raised to a great height to screen the view of the mean houses on the Quirinal.[46] This forum also had a mainly legal function.

South of the Forum Augusti and separated from it by a wide street – the Argiletum – which led from the Subura to the Forum Romanum Magnum, Vespasian built the Temple of Peace (dedicated in AD 75) around which he laid out the third forum, named after him. It is believed that it functioned as a public library and was the venue for literary discussion. The route space of the Argiletum, some 127 yards by 43, was shortly afterwards converted by Domitian and his successor Nerva (96–98) into the Forum Transitorium (later the Forum of Nerva).

The last and most magnificent of the imperial fora was built by Trajan between 112 and 114 from designs of Apollodorus of Damascus. It is the outstanding example of Roman civic design, both for its intrinsic architectural qualities and for the immense engineering works it required. To Zucker the Forum Traiani represents the definite triumph of the Roman spatial concept based on absolute axiality and symmetry.[47] In addition to doubling the area of the five existing fora, its construction on ground immediately to the north of the Forum Augusti enabled the problem of traffic circulation around the Capitoline to be resolved. Originally the Capitoline was not a completely isolated hill but was connected to the main Quirinal spur by a lower ridge. Access to the fora from the north was either across this ridge, by

Figure 3.12 – The imperial fora, related to the buildings and spaces of the Forum Romanum and the Palatine and Capitoline Hills. Key: 1, Forum Romanum Magnum (as Figure 3.10); 2, Basilica Julia; 3, House of the Vestal Virgins; 4, Temple of Venus and Rome; 5, Colosseum; 6, Basilica of Constantine; 7, Forum of Vespasian; 8, Forum of Nerva; 9, Forum of Augustus; 10, Forum of Trajan; 11, Forum of Caesar.

45. Quoted by J H Middleton in *The Remains of Ancient Rome*.
46. Carcopino, *Daily Life in Ancient Rome*. The Forum of Augustus is an outstanding example in ancient history of the effect of aggrandizement and aesthetic determinants.
47. P Zucker, *Town and Square*, 1959.

the steep and narrow Clivus Argentarius, or around the three sides of the Capitoline. Trajan and his architect-planner took the bold step of removing the ridge to create a level area about 200 yards in width between the hills, with a total length of about 720 feet. Private property covering nearly 10 acres was acquired for the site, from which about 1 million cubic yards of earth and rock were excavated and spread out-

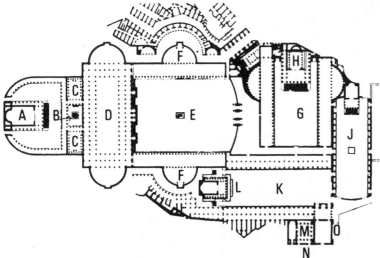

side the Porta Collina. The Column of Trajan, which was 124 feet in height without the statue, was erected 'to show posterity how high rose the mountain levelled to make room for the forum'.[48] This was the grandest historic instance of topographical adjustment.

Trajan's forum comprised five main parts: first, the *propylaia* – an entrance gateway in the form of a triumphal arch – which gave access from the adjoining Forum Augusti into the second part, the main forum area, 125 yards long by 110 wide, with double colonnades in front of twin hemicycles excavated into the hillsides and third, the Basilica Ulpia facing the forum on its longest side (also with hemicycles at each end). The fourth part, entered from the basilica, was a small court, only 76 feet wide by 52 long, in which still stands the 124 feet high Trajan's column, flanked at each end by library buildings. Fifth and last was the Temple of Trajan which stood in its own colonnaded space and closed the monumental group at its north-western end.

RECREATION

One of the major preoccupations facing the Roman emperors was to divert the anti-establishment potential of the urban mob which the handouts of food and money sustained. In the second century AD 175,000 people received public assistance from the city. If one accepts a figure of only three persons per family it is likely that, directly or indirectly, at least one-third and possibly one-half of the population of the city lived on public charity.[49] In addition to feeding the plebs, the authorities also had to amuse them during their holidays, which numbered 159 annually during the time of Claudius, rising to about 200 in the third century AD. A high proportion of the holidays were devoted to games staged at state expense and involving elaborate organization and building facilities. The largest recreation centre was the Circus Maximus, located in the valley between the Palatine and Aventine

Figure 3.13 – Detail plan of the sequence of imperial fora; the relationship to the Forum Romanum is given in Figure 3.12.

Key: A, the Temple of Trajan, which closed the monumental group at its north-western end; B, Trajan's Column, completed in AD 113 and traditionally taken as marking the depth of the excavation necessary to construct the forum; C, two libraries – the Bibliotheca Ulpia and the Templi Traiani; D, the Basilica Ulpia (also known as Trajan's Basilica), a magnificent hall 288 by 177 ft in area, surrounded by a double row of columns and with hemicycles at each end; E, the Forum of Trajan; F, hemicycles on each side of Trajan's forum serving as retaining walls against the Capitoline Hill to the south-west and the Quirinal Hill to the north-east; G, the Forum of Augustus; H, the Temple of Mars Ultor, behind which was raised a high stone wall to screen the view of mean houses clustered on the slopes of the Quirinal (Lanciani records that 'the irregular form of the wall at the back of the temple is accounted for by the circumstance that Augustus was unable to obtain a symmetrical area, as the owners of the nearest houses could not be induced to part with their property'); J, the Forum of Nerva, or Forum Transitorium, so called because it was traversed by the main thoroughfare of the Argiletum. Its long and narrow shape (400 by 128 ft) made it more like a handsomely decorated street than a formal public space; the small Temple of Minerva was located at its north-eastern end; K, the Forum of Caesar (also known as the Forum Julium) containing L, the Temple of Venus Genetrix, which was dedicated in 45 BC; M, the Curia; N, the Comitium; and O, the thoroughfare of the Argiletum. (These latter three parts of the plan are also located on the detail plan of the Forum Romanum Magnum shown in Figure 3.10.)

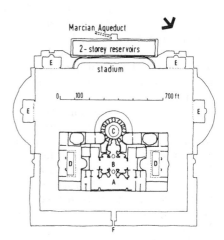

Figure 3.14 – Rome: outline of the Baths of Caracalla, built between AD 211 and 217, in the southern part of the city (see Figure 3.6). Accommodation was provided for more than 1,600 bathers, The main central building contained the various bathing halls as keyed: A, Frigidarium; B, Tepidarium; C, Calidarium; D, open colonnaded courtyards; E, lecture halls, libraries etc.

48. Lanciani, *The Ruins and Excavations of Ancient Rome*. The construction of this extraordinary sequence of buildings and spaces is the largest example in

Hills. It eventually attained dimensions of 654 by 218 yards and provided at least 255,000 seats, possibly as many as 385,000. The best known centre was the Colosseum (the Flavian Amphitheatre), completed in AD 80 on an imposing site at the south-east end of the Forum Romanum complex; here there were some 45,000 seats and 5,000 standing places. The three major theatres in Rome could accommodate about 50,000 spectators; in addition there were many smaller theatres.

Providing perhaps for more wholesome pastimes, a number of enormous baths were constructed in the city. They included those of Caracalla (officially designated the Thermae of Antonius), over 27 acres in extent, and those of Diocletian, 32 acres in area. In addition to every possible type of bath, these establishments included shops, stadia, rest rooms, libraries, museums, and numerous other facilities. As Carcopino put it, the Caesars had in fact shouldered the dual task of feeding and amusing Rome.[50] Here we have a restatement of Juvenal's famous indictment that the populace, who once bestowed commands, consulships, legions, and all else, now meddled no more and longed eagerly for just two things – bread and circuses.[51]

With this appropriate contemporary Roman conclusion we must turn from the city itself for description of a number of the most important provincial examples of imperial urban planning. The decline, fall and eventual resurgence of the city of Rome, which continued through to the end of the Renaissance period, is described later in Chapter 5.

Roman imperial urbanism

Chapter 3 concludes with a representative selection of examples of imperial urban planning, grouped by provinces as Italy, North Africa, Spain, Gaul and Germany, and Britain. Constantinople (Byzantium/ Istanbul) provides a separate epilogue to the Roman period. The basic urban planning principles adopted by the Romans have been summarized on pp. 57–8 related to Figure 3.3, with an introduction to the three main political classes of imperial towns – colonia, municipium and civitas. The extreme importance of colonia in the imperial planning 'policy', as it came to be consolidated, requires further elaboration.

COLONIA

As established earlier, the Romans avoided the need to maintain 'standing military presences' within their initially conquered imperial provinces, through the policy of local settlement of their own people, combined with 'Romanization through urbanization' of native tribesmen.[52] Although in its Latin derivation, colonia meant a group of coloni ('peasants', from colere, 'to cultivate'), the word took on legal precision, explains Salmon, as denoting 'a group of settlers established by the Roman state, collectively and with formal ceremony, in a specified locality to form a self-administering civic community'.[53]

A colonia was essentially a city-state entity comprising the urban nucleus and its immediately surrounding agricultural rural territorium. Although physically similar to the colonial city-states founded widely around the Mediterranean by the Greeks (described in Chapter 2) its foundation represented an invariable act of the Roman state implementing strategic imperial policy, rather than the haphazard enterprise for commercial gain of private Greek citizens (as also latterday European 'colonialists'). The royal (central government) bastides of the thirteenth century in Western Europe are a closer parallel in terms

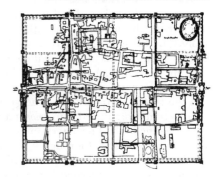

Figure 3.15 – Augusta Praetoria (Aosta): following the defeat in 25 BC of the local tribesmen, the Salassi, by a Roman force under Terentius Varro Murena, a *castra* was established on the site of their village in the valley of the Dora Baltea. This was strategically situated at the foot of the Great and Little St Bernard Passes, 25 miles north-west of Ivrea, which had been founded as a *colonia* before 100 BC. In 22 BC the Emperor Augustus upgraded the camp into a strongly fortified *colonia*, with the full title Augusta Praetoria Salassorum. It was settled by 3,000 time-expired legionaries from the *cohortes praetoria*, and their families.

The plan shows the limited extent of medieval organic-growth modification of the underlying Roman gridiron (which covered an area of about 100 acres) thereby evidencing continued urban status surviving the end of the empire. Six streets divided the town into sixteen main grid blocks which were further subdivided into *insulae* measuring about 90 × 70 yards. The arena was inside the unusually rectangular wall, in the north-west corner.

ancient history of man-made topographical adjustment for an architectural purpose. (The most important instance in history of natural-world topographical adjustment is Mexico City, described in Chapter 9, although this statement is best modified to refer to present-day cities, given the terminal impact of natural disaster on Pompeii, for which see pp. 71–2.)

49, 50. Carcopino, *Daily Life in Ancient Rome*.

51. This is the well-known quotation from Juvenal, *The Sixteen Satires, Satire X*. There are, of course, numerous parallels that can be drawn with late-twentieth-century 'social welfare' circumstances around the world, notably New York City, which is regularly reported as teetering on the brink of bankruptcy. In different policy directions, television now serves an international purpose of amusing 'the plebs' (as the dictionary sense of 'ordinary insignificant persons') and for the late-twentieth-century British, at least, there are also the diversions catered for by a plethora of garden centres and do-it-yourself home improvement emporia.

52. Collingwood, *The Cambridge Ancient History*.

53. Salmon, *Roman Colonization under the Republic*.

of origin, and leaping ahead in urban history, it is important to note that when, during the 1890s, the British 'garden-city' concept was evolved by Ebenezer Howard, as the lineal predecessor of the post-1946 British New Towns, it was self-contained, urban/rural historic city-state imagery that provided one of his main points of departure.[54]

As an alternative to colonization in the technical Roman sense, there was the *viritane* distribution (or *assignatio*) of conquered land to settlers who were not organized into *colonia* and who were under the jurisdiction of Rome itself. The determining consideration was that of need for local defence: only a thoroughly pacified region was deemed suitable for *viritane* distribution and for that reason, says Salmon, it follows that colonization often preceded it in a given region.

The actual foundation of a colony was a closely formalized procedure whereby three commissioners appointed for the purpose and invested with *imperium*, 'delimited the boundaries of the territory, assigned allotments to its settlers, adjudicated any disputes that broke out between them, or between them and the nearby natives, laid down the constitution for the new community, and appointed its first office-holders and priests'.[55] Surveyors on their staff divided up the territory on the basis of a centuriated grid comprising square *centuriae* of area 200 *iguera* (125 acres/50 hectares).[56] A 2-*iguera* plot was the traditional hereditary land-parcel of a Roman citizen. Divisions between *centuriae*, often as rural roads, were called *decumani* and *kardines*. The urban nucleus of the colony was similarly surveyed, on occasion in accordance with the rural land grid. (The modern counterpart to Roman centuriation was the western lands settlement grid, on which basis the expansion westwards of the USA was controlled from 1786: see Chapter 10.)

Given the role of colonies in the consolidation of Roman dominion over disputed regions, the task of the commissioners was not without risk. Salmon describes how those founding Placentia in 218 were captured by the Gauls and handed over to Hannibal's Carthaginian army.[57] Lewis Silkin, the British Minister of Town and Country Planning, may well have pondered his fate at local hands, when attempting in 1946 to explain to the infuriated burghers of Stevenage why they should be pleased to be made into a new town. In the event he escaped, lucky to have had only his car tyres deflated.[58]

Prefaced by the disclaimer, 'certainty about a town's status in the Roman Empire is not always possible', Salmon lists a total of 407 probable *coloniae*.[59] The earliest is Fidenae (Castel Giubileo) on a hill-top controlling the Tiber-crossing next above Rome, which is attributed to Romulus; although from AD 212 all provincial cities became Roman, thereby diminishing the practical significance of being a *colonia*, the title continued to enjoy great prestige and was conferred well into the fourth century.

ITALY

The Romans were masters of the peninsular by the time of the Emperor Augustus, their power initially imposed through, then lodged in, numerous imperial 'new towns'. There are too many to mention more than a representative selection; this section describes the origins and forms of five: Pompeii (arguably ancient history's most important remains); Ostia, Aosta, Verona, and Cosa. Others include Florentia (Florence – Chapter 4) and Mediolanum (Milan) and Augusta Taurinorum (Turin – Chapter 5).

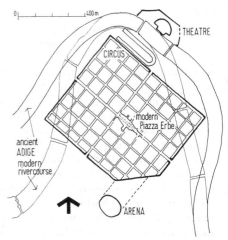

Figure 3.16 – Verona, one of the most attractive historic Italian cities (and the author's favourite); the beauty of its setting within (and now around) a tight river bend, derives, however, from the Roman engineer's eye for defensive potential when first locating a Latin colony in 69 BC to control the strategic crossing of the Adige. Reconstructed around 15 BC, the city was rewalled in AD 265, incorporating the arena as a 'citadel'. The theatre is across the river, on the steep valley slope.

54. For the bastides see Chapter 4; for the garden-city concept see E Howard, *Garden Cities of Tomorrow*, 1902 (reprinted 1946 with a preface by F J Osborn and an introductory essay by L Mumford). Howard's book was first published in 1898 under the title *Tomorrow: A Peaceful Path to Real Reform*, a convenient date for urban historians, right at the end of the nineteenth century, and which is to be regarded as the 'birthday' of the continuing British Garden City/New Towns movement.
55. Salmon, *Roman Colonization under the Republic*.
56. See Benevolo, *The History of the City*, 1980, Chapter 3, 'Rome: City and Worldwide Empire', and the extract from the Map of Italy published by the Instituto Geografico Militare showing the *centuriatio* for the land to the north and east of Imola.
57. Salmon, *Roman Colonization under the Republic*.
58. This may be an apocryphal story from the early days of the British New Towns programme; although it was vouched for by F J Osborn, who was at the centre of events at the time, in an interview with the author; see A E J Morris, 'New Towns: The Crucial Decade', *Building*, 29 November 1974; F J Osborn and A Whittick, *The New Towns*, 1977.
59. Salmon, *Roman Colonization under the Republic*.

POMPEII

Uniquely preserved beneath the ash and dust from Vesuvius (AD 79), Pompeii has provided archaeologists with a complete cross-section of urban form and social life of the time.[60] Whereas, in so many other examples of Roman planning, continued site occupation throughout centuries has made comprehensive excavation impossible, at Pompeii life was abruptly suspended and the site subsequently unoccupied.

Pompeii was originally founded as a Greek colonial city in the early sixth century BC. In form it is essentially late Hellenistic, having been rebuilt *c.* 200–100 BC, during which period Greek city building practice was gradually being superseded by Roman methods. Apart from the ultimate deficiency of its location Pompeii was well situated in an area alongside the Bay of Naples which was both beautiful and richly fertile.

The city in AD 79 was roughly oval in shape, about four-fifths of a mile long by two-fifths wide with an area of some 160 acres enclosed with a double wall. Population estimates vary between 25,000 and 30,000. The original Greek settlement centred on its old triangular

Figure 3.17 – Pompeii: general plan of the city, as revealed by the continuing excavation programme. Key: A, the main forum (Figure 3.18); B, the triangular forum, centre of the original Greek settlement; C, location of the two theatres; D, the arena. The detail housing district plan, as Figure 3.19, occupies ten major *insulae* in the western corner of the city (top left-hand corner). The House of Sallust (plan shown as Figure 3.20) is indicated in heavy outline at the southern end of the rectangular *insula* in this western corner (Region VI; Insula 2; Number 4).

Figure 3.18 – Pompeii: detail plan of the forum and surrounding buildings (north to the bottom right: the orientation of the forum has been moved in this plan clockwise through 90° from its position in Figure 3.17). Note how the two streets which approach the forum at its southern end (left) are discontinuous across the main central pedestrian space. Key (as Roman numerals): VII, Temple of Jupiter; seriously damaged during the earthquake of AD 62, it was still partially ruined in AD 79; IX, Temple of Apollo; XVI, public conveniences; XVII, market; XVIII, the basilica centre of Pompeii's commercial life; XIX, three civic halls, the central one probably the Tabularium (Public Archives Office); XX, Comitium, the location of public elections; XXI, Building of Eumachia, the most impressive construction of the forum, after the basilica; XXII, Temple of Lares; XXIII, covered market.

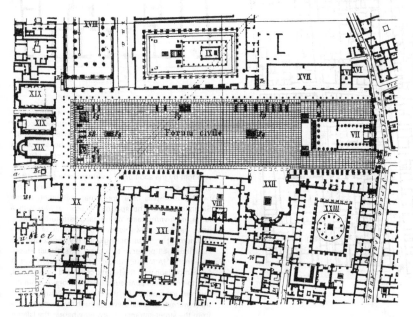

forum (agora) was in the south-east corner; the final area of the city, based on a freely interpreted gridiron form, was extended from here. There were eight gates into the city leading to well-paved main streets provided with raised pavements. Mercurio Street at 32 feet was the widest. Other main streets had widths of about 26 feet, while the minor ones, which served only to give access to the houses, varied between 18 and 12 feet wide.

The new forum was located roughly in the centre of the city, near to the harbour front. Enclosing a civic space of some 500 by 160 feet, it displayed carefully composed building relationships and unifying colonnades (Figure 3.18).[61] Anticipating modern practice by almost 2,000 years, and illustrating the traffic conditions in many Roman towns, the forum at Pompeii constituted a pedestrian precinct with gateways preventing vehicular access.[62] The city had two theatres seating 5,000 and 1,500 people, located near the old forum, and a magnificent oval

60. See A Maiuri (trans. W F McCormick), *Pompeii*, as the author's preferred guidebook to the excavations of Pompeii. For both Pompeii and Herculaneum, the near neighbour that suffered the same fate, see M Grant, *Cities of Vesuvius*, 1971.

61. The unifying colonnade constructed in front of the individual forum buildings represents that which can be acclaimed the earliest known instance in history of an 'urban beautification scheme'.

62. As far as I know, the Roman forum at Pompeii was also the first formally organized pedestrian precinct in urban history. In which respect there is further anticipation of social priorities in modern city traffic management, whereby, as revealed by their named tallies, certain sufficiently important persons were entitled to take their chariots into the forum, doubtless with their own reserved parking spaces.

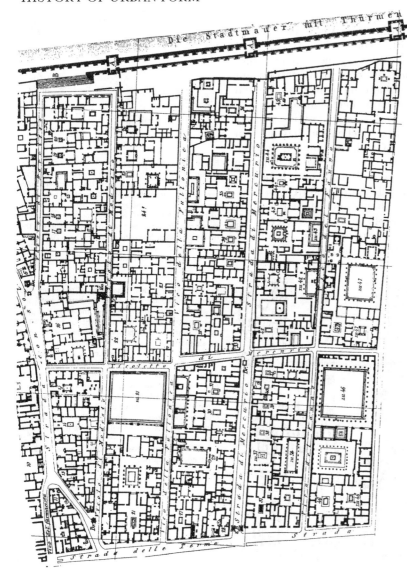

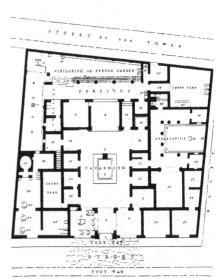

Figure 3.19 – Pompeii: detail plan of housing *insulae* in western corner of city. The House of Sallust (Figure 3.20) is keyed as No 16, forming the southern (bottom) end of the top left-hand *insula*. This beautiful archaeological plan bears close scrutiny for the extent of individual courtyard houses, large and small, and for the 'lock-up' shops alongside the streets.

Figure 3.20 – Pompeii: plan of the House of Sallust. (The orientation of this house in the north-west corner of the city, has been moved anti-clockwise through 90° from its position in Figure 3.19.) 'This house,' writes Amedo Maiuri, 'notwithstanding the presence of *tabernae* on either side of its entrance and the changes wrought by the addition of a *viridarium* in the peristyle and of a small *pistrinum* in the west wing, is one of the noblest examples of a pre-Roman habitation of the Samnite period. The spacious Tuscan *atrium* with a large *impluvium* basin and great characteristic doorways narrowing towards the top, the walls of the airy *tablinum*, of an *oecus* looking upon the portico and of a *cubiculum*, all preserve to a great extent their stucco revetments' (A Mairui, *Pompeii*). See Chapters 1 and 11 for comparison with the Islamic courtyard house types, and their Arabian peninsular predecessors. A main difference, exemplified by the House of Sallust, is that there were direct axial views through the Roman house from the street entrance doorway; whereas the Islamic private domestic interior is always screened from the public domain.

Thermopolium of Asellina, the most complete example of a bar for hot and cold drinks so far restored to us by Pompeii. The terracotta and bronze vessels and fittings, the hanging lamp, the petty cash, were all found in place. The heating vessel for hot drinks was found with its lid hermetically closed. The exotic female names written on the outside wall, 'Aegle', 'Maria', 'Smyrna', and even the nickname 'Asellina', leave us to suppose that the place owed its fame above all to the easy favours of the damsels who served in the bar and in the inn on the upper floor...
(Aemdeo Maiuri, Pompeii)

63. The courtyard houses at Pompeii include some of the grandest of their kind in history, as exemplified by the House of Sallust, Figure 3.20.

amphitheatre, some 450 feet in length and capable of holding some 20,000 people, in the eastern corner. With these three buildings alone able to accommodate almost the total population of the city, it is reasonable to assume that Pompeii functioned as a regional leisure and cultural centre. Near the forum to the north there were the two main public baths.

The forum area provided the main shopping and commercial facilities but flanking the main streets were many smaller shops and workshops. Although these uses formed part of the housing blocks, they generally stayed completely distinct from the dwellings themselves. The houses invariably followed the pattern whereby rooms faced into a central courtyard with only the entrance doors opening into the street.[63] These houses were on two, three and as many as five or six floors in the undulating southern part of the city.

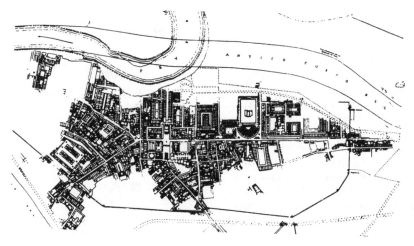

Figure 3.21 – Ostia, general plan of the port (north at the top). The ancient bed of the Tiber ran from east to west along the northern side of the town (Antico Corso, as plan); the modern bed approaches from the north before rejoining the ancient line. Lanciani, writing in 1897, noted that, 'Ficana, the oldest human station near the bar of the river, is now 12,000 metres inland and kingly Ostia 6,600 metres. The Torre di S. Michele, built in 1567 by Michelangelo on the edge of the sands, stands 2,000 metres away from the present shore. The average yearly increase of the coast at the Ostia mouth is 9.02 metres' (*Ruins and Excavations of Ancient Rome*).

The original *castra* is directly opposite the modern bend in the Tiber (immediately right of the fork in the *decumanus*, the main east–west street); the forum was the centrally located open space.

OSTIA

The traditional date for the foundation of Ostia, at the mouth of the Tiber, is during the reign of Ancus Marcius, fourth king of Rome.[64] However, the earliest remains so far discovered are of a 5½ acre *castra* built about 330 BC, in the angle formed by the Tiber and the Mediterranean. Ostia played an important role as a naval base during the Carthaginian Wars and became a busy commercial port from the third century BC when its growth and prosperity directly reflected the rise of Rome itself. Expansion required a new defensive perimeter in 80 BC, enclosing an area of approximately 160 acres, about the same as Pompeii.

Figure 3.22 – Cosa (Ansedonia): plan of the site on the coast of central Etruria, 85 miles north of Rome, which Salmon describes as 'the scene of but little human habitation either before its colonization under the Republic or after its abandonment under the Empire' (*Roman Colonization under the Republic*). 'Such a favourable combination of circumstances,' he explains, 'has made of Cosa "the very paradigm of a Latin colony".'

In 280 BC the Etruscan city-state of Vulci ceded the territory of Cosa to the Romans, who chose for the site of their new colony a truncated 350-ft-high hill, 'low enough to be accessible but high enough to be defensible', which was 500 yards from an excellent natural anchorage. The massive town wall (1,500 yards in length) follows the contour and encloses an area of 33 acres. The three gateways were in accordance with native Italic practice, whereas the eighteen square towers were an unusual defensive strengthening.

Inside the wall there was a pomerial roadway 15 feet wide, with the *insulae* formed by three *decumani* and seven *kardines*. There was no through *decumanus maximus* as such, but the *decumanus* from the north-west gateway to the forum served as one axis, and the *kardo* from the forum to the north-east gateway, the other. Both were about 20 feet wide. A third main route was the *kardo* (15 feet wide) linking the forum with the sacred precinct (the *arx*) in the highest, south-western corner of the town. Salmon informs us that public buildings occupied 25 per cent of the enclosed area and streets a further 15 per cent, the remainder was densely covered with predominantly single-storey housing of simple (non-atrium) designs.

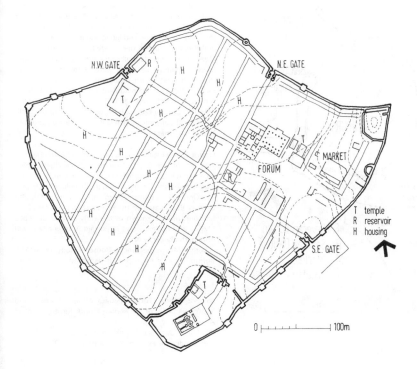

N.W. GATE   R   H   N.E. GATE

FORUM

MARKET

T   temple
R   reservoir
H   housing

S.E. GATE

0 ⊢—⊢—⊢—⊢—⊢—⊢—⊣ 100m

64. For Ostia see R Meiggs, *Roman Ostia*, 1960.

North Africa

This section covers present-day Egypt, Libya, Tunisia and Algeria, to the extents that their territories inland from the Mediterranean were part of the Roman Empire (see Figure 3.1 for locations). In AD 106, when the Empire had reached its furthest enduring limits, the provinces concerned were Aegyptus (30 BC), Cyrenaica (74 BC) and Africa (146 BC). (The western North African provinces of Mauretania Caesariensis and Mauretania Tingitana dated from AD 44.)[65]

Urban origins were various: in eastern North Africa there were Hellenistic settlements, headed by Alexandria; further west there were many well-established Punic towns – notably Carthago and Leptis Magna – and others of native origins, including Thugga. In addition to existing towns, which were extended and replanned to varying extents, there were the usual Roman new towns, planned either as replacements, or as legionary-settlement *colonia*, notably Thamugadi and Cuicul.

Olive oil, corn and other agricultural produce were exported to Rome in very considerable quantities; although the provinces saw only a small part of the profits, they were wealthy enough to sustain highly developed urban life-styles. The cities embodied all the usual imperial urban facilities and services, some at comparatively large scales: the amphitheatre at Thysdrus (modern El Jem, Tunisia) for example, was third largest in the Empire;[66] Carthago, as reconstructed, was supplied by an aqueduct more than 50 km in length; and the Temple dedicated in AD 166 as part of the marvellously surviving Capitol complex at Thugga (as illustrated) was one of the outstanding imperial works of architecture.

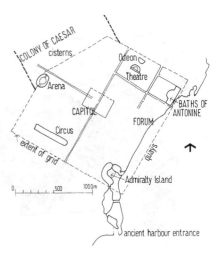

Figure 3.23 – Carthago, diagrammatic plan (after Alexandre Lezine, *Les Thermes d'Antonin a Carthage*) of the Roman city authorized by Julius Caesar as a *colonia* – most probably the first transmarine one – just before his assassination in 44 BC. The site was that of the historic Punic capital, totally destroyed by the Romans in 146 BC. (In 123 BC Gaius Sempronius Gracchus, plebian tribune, had received permission for a citizen-colony on the site, for 6,000 settlers, which was never implemented.)

Carthago subsequently owed much to the Emperor Augustus, who confirmed its status as a colony. The city was approximately square on plan, with three sides of about 1,800 metres. Within that area there was a regular gridiron on the gently sloping land. The northern corner (possibly the extent of Caesar's original proposal) had a differently orientated gridiron.

The *cardo maximus* and the *decumanus maximus* were both 12 metres wide, and the secondary *cardines* and *decumani* were 6 metres wide. There was a total of 120 *insulae*. The forum occupied the highest ground, in the centre of the city, and alongside the Mediterranean, in the eastern corner, there were the magnificent Antonine Baths, fourth largest in the Empire after the main three at Rome. The baths were supplied with water from a covered cistern, occupying an entire *insula*, supplied by the city's main aqueduct. The enormous circus provided seats for more than 80,000 spectators, and the amphitheatre ranked sixth largest in the Empire. (Little remains of Carthage, which is now a seaside suburb of Tunis: see Chapter 11.)

Figure 3.24 – Leptis Magna: the general plan. The city was one of the 'tri-polis' group of three cities, which gave the name to Tripolitania, and from which, in turn, was derived the name of modern Tripoli (see Chapter 11). Originally a Punic settlement, probably of the fifth century BC, located on the headland to the west of the harbour, little is known of the early history of Leptis Magna and the extensive important Roman city is still only partially excavated.

Leptis Magna became a Roman colony in AD 110, during Trajan's reign, and it benefited greatly under

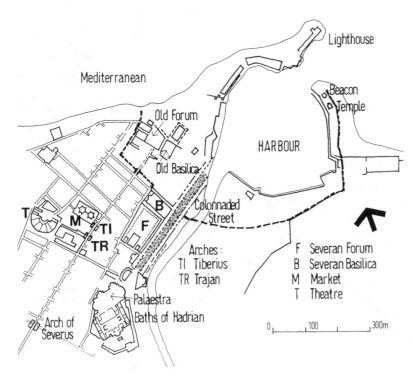

Arches:
TI Tiberius
TR Trajan

F Severan Forum
B Severan Basilica
M Market
T Theatre

65. For North Africa generally see Cornell and Matthews, *Atlas of the Roman World*; Raven, *Rome in Africa*, 1969.

66. For El Jem see S Tlatli, *Antique Cities in Tunisia* (Les Guides Ceres – English translation), 1971.

THUGGA

As modern Dougga, this most beautifully sited and almost miraculously preserved Roman town is located about 105 km south-west of Tunis near the small modern town of Teboursouk.[67] The readily defended hillside position overlooking a still richly fertile, extensive valley plain, had been recognized from early times as an organic growth settlement, which became a prosperous Punic city. Thugga survived the downfall of Carthage in 146 BC and became gradually romanized, eventually to gain formal status as a *municipium* early in the third century AD. Although acquiring several major works of architecture and urbanism, the city kept its haphazard street pattern; the effect, in combination, of age – with the vested site interests – and unsuitability of the hillside site.

Septimus Severus (193–211), a native of the city, who made possible the construction of a new forum and basilica, modernization of the port facilities, and the addition of a new, grandly colonnaded avenue leading from the harbour to the Baths of Hadrian. During the mid-fourth century the city came under threat from desert tribes and it suffered greatly under the Vandals. By the sixth century it had declined to the status of a small fishing village.

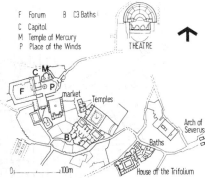

Figure 3.25 – Thugga (Dougga), plan of the city (the hillside sloping from north to south) showing the unaltered organic growth street layout of the original Punic settlement, and locations of the main Roman monuments and buildings.

Figure 3.26 – Thugga (Dougga), the superb Capitoline Temple (AD 166/67) seen across and above the organic growth streets. One of the most completely surviving major Roman temples, it is also probably the least well known. Only a short distance inland from the Tunisian 'costa del tourists', at Dougga it is still possible to walk the streets of an ancient Roman town where images of an urban past can be conjured up in comparative freedom from present day distractions. Or so it was, on both occasions that I have been there.

Interestingly, the portico and steps of the Capitoline Temple face south out across the town towards its agricultural hinterland and not into the Forum, which is alongside its blank western side. This may have been determined by symbolic religious considerations, or – and the more likely in combination – it was the result of topographical and historic constraints on an already long occupied hillside location. From the well preserved theatre on the highest ground there are sweeping views over the town and the magnificent landscape setting.

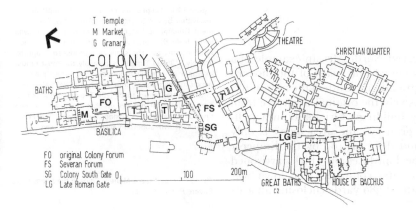

Figure 3.27 – Cuicul (Djemila): plan of the extensive two-part city which grew from its origin as a veteran colony, founded under Nerva (AD 96–98) as one of several along the main road from Sicca Veneria (El Kef) on the coast, through to Cirta (Constantine). The enlarged city was unbalanced by extensive later second-century growth to the south, with the bath complex and theatre; a situation resolved by constructing between the two parts the extensive new early third-century Severan forum. Note the contrast between the original grid plan and the more informal pattern of the 'new town'.

67. For Thugga (Dougga) see Tlatli, *Antique Cities in Tunisia*; for Thugga and also Thuburbo Majus, Mactar and Gightis, see A Mahjouhi, *Les Cités Romaines de Tunisie* (French).

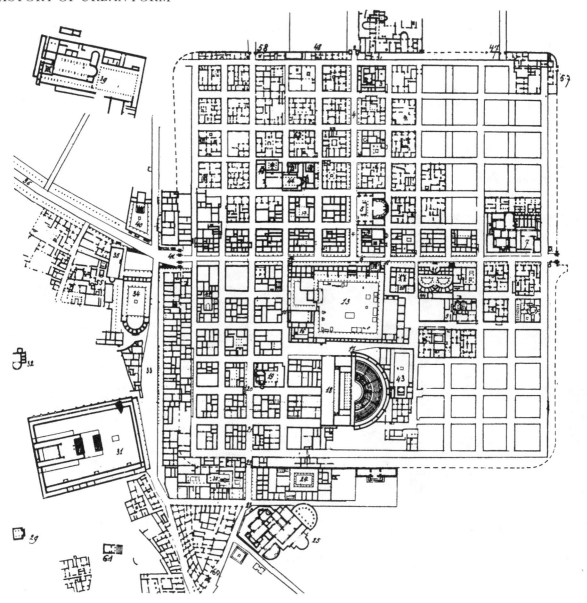

Figure 3.28 – Timgad: general plan of the settlement as excavated by the Direction des Antiquités, Gouvernement Général de l'Algérie. (Orientation is somewhat away from the cardinal points: the upper side faces north-north-west. The scale as reproduced is approximately 1:3500.) Timgad is possibly the most regular example of imperial gridiron-based urban planning, which makes more strange the inconsistent alignment of the central forum space (13). The plan departs from general imperial practice in that there are no continuous cross-axes. Outside the original perimeter organic growth suburban accretion can be seen.

TIMGAD

Timgad (Thamugadi) is a well-preserved North African example of a Roman legionary town.[68] Situated some 24 miles east of the modern town of Batna in Algeria, on flattish ground just south of the hills, it has been possible to excavate comprehensively the original town free from the complications of later, overlaying buildings. Originally in a well-watered, fertile part of the Roman 'granary', the area today is only sparsely cultivated. Timgad was founded in AD 100 by the Emperor Trajan for time-expired veterans of the Third Legion which garrisoned the nearby fortress of Lambeisis. As first completed the town was almost square on plan, with sides of about 380 yards enclosing an area

68. For Timgad see A Menen, *Cities in the Sand*, 1972; Raven, *Rome in Africa*.

of about 30 acres (Figure 3.28). Later suburban additions were added, mainly to the western side. The plan has a rigid gridiron pattern, formed by eleven streets in each direction which intersect to give square *insulae* with sides of 23 yards. There were originally four gates, three of which have been identified. The forum, about 160 by 145 feet, and the theatre, were both formed out of several *insulae*, and other public buildings occupied one *insula* each. Main streets were widened and flanked by colonnades and were well paved and drained.

Mumford considers Timgad to be 'an example of the Roman planning art in all its latter-day graces. Being a small town, like Priene, planned and built within a limited period, it has the same diagrammatic simplicity, unmarred by later displacements and renovations that busier towns subject to the pressures of growth would show'.[69] But the formal built-up pattern of Timgad, with virtually no incidental open space within its limits other than streets and public spaces, should be compared with the contemporary situation at Calleva Atrebatum (Silchester) in Britain. Here the native predeliction for informal groups of buildings in space had completely modified the imperial norm.

ALEXANDRIA

One of the greatest cities of the Roman world, Alexandria, had been planned for Alexander the Great, by Dinocrates of Rhodes, during the winter of 332–331 BC.[70] The propitious siting was on a strip of land between the Mediterranean and Lake Mareotis (Mariut), with the original harbour protected by offshore islands, the largest of which was the Island of Pharos, on which, in *c.* 280 BC, was constructed the lighthouse famed as one of the 'Seven Wonders of the World'. The Romans linked the consolidated islands to the mainland by a causeway, thereby creating the incomparable double harbour – the Porto Eunostos and the Magnum Portus. With time the causeway has been widened to form the present-day broad neck of land. During the Roman Empire, the city was second only to Rome itself (see also Chapter 11).

Figure 3.29 – Alexandria: diagrammatic plan of the late Roman city (after M Bey, *Memoire sur l'antique Alexandrie*, 1872, and others). The presumed orthodox Hellenistic gridiron of Dinocrates has been completely lost; although physical remains of the Roman city are sparse, two contemporary descriptions have survived:

Here Alexander noticed the harbour sheltered by nature with its magnificent emporium together with fields of grain spreading over the whole of Egypt, and the immense advantages of the giant River Nile. (Vitruvius, Book II, Introduction)

... the advantages of the city's site are various; for, first, the place is washed by two seas, on the north by the Aegyptian Sea, as it is called, and on the south by Lake Mareotis. This is filled by many canals from the Nile, so that the harbour on the lake was in fact richer than that on the sea. The shape of the area of the city is like a 'chlamys' (a Macedonian army cloak), the long sides of it are those that are washed by the two waters, and the short sides are the isthmuses.... The city as a whole is intersected by streets practicable for horse-riding and chariot-driving, and by two that are very broad ... which cut one another into two sections and at right angles. (Strabo XVII 1.6 ff)

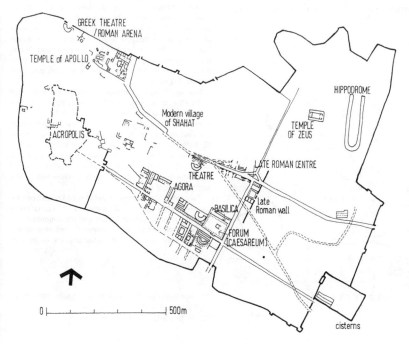

Figure 3.30 – Cyrene: the city was a Greek foundation of the seventh century BC, reputedly on the part of one Aristoles (King Battus), leader of a party of colonists from Thera. The site near the North African coastline some 800 km west of Alexandria, was chosen for its reliable water supply, as the legendary recommendation by local tribesmen: 'Here, O Greeks, ye may fitly dwell, for in this place there is a hole in the heavens' (R Goodchild, *Cyrene and Appollonia*).

The Romans gained control from 96 BC and, as with Alexandria, their continued building activities obliterated the Greek city, which is assumed to have been centred around the Temple of Apollo, near the spring and later sacred fountain of Apollo, on lower ground in the north-western corner of the enlarged city. From there, the valley road rising to the south-east, through the city, was one main axis; with a parallel main street along the south-western ridge through the regularly planned Roman city centre and leading up to the acropolis.

69. Mumford, *The City in History*.
70. For Alexandria see E M Forster, *Alexandria: A History and a Guide*, with introduction by L Durrell, new edn 1986.

Spain

It is one of the more ironic twists of history that Spain, the sixteenth- and seventeenth-century despoiler of a 'New World' in search of mineral wealth, had herself been a comparable, mystical treasure-land of the old world, visited from earliest times by Phoenician and Greek traders. Subsequent plunder, first by the Carthaginians and latterly the Romans, largely exhausted her once considerable deposits of gold, silver and copper.[71]

The Carthaginians succeeded the Greeks as traders along the Mediterranean coast; once established in Spain they recognized its potential as a mainland European base from which to mount military operations against Rome. In 223 BC Hasdrubal founded Carthago Nova (Cartagena) as the new Punic arsenal and naval base, with the Romans, across the River Ebro, based on Tarragona (rewalled in 218 BC) and Barcino (the future Barcelona – see Chapter 9). During the Second Punic War, when Hannibal was stalemated in northern Italy, a Roman expeditionary force under Publius Scipio captured Carthago Nova in 209 BC. Long-term systematic conquest of the Iberian peninsular was then required of the Romans, initially to consolidate defeat of the Carthaginians and latterly to secure themselves against uprisings and frontier incursions, a process that was to take some 200 years.

The provincial era in Spain dates from 205 BC when Roman territory was divided into two provinces of Hispania Citerior and Hispania Ulterior ('nearer' and 'further' Spain respectively). The Romans were quick to exploit their new Eldorado: estimated totals of 130,000 lb of silver and 4,000 lb of gold were sent back to Rome during the first ten years.[72]

With few exceptions, the sites of Roman cities in Spain have been in continuous occupation for some 2,000 years, during which time normal building activities and, for many, extensive reconstruction following devastation, has either erased the Roman period, or left its urban re-

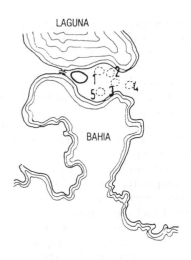

Figure 3.31 – Cartago Nova (Cartagena): diagrammatic location plan of the city on a readily defended peninsular at the back of a sheltered deep-water bay. An existing small port was greatly enlarged by the Carthaginians as their new arsenal and naval base, but after only fourteen years it fell to the Romans. Little is known of their city, other than traces of the forum identified beneath the modern Plaza de los Tres Reyes.

Figure 3.32 – Emporiae (modern Ampurias and the Greek Emporion): located on the Gulf de Rosas, southeast of Figueras in the north-eastern corner of Spain. The general site plan relates the only partially excavated extent of Emporiae to the two preceding Greek cities, the Neapolis (their 'new' city) and Palaipolis (their original settlement, now occupied by the village houses of Sant Marti d'Empuries).

Scipio's expeditionary force directed at Cartago Nova landed at Emporion, and in 195 BC the consul Cato chose it as his regional capital. The port prospered, enjoying taxation privileges and a degree of political autonomy. An Iberian village had grown up to the west of Romanized Emporion and this was taken by Julius Caesar in c 49 BC as the site for his new veteran's colony renamed Emporiae. The rectangular walled perimeter enclosing a formal orthodox gridiron measures 760 by 325 yards.

The forum (A) was approximately square with sides 74 by 68 yards enclosed by colonnaded porticos on the east, south and west, with a series of eight small temples along the northern side. The decumanus through the central southern gate leads across a market square (B) to the forum, around which it is diverted to the east as an unusual departure from imperial planning orthodoxy. (The even more exceptional diagonal street in from the south-western corner is unexplained by the archaeological record.)

Two extensive multi-courtyard houses have been excavated ('E' and 'F') revealing evidence of a prosperous Roman urban life-style; leisure and recreational activities were catered for by the arena ('C') and an enclosed palaestra for gymnastic and other physical exercising.

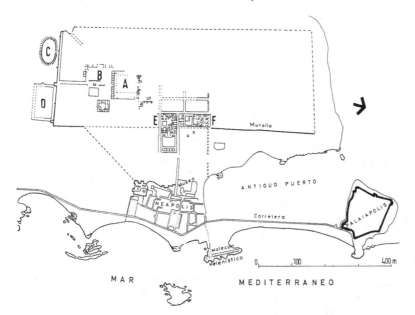

71. For Spain see S J Keay, Roman Spain, 1988; C H V Sutherland, The Romans in Spain.
72. Sutherland, The Romans in Spain.

mains more or less permanently concealed (often beneath greatly different Islamic urban forms: see Chapters 4 and 9). Emporiae (Ampurias)[73] and Italica (Santiponce) are two major exceptions which are still in process of excavation (see Figures 3.32 and 3.34). The extents of Roman nucleii of modern urban areas are known, however: where they originated as new towns, a characteristic grid plan can be presumed. Older cities of Celtic origin, taken over by the Romans, usually retained their essentially organic growth forms.

The Emperor Augustus (63 BC–AD 14) led campaigns in 26 and 25 BC against the remaining Celtiberian tribes in northern Cantabria and Asturia. Believing himself victorious, he founded Colonia Emerita for his time-expired veterans, only for bitter fighting to continue until 19 BC when all of Spain was finally conquered. Under Augustus, Hispania was reorganized into three provinces: a greatly enlarged Citerior (renamed Tarraconensis after its capital) and Ulterior divided into two parts (Baetica and Lusitania). The reorganization was of extreme long-term significance: 'the independence of modern Portugal', writes Sutherland, 'is a tribute to Augustus's shrewd wisdom'.[74]

New towns founded by Augustus include Caesaraugusta (Saragossa), a civilian colony occupying a position of great administrative importance on the River Ebro. In Tarraconensis he raised Barcino to *colonia* status and founded Libisosa and Ilici. Emerita (Merida), 'Queen of the province of Lusitania, if not all Spain,' is regarded by Sutherland as 'the clearest and finest example of the Augustan system.'[75] It was an extremely prosperous city; its lack of established urban traditions well compensated by architectural magnificence. It had an amphitheatre, a circus, a theatre built for Agrippa and restored under Hadrian, two aqueducts and associated reservoirs, and two bridges over the River Anas, the latter dating from Trajan's reign and still in use today (see Figure 3.33).

Under Augustus, in Baetica there were 175 urban communities, including nine *coloniae*, ten *municipia* and twenty-seven towns with Latin rights; Tarraconensis comprised twelve *coloniae*, thirteen *municipia* and fewer than twenty other privileged towns; while Lusitania out of a total of about forty-five urban settlements could show but five *coloniae*, one *municipium* and three towns with Latin rights.[76]

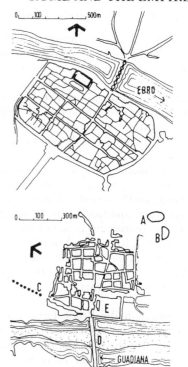

Figure 3.33 – The nucleii of modern Zaragoza (top) and Merida, respectively Roman Caesaraugusta and Emerita. Both present-day street plans reveal an underlying Roman grid form – Zaragoza to an unusual extent; as evidence of uninterrupted – or not seriously affected – urban status through from the end of the Empire to the present day.

Caesaraugusta was founded by the Emperor Augustus in 25 BC as a civilian *colonia*, near existing Salduba, alongside the River Ebro where it is joined by the Gallego and the Nuerva. It became an imperial administrative centre of great importance.

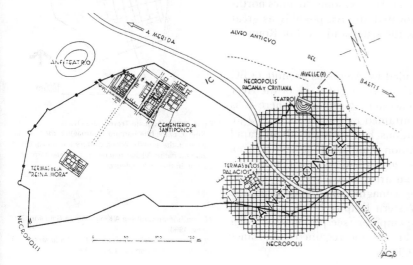

Figure 3.34 – Italica (Santiponce): the general outline of the Roman city, the south-eastern part of which is covered by modern Santiponce. Seville (Roman Hispalis) is some 5 miles (8 km) to the south-east.

73. For Ampurias see E R Perello, *Ampurias*, 1979; Keay, *Roman Spain*.
74, 75. Sutherland, *The Romans in Spain*; for Emerita see E G Gabriel and M A S Caballero, *Guia Breve de Merida*, 1987.
76. Salmon, *Roman Colonization under the Republic*.

Gaul, Germany and the Danube

The Romans had conquered southern Gaul (France) in 120 BC, creating the province of Gallia Narbonensis, its capital at Narbo (Narbonne) and extending up the Rhône valley to Lugdunum (Lyon) and beyond. Caesar's epic Conquest of Gaul between 58 and 51 BC established the imperial frontier along the River Rhine, with forays carried out further east and across the Channel into southern Britain. Lugdunum, at the confluence of the Rhône and the Saône, rose to be a great imperial city, the only one in Gaul to be honoured with full Roman citizenship.[77]

Of Roman cities in the present-day French part of Gaul, Colonia Nemausus (Nîmes) has the best-preserved monuments, notably its arena and the superb 'Maison Carrée' in the city, and the Pont du Gard aqueduct (and road bridge) some 20 km to the east. Lutetia (Paris) was only a comparatively minor Roman town (see Chapter 6) while Durocortorum (Reims) and Burdigala (Bordeaux) were provincial capitals. In present-day German Gaul, Moguntiacum (Mainz) and Colonia Agrippina (Cologne) were the respective capitals of Germania Superior and Germania Inferior, as created in AD 90 out of the previously military areas.

Although the Rhine formed a logical, reasonably easily defended frontier with the Germans, Augustus took the momentous decision to advance to the Elbe and the Danube, along which latter river, virtually from its source to the Black Sea, numerous legionary camps and other settlements were founded. These include Castra Regina (Regensburg – see Chapter 4); Vindobona (Vienna – Chapter 7); Aquincum (Budapest); and Singidunum (Belgrade). The especially vulnerable land frontier between the Rhine at Confluentes (Koblenz) and Castra Regina on the Danube was defended by numerous forts (the limes), before in the late Empire the Romans withdrew again behind the rivers.

AUGUSTODUNUM

Augustodunum (Autun) was established about 12 BC when Augustus, in the process of organizing the new territories in northern and central Gaul which had been conquered by Caesar, moved the Celtic tribesmen of the Eduens from their hilltop village of Bibracte, some 12 miles away, to a new location alongside the river Arroux, some 10 miles north-north-west of Lyon. Augustodunum covered an area possibly as great as 490 acres and became renowned as the sister and rival of Rome.

AUGUSTA TREVERORUM

Augusta Treverorum (Trier) was founded by Augustus around 14 BC as the 'new town' for the local Treveri, on the right bank of the river Moselle in north-east Gaul (modern western Germany, where it is renowned as 'the oldest town').[78] It attained *colonia* status in AD 50. Excellent road and river communications had brought about rapid growth and at that time its population numbered 50,000–60,000. Later, as refortified by Diocletian (285–305) it had become perhaps the largest city in Western Europe with an area of 704 acres. Diocletian made it the capital of Gaul, building a magnificent imperial palace, from which Constantine ruled the Western Empire before moving to Constantinople on becoming absolute emperor (see pp. 89–91). Trier survived repeated destruction by the Franks and regained importance as the seat of the Merovingian king, Clovis.

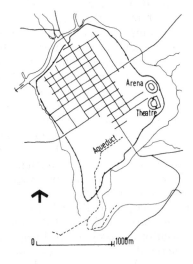

Figure 3.35 – Augustodunum (Autun): general plan.

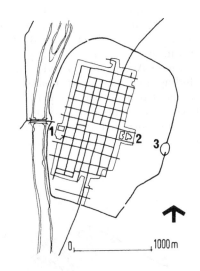

Figure 3.36 – Augusta Treverorum (Trier).
Key: 1, baths (the Barbara Thermae); 2, the Imperial Palace; 3, the superb second century arena capable of accommodating 30,000 spectators, and later forming a bastion in the city's defences.

77. For Roman Gaul see A King, *Roman Gaul and Germany*, 1990.

78. See E Zahn, *Trier: A Guide to the Monuments*; E M Wightman, *Trier and the Treveri*, 1970.

AUGUSTA RAURICORUM

Augusta Rauricorum (Augst) is located on the southern left bank of the Rhine, in modern Switzerland, some 11 km upstream from the centre of modern Bâle. It was founded in 44 BC as a military colony with the purpose of consolidating Caesar's conquest of Gaul and controlling a vulnerable, vitally important stretch of the Rhine frontier. Although initially slow to be equipped with the niceties of civilian provincial life, the town grew to become an important northern centre, as evidenced by the mid-second-century rebuilding of the theatre to provide some 8,000 seats. After third-century Germanic invasions, it would seem that the reduced population moved to the more readily defended riverside settlement of Castrum Rauracense (Kaiseraugst).

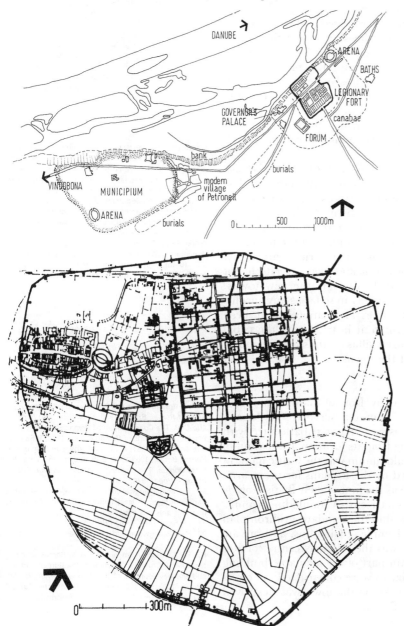

Figure 3.37 – Augusta Rauricorum (Augst): relating the castrum to the *colonia.*

Figure 3.38 – Carnuntum, founded by Tiberius (AD 14–37) as a legionary fortress camp on the west bank of the Danube some 35 km downstream from Vindobona (Vienna), is a clear example of the informal processes of imperial urban growth. Initially based on the military service-industry economy, and subsequently 'regularized' as a civilian *municipium.* The *canabae* that grew up around the fortress were under the jurisdiction of the legionary legate, and came to include a forum, baths and other 'urban' amenities.

The *municipium,* founded upstream later in the first century (now in part beneath the present-day village of Petronell) was a substantial town during the second century. In 193, Septimus Severus was there proclaimed emperor, but although Valentian (364–75) made it his base for a while, a contemporary writer (Ammianus Marcellinus) dismissed it as 'a deserted and unkempt town: but militarily convenient'.

Figure 3.39 – Aventicum (Avenches) showing the grid-iron structure of the vanished Roman town, within the extent of the outer defences, the ager, protecting the immediate agricultural land-holdings. The regularity of the Roman planning is in direct contrast to the organic growth form of the successor medieval town which became established closely to the west. The Roman arena, which could accommodate 12,000–15,000 spectators, survives at the eastern end of the present-day main street, and the theatre is also still to be seen, out in fields some distance to the south-east.

Roman Britain

The first invasion of Britain by Julius Caesar in 55 BC followed a five-year expansion period of the Roman Empire during which its frontier had been advanced from the Alps and the Cevennes to the shores of the Channel.[79] Troubles inside the Empire prevented first Julius, then Augustus from following up this reconnaissance, but contact between Rome and Britain had been established. In AD 43 Claudius invaded Britain and Roman legions methodically conquered England, Wales and southern Scotland. Before the conquest systematic town planning was unknown in Britain; the largest tribal capitals were probably no more than informal groupings of crude huts clustered together within earthwork defences.[80]

Following normal procedure Roman power lodged and consolidated itself by the building of towns, often on the sites of existing tribal villages. Four *coloniae* were recognized: three founded for demobilized veterans – Camulodunum (Colchester), Lindum (Lincoln) and Glevum (Gloucester) – and Eburacum (York), subsequently granted *colonia* status.[81] Verulamium (St Albans) is believed to have been the only *municipium*. Organization of the native Britons was on the basis of local authorities, *civitates*, which followed existing tribal divisions as far as possible, with a third class of town, the *civitas*, acting as cantonal capitals.

However, as A L F Rivet observes in his *Town and Country in Roman Britain*, whereas '*colonia* and *municipium* are precise Latin terms, and as applied to a town they defined its rank and the status of its citizens, the term "cantonal capital" is not precise, for it has no Latin equivalent. The reason for this is that to the Roman, reared in the traditions of the Mediterranean city state, the country was an adjunct of the town, not the town of the country. *Coloniae* and *municipia* had territory attributed to them, but the significant unit was the *colonia* and *municipium* itself, not the territory by which it was surrounded'. Outside the new towns, Celtic village communities were affected mainly in so far as any surplus of corn production was exported to Europe. The native Briton, even when Romanized, remained a countryman at heart,[82] with the tribal aristocracy preferring to live in Romanized villas established as country estates, generally within easy reach of the towns.

CAMULODUNUM

The ancient Celtic settlement replaced by Roman Camulodunum – modern Colchester – is one that would seem to deserve at least 'quasi-urban' recognition, in view of its status as a major tribal centre with considerable trading activities resulting from its location at the limit of navigation on the River Colne.[83] Camulodunum was the Celtic name of the capital of the Trinovantes, retained by the Romans for the nearby town which they founded in AD 49–50 as the first of their British *colonia* (known properly as Colonia Claudia Victricensis Camulodunensium). After ten years, completed buildings within the standard gridiron plan included a senate house, theatre, the Temple of Claudius, and private dwellings of varying size. The temple was the religious centre for Romans in Britain, and it received a main part of the available building resources. Perhaps as a result, no defences were constructed: an omission which proved disastrous when in AD 60 the town was destroyed during the short-lived uprising under Boadicea.

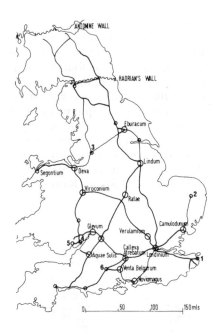

Figure 3.40 – Roman Britain: outline map relating major towns to the main Roman roads. Key: 1, Rutupiae (Richborough) one of three ports nearest to Gaul; 2, Venta Icenorum, tribal capital of the Iceni; 3, Mamueium (Manchester); 4, Venta Silurum (Caerwent); 5, Isca (Caerleon); 6, Sorviodunum (Old Sarum).

Roman place name equivalents: Londinium (London); Camulodunum (Colchester); Verulamium (St Albans); Ratae (Leicester); Lindum (Lincoln); Eburacum (York); Deva (Chester); Segontium (near Caernarvon); Viroconium (Wroxeter); Glevum (Gloucester); Aquae Sulis (Bath); Calleva Atrebatum (Silchester); Venta Belgarum (Winchester); Noviomagus (Chichester).

79. For Roman Britain see B Jones and D Mattingly, *An Atlas of Roman Britain*, 1990; G Webster (ed) *Fortress into City: The Consolidation of Roman Britain*, 1988; J Wacher, *The Towns of Roman Britain*, 1974; P Salway, *Roman Britain*, 1981; A L F Rivet, *Town and Country in Roman Britain*, 1958/1966; B W Cunliffe and T Rowley (eds) *Oppida: The Beginnings of Urbanisation in Barbarian Europe*, 1976.
80. The archaeological consensus now recognizes that Camulodunum, at least among Celtic tribal capitals, now warrants urban status.
81. I A Richmond, *Roman Britain*, 1955.
82. R G Collingwood and J N L Myres, *Roman Britain*, 1937.
83. For Camulodunum see M R Hull, *Roman Colchester*, 1958; P Crummy, 'Colchester: the Roman Fortress and the development of the Colonia', *Britannia*, vol. VIII: 65–105, 1978; R Dunnett, *The Trinovantes*, 1975; P J Drury, 'The Temple of Claudius at Colchester Reconsidered', *Britannia*, vol. XV: 7, 1985.

Figure 3.41 – Camulodum (Colchester): one of the earliest and most rectilinear Romano-British towns. Key: A, main gateway (from Londinium); F, forum; T, theatre; C, Temple of Claudius; B, baths.

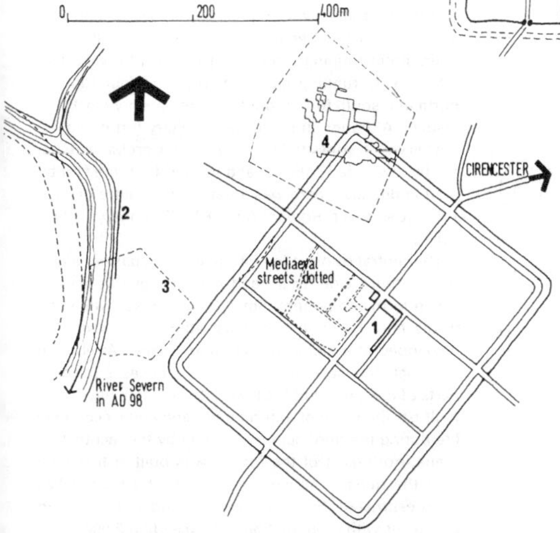

Figure 3.42 – Glevum (Gloucester) on the right bank of the Severn, had as its full title Colonia Nervia Glevensium and is therefore known to have been founded for veterans during the reign of Nerva AD 96–8) on a new site about half a mile downstream of an earlier legionary fortress of AD 50, probably that of the Twentieth Legion. Key: 1, Forum buildings; 2, Roman quayside; 3, first army fort; 4, Medieval Cathedral.

Collingwood and Myres in their *Roman Britain* consider that in order for the Romans to take the offensive in South Wales it was necessary to establish 'a legionary fortress on the left bank of the Severn at some point where it could be easily crossed, threatening the hostile right bank as Cologne or Mainz threatened the hostile right bank of the Rhine. The obvious position for such a fortress was Gloucester, which is still the lowest bridge on the Severn ...' Glevum covered an area of about 46 acres within a rectangular perimeter some 630 yards long by 440 yards wide.

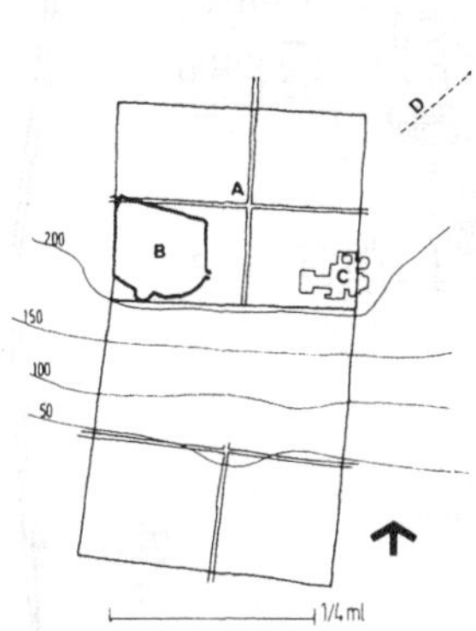

Figure 3.43 – Lindum (Lincoln). Key: A, the forum location; B, the Norman castle; C, the cathedral; D, known route of an aqueduct. Contours at 50 foot intervals; slope down from the original *castra* is very steep.

LINDUM

Lindum (Lincoln) was first established as a Roman fortress by troops of Legio IX Hispana in about AD 48. The area enclosed by a standard rectangular defensive perimeter was about 41 acres (16.6 hectares), situated on the edge of a limestone ridge some 200 feet above the river Witham, in a dominant, readily defended position. Subsequently the military abandoned Lindum in the late seventies when the Roman frontier was taken further north. In about AD 92 the old *castra* area was given new status as a *colonia* for veterans of Legio IX and by about AD 200 the lower market town, which had grown up between the *castra* and the river, was given its own defensive wall, enclosing a total area of some 97 acres.

Little is known of Lindum's street layout, or its buildings; continuity of site occupation, following only brief desertion after the end of Roman Britain, has left few known remains.[84] A partial system of sewers has been found with water brought in by aqueduct.

VERULAMIUM

Verulamium (St Albans) is generally assumed to have been the only *municipium* in Roman Britain and as such was established for the Catuvellauni, near to their own tribal centre. As first built, Verulamium seems to have been largely unplanned. However, it was a thriving commercial centre when burnt down by Boadicea's men in AD 60. The rebuilt Verulamium was considerably bigger than before, with an enclosed area of 140 acres, and included a great monumental forum and basilica of AD 79. Hadrian, however, seems to have been dissatisfied with the town, which was entirely rebuilt about 129–30, partly on and partly off the old site.[85]

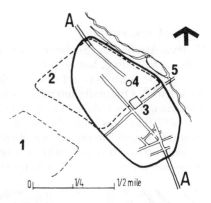

Figure 3.44 – Verulamium (St Albans). Key: 1, location of earlier tribal hill village; 2, first Romano-British city as rebuilt in AD 60 (dotted); 3, the forum; 4, theatre; 5, river ford; A, Watling Street. The perimeter of the second city is in heavy outline.

84. For Lindum see C Colyer, 'Excavations at Lincoln 1970–72: The Western Defences of the Lower Town', *Antiquaries Journal*, vol. 55, 1975; M J Jones, *The Defences of the Upper Roman Enclosure*, 1980.
85. For Verulamium see S S Frere, *Verulamium Excavations*, (i) 1972; (ii) 1983; (ii) 1984.

HISTORY OF URBAN FORM

VENTA SILURUM

Venta Silurum (Caerwent) was in South Wales, 10 miles from Newport and 5 miles from Chepstow. It dates from AD 75 and was founded to replace the Silurian tribal hill-fort situated on high ground a mile to the north-west.

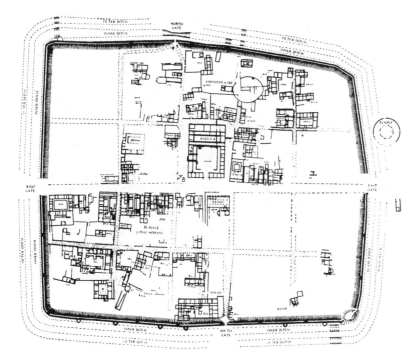

Figure 3.45 – Venta Silurum (Caerwent): general plan as excavated (north at the top). Scale: 1 inch to 450 feet (approx.) O E Craster has described how the new town had the customary rectangular Roman layout and covered an area of 44 acres. It was divided in half by the main street running through it from east to west; two other roads running east and west and four running north and south further divided it into twenty blocks, or *insulae*. Although in no sense a military fort it had defences which consisted of an earth bank probably revetted by a wooded palisade and fronted from the first by double ditches. There was a gate in the centre of each side. (*Caerwent Roman City*, HM Stationery Office, 1951).

The central colonnaded forum, with the basilica along the northern side, is a good example of this Roman urban form component. Small 'strip' houses with shop fronts line the main east–west street. Caerwent was developed at a higher density than Calleva Atrebatum and included several large 'patio' houses but some parts of the interior of the town appear not to have been built on. Evidence of a decline in Caerwent's economic life during the third century is given by the fact that an arena, north-east of the forum, was built at that time over the ruins of houses and on top of a road. It has been estimated that there were around 100 houses in Caerwent with a population not exceeding 2,000.

CALLEVA ATREBATUM

Calleva Atrebatum (Silchester), sited in Hampshire between Basingstoke and Reading, has been completely excavated (although each section of the town was returned to farmland after the findings were recorded).[86] Originally intended to have an ambitious area of around 200 acres, it was reduced to its irregular polygonal shape of about half that size. This was when defensive expediency required that a wall of economic length should be built. Richmond says that the town wall was an impressive structure, with a large fosse outside it and imposing semi-monumental gates. The main street was lined fairly thickly with

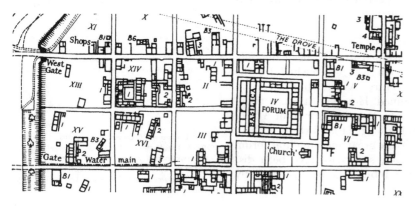

Figure 3.46 – Calleva Atrebatum, detail plan of the forum area, and six *insulae*; Nos V, VI, II, III, XIV, XVI, XIII, XV. The forum occupied the major part of a double *insula*, 130 yards (east–west) by 200 yards in size; the minor *insulae* are approximately 90 yards square. The forum and adjoining basilica together covered nearly two acres. The forum was an open square, 143 by 130 feet, lined by colonnades, shops and offices on three sides. The fourth side was closed by the basilica with a central hall 234 feet long by 58 feet wide. The baths were south-east of the forum on the side of a small valley in which 'the stream that flushed its drains still runs', (G C Boon, *Calleva Atrebatum; A New Guide*, Reading Museum in conjunction with the Calleva Museum Committee, 1967.) An intriguing, unexplained occurrence at Calleva Atrebatum was that the forum, baths and three temples were erected *before* the road system was established; the portico of the baths actually being removed to make way for the street (Boon).

86. For Calleva Atrebatum see G C Boon, *Silchester, the Roman Town of Calleva*, rev. edn, 1974.

shops and workshops. There were about twenty-five really large houses and something like the same number of small ones. To estimate accurately the population of such a place is difficult; surely at not less than 2,500 on the showing of the plan. But if, as seems very probable, many timber structures were missed this number may have been doubled or even trebled.[87] The most significant aspect of Silchester, in contrast to Timgad, as a hot-dry climate example, is the difference between the formal gridiron street pattern and the informality of the low-density building groupings within the grid blocks. The comparatively large proportion of open space between the buildings of Silchester has gained for it the title of 'Romano-British Garden City' and it exemplifies the difference between the introverted courtyard housing of the Mediterranean, and the outward orientated preference of damp-cold northerly climes.

LONDINIUM

Londinium (London) is believed to have been established in AD 43 as the Roman military base on the north bank of the Thames for the army's advance on Colchester.[88] During the uprising of AD 60 the base, and its adjoining civilian township, were burnt. Subsequently, as rebuilt, there was rapid assumption of the functions of provincial capital. 'No question of status', explains Professor Richmond, 'could prevent Londinium from becoming the natural centre for British trade and administration, once the Roman engineer picked it. If the first intention was to govern the province from Camulodunum, it is clear that within a generation the financial administration was using Londinium as its headquarters; whilst in the fourth century AD it was not only the seat of the provincial treasury but the residence of the civil governor who presided over the four divisions into which the province was then broken.'[89] Notwithstanding its *de facto* metropolitan role, Londinium, as far as is known, was never accorded an appropriately formal Roman urban status.

The importance of Londinium derived from its siting at the then lowest possible bridging point of the Thames, just downstream from what is believed to have been a non-tidal ford in Chelsea Reach. Here, Richmond emphasizes, 'the roads radiated from the bridgehead [and] the sea lanes converged from the Rhine, the Gallic coastal ports, and the North Sea, or by the channel route from Bordeaux, Spain, and the Mediterranean'.[90] Before the conquest, Celtic economic activity *vis-à-vis* Europe, such as existed, had been centred on Camulodunum, at which time, for all its incomparable strategic potential, the site of the future London was unoccupied, other than by minimal 'village' settlement. It is assumed that location at the intersection of several Celtic tribal boundaries may well have discouraged earlier proto-urban settlement.[91]

Two low gravel mounds on either side of the Walbrook – a small north bank tributary – provided the Romans with ample well-drained building land, with the comparatively sheltered Walbrook serving as the first Roman port of London, off the main tidal stream. A second small river – the Fleet – formed a natural western defensive barrier, and low-lying ground to the north and east could be easily fortified. (The caption to Figure 3.49 further explains what little is known of Londinium, including, most importantly, its presumed gridiron street layout.) The city expanded steadily and when walled, in the third century,

Figure 3.47 – Calleva Atrebatum (Silchester) – the 'Woodland Town' of the Atrebates – general plan, north at the top. (The greatest east–west width is almost exactly half a mile; from north to south the greatest width is 800 yards.) The diagonal route across the town, north-east of the forum, is a modern farm lane – 'the Drove' – superimposed on the Roman plan.

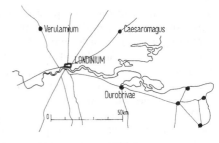

Figure 3.48 – Londinium: location map relating the main Roman roads to the Thames valley and estuary.

87. Richmond, *Roman Britain*.
88. For Roman London see R Morris, *Londonium in the Roman Empire*, 1982; G Milne, *The Port of Roman London*, 1985; T Brigham, 'The Late Roman Waterfront in London', *Britannia*, vol. XXI, 1990; J Morris, *Londinium: London in the Roman Empire*, 1982. Archaeological evidence to 1990 shows scattered Celtic pre-Roman occupation of isolated, small sites within the eventual extent of Londinium, but there is no evidence of even sizeable village settlement. Caution, as ever, is necessary in reaffirming that the Romans founded Londinium on unoccupied land (remembering that earlier editions had to note that 'the exact position of the Roman bridge over the Thames ... is still unknown' before it was positively located in 1981 – Martin Walker, 'The Romans: London Bridge found', *Guardian* (London and Manchester), 11 November 1981.
89, 90. Richmond, *Roman Britain*.
91. The location of the Thames crossing away from established tribal centres remains the accepted reason for the absence of pre-Roman settlement.

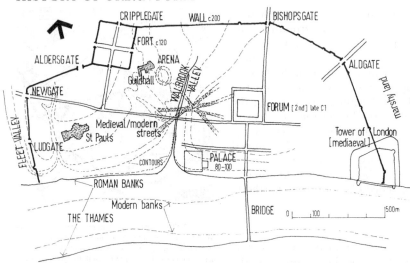

Figure 3.49 – Londinium: the sparse known Roman remains, within the wall, and showing the Roman river banks, in heavier line, with the medieval street pattern in lighter line. The contours of Londinium are in dotted line (after the *Geographical Magazine*).

The loss of even the major elements of the Roman street plan is emphasized in those towns where the Roman through routes do not always pass through opposed gates, but are for some reason diverted within the town to pass out of the walled area by a gate or gates not directly opposite the point of entry. Canterbury provides a case in point; the mediaeval and modern east–west streets now run diagonally across the grain of the Roman pattern.

London presents much greater difficulty. Almost the entire Roman street pattern has been lost, but there are some notable exceptions. The survival of the internal street plan of the Cripplegate Fort suggests that its defences exercised some control over the development of this area in the early middle ages and thus that the fort itself survived as a distinct and possibly still defensible enclosure into the Saxon period.

Elsewhere, lengths of the two most important streets of the Roman town have remained in use, while along intervening stretches of the same streets, the mediaeval and modern courses have wandered away from the earlier lines. It is remarkable that the great street markets of mediaeval London – Westcheap (Cheapside) and Eastcheap – both occupy approximately the line of Roman streets. There may be topographical controls here of which we are ignorant, but the possibility that some London streets did actually survive in continuous use with properties along them cannot be entirely ignored. They would provide the exception which emphasizes the general lack that has been stressed here of the demonstrable continuity of streets as built-up routes.

(D M Wilson, The Archaeology of Anglo-Saxon England, 1976)

it had an area of some 325 acres, making it one of the largest north of the Alps. In the ninth century the walls were reconstructed by King Alfred, for use against the Danes and still, in the medieval period, they formed the limits of the City of London (see Chapter 8).

Alfred's London, however, had by then effectively lost the original Roman gridiron street pattern, acquiring in its place an organic growth form which, with only comparatively minor modifications, has come through to the present day. As further described in Chapter 4, this was a characteristic fate of Romano-British cities, which were abandoned, more or less completely, when provincial law and order broke down following withdrawal of the legions early in the fifth century. In general, the 'open' undefended British urban centres were easy prey to invading Norsemen and others.[92]

When an abandoned urban settlement of Roman origin subsequently became re-established on its old site, the new street pattern was typically that of organic growth, overlaid on the original Roman gridiron; the controlling effect of which was lost, to greater or lesser extents. Conjectural description of what could have happened with Alfred's London exemplifies (by way of prelude to Chapter 4) the post-Roman urban re-establishment process.

LONDON 'VILLAGE'

Archaeological evidence of the early 1980s confirms extensive devastation by fire during the fifth century, when Londinium was sacked by the Vikings.[93] Even if completely burnt – and as a predominantly thatched-roof, timber-built city it was exceedingly vulnerable – there would still have been ruins substantial enough to shelter small groups of 'survivors', without there being sufficient movement activity of people, pack animals and wheeled traffic to maintain the old street pattern. Once lost, beneath rubble and refuse, and grass and trees, the Roman gridiron was largely gone forever. When economic activity in Saxon England resumed during the sixth century or so, London's pre-eminent locational advantages ensured its renewed dominant share. As above, the new medieval main street pattern came to embody much that was typical of the organic growth of an unplanned 'village'; in particular the apparently inexplicable routes radiating from what was

Figure 3.50 – 'London Village': the conjectural location on the eastern side of the assumed causeway (or bridge ?) over the Walbrook. See P Marsden, *The Roman Forum Site in London*, 1987, for an archaeological map which confirms the three Saxon routes radiating to the east from the Walbrook, disregarding the location of the Roman Forum'.

92. See T Dyson and J Schofield, 'Saxon London', in J Haslam (ed) *Anglo-Saxon Towns in Southern England*, 1984 which confirms that between 457 and 604 there is 'no historical evidence for the City of London'. Also M Biddle, 'Towns', in D M Wilson (ed) *The Archaeology of Anglo-Saxon England*, 1976.

93. For conclusive evidence of widespread destruction by fire see Dyson and Schofield, 'Saxon London'; Biddle, 'Towns'; P Marsden, *The Roman Forum Site in London*, 1987.

to become the modern nodal point between the Mansion House and the Bank of England, shown by Figure 3.50.

However (as established in Chapter 1), there is always the one, or a combination of determinants responsible for even the seemingly most baffling of organic growth patterns, on which basis an explanation can be offered for the apparently arbitrary nature of medieval London's primary street layout.

When early Saxon London began to grow, it is reasonable to believe that it did so as a renewed 'village' settlement, or perhaps the dominant of several, occupying a propitiously located part of the old city area. Where, then, is it most likely that this embryonic 'new London' would have been located? An answer is to be found by superimposing the eventual medieval street pattern on an accurate contour map of Roman Londinium;[94] whereby the author derives the personal theory that London renewed itself around the lowest crossing point of the Walbrook.

With its steepish valley sides, this stream would have constituted a swampy barrier to east–west movement along the north bank of the Thames. It is therefore reasonable to conjecture that London 'village' established itself at the lowest crossing point, on either side of the bridge or causeway, hence the medieval (and present-day) pattern of main streets radiating away from the bridgeheads.[95]

Roman Britain: Legionary fortresses and civil settlements

Numerous Romano-British cities comprise either wholly, or more usually in part, the area of an earlier legionary fortress or *castra*. It is vital not to confuse the essentially stereotyped military layout of a *castra* – where all the parts were usually in the same place – with the planning of civil towns, which while not of a particularly imaginative kind, nevertheless did admit of individual, topographically determined arrangements of the basic urban components in the form of variations on a simple planning theme. If the site of a *castra* was to be reused for a civil town, the military buildings and streets were usually completely cleared away to facilitate laying out the new town. Sometimes, as at Lincoln, the area occupied was that of the *castra* and the walls were retained; elsewhere, as at Cirencester, a larger area was required and the defences were also demolished. The *castra* – as with military camps throughout history – invariably generated market-place trade and the total conversion to civilian occupation frequently took the form of a planned military nucleus and an adjoining organic-growth 'town'. (See pp. 88–9 for Housesteads which, in its vulnerable frontier location by Hadrian's Wall, retained the contrasted *castra*/town two-part form.)

CORINIUM

Corinium (Cirencester) was founded around AD 43–44 in the land of the Dobunni, some 4 km south of their tribal capital at Bagendon.[96] The first settlement on the site was a typical large *castra*, about 30 acres (12 hectares) in area, capable of accommodating a half-legion of 3,000 infantry. This *castra* controlled Bagendon and was located at the vital strategic intersection of the Fosse Way, and Ermin and Akeman. It was abandoned in about AD 49 and replaced by a smaller *castra* of 6 acres built alongside for a cavalry garrison. Attracted by the trading needs of the military, the Dobunni gradually deserted Bagendon in favour of the new market town which grew up, organically, north-west of the fort.

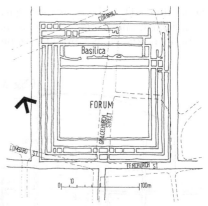

Figure 3.51 – Londinium: reconstructed plan of the Forum area (after P Marsden, *The Roman Forum Site in London*).

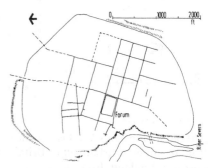

Figure 3.52 – Virconium (Wroxeter), initially the *castra* base for Legio XX, in succession to Legio XIV, until its abandonment by the military around AD 90, after which it was re-established as only a *civitas* capital for the local Cornovii – in contrast to the *colonia* status subsequently enjoyed by its former legionary counterparts of Colchester, Lincoln and Gloucester. Viroconium was located at a strategically important point where Watling Street meets the River Severn, not far to the east of the Welsh mountain massif, where it served from about AD 58 as the main base for the Roman conquest of Wales.

Viroconium is unusual among major Romano-British sites in that it remained unoccupied following the fall of the empire and possesses in its magnificent baths one of the best surviving examples of grand-manner Roman architecture in Britain. (John Wacher, in his definitive *The Towns of Roman Britain*, quotes the intriguing suggestion that the baths were the gift of the departing legionaries to the new civil town.) The baths occupy an entire *insula* adjoining the equally large forum. Wroxeter would seem to have been a prosperous market town with an area of some 180 acres within the late-second-century earthen ramparts which were augmented by a stone wall in the third century.

94. Marsden, 'Mapping the birth of London', *Geographical Magazine* (London), July 1980.
95. See Dyson and Schofield, 'Saxon London', for the outline street plan of Saxon London. Also M Walker, 'Londinium: Rome's grandiose frontier jewel', *Guardian* (London and Manchester), 12 December 1981.
96. For Corinium see J Wacher, *Corinium*.

Soon after AD 70 the permanent garrison moved away, the *castra* was demolished, and the land made available for a new *civitas* for the local tribesmen, known as Corinium Dobunnorum.

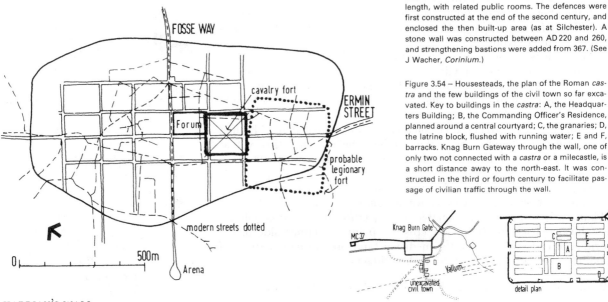

Figure 3.53 – Corinium, the Roman gridiron shown superimposed on the modern streets, is standard Imperial planning; note the large forum (in the angle of the two main streets), and the amphitheatre. The south-east side of the forum consisted of an aisled hall 84 m in length, with related public rooms. The defences were first constructed at the end of the second century, and enclosed the then built-up area (as at Silchester). A stone wall was constructed between AD 220 and 260, and strengthening bastions were added from 367. (See J Wacher, *Corinium*.)

Figure 3.54 – Housesteads, the plan of the Roman *castra* and the few buildings of the civil town so far excavated. Key to buildings in the *castra*: A, the Headquarters Building; B, the Commanding Officer's Residence, planned around a central courtyard; C, the granaries; D, the latrine block, flushed with running water; E and F, barracks. Knag Burn Gateway through the wall, one of only two not connected with a *castra* or a milecastle, is a short distance away to the north-east. It was constructed in the third or fourth century to facilitate passage of civilian traffic through the wall.

HADRIAN'S WALL

Although Hadrian's Wall is of little direct significance in an account of Roman imperial urbanism in Britain, its existence from AD 128 across northern England, from the River Tyne to the Solway Firth, forming a

Figure 3.55 – Hadrian's Wall, locating Housesteads Fort (see below) and two other towns of Roman origin described later in the book: Carlisle (Chapter 4) and Newcastle upon Tyne (Chapter 8).

continuous, impenetrable frontier barrier against invasion from the north, helped Romano-British towns to maintain freedom of growth, unrestrained by the need for individual urban defences, until the advent of unsettled times from the latter part of the second century.[97] (Directly comparable are the more recent politico-military frontier defensive systems in Europe, include the French fortresses of the second half of the nineteenth century, and the opposing French and German Maginot and Siegfried lines of the 1920s and 1930s. The former system enabled most major French cities – previously vulnerable – to dispense with their own individual defences; but see page 202 for note concerning readiness of the Parisian system as late as 1914.)

Roman domination of north-western England was first gained by the legions under governor Gnaeus Julius Agricola (78–85) and maintained, initially, through a chain of forts constructed between Carlisle and Corbridge along Stanegate military road. In AD 81 advances further north into central Scotland were consolidated behind a new line of forts constructed between the Clyde and the Forth, and the Stanegate line was abandoned. The Roman hold over central and southern

Figure 3.56 – The location of Constantinople (on the site of earlier Byzantium) at the south-western end of the Bosphorous, which links the Black Sea to the Sea of Marmora and the Mediterranean and which separates Europe from Asia.

97. For Hadrian's Wall see D J Breeze and B Dobson, *Hadrian's Wall*, 3rd edn, 1987; J C Bruce, *Handbook to the Roman Wall*, 13th edn, 1978; G D B Jones, *Hadrian's Wall from the Air*, 1976. For the Antonine Wall see A S Robertson, *The Antonine Wall*, 1979.

Scotland was precarious, however, and the rebellion of 117–118 forced their withdrawal back to the Stanegate line.

The Emperor Hadrian (117–138) came to Britain in AD 121–122 to restore order and established the northern boundary (*limes*) separating the Roman province from barbarian territory, and which has taken his name. His wall replaced Agricola's Stanegate fortresses, generally a short distance to the north.

HOUSESTEADS CASTRA AND CIVIL TOWN

Housesteads is located approximately halfway between Newcastle upon Tyne and Carlisle.[98] After construction of the *castra* between 125 and 128 a force of 1,000 infantry moved in, remaining for only some twelve years before being sent north to garrison the Antonine Wall, leaving behind only a 'care and maintenance' detachment. By 211 Housesteads had been reoccupied, this time by the 1,000-strong First Cohort of Tungrians.

There are in reality two Housesteads: first the *castra* itself, properly known as Roman *Vercovicium*, and illustrated in Figure 3.54; and second, the little-known and virtually unexcavated civil town which adjoins the southern wall of the *castra*, and which for our purposes is of much greater interest. It is also much bigger, covering an estimated 10 acres, about twice the area of the *castra*. While the *castra* is clearly of enormous historic importance as an extensively excavated Roman site, it was only ever under military occupation. The civil town, on the other hand, could have had 2,500 or more inhabitants, providing for the families of the legionaries, the various requirements of the military, and also, it is to be believed, serving as a market town for trading activities through the wall to the north. Only a cluster of shops and inns immediately outside the southern gate – a ribbon of development alongside the approach road – has yet been excavated, the landowner fearing greatly increased tourist incursions if the attractions of the town are added to those of the fort. Evidence so far gathered of the town point to an intriguing, characteristic combination of rigidly planned military form, and organic growth suburban civilian accretion. (See also such combination at Timgad, page 76.)

Epilogue: Constantinople

Constantinople (Istanbul), one of the world's greatest historic cities (after Paris, the author's close-second favourite) is the successor on the site to Byzantium, whose ancient history is given in the adjoining column.[99] After 1453, when the previously impenetrable defences were breached by the Turks, Constantinople became part of the Islamic world. This epilogue to the Roman Empire takes up the city's history in 324, when the Emperor Constantine arrived on the scene, and carries it through to the ninth-century apogee as capital of the Byzantine Empire. Within Islam, the city, renamed Istanbul in 1923, is most notable for its superb historic mosques, for which see Chapter 11.

The immediately preceding imperial context is summarized as follows: in AD 285 the Emperor Diocletian divided the Empire into four parts to improve its administration and defence, each with its own Caesar, taking the East for himself with his capital at Nicomedia.[100] Constantius ruled as Western Emperor until his death in 306 and his successor, Constantine (proclaimed at York, England) first consolidated his western power by defeating Maxentius in 312, and then

Byzantium

The site of the ancient Greek city of Byzantium, from which nucleus Constantinople was to grow, had been that of a Mycenaean settlement dating back to the 13th century BC, concerning which, however, very little is known. At the tip of a naturally fortified peninsular, controlling the western, Mediterranean end of the Bosphorous, the location had self-evident advantages for early settlement; or so it would appear today. Nevertheless, historical legend describes how Byzas, the city's eponymous founder, on consulting the Delphic oracle concerning the location of his proposed Megarian colony, was advised to settle 'opposite the land of the blind': a reference, apparently, to the inhabitants of Chalcedon, the earlier Greek colony which had opted for a greatly inferior position on the opposite, Asian side of the Bosphorous.

Byzantium occupied the first of the 'seven hills' over which Constantinople eventually came to extend. Cliffside defences on three sides required only a comparatively short length of land wall across the neck of the peninsular. After an eventful history, allied with one Greek power after another, without ever achieving great authority, Byzantium from about 150 BC became a free, tribute-paying city on the fringe of the Roman Empire.

This favourable status was withdrawn by Vespasian who formally incorporated the city into the Empire in AD 73. In 196, Byzantium paid the penalty for supporting an unsuccessful contender for imperial power: Septimus Severus, the victor, capturing it after nearly three years of siege and razing it to the ground. The location was far too important to be left unoccupied, so Severus re-fortified an extended area, renaming it Antoninia.

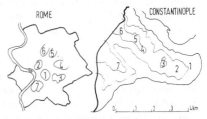

Figure 3.57 – Constantinople and Rome: a topographical comparison. Just as Rome's familiar description, the 'city of the seven hills', has been shown to be misleading (three of them are more like hillside spurs: Figure 3.6) so also is the same description as applied to Constantinople. Whereas the superb original acropolis location of Byzantium can be properly termed the 'First Hill', and the 'Seventh' has its separate identity; the other five are best regarded as areas of higher ground on the ridge which terminates as the First Hill.

The two diagrammatic topographical maps are drawn to the same scale. By and large, Constantinople was (as it remains) a much hillier city than Rome, with numerous steeply sloping minor roads and lanes leading off the more gently undulating main routes.

98. For Housesteads Fort see E Birley, *Housesteads Roman Fort*, DoE Guidebook, 1975; J F Johnston, *Hadrian's Wall*, 1977.

99. For Constantinople (Istanbul) see D T Rice, *Constantinople*, 1965 as an extensively illustrated history of the city; Sir Steven Runciman, 'Christian Constantinople', in A Toynbee (ed) *Cities of Destiny*, 1967.

100. Nicomedia (modern Izmit) was located in the far eastern corner of the Sea of Marmara. The other Caesars and their parts of the Empire were Maximian (Italy and Africa); Galerius (Illyricum and the Danube); Constantius (Britain, Gaul and Spain).

became absolute emperor when he captured Antoninia (Byzantium) from Licentius, his pagan eastern rival in 324.

As Emperor, Constantine assumed a formidable weight of responsibility; imperial prosperity and population were declining, while barbarian pressures along the frontiers continued to increase. Above all there was the problem of Rome itself where the accustomed life-styles imposed intolerable economic burdens, sustained only at great and increasingly disproportionate expense. After 313, when Christianity gained partial state recognition, Constantine resolved to create a new imperial capital away from threats of the present and distasteful religious reminders of the past. He recognized that a revived empire could but be centred in the eastern Mediterranean, where, after careful survey of alternatives, he opted for ancient Byzantium.

Constantine was in no doubt as to the required extent of his new city, and shrugging off incredulous courtiers' queries with the legendary reply – 'I shall still advance till He, the invisible guide who marches before me, thinks proper to stop' – before tracing out a boundary line, with ceremonial spear in hand, all of 2 miles west of the existing Severan Wall.[101] Divine guidance or not, it was the Emperor who held temporal purse-string power to ensure adequate population for the enlarged city. 'Presence of the Emperor and government,' explains Cecil Stewart, 'meant that a large part of the revenue of the Empire would automatically come to the new city. Many were attracted by a sense of duty; some came at the invitation of Constantine himself – an invitation which could not be distinguished from a command.' And as perhaps the most effective inducement, 'the wheat which had hitherto been sent from Egypt to Rome was diverted to the new capital, and provided for the free distribution of 80,000 loaves'.[102] First, however, he had to deal with a shortage of architects by commanding the immediate institution of new imperial schools of architecture!

Constantine's ramparts were completed by 330 and on 11 May he dedicated the city as the Christian capital of the East, and the declared rival of Rome in the West. From then on their fortunes were in marked contrast: Constantinople's prosperity increased, just as Rome's declined. By 410 – the year in which Rome fell to Alaric and the Goths – the imperial founder's optimism was justified; the area within his perimeter was fully occupied such that Themistius could write 'no longer are we cultivating more territory within our walls than we inhabit; the beauty of the city is not as heretofore scattered over it in patches but covers its whole area like a robe woven to the very fringe'.[103] (Although officially renamed 'New Rome', it was known from the outset as Constantinople.)

Although Constantine greatly extended and further embellished the city, its essential character was retained. The original acropolis, true to type, had become the city's religious precinct, with the Roman forum at its south-western end. Further in that direction, the Hippodrome had been given its essential form, later to be considerably extended. The embryonic Palace area had been established between it and the Sea of Marmora, on the south-east-facing slope. The route is known of the main street through to the forum from the gate on the Second Hill, where Constantine planted the column marking the centre of his new square; but otherwise little else has yet been revealed. The city was built on hills, eventually seven – as at Rome (Figure 3.57) – and the general form was that of a 'Y'-shaped main street system traversing five

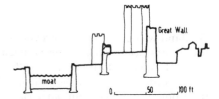

Figure 3.58 – Constantinople, section through the Land Walls. As completed by 413, the triple-wall system ran for 4½ miles (south–north) from the Sea of Marmora across to the Golden Horn. The inner wall was 40 feet high on the outside and 16 feet thick at the bottom. It was strengthened by ninety-six surviving towers about 60 ft high; there may have been a hundred of them before reconstruction of the northern end.

Figure 3.59 – Constantinople: the map published by the Society for the Diffusion of Useful Knowledge in 1840. The mid-nineteenth-century city is shown still tightly contained within the walls of Anthemius (or Theodosius) which were breached in 1453 by Turkish cannon, thereby requiring the development of artillery fortifications for other major European cities. Exceptionally, the defences of Constantinople were not similarly upgraded; neither is there any sign in this map of the line of Constantine's earlier walls of 330 (see Figure 3.60) which, had the city been extended during the seventeenth or eighteenth centuries, say, would probably have been retained as a boulevard main street.

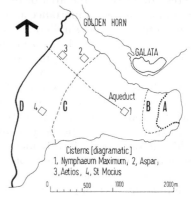

Figure 3.60 – Constantinople: an explanatory outline plan showing the city related to the Golden Horn, which provided water defences along its northern side, and the Sea of Marmora to the south. Key: A, the extent of Byzantium before its destruction in AD 196; B, the Wall of Septimus Severus; C, Constantine's fortifications of AD 330; D, the Walls of Anthemius (or Theodosius). See Figure 3.58 for a cross-section of these.

Very little is known of the detailed layout of Constantine's city. The line of its walls has been determined, and it is known that it comprised fourteen districts and covered seven hills, as did Rome, with the main street system in the form of a 'Y' traversing five main fora, the most important of which was the circular Forum of Constantine located at what had been the main gateway in the Severus wall.

101. See Runciman, 'Christian Constantinople'.
102. C Stewart, A Prospect of Cities, 1965. Brasilia, the new Federal Capital of Brazil, is a close modern parallel to Constantinople with the political decision to move the function of central government, whereby the host of Brazilian civil servants and their dependants also had little choice but to move along with their salaries – and their pensions.
103. Themistius, Oratio XVIII.

major fora. Other than the important, planned civic areas, it is most probable that the remainder of the city was a result of organic growth, *laissez-faire* development.

Under Theodosius II (408–450) the city was again expanded, within a new defensive system which was constructed about a mile further to the west of Constantine's wall. This system, which eventually took the form of three walls and a wide moat, is known variously after the emperor and the regent, Anthemius, who held the office of prefect (Figure 3.58). The Walls of Anthemius (or Theodosius) remained unrivalled and impregnable through to 1453 when they finally succumbed to Turkish cannon.[104] In addition to the land walls, there was also a complete circuit of sea walls.

The city had a major water supply problem which required the construction of a system of aqueducts for daily supply, and also the provision of numerous cisterns, several of vast proportions, for storage within the defensive perimeter during time of siege, for which see the adjoining column.[105] As a key part of the early system, the Emperor Valens in *c.* 375 added the aqueduct named after him, which survives above ground in large part, most impressively where it crosses the valley between the fourth and third hills. It led to a huge storage and distribution cistern, the *nymphaeum maximum*, on the third hill, near modern Beyazit Square. The aqueduct remained in use until the late nineteenth century.

104. See Chapter 5 for the significance of the fall of Constantinople to the Turks, and Z Celik, *The Remaking of Istanbul: A Portrait of an Ottoman City in the Nineteenth Century*, 1986.

105. Of all the Roman and Byzantine remains at Constantinople, the personally most fascinating are the open and underground water storage reservoirs. Even the smallest of the three surviving open reservoirs, the Cistern of Aetios on the Sixth Hill (*c* AD 421), readily accommodates a football stadium within its 224 m × 85 m area, and it was originally about 15 m deep. The largest is the Cistern of St Mocius on the Seventh Hill (*c* AD 500) at 170 m × 147 m; and the Cistern of Aspar on the Fifth Hill (*c* AD 470), which is 152 m square and 10 m deep, had housed an entire village until recently demolished for a 'modern improvement'.

Water was brought into the city by the Aqueduct of Valens, constructed from AD 375 by the Emperor Valens. It entered the city in pipes under the Wall near the Adrianople Gate, and ran along the ridges of the Sixth, Fifth and Fourth Hills before being taken on a high, impressively surviving, 1,000 m length aqueduct across the valley to the Third Hill. There was a major distribution cistern located near modern Beyazit Square, close by the modern equivalent. Often restored, the Aqueduct remained in use until the late nineteenth century. (See J Freely, *The Blue Guide to Istanbul*, 1991.)

4 – Medieval Towns

Classified on the basis of their origins there are five broad categories of towns in medieval Europe of the eleventh to fifteenth centuries.[1] Three categories are of *organic growth towns*:

1 towns of Roman origin – both those which may have retained urban status throughout the Dark Ages, albeit considerably reduced in size, and those which were deserted after the fall of the Empire, but which were re-established on their original sites
2 burgs (borough, burk, bourg, burgo), founded as fortified military bases and acquiring commercial functions later
3 towns that evolved as organic growth from village settlements.

The remaining two categories are of *planned new towns* which were established formally at a given moment in time, with full urban status, and with or without a predetermined plan:

4 bastide towns, founded in France, England and Wales
5 planted towns, founded throughout Europe generally.

The sequence is in approximate chronological order. Before the Romans, urban settlements in Europe are believed to have been few and far between. Chapter 3 has established that many of them were taken over by the Romans and redeveloped along planned lines. After the collapse of the Empire in the fifth century, urban life in Europe was greatly curtailed (in Britain it effectively disappeared) until, beginning in the tenth and eleventh centuries, political stability and resurging trade gave renewed life to many Roman foundations, converted burgs to commercially orientated towns, and instigated the slow process whereby a small proportion of villages were transmuted into towns. New towns were founded throughout the Middle Ages but the rate of creation was slow at each end of the period; the pronounced peak came during the thirteenth century.

Although the medieval period shows a two-dimensional contrast between organic growth and planned urban form as marked as during any other period, classification in terms of plans on the simple basis of either of these extremes is no longer possible. The route structures of the great majority of rebuilt Roman towns, and of many of those which had continued in being were as much the result of organic processes as determined by their original gridiron layouts. Similarly, numerous medieval new towns which started out from a plan subsequently underwent uncontrolled expansion and change. There are also a few examples of planned additions to organic growth towns. This chapter does

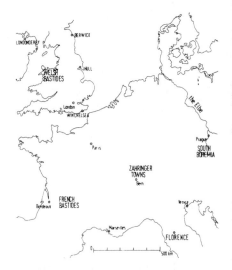

Figure 4.1 – Western Europe, showing the locations of six groups of planned new towns described in this chapter. Key: 1, the French bastides (detail map, Figure 4.37); 2, the Welsh bastides (detail map, Figure 4.51); 3, the Zähringer towns (detail map, Figure 4.64); 4, towns in East Germany and Poland, east of the Elbe; 5, towns founded by the Florentine Republic in the valley of the Arno (detail map, Figure 4.70); 6, towns in South Bohemia (Czechoslovakia).

The two English bastides, Winchelsea and Kingston upon Hull, are located individually. Londonderry, an English planted town in Ulster, is also shown.

1. Key references for this chapter are R H C Davis, *A History of Mediaeval Europe: From Constantine to Saint Louis*, rev. 2nd edn, 1987; H Pirenne, *Mediaeval Cities: Their Origins and the Revival of Trade*, 1969; C T Smith, *An Historical Geography of Western Europe before 1800*, 1967; R Latouche, *The Birth of Western Economy*, 1965; M W Beresford, *New Towns of the Middle Ages*, 1967; W G Hoskins, *The Making of the English Landscape*, 1978; P H Sawyer (ed.) *English Mediaeval Settlements*, 1979.

not attempt to present a complete description of medieval urban form and its determining background factors for all European countries. The medieval period was the formative era of town development in Europe: in many ways it is the most important part of this volume. Not only is the political, economic and social context extremely complex, but also the towns themselves present an almost infinite variety of forms. Nevertheless, several similarities in the context of urban development in European countries make possible a general consideration of a number of key factors: feudalism; the early medieval revival of commerce, related to the role played by the Church; the medieval industrial situation, and common features of urban form which are found in most parts of Europe. This shared background will comprise the introductory section to this chapter, to be followed by a description of towns in the five groups.

The origins of Rome and London are described in Chapter 3 and Paris in Chapter 6. To enable continuity of development to be properly established, the histories of these cities during the medieval period are included in later chapters dealing with development from the Renaissance onwards.[2] Chapter 7 includes the medieval periods of most of its city examples in other European countries. The medieval Islamic cities of central and southern Spain (Europe's uniquely different urban forms) are described on pp. 144–5.

RELIGION AND REVIVAL OF COMMERCE

In AD 395 the Roman Empire was divided into two parts. With Constantinople as its capital, the Eastern Empire prospered (see Chapter 3), but the already disintegrating Western Empire, under its capital Rome, lasted only until the beginning of the fifth century, by which date, writes Pirenne, 'all was over. The whole West was invaded. Roman provinces were transformed into Germanic kingdoms. The Vandals were installed in Aquitaine and in Spain, the Burgundians in the valley of the Rhône, the Ostrogoths in Italy'.[3]

Contrary to popular misconception the fall of the Empire in the West did not immediately result in a more or less complete breakdown of the Roman economic system. Although there was a continued commercial *raison d'être* for many Roman towns – in particular those in southern Gaul – the effect of the collapse of the Western Empire must not be exaggerated. The economic order of fifth- to eighth-century Gaul remained, as it always had been, founded on agriculture. Neither must we underestimate the role played by the Church in maintaining the continuity of urban life in many parts of Western Europe (although Britain was a notable exception to this). In general the Church had taken imperial administrative districts as the basis of its ecclesiastical organization, each diocese corresponding to a *civitas*. These districts were unaffected by the setting up of the new Germanic kingdoms such that 'from the beginning of the sixth century the word *civitas* took the special meaning of "episcopal city", the centre of the diocese'.[4] These cities also retained trading functions, varying from simple local produce markets to comparatively highly developed commercial activities.

Imperial economic unity, and the urban life it supported, survived the Germanic invasions largely because the Mediterranean remained open for trade. From early in the seventh century, however, the vital link was first constricted, then closed, by the rapid growth of Islam. Syria was taken from the Byzantines in 634–36 and progressive

In the fourteenth century the English town was still a rural and agricultural community, as well as a centre of industry and commerce. It had its stone wall or earth mount to protect it, distinguishing it from an open village. But outside lay the 'town field' unenclosed by hedges, where each citizen-farmer cultivated his own strips of cornland; and each grazed his cattle or sheep on the common pasture of the town, which usually lay along the river side as at Oxford and Cambridge. In 1388 it was laid down by Parliamentary Statute that in harvest time journeymen and apprentices should be called on to lay aside their crafts and should be compelled 'to cut, gather and bring in the corn'; mayors, bailiffs and constables of towns were to see this done (Stats. Of Realm II, 56.). In Norwich, the second city of the kingdom, the weavers, till long after this period, were conscripted every year to fetch home the harvest. Even London was no exception to the rule of a half rustic life. There was none of the rigid division between rural and urban which has prevailed since the Industrial Revolution. No Englishman then was ignorant of all country things, as the great majority of Englishmen are today.
(G M Trevelyan, English Social History)

The tremendous effect the invasion of Islam had upon Western Europe has not, perhaps, been fully appreciated.

Out of it arose a new and unparalleled situation, unlike anything that had gone before. Through the Phoenicians, the Greeks, and finally the Romans, Western Europe had always received the cultural stamp of the East. It had lived, as it were, by virtue of the Mediterranean; now for the first time it was forced to live by its own resources. The center of gravity, therefore on the shore of the Mediterranean, was shifted to the north. As a result the Frankish Empire, which had so far been playing only a minor rôle in the history of Europe, was to become the arbiter of Europe's destinies.

There is obviously more than mere coincidence in the simultaneity of the closing of the Mediterranean by Islam and the entry of the Carolingians on the scene. There is the distinct relation of cause and effect between the two. The Frankish Empire was fated to lay the foundations of the Europe of the Middle Ages. But the mission which it fulfilled had as an essential prior condition the overthrow of the traditional world-order. The Carolingians would never have been called upon to play the part they did if historical evolution had not been turned aside from its course and, so to speak 'de-Saxoned' by the Moslem invasion. Without Islam, the Frankish Empire would probably never have existed and Charlemagne, without Mahomet, would in inconceivable.
(H Pirenne, Mediaeval Cities, 1925)

2. See Rome (Chapter 5), Paris (Chapter 6) and London (Chapter 8). For other major European cities with particularly important medieval periods which are included in other chapters, see Venice (Chapter 5), Amsterdam, Prague and Vienna (Chapter 7) and Bristol and Edinburgh (Chapter 8). The contemporary Islamic cities of the Middle East are described in Chapter 11.
3. Pirenne, Mediaeval Cities.
4. Smith, An Historical Geography of Western Europe before 1800.

advances westward along the African coast brought Islam into Spain by 711. Its growth was contained only by the successful defence of Constantinople (713)[5] and Martel's victory in Spain (732).

Muslim control of the Mediterranean extended to continual piracy along its northern shores. Trade was completely disrupted; by the middle of the eighth century 'solitude reigned in the port of Marseilles. Her foster-mother, the sea, was shut off from her and the economic life of the inland regions which had been nourished through her intermediary was definitely extinguished. By the ninth century Provence, once the richest country of Gaul, had become the poorest'.[6] Already under sustained pressure from the south, the northern and western coasts of the Carolingian Empire and areas reaching for a considerable distance inland along navigable rivers also suffered increasingly from Danish and Norwegian raids. Their attention, writes Robert-Henri Bautier, 'was directed alternately towards the Continent and England. Plunder was the essential if not the sole object: only the ports, the main economic centres and the salt-works were affected. Every year from 834 to 837 Duurstede, the principal port for Frisian trade, was pillaged, as well as neighbouring ports, and also the island of Noirmoutier and the Thames estuary. Rouen was taken in 841, Quentovic (the chief Frankish port) and London in 842, Nantes was sacked in 843, Paris in 845, Bordeaux in 848'.[7]

Commerce of any significance, and certainly of a kind that had previously supported professional international merchant bodies, was impossible in these unsettled conditions. As a direct result urban life in Western Europe declined to its nadir by the end of the ninth century. Pirenne observes that 'an economy of exchange was substituted for an economy of consumption. Each demesne [great estate] in place of continuing to deal with the outside constituted from this time on a little world of its own ... the ninth century is the golden age of what we have called the closed domestic economy and which we might call, with more exactitude, the economy of no markets'.[8]

However, a commercial revival took root and steadily gained in strength as soon as the political climate became only relatively more stable during the early decades of the tenth century. Throughout Western Europe long-distance trading routes were reopened, notably those connecting with Venice and the north Italian trading communities, and by mid-century Flanders had become a comparable north-western focus. The resurgence of trade had a particularly early origin in Germany where, in 918, Conrad I granted market rights to Würzburg and, as Bautier notes, 'twenty-nine Acts granting market concessions have come down to us mainly dating from the third quarter of the tenth century'.[9]

The Church, which had preserved for Europe a semblance of civilized life during the Dark Ages, not only provided nuclei for many early medieval towns but also used its 'ubiquitous and persuasive influence to restrain the quick impulses which were at all times so close to the surface.'[10] In addition to the protection afforded by the bishops' walled cities and strongly defended monasteries, numerous burgs established in all parts of Europe as heavily fortified military and administrative centres also served to promote the tenth-century resurgence of trade. Again referring to Germany, C T Smith notes that, 'of about 120 towns identified in Germany in the eleventh century, about 40 were on the sites of bishops' seats, 20 were near monasteries, and no less than 60

In the Middle Ages the inhabitants of each town regarded themselves as a separate community, almost a separate race, at commercial war with the rest of the world. Only a freeman of the city could be bound apprentice; and the freedom of the city was but grudgingly bestowed on 'foreigners' or 'aliens', as the people of the neighbouring towns and villages were called. This idea of the civic community was now gradually yielding to the idea of the national community, and to the broader aspects of economic and racial policy introduced by the discovery of America and the struggle with Spain. But the old ideas still lingered.
(G M Trevelyan, England under the Stuarts)

A new class appeared on the edge of the feudal society: the merchants. Probably they originated among the landless men, escaped serfs, casual harvest labourers, beggars and outlaws. The bold and resourceful among them, the fair talkers, quick with languages, ready to fight or cheat, became chapmen or pedlars, carrying their wares to remote hamlets. They were paid in pennies and farthings and in portable local products, such as beeswax, rabbit fur, goose quills, and the sheepskins for making parchment. If they prospered, they could settle in a centre and hire others to tramp the forest paths. Such was the career of the English Godric of Finchale, who rose from peddling to be a ship-owner and entrepreneur, journeyed to Denmark and Rome, and ended by reforming and becoming a hermit and saint.

The merchants made the towns. They needed walls and wall-builders, warehouses and guards, artisans to manufacture their trade goods, caskmakers, cart builders, smiths, shipbuilders and sailors, soldiers and muleteers. They needed farmers and herdsmen outside the walls to feed them; and bakers, brewers and butchers within. They bought the privilege of self-government, substituting a money economy for one based on land, and thus they were likely to oppose the local lordling and become supporters of his distant superior, the king. Towns recruited manpower by offering freedom to any serf who would live within their walls for a year and a day.
(M Bishop, The Penguin Book of the Middle Ages, 1971)

5. The eventual fall of Constantinople (Chapter 3) gave Islam centuries of political dominion, with enduring religious influence, over much of south-eastern Europe, including Greece; it was only the successful defence of Vienna that subsequently kept the Turkish Muslim armies out of central Europe (see Chapter 7). There may be some urban historians who still doubt the importance of fortification as a determining influence on the form of cities and national or even international fortunes. The might-have-beens resulting from the fall of Vienna in 1529 or 1683 are beyond conjecture: suffice to note that had the hold of Islam not been quickly reversed, then (as in Spain) surviving historic nuclei of many central European cities (and perhaps further west) would not look the way they do today.

6. Pirenne, Mediaeval Cities.

7. R-H Bautier, The Economic Development of Mediaeval Europe, 1971.

8. Pirenne, Mediaeval Cities.

9. Bautier, The Economic Development of Mediaeval Europe.

10. J W Thompson and E N Johnson, An Introduction to Mediaeval Europe: 300–1500.

grew around royal foundations, including some 12 near the sites of royal palaces'.[11] Throughout Europe these towns, stimulated by long-distance trade, encouraged local trade through systems of market towns which developed naturally from village origins or were 'planted' in favourable new locations (see below for a further discussion).

By the eleventh century commerce was generally re-established but, to quote Pirenne, 'it was only in the twelfth century that, gradually but definitely, Western Europe was transformed. The economic development freed her from the traditional immobility to which a social organisation, depending solely on the relations of man to the soil, had condemned her'.[12] Urban expansion continued apace during the twelfth and thirteenth centuries and even the Black Death, which hit Europe generally between 1348 and 1378, did no more than produce temporary set-backs. Paradoxically its effect was greatest on rural populations. Chances of survival were greater in the country, but on the other hand the commercial advantages of living in towns were still very attractive. Thus there was a tendency to move towards the towns, so accelerating the problems of rural depopulation.

FEUDALISM

During the twelfth and thirteenth centuries, throughout Europe, feudalism was 'the basis of local government, of justice, of legislation, of the army and of all executive power. In this period ... all land is held from the king either mediately or directly. The king himself is a great landowner with demesnes (estates) scattered over the length and breadth of the realm; the revenues of these supply him with the larger part of his permanent income. The king is surrounded by a circle of tenants-in-chief, some of whom are bishops and abbots and ecclesiastical dignitaries of other kinds; the remainder are dukes, counts, barons, knights. All of these, laymen and churchmen alike, are bound to perform more or less specific services in return for their lands. ... These tenants-in-chief have on their estates a number of sub-tenants who are bound to them by similar contracts'. In a typical feudal state all members are grouped voluntarily or forcibly under the rule of persons higher up the feudal hierarchy, either as 'servile village-communities who give up perforce a large proportion of their working days to the cultivation of the lord's demesne, (or) ... the small freeholders who pay to this or that lord a rent in money, kind or services'.[13]

The feudal system encouraged slight, but highly significant improvements in agricultural techniques, making an increase in the rural food surplus available to urban settlements with their steadily growing proportions of non-agricultural specialists. (For the comparison with the 'urban revolutions' in Mesopotamia and the other first civilizations see Chapter 1, page 4). In the emerging towns the inhabitants are also subject to a lord or the king, but whilst 'some are only half-emancipated communities of serfs ... in others the burgesses have the status of small freeholders and in a minority, but a growing minority, of cases the burgesses have established the right to deal collectively with the lord, to be regarded as communes, or free cities'.[14] This latter factor was to be of profound importance in the rise of European urban society and the corresponding decline of medieval feudalism.

MEDIEVAL INDUSTRY IN ENGLAND

The two principal industries of medieval England were those for the

---

While the Greek and Roman thought that the happy life could only be lived in the city, the nascent civilisation of the Middle Ages was of the country not of the town. Its unit was the court and manor of the feudal landlord, the homesteads and farm buildings of his humbler tenants. There was neither the good government necessary for ordered town life, nor the commerce which made it economically possible for great hordes of men to dwell together in an urban area.
(T F Tout, Mediaeval Town Planning)

Throughout the Middle ages the manor was the unit of rural organization over the greater part of England. Manors existed in England long before the Norman Conquest, and, when that event took place, the manorial system of cultivation was already well established. It was not limited to this country; it was, in fact, to be found throughout central and western Europe. The history of its development in England is obscure, and is the subject of controversy. Some investigators have tried to show that the English manor was a development of the vill, an estate worked by slave labour in the time of the Roman Empire; others have professed to discover its origin in the German mark, an area owned and cultivated by a community of free men. At the present time most scholars hold that both Roman and Teutonic influences helped in the development of the medieval manorial system. The question cannot be regarded as settled; it is possible, and, indeed, probable, that some of the problems associated with the origin and early history of the manor will never be solved in a way which will meet with universal acceptance.

A manor was a large estate which consisted, usually, of a single village and an extent of land surrounding it. In many cases the manor was enclosed by a quickset hedge, known in Anglo-Saxon times as a tun, which served to mark its extent and to protect it. Such a hedge, kept in proper condition, would prove a formidable barrier to robbers or outlaws or wild animals, though it could not prevent the passage of an organised army. In the more fully occupied parts of the country manors were adjacent, and the hedge separated one from another; in the remoter regions there were large stretches of moor and wilderness which were part of no manor and which were inhabited only by the wolf and the boar, the robber and the outlaw.

Every manor had its lord, though a statement that the lord 'owned' the manor would give a false impression. There was no absolute ownership of land other than by the king, by whom or from whom all land was held. The lord of the manor was regarded as the 'tenant', the holder, rather than the absolute owner of his estate. But this tenancy was different from a tenancy of the present day. The lord, who might hold his manor either from the king or from some other lord who held it from the king, was secure in its possession, and he could not legally be deprived of it, unless, indeed, he committed treason.
(G W Southgate, English Economic History, 1943)

---

11. Smith, *An Historical Geography of Western Europe before 1800.*

12. Pirenne, *Mediaeval Cities.*

13, 14. Davis, *A History of Mediaeval Europe.*

production of woollen cloth and the smelting and working of iron. Timber, a most important raw material, was used both for general purpose construction and as the source of the pure charcoal fuel required for the iron industry. Coal was widely used only for domestic purposes.

Neither the location of medieval towns nor their form was significantly affected by industry. The appalling condition of roads, totally neglected since the Romans' departure, meant that the only way of transporting bulk materials for any distance was by water. As a result such 'heavy' industry as existed was decentralized, on a largely rural basis, in those districts where raw materials, e.g. iron ore and timber, were readily available. Production levels remained low until the end of the period when, with demand increasing, some small towns, notably Birmingham, started to become urban industrial centres. Long-established boroughs with vested guild interests, such as Coventry, frequently reacted against the introduction of new techniques thereby prejudicing their futures.[15]

Within towns and manufacturing villages there would have been no separate industrial premises of any great size. Industry would have been of a predominantly secondary nature, converting bar-iron, for instance, into a limited range of products, with the same premises often serving as a workshop, shop and the home of a proprietor. This characteristic European multiple use is in direct contrast to the separation of home and workplace which is usual in Islamic urban culture.

Wool

'From the twelfth to the nineteenth centuries the woollen industry was the premier English industry, and as such was largely responsible for the growth of the country's wealth.'[16] The history of the cloth industry goes back well before the Roman occupation. Although a uniform-producing centre for the Roman army is recorded at Winchester,[17] this is an exception to the rule that the spinning of wool fibre into yarn, to be woven into cloth, was a part-time rural 'cottage-industry' occupation.

Cloth production initially paralleled agricultural production, being for subsistence rather than for marketing. Most parts of the country were reasonably suited for sheep, and the wool industry was very widely scattered. From the twelfth century there was a surplus of raw wool and finished cloth and exports of both were being made to the Continent. Specialist cloth workers set themselves up in the market-town centres of the main wool producing districts, and by the reign of Henry I (1100–1135) weavers' guilds are recorded in London, Lincoln, Oxford and Winchester.

Coal

The Romans recognized some of the valuable properties of coal and organized its production from surface and shallow excavation workings in several parts of the country. During the Dark Ages there was hardly any demand for coal but within a few decades of the Norman Conquest several of the coalfields were again being worked. At first this was from shallow quarries, along the exposed surfaces of the coal seams, but later there were 'adit' mines following the seams underground on easy slopes. Shaft mining was also in early use, but depths were limited by the lack of pumping facilities until the Industrial Revolution.

*In 1427 there were 180 wool shops in Florence, and nearly twice this number 50 years later; but each one might produce on average less than 100 bolts of cloth a year. A 'shop' was more like an office, with a manager and his assistant, and perhaps a bookkeeper and one or two errand boys. Most of the operations – and there were 26 different stages in the production of wool cloth – were put out to pieceworkers whose status varied according to their job. At the bottom were the humble carders and combers, characterised by Archbishop Antonino as rowdy and foul-mouthed, who never saw their employer in person, and whose names never appeared on the company books. Forming a sort of aristocracy at the top were the dyers who sometimes even became partners in industrial ventures. In Genoa the silk dyers were the best organised working groups because they were not outworkers, but gathered together in shops; and it was a dyer, Paolo da Novi, whom the popular party elected as Doge in 1507.
(J Gage, Life in Italy at the Time of the Medici)*

*The great international commerce was in textile. The wool that was most prized came from England. Much of it went directly to Flanders for processing. The wool trade brought great prosperity to the English; the Lord High Chancellor in the House of Lords still sits on a woolsack, which symbolizes the nation's wealth. In both Flanders and England the wool was spun with spindle and distaff by women spinsters. Spinning was considered to be women's proper work; even at the world's beginning Adam delved and Eve span. The spinning-wheel, probably an Indian invention, did not appear in the West until the thirteenth century, and then came quickly into general use. Once spun, the yarn went to weavers, who worked at home or in shops in the towns. Often two weavers sat side by side at a broad double loom, weaving 'broadcloths'. Next, the woven wool went to fullers, or walkers, who stamped on the cloth, shrank and compacted it, and removed grease with fuller's earth. Water-powered fulling mills, with tireless hammers, eventually replaced the stamping walkers. (The importance of the cloth trade is evidence by the number of people named Weaver, Fuller, and Walker today; but for obvious reasons there are no Spinsters.) The material went then to the dyers; to fix the colours and give them brilliancy, alum was required, which came chiefly from the Aegean islands and was for centuries a Genoese monopoly. Finally the cloth was shipped to market by sea or in bulging bales on the backs of pack mules. Italy as well as Flanders had a great garment trade.
(M Bishop, The Penguin Book of the Middle Ages)*

15. Birmingham, which steadily overtook Coventry during the seventeenth and eighteenth centuries, had a population of 71,000 at the time of the first census of 1801 (compared to Coventry's 16,000). In 1831, when its population had doubled to 144,000, Birmingham was still in effect a feudal village, not receiving its Municipal Charter until 1838. Coventry over the same period increased to 27,000.

16, 17. L D Stamp and S H Beaver, The British Isles, 1941.

Coal was used basically for domestic heating purposes; London in particular constituted a growing market for the Northumberland and Durham coalfields, with coal being transported by sea down the east coast. Elsewhere only those towns with good navigable water links to coalfields could obtain it at economic prices. As workable faces were exploited further inland from water-fronts, expediency resulted in the earliest crude wooden tracks and 'railways'; along these, carts were hauled from the fifteenth century onwards. However, the use of coal for manufacturing purposes was strictly limited by the impurities it contained – until the Industrial Revolution found ways of converting it into a usable form.

Iron

The essential prerequisite for the production of workable iron from naturally occurring iron ore was the availability of wood; this needed to be converted into charcoal for the smelting process. For this reason the fairly widespread, but relatively small-scale, Celtic ironworking industry of the Iron Age and Roman occupation had concentrated by the Middle Ages in two zones of the country where ore was available in extensively wooded surroundings – the Forest of Dean and the Weald of Sussex and Kent. As early as 1282 there were sixty forges in the forest using the local ore and the industry continued to flourish for many centuries.[18] By the sixteenth and seventeenth centuries the Weald was the leading centre of production, with the first cast-iron guns made at Buxted in 1543. An acre of woodland provided charcoal for three tons of iron. In one week a sixteenth-century furnace could produce 20 tons of iron and the resulting demand on the nation's timber resources, both for ship-building and iron production, led to crisis legislation being put into effect under Elizabeth I to conserve the woodlands and secure supplies for the navy. This legislation forced the iron masters to develop alternative sources of iron ore along the wooded valleys of the Welsh borders, in Staffordshire, south Yorkshire and a few other areas.[19] Nevertheless shortage of fuel for smelting in the years before coal could be used resulted, by the early eighteenth century, in a situation whereby almost twice as much iron was being imported into this country, mainly from Sweden and Russia, as was being produced in our own furnaces. The marriage of coal and iron, from 1730 onwards, and the subsequent growth of their offspring, steel, made possible the major growth phase of the Industrial Revolution.

Medieval urban form

With the all important exception of Islamic Spain, medieval towns generally have similar social, economic and political contexts in most European countries.[20] They are also alike as regards most visual details: the same kind of local vernacular buildings make up both the formal gridirons of planned new towns and the informal uncontrolled layouts of their unplanned contemporaries. The component parts of the medieval town are normally the wall, with its towers and gates; streets and related circulation spaces; the market place, probably with a market hall and other commercial buildings; the church, usually standing in its own space; and the great mass of general town buildings and related private garden spaces.

The latter half of the seventeenth century had seen a decline in the manufacture of iron, and particularly of iron bar, in this country and the beginnings of a revival of this trade were associated with a migration of the iron-making centres from the Sussex Weald and the Forest of Dean, to the thick woodlands of the Welsh borders, the South Yorkshire and Derbyshire valleys, and to Furness. The search in all cases was for an abundant charcoal supply within reasonable reach of reserves of iron ore, and water power.

All these new areas had an ancient if modest tradition of ironworking in monastic forges, and many of the bloomery sites of the early sixteenth century became, in the seventeenth century, small furnace units. In the Midlands and the Welsh borders there were many men who described themselves as ironmongers, lock and nail makers, smiths and workers in small ironware, and from these many of the early Friends were drawn. The family of Lloyd of Dolobran had small forges which they leased from the landlord to whom they had been appropriated at the Dissolution of the Monasteries, and these lay in rich woodlands within reasonable reach of Welshpool and other centres on the Severn. For some years, into the early part of the eighteenth century, their pig iron or ore was purchased and brought up the river from Shropshire or Gloucestershire, but later they built a small furnace of their own. Other Friends had furnaces in many parts of the Midlands, and in the other new areas of Yorkshire and Furness. These furnaces were subservient to the forges where the iron was refined and converted to rod and bar iron, and so in the end were dependent upon the vagaries of the bar iron trade.

This trade suffered many checks in the eighteenth century, due to our changing relations with Sweden, and the incidence of taxation and protective tariffs, and the fortunes of the forges and furnaces fluctuated in sympathy.

(A Raistrick, Dynasty of Iron Founders)

18. G W Southgate, *English Economic History*, 1934,
19. A Raistrick, *Dynasty of Iron Founders: The Darbys and Coalbrookdale.*
20. For general references on European medieval urban form see J E Vance jr, *The Continuing City: Urban Morphology in Western Civilisation*, 1990; S E Rasmussen, *Towns and Buildings*, 1969; C Sitte, *City Planning According to Artistic Principles*, 1889 (1965, English edition); P Zucker, *Town and Square*, 1959.

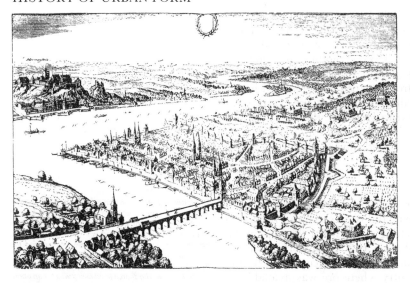

Figure 4.2 – Coblenz, an early seventeenth-century view of the city under siege. At this comparatively late date (many cities had already acquired complex Renaissance-period defensive systems) Coblenz has only its medieval wall and towers augmented by the beginnings of exterior earthworks. Note the clearly defined contrast between town and country – a medieval characteristic – and the bridgehead settlement which has become established across the Moselle. A typical hilltop castle can be seen on the far side of the Rhine. Coblenz was founded by the Romans who erected a castellum, named Confluentes, to control the crossing of the Moselle on the important left-bank route alongside the Rhine. The map published in the 1926 edition of Baedeker's The Rhine clearly shows the extent of that original settlement around the right-bank Moselle bridgehead, as revealed by the concentric sequence of streets on the line of that defensive wall. The later complex fortifications were not removed until 1890. The Moselle Bridge erected in 1344, and restored in 1440, crosses the river in fourteen arches. Only in the eighteenth century was there significant expansion towards the Rhine, which was not bridged until the mid-nineteenth century.

THE WALL

From the list of England's largest towns in the poll tax of 1377, it appears that 'only Boston, among the first ten, lacked substantial defences and even then there was a ditch. All but three of the forty largest towns and cities were defended'.[21]

One clear distinction, at least, can be drawn between English and continental European town defences. By the fourteenth century the military significance of the English town wall became greatly reduced because of the state of peace within the island. As a result the wall subsequently served mainly as a customs barrier, protecting trading interests of the townsmen and enabling tolls to be levied on all goods passing through the gates. Norwich is an instance of this diminished military significance. Given its charter by Richard I in 1194 (100 years after the foundation of the cathedral) its city wall, 'which replaced the original ditch and bank, was commenced towards the end of the thirteenth century and finished in 1343. ... It was built of flint and, in comparison with examples elsewhere, had no great strength and was probably not intended as a major fortification'.[22] Nevertheless, over 2 miles in length, 20 feet high and 5 feet thick, it was no mean undertaking.

On the continent, however, the wall did retain its primary military function (in addition to its use as a customs barrier). As Chapters 5–7 will show, it was to assume extremely complex and costly characteristics during the Renaissance period – so much so that city defences became probably the most important determinant of urban form. Because larger, prosperous continental cities found it essential to maintain strong defences, horizontal growth could not be a continuous process; it had to take place in stages, each of which was normally preceded by the construction of a new wall, although previously undefended 'suburbs' were also frequently included. Typically, the new wall completely surrounded the city, the distance out from the previous perimeter representing a careful compromise between short-term investment considerations and the need to enclose enough land for future expansion. Walls were also sometimes built to enclose discontinuous new suburbs. Either way this is in direct contrast to the situation in

M. Pirenne, writing more particularly of the Low Countries, tells us that up to the close of the Middle Ages a sum never falling short of five-eighths of the communal budget was expended on purposes connected with the maintenance of the walls and the provision of instruments of war. In Italy, despite the fact that she was now fast securing for herself the leadership of the world in craftsmanship and international commerce, the warfare of city with city was almost perpetual. Cities would fight about diocesan boundaries and feudal rights, over tolls and markets, for the extension of their powers over the 'contado' or surrounding country, or in pursuance of the long-inherited feuds of the nobles within their walls. (H A L Fisher, A History of Europe)

A wall girding a town was an earlier method of defence than a castle, and a much more popular one; while the hated castle was the visible symbol of subjection, the wall around the town reminded the burgesses of their rights as citizens and their community of interests; they looked on their allotted share in its defence as a privilege, no less than a duty, and the townsmen of the open towns eagerly applied for and warmly welcomed the right to protect themselves with a wall.
(A Harvey, The Castles and Walled Towns of England, 1911)

21. M W Beresford and J K S St Joseph, Mediaeval England: An Aerial Survey, 1958.
22. City of Norwich Plan: 1945; see also M D Lobel (ed.) The British Atlas of Historic Towns, Volume Two, 1975.

England where there are comparatively few instances of post-medieval defence work (although see Chapter 8, pages 280–5, for five later systems). Among these later systems were Civil War fortifications, and 'Palmerston's Folly' – the extremely costly, and, as it turned out, effectively pointless system of forts constructed between 1857 and 1868 to defend Portsmouth against the French.[23]

Florence is a clear example of the European concentric-ring type of growth, her two medieval walls of the late twelfth and early fourteenth centuries enclosed the original Roman nucleus (Figure 4.3). The tremendous growth of this city between 1172 and about 1340 involved an increase in area from 197 acres to 1,556 acres. The second medieval circle 'cost about 6,000 L a year in the first years of the fourteenth century and in 1324 nearly 20,000 L was spent on it in five months, which represented roughly a quarter of the commune's total expenditure'.[24] The growth of Paris required perhaps the greatest number of walls: that of AD 360, probably enclosing the Ile de la Cité; the second, of 1180, on both banks of the Seine; the wall of 1370, built on the right bank only (extended 1610–43); the one of 1784–91 (essentially a customs barrier); and lastly that of 1841–45, now the line of the Boulevard Periferique (Figure 6.13). Cologne (Figure 5.6) is one of the best illustrations of a town combining suburban accretion with concentric ring addition (the latter was on the west bank only of the Rhine).

STREETS

All medieval towns contained a space, if not several, which acted as a market – this is explained below. However, as Howard Saalman stresses, 'the existence of these specialised spaces dedicated to trade should not blind us to a basic fact: the *entire* mediaeval city was a market. Trade and production went on in all parts of the city; in open spaces and closed spaces; public spaces and private spaces'.[25] As a result, although frequently little more than narrow, irregular lanes in organic growth towns, main thoroughfares leading to the gates from the centre were as much linear extensions of the market place as communication routes, and the notion of a 'traffic network' was as absent as constant wheeled traffic itself.[26] Street frontage was therefore a valuable commercial asset, especially near the gates and market place, and its continuous development was normal. Later it also became usual for narrow passageways to be formed off the streets, providing access to new minor street and court development of back gardens. The City of London is one of the best examples of this type of internal elaboration.

Movement in medieval towns was very largely on foot; wheeled traffic reached significant proportions only late in the period and transport of goods was mainly by pack-animal. Street paving commenced early in the period; Paris 1185, Florence 1235, Lübeck 1310 – indeed by 1339 all of Florence was paved.[27] Throughout the Middle Ages there was a tendency for buildings to encroach even further on to streets (including bridges) and into public open spaces. Attempts to regulate this gradual strangulation met with little success. Upper floors projected still further out over the street until eventually it was literally possible to shake hands between opposite windows. Thus the medieval city acquired its traditional street scene – here was informality, 'romance', repeated visual surprise. Above all it was *apparently* accidental, although there was collective action more frequently than might be supposed.

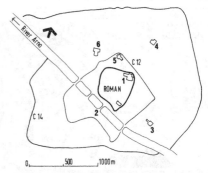

Figure 4.3 – Florence: showing the two rings of medieval defences around the Roman nucleus. Key: 1, Cathedral; 2, Ponte Vecchio; 3, S. Croce; 4, S. Marco; 5, S. Lorenzo; 6, S. Maria Novella; C12, twelfth-century fortifications; C14, fourteenth-century fortifications. As the 'birthplace' of the Renaissance in architecture and urban design, Florence is also considered in Chapter 5 (with references). At this medieval stage in its history, Florence exemplifies the characteristic of considerable areas of open, unbuilt-on land within the fortified perimeter. The line of the fourteenth-century wall, as this outline map shows, was subsequently consolidated by the zone of artillery fortification, which, as late as the first part of the nineteenth century, still included extensive open areas; for which see the SDUK Map of 1835, in M C Branch, *Comparative Urban Design*, 1978.

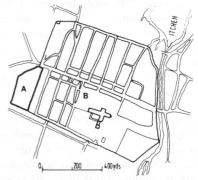

Figure 4.4 – Winchester: the medieval plan. The influence of the original gridiron structure of Roman Venta Belgarum has resulted in an uncharacteristically regular medieval street pattern. Key: A, the castle; B, the market place.

For early medieval Winchester see M Biddle, 'Towns', in D M Wilson (ed.) *The Archaeology of Anglo-Saxon England*, 1976; N Pevsner and D Lloyd, *The Buildings of England: Hampshire*, 1967/1990.

23. S A Balfour, *Portsmouth*, 1970; A Temple Patterson, *Palmerstone's Folly: The Portsdown and Spithead Forts*, 1967; see also Chapter 8. (The general references are C Duffy, *Siege Warfare, Volume I – The Fortress in the Early Modern World 1494–1660; Volume II – The Fortress in the Age of Vauban and Frederick the Great 1660–1789*, 1985.

24. D Waley, *The Italian City-Republic*, 1978.

25. H Saalman, *Mediaeval Cities*, 1968; see Chapter 11 for the contrasted form of market activity in the historic Islamic city.

26. L Mumford, *The City in History*, 1961; see Chapter 11 for analytical descriptions of the streets and lanes in the contemporary Islamic city.

27. Waley, *The Italian City-Republics*.

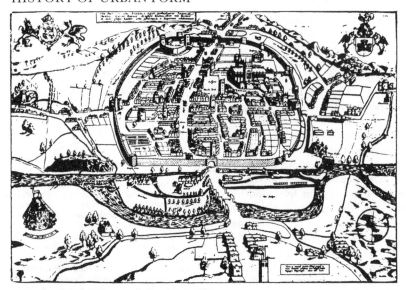

Figure 4.5 – Exeter: the city depicted in Hooker's plan of 1587. Even allowing for artistic licence, the wall, strengthened only by infrequent towers, can be seen to be of negligible military significance; there is also no sign of urban overcrowding at this late sixteenth-century date. For Exeter see B Cherry and N Pevsner, *The Buildings of England, Devon*, 2nd edn, 1987.

Exeter is not only a transit city for the south-west of England, but is itself in a state of transition. Ancient as York or Coventry, it is nevertheless losing, one after another, its characteristic signs of primordial life, and assuming a new character. Whether it be also, and at the same time, emerging from monkish superstition and its attendant darkness, it is not for a mere chance passenger to determine.

In its exterior, the city of Exeter, from being in a transitive state, offers some peculiar and startling contrasts as well as features. No one, for instance, can walk along its main street, the ancient way of the place from north-east to south-west, without having his attention attracted, first, by the buildings of a Gothic age, and then by those imitated from the Greeks and the Romans, either opposed to each other, or rising side by side – the one marking the days of yore, the others those that are even now passing. High-street and Fore-street in Exeter, to which I am thus alluding, are perhaps in those respects as interesting as High-street, and Cannongate, and Princes-street in Edinburgh would be, if the respective elements of architecture of those separate loacalities were mingled perchance together to form but one successive line of buildings as in Exeter. Here we have, for instance, the imposing elevation of the free Grammar-school, with its few yet grand Gothic windows and a fine gateway on the one side of High-street, contrasting with the Corinthian and pretending front of the West of England Insurance Office on the other side.

One of the popular misconceptions concerning medieval towns is that inside the wall the usual situation was one of 'over-crowding and muddle – picturesque owing to architectural treatment, but insanitary'.[28] Possibly because of the accepted image of narrow, continuously built-up streets, it is assumed that away from the two main public spaces – those containing the market and the church – there was only strictly limited private open space, and that development was uniformly dense throughout. In a small number of instances this may well have been the case, but, as Mumford points out, 'the typical mediaeval town was nearer to what we should now call a village or a country town than to a crowded modern trading centre. Many of the mediaeval towns that were arrested in their growth before the nineteenth century still show gardens and orchards in the heart of the community'.[29] To Sir Patrick Abercrombie, the little town of Furnes, in Flanders, epitomized 'the mediaeval conception of a business town, with its noble central place, its range of public buildings, including a cathedral, a great town church, town hall, law courts, etc., its houses continuously lining the streets, economically making use of every foot of frontage, but backed by ample gardens'[30] (Figure 4.6).

Overcrowding in continental towns more generally occurred in the late medieval and Renaissance periods – from growth constrained by inflexible fortification systems. In Britain it particularly accompanied the Industrial Revolution: existing gardens were developed for the new working-class housing, which, in the absence of mass transit systems had to be within reasonable walking distance of the workplace (in effect, a negative urban mobility determinant). Sanitary conditions are closely related to density. Although medieval towns had only rudimentary refuse disposal arrangements and water supply was a continual problem – particularly in hill-towns – it must not be assumed that disease was necessarily an everyday accompaniment to urban life. Mumford neatly disposes of this misconception: 'we must remember that practices that are quite innocuous in a small population surrounded by plenty of open land, become filthy when the same number of people crowd together on a single street ... in all probability the

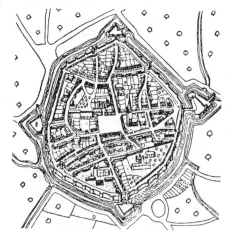

Figure 4.6 – Furnes, in Flanders: a view of about 1590. See D Nicholas, *Mediaeval Flanders*, 1992

28. Sir Patrick Abercrombie, *Town and Country Planning*, 1959.
29. Mumford, *The City in History*.
30. Abercrombie, *Town and Country Planning*.

early mediaeval village or town enjoyed healthier conditions, for all the crudeness of sanitary accommodation inside and outside the house, than its more prosperous sixteenth-century successor'.[31]

THE MARKET PLACE

Marketing – the *raison d'être* of medieval towns – was accommodated in a number of basic ways. Two types are common in both planned and organic growth towns: first, where the market occupies a square to itself, normally located at or near the centre; second, where it is located at a widening of the main street. Paul Zucker notes two further types of market place in organic growth towns: as lateral expansions of the main street; and as squares at the town gate.[32] In planned towns laid out with a regular gridiron structure the market square is usual. Its general form here is that of a void within the grid, bounded by streets on all four sides. (Important exceptions are some of the Welsh bastides, where the market is located in front of the castle.) On the continent it was usual for the surrounding buildings to be of the same height, and unified at ground level by arcades under which the streets frequently continued alongside the square; Monpazier – the archetypal French bastide – New Brandenburg and Ceske Budejovice (see Figure 4.90) exemplify this type of square. Although arcades would have suited the climate, they were not part of British tradition. Typically, most squares contain market halls, sometimes on two floors but instances of the town church facing into the market place are rare. The market street was much less frequently incorporated into planned towns and never in bastides. Its most important such use was in the Zähringer new towns of Switzerland and southern Germany, where it characteristically ran the length of the town between the gates. Rottweil and Berne (the latter illustrated in Figure 4.65) are important examples.

In unplanned towns both the market square and the market street defy precise description: no two layouts were alike, each had its own distinct spatial character. Many are still outstanding in Europe's cultural heritage. There are a few instances in Roman foundations of the market occupying the site of the old forum area, but usually a central position resulted from an original location within the village from which the town had gradually developed. 'More often than not ... the market place would be an irregular figure, sometimes triangular, sometimes many-sided or oval, now saw-toothed, now curved, seemingly arbitrary in

Figure 4.7 – Munich: north at top; the market – Marienplatz – a lateral expansion of the main street. For the form of this medieval nucleus in the context of the later mid-nineteenth-century city, see Branch, *Comparative Urban Design*, for the SDUK Map of 1840. A view of Munich in 1586 is included in J Goss, *Braun and Hogenberg's The City Maps of Europe: A Selection of 16th Century Town Plans and Views*, 1991.

Figure 4.8 – Examples of medieval civic spaces in seven cities; in several instances two or more squares form a sequence of spaces. (In addition, a number of these spaces were remodelled during the Renaissance period.) Key: a, Lucca, a complex sequence of spaces, reading from left to right; Piazza Grande (Napoleone), Piazza del Giglio, Piazza S. Giovanni, Piazza S. Martino, Piazza Antelminelli; b, Bruges, left, the Grand Place, right, Place du Bourg; c, Rothenburg, the market place in front of the Rathaus; d, Nuremburg, the Hauptmarkt in front of the Frauenkirche; e, Arras, Petite Place (top left) and the Grande Place; f, Perugia; g, Todi, the square in front of the cathedral.

See P Zucker, *Town and Square*, 1959, and C Sitte, *City Planning According to Artistic Principles* (translated by G R Collins and C C Collins), for detailed consideration of medieval civic space.

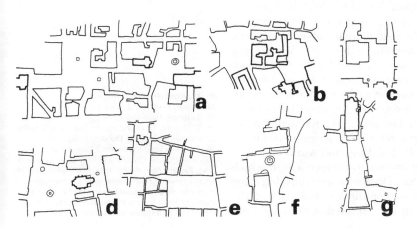

31. Mumford, *The City in History*.
32. Zucker, *Town and Square*; see also M Webb, *The City Square*, 1990.

shape because the needs of the surrounding buildings came first and determined the disposition of the open space.'[33] Several squares are shown in outline in Figure 4.8. In addition to their commercial function many squares were large enough for public gatherings. 'In towns which have evolved naturally from former villages, trading posts, etc., the main thoroughfare automatically becomes the market square since traffic is the vital element in the growth of the town.'[34] The street market is common to all European countries; the best examples are perhaps those in Germany, Austria and Switzerland. Market places possibly developed here as lateral extensions of the main street, in order to free their activities from pedestrian and vehicle movement through the town. Corporate action would usually have been necessary to clear away existing buildings. Immediately inside the town gate was another logical place for trade to develop. Here, 'a peasant with produce to market has at last reached his goal. He is inside the city, in the market ... his cart is heavy, why bother to move it further?'[35] The latter type of market, however, rarely became the most important in the town. (It should be noted that the market square is not part of Islamic urban tradition.)

CHURCH SQUARE

The space before the church – the medieval *parvis* – should not be confused with the burial ground where the latter adjoins the church, as is usual in Britain. 'It was on the *parvis* that the faithful gathered before and after the service; here they listened to occasional outdoor sermons and here processions passed. Here in front of the west portals of the church, mystery plays were performed from the twelfth century on. Here people from out-of-town left their horses and soon stalls of various kinds were set up. None the less, the *parvis* was never intended to compete with the market square.'[36] The *parvis* – or, in Britain, the burial ground – meant that churches were generally located within their own space. As this frequently adjoined the market square, a two-part nucleus is a typical characteristic of medieval towns, both planned and unplanned.

To conclude this general introduction to medieval urban form we must briefly consider the extent to which its development was subject to control: how far was it pursued as a conscious effort to achieve order and beauty? While Zucker's statement that, except in the comparatively few planned towns, the organization of a town as a whole was neither understood nor desired by the builders of the Middle Ages,[37] is a fair summary of the situation, we must be careful not to presuppose that there was absolutely no concern for spatial organization or aesthetic unity. This is a neglected area of documentary research. We are not concerned with those collective decisions which had to be taken from time to time – width of streets, maintenance or extension of the wall, rudimentary health measures, etc. These are well enough documented; but unfortunately there are only a few recorded instances of aesthetic awareness. Most of these concern Italian cities. Daniel Waley notes that 'in the thirteenth century Bologna retained a series of architects whose task it was to supervise all public buildings and works' and that 'the Sienese in 1309 asked their Dominicans to remove a wall which partially hid the grandiose church of San Domenico and thus detracted from the dominant architectural feature of the western part of the city'.[38] It is also known that following the completion of Siena's im-

Market means little to the average citizen today, but in this period it was the centre of his week, the day he could take his goods to sell and buy himself what the market could provide. Retail trade was conducted very largely in the market, for a mediaeval shop was less a store than a workshop. Traders kept no stocks of made-up goods, but sat in their shops and made what was ordered of them. Shops were generally very small, often no more than 6 feet wide. Osney Abbey built the Golden Cross Inn towards the end of the twelfth century. It sold the inn but kept the ground floor, consisting of 4 shops, each measuring about 6 by 15 feet.
(D M Stenton, English Society in the Early Middle Ages)

The church and its teachings pervaded man's entire life. One could not strike a bargain, cut a finger, or lose a farm tool without invoking celestial favour. One was seldom out of sight of a church tower, out of hearing of a consecrated bell. It is estimated that in England there was a church for every forty or fifty households. The visitor today marvels at seeing an enormous, beautiful, ancient church, usually empty, alas, in every East Anglian hamlet. According to an eleventh-century chronicler, the world was 'clothed in a white garment of churches'. Love and pride built the parish churches, and love and pride found a transfiguration, in the cathedrals. All men contributed their building, giving labour as well as money, even harnessing themselves to the supply wagons. The ecclesiastical fund-raisers were astute; they knew that by offering a little they could obtain much; thus they rewarded those who contributed generously towards the construction with indulgences or freed them from the strict observance of certain church laws. The 'Butter Tower' of Rouen Cathedral was built largely with funds received for permission to eat butter during Lent.
(M Bishop, The Penguin Book of the Middle Ages)

33. Mumford, *The City in History.*
34. Zucker, *Town and Square.*
35. Saalman, *Mediaeval Cities.*
36, 37. Zucker, *Town and Square* and Webb, *The City Square;* see also Vance, *The Continuing City.*
38. For Siena see J Keates and C Waite (photography), *Tuscany,* 1988; G Grazzini and G A Rossi (photography), *Tuscany from the Air,* 1991. Revisited in 1990, after an interval of nearly a quarter of a century, Siena is an outstandingly successful example of a major European city which has assimilated the late twentieth century within its historical urban fabric.

pressive Town Hall in 1310 (the tower was not finished until 1344) it was decreed that other buildings on the Piazza del Campo should have similar windows.

This fourteenth-century concern for visual order would seem to be a specifically Italian trait, presaging the emergence of the Renaissance and with it four centuries of more disciplined urban design. It also provides evidence of the continuing influence from the Roman past – one which included the visual unification of the forum at Pompeii by the construction of a colonnaded arcade in front of the disparate individual buildings (see page 71).

Figure 4.9 – Siena, the Piazza del Campo, dominated by the Torre del Mangia (1338–49). Daniel Waley observes that 'this tower had to be particularly imposing so that it rose above the cathedral. The curious site of the Sienese *palazzo*, on low ground where huge foundations were required, was dictated by the wish to choose neutral territory between the three hills on which the city lies. The result was a triumph and indeed Siena had benefited from postponing for so long the erection of a palace (*The Italian City-Republics*).

The paving of the central space is regulated by radii marked out from in front of the town hall, facing which are the uniform-height medieval palaces.

HISTORY OF URBAN FORM

Organic growth towns

Although when the Middle Ages opened towns were few and far between, the essential fact remains that by the eleventh and twelfth centuries practically all the settlements which were subsequently to evolve into towns were established on their sites. Either they were the vestigial remains of Roman foundations, about to take on a new lease of life; or they were burgs, built in the ninth century as fortified bases and acquiring commercial functions later; or again they were agricultural, subsistence-economy village settlements, able to exploit geographical advantages and promote themselves from village to urban status. These are the organic-growth towns of Europe and they constitute the great majority of medieval towns, although in some few regions planned towns are in a local majority – notably in central Europe east of the Elbe.

ROMAN ORIGINS

Any apparent contradiction presented by such a classification, whereby surviving Roman *urbes* became medieval organic-growth towns, is explained by the fact that, with but a few exceptions, the original Roman gridiron structure was lost during the decades, often centuries, when the town was deserted, or reduced to occupying only a small part of its earlier area. When rebuilding eventually took place it usually constituted unplanned organic growth of the medieval norm, the gridiron being ineffective. (This process is exemplified by the City of London, as described in Chapter 3.)

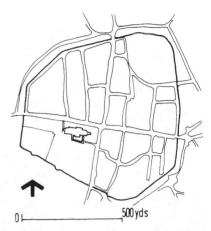

Figure 4.10 – Chichester, the medieval plan. Founded as Noviomagus Regensium, this site, in common with other Romano-British settlements, is believed to have been deserted for a considerable period of time following the Roman evacuation. As re-established however, the main axial crossing of the *cardo* and *decumanus* formed the basis of the layout; other minor streets have also followed the original gridiron to varying extents. Chichester exemplifies the characteristic retention of the main cross axes of a Roman plan where they were originally both continuous across the town.

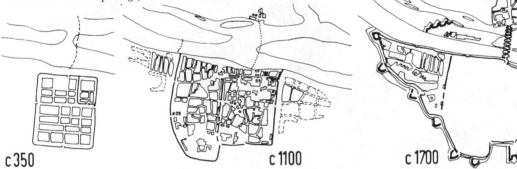

c 350   c 1100   c 1700

A distinction should be made between the fate suffered by Romano-British towns, following the final withdrawal of the Roman legions in AD 407, and the changed circumstances of their continental contemporaries after the fall of the Roman Empire. During their occupation of Britain the Romans had imposed only a thinly spread urban culture on the essentially rural civilization of the native Celtic tribes. The Romano-British towns were small, and even if London could boast some 15,000 inhabitants most towns barely averaged a tenth of that number.[39] The invaders of the fifth century must have regarded the towns and villas as sources of plunder and objects of destruction. Celtic villagers had little to lose and for them the end of Roman rule was no overwhelming calamity.[40] Perhaps, too, it was no more than an enforced return to their old way of life for the urbanized population. Lacking organized military resistance, they would have to take to the woods and hills in the face of invasion. The result was a temporary end to urban life in Britain. Some towns, notably Silchester, remained unoccupied long enough to disappear completely and it seems most prob-

Figure 4.11 – Regensburg, three stages in the development of the city from its Roman frontier *castra* origin on the southern bank of the Danube. Although not completely deserted following the withdrawal of the legions, the layout of 1100 within the *castra* perimeter shows little sign of the original gridiron. This sequence of stages in the growth of a strategically situated town also illustrates the effect of fortifications as an urban form determinant.

39. London's temporary loss of urban status is characteristic of Romano-British towns (see Chapter 3); similarly its subsequent re-establishment along organic growth lines, its original gridiron plan buried beneath the rubble.
40, 41. M Sayles, *Mediaeval Foundations of England*, 1948; see also M Biddle, 'Towns', in D M Wilson (ed) *The Archaeology of Anglo-Saxon England*, 1976.

able that all the Roman sites were deserted for lengthy periods from 457 onwards. The advantages of their locations, and the existence of ruined buildings and defences which could be repaired, subsequently attracted back a proportion of the old inhabitants and may have served as a ready-made base for immigrants into the district. But the original urban patterns, once disrupted, were never completely reinstated.

For a number of reasons the fate suffered by the Romano-British towns was worse than that endured by continental towns. Britain was the least Romanized of the provinces, it had been the last major ac-quisition and it was furthest from the origins of the imposed Mediterra-nean culture. The Barbarian invaders of Britain, in their turn, were much less affected by Roman influences than the new rulers of Western Europe generally. The latter had learned to appreciate the high civilization into which they entered and sought in their clumsy way to copy it.[41]

BURGS

The anarchy of the ninth century, which had compelled bishops to reinstate the walls of their Roman *civitates* and turned monasteries into religious citadels, also directly resulted in the emergence of a new class of town in Western Europe. On the continent the Carolingian Empire, disintegrating in the face of continual Scandinavian invasions, 'was parcelled out, after the middle of the ninth century, into a number of territories subject to as many local dynasties, and attached to the Crown only by the fragile bond of feudal homage'.[42] Defences, as much against rival princes as against invading forces, were an essential re-quirement and castles were therefore built in many localities. As de-scribed by Thompson and Johnson 'these were not yet the great stone structures of the thirteenth century and later, but only rude, palisaded wooden blockhouses, often erected in a hurry to meet some crisis of invasion ... countless burgs were built by dukes, counts, and mar-graves in the ninth and early tenth centuries for protection against Norsemen, Magyars, Slavs and Saracens, especially in Saxony and on the eastern German frontier and in England, where many boroughs (burgs) served for defence against the Danes'.[43]

David Wilson confirms, on the basis of archaeological and historical sources, that in England, even before the Viking invasion, 'some towns were flourishing centres of trade and mutual protection. Such towns were Rochester, Canterbury, Carlisle, Thetford, Winchester, and, of course, London ... nevertheless, the great development in English town life came in and after the reign of Alfred in the late ninth century. The need to protect the country against Danish attack led to the found-ation of a series of fortified boroughs'.[44]

The Danes first appeared in England in 793; the Anglo-Saxon Chronicle records that on 8 June the ravages of heathen men miserably destroyed God's church on Lindisfarne, with plunder and slaughter.[45] The essentially agricultural communities of England were not orga-nized to resist these incursions, which by the middle of the ninth cen-tury had escalated into large-scale military operations. As a result Mercia and Northumbria were lost to the invaders and Wessex was seriously threatened. King Alfred (871–99) was responsible for first containing the Danish advance and then defeating them in 878 at the battle of Edington. Dorothy Whitelock describes how the defensive strategy was based on his burghal system – the provision of fortified

If towns on sites once Roman were richer and larger than towns on newer sites, as York was richer than Norwich and Lincoln than Ipswich, it was because of present circumstances, not an inheritance from their remote past. When William I became king there were in England boroughs which would never develop into towns in the modern sense, places little more in size than villages, and there were villages that would in time become towns.

Bedwyn in Wiltshire cannot now be called anything but a village, but it was a borough in 1086 when Doomsday Book was compiled; so were Bruton and Langport in Somerset. It was not size or wealth that made a borough, but an act of royal will. The larger English boroughs at the beginning of this period had all grown up under royal protection and control as an in-tegral part of the royal policy for the development of national resources and national defence.

(D M Stenton, English Society in the Early Middle Ages)

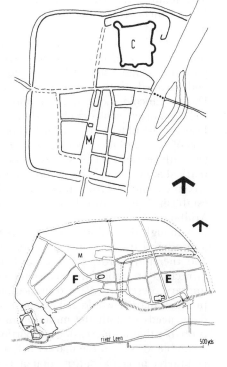

Figure 4.12 – Wallingford (top) and Nottingham: two examples of English burgs. Key: C, castle; M, market place.

At Nottingham, after the Norman Conquest a new French borough (F) was established adjoining the area of the original English burg (E). For Nottingham see M D Lobel (ed.) *The British Atlas of Historic Towns, Volume One*, 1969.

42. Pirenne, *Mediaeval Cities*.
43. Thompson and Johnson, *An Introduction to Mediaeval Europe, 300–1500*.
44. D Wilson, *The Anglo-Saxons*, 1976.
45. D Whitelock, *The Beginnings of English Society*, 1968.

centres which could protect a tract of country against enemy attack, a policy continued under his son Edward from whose reign comes a document known as the Burghal Hidage, which gives the names of the boroughs thus formed under West Saxon rule.[46] According to Sir Frank Stenton there were thirty-one of these strongholds in Wessex during Alfred's reign, so located that no village was more than twenty miles away from safety.[47]

Alfredian boroughs were of various origins. Many were based on existing settlements which had been given new or strengthened walls; others were new foundations. Oxford and Wallingford, each founded on eight yardlands of land, seem to belong to this class, and the account in Domesday Book suggests that population has been attracted to them by favourable conditions of tenure.[48] The boroughs created by the kings had their fates determined by economic circumstances. The borough of Sceaftesage is one of those that disappeared and it has been only tentatively identified as Sashes, on the Thames near Cookham. Many others prospered only to the extent of becoming well-established market and country towns.

The Danes also founded boroughs as bases for their offensive operations and after the treaty of 878, which recognized the Danelaw as their part of England, they settled peacefully around them as social, market and fortified centres. The five most important Danish boroughs – Darby, Lincoln, Leicester, Stamford and Nottingham – constituted a well-organized confederacy and the English counter-offensive under Edward the Elder (901–25) turned on their reduction. With the exception of Stamford these boroughs were confirmed as county towns in the eleventh century. Darby shows how the reconquest of the Danelaw was marked at every point by the creation of a stronghold. From the Anglo-Saxon Chronicle a list can be compiled of just over a score of boroughs established by Edward the Elder to secure his conquests in the Midlands.[49]

Although the majority of burgs in Europe were essentially military and administrative centres, without significant commercial activities and therefore hardly to be accorded urban status, they constituted attractive 'pre-urban' nuclei around which many towns developed. The military presence generated immediate service industry activity and a produce market would have soon become established to provide for daily needs of both the military elite and the serf community. It is probable that an increasing proportion of local craft products became available for sale in the market before the general revival in European commerce brought confirmation of urban status – a revival encouraged first by the packs of itinerant pedlars and subsequently by the wagons of increasingly prosperous merchants.

Civil trading communities which grew up outside the walls of burgs were frequently called faubourgs (Latin *foris burgum* – 'outside the burg') or suburbs (Latin 'close to the *urbs*'). As they became established, perhaps even completely surrounding the burg, they required defensive walls of their own. There is no evidence that these simple two-part burgs which are found in all parts of Western Europe, were other than the result of organic growth. Some burgs, however, had more complex origins and were possibly planned in part. Magdeburg (Figure 4.13) is taken by C T Smith as an example of a multi-nuclear burg, comprising a Carolongian castle-site of 805 (with a monastery added in 937); the cathedral of 968 and its related buildings; the

The status of a borough was no assurance of prosperity and it was easy even for an ancient borough to lose ground to a new competitor. When late in the Middle Ages a bridge was built over the Thames at Abingdon it diverted the London to Gloucester road and drew trade from the borough of Wallingford, in early days the chief borough of Berkshire. But long before this, between Abingdon on the one hand and Reading on the other, where Henry I's new abbey brought trade to the town, Wallingford was losing ground. Even today it has spread little beyond its Anglo-Saxon defences.

Many of the boroughs founded by magnates did not prosper, either because of too close competition or because the site was unsuitable to a town. In Lancashire only 4 of the 23 boroughs founded there between 1066 and 1371 survived the Middle Ages as boroughs. Even Manchester and Warrington failed to keep their burghal status continuously into the modern age.

(D M Stenton, English Society in the Early Middle Ages)

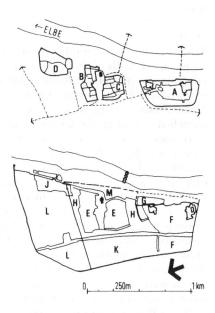

Figure 4.13 – Magdeburg: two stages in the development of this complex multi-nuclei settlement located on high ground on the west bank of the Elbe.

Above (c. AD 1000). Key A, Carolingian castle-site of 805 (on the probable site of a Saxon lord's house) and the monastery of St Moritz (937); B, cathedral of 968, and the early unplanned Ottonian town; C, planned Ottonian town of c. 1200 with a market place; D, the count's burg.

Below (c. AD 1250). Key: E, Ottonian town; F, the immunity of the cathedral; G, monastic land; H, infilling development of the eleventh and twelfth centuries; J, part of the count's burg, used for church buildings from the thirteenth century; K, organic growth extension of 1152–1192; L, planned new district of 1213–1236; M, twelfth- and thirteenth-century riverside development.

46. Whitelock, *The Beginnings of English Society*.
47. Sir Frank Stenton, *Anglo-Saxon England*, 1947.
48. Whitelock, *The Beginnings of English Society*.
49. H C Darby, *An Historical Geography of England before 1800*, 1961.

count's *burg*, further downstream; and, between cathedral and castle, a regularly laid-out market district of the early thirteenth century.[50]

VILLAGE SETTLEMENT

By definition, therefore, organic growth towns of the Middle Ages developed from, or were based on, village settlements. Throughout Europe, the location of these towns and, to a very great extent, their urban form, were determined by the preceding slow, accumulative processes of village settlement. Thus it is first necessary to consider the circumstances of their village origins. England provides an instructive illustration of these processes.

With relatively few exceptions the 13,000 or so English villages were in existence on their present sites by the time of the Domesday Book survey of 1086, although in many instances these could have been only the embryonic nuclei of later medieval settlements. England became a country of villages as a result of the Anglo-Saxon settlement between the fifth and tenth centuries. By the time of the first Anglo-Saxon villages much of the earlier localized, cleared and tamed landscape had reverted to its natural state. W G Hoskins, in his admirable *The Making of the English Landscape*, records that in certain favoured regions, like the Cotswolds and north Oxfordshire, the Anglo-Saxons may have entered a fairly civilized landscape, but in general they had literally to start from scratch.[51] The great majority of the English settlers faced a virgin country of damp oak/ash forest, or beech forest on or near the chalk; what was not thickly wooded was likely to be cold, high, mist-wrapped moorland or water-logged, wet, heath, drowned marshes and estuary saltings, or sterile, thin-soiled, dry heath – hardly the sort of land that gave itself to cultivation.

This situation facing the Anglo-Saxon village settlers was the first of four distinct phases in the continuing evolution of the English landscape: first, more or less wild, inherited landscape; second, the result of village settlement, which, over most of the country, saw the forest largely cleared for the typical open-field farming system; third, the sub-division of the large fields into the 'traditional', small hedgerow-enclosed fields during the enclosures of the sixteenth–eighteenth centuries; and lastly the continuing process of recent years whereby the small fields are being formed into larger areas suitable for mechanical cultivation, and the hedgerows up-rooted.

At the time when a village was founded or inherited, it is to be expected that the settlers had only a vague conception of the ultimate extent of its fields and the quality of land which lay below still uncleared forest. The distance at which neighbours (or, for that matter, daughter settlements) came to be tolerated could not have been based on any calculation of 'adequate' agricultural territory. The field-area was but a tiny island in a sea of uncleared land, and future population requirements are unlikely to have concerned the small groups of settlers who lived hand to mouth. Yet, as W F Grimes has written of Northamptonshire villages, the whole has the semblance of a consciously planned and co-ordinated allocation of the land best suited for primary settlement.[52] In two sample areas in England the distance from each village church to the nearest adjacent church has been measured. The uniformity is remarkable. In Northamptonshire fifty-three villages recorded in Domesday Book give an average distance between villages of 1.2 miles: thirty of the villages having their neighbours between 1.0 and

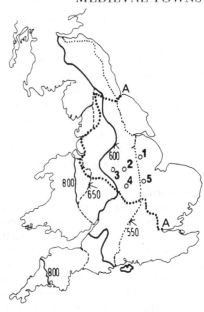

Figure 4.14 – Map of England showing successive stages in the extent of the Anglo-Saxon settlement at 550, 600, 650 and 800. This map also shows in heavy dotted line (A–A) the extent of the Danelaw and locates the five boroughs referred to on page 106 as: 1, Lincoln; 2, Nottingham; 3, Derby; 4, Leicester; 5, Stamford.

50. Smith, *A Historical Geography of Western Europe before 1800*.
51. Thoroughly re-read after an interval of years, *The Making of the English Landscape* remains one of those deceptively definitive works for which even an equal can scarcely be imagined.
52. Quoted in Hoskins, *The Making of the English Landscape*.

1.2 miles away. Similarly in an area of Huntingdonshire, thirty-seven villages dating from at least Domesday Book are an average of 0.95 miles apart.

These two areas are of undulating open country without natural barriers to an even spread of colonization in all directions. In a typical river valley, the Avon between Salisbury and Pewsey, there are twenty villages spaced on average 0.89 miles apart, although in several instances the distance over the ridge to the next village is considerably more. Beresford and St Joseph conclude therefore that where the physical conditions are uniform, or nearly uniform, settlements have been tolerated at about the same distance from each other with a strong preference, in the three districts examined, for having neighbours about a mile away.[53]

Although our concern is not with the plans of medieval English villages as such, a brief consideration of their essential characteristics is necessary. Through to recent times the form of those towns which have developed from village origins has been largely determined by the pre-urban cadastre,[54] in combination with topographical considerations. Although every one of the English villages had a unique layout,[55] uniquely formed to meet the requirements of its location, it is possible to classify them under three broad headings: first, enclosed villages (also known as nucleated or squared); second, linear villages (also known as street or roadside villages); third, dispersed or disintegrated villages. The first two headings are generally accepted, notably by Thomas Sharp, author of *The Anatomy of the English Village*, and joint secretary of the 1942 *Report of the Scott Committee on Land Utilisation in Rural Areas*. A third group, not allowed for by other authorities, is the dispersed or disintegrated village, as referred to by W G Hoskins.[56]

Illustrations of enclosed and linear villages are shown adjoining this page together with a further description in the captions. It must be stressed however that only a small minority of the total number of villages had such simple clarity of form; most combined one or other, perhaps even both characteristics, with other non-conforming elements. The dispersed village normally had no coherent form; houses were dotted about singly or in groups of two or three, linked by a network of paths and lines, but nevertheless they constituted recognizable social units. Such villages were generally off the main routes and few developed into towns. The conjectural form of a typical village in about AD 900 is shown in Figure 4.21; large open fields, collectively worked, surround the built-nucleus, within a progressively enlarged area of cleared woodland.

Even for the Anglo-Saxons, the open-field village was not the only form of human settlement. 'No single type of settlement,' says Sir Frank Stenton, 'can ever have prevailed throughout the whole, or even most of southern England.' On heavy lands and indeed wherever there was a prospect of a steady return to co-operative agriculture, ceorls tended to live together in villages. But as late as the eighth century life for perhaps a quarter of the English people was a struggle for existence against unprofitable soil and scrubland vegetation which would spread again over cultivated fields on any slackening of effort. It was by individual enterprise that these poor lands had been brought into cultivation and innumerable isolated farmsteads bearing Anglo-Saxon names remain as memorials.[57]

The main phase of farmstead building, however, came much more

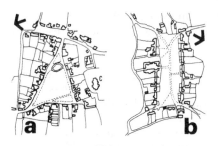

Figure 4.15 – Two enclosed villages with traditional, grassed village greens: a, Writtle in Essex, with a large triangular green with a pond in its eastern corner; b, Milburn in Westmorland.

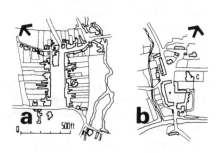

Figure 4.16 – Two enclosed villages with unusual, hard-surfaced 'urban' central spaces; a, Wickham in Hampshire between Southampton and Portsmouth; b, Blanchland in Northumberland.

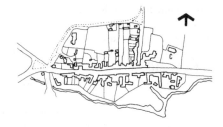

Figure 4.17 – Wycombe in Buckinghamshire: a clear example of linear village form.

An essential characteristic of both village types is the long back garden, frequently amounting in effect to a small-holding, which was attached to each dwelling, and which in many instances was approached from the rear by an access lane. The back lanes were generally upgraded into roads providing access to new houses during early stages in the expansion of a village. In small market towns to this day many of the back-gardens of the original cottages have yet to be developed. In towns which prospered and continued to expand this space was usually built over by the advent of the Industrial Revolution.

53. Beresford and St Joseph, *Mediaeval England: An Aerial Survey*.

54. Although not spelt out as such, the phrase (pre-urban) 'route structures and property boundaries' refers to the pre-urban cadastre. Of urban form determinants throughout history, it is the pre-urban cadastre that had most widespread effect. See D Ward, 'The Pre-Urban Cadastre and the Urban Pattern of Leeds',

recently. During the Parliamentary Enclosure Movement – the third phase in the creation of the landscape (c. 1750–1850) – the remaining medieval open fields were converted into the modern chequer-board pattern of small, squarish fields enclosed by hedgerows of hawthorn.[58] Only in a relatively few instances could the enclosure commissioners establish the lines of the new fields such that direct access could be given for those owners whose original homestead was in the village itself.

As a result new farmsteads were required out in the fields away from the village, but according to Hoskins the total number built was very small; one would guess at not more than half a dozen in the average parish, often fewer than that. He adds that, whatever the actual figures, the number of Georgian farmers in any parish was generally only a fraction of the number there had been in the medieval or Tudor village.[59]

URBAN ORIGINS

We must now consider two interrelated questions. When, during the Middle Ages, did a village attain town status? And why did only selected villages develop into towns, perhaps engulfing other villages, while their neighbours retained their original form and agricultural function through to the private-transit revolution of recent decades? To generalize in answering the first question we can say that a medieval village became a 'town' when it acquired the secondary function of a local trading centre, with probably also some small-scale specialist industry, with a proportion of its inhabitants spending some of their time on these non-agricultural pursuits. As the 'town' gains in strength, developing trade and meeting demands for its products, the proportion of non-agricultural specialists rises, and their involvement with agriculture diminishes. But – and this fact must be kept clearly in mind – only a small minority of the inhabitants would lose all contact with the land and a significant proportion of the day-to-day agricultural requirements of the town would be met by its own production. In addition, and this point is stressed, the great majority of towns were of very small size by modern standards, and until the later Middle Ages many were not much bigger than their village neighbours.

Certainly mere size, either of area or population, is not a safe criterion to apply in assessing town status, G W Southgate, in his *English Economic History*, notes that about eighty towns are mentioned in Domesday; only some forty of these had, in 1377, populations of over 1,000. Doris Stenton, in her *English Society in the Early Middle Ages*, also tries to estimate the size of Domesday Book towns, noting 'that it is possible to gather an impression of the immediate results of the Conquest on the thriving communities of Saxon days; most towns seem to have declined in population since the Conquest'. Neither London nor Winchester is described in Domesday but Lady Stenton attributes populations of 8,000 plus to York, more than 6,000 to Lincoln and around that figure to Norwich, to Thetford about 5,000 and Ipswich 3,000. These are the exceptionally large early medieval towns.

Simple definitions of medieval urban status must be questioned, but attempts to cover all the criteria tend to become unwieldy like Professor Hofer's statement that 'a medieval town is the result of the interrelationship of the following six aspects: economic structure (market handicrafts, trade); social structure (craftsmen, merchants, clergy, aristocracy);

The interdependence of city and countryside was not merely the consequence of landowning by citizens. The essential function of the great majority of towns was as the principal market centre for local commodities. Most towns were probably mainly dependent on their own rural territories for grain, wine, meat, cheese, vegetables and fruit, a majority even for their hides and wool, a great many too for their oil and fish. Those cities, such as Genoa and Florence, which became so large that they could not find sufficient cereals in their own vicinity, were quite exceptional. And the commodities that most towns had to import from further afield – salt, iron, perhaps building-stone – were also the exceptional ones. Its position as the centre of roads, and often waterways, for receiving and marketing wares is the key to the economic life of almost every city except the greatest nuclei of international commerce.
(D Waley, The Italian City-Republics)

Tax lists and muster-rolls show that most Tudor towns were not only more densely populated but also wealthier than their village neighbours. The concentration of economic power had come through workshops and counters rather than by ploughs and animals. Fertility of soil explain why one Elizabethan village was four times the size of another, but not why a town could be forty times the size of a village.

The protection afforded by walls and the freedom from feudal obligations had provided a climate in which trade and crafts could flourish more luxuriously than in a village, and even when privilege hardened into jealous restriction of competition there were still enough advantages in town life to prevent every townsman and every occupation from fleeing countrywards.
(M W Beresford and J St Joseph, Mediaeval England – an Aerial Survey)

Annals of the Association of American Geographers, vol 52, 1962.
55. Generally, now, when considering organic growth urban form I use the term 'layout' (or 'arrangement'), rather than 'plan', for example 'each English village had a unique layout'. In this case the original first edition wording has been retained in the text in order to emphasize this point.
56. Hoskins, The Making of the English Landscape.
57. Stenton, Anglo-Saxon England.
58. In many parts of England, the pattern of small hedgerow enclosed fields, which suited eighteenth- and nineteenth-century farming methods, has been converted back to large open fields in response to later twentieth-century mechanized farming economies.
59. Hoskins, The Making of the English Landscape.

Figures 4.18, 4.19 and 4.20 – Three aerial photographs of English villages which, if they could have been taken 100 or 150 years ago, might have represented towns in the making. Today, in the context of British town and country planning legislation, it is extremely unlikely that such transitions from rural to urban status could take place (at least during the foreseeable future). Nevertheless, in the absence of aerial photographs of villages of the past which did evolve as organic growth towns, these three views can serve as examples of possible embryonic towns, showing how the form of a future town would be largely determined by factors inherited from its rural past.

The main 'form determinants' of organic growth in England (and generally also throughout Western Europe) were land ownership or property boundaries and regional routes. Together with natural topographical constraints, these two man-made factors underlie the forms of virtually all organic growth towns.

Expansion of a village (or, in turn, of the town it became) was on an incremental basis, field by field if the land-holdings were small enough, or as parts of larger fields. The photograph of Appleton-le-Moors (facing page) shows an unusually regular, small-scale field pattern which could have been expected to determine an incrementally regular town. The sharply contrasted winding line of the road away across the fields would, however, have been retained as one of the town's main streets.

Facing page: Appleton-le-Moors, Yorkshire – a 'street' village with the bounding back lanes clearly delineated, and with the long and narrow village land-holdings also clearly shown. (The regular field pattern mentioned above is a result of the re-apportionment at the later stages of the Enclosure Movement.)

Above: Blanchland, Northumberland – an 'enclosed' village with unusually consistent urban spatial characteristics, as seen from within – the result of its 'planned' origins in the mid-eighteenth century, when a long-deserted monastery site was partially rebuilt.

Left: Warkworth, Northumberland – a twin-nucleus village, dominated by its superb castle of Norman origin, and with its equally impressive church at the far end of the main village 'street'. (Growth to urban status would, in this instance, have been greatly complicated by the location in the bend of the river.)

I have retained the original first edition sense of these captions in respect of the main 'form determinants' of organic growth from village origins in England. The third edition revision is that I now combine 'land ownership or property boundaries and regional routes' under the one heading as the pre-urban cadastre.

physical structure (town plan, public buildings, fortifications); legal personality (constitution, legal organs, districts); situation (land and waterways, bridge, halting places, reloading places); and political vitality'.[60] For Professor Hofer one or two of these aspects may be slight or absent but a vital strong town is created when all six are present and equally developed; if there are only two or three of these elements, the town remains small, reverts to village status or even disappears. Hofer rejects attempts at statistical definitions of a town, dismissing as 'absurd' the suggestion that a dividing line between a village and a town could be placed at 10,000 inhabitants.

To generalize further it is safe to assume that if the trading activity was established, then the other aspects would most probably follow, given a favourable economic climate, but if there was no trade then there could be no town. The development of trading and industrial functions did not at first greatly affect the traditional relationship between the lord of the manor and the inhabitants of the town. 'The burgesses of a town, however, formed a larger and wealthier body of men than the serfs in the rural manor, and they were often able to extort privileges from the necessities of their lord.'[61] The ambition of the townsmen was to achieve as great a control of their interests as possible, and to ensure protection against competition in their trading activities. Towns therefore petitioned for a borough charter giving 'corporate' status (normally obtained by purchase) which would ensure significant privileges for the inhabitants. Although there was no consistently applied formula on which borough charters were drawn up, there was a common basis of rights which were conferred. One of the most sought-after benefits was the right to hold a weekly market, supplemented if possible by one or more annual fairs. As implied by the definition of 'town' an informal, unrecognized market of some kind existed in all medieval towns, but the establishment of a market was not in itself sufficient to constitute a borough, and the distinction between 'corporate towns' and 'market towns' must be kept in mind.[62]

The second question as to why only certain villages attained urban status becomes, in effect, a question as to how such villages acquired trading functions. Part of the answer has been given earlier when considering the general revival of European commerce,[63] some villages became trading centres because they were conveniently located on through-routes, attracting custom to natural stopping places. Many others (and they most probably amount to a considerable majority of English market towns) became the dominant of the eight to twelve *vills* which together constituted a *hundred*, most probably adding commercial functions to their administrative roles. As Stamp and Beaver explain,[64] 'the market towns of mediaeval England were closely spaced. The visit to the market town had to be made on foot ... it is clear in the more settled rural parts of England that between seven and ten miles was regarded as the proper distance between marketing centres. Indeed there is an old law in existence which makes it illegal to establish a market within 6⅔ miles of an existing legal market'.[65]

Organic growth towns: England

Eight examples are illustrated, each with description of a reproduction of an historic map showing the medieval form. Readers are referred, in particular, to the series of maps published by John Speed in 1611.

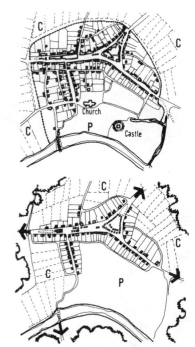

Figure 4.21 – Diagrammatic illustration of two stages in the transmutation of a village into a market town: below, c. 900; above, c. 1100. Key C, common fields; P, pasture.

These sketch plans are based on conjectural stages in the growth of Hereford, as illustrated in Cecil Stewart's *A Prospect of Cities*. But Stewart, in missing this town's vital growth phases as a burg (as revealed in Volume One of *Historic Towns*, edited by Mrs Lobel), has compromised the validity of his argument that trade alone made Hereford. Nevertheless, as abstract diagrams they admirably illustrate the general process of organic growth change from a riverside trading village, conveniently located by a ford, with temporary market stalls set up alongside through-routes, to a fully fledged market town.

The plan of c. 900 also shows the entirely typical relationship of the built-nucleus to its surrounding fields, enclosed in turn by as yet uncleared woodland. (By c. 1100 the limit of cultivation had been pushed farther out.) See Hereford in M D Lobel (ed.) *The British Atlas of Historic Towns, Volume One*, 1969.

60. P Hofer, *The Zähringer New Towns*, Exhibition Catalogue, Department of Architecture, Swiss Federal Institute of Technology.

Other urban definitions of a specifically medieval nature include M W Beresford: 'any place that passes one of the following tests: had it a borough charter? did it have burgages? was it called a burgus in the Assize Rolls, or was it separately taxed as a Borough? did it send members to any mediaeval Parliament?' (*New Towns of the Middle Ages*).

Reduced to just the one criterion, this book is based on the belief that if there was no trade then there could be no town (A E J Morris).

61, 62 G W Southgate, *English Economic History*, 1934.

63. See also the introductory Locational Determinants section in Chapter 1.

64. L D Stamp and S H Beaver, *The British Isles*.

65. See also R E Dickinson, *Geography* XVII, March 1932. This offers another estimate, based on a study of East Anglia, of the maximum range of influence of the mediaeval market town as 6 miles.

Although not cartographic records, these maps are attractively presented, accurate representations of the early seventeenth-century forms.

OXFORD AND CAMBRIDGE

The real importance of both Oxford and Cambridge as significant examples of historic urban form has been largely disregarded in other comparable histories, which tend to concentrate on the unique townscape qualities of their individual and collective college building groups, without adequate description of their historic urban contexts. In their different ways each city is an excellent example of characteristic origin and historic form to which emphasis is given on these pages. Summary histories of the two universities are given in the captions to the respective plans published by *The Weekly Dispatch* in the 1860s.

Oxford: the city

Whereas the university predates by some decades its great traditional rival, the city of Oxford itself is of considerably more recent origin.[66] The first mention of its name, as an important 'oxen-ford' across the River Thames at Hinksey, appears in the Anglo-Saxon Chronicle in 912, when it was also recorded: 'King Edward took possession of London and of Oxford and of all the lands that owed obedience thereto.' The town's strategic location, combined with fortification within natural moats, prompted the decision to make Oxford one of the system of English burghal strongholds in face of renewed Danish attacks. It is reasonable to assume that under King Edward the town was refounded on the basis of a gridiron street layout, as evidenced by a residual rectilinear pattern in the centre of modern Oxford (Figure 4.22). In addition, an account in Domesday Book of citizens still enjoying favourable conditions of tenure suggests their having been attracted into a new royal town – see below for this incentive as part of bastide planning.

Oxford eventually did fall to the Danes in 1009 and was plundered and burned, recovering only slowly through to 1086 when it was recorded that for 243 houses paying taxes, there were 478 'so waste and destroyed that they cannot pay the geld'.

Figure 4.22 – Oxford: the burgh, conjectural extent and street pattern based on the modern layout.

Figure 4.23 – Oxford in about 1250 with the Norman castle and the reconstructed circuit of walls. Under the Normans a powerful castle was constructed and in 1226 the circuit of town walls was rebuilt. A six-day annual fair was awarded by Henry I in 1122, and a town charter from Henry II followed in 1161. Once again commercial prosperity was transient, although decline in the one aspect of Oxford's life was to be offset by the growth of the university, a main reason for which was the existence of a nucleus of teachers in monastic schools, notably at St Frideswide's Priory – later the site of Corpus Christi and Merton Colleges. (This factor also helped establish the University of Cambridge.)

Figure 4.24 – Oxford: the university as shown by the central section of the *Weekly Dispatch* map of the mid-1850s. The first twelfth-century students received their instruction in their tutors' homes and made their own living arrangements. There were no separate colleges and all students were members of the university. Rivals for precedence among the Colleges involve Merton (O), Balliol (D) and University College (N), which all claim to be the oldest founded at dates set variously between 1249 and 1263. Exeter (G), Oriel (Q) and Queen's (L) followed during the fourteenth century.

Monastic houses in both Oxford and Cambridge not only influenced the establishment of a communal academic way of life, but also provided the architectural form of buildings grouped around courtyards which was adopted as the characteristic arrangement of the colleges. University expansion entailed considerable building demolition within the medieval walls, thereby accelerating the city's commercial decline. New College of 1379, for example, was built on the sites of more than 30 houses and the garden of Merton College took the land of ten or twelve others.

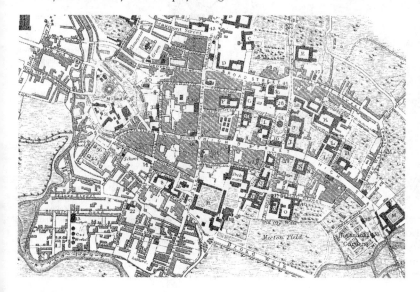

66. For Oxford see D Defoe, *A Tour Through the Whole Island of Great Britain, 1724* (1971 edition); C Duffy, *Siege Warfare: The Fortress in the Early Modern World 1494–1660*, for the Plan of the Oxford Fortifications, as appended Figure 4.24.

Cambridge: the city

Local Belgic tribesmen first recognized the trading potential of the place where the main east–west overland route crossed the River Cam at its lowest fording point.[67] They settled around present-day Castle Hill from where they controlled the ford near Magdalene Bridge. Later, around AD 70 the Romans replaced the native village by a fortified military town which survived into the fifth century. As rebuilt by the Danes from 875 the name was changed to Granta Bridge, from which derives the modern form.

Growth of trade in the tenth century resulted in the establishment of a right-bank river port above the bridge, between High Street (Trumpington Street) and the Cam. By 1066 it is probable that the river port had become the dominant of the two settlements. Two years later William the Conqueror constructed a motte and bailey castle within the old left-bank town, demolishing some thirty dwellings in the process (Figure 4.26). In the Hundred Rolls of 1279 there were three parishes recorded on the left bank compared with fourteen across the river. Cambridge was the county town and with upwards of 2,500 inhabitants it ranked as one of England's most important urban centres. Commercial prosperity did not last and during the fourteenth century, coincident with the rise of the university, the town's trading role gradually diminished, possibly because of silting-up of the river route by way of the Ouse to the sea at King's Lynn. In addition to the existence of monastic schools, a second factor which helped the growth of the university was availability of vacant buildings in the old river port area. Nevertheless residual river front commerce remained through until the mid-fifteenth century when, in order to build his King's College, Henry VI was obliged to 'compulsorily' acquire and demolish a tightly packed area of medieval lanes and buildings and their river wharves.

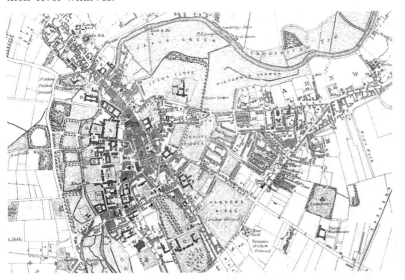

YORK

Although no longer second in size to London among English cities – an eminence which York enjoyed from its Roman origins through to late medieval times – the city is nevertheless still unrivalled in several respects as a second, northern capital.[68] By avoiding nineteenth-century

Figure 4.25 – Cambridge: the Roman military town on the left bank of the Cam; evidence for an early predecessor of Magdalene Bridge is inconclusive.

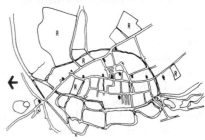

Figure 4.26 – Cambridge: the right-bank river port showing the sites of churches; the land occupied by religious houses; and the typical pattern of narrow lanes leading through to the riverside wharves. (Also the Norman castle and the subsidiary settlement remaining on the left bank of the Cam.)

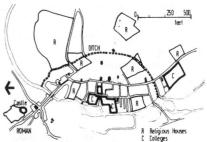

Figure 4.27 – Cambridge: the extent of the late-thirteenth-century university (both within the old town area, bounded by the Ditch) and the early colleges that were located outside.

Figure 4.28 – Cambridge: the comparable central section of the *Weekly Dispatch* map. The colleges have completed their taking-over of the land between Trumpington Street and the Cam, with the creation of the marvellous sequence of riverside gardens known as the 'Backs' (or Back Greens) between St John's College (B) in the north and Queens' College (K) in the south, just outside the limit of the medieval city. The first college was Peterhouse (P) of 1284, to the south of the city; followed by Michaelhouse of 1324, which had its own wharf, and which was merged with King's Hall into Henry VIII's new foundation of Trinity College (E) in 1546. Other early colleges include Clare (H), King's Hall, Pembroke (O), Gonville and Caius (G), and Corpus Christi (M).

67. For Cambridge see M D Lobel (ed.) *The British Atlas of Historic Towns, Volume One*, 1969; J Evelyn, *The*

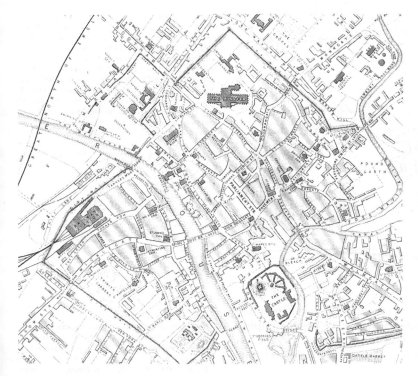

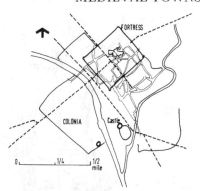

Figure 4.29 – Eburacum (York), the extent of the *colonia* and the *castra* related to the modern streets, which differ from those of the mid-nineteenth-century Tallis map in several respects. The *castra* was about 50 acres in area, its sides 550 yards long on the north-west/south-east, and 468 yards on the other two. The first defences of AD 71 were earthen banks with timber palisading; stone walls were constructed early in the second century with later rebuilding and strengthening. There were wharves alongside the main stream of the Ouse. The main building in the *castra* was the Commander's headquarters (*praetorium*) directly above which the first minster church was built.

The *castra* initially provided accommodation for the 5,000 or so legionaries of the IX Hispana Legion and the combination of military base and civilian market town which grew up along the main approach road from the south, across the Ouse, is typical of the process of Romano-British urban settlement described generally in Chapter 3. Eburacum was the scene of the Imperial Court of the Emperor Severus in 208–211 (where he died) and it was officially recognized as a *colonia* and capital of Britannia Inferior in about 213. Eburacum was rebuilt in *c.* 300 after destruction in an uprising and in 306 Constantine was proclaimed Emperor there, in succession to Constantius.

Figure 4.30 – York, the central section of the Tallis map of about 1850, showing the circuit of medieval walls beyond which at that late date there was still only the beginnings of modern suburban growth. Within the walls the old street pattern was little changed and the four main gateways of Norman times – Micklegate, Monk, Walgate and Bootham Bars – controlled access to city then, as indeed they still largely do today.

The main Norman Castle was at the confluence of the Ouse and its minor tributary the Foss; the motte was later strengthened by the stone-built Clifford's Tower, and the bailey court was enclosed by the Assize Courts and County Gaols. The second, minor castle was directly across the Ouse and survives only as the rudimentary Baile Hill. The famous Shambles – the city's best surviving example of medieval street-scape – is the unnamed lane between the Pavement and the northern end of Church Street (King's Square, where the Danes had their palace).

commercial and industrial growth and prosperity, the city's subsequent loss in terms of population, extent and superficial Victorian grandeur is history's gain. York alone of medieval towns retains not only the full extent of her medieval walls, but also much of the close-grained, characteristically small-scale complexity of medieval urban form.

The city was founded by the Romans in AD 71 as the strategic legionary fortress base for their conquest of Celtic Brigantia. The location was at the limit of tidal water on the River Ouse, where the *castra* commanded both the main north–south land route and also access to the important inland waterways of the Ouse and its tributaries. Figure 4.29 gives the location of the *castra* related to the modern streets.

After the Roman withdrawal Eburacum was deserted and largely destroyed but the walls were still in good repair when the Angles took possession around 560. As re-established, the original gridiron plan was replaced by a typical organic growth street pattern, although a remnant of the primary cross-axes can be detected in the Tallis map (Figure 4.30), The first church on the site of the famous Minster was erected in 627. By then known as Eoforwic, the town was captured by the Danes in 876 and their royal palace was built at the northern end of Church Street. In turn renamed Jorwik, the town prospered under the Danes until taken by the English under King Athelstan in 926. The Normans confirmed York as their northern capital, after initial local rebellion, and built two motte and bailey castles to ensure their dominance. They also rebuilt the circuit of stone walls. The Domesday population of about 4,150 and fourth ranking, had increased to almost 11,000 by 1377 and second place after London with 30,000. Not taking part in the early Industrial Revolution, York had a population of only 16,846 by 1801 and 36,000 in 1851, by which date the railways were bringing belated suburban expansion.

Diary, Volume One, 31 August 1654 (1907 edition).
68. For York see P Nuttgens, *York: The Continuing City*, 1976; *Rock's Views of York: 1852–1861*, Leeds, 1971; Defoe, *A Tour Through the Whole Island of Great Britain*, 1724.

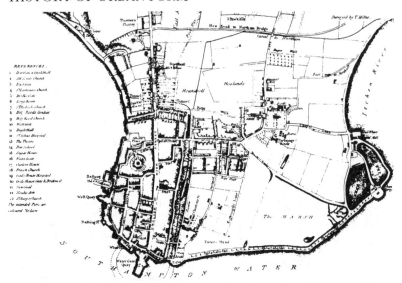

REFERENCES.
1 Bar·in & Guildhall
2 All Saints Church
3 East Gate
4 St Laurence Church
5 Bridle Well
6 Long River
7 St Michaels Church
8 Hol. Roode Conduit
9 Holy Rood Church
10 West Gate
11 Bayly Hall
12 St Julians Hospital
13 The Theatre
14 Free School
15 Square House
16 Water Gate
17 Custom House
18 Gods House Hospital
19 Gods House Gate & Bridewell
20 Nunnerie
21 Noahs Ark
22 St Marys Church
The extended Part are coloured Yellow

SOUTHAMPTON

The medieval nucleus of the future city was the third (possibly even the fourth) riverside settlement to be founded at the head of Southampton Water.[69] The known predecessors were Roman Clausentum and Saxon Hamwich (or Hantune), on the eastern and western banks, respect-

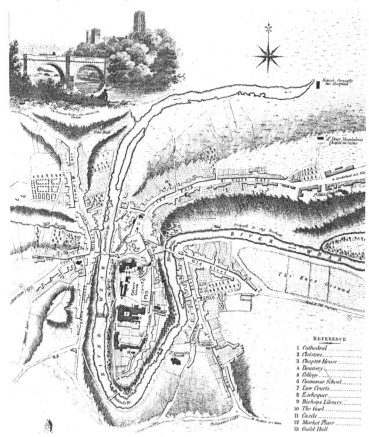

REFERENCE
1 Cathedral
2 Cloisters
3 Chapter House
4 Deanery
5 College
6 Grammar School
7 Law Courts
8 Exchequer
9 Bishops Library
10 The Gaol
11 Castle
12 Market Place
13 Guild Hall

Figure 4.31 – Southampton in 1791: still essentially contained within the circuit of the medieval walls but with future nineteenth-century growth presaged by 'ribbon' development of the main London road leading up to the northern Bargate, and more solidly established ribbon and infill to the east between the town gate and the mother church of St Mary, originally founded in the otherwise long since deserted Saxon town of Hamwic.

The Normans consolidated the port as a main link with France, controlled by an impressive castle, and medieval prosperity followed. In 1791 the town was enjoying transient popularity as a seaside resort (note the western shore bathing houses) shortly to be lost to Bournemouth and elsewhere, as the port activities rapidly increased.

Figure 4.32 – Southampton: the locations of the earlier settlements related to the medieval nucleus, and showing also how the coastline has been radically changed by the construction of the nineteenth- and twentieth-century docks. The Old Docks to the south, extending the form of the peninsular, were created between 1842 (following the arrival of the railway from London in 1841), and 1890–1911. The New Docks to the west, alongside the River Test, were built in the 1920s and early 1930s together with an extensive land-fill programme.

Figure 4.33 – Durham: centred within an acute bend of the River Wear, is perhaps the most dramatically situated historic city in England. As an example of organic growth in its most romantically intriguing forms the city is rivalled by few other such places in Europe (one of which, Cesky Krumlov, in South Bohemia Czechoslovakia, is described in page 151). The British Atlas map of 1804, from which the illustration is extracted, shows the practically unchanged medieval nucleus of the city not only as it was earlier, but also as it has been retained through to the present day.

It is likely that the defensive potential of the 70 ft high rock above the river, and command of an important crossing, proved attractive to Celtic or Romano-British settlers, before Bishop Ealdune founded a church there in AD 995 and transferred his see from Lindisfarne. Known at first as Dunholme, a small walled town grew up to the north and east of the cathedral protected by the precipitous river valley on all but the narrow northern side. The Normans confirmed the city's religious status, rebuilding the cathedral as one of the finest examples of their architecture. They also constructed an equally impressive castle at the northern end where it controlled the bridges and dominated the emergent northern suburb with its important market place. For Durham see N Pevsner, *The Buildings of England: Durham*, 2nd edn revision, E Williamson.

69. For Southampton see C Platt, *Mediaeval Southampton: The Port and Trading Community AD 1000–1600*, 1973; A Temple Patterson, *History of Southampton 1700–1914*, 1966; A B Granville, *The Spas of England, Volume 2*, 1841 (1971 edition).

ively, of the River Itchen, with a possible, later Saxon forerunner across the peninsular alongside the western River Test (Figure 4.31). Before the Norman Conquest, however, Southampton – already a royal borough by charter of 962 – was firmly established in the south-western corner of the peninsular.

Figure 4.34 – Chester: John Speed's map of 1611 showing the city in the bend of the river with the Norman castle and bridge at the southern end. Although a pictorial map this is an accurate representation of the early seventeenth-century city with organic growth suburban ribbon development extending outwards from the main gates, and with the beginnings of a bridgehead across the River Dee.

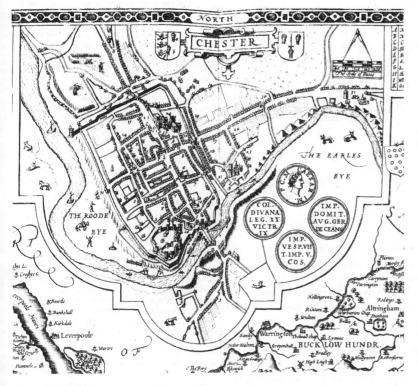

CHESTER

Chester was founded by the Romans in about AD 60 as a strategically vital legionary fort at the head of the estuary of the River Dee.[70] (The Roman name was Deva – the camp on the Dee.) It was an important harbour and served as the northern base for the conquest of Wales. Deva covered approximately 59 acres within a rectangular perimeter of north–south sides of 1,930 feet. The usual civil market town established itself on either side of the main road to York outside the fort's Eastgate. A permanent stone wall with towers at about 200 feet intervals was constructed at the end of the first century and, as repaired by the Saxons in 905, it subsequently also formed the defences of the later medieval city of Chester (which name derived from the Saxon *ceastre*, or camp). Possibly because of the permanence of the Roman masonry of buildings fronting Deva's main grid streets, Chester has retained the original Roman primary cross-axes. The Normans built a major castle and a first bridge over the Dee and granted important commercial privileges which were consolidated during the medieval period, from which period date the Rows, the city's unique system of covered walkways fronting shops at first-floor level.

As I am now at Chester, 'tis proper to say something of it, being a city well worth describing: Chester has four things very remarkable in it. 1. Its walls, which are very firm, beautiful, and in good repair. 2. The castle, which is also kept up, and has a garrison always in it. 3. The cathedral. 4. The River Dee, and 5. the bridge over it.

It is a very ancient city, and to this day, the buildings are very old; nor do the Rows as they call them, add anything, in my opinion, to the beauty of the city; but just the contrary, they serve to make the city look both old and ugly. These Rows are certain long galleries, up one pair of stairs, which run along the side of the streets, before all the houses, though joined to them, and as is pretended, they are to keep the people dry in walking along. This they do indeed effectually, but then they take away all the view of the house from the street, nor can a stranger, that was to ride through Chester, see any shops in the city; besides, they make the shops themselves dark, and the way in them is dark, dirty, and uneven.

The best ornament of the city, is, that the streets are very broad and fair, and run through the whole city in straight lines, crossing in the middle of the city, as at Chichester. The walls as I have said, are in very good repair, and it is a very pleasant walk round the city, upon the walls, and within the battlements, from whence you may see the country round; and particularly on the side of the Roodee, which is a fine large low green, on the bank of the Dee.

There are 11 parishes in this city, and very good churches to them, and it is the largest city in all this side of England that is so remote from London. When I was formerly at this city about the year 1690, they had no water to supply their ordinary occasions, but what was carried from the River Dee upon horses, in great leather vessels, like a pair of bakers' paniers; just the very same for shape and use, as they have to this day in the streets of Constantinople, and at Belgrade, in Hungary; to carry water about the streets to sell, for the people to drink. But at my coming there this time, I found a very good water-house in the river, and the city plentifully supplied by pipes, just as London is from the Thames; though some parts of Chester stands very high from the river.

This county, however remote from London, is one of those which contributes most to its support, as well as to several other parts of England, and that is by its excellent cheese, which they make here in such quantities, and so exceeding good, that as I am told from very good authority, the city of London only take off 14000 ton every year; besides 8000 ton which they say goes every year down the Rivers Severn and Trent, the former to Bristol, and the latter to York including all the towns on both these large rivers.

(D Defoe, A Tour through the Whole Island of Great Britain. 1724–26. (Letter 7.))

70. For Chester see B Harris, *Chester*, 1979; B C A Windle, *Chester: An Historical and Topographical Account of the City*, 1903.

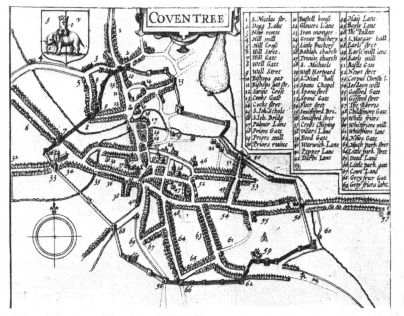

The list of labels in the Coventry map reads:

1 S.Nicolas ftr.	24 Baftell houfe	47 Haie Lane
2 Digg Lane	25 Gloucrs Lane	48 Bayly Lane
3 New rents	26 Iron monger	49 The Pallace
4 Hill mill	27 Great Buchery	50 S.Margar hall
5 Hill Crofs	28 Little Buchery	51 Earls ftret
6 Hill Street	29 Bablak church	52 Earle mill lane
7 Hill Gate	30 Trinite church	53 Earle mill
8 Well Gate	31 S. Michaels	54 Bafile Gate
9 Well Stret	32 Weft Hortyard	55 Corpus Chrifti l.
10 Bifhops gat	33 S.Nicol. hall	56 Iordayn will
11 Bufhops gat ftr.	34 Spone Chapel	57 Gofford Gate
12 Swine Crofs	35 Sponeftret	58 Gofford ftret
13 Cooke Gate	36 Spone Gate	59 The Barres
14 Cooke ftret	37 Flect ftret	60 Childmore Gate
15 S.Ioh.Schole	38 Smithford Brs.	61 White friers
16 S.Ioh.Bridg	39 Smitford ftret	62 Whitfriers mill
17 Priors Gate	40 Viilers Lane	63 Whitfriers Lane
18 Priors mill	41 Brod Gate	64 New Gate
19 Priors ruine	42 Warwick Lane	65 Much park ftret
	43 Peppor Lane	66 Little park. ftret
	44 Darbi Lane	67 Dead Lane
		68 Little park gate
		69 Cowt Lane
		70 Grey frier Gat
		71 Grey friers lane

Figure 4.35 – Coventry: John Speed's map of 1611. Within the period of this volume, Coventry was not an exceptionally important town. It was neither particularly ancient (the earliest definite dating is that of the 1043 founding of a Benedictine monastery by the Earl of Mercia and his wife, Lady Godiva); nor was it of specially significant form, although its unplanned 1611 street pattern epitomizes the organic growth type of town. It was a sizeable town, numbering some 5,000 adult inhabitants in 1377 but the total population in 1520 was no more than about 6,600.

The city wall was an excellent example of a trade protection barrier, constructed between 1355 and 1400 when risk of military plundering was negligible but fear of unauthorized commercial infiltration was ever-present.

The importance of Coventry comes later and derives from its strength as a major midlands' industrial manufacturing centre. This came about first and paradoxically, because its entrenched craft guilds kept out the new mechanical technology which gave Birmingham, till then of little importance, its nineteenth-century dominance; and second, during the Second World War, when its own belated industrial prominence attracted Nazi bombing which destroyed much of the area of the Speed map, thereby affording the city the opportunity to replan its central area, with results which must await a further volume. See Coventry in M D Lobel (ed.) *The British Atlas of Historic Towns, Volume Two*, 1975; K Richardson, *Coventry: Past into Present*, 1987.

CARLISLE

When the Romans first campaigned through the region in AD 71–74 the site of the future city, on high ground commanding the lowest crossing of the River Eden, was occupied in part by a tribal settlement of the Cymry.[71] As their key base in the north-west, the Romans constructed a fortress immediately across the Eden known as Stanwix, the name of a present-day suburb of the city. The Celtic village was transformed into the walled Roman town of Luguvallium which initially served the fortress but which subsequently developed into a prosperous market place with considerable trade northwards through Hadrian's Wall (see Figure 3.55). The Normans captured Carlisle from the Scots in 1092 from which time through into the seventeenth century it remained a much contested frontier town.

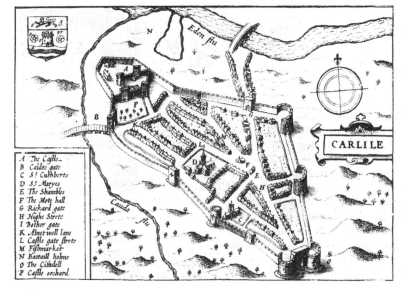

A The Caftle
B Caldoe gate
C S.t Cuthberts
D S.t Maryes
E The Shambles
F The Mote hall
G Rickard gate
H Highe Strete
I Bother gate
K Abnet well lane
L Caftle gate ftrete
M Fifhmarket
N Baftaill holme
O The Citadell
P Caftle orchard

I return to Carlisle: the city is strong, but small, the buildings old, but the streets fair; the great church is a venerable old pile, it seems to have been built at twice, or, as it were, rebuilt, the upper part being much more modern than the lower. King Henry VIII fortified this city against the Scots, and built an additional castle to it on the east side; there is indeed another castle on the west, part of the town rounds the sea, as the wall rounds the whole, is very firm and strong. But Carlisle is strong by situation, being almost surrounded with rivers.

Here is a bridge over the Eden, which soon lets you into Scotland; for the limits are not above eight miles off, or there-about. The south part of Scotland on this side, coming at least fifty miles farther into England, than at Berwick. There is not a great deal of trade here either by sea or land, it being a mere frontier. On the other side the Eden we saw the Picts Wall, of which I have spoken already, and some remains of it are to be seen farther west.

(D Defoe, *A Tour through the Whole Island of Great Britain*, 1724–26. (Letter 10.))

Figure 4.36 – Carlisle: John Speed's map of 1611, the castle in the north-west corner dominating the river Eden, and the citadel in the south-east controlling the main road approach from the south. There is suburban ribbon development alongside the road to Scotland across the bridge. The Roman grid (if one existed) has disappeared and the main form determinant is the distinctive 'Y'-shaped main street pattern.

71. For Carlisle see E A Gutkind, *International History of City Development*, Volume VI, *Urban Development in Western Europe: The Netherlands and Great Britain*, 1971.

New towns

The other category of medieval town – those settlements founded at a given moment in time with instant urban status – is divided into two sub-types: first, the 'bastides' which, as defined below, were built according to a predetermined plan; second, the various forms of 'planted towns', only a proportion of which were planned. The term 'bastide' has been misapplied by some urban historians to give this generic status to all the medieval new towns. Such a misuse of a valid and valuable term tends either to imply that all the new medieval towns, in all European countries, were bastides, or results in the completely erroneous impression that the *only* medieval new towns are those, in a limited number of countries, which are recognized as being bastides. Professor Beresford employs 'bastide' with reference to foundations in France, and 'new town' or 'planted town' for the foundations in England and Wales. He notes that in France, 'bastida' in its Latin form occurs in almost every foundation charter, whereas in England the word was not employed.[72]

Limitation of 'bastide' to France seems unduly restricted, bearing in mind its traditional application to examples in England and Wales and the foundation by Edward I of towns in all three countries. A preferred definition is that bastides are the planned new towns of the thirteenth century, built in France – predominantly in the south-west of the country – by the French generally and by Edward I; and also those foundations of Edward I in England and Wales. With all these examples the term 'planned' is employed in its complete sense of the foundation, at a given moment in time, of a new settlement with full urban status and with a predetermined town plan.

'Planted town' is accepted as a term for all other medieval new towns, either with or without a predetermined plan. (The thirteenth century foundations of, for example, the Teutonic knights in the eastern part of Germany, cannot therefore, still be strictly classified as bastides.)[73]

THE BASTIDES

The French, English and Welsh bastides have much in common; nevertheless there are several significant differences in form and function which establish distinct national characteristics. The three main principles followed in the planning of all bastides were first that they were new urban foundations started with predetermined plan forms; second that the gridiron system of rectilinear plot subdivision formed the basis of their layout; and lastly that the main inducement to settle in a bastide was the grant of a house plot within the town together with farming land in the vicinity and other economic privileges.

These principles must be qualified, however. Not all bastides were built on new sites; many were based on existing village settlements redeveloped along planned lines. Although a gridiron of streets, forming rectangular house plots, is present by definition in all bastides, there is no standard plan and in many examples the grid, often considerably distorted, is used only for a part of the town.[74] As with the majority of applications in preceding historical periods it is a grid of convenience – the quickest and most equitable way of laying out a town on a new site. The granting of farming land outside the town was essential in that in common with all early medieval towns, even the

During the Middle Ages, understood here as being the period from the ninth to the fifteenth century – in other words from the beginning of the Romanesque to the end of the Gothic style – the concept of a town was so entirely different from the Greek idea of a polis or the Roman concept of an urbs that the problem of town and square must be approached from new angles, both sociologically and visually.

Except for the bastides in France and England and the foundations of the Teutonic Knights, the organisation of a town as a whole was neither understood nor desired by the builders of the Middle Ages. Even in those medieval towns which were of Roman origin, any changes or additions were interpolated without reference to the ancient general plan.
(P Zucker, Town and Square)

One of the most interesting aspects of the urban geography of mediaeval France is the creation of bastides in the south and particularly in the south-west. The recent study of Dr Deffontaines ('Les Hommes et leurs Travaux ... Moyenne Garonne') of the historical geography of the middle Garonne region, where the building of the bastides in the thirteenth and fourteenth centuries was most actively undertaken, illustrates effectively this phase of French urban development. The middle Garonne region was a frontier district which included lands of the French king, of his vassal the Count of Toulouse and of the English king in Aquitaine, and it had suffered much depopulation in the course of Anglo-French wars and in the Albigensian religious war. The bastides were usually (though not invariably) newly created settlements and, what is more, as in the case of Roman cities, they were laid out according to a definite geometric plan. The plans were very often rectangular, but other shapes were adopted in conformity with the topography of the site, where, for example, the town was strung out along a valley or seated on the top of a spur of a plateau. Many of the bastides marked the advance of agriculture at the expense of the forest, as was the case in the molasse country south of the middle Garonne.
(W G East, An Historical Geography of Europe)

72. Beresford, *New Towns of the Middle Ages*, is the most important reference on this topic; but see also Vance, *The Continuing City*; and F Divorne, B Gendre, B Lavergne and P Panerai, *Les Bastides d'Aquitaine, du Bas Languedoc et du Bearn*, 1985.

73. Salisbury (New Sarum), an English ecclesiastical foundation of 1220, widely referred to as a 'bastide', is the main British misapplication of the term. Whatever the terminology, it is stressed that the European medieval new towns embodied broadly the same urban characteristics.

74. An essential characteristic of a bastide gridiron plan is that it was an urban land allocation document, drawn up on a basis of predominantly rectangular, finite individual urban building plots, rather than, by way of contrast, serving as a larger scale network of main streets dividing an urban area into residential districts that were distributed, by separate agreement, between a number of families. The latter method, described in Chapter 1, would now appear to have been the custom of Egyptian and Harappan city planners.

largest, bastides were primarily agricultural communities with only a small proportion of their inhabitants exclusively involved in non-farming activities.

The great majority of bastides were built by the royal central authority, either to impose itself over dissident parts of its territory, or to extend its domain. It generally paid for and organized the town defences, present from the start in most bastides, and controlled the layout of the town. New tenants generally held their land direct from the Crown: in return for the valued status of freeman and other privileges they were usually committed to some form of part-time militia service.[75]

Apart from its importance as a fortified militia garrison, the French bastide was also a source of primary agricultural production and the local market centre for trading. The Welsh bastides, on the other hand, were initially intended as impregnable bases for regular army garrisons and had only subsidiary trading functions. Despite their military function however, the French bastides and the two English examples, Winchelsea and Kingston upon Hull, were planned without any form of inner citadel.

As well as a contrast in the roles played by the French and Welsh bastides, there was also a difference in the ways in which they were populated. Whereas the Welsh bastides contained a proportion of English immigrant families, brought in as part of a settlement policy, in the case of the French bastides, as C T Smith suggests: 'they often involved no more than a regrouping of population from hamlets to nucleated villages and small towns'.[76]

With few exceptions bastides were fortified, but there is a marked contrast between the formidable castle towns of Edward I in North Wales, which could and did successfully resist organized military assault, and the town walls of most French examples, which gave their inhabitants little security other than from small-scale localized attack. Fortifying a gridiron town created planning problems around the perimeter. Combining the rectilinear street system with a circular wall – which enclosed maximum area for a given length and in addition offered the best means of defence against medieval assault tactics – created odd-shaped plots around the perimeter. Economic and military considerations frequently prevailed and a number of French examples have more or less circular walls – notably Monflanquin (Figure 4.41) and Sauveterre-de-Guyenne. In east Germany, where strong fortifications were essential, the planted towns were normally enclosed within circular walls. New Brandenburg (Figure 4.69) is a good example. Czeske Budejovice, further south in Bohemia (Figure 4.93) also follows this pattern. Other French examples, however, were given rectilinear walls, including Aigues-Mortes and Monpazier (Figures 4.39 and 4.44). In England the two bastides are compromise solutions with only their corners rounded off. The later Welsh bastides, as the most advanced examples of military engineering of the time, had the emphasis placed on their massive castles, but the civil towns were also well protected within similar compromise defences.[77] Neither English nor French bastides were given castles, although the latter sometimes had strongly constructed churches which served as emergency citadels.

THE BASTIDES: FRANCE

In the early years of the thirteenth century, partly on the pretext of a

Figure 4.37 – South-western France: showing the location of Aigues-Mortes and Carcassonne related to major towns and to a group of bastides of both English and French origin, situated between the Dordogne and Lot rivers, within a circle of 25 km radius centred on Villereal.

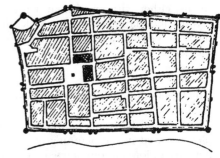

Figure 4.38 – Aigues-Mortes: general plan (north at top), alignment approximately as in Figure 4.39.

75. Land grants in return for military service made excellent economic sense: the king was spared the heavy expenses of permanent, regular garrison forces and yet, for a very reasonable capital investment, maintained an adequate local military presence.
76. Smith, *An Historical Geography of Western Europe before 1800.*
77. See the various excellent illustrated guides prepared by the (late) Ministry of Public Building and Works – now part of the Department of the Environment – and obtainable from HM Stationery Office, London, and at other addresses. Notably, Beaumaris Castle; Harlech Castle; Caernarvon Castle; Conway Castle; Rhuddlan Castle and Flint Castle.

religious crusade against the Albigensian heretics, the Languedoc region was conquered by the northern rulers after a bloody struggle which left it a land depopulated and exhausted by war, rich in resources, and sullenly hostile to its conquerors, ready for the victor to work his will on.[78] Like Edward I, later in the same century, the new rulers of the south relied on new settlements as a key factor in establishing their authority, building some hundreds of them strewn so thickly over the map that only a small proportion became real towns. St Louis, King of France, with his brother Alfonse, Count of Poitiers and son-in-law and successor to the last native Count of Toulouse, were the

Figure 4.39 – Aigues-Mortes: aerial plan view. The Mediterranean is south of the town, separated from it by salt marsh and industrial salt-pans, across which runs the canal, on the left of the picture, giving access for small coastal vessels. The original coastline runs diagonally from left to bottom right, marked by the edge of the vineyard field pattern. The field/vineyard boundaries and the regional routes are another characteristic example of the pre-urban cadastre.

78. T F Tout, *Mediaeval Town Planning: A Lecture*, 1917 – an elderly reference but still with background value.

most active bastide builders and if the great king's bastides were the more enduring and important those of Alfonse were by far the more numerous.[79]

Several general principles were followed in the design of the French bastides, which made them differ considerably from their Welsh counterparts. As far as possible a square or rectangular area, enclosed by a defensive wall, was laid out with equal-sized house-building plots. But although this was the standard approach there were bastides with eccentric outlines, as for example Sauveterre-de-Guyenne, shaped almost like a pear. The bastides were invariably protected by a wall and a ditch but, in direct contrast to the Welsh practice, only in a very few instances by a castle. Generally only the site and its defences were provided by the founder; tenants were responsible individually for their houses, and collectively for the town hall and the church – the two main public buildings in the town. The town hall often took the form of a two-storeyed building, with the ground floor used as covered market accommodation: it stood in the main square of the town, at the meeting point of the main streets leading from the entrance gates. Alongside the square the streets were usually within arcades formed beneath the first floor of the buildings.[80] The church, often strongly built so that it could function as an emergency inner citadel, was generally located in a separate but frequently adjoining square. Within the town each settler received a house plot, although there is evidence that important people often had several allotments assigned to them. Each house plot carried with it the condition that it should be built on within a stipulated period and that the house was to cover the entire street frontage. However, it would be wrong to consider this as an aesthetic requirement. Rather, it seemed to stem from considerations of defence.

AIGUES-MORTES

As a result of the Albigensian crusade the territory of the king of France extended to the Mediterranean for the first time. Louis IX (Saint Louis) could then build himself a base for his first crusade to the Holy Land and in 1240 work was started on the new town and port of Aigues-Mortes. Progress by 1248 was such that it was able to serve as the gathering point for the Seventh Crusade, led by Louis IX. In 1272 Phillipe le Hardi brought Simone Boccanegra from Genoa to add the massive defensive wall around the town. This wall, which still exists in its original form,[81] is about 35 feet high and strengthened by fifteen engaged towers; it encloses an area some 650 yards long by 300 yards wide. Inside the wall the town is laid out to a slightly distorted gridiron. The main town square is located about one-third of the way across from the western side; around it are grouped the main public buildings (Figure 4.38). At the north-west corner of the town the Tour de Constance serves as an inner citadel, with a 'lighthouse' beacon on its roof. Aigues-Mortes suffered a similar fate to Winchelsea (see below): before the end of the fourteenth century the channel linking the port to the Mediterranean silted up, leaving it marooned inland with three miles of salt marsh between it and the sea.

CARCASSONNE

There are two towns of Carcassonne – the old original *cité* on its hilltop, east of the River Aude, and the *ville-basse* on the lower ground across the river. Despite being heavily restored by Viollet-le-Duc and others dur-

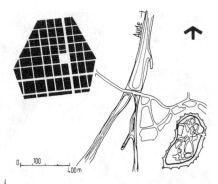

Figure 4.40 – Carcassonne: the bastide *ville-basse* related to the original hilltop *cité*, on the other side of the Aude. Conveniently side by side across the river, the twin nucleii of modern Carcassonne exemplify contrasts between organic growth and planned urban form. However, while the hilltop site is still largely in an edge-of-town location on the south and east, the bastide *ville-basse* is now at the centre of an extensive urban area. Pleasingly it has retained much of its history in a natural way, compared to its self-conscious parent.

Figure 4.41 – Monflanquin: detail plan, giving modern plot boundaries within the area of the bastide and showing the surrounding suburban development.

79. Tout, *Mediaeval Town Planning*.

80. Continuous street frontage development is also a characteristic of traditional Islamic urban form, with its self-policing benefits, for which see Chapter 11.

81. The wall and town of Aigues-Mortes, in happy contrast to Carcassonne's, have not suffered obvious restoration and, as yet, have not been unduly spoiled by the considerable numbers of tourists it attracts.

I wrote not 'unduly spoiled' when visiting in the early 1970s. Both 'unduly' and 'spoiled' are relative terms; somewhere that has been spoiled for me as a result of pressure of visitor numbers may well have been turned into an attraction for others – possibly, there's no way of knowing, the considerable majority? However, if it is the past that is the reason for visiting, best do so out of season: an observation that holds true for Aigues-Mortes, just as for Monpazier and Eymet, which were revisited in the summer of 1991. Which makes it even more pleasing to be able to note that there are still bastide towns, such as Villereal, Castillonnes and Monsegur, which are relatively unspoiled.

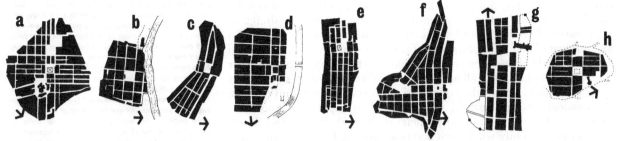

ing the last century, the *cité* exists as a unique example of a medieval fortified town complete with two rows of ramparts, church and castle. The preservation of the *cité* is indirectly due to the action of Louis IX in expelling its inhabitants in 1240 and subsequently permitting them to found the new town across the Aude. From then on the *cité* was used mainly for military purposes, while the *ville-basse*, surviving destruction by the English in 1355, steadily grew in size and importance as the chief town of the department of the Aude – one of very few bastides to have flourished to such an extent.

MONPAZIER

Monpazier is situated on the northern slopes of the valley of the upper Dropt.[82] It was started in 1284 as a part of the system of defended towns which Edward I established to protect his territory in Gascony against French attack from the east, and to consolidate his authority over the district.[83] Edward had been resident in Gascony since 1279, reorganizing his administration there, and in 1284 negotiations were completed for the site of Monpazier, the landowner, the Duc de Biron, turning it over in return for an equal share in all revenues. Having selected the location Edward next wrote back to England asking for four men who knew how to arrange and order a new town 'in the manner most beneficial to us and the merchants'. It is not recorded how the request was answered, but the resulting layout is a classic of bastide planning in its regularity and relationships between the elements of the plan.

The town was laid out to the module of the standard house plot; these each had a frontage of 24 feet on to the 20 foot wide main streets, and a depth of 72 feet. Lanes 6 feet wide separated the plots at the rear. These house plots were grouped into twenty blocks, one of which was used for the market place, with part of a second (linked through at the corners) containing the fortified church. The market square contained the town hall and the well. Three streets ran through the town along the length of the ridge site, two of them being interrupted by the square; four further streets ran across the town, two of them forming the other sides of the arcaded central square. These streets passed through the ten gates in the wall. The latter had been completed by 1290; the church was added in the fourteenth century.

Each settler was given a house plot in the town, an 'allotment' close to the walls, and farmland in the vicinity. The house had to be completed within two years, to a building line along the main streets. A ten-inch gap between adjoining houses formed a fire-stop and contained the open sewer over which the latrines were corbelled. The sewers were taken down the ridge to the allotments.

In addition to these three French bastides, the plans of eight others are given in Figure 4.42. A main reason why, until recently, the French

Figure 4.42 – Eight French bastide plans illustrating their essential 'variations-on-a-theme' character. Key: a, Villeréal; b, Lalinde; c, Castillonnes; d, Eymet; e, Villefranche-du-Périgord; f, Domme; g, Beaumont; h, Monflanquin.

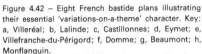

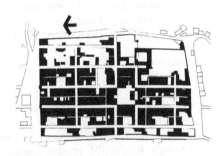

Figure 4.43 – Monpazier: present-day plan. There is reason to believe that the south-east quarter was never fully developed.

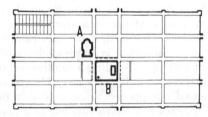

Figure 4.44 – Monpazier: stylized plan based on over-regularized versions of the actual layout, as appearing in other urban histories. Key: A, church; B, market square with market hall and pump. The street-frontage outward-oriented residential gridiron planning (at top left corner) is the antithesis of Islamic urban form, for which see Figure 11.32, chapter 11 generally, and also the aerial photograph of Erbil in Figure 1.10.

82. See Vance, *The Continuing City*, for further illustrated description of Monpazier.
83. Twenty years on, the French bastides are variously in the tourist guides; none more so than Eymet on the Dordogne with enough British settlers to support a cricket club. However, revisited in the summer of 1991 a group of bastides centred on Monsegur proved to be still comparatively off the beaten track.

bastides were largely ignored in the guide books was the misconception that because of their planned origins they were monotonously similar in appearance. Professor Tout has much to answer for in this respect. In his authoritative if dated *Mediaeval Town Planning* he wrote, 'when you have sampled half-a-dozen or so, you have no real need to pursue your travels any further, since all are very much alike'. In reality one of their main attractions is that they offer variations on a theme, and (to take the musical analogy further) it is essential to experience numerous variations to understand fully their individual forms and inter-relationships.[84] Only when each one is appreciated in the context of a number of others can its essential *individuality* be understood. Quite apart from the rebuilding in the last 700 years, which has increased the superficial differences between each bastide, the limited vocabulary of planning components was, in any case, adapted to each site in such a way that individuality of form was established from the outset. Time has enhanced the variations, with many sympathetic relationships of medieval and minor Renaissance vernacular buildings.

THE BASTIDES: ENGLAND

Only two bastides were built in England – Winchelsea and Kingston upon Hull. Both were replacement harbours and present a fascinating difference in urban fortune – decline back to village status at Winchelsea when its *raison d'être* was removed and economic prosperity at Hull, as a major east coast port and industrial centre.

Although Edward I is also known to have been involved in the planning or rebuilding of two other places which might qualify as bastides, evidence concerning his precise involvement at either is scanty. In their fascinating *Mediaeval England: An Aerial Survey* Beresford and St Joseph describe proposals by Edward I for the foundation of Newton on the shores of Poole Harbour in Dorset, but there is no physical evidence of a start having been made. The second place, Berwick-on-Tweed, came into Edward's possession at the end of the century and there is ample evidence of the procedure he adopted for its rebuilding – including a 'conference' of town planners and others held at Harwich in January 1297. However, it is not known for certain what actually happened. Most probably Berwick, which was chartered in 1302, resembled Caernarvon as a castle town intended to control the key river crossing over the Tweed.

The more important period in Berwick's history is the refortification of the town in the mid-sixteenth century (an account of which is given in Chapter 8) related to comparable works at Portsmouth, Hull (see also below in this chapter) and Plymouth. As bastides, neither Winchelsea nor Hull was strongly fortified: the former suffered accordingly from French assaults and eventually fell victim, paradoxically, to the sea's retreat.

WINCHELSEA

Old Winchelsea, one of the Cinque Ports and a key defence station against French coastal incursions, had been seriously threatened by the sea since the beginning of the thirteenth century.[85] It was located precariously on a crumbling low cliff line on the east bank of the estuary of the Brede and suffered extensive storm damage in 1244 before succumbing, inevitably, to inundation in 1287. Edward I had foreseen this disaster and planned accordingly. In 1280 he directed his steward to

Since his boyhood Edward I had been familiar with such town-planning as there was in his time in western Europe. The royal hunting-lodge or palace of Clarendon overlooked the new town of Salisbury, which had been laid out around the new cathedral and in 1227 was given the liberties of Winchester. Edward must have been there often. Later, as lord of the town and castle of Bristol, he would have been aware of the engineering feats done there between 1240 and 1250, when the Avon was diverted into its new cut to provide for town extension and Redcliffe and the manor of the Templars were enclosed by wall and ditch ... As Duke of Aquitaine he knew all about the extensive foundations of bastides in Languedoc by Saint Louis, Alphonse of Poitiers, and other lords, and had encouraged the extension of the movement to Gascony in the time of his seneschal John de Grilly.
(T F Tout, Mediaeval Town Planning)

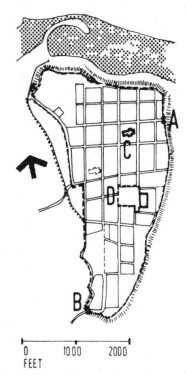

0   1000   2000
FEET

Figure 4.45 – Winchelsea as originally laid out. The river Brede, with its quayside providing a total of eighty-one harbour plots, was at the bottom of the cliff at the northern end of the town. The long eastern side facing the English Channel was similarly protected by the cliff line. Key: A, the Strand Gate (still existing); B, the South (New) Gate, also still in existence but seemingly isolated from any urban function across a lane in the middle of fields; C, St Thomas's Church; D, the market place, long since disappeared.

84. For Edward I, known as the 'Town-builder', see M W Beresford, *New Towns of the Middle Ages*, 1967.
85. For Winchelsea, which lacks definitive individual description, see Beresford, *New Towns of the Middle Ages*; A E J Morris, 'Cities Built for a King', *Geographical Magazine*, November 1975.

Figure 4.46 – Winchelsea: the aerial view from the south showing the surviving fragment of the old town, with agricultural land taking the place of the once busy harbour at the top of the photograph.

obtain, by purchase or exchange, land at Iham suitable for the new town.[86] The chosen location was a wooden plateau some 3 miles north-west of Old Winchelsea, above the level of the marshes, with the River Brede forming two sides of the site. In the following year Edward nominated his three commissioners to develop the new town for 'the barons and good men of Winchelsea'. They were Stephen of Penshurst, a warden of the Cinque Ports; Henry le Waleys, a London merchant and sometime mayor of both London and Bordeaux; and Itier of Angoulême, an experienced Gascon bastide planner. Penshurst represented local political interest, le Waleys those of commerce, in particular the Bordeaux wine trade for which Winchelsea was the main port of entry. Itier of Angoulême was appointed to act as their professional adviser.

Progress was virtually non-existent at first. Prior to Old Winchelsea's final destruction the merchants were reluctant to leave, even after the king had begun to lay out the new site to receive them.

In 1284 Gregory of Rokesley, Mayor of London, was added to a reconstituted commission which was to plan and give directions for the necessary streets and lanes, for places suitable for a market, and for two churches to be dedicated to St Thomas of Canterbury and St Giles, the patron saints of the two parishes in the old town.[87] In addition, building plots were to be assigned according to the requirements of the population of the old town. Still they refused to move, both from natural reluctance to break with their past and because of their misgivings at the absolute control which Edward I would have over the new town. Edward at this time was in France; growing impatient at all this delay in implementing his desires, he instructed John Kirkby, Bishop of Ely, and his treasurer, to take personal charge of the work. At around the

Described in various charters and records as the 'Ancient Towns' Rye and Winchelsea acquired their status as quasihead ports in the Charter of 1278, which was unusual as embracing the five ports and the two 'ancient' towns jointly. At that time they fulfilled a role which made them important members of the Confederation, but their history goes much further back. It goes back to the days when the Romans made use of the forest ridge running through Battle to Netherfield and Uckfield as the most direct route for the movement of iron recovered at Battle and Sedlescombe.

Queen Elizabeth visited Rye in 1573 and named the town Rye Royal. If this gives Rye a special status, Winchelsea can claim to have witnessed the birth of Methodism for in 1773 John Wesley came to Winchelsea and opened a little chapel in which, today, can be found the pulpit from which he preached. It was to Winchelsea that he returned in 1790 to preach his last open air sermon under a large tree beside the church a few nights before he died. The spot is marked for the visitor by a notice at the foot of an ash tree which replaced the original one, blown down in a storm.

The church of St. Thomas the Martyr is partly a ruin, but from what is left it is possible to imagine the splendour of the great building when it was first dedicated. The oak timbers of the roof came from the great forest of Anderida, while Caen stone was brought from Normandy.

(E Hinings, History, People and Places of the Cinque Ports, 1975)

86. Tout, *Mediaeval Town Planning*.

time of this appointment, there occurred the storms and coastal flooding which finally compelled the move to the new site. Advance knowledge of the decision taken in 1288 to hand over the site of the town to the 'barons of Winchelsea', with the King himself keeping only some 10 acres may perhaps have encouraged the move.

On the level plateau site the new town was laid out to the normal bastide/gridiron pattern, with streets forming a total of 39 quarters varying in size from one to three acres. Beresford and St Joseph record that 87½ acres were available for the 611 houses. In addition to these there were also 79 houses on the slopes north and west of the town overlooking Iham marshes.[88] At first there were no proposals for a town wall but by 1297 New Winchelsea was so firmly established that it could accommodate the court and army which Edward I assembled there for transit to France. In 1295 permission was given to raise a tax on shipping to pay for defences, required because of French threats. Further works followed in 1321–8, although it is considered that the full perimeter was never completed. Later, in 1415, another wall was started for a portion only of the town but this again was not finished.

For a while the new town prospered but vacant holdings began to appear before the mid-fourteenth century. The drift away resulted in part from French coastal raids, and also from the clear signs of the silting-up process of the Brede, which eventually left New Winchelsea high and dry some distance inland. By 1369 there were at least 377 houses, but in 1575 the total was put at only around 60, and silting up of the harbour was complete; with its dual naval and commercial port function removed New Winchelsea had slowly faded away. Today, some twenty-seven of the original thirty-nine quarters are grass covered. The streets are recognizable, if at all, only as lines of hedges or depressions in the fields. One of the churches has long since disappeared and the remaining church exists only as a fragment of the original building. Quarter No. 15, where in 1292 twenty-five houses are recorded on two and eleven-sixteenths acres, is now the local cricket ground. The Strand Gate still stands in the north-east corner of the remaining occupied area, on top of the slope down to the old harbour, but the South Gate, at the far end of the town, is now isolated in the middle of the fields. Defoe wrote of Winchelsea as rather the skeleton of an ancient city than a real town.[89]

KINGSTON UPON HULL

The essential difference in terms of urban fortunes between Kingston upon Hull (or Hull as familiarly known) and its late-thirteenth-century contemporary on the south coast, Winchelsea, has been established above.[90] Winchelsea's rise to fame and prosperity, albeit transitory, was comparatively more rapid: its decline and fall equally so. Less is known of the origin and early history of Hull; its more interesting and significant period (introduced in Chapter 8) is that from 1750 to 1850 when the basis of its supremacy as an east coast port was laid.

The first record of a settlement with a harbour on the eventual site of Edward I's Kingston is that of 1193 when reference is made to the wool collection for Richard I's ransom at the 'port of Hull'. The increasing wool trade of the Abbey of Meaux, about 7 miles to the north, had led to the creation of Wyke-upon-Hull in the upstream angle formed by the River Humber and its sheltered north bank tidal tributary, the Hull. Little is known of the streets and buildings of Wyke, although it was

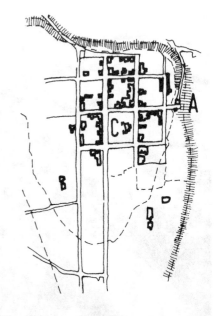

Figure 4.47 – Winchelsea: a tracing from the Ordnance Survey map of the present-day town which has contracted to occupy only the north-eastern corner of the original thirteenth-century foundation. Key: A, the Strand Gate; C, St Thomas's Church (as Figure 4.45).

Here I stayed [at Rye] till the 10th with no small impatience, when I walked over to survey the ruins of Winchelsea, the ancient cinq-port, which by the remains and ruins of ancient streets and public structures, discovers it to have been formerly a considerable and large city. There are to be seen vast caves and vaults, walls and towers, ruins of monastries and of a sumptuous church, in which are some handsome monuments, especially of the Templars, buried just in the manner of those in the Temple at London. This place being now all in rubbish, and a few despicable hovels and cottages only standing, hath yet a mayor. The sea, which formerly rendered it a rich and commodius port, has now forsaken it.
(J Evelyn, Diary)

87. See Vance, *The Continuing City*, for the administrative procedures involved in setting up a bastide (planted town).
88. Beresford and St Joseph, *Mediaeval England: An Aerial Survey*.
89. Defoe, *A Tour through the whole Island of Great Britain*.
90. For Hull see E Gillett and K A MacMahon, *A History of Hull*, 1980; Morris, 'Cities Built for a King'; see also Chapter 8.

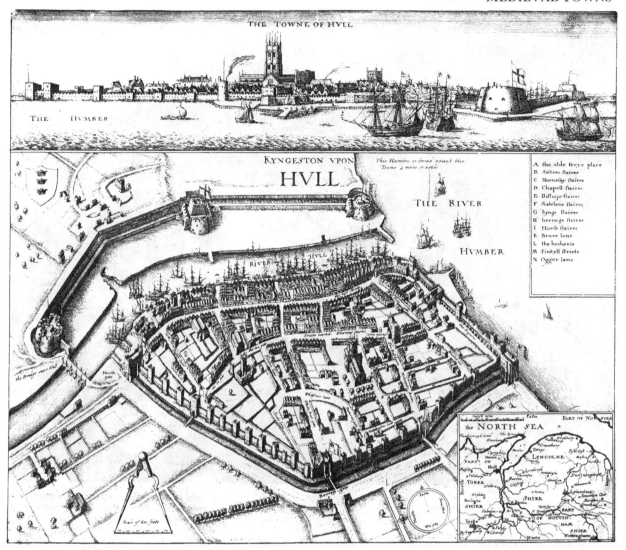

clearly a place of some consequence, paying the sixth largest tax contribution of the east and south coast ports to King John in 1203–5, and the third largest payment to royal customs revenues from 1279 to 1290, exceeded only by London and Boston. In 1279 Meaux Abbey was granted the right to a fifteen-day annual fair and a weekly market. Edward I's interest in Wyke as a potential replacement for Ravenser as the royal base for military and naval operation in the north of England probably dates from 1292, when he is known to have been in the neighbourhood on two occasions. A valuation of Wyke was ordered in November 1292, carried out in January 1293 and in March Edward took possession of Wyke and its neighbour, Myton Grange.

Hull quickly assumed its strategic role and served as a base for Edward's Scottish campaign of 1297. A royal charter was granted in 1299, confirming borough status with a twice weekly market and a thirty-day annual fair. It seems likely that these valuable trading concessions were granted in order to encourage people to move to Hull.

Figure 4.48 – Kingston upon Hull: Wenceslaus Hollar's famous engraving of 1665, seen from the north-west, the same viewpoint as the modern aerial photograph as Figure 4.49. The medieval wall and towers, within the moat, tightly confined the town around the land side. Across the River Hull can be seen the mid-sixteenth-century artillery fortifications added by Henry VIII, comprising the central castle and two flanking blockhouses linked by a wall and a moat. (See later map of 1817, as Figure 8.36, which shows the citadel which took the place of Henry VIII's defences; and also the first two inner docks which were in part formed from the old medieval moat.)

The harbour in 1665 comprised both common staithes and private quays alongside the River Hull (or 'Old Harbour' as it became known). The characteristic medieval waterside pattern of long narrow warehouse/residential plots, separated by frequent narrow lanes (or staithes) has been extensively retained through to the present day.

Figure 4.49 – Hull: the aerial view from the north-west which directly compares with Hollar's record of the 1665 scene.

There were perhaps only sixty families living there in 1293 and even after 1300 Edward constantly had to urge the letting of empty plots to enable the town's commercial survival.

The first fortification of 1321–4 is an indication of growing importance reflected also in the 1377 Poll Tax return of 693 families, one-sixth of whom kept servants. By the middle of the sixteenth century Hull's strategic importance demanded a new, up-to-date defensive system, and following a visit in 1541 Henry VIII authorized the construction of a castle and two blockhouses, linked by walls and a moat, on the eastern, undeveloped side of the River Hull (see Figure 4.48). Resurvey of the defences in 1680 found them to be inadequate in face of possible artillery assault and an extensive citadel was constructed on the eastern side of the Hull. The importance of this citadel as a rare instance of seventeenth-century urban fortification in Britain, and its effect on the subsequent growth of Hull, are described in Chapter 8.

By the seventeenth century Hull had become a prosperous port, and Hollar's famous imaginary aerial view records the densely developed waterside.

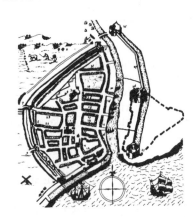

Figure 4.50 – Hull: as recorded in pictorial form by John Speed in 1611, showing the Castle (26), the Blockhouses (25) and the Fortification (27) on the eastern side of the River Hull, with the approximate extent of the later citadel added in dotted line.

THE BASTIDES: WALES

In Wales Edward I was responsible for a total of ten bastides, built in three phases following successive military campaigns against the Welsh. Three bastides resulted from the end of a period of hostilities in 1277 – Flint, Rhuddlan and Aberystwyth; five were under construction at the same time in 1283 after the defeat of Llewellyn – Caernarvon, Conway, Criccieth, Bere and Harlech; and the last two – Beaumaris and Bala – followed the uprisings of 1294. As with the earlier bastides in Aquitaine their origin was primarily military; second to this was the economic motive, emphasized by the desire of the Englishman, already rather a 'superior person', to teach 'civility' to the 'wild Welsh' and to direct them, not necessarily too gently, in the right way.[91]

Each of the Welsh bastides was attached to a castle as the base of a regular army garrison; this was in direct contrast to French examples where the military function was limited to the part-time performance of the citizen militia. They were granted borough status, with the constable of the castle also serving as ex-officio mayor of the civil town. Application for burgess-ship of the towns was considered only from free Englishmen – Welshmen and Jews were not admitted – with the usual inducements offered to the new settlers of a house plot within the walls and farmland outside, a monopoly of the commerce of the district and as many privileges as were compatible with the military unity of the borough. The success of Edward I's military campaigns in North Wales was due to his combined use of naval and military power to gain control of the coastal plains, leaving the Welsh insurgents to starve in the mountains. With two exceptions his bastides were concentrated at strategic locations along the coastline, controlling key river crossings and navigable inlets.

FLINT

The first of Edward's Welsh bastides was Flint, started in July 1277. It was an entirely new foundation, with the castle and the civil town planned separately, in marked contrast to the integrated form of later examples. The town was laid out to a regular gridiron plan with the main street connecting it to the castle gate.

The extreme importance given by Edward to the rapid establishment of the chain of castle towns is shown by the presence in his army, towards the end of July and throughout August 1277, of some 950 dykers, 330 carpenters, 200 masons and 320 woodcutters employed on the site work and construction of the castle and timber palisade of Flint.[92] This is a sizeable enough work-force by present-day standards and its advance organization is clear proof of Edward's understanding of the strategic role of the castles and the associated towns. In subsequent centuries, however, as Flint's military function diminished so did the town, and on Speed's map of 1610 it has almost disappeared, with a few houses shown dispersed within the town boundary. In 1652 a visitor could record that the town had no saddler, tailor, weaver, brewer, baker, butcher, button-maker; there was not so much as the sign of an ale-house.[93]

Two hundred years later Flint was expanding again under the influence of the Industrial Revolution and the original gridiron street pattern became lined with standardized 'by-law' housing. In the convenient gap between the castle and the town the railway, the bringer of prosperity, had neatly slipped in.

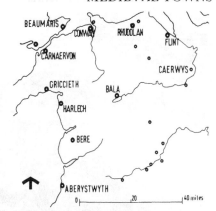

Figure 4.51 – North-west Wales: showing the locations of Edward I's bastides, as named individually, and other planted towns. (Caerwys was given its charter in 1290 but it was not military in character; Bala was not an Edwardian bastide as such.)

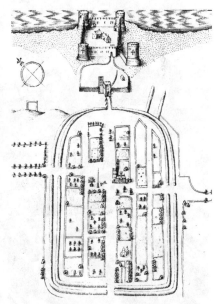

Figure 4.52 – Flint: depicted in 1610 by John Speed in his *Theatre of the Empire of Great Britaine*. The castle, located on a sandstone outcrop about half way between Chester and Rhuddlan, commanded the estuary of the River Dee.

91. For general references on medieval architecture and urbanism in Wales, see J B Hilling, *The Historic Architecture of Wales*, 1975; H M Colvin (ed.) *The King's Works in Wales 1227–1330*, 1974, from *The History of the King's Works, Volume 1, The Middle Ages*, 1963; H Carter, *The Towns of Wales*, 1965.
92. H W Dove, *The Annals of Flint*.
93. See C Stewart, *A Prospect of Cities*, 1965. Figure Survey Map Sheet IX. This map shows the nineteenth-century reinstatement of empty spaces within the town grid as revealed by Speed in 1610, but subsequent 'slum-clearance' redevelopment has replaced many of the terraced industrial cottages with anonymous tower

Figure 4.53 – Caernarvon: aerial view from the north-east looking across the clearly defined area of the bastide (the castle to the left) to the open country on the far side of the estuary of the River Seiont where it flows into the Menai Strait (right). The regularity of the bastide layout, within the surviving original town wall, is in marked contrast to subsequent uncontrolled 'suburban' additions.

CAERNARVON

The strategic military importance of the estuary of the River Seiont, where it flows into the Menai Strait, had been recognized by the Romans in building their legionary fortress of Segontium, as described in the caption to Figure 4.55. Subsequently little is known of the history of the area until in about 1090 Earl Hugh, the first Norman Earl of Chester, constructed a motte and bailey castle on the previously unoccupied peninsular formed by the Seiont, the Menai Strait and the Cadnant Brook.[94] Improved by the addition of a stone hall and other permanent buildings, this first Caernarvon Castle was used on occasion by the Welsh princes, and was served by a prosperous royal *maenor* (manor).

In December 1282 Llewellyn was defeated and killed by Edward I's army and in the January the English capture of Dolwyddelan castle

flats and broken up the clarity of the ancient pattern. However, the massive textile mill shown by Ordnance Survey as adjoining the north-western wall of the castle has been demolished and sympathetic restoration of its towers and walls and the surrounding area is in progress.

94. For Caernarvon see M D Lobel (ed.) *The British Atlas of Historic Towns – Volume One*, 1969; G C Boon, *Segontium Roman Fort*, HM Stationery Office, 1966.

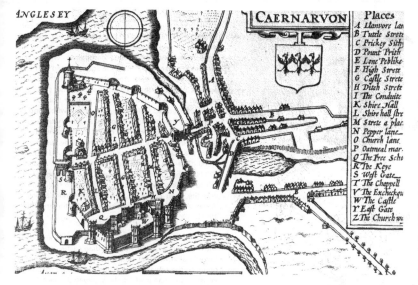

Figure 4.54 – Caernarvon in 1610, after Speed. The alignment is the same as Figure 4.55. The outer wall shown by Speed along the western side of the town has disappeared; now only a narrow quayside between the castle and the town's western gate, and a narrow sea-walk further north remain.

Caernarvon castle was designed by James of St George (c. 1235–1308) a master mason and military engineer of genius who came over from Savoy to work for Edward I on the Welsh castles. Caernarvon and its contemporaries represented the most advanced military architecture of the age, with Caernarvon itself believed to have been influenced by the great defensive system of Constantinople.

Caernarvon Castle, begun in 1283 and still unfinished when work on it stopped in about 1330, is one of the most striking buildings the Middle Ages have left to us. In North Wales it is one of a group of imposing castles that still remain in a state of some completeness, but it stands apart from the others in its sheer scale and majesty and degree of architectural finish, different from yet challenging comparison with the great Crusader castles of Syria and Palestine, with Avignon and Agues Mortes, or with the enchanting castles made famous by the gaily coloured miniatures of the Très Riches Heures of the Duc de Berry. Small wonder, than, that Caernarvon should have caught the imagination of writers and painters through the ages. It was a seventeenth-century author who wrote of it, 'I have seen many gallant fabrics and fortifications, but for compactness and completeness of Caernarvon I never yet saw a parallel; and it is by Art and Nature so fitted and seated, that it stands impregnable.' And Dr. Johnson, when he saw it over a hundred years later, in 1774, wrote in his diary that its stupendous magnitude and strength surpassed his ideas, for, as he said, he did not think there had been such buildings. From the early eighteenth century onwards the castle has been a favourite subject of the topographical artists and engravers; Samuel and Nathaniel Buck, John Boydell, William Pars, Richard Wilson, Moses Griffith, J. M. W. Turner, Thomas Girtin, Peter de Wint, Joseph Buckler and David Cox, and, in our own day, Dennis Flanders and John Piper, have all drawn or painted it.
(A J Taylor, Caernarvon Castle, 1969)

sealed the victory in north-west Wales. Edward I took up residence at Conway in March 1283 where work started immediately on its new castle and town walls. The importance of Caernarvon was likewise confirmed from early June with commencement of work on a new castle, embodying much of the fabric of its Norman predecessor, and the bastide with its town wall and gates. In addition a stone quay was constructed in order to handle the enormous quantities of building materials. Two closely related events of 1284 determined Caernarvon's future status; first, the Statute of Rhuddlan in March which established the three shires of North Wales with the town as their administrative capital; and second, on 25 April, the birth at the unfinished castle of Edward of Caernarvon, the king's heir from later that year, to be created Prince of Wales in 1301.

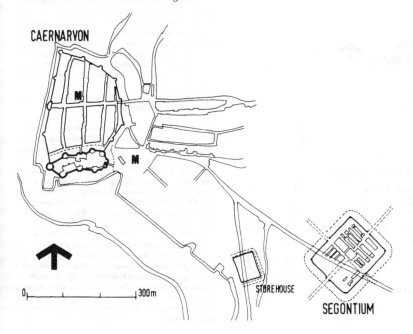

Figure 4.55 – Caernarvon related to Segontium. G C Boon has written that 'the fort at Caernarvon was probably founded in AD 78, when the Romans conquered this part of Wales. It was an important bastion of imperial power in the west, and for the greater part of three centuries was garrisoned by a cohort of auxiliary troops.'

Segontium was some 710 yards south-east of Caernarvon and faced south-west towards Lleyn. 'Its ramparts, girt originally by a double ditch, were marked out – imperfectly – as a rectangle with rounded corners, like a playing-card, 550 feet by 470 feet. There were four gateways, one in the middle of either short side, and one just south-west of the central point of either long side. Within, the layout was of a standard pattern which the authorities diversified only as far as the needs of different types of regiment required. A clear space (inter-vallum) running round the foot of the rampart delimited the closely-packed buildings, which were arranged in three lateral blocks with the administrative buildings in the middle. Like other forts of its date, Segontium was built at first in earth-work and timber; the stone structures which we see today are later' (G C Boon, Segontium Roman Fort, HM Stationery Office, 1966).

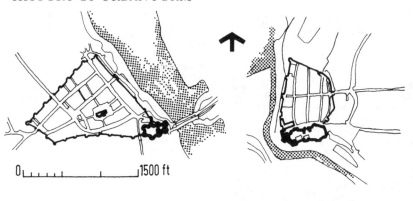

0 |——————| 1500 ft

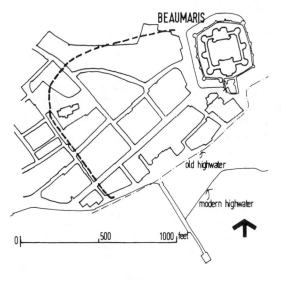

BEAUMARIS

old highwater

modern highwater

0 |———————|————————| feet
        500   1000

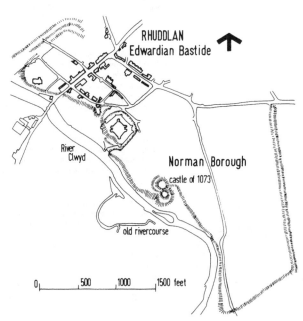

RHUDDLAN
Edwardian Bastide

River
Clwyd

Norman Borough

castle of 1073

old rivercourse

0 |———|———|———| 1500 feet
     500   1000

Figure 4.56 – Left, Conway, drawn to the same scale as Caernarvon (right). From a starting date in 1283 construction of the castle and town walls of Conway was urgently carried forward; it was garrisoned after only two years and effectively completed within five (accounts ceased by 1292). 'The spot selected for headquarters was a broad precipitous rock at the southeastern angle of the existing hamlet, one side of it washed by the Conway river and another bounded by the Gyffin brook. Chief designer of the whole monument was James of St George, an international celebrity from Savoy, where Edward could have studied his methods. At the peak of activity, in the summer of 1285, the craft and labouring force employed reached 1,500. The bill is reckoned to have come close to £20,000, and to get a comparable figure for today it is fair to multiply by a hundred, making two million. This was the most costly of all Edwardian foundations in Wales (A Phillips, *Conway Castle*, HM Stationery Office, 1961).

Figure 4.57 – Beaumaris: although primarily and justly known for its superlative Edwardian castle, Beaumaris town is also of great interest as a bastide, owing its origin to the replacement of the erstwhile prosperous Welsh royal manor of Llanfaes, situated about 1 mile away to the north along the Anglesey coast. Llanfaes had grown up around a Franciscan friary founded by Llewellyn the Great in 1237, and before 1295 (when it was supplanted by Beaumaris) it had been the main port and commercial centre of Anglesey.

These mountains are indeed so like the Alps, that except the language of the people, one could hardly avoid thinking he is passing from Grenoble to Susa, or rather passing the country of the Grisons. The lakes also, which are so numerous here, make the similitude the greater, nor are the fables which the country people tell of these lakes, much unlike the stories which we meet with among the Switzers, of the famous lakes in their country.

There is nothing of note to be seen in the Isle of Anglesea but the town, and the castle of Beaumaris, which was also built by King Edward I and called Beau-Marsh, or the Fine Plain; for here the country is very level and plain, and the land is fruitful and pleasant. As we went to Holly Head, by the S. part of the island from Newborough, and came back through the middle to Beaumaris, we saw the whole extent of it, and indeed, it is a much pleasanter country, than any part of N. Wales, that we had yet seen; and particularly is very fruitful for corn and cattle.
(D Defoe, A Tour Through the Whole Island of Great Britain, 1724–26. (Letter 6.))

Figure 4.58 – Rhuddlan: the plan redrawn from that included in *Rhuddlan Castle* by A J Taylor (published by HMSO), shows the extent of the Anglo-Norman town, with its motte and bailey castle; and also that of the Edwardian bastide with its new castle, work on which was commenced in 1277. Rhuddlan was an important eighth-century settlement controlling the lowest crossing of the River Clwyd – the natural barrier before the mountainous heart of Wales. The Norman castle was established in 1073, following which Rhuddlan was a frequently contested frontier town. The fine Edwardian castle and the planning of the bastide are of greatest relevance, but the feat of civil engineering that straightened the meandering Clwyd, providing for a sea-going shipping harbour below Rhuddlan bridge, is also of exceptional importance. Three hundred diggers were brought over from the English Fenlands for the excavation of the 'new cut', and in mid-September 1277 it was recorded that a total of 968 *fossatores* were at work.

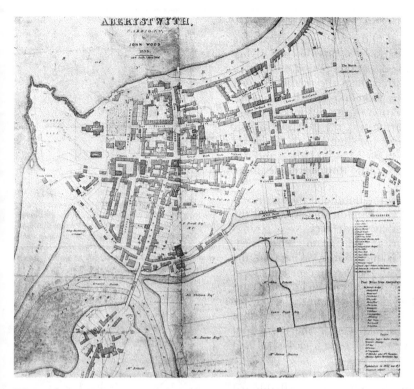

ABERYSTWYTH.

Figure 4.59 – Left, Aberystwyth, as John Wood's map of 1834. Under Edward I's brother, Edmund of Lancaster, work was started in July 1277 on a castle near the mouth of the river Rheidol. This castle replaced Henry I's earlier castle at nearby Aberystwyth and also gradually assumed its name: a process assisted by the River Ystwyth conveniently shifting its course nearer to the new castle. Little remains of the Edwardian castle and Aberystwyth is now best known as an attractive seaside resort and Welsh university city. Nevertheless, there is still a typically close-grained medieval appearance to the old bastide streets east of the castle, and the nineteenth-century map shows the extent to which the original gridiron layout has been retained. See H Carter, *The Study of Urban Geography*, 1975, for a growth diagram of the town to the end of the nineteenth century.

The founders of towns were not always passive spectators of economic change, waiting for an opportunity to be presented to them as the result of political and economic action far outside their own control. The urban potential of a site was sometimes the result of a seigneur's own actions. It has been seen already that the construction of a new bridge and the diversion of roads could be turned to the seigneur's advantage. The intensive development of their rural estates could not help but improve the chances of any plantation whose modest aim was to be a local market centre for the products of the fields. The development of quarries and mines by a seigneur would also improve the fortunes of local markets and ports.

Besides the general encouragement of local economic development the seigneur might spend money on constructional work that had the indirect result of encouraging town development ... building of new bridges in Essex, Lincolnshire and Yorkshire encouraged the diversion of roads and the creation of Chelmsford, New Sleaford and Boroughbridge. There were other building ventures where the seigneur had no direct intention of fostering trade or towns, yet where the natural consequence of building was the enhancement of a site for a new town. The residence of a seigneur could not help being a place of public resort. A new residence for a great seigneur that was built away from an existing village promoted a concourse of social visitors, tenants, suitors at courts, knights of the Honour and perhaps even the king himself; and with the visitors came servants, retinues and spectators. The suddenly inflated population had to be housed and fed, and if the seigneur was regularly in residence there were regular opportunities for traders and craftsmen established at his gate.

These magnetic forces were strong around the residences of bishops and abbots and even stronger around the residences of the king and the royal family, and they can be detected – though less strongly – around the manor houses of the smallest seigneur. The forces of attraction were strengthened if the seigneur's residence was fortified and could offer a place of refuge in time of war.

(M W Beresford, New Towns of the Middle Ages, 1967)

Planted towns

ENGLAND

In addition to Winchelsea, Kingston upon Hull, and the ten Welsh castle towns, Professor Beresford lists a total of 120 other medieval urban foundations which he classes as planted towns.[95] Several of these have similar strategic, politico-military origins to the Edward I bastides, notably Portsmouth (1194) and Liverpool (1207).[96] The great majority of the planted towns were, however, founded for a commercial function resulting from the general development of trading activity after the Dark Ages. It has been observed that during the Middle Ages it was the routes which made the towns and that subsequently, particularly during and after the Industrial Revolution, it is the towns which have made the routes. This locational determinant is particularly relevant to the creation of planted towns. The essential prerequisite for a viable foundation was a roadside or a riverside location – assuming that the need for a town in the general locality had been established. Beresford points out that the choice of site for any particular *seigneur* (landowner) was correspondingly limited; if an important artery of road or water touched only the edge of his estate there was no choice at all but to lay out the town at the point of tangency. Many old-established roads, notably the Roman system, were taken by the Anglo-Saxons as boundaries between villages (later parishes) and, as a result, a number of planted towns was formed out of two adjoining parishes – Professor Beresford lists Royston, Newmarket, Wokingham, Bocastle, Mitchell and Maidenhead. Intersections of roads and river-crossing points were natural first choices, if the estate was so favoured. A number of landowners were hence able to establish new harbours for imports and exports – notably on the east and south coasts.

---

95. Beresford, *New Towns of the Middle Ages*.
96. For the origins and later histories of Portsmouth and Liverpool see Chapter 8.

SALISBURY

The original settlement on the hill site of Old Sarum dates from the early Iron Age, though permanent occupation was possibly not earlier than the first century AD.[97] The earthworks of this period enclose an area of some 30 acres (Figure 4.60). Within them the medieval town sheltered, in all probability preceded by Romano-British Sorviodunum. Actual evidence for Roman occupation of the hill fort is slight, but in the immediate vicinity of Old Sarum was the junction of four Roman roads – those to Winchester, Exeter (Ackling Dyke), Silchester and the Severn, by way of the lead mines in the Mendip Hills – and it is reasonable to presume that, initially at least, the Romans made use of the substantial existing defences.

The hilltop was probably deserted from the middle of the sixth century AD, when the Britons were forced to abandon the site, until the time of King Alfred, when it is recorded that the defences were strengthened. In 960 King Edgar held his court at Old Sarum and in 1003, while the nearby valley town of Wilton was burned in a Viking raid, the strength of Old Sarum ensured its survival. (It also assumed the local mint function from Wilton.) William the Conqueror heightened and broadened the outer fortifications and levelled 6 acres in the centre for the 60-foot high motte of his new castle. By decree of the Council of London in 1075 the seat of the amalgamated sees of Sherborne and Ramsbury was transferred to Salisbury, and by 1092 a cathedral had been built.[98]

During the next 120 years or so this crowding together of church, soldiers and townspeople occasioned much conflict, and probably led more than anything to Bishop Richard Poore's decision to remove his cathedral church down from the hilltop to a new location on the river plain. The power struggle between the governors of the castle and the bishops, who, in spite of a titular lordship of the manor, had little but spiritual authority and felt themselves to be prisoners in their own cathedral[99] was, however, only one of a number of reasons for moving. Certain others were noted by Pope Honorious in his consent to the bishop's proposals: 'Situated within a castle, the church is subject to such inconvenience that the clergy cannot stay there without danger to their persons. The church is exposed to such winds that those celebrating the divine offices can hardly hear each other speak. The fabric is so ruinous that it is a constant danger to the congregation, which has dwindled to the extent that it is hardly able to provide for the repair of the roofs, which are constantly damaged by the winds. Water is so scarce that it has to be brought at a high price, and access to it is not to be had without the governor's permission. People wishing to visit the cathedral are often prevented by guards from the garrison. Housing is insufficient for the clergy, who are therefore forced to buy houses from laymen. The whiteness of the chalk causes blindness.'[100] Beresford and St Joseph note that the abandonment of the medieval town between 1220 and 1227 is one of the most curious episodes in settlement history, so sudden that the chronicle story might be suspected of over-dramatization to point the moral of a modern Sodom and Gomorrah. At the time of its desertion the earthwork contained a castle, a cathedral and the houses and streets of a great medieval town. Ruins of the castle remain, but the cathedral was buried until excavations conducted between 1909 and 1915 by the Society of Antiquarians laid its foundations bare. Very little of the houses and streets has yet been

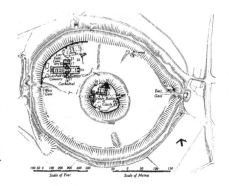

Figure 4.60 – Old Sarum as partially excavated showing the location of the cathedral (1092) in the north-west corner related to the 60-ft high central *motte* of William the Conqueror's castle. Within the impressive earthworks of the outer fortifications most of the remainder of the city would have consisted of tightly packed, organic-growth housing.

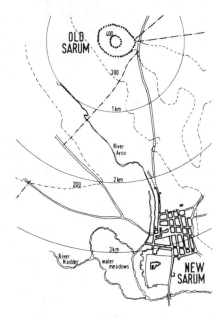

Figure 4.61 – The relationship between Old Sarum and New Sarum (Salisbury) showing the extent of the city's present-day built-up area, the contour lines and also the Roman roads at Old Sarum.

97. For Salisbury see H de S Shortt, *Old Sarum*, HM Stationery Office; and Lobel, *The British Atlas of Historic Towns*.
98. Beresford, *New Towns of the Middle Ages*.
99, 100. Shortt, *Old Sarum*.

uncovered.[101] Before 1220, in which year the new Salisbury cathedral was commenced (it was to be the only English cathedral effectively started and finished in the same Gothic period), there is considerable evidence of a move away from Old Sarum. In 1187 a 'New Salisbury' is recorded in the pipe rolls on the bishop's rich meadow land alongside the River Avon. In contrast to the English and Welsh bastides the foundation of New Salisbury was based on purely ecclesiastical and economic motives. It was necessary to find room and comfort for clergy and traders in a well-planned city of the plain. The unimportant castle could safely remain on the hill. Salisbury Cathedral was built on its magnificent close and stood, together with the Bishop's Palace, in the centre of a broad bend in the River Avon; the new city was laid out to a somewhat irregular gridiron to the north and north-east (Figure 4.62). From its inception New Salisbury was intended to attract the remaining inhabitants of Old Sarum, and to take away the trade of the flourishing borough of Wilton some 3 miles to the west.[102] These objectives were finally achieved when the great western road was diverted to it from Old Sarum. The new city was not fortified, although walls had been authorized. A ditch fed by the Avon provided adequate defence for the inhabitants. The market square was located in the centre of the city with the town hall in the south-east corner and the main parish church in the south-west corner. By 1227 the city had received its charter and the cathedral was in use. Old Sarum was deserted with the exception of the castle garrison and in 1331 a licence was granted for the use of the stone from the old cathedral for building purposes in the new city. The castle remained in use until the late fourteenth century, but the poll tax collectors of 1377 found only ten persons over sixteen years of age in Old Sarum. Nevertheless it continued to send its burghal representatives to Parliament until the Reform Bill of 1832.[103]

PLANTED TOWNS: THE ZÄHRINGER FOUNDATIONS

In the twelfth century the Dukes of Zähringen created a dynastic state on both sides of the Rhine, in what is now Switzerland and southern Germany. Expansion of the Zähringer lands and the securing of their frontiers was based on the establishment of towns, castles and monasteries.[104] Offenburg, as the first of the Zähringer towns, was followed by a group of towns founded after 1122, on the right bank of the Rhine – Freiburg im Breisgau, Villingen and Rottweil. In all twelve towns were created, with three others strongly influenced by the dukes. It has been noted that 'the success of these twelfth-century foundations gave rise to a veritable new town boom in the thirteenth century, with every petty feudal lord gambling his financial future on new foundations ... most of which failed, partly because of whimsical location and lack of hinterland, but primarily because of insufficient initial urban mass. What sufficed in the sparsely urbanised landscape of the twelfth century did not suffice in the relatively saturated urban region of the thirteenth, since the new creations now had to compete with the earlier foundations grown to maturity'.[105] Eight basic elements governed the layout of the Zähringer towns in their fully developed state at the end of the twelfth century: (1) a market thoroughfare, 75 to 100 feet wide running the full length of the town between the gates; (2) the absence of other interior spaces; (3) the use of the homestead (area) as a planning module and as (4) the taxation unit; (5) an orthogonal geometry (gridiron) basis of plan, in harmonic proportions of 2:3 and 3:5; (6) location

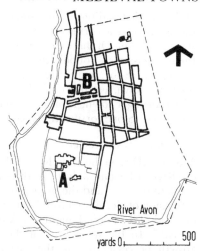

Figure 4.62 – New Sarum (Salisbury) as laid out from 1220 on the Bishop of Salisbury's meadows below the old hilltop location. Key: A, the cathedral; B, the market place. The line of the rudimentary defensive perimeter is shown dotted. New Sarum's grid structure is much less rigid than, for example, Winchelsea, but its comparative regularity has proved well suited to mid-twentieth-century city centre regeneration.

Figure 4.63 – Londonderry: general plan of this planted town in Ulster (Northern Ireland). After the province of Ulster was acquired by the English Crown at the beginning of the seventeenth century, the plans drawn up for the creation of the six counties involved the foundations of twenty-three new towns.

Two of these towns – Derry and Coleraine – were planned about 1611 by the Irish Society – a colonizing company established by the City of London. Londonderry's plan was based on main cross-axes with a large building in the centre of the main square providing market, town hall and prison facilities. For further reading G Camblin, *The Town in Ulster*. For Londonderry see the detail consideration in J S Curl, *The Londonderry Plantation 1609–1914*, 1986.

101. Beresford, *New Towns of the Middle Ages*.
102. Tout, *Mediaeval Town Planning*.
103. Modern suburban Salisbury has extended up to the southern side of Old Sarum (Figure 4.61), but to the north and west there is still the extensive open undulating landscape of Salisbury Plain.
104. For the Zähringer Towns see P Hofer, *The Zähringer New Towns*; E A Gutkind, *International History of City Development*, Volume II, *Urban Development in the Alpine and Scandinavian Countries*, 1965.
105. Rasmussen, *Towns and Buildings*.

of the public buildings away from the main market-street; (7) placing of the fortress at a corner or on a side wall; and (8) construction of a sewage system. Of these elemental 'laws' the most important by far was the market-street, not only the *raison d'être* of the town but also the point of departure for the entire plan. Contrary to what was usual in a medieval town, a strong encircling stone wall was not a component part of a Zähringer town; it probably had no more than a timber palisade and moat. Stone fortifications were added after the Zähringer period.

PLANTED TOWNS: EASTERN GERMANY

Around AD 1200 the Holy Roman Empire had about 250 towns west of the River Elbe and only 10 to the east. Two centuries later there were 1,500 to the west and the same number to the east: the result of eastern expansion of Germany brought about by land shortage and the crusading zeal of Teutonic Knights, seeking to establish Christianity in new areas.[106] In the west most of the towns had developed by organic growth processes but in the east almost all were new foundations, generally established as 'colonies' by towns in the west.

The best known example of such a thirteenth-century German colonial town is New Brandenburg. Rasmussen records that 'by letters

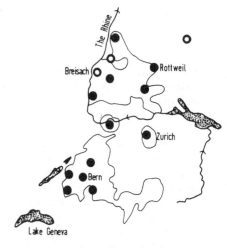

Figure 4.64 – The Zähringer towns, locations related to the Rhine and Lake Geneva. Towns shown as solid circles were founded by four Dukes of Zähringen between 1122 and 1218; open circles denote the towns founded in the Zähringer conception.

Figure 4.65 – Berne: an aerial photograph showing the distinctive gridiron form of the original Zähringer town as the nucleus of the modern city, on its exceptionally beautiful location in the bend of the river.

106. Rasmussen, *Towns and Buildings*; for Central and Eastern Europe see also E A Gutkind, *International History of City Development*, Volume I, *Urban Development in Central Europe*, 1964.

patent, dated January 4th 1248, Markgraf Johann of Brandenburg authorised a certain knight, Sir Herebord, to build the town of New Brandenburg'. The purpose of the town was twofold. In addition to serving as part of the strategy of German expansion, it also followed the Greek city state precedent of founding a new town to take the overspill population from the over-crowded farmlands of the original town. As such there was no difficulty in attracting population to the settlements. 'At home in Brandenburg the younger sons of peasantry had no prospects of ever getting their own farms. They would have to be farmhands all their lives. In the new town they would be given land as such a colony was always started as a great agricultural undertaking. The newcomers settled in an untilled district thinly populated by a scarcely civilised people.'[107] New Brandenburg was laid out to a regular gridiron, each inhabitant being allocated land inside and outside the wall.

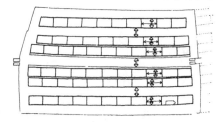

Figure 4.66 – Berne (north at the top): founded around 1190–91 by Duke Berthold V on an elevated site surrounded on three sides by the River Aare. The diagrammatic layout shows the arrangement of the 64 homesteads, each originally with a frontage of 100 feet and depth of 60 feet but later divided up into narrow-frontage strip lots, and the broad east–west market square.

Figure 4.67 – Berne: a characteristic street scene within the old Zähringer town illustrating the use of arcades containing the pedestrian footpaths alongside the main streets.

Figure 4.68 – Breisach: founded in 1185 on a steep hill above the Rhine with a long but obscure history of pre-Roman and Roman settlement.

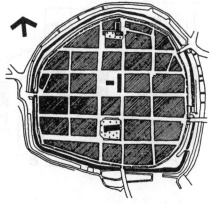

Figure 4.69 – New Brandenburg: founded in January 1248 as a typical agricultural settlement in eastern Germany. The roughly circular form resulted from the fact that New Brandenburg's defensive requirements were greater than those of the generally more rectangular French bastides and Florentine new towns (see below).

107. Rasmussen, *Towns and Buildings*.

137

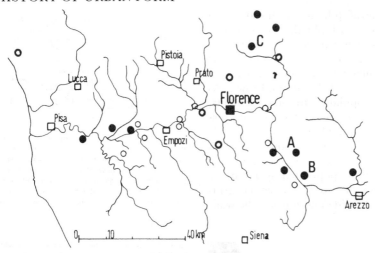

Figure 4.70 – Outline map of the valley of the Arno, showing the location of Florence and other major cities. *Terre murata* are denoted by solid circles; other towns as open circles. Key: A, Castelfranco di Sopra; B, Terranuova; C, Scarperia.

The Florentine new towns were works of art in the literal sense that their designers were artists. We do not know this about any other medieval new-town project. In part this is the result of the surviving documentation. For most new towns only the official acts of foundations are preserved. These documents assign authority and grant legal privilege; they say almost nothing about design and construction. The Florentine archives preserve both this and another level of record; in addition to legislative documents, the deliberations and financial accounts of planning committees and builders chronicle the city's activities as a founder of towns. The daily records identify the designs of the new towns as mason-architects from the circle of builders active in the city's public projects. Careful examination of the plans reveals their contribution. In an age in which geometry was considered the essence of art, the indispensable theoretical base for all design, the geometrically generated proportions of the new-town plans are a sure sign of the participation of professionals ...
(D Friedman, Florentine New Towns: Urban Design in the Late Middle Ages, 1988)

PLANTED TOWNS: THE FLORENCE REPUBLIC

The city of Florence has been briefly described earlier in this chapter with reference to its medieval defensive systems; the city is also returned to in Chapter 5 when its originating role in the Italian Renaissance is described together with an illustration of its Renaissance urban form.[108] The Florentine Republic was also notable for a town-building programme which was carried out around 1280 to 1310 in the main and tributary valleys of the River Arno. As Gutkind observes, 'this was not merely a political move directed against the feudal lords of the *contado* but an important step towards the unification of the city and surrounding country, one of the goals of every Italian major municipal administration'.[109]

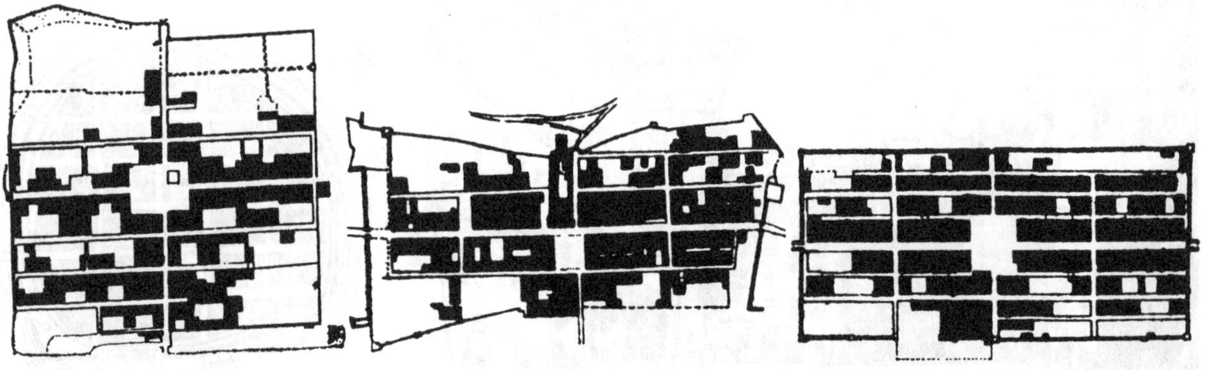

The policy of thirteenth-century Florence was to control its territory with castles and citadels in subject communities; from 1284, however, with the foundation of the first two *terre murata* (walled new towns), Santa Groce and Castelfranco di Sotto, the Republic's new policy involved the creation of upwards of twelve new towns and associated roads and bridge works (Figure 4.71). From the point of view of their physical form and their socio-economic basis, *terre murata* are similar to French bastides.

Figure 4.71 – Three typical *terre murata*. From left to right: Castelfranco di Sopra; Scarperia, one of a group controlling the route to Bologna; Terranuova, guarding the valley leading to Arezzo.

108. For the Florentine New Towns see D Friedman, *Florentine New Towns: Urban Design in the Late Middle Ages*, 1988; E A Gutkind, *International History of City Development*, Volume IV, *Urban Development in Southern Europe: Italy and Greece*, 1969.
109. Gutkind, *Urban Development in Southern Europe: Italy and Greece*.

The Netherlands

Few readers will be without some knowledge of the importance of water in Dutch history, whether that water be the North Sea or the major European rivers, or related to the exceptionally high-water tables. The compelling if improbable image of a boy holding his finger in a vulnerable dam must have remained with many from school-days past. However, as Renier cautions: 'One can exaggerate the influence of geography upon the history of the Dutch, and far too much is said about climate by those who study comparative national character.' Nevertheless, he adds, 'the fact that the Dutch fought the sea and fought their rivers, and are fighting them still, did put its mark upon their character. Ever since the Romans taught them to construct dikes, the people of the Low Countries have built defenses against the sea and the rivers and have continued to wrest portions of territory from the waters. Every man had to be a soldier in this war; every soldier was permanently on sentry duty'.[110]

Where the physical characteristics of settlement in The Netherlands differ from those of Western Europe generally (and indeed is still so with the reclamation of new land) is in not only the 'planning' of an unusually large proportion of the original villages, but also the subsequent extensions of many of those that grew into towns. The only truly 'unplanned' villages were those which were established on existing high ground, and the only organic growth consonant with Western European tradition was that which was possible before all the available agricultural land had been taken up. The historic nuclei of present-day Haarlem and Leiden are outstanding examples of organic growth, as clearly revealed by old maps, and they are happily still there to be visited.[111]

Villages founded in connection with land reclamation works – or where the cause and effect were reversed – ordinarily involved the construction of one or more dikes for land drainage and the creation of a safe land route, with adjoining building plots, raised above the high-water level. The simplest form was that of a canal lined by buildings on one side only. Variants included land routes and buildings raised above both sides of the canal; a cross form, where the dike intersects an existing land route, for which Burke cites Sloten as a clear example;[112] and a combination of dikes and a dam, whereby the main flow of the river is diverted along both sides of an enclosed 'harbour'. The latter method required the greatest resource investment and was accordingly employed very rarely; by far its most important use was in the creation of Amsterdam (for which see Chapter 7).

The resultant village, or eventual urban form, was not necessarily 'planned' to the extent, for example, of a thirteenth-century bastide in south-western France, where the market square and the layout of streets and building plots were variously predetermined. Nevertheless neither was the form the result of unplanned processes; rather it was a uniquely Dutch modification of organic growth, controlled along lines determined by the all-important 'water factor'.[113]

Of greater consequence than the shaping of urban nuclei – because of the comparatively extensive areas involved – was need to control the reclamation and use of new suburban land. This brought the 'water factor' into even greater prominence during the urban expansion phases of the fourteenth to eighteenth centuries. Not only were new

Figure 4.72 – The map of The Netherlands locating the principal cities and the main rivers.

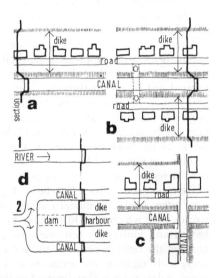

Figure 4.73 – Diagrammatic layouts of four basic types of Dutch urban settlements.
a — canal with dike;
b — dikes on both sides;
c — road and canal crossing;
d — dikes and dam

110. G J Renier, The Dutch Nation, 1944. (I knew Gustav Renier when I was a student at University College London; his work is my preferred background reference.) See also A E J Morris, 'Historical Roots of City Planning', in A K Dutt and F J Costa (eds) Public Planning in the Netherlands, 1985; G L Burke, The Making of Dutch Towns, 1956.
111. Dutch urban cartography of the sixteenth and seventeenth centuries has provided a fascinating, detailed record of the period. Jacob van Deventer (c. 1510–75) mapped more than 200 towns in the Low Countries between 1557 and 1575 for Philip II of Spain; Braun and Hogenberg published the six volumes of their Civitates Orbis Terrarum between 1572 and 1618; Johannes Blaeu (1596–1673) published two volumes of Dutch maps in about 1848, as part of his incomplete

canals required in order to drain land and provide excavated soil for land formation, but they were also necessary in order to provide water-transport access up to a considerable proportion of the buildings which were to serve both as the residences and the warehouses and workshops of burgher families. Moreover, the outermost canals characteristically performed the additional function of a 'wet moat' within the defensive system of that time.[114] The Dutch term for the 'water towns' is *grachten-staden*, from the *grachten* (drainage canals) which were excavated for either the foundation or extension of towns.

In numerous instances, the plan of a new urban district was determined by a pre-urban cadastre of ditches (*sloten*) used for draining the adjoining agricultural land over which expansion was to take place. Typically, these had been dug as parallel straight lines, a uniform distance apart, and when widened and deepened, a proportion were upgraded as *grachten* serving a new urban district. The regular *sloten* are the uniquely Dutch counterpart of the haphazard English field hedge-row patterns,[115] and as is to be expected, these two fundamentally different types of pre-urban cadastre resulted in the creation of fundamentally different urban forms. The map of Delft exemplifies the Dutch effect.

Vulnerable Dutch cities requiring fortification were secure within simple brick walls supplementing their natural water defences, or easily excavated wet ditches. Fortification of a community was ordinarily carried out by its inhabitants – yet another instance of the necessary acceptance of planning by the Dutch. At Naarden, a small town famously refortified later, an ordinance of 3 May 1442 required that 'no one can be accepted as a citizen without making a donation of 20,000 bricks to the town to be paid within two years'.[116]

The acquisition of land by a community for the construction of its new defences, or the extension of an existing system, involved its citizens in 'compulsory purchase' legislation, a process which is the legal prerequisite for modern urban and rural planning in the Western world, with which the Dutch have been reconciled ever since the thirteenth and fourteenth centuries. In 1271 the town of Dordrecht was granted the right to acquire land for a new moat, regardless of existing uses, with 'right of free access to and use of land needed for this purpose'. A fair price was to be paid to the owners of the land, which, it is to be believed, 'did not considerably exceed its *agricultural value*' (added italics).[117] There is a record of 1386 from Leiden whereby the granting (by the feudal lord) of the right to compulsorily acquire land for extended fortifications, instructed that 'compensation fixed by the Aldermen, should be paid'; at not much more than agricultural value.

In addition to recognition of need to pay fair compensation for private property acquired in the common interests, Dutch city councils also evolved a complementary system whereby 'betterment' payments should be made towards the cost of an urban improvement by those neighbouring owners who would benefit from the increased value of their property. Alkmaar exemplifies this process, when in 1558 the cost of compulsorily acquiring seven houses for the area of the new market square was apportioned between 119 landowners in the nearby streets who were deemed to gain from the development.

The characteristic basic component of historic Dutch urban form is the individual burgher's house, and it was the organization of these houses as a 'harmonious whole' that prompted Rasmussen to write of

GEESTGRONDSTAD   GRACHTENSTAD

Figure 4.74 – Alkmaar in 1597, from the map published by Cornelius Drebbel. This smaller city, located 30 km north-west of Amsterdam and 8 km inland from the North Sea, beautifully exemplifies that characteristic process of historic Dutch urban development whereby a planned *grachtenstad* extension was added to an essentially unplanned *geestgrondstad* nucleus.

The *geestgrondstad*, which had grown up on the eastern end of a sandspit, is known to have had a market as early as 1134. A charter was granted by Count William II in 1254. The *geestgrondstad* has a more uniform street pattern than is usual with early medieval unplanned urban form, with the Langestraat, an impressive straight main street, linking the Grootekerk, and its monasteries, with the quay of the early town on the Mient, to the east: this regularity of layout is perhaps suggestive of a controlled origin?

By the early sixteenth century – the exact year is unknown – the related needs for more residential land and improved port facilities, had resulted in a city council decision to develop a main *grachtenstad* extension to the east of the Mient, with additional new area south of the Oude Gracht. A new moat and rampart defensive perimeter was completed, it is believed, by the 1560s. The new market square of 1558, mentioned before as an early example of Dutch increment-value tax ('betterment' in British planning practice) is also on the sketch map. The surrounding marshland was drained in the last decade of the seventeenth century and Alkmaar became a prosperous market centre, notable for its export trade in cheese.

Stedenatlas van de Vereenigde Nederlanden.
112. Burke, *The Making of Dutch Towns.*
113. See Chapter 11 for the comparison with the uniquely Islamic control of organic growth urban form.
114. See C Duffy, *Siege Warfare, Volume 2: The Fortress in the Modern World 1494–1660,* 1979.
115. See the aerial photograph of Appleton-le-Moors for the comparable field-boundary landscape context of an English village (Figure 4.18, Chapter 4); see also A M Lambert, *The Making of the Dutch Landscapes,* 1985.
116. Quoted in E A Gutkind, *International History of City Development, Volume VI, Urban Development in Western Europe: The Netherlands and Great Britain,* 1971.
117. Quoted in E A Gutkind, *Urban Development in Western Europe: The Netherlands and Great Britain.*

Amsterdam as 'one great composition'.[118] In this respect Amsterdam must be regarded as the outstanding Dutch city, if only because of the exceptional extent of its historic central districts. Size apart, however, Amsterdam is but one of several claimants as the most beautiful Dutch city, and there are numerous more modern historic centres which have their own particular attractions as variations on the uniquely national townscape theme.[119]

The traditional urban scene is a blend of buildings and open waterside spaces. The great majority of Dutch buildings originated as multipurpose combinations of residence and warehouse (and also workshop, if the owner was involved in small-scale domestic industry), and the open space was the canals and their quayside required to provide them with the vital waterway access. Fortunately, indeed in effect miraculously, as compared with the commercial and wartime devastation of urban history in other countries, the many changes of use from original purpose have been accommodated behind the historic façades with remarkably little effect on external appearances. Similarly, most of the canals have not been filled in, as could well have been a fate in other countries. The greatest change that they have seen, intrusive as it is, has been use of the one-time quaysides as open-air car parks.

Each building fronting on to a canal had its right of access to that waterway and quayside. Minor building plots fronting cross-streets were on literal 'backwaters'. A description of an average merchant's house in Amsterdam is provided by Cotterell: 'it would have a cellar, a ground floor where he and his family lived and slept, and an upstairs apartment which was probably his workroom but possibly included a bedroom, and above that an attic beneath the gable which was his private warehouse – goods went up and down by pulley. The street, whether it faced the canal or another row of houses, was an extension of his house. In fact it was like a large room belonging to all the inhabitants: They sat out in it, worked in it, and played in it'.[120]

Nijmegen: a study in urban origins

Nijmegen in Holland is situated some 90 kilometres south-east of Amsterdam on the southern bank of the River Waal (the Dutch name for the German Rhine). The city's historic development is an important example both of multi-nuclear origins and also the effect of successive fortification systems as urban form determinants. The name is of Gallo-Roman origin, derived from *Noviomagus*, or 'new market', and the town itself dates from a Roman settlement established where a major military road crossed the Waal by ferry. Subsequently the crossing retained its strategic importance for Charlemagne to ensure control through an imposing castle constructed on the heights above the ferry. Beneath its walls a trading town became established, at first unfortified until required by twelfth- and thirteenth-century commercial prosperity.

Expansion took place initially to the west, where an extensive new district was developed during the first part of the fourteenth century, and also up to castle walls on the east. These areas were enclosed within a second circumvallation of 1334–56, which was later extended down to the riverside when the lower-lying land was used for the river port. The extensive re-fortification completed in 1468 was still of the medieval wall and ditch type and it was incorporated as the inner ramparts of the complex early eighteenth-century artillery works.

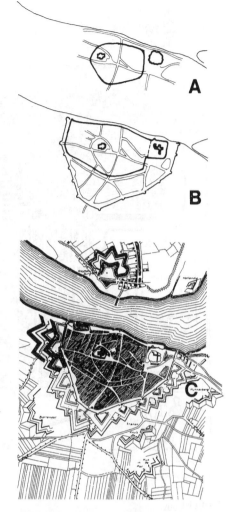

Figure 4.75 – Nijmegen: three sequential maps showing the main phases in the city's growth delineated by successive defensive systems:

(A) the castle commanding the ferry crossing and the lower *burgus* and *portus* joined together as the *oppidum* within the mid-thirteenth-century wall.

(B) the extended city of the mid-fifteenth century and the walls which also enclosed the castle; and the line of the refortification completed in 1468 which ambitiously provided adequate land through to the eighteenth century.

(C) the early eighteenth-century complex of artillery fortifications designed by Menno van Coehoorn, which were added outside the line of the 1468 wall. There is a separate strongly fortified bridgehead settlement across the Waal and extensive defensive outworks on the southern side. The streets marking the lines of earlier walls are clearly delineated.

118. Rasmussen, *Towns and Buildings*.
119. G Cotterell, *Amsterdam: The Life of a City*, 1973.
120. See Gutkind, *Urban Development in Western Europe: The Netherlands and Great Britain*, for extensive illustration of Dutch cities of the period.

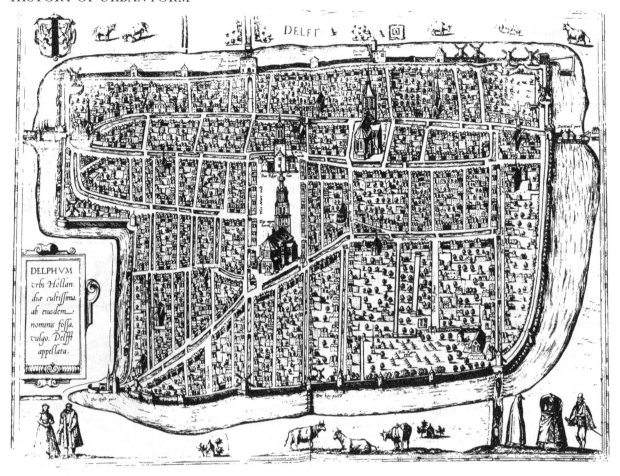

Figure 4.76 – Delft in 1650: the map published by Joannes Blaue.

DELFT

The name derives from a shallow ditch (delve) that had been excavated, perhaps as early as the ninth century, to connect the rivers Schie and Vliet. The first settlement was on the eastern bank of the Schie. A moat and walls were recorded in 1071 and a charter was granted in 1246. Delft is an excellent example of a grachtenstad; its form determined by three main waterways: the Oude Delft; the Nieuwe Delft – some sixty metres to the east; and a minor stream, parallel in part to the Oude; together with the regular east/west alignment of the surrounding polder ditches. (The role of minor drainage ditches as a characteristically Dutch urban-form determinant has been described above.) At the beginning of the fourteenth century the urban nucleus was approximately rectangular in outline, contained between the Oude Delft and the parallel stream, and divided lengthways by the Nieuwe Delft. Expansion at mid-century to the east and south-east, beyond the Verversdijk, was controlled on the 'planned' basis of deepening and widening the existing polder ditches into new commercial-frontage grachten. Similar extensions to the north and west which were carried out at the end of the fourteenth century made Delft the third largest Dutch city at that time.

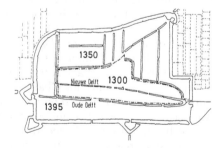

Figure 4.77 – A diagram locating the Oude and Nieuwe Delft and the fourteenth-century growth stages. The dotted pattern at top right and left exemplifies the uniquely characteristic Dutch pre-urban cadastre of land-drainage sloten.

(Map has north at top; diagram north at bottom.)

LEIDEN

The location was at the confluence of three rivers: the old Rhine, the New Rhine (the Vliet), and a minor stream, the Mare. Early village settlement is believed to have taken place on naturally higher ground between the Rhine branches, and there is an unsubstantiated possibility of an earlier Roman trading-post at that location. Summarized, this complex example of Dutch urban form was successively a burcht, a dike-town, and a grachtenstad. By the later twelfth century a burcht was well established on that higher ground, with a huddle of cottages comprising the trading village. Early in the thirteenth century a first extension was developed on the left bank of the New Rhine using the dike for the Breestraat, a curving street some 700 metres in length, and the lower, protected land for the Gravensteen – a residence of the Counts of Holland. A moat and a wall were constructed from 1266 along the line of present-day Rapenburg.

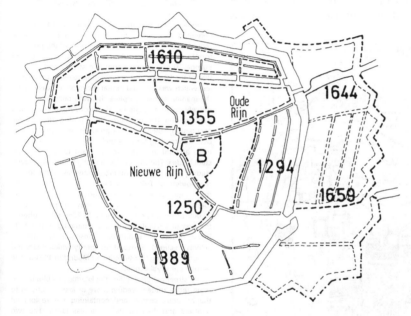

Figure 4.78 – Leiden, at mid-seventeenth century: abstracted from the map published by Joannes Blaeu, with the dotted outline of subsequent seventeenth-century extensions. Blaeu's map shows a defensive perimeter approximating in outline to the ideal circular shape, enclosing an essentially 'formless' city, lacking any centrally unifying element, which had grown up by the addition of a number of successive new districts.

From 1294, in response to rapidly increasing prosperity based on textile trade, a *grachenstad* extension was planned to the south of the *burcht* (B), making use of the old *sloten* (drainage ditches) as the basis of four approximately parallel canals, the outer one of which, the Heeren Gracht, formed the moat of the Blaeu map. Two fourteenth-century extensions are shown which, with that of the early seventeenth century, comprised Blaeu's Leiden. With a population exceeding 10,000, Leiden by 1500 had become the third largest city in the Netherlands. Later still, between 1644 and 1659, three further new districts were added across the southern side of the city. The total urban area was then approximately 170 hectares. See also J Goss, *Braun and Hogenberg's The City Maps of Europe: A Selection of 16th Century Town Plans and Views*, 1991.

HAARLEM

The city has its origin in the choice by the Counts of Holland of marginally higher ground enclosed within a bend in the river Spaarne as the site of their residence, the Bakenesse, which they occupied from the eleventh to thirteenth centuries. The service-town which established itself alongside the castle was defended by a ditch, the Bakenessegracht, cut across the river bend. A charter was granted by Count Willem II in 1245. The first extension was to the west of the Bakenessegracht, centered on a new market square known as t'Sand, denoting its position on a dry ridge. A new wall and moat, the Oude Gracht, enclosed an area of about 90 hectares in 1250. In 1335 and 1360, two geestgrond districts were added across the Spaarne to the east, and then at the beginning of the fifteenth century there followed the extensive new area to the west and south of the Oude Gracht (which became a commercial canal), enclosed within the wall and moat of the Raamsingel and Zijlsingel. The total area was then nearly 365 hectares.

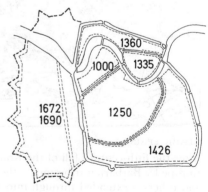

Figure 4.79 – Haarlem: the structural form of the city in 1647, abstracted from the map engraved by F de Wit, and published by Pieter Wils. The city's historic nucleus was completed during the last decades of the seventeenth century when, from 1672, a further 200 or so hectares were laid out according to a plan prepared by Salomon de Bray in 1644.

Spain: the eighth to thirteenth centuries

This description of Spanish medieval Islamic cities follows on from Chapter 1's introductory description of the general characteristics of Islamic urban form, as derived from the Mesopotamian organic growth precedent. Spain was divided during the period between the Christian north and the Islamic south; the boundary being pushed back further south in stages, after the initial Muslim conquest was stemmed (as shown by Figure 4.80).[121] An account of the *reconquista* and its formative effect on Spanish imperial policies and Latin American urban form introduces Chapter 9.

A clear distinction must be made between northern forms of settlement, and those which became established in Islamic Spain. Whereas usual European patterns are found in the north, with some Spanish modification in detail; the Islamic cities of southern and central Spain are the uniquely different cities of Western Europe; their characteristics those of eastern Arabian tradition: in direct contrast to those of the Christian north.[122]

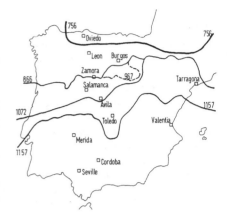

Figure 4.80 – The Iberian Peninsular: a map showing the successive phases of the *reconquista*, and locating the cities described in this medieval Spanish section.

Figure 4.81 – Toledo: the old city on its superb, naturally defended location in the bend of the River Tagus, requiring only the wall across the neck of the promontory to make it one of the most impregnable cities of Europe. The enduring pattern of streets and lanes is that of the Islamic city of the eighth to eleventh centuries, which was unusual in that it comprised essentially the madina as the complete city, with only the comparatively small suburbs (*arrabals*) of Antequeruela and Santiago on the northern side. During the Islamic occupation, Toledo became an exceedingly prosperous city; outstanding for its woollen and silk textile industries, and the manufacture of finest quality Toledan steel blades. The city's large Jewish community played leading roles in the commercial activities, and after the *reconquista* of Toledo by the Castilian army in 1085 their importance increased. During the reign of the tolerant Ferdinand III (1217–52) they comprised some 12,000 of the city's inhabitants. In 1355 they suffered the first of several pogroms, culminating in the 1492 Royal decree of expulsion. Consequently Toledo's activity declined; a loss of prosperity which increased rapidly following the selection of Madrid in 1561 as the permanent royal capital.

The street pattern typifies the Islamic combination of main radial routes, leading more or less directly in to the religious centre, and containing the residential quarters and their cul-de-sac access lanes. The two characteristically dominant buildings are the Cathedral, constructed from the early thirteenth century on the site of the Mezquita Mayor, and the Alcazar, reconstructed after Civil War devastation in 1936 to its appearance during the time of Charles V, when it was an imperial residence. There was neither the space, nor a sufficiently level area, to create a formal Plaza Mayor, and that role is performed by the irregularly triangular Plaza del Zocodover, also reconstructed after Civil War damage.

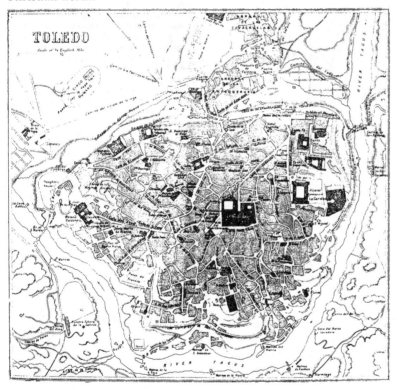

ISLAMIC SPAIN

After the fifth-century fall of the Roman Empire, conquest of Western Europe by Barbarian tribesmen – the Visigoths, Vandals, Suevi, Alans and others – extended through into the Iberian peninsular. Visigoths (West Goths) eventually ruled in Spain, taking over the major Roman cities but characteristically at reduced scales of economic activity. By the eighth century the Visigoths were in weakened disarray and Arab and Berber armies invaded from North Africa. Referred to henceforth as the Moors, a first plundering band of 500 men made a successful sortie in 710 to be followed the next year by an army of some 5,000. In

121. For the most relevant general background history see J H Elliott, *Imperial Spain 1469–1716*, 1983, Chapter 2, which sets a context for the Spanish imperial enterprise.
122. In addition to the formative Islamic influences in southern and central Spain, there are also the comparable towns in south-eastern Europe – notably Greece – which were part of the Islamic world from the 1450s, after the fall of Constantinople, through to the early nineteenth century, and those of European Turkey, in particular Istanbul (Constantinople) which are still within Islam (see Chapters 2 and 3 respectively).

711 in battle on the River Guadalete, the Visigoth forces were routed and an easy conquest of southern and central Spain proceeded apace. Cordoba and Malaga were lost in that year; Sevilla in 712; Valencia, Gerona, Zaragoza and Lugo by 714. Barcino (Barcelona) fell in 713 but was regained in 801.[123]

Cordoba remained a Muslim city until reconquered in 1236, Sevilla until 1248. Toledo had been regained in 1085, Saragossa in 1118, but Granada, a comparatively isolated mountainous enclave, was tolerated until 1492, on the eve of the Spanish Latin American Empire. Centuries of change left indelible Islamic marks on the cities concerned, greatly to the enduring benefit of European urban culture.[124]

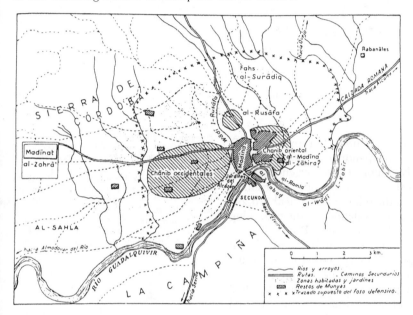

CORDOBA

The favourable location of Roman Cordoba had seen its promotion to capital of Baetia province, an advantage recognized by the Visigoths who made Kordhoba the religious centre of their Christian Kingdom.[125] The city was taken by the Moors in the first year of their conquest and steadily gained importance, first as a regional centre and then as the capital of Islamic Spain. It was at its peak under the Omayyas dynasty in the tenth century and could rival the prosperity and splendours of Baghdad and Constantinople; while the rest of Western Europe languished back at the level of semi-barbarian villages. By then, it would seem, the Roman gridiron had been completely overlaid by the city's characteristic Islamic Spanish form.[126] In the early eleventh century Cordoba was the centre of Moorish civil wars, suffering damage from which it did not recover. Although remaining the cultural centre of the western Islamic empire, its influence continued to wane until it was finally extinguished in 1236 when Ferdinand III of Castile regained the city for the Christian north. The outline map of Cordoba in the tenth century shows the Medina on the right bank of the Guadalquivir enclosed within extensive suburban districts. The left bank bridgehead suburb of Secunda in the bend of the river had been razed in 818 and left in ruins as a necropolis.

Figure 4.82 – Cordoba: the outline map of Cordoba in the tenth century shows the Medina on the right bank of the Guadalguivir enclosed within extensive suburban districts. The left bank bridgehead suburb of Secunda in the bend of the river had been razed in 818 and left in ruins as a necropolis. It is believed that the Medina, which had a walled circumference of about 2½ miles, occupied much the same area as that of the Roman city. A main street from the north ran through the Medina to the Roman bridge at the Bab al-Quantara (Puerta del Puente), the most important of the total of seven gates. Inside the Medina the two main buildings were the Mequita mayor (the cathedral Mosque) and, on the downstream side of the bridge, the Alcazar.

The Mosque occupies the site of the Visigothic Christian Cathedral of St Vincent, and was enlarged subsequent to its initial completion in c. 790. The Christians had been allowed to keep St Vincent as their Cathedral until 747, after which date they had to surrender half of its area as a Muslim mosque. The Moors soon found that space too limited and they purchased the remaining half to complete the site of their new congregational mosque.

The Alcazar also had become inadequate for the requirements of the Court of the Caliphs by the tenth century, and Abd ar-Rahman III built the magnificent new palace-city of Madinat al-Zahra on a site overlooking the river valley some 7 km west of the centre of Cordoba. This Islamic precedent of the seventeenth-century Versailles, required some twenty-five years to build and absorbed one-third of the state's revenues. Its total area of 279 acres contained numerous private and public buildings, courtyards and parks, and latterly its own commercial and market quarter, before its complete destruction around 1010 in the Civil War.

Much confused, I left the Great Mosque. In many respects it was unsatisfying; its Christian and Muslim halves were uneasy with each other and failed to attain that harmony one finds in Sancta Sophia. There is here a sense of imbalance and restlessness, as if the Muslim component of Spanish life had accepted its role of submission and were trying to escape; or as if the Christian component were not content with its conquest and were endeavouring to suppress even further the Moorish. In such circumstances the Muslim explodes to the surface through weakened fissures. There is in much of Spain this contradiction: it is a Christian country but one with suppressed Muslim influences that crop out at unforeseen points; it is a victorious country that expelled the defeated Muslims from all places except the human heart; it is a land which tried to extirpate all memory of the Muslims but which lived on to mourn their passing; and it is a civilization which believed that it triumphed when it won the last.
(J A Michener, Iberia, 1968)

123. For the conquest by the Moors see Davis, *A History of Mediaeval Europe*; F McGraw Donner, *The Early Islamic Conquests*, 1982.

124. For the *reconquista* see Elliott, *Imperial Spain 1469–1716*; J H Parry, *The Spanish Seaborne Empire*, 1966/1977; see also Chapter 9.

125. For Cordoba see A J Arbery, 'Muslim Cordoba', in A Toynbee (ed.) *Cities of Destiny*, 1967; and R Ford, *Murray's Handbook for Travellers in Spain*, volume II, 1898.

126. Through to the late 1991 date of revision, virtually nothing is known of the (presumed) gridiron plan of Roman Cordoba. The historic right bank nucleus has largely retained its Islamic form.

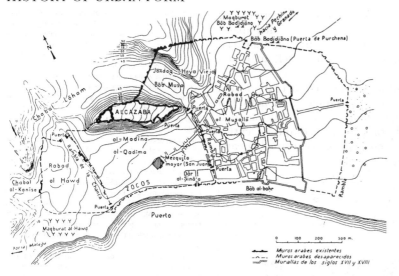

Figure 4.83 – Cordoba: a detailed present-day sketch plan of the north-bank bridgehead area relating the Mezquita (M) to the Alcazar (A) and the surviving Islamic lanes and housing access culs-de-sac. The orientation is the same as Figure 4.82 (see also Figure 9.4 showing the late nineteenth- and early twentieth-century gridiron residential district developed on the site of the abandoned western Islamic suburb).

Figure 4.84 – Almeria: the general form of the Islamic city, relating the eastern and western suburbs (the Rabad al-Musalla and the Rabad al-Hawd, respectively) to the Madina al-Qadima on the generally sloping ground between the hilltop Alcazaba and the sea. This plan shows the contracted extent of the seventeenth and eighteenth century within the bastioned eastern defences of that period.

CHRISTIAN CITIES

In those northern parts of Spain which remained free from Islamic domination, or early regained their independence, the nature of medieval urban settlement more closely conformed to the general western European characteristics than was the case in the conquered south. There are broad differences, however, in the relative numerical proportions of the types of medieval urban origins, as defined at the beginning of the chapter.[127] The comparatively few northern Roman towns underwent usual grid modification, while the limited agricultural potential of the northern regions supported fewer market towns transmuted from village origins. Conversely mountainous terrain and centuries of anarchy resulted in a higher proportion of strongly fortified military-origin burgs. The foundation of new towns was concentrated during and following the *reconquista* in central and southern Spain, where they played important strategic roles in previously Islamic regions.[128]

Organic growth – Roman origins

Not all the northern Roman towns had been given a regular gridiron plan; a proportion would have their pre-existing unplanned forms tidied-up to varying extents. In the former case, when urban regeneration followed resurgence in economic activity, the Roman street pattern lost much of its rectilinear uniformity such that by the thirteenth to fourteenth centuries all towns of Roman origin had acquired broadly similar characteristics.

Leon is an example of a Roman *urbs legionis* (from which its modern name derives) which comprised a regular gridiron legionary fortress and one or more extra-mural settlements which grew up, unplanned, for Roman civilian and other inhabitants.[129] In the mid-sixth century, when the occupied area would have been within the line of the old walls, Leon was captured by the Visigoths. In 717 it fell to the Moors but was retaken in 742, being devastated in the process. Rebuilt, it was made capital of the kingdom of Leon, Asturias, and Galicia. From 996 to 1002 it was again in possession of the Moors, suffering yet more destruction. Alfonso V re-established Leon as a residence city in the

127. See E A Gutkind, *International History of City Development*, Volume III, *Urban Development in Southern Europe: Spain and Portugal*, 1967.

128. See Elliott, *Imperial Spain 1469–1716*; see also Chapter 9, where the Spanish domestic experience in new town planning was carried over into their New World dominions.

129. For Leon see Gutkind, *Urban Development in Southern Europe: Spain and Portugal*, 1967; R Ford, *Murray's Handbook for Travellers in Spain*, 9th rev. edn, 1898.

early eleventh century, a status which it held through until 1230. It was awarded a *fuero* in 1020 and received other privileges; yet its area at first was only some 50 acres, before a large southern suburb was incorporated towards the end of the thirteenth century. After all the devastation it had endured it is somewhat surprising to find the Roman gridiron remaining to the extent shown by the plan of eleventh-century Leon.

Organic growth – village origins

Towns promoted from village status by their increased economic activity were comparatively rare in northern Spain. Such opportunity was controlled by limited agricultural potential and the existence of an adequate, albeit dispersed, urban settlement pattern. Gutkind describes a type of town established by combination of neighbouring rural settlements, giving Salamanca and Soria as examples of such 'urban' agglomerations with originally enclosed areas respectively of 272 acres and 247 acres.[130] Salamanca, which is centred on three low hills alongside the River Tormes, had had a long history of Celtic, Roman and Moorish occupation before, about 1102, its repopulation for the Christians by Count Raymond de Bourgogne. Five distinct ethnic settlements were established which were unified within the one defensive system under Alfonso VII from 1147. Early in the fourteenth century the wall was extended and the population including students of the first Spanish University (founded in 1230) is estimated as high as 50,000.

Organic growth – Burgos

Burgos exemplifies this type of new settlement. Founded, it is believed, in 884 by Count Diego Rodriguez Porcelos, the castle controlled the crossing of the River Arlanzon and the town became established beneath it, alongside the river. The main streets within the walls follow the contour lines; cross-access is by steep lanes and steps. By the fourteenth century Burgos had become one of the wealthiest cities in Spain and after the union of Aragon and Castile in 1574 it was one of the capital cities until Madrid was made *unica corte* in 1560 (Figure 4.86). Vitoria is a smaller counterpart, founded on a hilltop in 581, which received its charter in 1181 and is notable for the concentric uniformity of its hillside streets below and around the original nucleus (Figure 4.87).

Planned 'new towns'

Numerous systematically planned urban settlements were created in Spain by victorious Christian rulers, either as strategically located military bases, needed to hold still disputed territorial gains, or as a main basis of the interior repopulation that followed the *reconquista*. While these towns are of importance in themselves, as further variations on the theme of European medieval new town planning, their main significance is that of providing a determining influence on the colonial settlement policies adopted by the Spanish conquistadores in Latin America, as described in Chapter 9.

Santa Fe, one of the most regularly planned examples, was constructed as the royal military base from which the final assault on nearby Granada was conducted. The year was 1492, when Columbus set sail for his unimagined 'new world' and it was most probably to Santa Fe that he was summoned to report on his return the following year.[131] The town is described in Chapter 9.

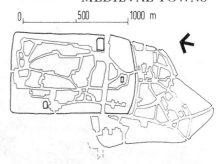

Figure 4.85 – Leon: the street plan of the early eleventh century within the extent of the Roman wall (in heavy line) showing the survival of the main west–east axis through the city from the bridge over the Bernesga, and the remnants of other gridiron streets. The Plaza Mayor occupies an area immediately outside the old wall, within the large suburban addition of the late thirteenth century.

Figure 4.86 – Burgos: the burg on its impressive hilltop, commanding the crossing of the River Arlanzon, and the distinctive organic growth form of the lower town, within the line of the medieval wall, its street pattern determined by the contours sloping down to the south-east.

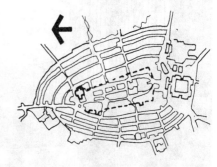

Figure 4.87 – Vitoria: the street plan of the old medieval city which developed outwards from the hilltop nucleus, following the contours to a remarkable extent. The Plaza de Espana is a characteristic example of an enclosed eighteenth-century plaza mayor; see also Chapter 9.

130. Gutkind, *Urban Development in Southern Europe: Spain and Portugal.*
131. For Santa Fe see Gutkind, *Urban Development in Southern Europe: Spain and Portugal*; J W Reps, *The Making of Urban America*; see also Chapter 9.

Figure 4.88 – Avila, Spain, situated 90 km west-north-west of Madrid, halfway to Salamanca. As with the old Cite at Carcassonne (Figure 4.40), this best preserved Spanish medieval walled city appears too good to be true; yet, on close aquaintance there is little of the self-conscious tidiness which reveal Carcassonne to be essentially a nineteenth-century reconstruction.

Richard Ford in Murray's *Handbook for Travellers in Spain* of 1898 found Avila to be 'undoubtably one of the most picturesque towns in Spain ... its granite walls are perfect; they were begun in 1090 and are 40 ft high and 12 ft thick, and there are no less than 86 towers and 10 gateways. Before the use of artillery the city must have been impregnable, for every point commands the plain below the hill on which it is built and even the grand Cathedral is half church, half fortress.' James A. Mitchener writing in his *Iberia*, 'went north over the Grados mountains to the walled city of Avila, judged by most people to be the finest mediaeval remnant in Spain ... the gates look as if horsemen might clatter out through the portcullis ...

Figure 4.89 – Alberca, Spain, still not much more than a large village, situated 80 km south-west of Salamanca. The high-noon photograph of the little central Plaza epitomises the mediaeval urban character of small-town inland Christian Spain. All that is missing is the heat and the silence of deserted open spaces. Ford describes Alberca in course of an equestrian excursion from Plasencia to Ciudad Rodrigo, for which he cautions 'attend to the provend and take a local guide.' Passing through Alberca he was clearly unimpressed, noting: 'this dingy hamlet is composed of prison-like houses built of granite. Its situation is, however, extremely beautiful ...'

South Bohemia

This concluding section to Chapter 4 takes the form of an illustrated supplement and presents four virtually unknown towns in South Bohemia. Recent history has served to keep this most beautiful part of Czechoslovakia off the tourist map (although it is nearer to Calais than the French Riviera) and except for a few minor effects of military and industrial action, these towns have come through from medieval to modern times virtually unscathed.[132]

CESKE BUDEJOVICE

Ceske Budejovice has a present-day population of around 75,000 and is the main town of South Bohemia.[133] It is located about 90 miles south of Prague, on the east bank of the Vltava, where it is joined by its tributary the Malse. Although it first felt the impact of the Industrial

Figure 4.90 – Ceske Budejovice: a nineteenth-century photographic panorama of three sides of the magnificent central square, seen from the middle of the southern side. (See Figure 4.91, which illustrates the north-eastern corner in greater detail.)

Figure 4.91 – Ceske Budejovice: the north-eastern corner of the central square showing the arcaded building form. Inevitably the square is used as the town's main car park and is clear only on special occasions. The proposals for conservation and rehabilitation include provision of parking spaces outside the square. Compared with the nineteenth-century view the appearance of the square has hardly changed; moreover there is no need for it to do so in order to readily accommodate individual modern shop-fronts behind the arcade screen.

Revolution as early as the 1790s, Ceske Budejovice has expanded on the basis of new industrial suburbs, distinct from the old historic city, which has changed very little over the last 200 years. Ceske Budejovice was founded in 1265 as a royal town by the King of Bohemia, Premsyl Otakar II (1253–78). Known as the gold and iron king by his contemporaries, Premsyl II greatly enhanced the already powerful position held by the Czech state in Central Europe, extending its territory in the south practically to the Adriatic, and in the north exerting pressure on Poland by building outposts on the Baltic, including Königsberg (present-day Kaliningrad). The founding of Ceske Budejovice was the result of two strategic considerations – first, to enable the natural resources of the district to be more fully exploited; second, to consolidate the authority of the Czech crown over South Bohemia in the face of potential aggrandizement by the Austrians across the not very formidable mountain chain of Sumava and Nove Hrady to the south. Premsyl II organized two colonization programmes in South Bohemia. The other was centred on a new monastery at Zlata Koruna, to which extensive lands were granted, some 14 miles further south, just off the main road to Cesky Krumlov. Premsyl II selected as an agent for this work the Chevalier Hirzo, who was already a new settlement planner of some experience.

At Ceske Budejovice, Hirzo selected the site of the existing small village of Budejovice, situated among the marshes at the confluence of the multi-streamed Vltava and Malse. The main factor determining the plan of the new town was the need for it to be strongly fortified. For this purpose the site was ideal, and with relatively little effort an approx-

Figure 4.92 – Ceske Budejovice in the mid-eighteenth century (north to the right). After some 500 years of urban existence there are still only the beginnings of ribbon development along main routes from the gates. The Vltava and the Malse are immediately to the south and west.

Figure 4.93 – Ceske Budejovice: the modern conservation, rehabilitation and development plan prepared with great sympathy and imagination by the Czechoslovakian Institute for Historic Buildings and Monuments, in Prague, and associated local architects and planners. The historic nucleus, with its magnificent arcaded town square, is separated from the more recent suburban developments by the landscaped zone of the former fortifications containing main traffic routes around the core. A combination of historical accident and an enlightened modern planning programme has ensured the survival, largely unscathed, of the old town, although it is now surrounded by extensive housing districts and related industrial zones.

The primary locational, and several of the morphological determinants have been given in the main text. Others include climate – responsible for the steeply pitched roofs and ground level arcading; and aesthetic 'control' resulting in an unusually consistent sequence of arcade bays surrounding the central square, with their upper floors creating a series of attractive variations on the local vernacular construction theme.

In addition to the climate defence role of the arcades – which prompts the question why was such a natural response never a characteristic of mediaeval English urban form? – there is a further invaluable benefit to presentday urban designers whereby it is possible to insert modern shopfronts, and make other changes behind their unaffected solid architectural rhythm, without introducing immediately apparent discordant notes around the superb central space. And, at night, the intensity of lighting within the arcades further enhances their form and that of the square itself.

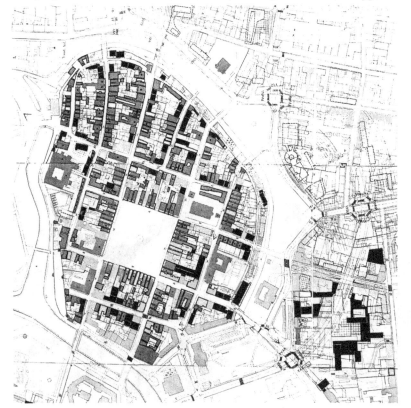

132. South Bohemia, as Czechoslovakia generally, was spared the street-fighting and aerial bombardment of the Second World War. Subsequently unlike in affluent Western Europe the comparatively backward communist-bloc economy did not finance discordant urban change: but it remains to be seen if 'democratised' city planning controls can continue to retain the best of the past.

133. J Pavel, *Ceske Budejovice* (photographs by V Fyman and V Hyhlik), Prague, 1965 (Czechoslovakian); E A Gutkind (ed. Gabriele Gutkind), *Urban Development in East-central Europe: Poland, Czechoslovakia and Hungary*, 1972.

imately oval-shaped area was raised above water level (Figure 4.93). A new canal served as a moat around the eastern and northern sides of the town, where it was not protected by the rivers. The main regional routes approaching the town determined the location of the three gates – Prague and Pisek, to the north; Trebon and Vienna to the east; and Linz, via Cesky Krumlov, to the south. Within the town the plan was dominated by the central square; each side measured 650 feet with the streets and building plots arranged on a regular gridiron basis, modified only slightly in the northern corner. The plan of Ceske Budejovice, one of a series prepared for its conservation and renewal, exemplifies the narrow street frontage, deep individual building plots which are a characteristic of the European medieval town. Work on draining and laying out the site started in 1263, and by 10 March 1265 Hirzo was able to report to the king that the town was effectively ready for occupation. Ceske Budejovice was populated from the surrounding countryside and further afield by extensive 'advertising'.

CESKY KRUMLOV

Cesky Krumlov is a three-part castle town intricately entwined with the river Vltava in a sharply defined valley, which is unexpectedly encountered in the otherwise gently undulating countryside that ascends to the mountains of the Austrian frontier.[134] With Durham in England (Figure 4.33), Cesky Krumlov is a pre-eminent European example of the effect of topography on urban form. From its earliest history, dating back well before the first recorded mention of the castle in 1235, the site of Cesky Krumlov has great political significance, controlling the river route from the south through to the Elbe, which the Vltava joins north of Prague. The Vltava takes the form of a triple

Figure 4.94 – Ceske Krumlov: a painting of the seventeenth century showing the lower, civil town in the foreground with the bridge across the Vltava to the newer bridgehead suburb, and the dramatically located castle of the Rozmberks on the escarpment beyond. The romantic informality of the building ensemble reflects the archetypal organic growth of the underlying urban form.

134. For Cesky Krumlov see V Mencl (ed.) *Czesky Krumlov*; E Samankova and J Vondra, *Cesky Krumlov*.

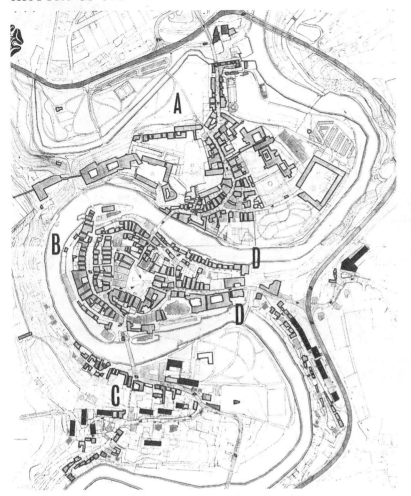

Figure 4.95 – Cesky Krumlov: north at the top, the river Vltava flowing south to north. Key: A, the castle and its associated town; B, the lower, 'civil' town around its market square, surrounded almost on all sides by the Vltava; C, later development across the river; D, the artificial moat cut across the narrow land approach to the lower town.

Only about 20 km south-west of Ceske Budejovice, Cesky Krumlov is the archetypal mediaeval organic growth town in east central Europe; the equivalent of Durham as a British representative, and Alberca as a small town in the Christian north of Spain (see Figure 4.89). Together, these two most attractive towns provide a conveniently related way of comparing and contrasting the two types of mediaeval urban form. The main text deals with the most important locational and morphological determinants, and the colour photograph on the front cover, taken just down river from the viewpoint of Figure 4.94 (opposite 'B' on the plan) further illustrates this most fascinating of mediaeval urban survivors. The use of soft-toned pastel-colour elevational painting is a characteristic of the local vernacular, for which see the facing caption to Figure 4.98.

Just as a photograph of Alberca can but fail to register the oppressive heat and the silence of summer noontime; neither can the colour view of Cesky Krumlov do more than hint at the crisp winter air, or serve to remind of a further urban dimension – that of the distinctive smell of acrid soft-coal smoke drifting through the valley. Similarly the photograph cannot possibly do justice to the equally attractive South Bohemian countryside, here in its winter guise.

bend creating three well-defined areas of land, the middle one of which is virtually a natural island. This area was in fact surrounded by water as a result of a moat excavated across it as part of the medieval defences. The artificial island so formed became the site of the lower civil market town of Cesky Krumlov developing, in the thirteenth century, from an earlier bridgehead settlement.

Immediately across the river to the north, on a high cliff, the powerful Rozmberk family consolidated, from the thirteenth century onwards, the earlier hill fortress as one of the most dramatically sited, urban-related castles of history. Below the castle and filling in the lower ground within the northernmost bend of the river, the castle town forms the third part of Cesky Krumlov. The existing development within the southern river bend is of more recent origin. The castle was largely rebuilt around the end of the sixteenth century, and further work was carried out in the eighteenth century in the Baroque and Rococo styles. Along the top of the cliff, to the west of the castle, formal gardens, now much overgrown, were laid out. The lower town comprised medieval and Renaissance buildings, intermixed with hardly a discordant note – either from the past or, fortunately, from more recent times.

Figure 4.96 – Tabor: general plan of the old Hussite town on its dramatic hilltop site (north at the top).

Figure 4.97 – Tabor: aerial view of the market square from the north-east, illustrating most of the characteristics of medieval organic growth form – the unevenly defined perimeter of the square which may have been regularly laid out but which tolerated gradual encroachment; the uneven roof lines and variety of building elevations which none the less add up to a consistent, harmonious ensemble; and the narrow meandering streets leading away into the town. Although not clearly shown in this photograph, the square slopes markedly, adding a further dimension to the scene.

TABOR

Tabor, on a magnificent hilltop location, is of relatively recent origin. It was founded at the beginning of the fifteenth century, around 1420, as a revolutionary centre of the Hussite movement. The Hussites, started by John Hus, the Rector of Charles University in Prague, were originally outspoken critics of the power of the Church in the Czech state before turning their attention to the repressive system of feudalism in general.

Hus himself was forced to leave Prague in 1412 to live at Kozi Hradek in South Bohemia, where he continued to preach against the papacy, with the support of the local nobility. In 1415 Hus was burned at the stake, resolutely refusing to recant. After his death the movement gained strength throughout the Czech lands. Prague was taken, and

Figure 4.98 – Tabor: a detail view of the buildings forming the western side of the square. Typifying the construction of South Bohemian towns, Tabor is deceptively solidly built. The local vernacular of this part of eastern Central Europe is that of comparatively rough stone masonry or brickwork rendered over as the base for applied architectural decoration and soft-toned pastel-coloured weather-resistant painting. As illustrated, the former generally embodies a low-key 'classical' informality with visual emphasis provided by decorative roof gable ends. The frequent use of colour adds greatly to the winter scene, providing pleasing contrasts to snowy foregrounds, whether pristine white or muddy grey. Unlike the southern European and Mediterranean reliance on whitewash for sunlight reflection; white coating is not in traditional use in northern Europe for protection and decoration. Not least because atmospheric soft-coal smoke pollution quickly turns it shades of grey. Without the visual relief provided by coloured buildings (as Cesky Krumlov on the cover), beneath damp grey skies the wintertime eastern European urban scene often can but assume depressingly dreary characteristics.

'revolutionary' authority was established in most Czech towns. Near to Kozi Hradek, where Hus had stayed, the new town of Tabor was created on the basis of a totally new kind of society: feudal privileges were abolished, all people were declared equal, property was held in common and both the administrative and military leaders were elected by popular vote.

Tabor can have changed little since the middle of the sixteenth century: those changes that occurred have resulted only in the substitution of Renaissance elevations for earlier medieval ones. Although new housing districts have been added in recent years, they have not blurred the clearly defined outline of the old city on its hill above the broad expanse of the Jordan Lake.

TELC

This marvellous little town, set among lakes in typically undulating South Bohemian countryside some 55 miles to the east of Ceske Budejovice, is without doubt one of the most beautiful towns in all of Europe, although it still remains one of the least known.[135] Telc had its origins in a moated fortress built towards the end of the thirteenth century and, during the succeeding one hundred years or so, the incomparable town square grew up either side of the approach route from the south-east. The unusual, elongated form of this square was determined by the narrowness of the neck of land in front of the fortress, bounded by lakes to the north-east and south-west, and the division of the approach route into two branches at its far end. The old town therefore took the form of a hollow shell, consisting only of the single row of burghers' houses fronting the square on each side.

The fortress was remodelled as a Renaissance palace in the latter part of the sixteenth century. The town itself was destroyed by fire on several occasions: a final rebuilding of the eighteenth century is essentially the Telc of today, although since 1945 a sympathetically conceived conservation programme has been carried out.

Figure 4.99 – Telc: view looking south-east from the palace, along the square, past the twin towers of St Mary's Church, and showing the lake to the north-east. The fifty or so burghers' houses around the square are connected by continuous arcades providing access to ground-floor commercial premises. Each of the houses has an individual elevation design, notably the gable profiles, yet the total effect is that of entirely successful, subtly expressed variations on a theme.

Telc is the author's favourite of these four South Bohemian towns – and there are others for which space could not be found: notably Jihlava and Prachatice, with Jicin of several elsewhere in Czechoslovakia. Telc is the first place to which I direct unfamiliar visitors to South Bohemia, while stressing that the others are conveniently close at hand. Although personally emotive intangibles play a main part in reaching such a conclu-

135. See V Mencl, *Mesta Hrady a Zamky*; E Vasiliak, *Nad Ceskoslovensken*, for aerial photography of Telc and numerous other historic towns.

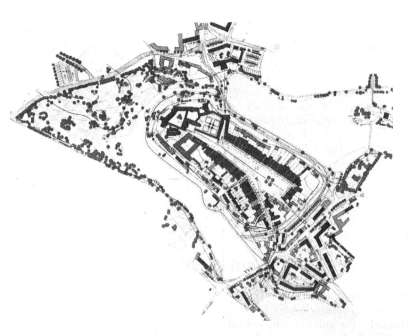

sion, Telc is favoured because it combines historic aspects of organic growth and planned morphology to an extent which is not only extremely unusual, but which also points highly significant lessons for modern urban designers seeking to give human faces to necessarily standardized buildings . . . and who all too often end up with shallow cosmetics.

Who today, for instance, when charged with designing the basically repetitive sides of an urban space, would think of a series of variations on a three-bay upper floor theme, flowing over and above – to use a musical allegory – a solid rhythmic arcade base? And yet that is exactly what an unknown urban designer created for Telc when it was rebuilt after fire in the eighteenth century. (However, it is another matter to attempt to recreate, overnight as it were, history's centuries-old subtle shaping of the square itself.)

Figure 4.100 – Telc: general plan of the town (north at the top), with the palace and church occupying most of the north-western corner. The town was protected by a wall constructed across its south-eastern end, between the two lakes.

Figure 4.101 – Telc: the burghers' houses forming the north-eastern side of the square, showing the continuous arcade at ground level and the different, yet perfectly compatible, upper-floor elevational designs.

Medieval urban populations

This chapter has established that during the Middle Ages in Western
Europe the great majority of people lived in agricultural villages. Unfortunately, at neither end of the period is it possible to give accurate
figures for the proportions of the total populations that lived in the
country and in the towns. Reasons for this include the problem of
defining urban status and the need to interpret such population records
as are available. However J C Russell found it possible to arrive at
reasonably convincing estimates, both for the English national population and for the numbers of inhabitants in boroughs, based respectively
on Domesday Book and the poll tax.[136]

Russell calculated total populations of 1,099,766 in 1086 and of
2,232,375 in 1377, based on an average 'household' figure of 3.5 (considerably less than assumptions made by other historians). He
accepted for his purpose that settlements with more than 400 inhabitants could be classed as boroughs.

If there were eighty settlements with urban status in 1086 and if the
average number of their inhabitants was 1,250, then the total 'urban'
population of 100,000 would have been something under 10 per cent of
the national total. By 1377 this proportion would certainly have risen to
between 10 and 15 per cent but, unfortunately, the poll tax evidence on
borough populations was incomplete. Russell also concluded that it
was probable that a smaller proportion lived in the boroughs in 1545
than in 1377. (See the conclusion to Chapter 8 for the effect on population location of the Industrial Revolution in the eighteenth and
nineteenth centuries.)

136. J C Russell, *British Mediaeval Population*, Albu-
querque, 1948: the basis on which Russell calculated
his medieval populations has been criticized (possibly
because of the lack of rounding-off?) but his conclusions still provide valuable comparative data. The table
below gives Russell's population figures for the ten
largest boroughs in 1086 and their comparable 1377
totals, together with the factor of change (not percentage) and the corresponding factor for the counties in
which they were situated.

Town	Population 1086	1377	Factor of change	Factor of change for county	
London	17,850	34,971	1.96	Middlesex	2.02
Winchester	6,000	1,440	0.24	Hampshire	1.44
Norwich	4,445	5,928	1.33	Norfolk	1.54
York	4,134	10,872	2.63	Yorkshire	6.88
Lincoln	3,560	5,354	1.50	Lincolnshire	1.58
Bristol	2,310	9,518	4.12	Gloucestershire	2.01
Gloucester	2,146	3,358	1.56	Gloucestershire	2.01
Cambridge	1,960	2,853	1.46	Cambridgeshire	2.31
Hereford	1,689	2,854	1.69	Herefordshire	1.26
Canterbury	1,610	3,861	2.40	Kent	1.84

5 – The Renaissance: Italy sets a Pattern

The Renaissance period in urban history is taken as extending from its commencement in Italy, at the beginning of the fifteenth century, until the end of the eighteenth century. 'Indeed,' as Sir Patrick Abercrombie observes, 'it might be placed a little later at each end, for Bacon's dictum that men come to build stately, sooner than to garden finely, holds good also of site planning, which does not make its appearance until the Renaissance is well advanced ... and continues to lap over into the nineteenth century.'[1] It is important to bear in mind that Renaissance urbanism spread slowly from Italy to other European countries, taking some seventy-five years to reach France and a further eighty-five years to become established in England. Renaissance architecture – the essential precursor of urbanism – takes over from the Gothic as the momentum of that style wanes. Never strongly established in Italy, Gothic architecture was at its fifteenth-century peak in England at a time when the Renaissance was fully under way in both Florence and Rome. In turn the Renaissance flowered and died. In its final decadent phase it was to be overwhelmed, initially in Britain, by the irresistible, uncontrollable onslaught of the Industrial Revolution's urban expansion.

The term Renaissance means, literally, rebirth: a revival of interest in the classical art forms of ancient Rome and Greece, and their use as the inspiration of European painting, sculpture, architecture and urbanism. The competition-winning designs of Lorenzo Ghiberti for new bronze baptistery doors at Florence Cathedral in 1401 are generally taken as the first sign of the Renaissance in the plastic arts. The first architecture is similarly seen to be the Foundling Hospital designed by Filippo Brunelleschi, also in Florence, which was started in 1419. The earliest Renaissance urbanism – the conscious arrangement of buildings into a predetermined form – is considered to be the Via Nuova in Genoa, of 1470.[2] The development of the Renaissance in the plastic arts is closely linked with the growth of literary and scientific humanism. In predating them by about a century, this established an intellectual context favourable to a successful revolt against reactionary medieval mysticism. In this the leading writers are Dante (1265–1321), Petrarch (1304–74) and Boccaccio (1313–75). Following their lead came many scientists and voyagers whose work extended the knowledge of the physical world.

The Renaissance originated in Florence where, as Nikolaus Pevsner observes, a particular social situation coincided with a particular nature of country and people, and a particular historical tradition.[3] The social situation is that of an immensely rich and powerful city

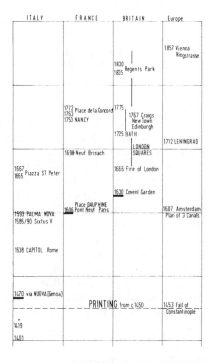

Figure 5.1 – Time chart showing the duration of the Renaissance in Italy, France and Britain with dates of key buildings and examples of urbanism in these three countries and, as a fourth column, other European countries.

1. Sir Patrick Abercrombie, *Town and Country Planning*, 1959. I am still (1992) a great admirer of this valuable, 'little' book but see also N Pevsner, *An Outline of European Architecture*, 7th edn, 1963; R Wittkower, *Art and Architecture in Italy 1600–1750*, 1973.
2. Contemporary with the Via Nuova at Genoa, possibly preceding it as that which can be regarded as the first Renaissance urbanism, there is the Piazza in front of the Cathedral at Pienza, near Siena, designed by Bernardo Rossellino in about 1460. See L Benevolo, *The History of the City*, 1980.
3. N Pevsner, *An Outline of European Architecture*, 1963; see also J H Plumb (ed.) *The Horizon Book of the Renaissance*, 1961.

state, in which leading families were active patrons of the arts. For these wealthy merchants the religious conventions of medievalism had little attraction, 'they tended to worldly ideals, not to the transcendental; to the active, not to meditation; to clarity not to the obscure'.[4] Further- more Italy had never taken to the essentially northern, Gothic style of architecture with any enthusiasm, and it was a country with innumer- able impressive, if ruined, Roman remains. It seems inevitable that in Florence 'the clear, proud and worldly spirit of Roman antiquity should be rediscovered . . . that its attitude to physical beauty in the fine arts and beauty of proportion in architecture found an echo'.[5]

From Florence the new architectural style quickly spread throughout Italy and by the end of the fifteenth century it had become firmly established in Rome. Although the term Renaissance is used in this urban history for the entire period, architectural history usually divides it into phases: Early Renaissance (1420–1500); Late Renaissance (1500–1600); Baroque (1600–1750); and Rococo or Neo-classical (1750–1900). Of these phases, the Baroque is the only term with rele- vance to urban history, as explained below. (The time-chart as Figure 5.1 shows the key dates for Renaissance urbanism in Italy, France, Britain and elsewhere in Europe.)

A key factor in the spread of the Renaissance was the development of printing, which following, it seems, some obscure preludes in The Netherlands, was used at Mainz for reproducing books around 1450 by John Gutenberg.[6] In architecture and urbanism two significant in- fluences resulted from the 'discovery' in about 1412, and printed pub- lication in 1521, of the writings of Vitruvius – an architect practising in the Rome of Augustus, and from the influx of Greek scholars and artists into Italy from Constantinople when it was captured by the Turks in 1453. At a time when architects were making detailed studies of the surviving Roman Buildings, on which to base their designs, the *De Architectura* of Vitruvius had a mystical significance far beyond its real value to architects. The influence of the Constantinople émigrés has a modern parallel with that of the architects and designers who, forced to leave Nazi Germany in the 1930s, worked elsewhere in Europe and the USA.[7]

Renaissance urbanism

This section serves as an introduction to Renaissance urbanism in European countries generally, since as Abercrombie writes, during all this period, in spite of the marked changes of architectural style from Bramante to Adam, the planning continued to be practised on nearly similar lines.[8] The Renaissance coincided with marked increases in the extent and population of European cities. London's population grew from about 50,000 in 1530 to 225,000 by the early 1600s, although her boundaries were constrained until later on in that century. Berlin's population expanded fivefold between the mid-fifteenth and early seventeenth centuries. Rome – probably presenting the most marked change – grew from an estimated 17,000 in the 1370s to about 124,000 by 1650, yet this was still perhaps but one-tenth of her former maxi- mum total.

Mainly because of size there were few opportunities for comprehen- sive redevelopment. Destruction by fire – the greatest scourge of pre- nineteenth century towns – or military action, as on the continent,

The historian Guicciardini, writing amid the troubles of the sixteenth century, thus describes the Florence of Lorenzo dei Medici in which his childhood was spent: 'The city was in perfect peace, the leading citizens were united, and their authority was so great that none dared to oppose them. The people were entertained daily with pageants and festivals; the food supply was abundant and all trades flourished. Talented and able men were assisted in their career by the recognition given to art and letters. While tranquillity reigned within her walls, externally the city enjoyed high honours and renown.' All the features upon which Guicciardini dwells – the internal peace and prosperity of Florence, her artistic and literary pre-eminence, and the prestige which she enjoyed throughout Italy – were due in large measure to the direction of her affairs by the Medici.
(A Hearder & D P Waley, A Short History of Italy)

As part of his research Toscanelli wished to measure the sun's noon height. The higher he could place the upright of his gnomon, the longer the shadow, and the more accurate would be his calculations. In 1468 he obtained permission to construct a gnomon in the cathedral, using as his upright a column of the lantern of Brunelleschi's dome: it was typical of Florentine broadmindedness to allow a scientist to conduct ex- periments in the house of God. Toscanelli installed a brass plate on the cathedral floor near the new sacristy. By measuring the shadow on this plate he was able to calculate the sun's meridian altitude, and thence plot its relation to the earth over the months. One of the dis- coveries he made was that the equinox fell twenty min- utes earlier than would follow from tables based on the Ptolemaic system This small discovery has an intrinsic and also symbolic importance. It shows a scientist, armed with Platonic hypotheses, finding that his observations disagree with Ptolemy, the accepted authority in the field; and this was to be the pattern of future developments.
(V Cronin, The Florentine Renaissance)

4, 5. N Pevsner, *An Outline of European Architecture*, 1963; see also J H Plumb (ed.) *The Horizon Book of the Renaissance*, 1961.
6. C W Previte-Orton, *The Twelfth Century to the Re- naissance*, 1952.
7. Notably Walter Gropius, Mies van der Rohe and Marcel Breuer, whose combined influence, with that of other expatriates, has had a continuing and widespread effect on the course of architecture in the United States; their influence, however, did not extend through into significant-scale city planning, either in practical or in- fluential theoretical ways.
8. Abercrombie, *Town and Country Planning*.

seldom necessitated complete rebuilding and when a chance was presented, as happened after the 1666 Fire of London, there was neither the political will to rebuild to a new plan, nor the bureaucratic means to achieve such an end. Furthermore there was no demand for new commercially orientated urban settlements. Europe generally was adequately and in part even over-provided with such towns. Only in the closing decades of the Renaissance did industry become a significant generator of urban settlement. The relatively few new foundations of the fifteenth to eighteenth centuries were therefore primarily either of a strategic military origin, e.g. Palma Nova in Italy, Neuf Brisach as a Vauban example in France, and Christiansand in Norway – or the result of autocratic rule, e.g. Richelieu and Versailles in France and Karlsruhe in Germany.[9] St Petersburg, the only example of a major city founded during the Renaissance period, combines both origins.[10] Renaissance urbanism was therefore effectively limited either to the expansion of existing urban areas, or to their redevelopment in part. Furthermore, as explained below, there was only limited activity in other than the main cities.

For the purposes of this introductory study five broad areas of Renaissance urban planning can be distinguished: fortification systems; regeneration of parts of cities by the creation of new public spaces and related streets; restructuring of existing cities by the construction of new main-street systems which, extended as regional routes, frequently generated further growth; the addition of extensive new districts, normally for residential purposes; and the layout of a limited number of new towns. The construction of fortification systems had a primary determining effect on the form and social conditions of many continental cities (as also did the absence of this constraint in Britain). The role of fortifications is therefore considered separately, later in this chapter.

Renaissance urbanists had three main design components at their disposal: first, the primary straight street; second, gridiron-based districts; third, enclosed spaces (squares, piazzas and *places*). Writing about their use in general, Abercrombie remarks that these components 'were sometimes fused together to make a composite plan, but more often were found somewhat disjointedly used, as though the designer had now been under one influence, now under another'.[11] Before briefly describing the application of these components in general, and the related key examples of urban planning activity (regeneration, restructuring, expansion and new site layout), we must first establish the differences between early Renaissance and Baroque architecture (as defined earlier), the essential characteristics of Baroque urbanism, and briefly consider several significant social and political background factors.

RENAISSANCE AND BAROQUE

During the Renaissance period of urban history the aesthetic determination of spatial design and that of the enveloping architecture was more closely integrated than at any other time. This holds true equally for the early and Baroque phases, when generally applied rules of proportion governing the plans, three-dimensional massing and detailed elevational design of buildings were extended outwards for the organization of urban space.[12] Contrasting Renaissance and medieval urbanism, Zucker writes that 'from the fifteenth century on, architectural design, aesthetic theory, and the principles of city planning are

---

Pitirim Sorokin (in his work Society, Culture and Personality) *traces the period of emergence of 355 great cities, those having more than 100,000 population.*

Period of Emergence	Number	%
BC to 5th century AD	67	18.8
6th to 10th century	69	19.4
11th to 15th century	75	21.2
16th to 20th century	144	40.5
Total	355	99.9

Forty per cent of the cities he studied developed in the sixteenth to the twentieth centuries, and another twenty-one per cent in the eleventh to the sixteenth centuries. Sorokin's data indicate that great cultural changes had to take place before great cities could be built and before people could survive within them. Such changes also had to be made before the world's population could increase substantially and before it could have a somewhat steady rate of growth.
(W E Cole, Urban Society)

In the old medieval scheme, the city grew horizontally: fortifications were vertical. In the baroque order, the city, confined by its fortifications, could only grow upward in tall tenements, after filling in its rear gardens: it was the fortifications that continued to expand, the more because the military engineers had discovered after a little experience that cannon fire with non-explosive projectiles can be countered best, not by stone or brick, but by a yielding substance like the earth: so the outworks counted for more than the traditional rampart, bastion, and moat. Whereas in earlier baroque fortifications, the distance from the bottom of the talus to the outside of the glacis was 260 feet, in Vauban's classic fort of Neuf-Brisach it was 702 feet. This unusable perimeter was not merely wasteful of precious urban land: it was a spatial obstacle to reaching the open country easily for a breath of fresh air. Thus this horizontal expansion was an organic expression of both the wastefulness and the indifference to health that characterized the whole regime.
(L Mumford, The City in History)

---

9. In contrast to the lack of means in London (see Chapter 8), both Pope Sixtus V in Rome (see below) and Louis XIV in Paris (see Chapter 6) were able, in their different way, to realize their city planning ideas, the former in order to enhance the status of the Roman Catholic Church and the latter to the glory of the French monarchy.

10. See Chapter 7 for Karlsruhe and for St Petersburg, which has now regained its old name after the decades as Leningrad.

11. Abercrombie, *Town and Country Planning.*

12. See E Bacon, *Design of Cities*, rev. edn 1974.

directed by identical ideas, foremost among them the desire for discipline and order in contrast to the relative irregularity and dispersion of Gothic space'.[13] The characteristic informality of medieval (Gothic) space, even when developed from a regular plan, resulted in the picturesque effect of Gothic architecture's asymmetrical massing, punctuated skylines and frequently intricate detailing. Renaissance architecture on the other hand rejected asymmetrical informality for a classical sense of balance and regularity: emphasis was placed on the horizontal instead of the vertical.[14] 'Gothic architecture,' observes R Furneaux Jordan, 'was born in France and, however many palaces or castles were built, was primarily ecclesiastical. The Renaissance was born in Italy and, however many churches were built, was primarily royal and mercantile – especially north of the Alps,'[15] a distinction which is returned to below when considering the nature of Baroque urbanism.

For a simple, straightforward summary of the essential differences between early Renaissance and Baroque architecture we can do no better than to refer to an art-historian, Heinrich Wölfflin, who wrote: 'In contrast to Renaissance art, which sought permanence and repose in everything, the baroque had from the first a definite *sense of direction*.' Wölfflin adds that 'Renaissance art is the art of calm and beauty ... its creations are perfect: they reveal nothing forced or inhibited, uneasy or agitated ... we are surely not mistaken in seeing in this heavenly calm and content the highest expression of the artistic spirit of that age.... Baroque aims at a different effect. It wants to carry us away with the force of its impact, immediate and overwhelming. Its impact on us is intended to be only momentary, while that of the Renaissance is slower and quieter, but more enduring, making us want to linger for ever in its presence. This momentary impact of baroque is powerful, but soon leaves us with a certain sense of desolation.'[16]

Wölfflin's profound observations are directly applicable to urbanism. Early Renaissance spatial organization aspired to a quiet, self-contained balance: the result is essentially *limited* space at rest. The Piazza Annunziata in Florence is perhaps the clearest example of this philosophy (see Figure 5.21). By comparison Baroque urbanism either strived for an illusion of *infinite* space, when contained within small-scale limits (e.g., the Piazza Navona in Rome – see Figure 5.24); or, for reasons explained below, was able to achieve effectively infinite perspectives. But what is more, although personal reactions will vary, experience of such vast urban perspectives can also readily engender 'a sense of desolation' once their grandiose impact was dispelled.

Where 'infinite' perspectives and the grand scale of the Baroque were achieved they were possible only as a result of the immense, centralized, autocratic powers which came to be vested in heads of certain European states. These were welded together out of the numerous medieval communities, founded on local authority, and personal aggrandizement came to replace collective interest in a number of instances. Absolute rulers acquired the political power and economic means to instigate and implement complex planning programmes on hitherto unheard-of scales: most notably those of Louis XIV and XV at Versailles, Peter the Great at St Petersburg, and with different objectives, Sixtus V in Rome.[17] At a correspondingly reduced scale, other lesser rulers transfomed their capitals to create an urban scenery appropriate to the grandeur of their activities and availability of re-

The functional and aesthetic aims of the town planners were also becoming clear. In so far as they were not military, they were closely connected with each other. The city was meant to impress, firstly by its layout, in which its different parts and subordinate centres were to be connected by straight avenues, very much like the formal Italian gardens which were just beginning to be imitated beyond the Alps. Secondly, the city was meant to impress by the magnificent façades of its churches and palaces, and by elaborate fountains. Thirdly, and perhaps most important, it was meant to impress by monumental perspectives. The architects and town planners had learned this from the Renaissance and mannerist painters whose idealized architectural compositions they now began to translate from canvas into stone. To heighten the dramatic effect of perspective Sixtus V set up obelisks in front of St Peter's and in the Piazza del Popolo. Where the Renaissance statue had been related to a building – Verrocchio's Colleone in Venice, for example, or even much later Cellini's Perseus in Florence – the mannerist and the baroque statue was moved into the centre of a square, related no longer to a building but to a view. The possibilities of this new fashion for the glorification of the subject of such monuments were not lost on kings and princes. The baroque towns, as they began to be planned in the sixteenth century in Italy and developed over much of Europe in the seventeenth and eighteenth, became part of the deliberately dramatic and theatrical appeal of absolutist monarchy. Just as the new baroque style of church decoration developed a deliberate popular appeal by making the interior of the church, and especially the high altar, into a kind of stage where mass was celebrated as a theatrical performance for an audience-like congregation, so the baroque city became a huge theatrical setting for the display of the court, the princes of the church, the nobility and other rich and powerful persons. It was the visual aspect of the political and social change from the city state, with its free citizens, to the capital of the absolute monarch, with its court and its subject inhabitants.
(H G Koenigsberger and G L Mosse, Europe in the Sixteenth Century)

---

13. P Zucker, *Town and Square*, 1959.
14. Summarized, the aesthetic determinant of Renaissance city planning embodied and extended the principle of Renaissance architecture. Balance – usually involving symmetry about one or more axes, regularity in elevational design – in particular the heights of individual buildings as viewed in perspective; an emphasis on focal points to terminate distant vistas; and conformity of construction materials for walls, usually stone or stucco substitutes, and roofs, the last of diminished significance behind building cornices.

Outstanding examples include the eighteenth-century developments in Bath, the London Squares (Chapter 8) and parts of Rome and Paris (described below and in Chapter 6 respectively). (See Abercrombie, *Town and Country Planning*; R Furneaux Jordan, *A Concise History of Western Architecture*, 1969; Bacon, *Design of Cities*; Zucker, *Town and Square*.)
15. Fourneaux Jordan, *A Concise History of Western Architecture*.
16. H Wolfflin, *Renaissance and Baroque*, 1966.
17. Rivalry between rulers played its part: outside our period, Queen Victoria is reputed to have envied Napoleon III's achievements in Paris – without anything much happening in London; and there are other similar instances.

sources. (Examples of such urbanism are described at appropriate places in the succeeding chapters.)

The British, however, were always careful to constrain their monarch's power over both their national capital and the country's purse strings. Accordingly Britain was effectively untouched by Baroque urbanism; and for this reason as much as any, it has remained a monarchy long after the halls of Versailles and St Petersburg echoed with new, revolutionary rulers.[18] It must be stressed that not only Baroque but also virtually all of Renaissance urbanism was created for minority sections of society, varying in extremes from Versailles to the unpretentious but none the less privileged squares and streets of Georgian London and other similar developments.[19]

Throughout the Renaissance period several dominant aesthetic considerations determined general attitudes to urbanization in all countries. First, there was a preoccupation with symmetry, the organization of parts of a planning programme to make a balanced composition about one or more axial lines. This was sometimes carried to ridiculous extremes as in the Piazza del Popolo in Rome, where the placing of identical churches on either side of the central street led Abercrombie to observe that 'churches are the last thing, ordinarily, to be produced in pairs, like china vases'.[20]

Second, great importance was attached to the closing of vistas by the careful placing of monumental buildings, obelisks or suitably imposing statues, at the ends of long, straight streets. Third, individual buildings were integrated into a single, coherent, architectural ensemble, preferably through repetition of a basic elevational design. Fourth, perspective theory was 'one of the constituent facts in the history of art, the unchallenged canon to which every artistic representation had to conform'.[21]

Before examining the practical results of these design considerations, as demonstrated by examples of Renaissance urbanism in various Italian cities, it is necessary to consider in general terms the main components of Renaissance planning: the primary straight street, gridiron based districts, and enclosed spaces.

THE PRIMARY STRAIGHT STREET

In the form of main routes 'emancipated from being mere access to a building plot on the one hand and an urban extention of the national highway on the other',[22] the primary straight street is a Renaissance innovation. In the majority of instances it still provided the approach to buildings, and frequently had direct connections with regional routes, but its main function was to facilitate movement, increasingly by carriage, between parts of the city. Rome and Paris are unique, certainly among major cities, in acquiring primary street systems as the result of comprehensive restructuring: Rome during the Renaissance proper (as described later in this chapter) and Paris during the 1850s and 1860s. On the other hand such new streets as London did acquire were largely unrelated; the best example is Regent Street (Figure 8.22), cut between Soho and Mayfair in the early nineteenth century, to link the St James's area with the Regent's Park developments by means of a suitably imposing route.

The clearest examples of the primary straight street as a generative element, determining growth of existing cities, are the Champs Elysées and the avenue, Unter den Linden: both remarkably similar, royal

The avenue is the most important symbol and the main fact about the baroque city. Not always was it possible to design a whole new city in the baroque mode; but in the layout of half a dozen new avenues, or in a new quarter, its character could be re-defined. In the linear evolution of the city plan, the movement of wheeled vehicles played a critical part; and the general geometrizing of space, so characteristic of the period, would have been altogether functionless had it not facilitated the movement of traffic and transport, at the same time that it served as an expression of the dominant sense of life. It was during the sixteenth century that carts and wagons came into more general use within cities. This was partly the result of technical improvements that replaced the old-fashioned solid wheel with one built of separate parts, hub, rim, spoke, and added a fifth wheel to facilitate turning.

The introduction of wheeled vehicles was resisted, precisely as that of the railroad was resisted three centuries later. Plainly the streets of the medieval city were not adapted either in size or in articulation to such traffic. In England, Thomas tells us, vigorous protests were made, and it was asserted that if brewers' carts were permitted in the streets the pavement could not be maintained; while in France, parliament begged the king in 1563 to prohibit vehicles from the streets of Paris – and the same impulse even showed itself once more in the eighteenth century. Nevertheless, the new spirit in society was on the side of rapid transportation. The hastening of movement and the conquest of space, the feverish desire to 'get somewhere', were manifestations of the pervasive will-to-power.
(L Mumford, The City in History)

18. The presidents of the United States have been similarly constrained in their city planning involvements, such as they have been, in their capital Washington, DC (see the concluding section to Chapter 10). I would add that if John F Kennedy was initially concerned to enhance the city (as legend would suggest) he rapidly became embroiled in weightier matters of state, as also his successors.

19. For illustration of the living conditions of Voltaire, a not so ordinary person living at Versailles, see an extract from Cobban's A History of Modern France, quoted on page 211.

20. Sir Patrick Abercrombie, 'Era of Architectural Town Planning', Town Planning Review, vol V, 1914.

21. S Giedion, Space, Time and Architecture, 1961.

22. Abercrombie, Town and Country Planning; see also Bacon, Design of Cities, for whom the primary straight street has extra-special significance as an urban form determinant.

routes west from palaces in Paris and Berlin respectively.[23] New urban plans, either based on primary straight streets or incorporating them as major elements, include Versailles, Karlsruhe and St Petersburg in Europe; and Washington, most important among the examples in the USA. Sir Christopher Wren's unrealized proposals for rebuilding the City of London after the fire of 1666 were also based on the use of such main streets.

In addition to effecting changes in its function, the Renaissance also introduced the aesthetic concept of the street as an architectural whole. Although at first it is clear, from Alberti's contemporary writings, that streets could be considered to consist of individual building elevations – best appreciated from curved approaches – as the period progressed architectural uniformity became *de rigueur*. 'From the end of the fifteenth century,' Zucker observes, 'three-dimensional distinctness corresponded to structural clarity. Definite laws and rules directed the limits of space and volume. Purity of stereometric form was in itself considered beautiful.'[24] Perspective effects were emphasized by the location of terminal features, both architectural and sculptural, in the form of statues, fountains and obelisks (notably in Rome). 'The monument at the end is the recompense, as it were, for walking along a straight road (devoid of the surprises and romantic charm of the twisting streets) and economics are met by keeping the fronting buildings plain so as to enhance the climax – private simplicity and public magnificence.'[25]

THE GRIDIRON

During the Renaissance period in Europe three main uses are made of the gridiron, history's oldest known urban form determinant: first, and by far the most widespread, as the basis of residential districts added to existing urban areas; second, for the entire layout of a limited number of new towns; third, in combination with a primary street system, for the layout of other new urban areas.[26] Because of the comparatively greater area that they covered, gridiron districts, in contrast to primary streets and enclosed-space components, are seldom found in redeveloped parts of Renaissance cities. In addition to being efficient and producing an equality of land subdivision – the same reasons for its use in preceding historical periods – the gridiron also conformed to the Renaissance ideal of aesthetic uniformity, even if the resulting townscape all to frequently reveals this to be mere monotony. Camillo Sitte, writing of Mannheim, a major new gridiron-based town (see Chapter 7) refers to the rule that all streets intersect perpendicularly.[27] However, although its Renaissance applications may have been unimaginative, the results generally had urban qualities, notably spaciousness, which were to be sadly lacking in the inhuman gridiron 'by-law' housing of the Industrial Revolution.[28]

New housing districts were most frequently either bounded by primary routes or divided into sections by their inclusion, with the gridiron streets themselves normally having only a minor, access nature. Those districts which may have escaped Sitte's opprobrium would most probably have done so only because their layouts embraced relieving natural landscape features, or were planned, usually in the form of squares, by the more sensitive urbanists. London's Mayfair, the area between Oxford Street and the New Road (Marylebone Road) (Figure 8.15), and later Bloomsbury developments contain examples of landscaped

23. In Western Europe the direction of prevailing weather systems is generally from the west and southwest, with the effect of blowing low-level atmospheric pollution of soft-coal domestic fires and industrial furnaces over the eastern parts of town. Here we have a little-remarked instance of the climatic determinant of urban form whereby those that could afford to do so lived on the windward, western sides of town.

In addition to Paris and Berlin, London is a third capital city whose moneyed-class residential districts were located to the west, although (as explained in Chapter 8) there were other contributory factors.

24. Zucker, *Town and Square*.

25. Abercrombie, *Town and Country Planning*.

26. During the European Renaissance period, particularly so in Italy, there was considerable interest in circular city planning. Most examples were theoretical 'paper' exercises, but a few were built, notably Palma Nova, in northern Italy (Figure 5.18). See N J Johnston, *Cities in the Round*, 1983.

27. C Sitte, *City Planning According to Artistic Principles* (translated by G R Collins and C Collins, 1889, 1965 English edition).

28. The earliest 'working-class' housing in the emerging industrial towns usually consisted of organic growth, small-scale additions; later, but only in response to the dictates of ruthless construction economy, the gridiron was used as the basis of large housing areas. Gridiron 'by-law' housing layouts, giving minimum health standards of light, air, and drainage date from the mid-nineteenth century. (See A E J Morris, 'Unplanned Towns', *Official Architecture and Planning*, April 1971; and 'Philanthropic Housing', *Official Architecture and Planning*, August 1971.) The extent to which the nature of a later nineteenth-century industrial housing gridiron district should be regarded as 'inhuman' raises complex questions, notably that of extent, as it determined the distance from open country.

squares which mitigate the effects of otherwise more or less straightforward gridiron planning. Craig's New Town, the major later eighteenth-century addition to Edinburgh, was saved from mediocrity by its two squares, its (originally) limited extent, and the open-sided nature of the two long boundary streets – most famously Princes Street, with its incomparable views south across the valley to the old city. Berlin, as a third major example, added less interestingly planned districts on either side of the Unter den Linden in the late seventeenth century.

ENCLOSED SPACE

There are semantic difficulties raised by this aspect of Renaissance urbanism. They have been resolved here by using the English word 'square' for enclosed spaces in Britain and other European countries, with the exceptions of Italy and France for which 'piazza' and 'place' respectively are used. On the basis of their urban mobility functions Renaissance urban spaces can be grouped under three broad headings: first, traffic space, forming part of the main urban route system and used by both pedestrians and horsedrawn vehicles; second, residential space, intended for local access traffic only and with a predominantly pedestrian recreational purpose; third, pedestrian space, from which wheeled traffic was normally excluded. In addition to the above physical uses, Renaissance spaces frequently served aesthetic and aggrandizement purposes, either as a setting for a statue or a monument, or as a forecourt in front of an important building; and there are also the uniquely Spanish *plazas mayores* described in Chapter 9.[29]

Spatial enclosure was effected with three main types of buildings: first, civic or religious architecture; second, residential buildings, usually in terrace form; third, market and related commercial buildings. Renaissance urbanists also defined space by the use of architectural landscape elements, e.g. colonnades, screens and terraces, and by various forms of tree and shrub planting. These ways of enclosing space were often used in combination and in a number of instances existing buildings and natural features were incorporated into the design.

ENCLOSED SPACE: TRAFFIC SPACE

Before nineteenth-century increases in urban traffic there were few instances of formally designed spaces at intersections of main streets.[30] Most were located on the urban perimeter – e.g. the Piazza del Popolo in Rome, and the three squares on the west side of Berlin: Potsdamer, Leipziger and Pariser Platz. The Place de la Concorde, which also performed this traffic function at the eastern end of the Champs Elysées on the edge of central Paris, was a unique form of space, combining civic buildings along its northern side and landscape elements on the other three sides. In addition to resolving the junction of routes, it also served as the setting for a statue of Louis XV. Westward along the Champs Elysées, the Place de l'Etoile around the Arc de Triomphe, the epitome of traffic space, was not completed until the middle of the nineteenth century. Paris also had the Place des Victoires, surrounded by impressive residences and containing a statue of Louis XIV. A number of theoretical traffic spaces for Paris were recorded on a map of the city published by Patte in 1765, for which see Figure 6.14.

The main purpose of these proposals was to provide a setting for a statue of the king. London, and British cities generally, were in contrast hardly affected by this aspect of Renaissance urbanism: Nash's layout

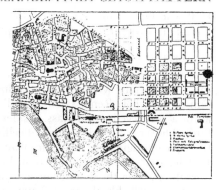

Figure 5.2 – Ystad in southern Sweden, a small fishing town that became internationally famous in the later Middle Ages for its herring catches. The rudimentary gridiron of the new housing district is in stark contrast to the organic growth form of the original medieval nucleus. This is an example of the gridiron used merely as an end in itself: that of expedient land-subdivision, which cannot be accorded the status of 'city planning'. (See K Astrom, *City Planning in Sweden*, 1967.)

In one place, however, baroque planning rose above its political and military premises; here it created a form independent of the purposes of the palace. This was in the conception of the residential square. The open square had never disappeared; but by the same token it had never, even in the Middle Ages, been used entirely for residential purposes, if only because the counting house and the shop were then part of the home. But in the seventeenth century, it reappeared in a new guise, or rather, it now performed a new urban purpose, that of bringing together, in full view of each other, a group of residences occupied by people of the same general calling and position. Dr Mario Labo is right in regarding the Strada Nuova in Genoa as more of a quarter than a street; but the new squares gave a fresh definition to this kind of class grouping.

In the older type of city, particularly on the Continent, the rich and the poor, the great and the humble had often mingled in the same quarter, and in Paris for instance, they long continued to occupy the same buildings, the wealthier on the ground floor, the poorest in the attic, five or six storeys above. But now, beginning, it would seem, with the establishment of Gray's Inn in London in 1600, a new kind of square was formed: an open space surrounded solely by dwelling-houses, without shops or public buildings, except perhaps a church. Gray's Inn indeed was a transitional form between the medieval walled enclosure, with inner gardens, dedicated to a convent or a great lord's mansion, and the square, walled in only by its own houses conceived as part of the new street pattern.
(L Mumford, The City in History)

---

29. Here we have instances of royal aggrandizement (see Chapter 6) and aesthetic considerations in combination as a primary form determinant.
30. Increase in the scale of urban traffic from pedestrian or horseback, to carts and private or hire carriages, required appropriately spacious junctions. As focal points in the city route-system, other determinants then came into effect.

of Piccadilly Circus as a traffic space to accommodate a change in direction of Regent Street is an exceptional example.[31]

ENCLOSED SPACE: RESIDENTIAL SPACE

The creation of an enclosure 'with no more monumental object than that of uniformity within itself, is perhaps the most attractive contribution of the whole Renaissance period'.[32] Such enclosures were almost all of a privileged social class, residential nature; wheeled traffic was limited to serving the individual dwellings. In Paris, where the residential space originated with the Place des Vosges (originally Place Royale, 1605–12) and elsewhere in France, such spaces were also frequently used as the setting for a royal statue. The first of London's squares – probably the most famous examples of this kind of space – was Covent Garden (1630). London's expansion to the west was largely based on a combination of residential squares and gridiron streets. The squares – usually containing a planted central area – provided a basis for urban family life which is held in the highest esteem by mid-twentieth-century planners faced with mass housing problems. It must not be overlooked, however, that only a small minority of urban homes had such an advantageous situation. By the time nineteenth-century London squares in Bloomsbury and Belgravia, and the comparable larger scaled development of Regent's Park (1810–30) were under way the Industrial Revolution had already created great tracts of effectively uncontrolled, high-density housing in other parts of London and primary manufacturing centres, lacking almost all of even the minimum basic necessities of life. Residential spaces are a characteristic of *controlled* seventeenth- and eighteenth-century urban growth in Britain;[33] few cities and towns of any size were without at least one pleasantly unpretentious square, those at Bath and Edinburgh being especially distinguished by unusual qualities of spatial organization and architectural attention to detail.

ENCLOSED SPACE: PEDESTRIAN SPACE

A number of extremely important enclosed spaces were either completely closed to wheeled traffic, or arranged so that pedestrians were not unduly affected there – e.g. wheeled traffic was not continuous across the space or was restricted to one side only.[34] The majority of these spaces served as forecourts or public assembly areas in front of important civic, religious and royal buildings. The most important examples are Italian, two of which are in Rome: the Piazza of St Peter's, where the east front of the church dominates the colonnade-enclosed space; and Michelangelo's Capitoline Piazza where identically designed buildings form the sides of the forecourt to the Palace of the Senators, the fourth side consisting of a monumental flight of steps up the hillside. Venice has the incomparable Piazza of St Mark, where the enclosing buildings had civic, commercial and religious functions. Venice, as a unique water-orientated city with only pedestrian land traffic, contains several spaces of great beauty. Elsewhere in Europe there were few opportunities to create forecourt spaces for pedestrians only.

Urban designers

By the time that Renaissance attitudes and style had been firmly established, the new technique of printing enabled new designs and theories

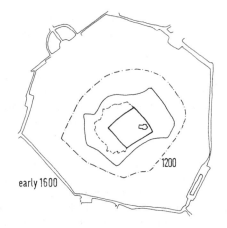

Figure 5.3 – Milan: concentric rings of fortification around the Roman nucleus, the outermost, that of the early seventeenth century, was used for the ring road system. (See M C Branch, *Comparative Urban Design*, 1978, for the 1832 SDUK Map of Milan; E Noyes, *The Story of Milan* (Mediaeval Towns series), 1908; P Sica, *Storia dell 'Urbanistica L'Ottocentro*, 4th edn, 1991.)

Milan seems to have held from the first the chief position among the cities of Lombardy. In the early centuries of our era it was hardly less important in the North of Italy than Rome was in the South. The line of the Po, cutting across the peninsula, or perhaps more correctly, the Apennine chain, originally divided Italy ethnologically and politically, a division which still endures to some degree in the character and sentiments of the respective inhabitants on either side. The Insubri, who drove out the Etruscans and settled in Lomardy about the sixth century (B.C.), were a race of Gallic origin. They had no ties of blood with the Romans, who subjugated them later, and their country – called by the conquerors, Cisalpine Gaul – was as much a foreign province of the Latin dominion as the Gaul beyond the Alps.
(E Noyes, The Story of Milan, 1908)

31. See Chapter 8 for the section describing Regent Street and Regent's Park, which it linked to Whitehall as a prestigious route.
32. Zucker, Town and Square.
33. I have used the term 'controlled' urban growth rather than 'planned', because the layouts of these British residential developments, did not have to observe requirements of any preconceived 'master plan' for the entire urban area.
34. See B Rudofsky, Streets for People, 1969.
35. Availability of books also had the effect of encouraging the 'Grand Tour', whereby wealthy aesthetes travelled to see for themselves the important foreign buildings and examples of urban design. John Evelyn was one such mid-seventeenth century traveller and amateur urbanist, who visited Rome, among many other cities, where his findings were recorded in The Diary, from which a number of extracts are taken. (See also J Boswell, Boswell on the Grand Tour: Italy, Corsica and France, ed. F Brady and F A Pottle, 1955.)

to be communicated internationally; it was no longer necessary to turn ideas into buildings to demonstrate architectural intentions and to influence others. Urban designers of the Renaissance were presented, for the first time in history, with the possibilities of making their theories and experience available to others on a wide scale. This outlet came at a fortunate time. Italy had inherited from her imperial Roman past more towns than were needed, with the result that only two significant examples of new-site planning were actually built; however, there are countless examples of military engineering theory applied to the fortifications of existing towns.

From the fifteenth century onwards there was a succession of published works dealing with the theory of architecture, urban design and military engineering. Before describing the ideas of individual theorists and main features of a major completed example, Palma Nova, a general description of the role of fortification (a primary urban design determinant) is required. This sets the context for the theoretical and practical work.[35]

FORTIFICATION

The role of fortification as an urban form determinant has been largely neglected by urban historians.[36] In 1529, and again in 1683, Vienna's defences were all that stood between the Turks and the hinterland of Western Europe. As recently as 1914 the defences of the city of Paris, which had previously successfully resisted the Prussians in 1871, were put in readiness to withstand the German advances from the Marne. At both Vienna and Paris, as well as at countless other large and small continental European cities, need to ensure efficient defence against attack was a major reason for high-density urban life based on relatively high-rise apartments. When successive rings of defence became obsolete, the land made available provided the opportunity to create inner-ring boulevards. Crammed within their fortified girdles, for ever increasing in population and density, the typical continental European city of the fourteenth to mid-nineteenth centuries could expand only upwards. This is in direct contrast to Britain, whose island location with much more settled internal conditions, helped to encourage a tradition of horizontal growth. Where towns in Britain did keep their walls, this was more in the interest of commerce than military necessity (in many cases they remained as such until the fifteenth and sixteenth centuries – see pages 98–9). The British therefore tended to evolve their own 'half-acre and a cow' philosophy – an essentially anti-urban attitude which has been maintained as the 'suburban-semi with a garden' preference, right up to the present day.

If European Renaissance cities, and those of later periods, could have been defended at lower cost, boundary expansion would have been more feasible and it is probable that growth would have been much more horizontal in character. With the military developments of the fourteenth and fifteenth centuries this was not possible. The factor which most upset the balance between defenders and attackers was the perfection of gunpowder and the cannon. Sir Reginald Blomfield writes that in medieval days if the walls of the castle were strong enough and high enough, and there was a good moat around it filled with water, the castle was impregnable except by starvation or treachery.[37] Such defensive systems often drained resources but essentially they had to follow only one dimension – the vertical; they could be extended to take

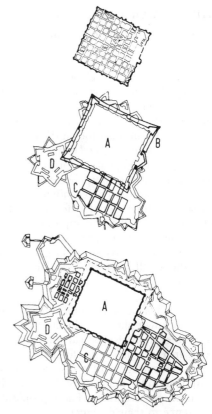

Figure 5.4 – Turin: three stages in the development of the city from its days as a Roman colony to the late seventeenth century, showing the extent and complexity of the Renaissance fortifications in marked contrast to the simple two-dimensional walled perimeter of Roman and medieval periods.

Above: The plan of the original Roman city of Augusta Taurinorum with the street pattern modified during the later Middle Ages.

Centre: Turin at the beginning of the seventeenth century with the Roman core (A) surrounded by an earlier wall of the sixteenth century (B); the new areas are (C), a Renaissance extension, and (D), the citadel.

Below: Turin at the end of the seventeenth century with further extensions to the north-west and south-east and a further ring of fortifications.

(See Sica, *Storia dell'Urbanistica L'Ottocentro*, for an excellent series of maps; Branch, *Comparative Urban Design*, for the 1833 SDUK Map of Turin; and S E Rasmussen, *Towns and Buildings*, 1969, on which these three sequential plans have been based.)

36. As a European urban historian I place great emphasis on the role of fortification as an urban form determinant, a view not shared by all my contemporaries. One European social historian, when reviewing the second edition of this book, even queried its relevance, which was surprising, given the evident socio-political differences between fortified continental European societies, evolved within their girdles, and the unconstrained British counterparts. When looking abroad from their own wide-open backyards, American historians are, perhaps, to be excused their comparative disregard for the role of fortification.

37. Sir Reginald Blomfield, *Sebastien le Prestre de Vauban 1633–1707*, 1971.

in new districts, without compromising the overall strength of the system.

The cannon changed all this. Its use by the Turks when they overwhelmed the city of Constantinople in 1453 led to a new era in the history of military fortification. After resisting the power of Islam for over 700 years the city's triple defensive wall system succumbed to a monster cannon capable of firing projectiles exceeding 800 pounds in weight.[38] From this time on the creation of adequate defences required ever increasing *horizontal* distances to be left between the city perimeter and the outer edge of the fortifications. In addition to this extra space, fortifications themselves became increasingly complex, involving intricate systems of inter-dependent bastions and forts. Once established, these two-dimensional defences in depth could be extended only at enormous cost, thereby imposing ever higher densities on the city. The diagrams of Turin's growth, based on originals in Rasmussen's *Towns and Buildings*, clearly show the effect of fortifications on its development. For the small town of Neuf Brisach, a Vauban example in eastern France, the width of the defensive zone exceeds 700 feet (see Chapter 6).

The fall of Constantinople had two immediate effects on the development of town planning during the period of the Renaissance in Italy. The migration of classical scholars to Italy during the formative Renaissance era has been described earlier in this chapter. The second effect, encouragement of the science of military engineering, was exemplified in the majority of the 'ideal city' schemes produced by Italian Renaissance urbanists. These proposals are described separately, with reference to their designers.

Although at the time of writing (1992) there is still no specialist work which deals fully with fortification and its determining effects on urban morphology and the social history of cities, there is the two-volume

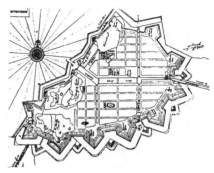

Figure 5.5 – Gothenburg: plan as founded in 1630 with a characteristically complex defensive perimeter. The city was laid out by a Dutch engineer as a 'Venice of the North'. Astrom writes that Gustavus Adolphus became a diligent founder of towns. His greatest achievement was to lay out Goteborg at its present site, to combine a strong border fortress with a thriving port in the Dutch manner. (See Astrom, *City Planning in Sweden*; C Duffy, *The Fortress in the Age of Vauban and Frederick the Great 1660–1789*, 1985.)

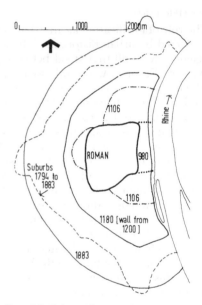

Figure 5.6 – Cologne: diagrammatic plan showing the city's growth from a Roman *colonia* to the nineteenth century as determined by successive defensive systems. The *colonia* was a defensive left-bank-only settlement: the river was too wide to be bridged before the nineteenth century. The first expansion of the Roman nucleus was the Rheinvorstadt of 980, between it and the river. In 1106 three suburban *faubourg* areas were included and in 1180 the line of the medieval wall was established.

Figure 5.7 – Cologne: view of 1646 within the defensive perimeter of 1180, as up-dated in part by Renaissance fortification works. This encircling zone of land was subsequently used for the city's inner ring road system.

38. It is one of the more intriguing ironies of urban history to know that the cannon used so successfully by the Turkish gunners had previously been offered by its makers to the city's defenders ... only to be rejected as too expensive.

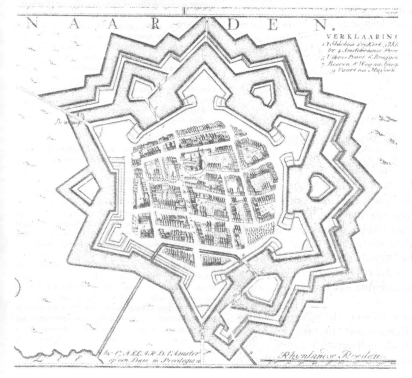

Figure 5.8 – Naarden: east of Amsterdam on the southern shore of the Zuider Zee, the epitome of a small, strongly fortified town of the late seventeenth century. The original fishing village was destroyed by fire in 1350 and Naarden was rebuilt by Count Willem V as a *nieuwestad* some 1,000 yards further inland, linked to the Zuider Zee by a canal.

During the seventeenth century Naarden became a vital strategic location and between 1673 and 1685 earlier walls were replaced 'by mighty installations in the Vauban style, comprising six great bastions and ravelins set in wide moats, protected gun emplacements and enfilade firing points and a vast network of covered routes and passages connecting armouries and ammunition stores to firing positions. The super-structure was dismantled during the nineteenth century, but the outlines of rampart and moat, as seen from the air, still present a most formidable appearance' (G L Burke, *The Making of Dutch Towns*).

See Figures 5.19, Palma Nova, and 6.29, Neuf Brisach for two other aerial views of fortress towns; see also Duffy, *The Fortress in the Age of Vauban and Frederick the Great 1660–1789*, for ground-level photographs of Naarden.

Figure 5.9 – Naarden: the layout of the third system of fortifications constructed between 1673 and 1685 to plans prepared jointly by the military engineer, Paen, and the architect, Dortsman.

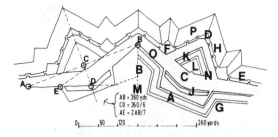

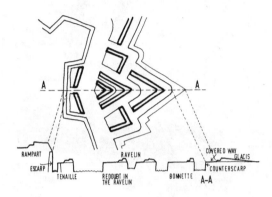

Figure 5.10 – Ground plan (or 'trace') of a typical fortification system, showing the basic geometry and illustrating terminology. Key: A, curtain; B, bastion; C, ditch; D, place of arms; E, covered way; F, salient place of arms; G, shoulder; H, traverses; J, tenaille; K, ravelin; M, gorge of bastion; N, embrasures; O, parapet; P, glacis.

The trace and cross-section (at proportionately greater scale) of an alternative system.

If all the various elements of Vauban's designs were deployed, an enemy would be faced by a succession of daunting obstacles. Confronted today by a full-blown example of 'Vaubanism' an observer might well be confused and puzzled. What was the justification for such intricacy? In truth, the extravagances associated with the Vauban school of military engineering are really those of his later followers, whose obsession with their own interpretation of his ideas led them to produce architecturally immaculate schemes which increasingly flew in the face of military realities. Their pursuit of perfection encouraged the design of symmetrical, idealized enceintes resembling those of the early Italian school. The fixation on the virtues of enfilading fire and defence in depth meant that the importance of frontal fire was neglected; indeed, the configuration of some defensive structures severely discouraged it. These complex linear fortifications were expensive to build, maintain and garrison, and yet their proponents continued to dominate military engineering for well over a century after Vauban's death.
(I Hogg, The History of Fortification, 1981)

Being small and insecure (London, a Leviathan among our English cities, can hardly have numbered more than twenty thousand inhabitants), mediaeval towns were everywhere fortified and organized for defence. M. Pirenne, writing more particularly of the Low Countries, tells us that up to the close of the middle ages a sum never falling short of five-eighths of the communal budget was expended on purposes connected with the maintenance of the walls and the provision of instruments of war. In Italy, despite the fact that she was now fast securing for herself the leadership of the world in craftsmanship and international commerce, the warfare of city with city was almost perpetual. Cities would fight about diocesan boundaries and feudal rights, over tolls and markets, for the extension of their powers over the contado or surrounding country, or in pursuance of the long-inherited feuds of the nobles within their walls.
H A L Fisher, A History of Europe, Volume 1: From the Earliest Times to 1713, 1936)

work by Christopher Duffy on the subject of Siege Warfare for the periods 1494–1660 and 1660–1789.[39] The accompanying illustrated glossary of terms (Figure 5.10) has been derived from these books. Examples of fully developed applications of military engineering science to urban defensive systems include the following: Nancy, Neuf-Brisach, Toulon and Le Havre (Chapter 6); Antwerp and Copenhagen, (Chaper 7); Portsmouth, Devonport and Berwick-on-Tweed (Chapter 8); Havanna and Manila (Chapter 9).

VITRUVIUS

Marcus Vitruvius Pollio, as mentioned briefly earlier in this chapter, was an architect working in Rome at the time of Augustus – the emperor who, it is claimed, found Rome a city of brick and left it one of marble. Towards the end of a seemingly uneventful life Vitruvius wrote several essays on the theory and techniques of architecture and on related aspects of town planning and civic engineering. These, in effect a textbook of classical Roman practice, are known collectively as the *De Architectura* of Vitruvius. Its 'discovery' as such about 1412–14 added considerably to the momentum of the Renaissance in architecture, urbanism and the arts generally. Its influence was again increased when the first of numerous Italian editions came out in printed form in 1521.[40]

De Architectura's early acclaim has not withstood the passage of time. Revered through much of the early Renaissance literally as a civic design gospel (an attitude not entirely dispelled), this work is now regarded only as that of an obscure and freely misinterpreted Roman

39. See also C Duffy, Siege Warfare, Volume I, The Fortress in the Early Modern World 1494–1660, 1979; Volume II, The Fortress in the Age of Vauban and Frederick the Great, 1985.
40. Contrary to the impression given by some urban historians, the Vitruvian manuscript, in common with other ancient works, was never lost at all. What happened around 1412–14 was that Poggio Bracciolini (apostolic secretary at the Council of Constance) drew attention to its existence at the monastery of St Gallen. Sem Dresden has written of such texts that 'although they had been virtually ignored during the earlier Middle Ages, they had nevertheless been preserved during these hundreds of years.... In our desire to credit the Renaissance with its due we should not completely strip the Middle Ages of theirs. The difference is rather that the humanists showed a fresh and, one might say, unprejudiced interest in ancient texts ... the manuscript of the poems of Catullus, which came to light as early as 1295, could be considered to be the first of these' (Humanism in the Renaissance).

authority on architecture.[41] The parts of the *De Architectura* of relevance in urban history are the fourth to the seventh chapters of the first book. In these chapters Vitruvius outlines fundamental considerations to be observed in designing towns and describes the features of a city designed in circular form. His ideas were not, however, illustrated by an actual plan. So far as is known this is a form never used in practice by the Romans for any of the countless military camps and towns they established throughout the empire. Vitruvius himself was therefore advocating a theoretical, ideal city plan.[42]

The interpretations of Vitruvian theory are generally given a radial-concentric form, enclosed within an octagonal defensive wall; eight radial streets lead out to the angle towers rather than to the gateways in the centre of four of the sides (Figure 5.11). Vitruvius advised this approach to avoid adverse winds. This was a curiously romantic idea which would not have worked as such in practice, although it would have had valuable defensive military advantages in that the gateways did not give direct access to the centre of the town. The main forum area was to be at the centre, enclosed within an octagonal space with eight secondary open spaces in the middle of each of the sectors. A primary consideration was the foundation of regularly shaped residential blocks. Vitruvius also wrote at length on the factors to be observed when siting towns.

ALBERTI

Leon Battista Alberti (1404–72) was born in Genoa, spent his childhood in Venice, and studied at the University of Bologna. Between 1432 and his retirement in 1464 he was employed as a secretary in the papal chancery, at first with general duties but subsequently concentrating on architectural and town-planning matters. In 1434 he accompanied the court of Eugenius IV to Florence, where he encountered Renaissance architecture for the first time and met, among other designers, Brunelleschi and Donatello. Systematic study of those buildings of ancient Rome that were still well preserved, consolidated his special interests and led, in 1447, to Nicholas V appointing him as his architectural adviser. Together they undertook the first of the grand projects which were ultimately to restore the visual majesty of the eternal city.[43] Alberti put the Aqua Virgine back into operation, taking water to fountains in the Piazza del Trivio, and initiated reconstruction of the crumbling ancient fabric of St Peter's (eventually to be demolished in 1505, making way for the present church) and the Vatican Palace. He was also the architect of several extremely important early Renaissance buildings – notably the Ruccellai Palace and S Maria Novella at Florence and S Andrea at Mantua.

In the urban planning field, however, his bold design for a new Borgo Leonino, the quarter that runs from St Peter's to the Castel Sant' Angelo, is significant as one of the earliest examples of the geometric spatial plans of the Renaissance, although it was only partially carried out. Here, 'Alberti wanted to have a plaza at both ends of the long rectangular area, connected by three broad avenues, and the entire scheme was to be given formal emphasis by the great obelisk which he wanted to set in the centre of one of the plazas, in front of St Peter's.'[44] This latter proposal was perhaps the inspiration of Sixtus V's programme to set up a system of focal point obelisks, one of which was erected in front of the still incomplete new St Peter's in 1589.

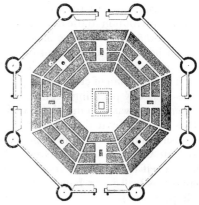

Figure 5.11 – Vitruvius: ideal city plan, as described but not drawn in his writings of the early first century AD. The Vitruvius 'Roman city plan' (interpreted as may be) differs in too many respects from accepted Roman imperial city planning practice to be other than extremely suspect in respect of the author's intentions. Whoever drew up this interpretation of Vitruvius's writings could well have done so in order to lend credence to emergent Renaissance theories.

On Pope Nicholas V
Both his predecessors, Martin V and Eugenius IV, the austere, dignified and extremely tall son of a rich merchant, had done what they could to restore the ravaged city. Pope Martin had revived the ancient office of Overseer of the public thoroughfares with a view to clearing away the rubbish and filth that filled the streets and poisoned the air. He had restored several churches and other public buildings; reconstructed the aqueducts which were in such a ruinous condition that many citizens had no idea what their original purpose had been; and, after rebuilding the Acqua Vergine, he had erected a fountain facing the Piazza dei Crociferi which was to be transformed in the eighteenth century into one of Rome's most celebrated sights, the Trevi Fountain. Pope Martin had also summoned to Rome the great Tuscan master, Masaccio, and had brought from Ostia the relics of St Augustine's mother, St Monica, whose tomb can now be seen in the church of S. Agostino, yet when Pope Eugenius had returned to Rome in 1443, having been driven from it after quarrelling with his predecessor's family, the Colonna, the city was still in the most parlous condition; S. Maria in Domnica and S. Pancrazio both remained on the verge of collapse; S. Stefano had no roof, and many other churches were in as bad or worse a state. Several lanes in the Borgo were avoided by the prudent citizen because of the ever-present danger of tumbling masonry. The streets, filthy

41. Pevsner, *An Outline of European Architecture.*
42. At time of writing (1992) I have found no reason to change my view that Vitruvius has little real significance in urban history.
43. J Gadol, *Leon Battista Alberti: Universal Man of the Early Renaissance,* 1969; see also M Jarzombek, *On Leon Baptista Alberti: His Literary and Aesthetic Theories,* 1990.
44. Gadol, *Leon Battista Alberti.* See Figure 5.29 for the Piazza of St Peter's, and Figure 5.32, which shows it was isolated from the city through until the 1930s. Alberti's proposal would have ensured a suitably grand setting for St Peter's.

Important as were these completed and theoretical designs. Alberti's main contribution to the development of the Early Renaissance was his *De Re Aedificatoria* – twelve books on architecture and related matters – which he presented in manuscript to Nicholas V in 1452. This work was published posthumously in 1485, and established Alberti as the 'first theoretician of city planning in the Renaissance; with his treatises conscious city planning begins'.[45] Joan Gadol considers that his *De Re Aedificatoria* was prompted by deficiencies in Vitruvius's writings.[46] This is substantiated by Alberti's scathing denunciation of Vitruvius, accusing him of writing in such a manner that to the Latins he seemed to write Greek, and to the Greeks, Latin, though it was plain from the book itself that he wrote neither Greek nor Latin, and that he might almost as well never have written at all (*De Re Aedificatoria*, Book Six, Chapter One).

Alberti recorded his architectural philosophy more completely than his urban design concepts. He included neither plans of ideal cities nor examples of urbanism, but he discusses at length many aspects of city planning which were to be more fully developed by later theoreticians, notably his concept of a centralized square with radiating streets which remained the crystallization of theoretical thought and was not realized until more than one and a half centuries later.[47] To enhance the majesty of important cities and to facilitate movement, particularly of soldiers, Alberti preferred wide, straight streets. But although they were clearly out of keeping with early Renaissance planning principles, he also saw advantages in winding medieval streets; these minimized the effects of climatic extremes, assisted internal defence, and allowed each of a sequence of individual buildings to be clearly presented to view.

It was inevitable that Alberti's writings should reflect his role as a bridge between the Middle Ages and the Renaissance: he was, as Mumford observes, in many ways a typical medieval urbanist.[48] Nevertheless, in other respects he was well ahead of his time, notably in advocating that the front and whole body of the house should be perfectly well lighted, and that it be open to receive a great deal of light and sun, and a sufficient quantity of wholesome air. It is also possible to see the germ of twentieth-century suburbia in his observation that there is a great deal of satisfaction in a convenient retreat near the town, where people are at liberty to do just what they please.[49]

AVERLINO AND SOME OTHERS

Antonio Averlino (1404–72), known under his adopted name of Antonio Filarete, has the credit for producing the first fully planned ideal city of the Renaissance.[50] This was described, and illustrated by a plan, in his *Trattato d'Architettura*, written between 1457 and 1464 but not published, and then only in part, until the nineteenth century. Its influence was, however, spread throughout Europe in numerous manuscript copies. Filarete's city was named Sforzinda, after his patron Francesco Sforza. The basis of the plan consists of two squares overlaid on each other to create an octagon within a circular perimeter (Figure 5.13). From the centre of the city sixteen radial routes, one of which takes the form of an aqueduct, lead out to the perimeter. An intermediate ring road links secondary squares, sited at intersections with the radial routes; these have alternating market and church location functions. The central area of the city includes three separate squares, the most important of which contains the cathedral and the ruler's palace,

as ever, still resembled those of a country village in which cattle, sheep and goats, driven by their owners in long country capes and knee-boots, wandered from wall to wall.
(C Hibbert, Rome: the Biography of a City, 1985)

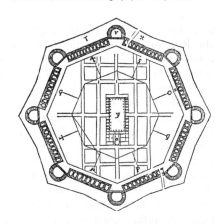

Figure 5.12 – Danieli Barbaro: ideal city plan from his *Commentary on Vitruvius* (1567). Neither Barbaro nor Filarete (see Figure 5.13) were other than theoretical book illustrations and they are not to be taken seriously in practical city planning terms. Recent (printed) urban history includes many fantastic or futuristic theoretical plans: these two, and other, Renaissance ideal city plans are early examples.

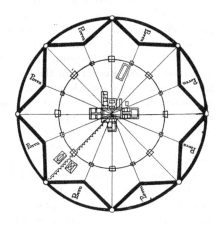

Figure 5.13 – Antonio Filarete: Sforzinda, the ideal city designed to illustrate points in his *Trattato d'Architettura* of 1457–64 (but not printed until the nineteenth century).

45. Zucker, *Town and Square*. Zucker's statement has been retained in the sense that Alberti is to be regarded as the first theoretician of the Renaissance, but (see Note 2) he was not the designer of that which can be regarded as the first Renaissance urbanism.
46. Gadol, *Leon Battista Alberti*.
47. Zucker, *Town and Square*.
48, 49. L Mumford, *The City in History*, 1966/1991.
50. Zucker, *Town and Square*.

the two lesser squares being provided for the market and the merchants. This arrangement is much nearer to Roman forum-planning than to the highly centralized nuclei of the later Renaissance ideal cities.

Francesco di Giorgio Martini (1429–1502) wrote his *Trattato d'Architettura* in 1495. In this work the fifth book is concerned with fortifications and some authorities regard it as marking the beginning of the new approach to military engineering required by rapid developments in offensive artillery. Sir Reginald Blomfield disagrees in no uncertain terms, however; he argues that, 'the true forerunner of Vauban was in fact Michele San-Michele of Verona (1484–1558) who was responsible, among many other works, for the fortifications of Verona and Padua'.[51] Martini was the most prolific designer of ideal cities; included in his work are examples of both formally centralized designs and those more freely adapted to sites.

Pietro Cataneo published his *Four Books on Architecture* in Venice in 1554. This work includes a large number of ideal city plans based on the regular polygon, including some with a separate citadel for the ruler of the city. This approach was to be used some fifty years later for the design of Mannheim in 1606 (see Chapter 7).

Buonaiuto Lorini's *Delle Fortificatione Libri Cinque*, also published in Venice, came out in 1592. It reflects the way in which, by this time, the depth of the defensive zone was steadily increasing to keep besieging artillery away from the edge of the city.

Not only did the ideal cities of the early Renaissance have little immediate influence on European urban form, other than improving the design of fortifications, but also they failed to have any effect on European town planning abroad, in particular in North America where it was the orthogonal grid geometry of the *colonia* and bastide which was favoured, not the stellar geometry of an ideal city. If the American Indians had had artillery, however, urban history could well have taken a different direction (see Chapter 10). Meanwhile, back in Europe, E A Gutkind considers that 'it is characteristic that not a single plan of an ideal city contains even the slightest indication of the arrangement of the houses. Only streets, squares and walls are shown ... the designers had the doubtful privilege of being the unintentional originators of the cult of the street, which finally led to the empty plans of drawing-board architects'.[52]

LEONARDO DA VINCI

Leonardo da Vinci (1452–1519) included among his theoretical work a wide range of proposals for architectural and planning schemes: this latter work involved both detailed town-planning considerations and regional development plans.[53] There is, however, no conclusive evidence of completed work in this field, other than the addition of fortifications to several existing towns. Leonardo acted as consultant, on military engineering, road-making and canal building, to Ludovico Sforza, the ruler of Milan.

The plague in 1484–5, which resulted in the death of a large proportion of the population of Milan, was recognized by Leonardo to be a direct result of the overcrowded and insanitary conditions within the city. Leonardo advised the duke to rebuild the city at a lower density and accommodate the resulting 'overspill' population in a total of ten new towns, designed for populations of 30,000. Each of the new towns

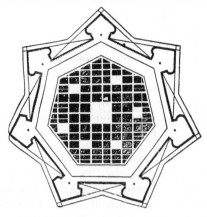

Figure 5.14 – Pietro Cataneo: ideal city plan from his *L'Architettura* of 1554, based on the gridiron with an oddly asymmetrical arrangement of the six minor squares. Most ideal city theoreticians were unable to reconcile conflicts of the essential geometry, whereby a circular or faceted polygonal defensive perimeter was required to enclose a gridiron or sensibly regular street pattern. Around the edges, where Cataneo's street grid meets the fortification zone, there are uselessly small and irregular building plots. (See also Scamozzi's ideal plan as Figure 5.17.)

A radial main street system, exemplified by Buonaiuto Lorini (Figure 5.15) and Scamozzi (or whoever) at Palma Nova (Figure 5.18), was the most economic compromise solution.

Figure 5.15 – Buonaiuto Lorini: ideal city plan published in his *Delle Fortificatione Libri Cinque* (1592), an intricately devised defensive system with three most insignificant entrances (bottom, top left and top right). Shows radial street planning with the church facing onto the central piazza.

51. Blomfield, *Sèbastien le Prestre de Vauban 1633–1707*.

52. E A Gutkind, *International History of City Development – Southern Europe: Italy and Greece*, 1969.

53. For Leonardo da Vinci see M Kemp, *The Marvellous Works of Nature and Man*, 1981; J Bronowski, 'Leonardo da Vinci', in J H Plumb (ed.) *The Pelican Book of the Renaissance*, 1982.

was to have 5,000 houses. (This figure of 30,000 is remarkably close to that of 32,000 recommended by Ebenezer Howard at the end of the nineteenth century as the basis for his garden city proposals.[54]) In his ideal city Leonardo was also centuries in advance of his time when he advocated multi-level separation of vehicular and pedestrian traffic, with special routes reserved for the heaviest goods traffic. In regional planning matters it was he who, though still only a youth, first suggested the formation of a canal from Pisa to Florence, by means of changes to be effected in the course of the River Arno.

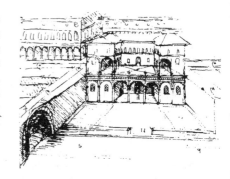

Figure 5.16 – Leonardo da Vinci: prophetic sketches for a multi-level city. (The original drawing is in the Bibliotheque de l'Institute de France, Paris; reproduced in M Kemp, *The Marvellous Works of Nature and Man*, 1981.)

SCAMOZZI AND PALMA NOVA

Vincenzo Scamozzi (1552–1616) was exceptional among Italian Renaissance urban theorists in that his ideas were actually realized in practice. He is usually credited with building the small fortified town of Palma Nova which was started in 1593, although some writers have

Figure 5.17 – Scamozzi: ideal city plan from his *L'Idea dell'Architettura Universale* (1615), notable for its use of the gridiron for internal layout, the hierarchy of squares and the introduction of a canal waterway system. (But note the ill-shaped small building blocks at the perimeter.)

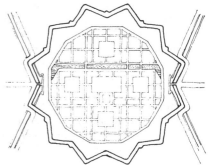

Figure 5.18 – Palma Nova (1593) the plan for which is generally attributed to Scamozzi. Adequate architectural detail is given in this reproduction for it to be seen that the regularity of the plan has not been complemented by standard building elevations. See also Philippeville (Figure 6.3), a Spanish frontier fortress town, with a radial street system.

queried his authorship of the plan. In 1615, the year before his death, he published in Venice a comprehensive ten-volume work, *L'Idea dell'Architettura Universale*, in which he recorded his architectural experiences. Nine chapters of the second book are concerned specifically with the subject of military fortifications based on a comprehensive study of outstanding western European examples. In his treatise Scamozzi includes a detailed plan of a theoretical fortified city which has features in common with Palma Nova, except for one basic differ-

---

54. See E Howard, *Garden Cities of Tomorrow*, 1904; the closely comparable 'new town' populations can but be entirely coincidental.

ence – its street system is organized as a gridiron within the defensive perimeter (Figure 5.17).

The function of Palma Nova was to be a fortified garrison outpost of Venice's defences. Its perimeter is a nine-sided polygon and its central square a regular hexagon.[55] These shapes are resolved into an integrated pattern by a complex arrangement of radial streets; six lead out from the centre to an angle of the wall, or, alternatively, to the centre of a side. Additionally twelve radial streets start from the innermost ring of three concentric streets. Main civic buildings are grouped around the central square. Six secondary squares are formed in the centres of house blocks. Palma Nova exists today as a quiet country town; fortunately it has retained the basis of its original plan, unobscured by subsequent developments.

Figure 5.19 – Palma Nova from the air, showing the great extent to which the town has retained its original plan and the comparatively vast area of the encircling fortification system. The pattern of field boundaries and main and local roads is that of another version of the pre-urban cadastre. (See also the aerial photograph of Neuf Brisach, a Vauban fortress town of the late seventeenth and early eighteenth centuries, Figure 6.29.)

55. For detailed plans of Palma Nova, see L di Sopra, *Palmanova: Analisi di una citta-fortezza*, 1983.

Florence

The city originated round about 200 BC when a new settlement was established where the Via Cassia, extended north from Arezzo to Bologna, crossed the Arno.[56] It was named Florentia and received brief status as a *municipium* in 90 BC before being completely destroyed by Sulla in the civil war of 82 BC. Florentia was rebuilt as a Roman colony a short distance downstream, occupying an area of 79.1 acres within the first of a sequence of defensive perimeters. (See page 99 for a description of the city's medieval town-building programme.)

Florence, for all its originating role during the early years of the Renaissance, remained essentially a medieval city, its magnificent Renaissance *palazzi* unrelated, and in most instances, uncomplemented by suitable spatial settings. Of the city's two most important public spaces, one – the Piazza della Signoria (the civic centre for more than 600 years) – is a medieval creation, its spatial anomalies only partially resolved by Renaissance furnishing (Figure 5.20). The Piazza Annunziata, on the other hand, is of great significance as a work of Renaissance urbanism.

Brunelleschi's Foundling Hospital of 1419–24, as completed, presented its beautiful arched arcade to the unresolved space in front of the Church of Santissima Annunziata, but it was to set the pattern for the eventual enclosure of the Piazza Annunziata – a space which had the dual function of providing an entrance forecourt to the church, and of terminating a long vista from the cathedral, on the axis of linking cathedral with state and church (Figure 5.21).

Following the construction of the Foundling Hospital, Michelozzo in 1454 designed a one-bay entrance porch to the church in harmony with Brunelleschi's arcade. This porch was subsequently enlarged by Giovanni Caccini between 1601 and 1604 into an entrance colonnade running the length of the north-western side of the square. A third arcaded side to the square, facing the Foundling Hospital, was designed by the architects Antonio da Sangallo the Elder and Baccio d'Agnola in 1516. These arcades serve to unify a number of disparate buildings into one spatial unity, in a manner reminiscent of the colonnades around the forum at Pompeii. Paul Zucker considers that in contrast to the medieval period, where an arcade belonged to an individual building, in the Renaissance arcades expanded the space of the square, integrating volume of structure and spatial void.[57]

Edmund Bacon in an extremely important section of his *Design of Cities* bases his 'principle of the second man' on the Piazza Annunziata, stating that 'any really great work has within it seminal forces capable of influencing subsequent development around it, and often in ways unconceived by its creator. The great beauty and elegance of Brunelleschi's arcade of the Foundling Hospital ... found expression elsewhere in the piazza, whether or not Brunelleschi intended this to be so'. The crucial decision, was that of Sangallo 'to overcome his urge toward self-expression and follow, almost to the letter, the design of the then eighty-nine-year-old building of Brunelleschi. This design set the form of the Piazza Annunziata and established, in the Renaissance train of thought, the concept of a space created by several buildings designed in relation to one another. From this the "principle of the second man" can be formulated: it is the second man who determines whether the creation of the first man will be carried forward or destroyed'.[58]

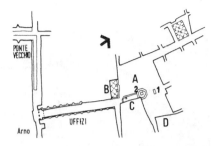

Figure 5.20 – Florence: the plan of the Piazza della Signoria related to the Uffizi and the River Arno and the Ponte Vecchio. Key: A, Piazza della Signoria; B, Loggia de Lanzi; C, Palazzo Vecchio; D, Palazzo della Tribunale di Mercanzia; 1, equestrian statue of Cosimo I; 2, statue of Neptune. See E Bacon, *Design of Cities*, 1974, for a diagram of the entire city centre.

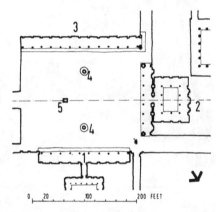

Figure 5.21 – Florence: plan of the Piazza Annunziata. Key: 1, the Foundling Hospital of 1419–24 (the first Renaissance building; 2, the Church of Santissima Annunziata; 3, Hall of the Servi di Santa Maria Brotherhood; 4, fountains; 5, equestrian statue of 1608 on the main central axis of the piazza.

Below the Ponte alle Grazie comes the Ponte Vecchio, the Bridge par excellence; il ponte, or il passo d'Arno, as Dante calls it. More than a mere bridge over a river, this Ponte Vecchio is a link in the chain binding Florence to the Eternal City. A Roman bridge stood here of old, and a Roman road may be said to have run across it; it heard the tramp of Roman legionaries, and shook beneath the horses of Totila's Gothic cavalry. This Roman bridge possibly lasted down to the great inundation of 1333. The present structure, erected by Taddeo Gaddi after 1360, with its exquisite framed pictures of the river and city in the centre, is one of the most characteristic bits of old Florence still remaining. The shops of goldsmiths and jewellers were originally established here in the days of Cosimo I., for whom Giorgio Vasari built the gallery that runs above to connect the two Grand Ducal Palaces. Connecting the Porta Romana with the heart of the city, the bridge has witnessed most of the great

56. For Florence see E G Gardner, *The Story of Florence*, 1908; P Sica, *Storia dell'Urbanistica L'Ottocentro*, 4th edn, 1991; M C Branch, *Comparative Urban Design*, 1978, for the 1835 SDUK Map of Florence.

57. Zucker, *Town and Square*.

58. Bacon, *Design of Cities*.

Rome

DECLINE AND FALL

Norwood Young in his book *The Story of Rome*, on which this section relies to a considerable extent, considers that 'if the strictly mediaeval history of Rome closes with Boniface VIII in 1303 and the period of desertion and decay comes to an end on the arrival of Martin V in 1420, the era of new life, of renaissance, begins with Nicholas V in 1447'.[59] Before concentrating on the planning achievements of the Renaissance popes during the century and a half from Nicholas V to Sixtus V, we must briefly take the history of the city through from its ancient zenith under the Caesars in the second century AD, to the deserted, desolated ruins which were the inheritance of the Renaissance.

Ancient Rome at the summit of its fortunes, in particular during the reign of the Emperor Trajan (AD 98–117) has been described in Chapter 3. From this time on, the fortunes of both the Roman Empire and its original capital city were on the wane. The unprecedented flooding of the city in AD 162 and the subsequent famine and plague seem to mark the beginning of the end. In AD 334 the first Christian emperor, Constantine, recognized this decline by moving the capital of the Empire from Rome to his new city of Constantinople (see end of Chapter 3).

The physical destruction of ancient Rome starts with the pillage of the city by the barbarians under Alaric the Visigoth in 410, after he had been bought off two years previously. Subsequent devastation followed the incursions of Genseric the Vandal in 476; he governed Italy from Ravenna while Rome was controlled only by a local prefect. A brief revival followed under Theodoric the Ostrogoth who took the city in 483, at the invitation of the Eastern Emperor, and repaired the aqueducts, the walls and many public buildings.

In 546 the Goths again besieged and captured Rome, this time under Totila who, lacking military strength sufficient to garrison the city while campaigning elsewhere in Italy, was determined to destroy the town entirely, to raze it to the ground, to make it a pasturage for cattle.[60] Totila was diverted from this extreme course, but as a secondary measure forced the entire population to abandon the city which even at this time was almost as magnificent as ever. Some temples and walls had been damaged and many statues taken away, but the city still contained numerous immense marble buildings.[61]

The first pope to be the practical lord of Rome in municipal affairs was Gregory I (590–604). 'In ecclesiastical matters he was undisputed head of Western Europe and in temporal matters a rival of the Eastern Emperor.'[62] In his first sermon in St Peter's he painted a dismal picture of the world laid waste and the devastation of the city of Rome. In 663, however, Constans II, the first Eastern Emperor to visit the city since the fall of the Western Empire, found it worth while removing from the roof of the Pantheon the bronze tiles and all the remaining bronze statues, with the exception of that of Marcus Aurelius which was thought to be of Constantine. (This is the statue which was used in 1538 as the focal point of Michelangelo's new Capitoline Piazza.)

From 800, as the generally acknowledged spiritual centre of the Holy Roman Empire, Rome again became for a while a great city and a centre of pilgrimage in Western Europe. Following further attacks, this time by the Saracens, Leo IV in 852 built the first fortifications of the

pageants and processions in Florentine history. Popes and Emperors have crossed it in state; Florentine generals, or hireling condottieri, at the head of their victorious troops; the Piagnoni, bearing the miraculous Madonna of the Impruneta to save the city from famine and pestilence; and Savonarola's new Cyrus, Charles VIII., as conqueror, with lance levelled.
(E G Gardner, The Story of Florence, 1908)

Probably neither rebel generals nor barbarians on the frontier would have much endangered the Empire if it had not been steadily weakening within. Great nations break down more often from internal than external causes. That the resources of the Empire were wasted for centuries is only too clear. The pillage of the provinces in the last century of the Republic was shameless; and whatever recovery was made under the Empire, certain wounds cannot be healed. One Roman conqueror, a man of high character and repute, sold one hundred and fifty thousand inhabitants of Epirus as slaves. Sulla's furious vengeance on Asia will be recalled. The money that rolled up in masses to Rome was in large part wasted, as the wealth of Alexander's successor was. A mania for building huge villas, baths, arches, temples; a still more wasteful mania for beast-shows, in each of which the public slaughter of the animals (fetched with extreme cost from the ends of the earth); a universal desire to wear silk and to have spiced food – all these things and others wasted labour.

Lately the point has been well made that there was a new shortage of labour. For some two centuries, down to Actium, Eastern and Northern wars had meant a ceaseless supply of slaves in Italy. Agriculture and manufacture were carried on by slaves – a bad thing in itself, for slave labour is reluctant and careless labour, wasteful of material; there was no improvement in tools or methods (and improvement was needed); slaves generally left no offspring, and they crowded out free labour with its families. This contributed to decline in population, while land went out of cultivation. Italy began to repeat the experience of Greece.
(T R Glover, The Ancient World)

59. N Young, *The Story of Rome* (Mediaeval Towns series, 1901), rev. edn 1953, still a valuable history of the city); see also R Krautheimer, *Rome: Profile of a City 312–1308*, 1980; C Hibbert, *Rome: The Biography of a City*, 1985.
60. Young, *The Story of Rome*; for other less fortunate captured cities, reference is made to Carthage, which was completely razed by the ancient Romans (Chapter 3); see also Warsaw, which was systematically devastated, if not completely destroyed, to Hitler's orders during the Second World War (Chapter 7).
61, 62. Young, *The Story of Rome*.

Vatican quarter. The most destructive of all the sackings of Rome took place in 1084 when the Norman allies of Gregory VII set the city on fire with the results that the Field of Mars was swept nearly bare and the region between the Lateran and the Colosseum was utterly destroyed. The Caelian and Aventine Hills have never since returned to their former populous condition.[63]

In 1065, the Turks captured Jerusalem and the ensuing Crusades to liberate the Holy Land gave the popes their last chance to unite Christendom in one cause under their banners. The First Crusade, sponsored by Urban II and led by Godfrey de Bouillon, ended with the recapture of Jerusalem in 1099. Later crusades, including the two led by the French King Louis IX, for which he built the bastide port of Aigues-Mortes as a point of departure (see Chapter 4), were progressively less successful. When the crusades came to an end with the fall of Acre in 1292 Boniface VIII hit upon the plan of using the Jubilee Year as a means of stimulating Christian enthusiasm and assisting papal finances by the sale of indulgences.[64] Previously indulgences – pardons for all sins against the Church – were granted to participants in the crusades and their financial supporters. The Jubilee of 1300 was an immense success. Norwood Young recounts that all Europe responded in a general contagion of religious zeal. The roads in the remotest parts of Germany, Hungary and Britain swarmed with pilgrims on the march to Rome. It was estimated that there was a traffic of 30,000 pilgrims in and out of the city daily, and that 2 million had entered Rome in the year. The offerings were gathered in at the altars with long rakes, the copper coins alone giving a value of 50,000 gold florins.[65]

STRATEGIC REPLANNING

The jubilees were at first intended to mark the beginning of a new century but later were held every twenty-five years. To qualify for their indulgence, pilgrims to the city had to visit certain specified churches and from 1300 onwards the papal financial policy relied on the income from the pilgrims. Related papal plans for the rebuilding of Rome were to a considerable extent based on facilitating the movement of pilgrimage groups between the seven main churches. Providing suitable accommodation for them and policing the city for their safety were also primary considerations.[66]

Important as it was to encourage the pilgrimage industry, Rome had of course a wider role to play in promoting the general authority of the Church. This policy was clearly reaffirmed by Nicholas V (1447–55) at the beginning of the Renaissance period in Rome. 'To create solid and stable convictions in the minds of the uncultured masses there must be something that appeals to the eye: a popular faith, sustained only by doctrines, will never be anything but feeble and vacillating. But if the authority of the Holy See were visibly displayed in majestic buildings, imperishable memorials and witnesses seemingly planted by the hand of God himself, belief would grow and strengthen like a tradition from one generation to another, and all the world would accept and revere it. Noble edifices combining taste and beauty with imposing proportions would immensely conduce to the exaltation of the chair of St Peter.'[67]

In itself this was not original thinking. Several popes between Boniface VIII and Nicholas V had had visions of a revival of the grandeur that was Rome but such sweeping planning programmes presuppose a settled, ordered community. Instead, the period between the papal

To a very large degree the first crusade owed its success to the spirit of the men who went on it. The common soldiers had been stirred in a way which would brook on halting; it was they who forced their commanders to lead them on to Jerusalem...

The Church had long been seeking to channel the vigour of such men into activities more pleasing to God than the endless feuding of the nobility. She had given her blessing to wars against the heathen, and especially had sought to organise soldiers for service in Spain against the Saracens.... Churchmen preached the crusade, they recorded the vows of the future crusaders, they collected and administered the alms of the faithful which helped to finance it. Acting through her diocesan bishops, the Church took into her special protection the lands and property of all who left for Palestine. Without the unitary papal system of administration behind it, the crusade could never have been organised as a large scale enterprise.

The crusade demonstrated, indeed, what a real unifying force in Christendom the Roman Church had become. To the canonists, the crusade became known as the 'Roman war', 'because Rome is the head and mother of our Faith'.
(M Keen, A History of Medieval Europe)

Rome in 1300
The profits made by the people of Rome that year were incalculable. One visitor saw so large a party of pilgrims depart on Christmas Eve that no one could count the numbers. The Romans reckon,' he continued, 'that altogether they have had two millions of men and women. I frequently saw both sexes trodden underfoot, and it was sometimes with difficulty that I escaped the same danger myself.' Day and night the streets were packed with people, lining up to pass through the churches; to see the shrines and the most famous relics; to gaze with reverence upon the handkerchief with which St Veronica wiped the sweat from Christ's face on his way to Calvary and which still bore the image of his features; to throw coins upon the altar of S. Paolo fuori le Mura, where two priests remained constantly on duty with rakes in hand to gather up the scattered offerings; to buy the relics, amulets, mementoes and pictures of saints whose sale brought such profits to the street-vendors of Rome.
(C Hibbert, Rome: the Biography of a City, 1986)

63, 64, 65. Young, The Story of Rome.
66. As a modern parallel, the Saudi Arabian Government has carried through a major series of improvements of the facilities provided for the movement and accommodation of pilgrims to the Islamic Holy City of Makkah.
67. Young, The Story of Rome. The last sentence is a straightforward statement of the aggrandizement determinant.

move to Avignon in 1309 and the end of the great schism with the election of Martin V in 1417 was a time of uncontrolled disorder in Rome and the city was a prey to anarchy.[68] Instead of being occupied with building the city of the future, all energies were consumed in continual power struggles among the leading family groups, and between them and the papal authority. In 1329 an earthquake severely damaged many of the ancient buildings, adding to the desolation, and in 1347 the Black Death took its toll of the already greatly reduced population. The Jubilee of 1350 represented a brief interlude of prosperity with 1.2 million pilgrims arriving in Rome between Christmas and Easter and 800,000 more at Whitsun. The condition of the city, however, was appalling. Petrarch was one of the visitors and he observed that the houses were overthrown, the walls come to the ground, the temples fallen, the sanctuaries perished, and the laws trodden underfoot. The Lateran lay on the ground, and the 'Mother of all Churches' stood without a roof, exposed to wind and rain.[69]

THE RETURN FROM AVIGNON

Gregory XI (1370–78) brought the papacy back to Rome from Avignon, after the 68 years of absence, returning to the hollow shell of a city in which perhaps only 17,000 occupied the area inside the Aurelian Wall where upwards of 1.5 million had lived under the Caesars. Gregory's death in 1378 led to a yet greater schism in the church, between the elected Roman successor and the rival pope chosen by Naples, Savoy, France and Spain.

For the next forty years Roman history is a chronicle of intrigue and revolution, ending only with the election of Martin V (1417–31); this re-established settled, orderly government for the first time since Boniface VIII. By then Rome had reached the nadir of her fortunes. The Tiber river defences were broken down – in 1420 flood water lapped the high altar in the Pantheon – the drainage system was silted up, the aqueducts were long since destroyed. All of the inherent deficiencies of the site had reasserted themselves and disease was rampant. The ancient Christian churches were dangerously decayed and from the account of Bracciolini, a Florentine visitor of 1420, it is evident that classical Rome had already been almost entirely destroyed.[70]

Martin V achieved little in the way of the practical renewal of the city, other than starting to repair the churches, but his pontificate established the social and economic conditions essential for the work of future popes, although his immediate successor Eugenius IV (1431–47) was forced to leave the city in 1434 when a republic was once more established. During his reign the arrival of the Renaissance in Rome is marked by the new bronze gates for St Peter's, designed by Antonio Filarete.

Nicholas V, as recorded earlier, was determined to increase the prestige of the Church by the grandeur of a new Rome. The Jubilee of 1450 produced the revenue for a programme of church repairs and in addition the city's defences were reconstructed. There is little evidence of any general plan for rebuilding but key sites were made available for new buildings at nominal rent. Nicholas V also played an extremely important role in the development of the Renaissance, in Rome and in Italy generally, in extending a generous hospitality to the émigré intelligentsia from Constantinople, particularly after it fell to the Turks in 1453.

Figure 5.22 – Rome: approximate extent of the medieval city. Destruction or blockage of the aqueducts serving the healthier higher districts had long since rendered them uninhabitable: Rome's inhabitants, such as they were, were concentrated on to low-lying ground in the bend of the Tiber opposite St Peter's. Here for the most part they found shelter in the substantial remains of imperial monuments, continually subject to flooding and disease. Medieval Rome, on the flat ground in the bend of the Tiber, has retained to the present day a closely informal street pattern in direct contrast to the spacious additions of the Renaissance and Baroque centuries. See the extract from Nolli's Map of 1748 (Figure 5.24) and the concluding map of Rome in 1830 (Figure 5.32).

68, 69. Young, *The Story of Rome*.
70. The same Poggio Bracciolini who had drawn attention to the Vitruvian manuscript in 1412–14. Norwood Young also quotes an English visitor to Rome of the same time exclaiming, 'O God, how pitiable is Rome! Once she was filled with great nobles and palaces; now with huts, thieves, wolves, beggars and vermin, with waste places, and the Romans themselves tear each other to pieces.'

At this time the inhabited part of Rome was concentrated largely on the low-lying Campus Martius, within the broad bend in the Tiber opposite the Castel Sant'Angelo; this latter served the popes as treasure house, prison and place of refuge in times of invasion or revolt (Figure 5.22). Although of secondary importance to the later system of main streets linking the seven churches of Rome, from Nicholas V onwards, the bridgehead across from the castle was established as the focal point of a network of routes through the medieval city. Bufaldini's map of 1551 shows this bridgehead as the Forum Pontis, later to be renamed the Piazza di Ponte. Leading from it, five streets were pieced together out of the irregular lanes (Figure 5.23); the most important, the Via Papalis, was subsequently widened and extended as the Corso Vittorio Emanuele. In an attempt to persuade people to move back on to the deserted higher parts of the city, Nicholas V in a Bull of March 1447 exempted the Quirinal, Viminal and Esquiline Hills from taxes, and reconstructed the Aqua Virgine to give the city a supply of 63,000 cubic metres of fresh water daily.

After Nicholas V three popes, Claistus III, Pius II and Paul II, contributed nothing of note and even failed to maintain settled conditions in the city. Sixtus IV (1471–84), however, took up the process of reconstruction and carried out many improvements both in preparation for the Jubilee of 1475 and, subsequently, with its proceeds. Before the arrival of the pilgrims he converted the old Ponte Rotto into the Ponte Sisto and, in order to avoid the recurrence of an earlier pedestrian crush in the narrow streets in which over 200 people had died, he introduced one-way routeing; the crowd was made to use the Ponte Sant'Angelo in going to St Peter's and the Ponte Sisto for their return across the Tiber. Also in preparation for the jubilee, he rebuilt the great Santo Spirito Hospital, repaired the Aqua Virgine, and restored the Trevi Fountain. A direct network of main streets linking the city's major churches was also started and, in an edict of 1480, he commanded that all building projections and street obstructions be cleared away.[71]

Sixtus IV was also an extremely able planning administrator, with views considerably in advance of his time in that he recognized that private interests should be subordinate to the public good. He insisted that although compensation for private loss must be fairly administered, there also existed on the other hand what became known later as 'betterment' and this ought to be also taken into account. A commission was established to consider the effect of property boundary adjustments brought about by the street improvements.

Alexander VI (1492–1503) restored the Castle of St Angelo and connected it to the Vatican by a new street, the Borgo Nuovo. His successor Julius II (1503–13) was mainly preoccupied with the problem of the Basilica of St Peter and took the final decision to demolish the old church rather than attempt to repair its decayed fabric. In April 1506 the foundation stone was laid of the present church to the designs of Bramante, who lived only to see the completion of the four main piers and the arches under the dome. A succession of famous Renaissance artists subsequently worked on St Peter's before its completion in 1626, including Raphael and Michelangelo, who built the magnificent dome and the supporting drum. But illustrative of the continuing disregard of the Romans for their ancient architectural heritage, Norwood Young comments that the new St Peter's was made entirely, even to the

Rome, Paris and London – the most important foci of western civilization – created the prototypes of the large cities of today. Rome's contribution came first. It was the work of the Popes of the sixteenth century, systematized and epitomized by Sixtus V. Under his initiative the limited, wall-girdled, star-shaped City of the Renaissance was converted into the City of Baroque, with those boldly drawn traffic lines which still form the warp and weft of the modern city.

Yet, by the beginning of the Renaissance, Rome was a desolate city, at the end of a millennium of decline. When the Popes moved from the Lateran to the Vatican they preferred to build a new suburb, and so the Borgo Nuovo gradually grew round the Basilica of St Peter. Thus, about 1500, when the Popes began to rebuild medieval Rome in earnest, the Popes became the greatest builders in the world. Neither they nor their leading architects and planners were ever Roman by birth or upbringing. Julius II, a Rovere from Urbino, and Leo X, a Medici from Florence, called in their close compatriots – Bramante and Raphael from Urbino and Michelangelo from Florence – to carry out their grandiose schemes: and so it continued even in the time of Baroque Rome.
(S Giedion, 'Sixtus V and the Planning of Baroque Rome', Architectural Review, April 1952)

For Rome it was the century which transformed a relatively small medieval town of barely 20,000 inhabitants into a city of over 100,000, and the capital, not just of the Papal States, but of Catholic Christendom. It changed the name of the 'campo vaccino', the cow field, back to Forum Romanum, as an archaeological site and a tourist attraction. In 1526, just before Rome was sacked by the imperial armies, it had already 236 hotels and inns; at the end of the century, there were at least 360, plus innumerable furnished houses and rooms to let. Thus it was that during the jubilee year of 1600, Rome could accommodate over half a million pilgrims.

Everything was geared to this: there was one wineshop for every 174 inhabitants; there were hundreds of tailors, goldsmiths, manufacturers of 'objets d'art' and religious souvenirs and, inevitably, street vendors. Yet all attempts to introduce a textile manufacturing industry into the city failed. Rome lived on the contributions of Catholic Europe, on the income of ecclesiastical offices and church lands, spent by their owners in the Eternal City; but, above all, it lived on its visitors: a city of beggars and prostitutes, of devoted clergy, pious pilgrims and indifferent tourists, of nobles and princes of the church, displaying their wealth and ruining their fortunes by sumptuous buildings and princely dowries to their daughters and nieces. Two things were necessary for success in Rome said St Carlo Borromeo, to love God and to own a carriage.
(H G Koenigsberger and G L Mosse, Europe in the Sixteenth Century)

---

71. S Giedion, 'Sixtus V and the Planning of Baroque Rome', Architectural Review, April 1952.

mortar, of materials taken from the ruins of classical monuments.[72] Julius II also laid out two new streets on either side of the Tiber – the Via Giulia on the left bank (downstream from the Piazza di Ponte) and the parallel Via Lungara on the right bank.

The next two popes, Leo X (1513–21) and Paul III (1534–49), were responsible respectively for the Via Ripetta and the Via Babuino, symmetrically laid out on either side of the Strada del Corso – the city's main approach from the north, by way of the Piazza del Popolo (see below and Figure 5.28). Paul III also continued redevelopment around the Piazza di Ponte and laid out the line of the Via Trinitatis, extending a route from the piazza towards the church of Sta Trinità dei Monti. In 1561 Michelangelo built his magnificent Porta Pia in the old Aurelian Wall for Pius IV, who connected it back to the deserted Quirinal Hill by the Strada Pia (now the Via del Quirinal); he also built the Via XX Settembre, from 1555 to 1567.

SIXTUS V

Between the papacies of Sixtus IV and Sixtus V there is a period of almost exactly a century, during which the city was steadily improved. The major achievement of the Capitol Piazza was carried through and redevelopment of the Piazza del Popolo (see below) was commenced. But the final form of Renaissance Rome owes most to the five-year pontificate of Sixtus V (1585–90). In this relatively short time he carried out an extensive programme of works, mainly with the architect-planner Domenico Fontana as his executant adviser. The programme was based on three main objectives: first, to repopulate the hills of Rome by providing them with the direct water supply lacking since the cutting of the ancient aqueducts; second, to integrate into one main street system the various works of his predecessors by connecting the main churches and other key points in the city; lastly, to create an aesthetic unity out of the often disparate buildings forming the streets and public spaces.[73] Sixtus V was 64 when elevated to the papal throne. He died of malaria, in his unfinished Quirinal Palace, just five years and four months later, during which time nowhere is his race with death more apparent than in the incredible rapidity with which he carried through his building programme; again and again Fontana remarks that nothing could be accomplished quickly enough to please his 'beloved lord'.[74]

It is clear that before becoming pope, Sixtus V had given considerable thought to Rome's planning problems. Immediately following his accession work was under way on the Aqua Felice; the Strada Felice, linking Santa Croce in Gerusalemme to Sta Trinità dei Monti, was commenced and completed during the first year; the obelisk, later to be erected in front of St Peter's, was transported to the site; the Lateran Palace and Basilica were under construction; and some 2,000 workers were put to draining the Pontine marshes. Although other popes had restored water supplies to the lower parts of the city they had failed to do so for the hilly districts. To bring water to the Quirinal, Viminal and Esquiline Hills, Sixus V constructed the Aqua Felice between 1585–89 (so called because Sixtus's name was originally Felice Peretti). This aqueduct was created in part from the ancient Aqua Marcia and Aqua Claudia; it was about 16 miles in length and because of a severely limited fall from its source involved 7 miles of high aqueduct and 7 miles of tunnel. It supplied over 4 million gallons a day.[75]

Sentence of death on the monuments of the Forum and of the Sacra Via was passsed on July 22nd, 1540. By a brief of Paul III 'the privilege of excavating or giving permission to excavate is taken away from the Capitoline or Apostolic chambers, from the "magistrates of streets", from ecclesiastical dignitaries etc, and given exclusively to the "deputies" for the Fabbrica di S Pietro. The pope gives them full liberty to search for ancient marbles wherever they please within and outside the walls, to remove them from antique buildings, to pull these buildings to pieces if necessary; he orders that no marbles can be sold by private owners without the consent of the Fabbrica, under the penalty of excommunication "latae senteniae", of the wrath of the pope, and of a fine of 1,000 ducats'. No pen can describe the ravages committed by the Fabbrica in the course of the last sixty years of the sixteenth century. The excesses roused the execration of the citizens, but to no purpose; on May 17th, 1580, the 'conservatori' made an indignant protest to the town-council, when a portion of the palace of the Caesars had fallen in consequence of its having been undermined by the searchers for marble. A deputation was sent to Gregory XIII to ask for the revocation of all licences ('ad perquirendos lapides etiam pro usu fabricae Principis apostolorum'). We may imagine what answer was given to the protests of the city when we learn that by a brief of Clement VIII, dated July 23rd, 1598, the archaeological jurisdiction of the Fabbrica was extended over the remains of Ostia and Porto. The Forum Romanum was swept by a band of devastators from 1540 to 1549; they began by removing the marble steps and the marble coating of Faustina's temple (1540), then they attacked what was left standing of the arch of Fabius (1540). Between 1546 and 1547 the temple of Julius Caesar, the Regia, with the 'fasti consulares et trimphales', fell under their hammer.

The steps and foundations of the temple of Castor and Pollux were next burnt into lime or given up to the stonecutters, together with the arch of Augustus. The temple of Vesta, the Augusteum, and the shrine of Vortumnus, at the corner of the Vicus Tuscus, met with the same fate in 1549.

(R Lanciani, The Ruins and Excavations of Ancient Rome)

72. Young, The Story of Rome; see also R Lanciani, The Ruins and Excavations of Ancient Rome, 1897/1968; R Krautheimer, The Rome of Alexander VII 1655–1667, 1985.

73. For a reassessment of the achievements of Sixtus V, see T A Marder, 'Sixtus V and the Quirinal', Journal of the Society of Architectural Historians, vol XXXVII, December 1978, who stresses that the Pope had no 'master plan' per se, and that individual improvement projects usually proceeded by fits and starts.

74. S Giedion, Space, Time and Architecture, 1961.

75. Marder, 'Sixtus V and the Quirinal', regards the Aqua Felice as the main urban planning achievement of Sixtus V, on which there were employed as many as 4,000 daily workers.

THE ROMAN STREET SYSTEM

As integrated by Sixtus V into a comprehensive movement system, the principal streets provided, as one of their major functions, a link between the seven pilgrimage churches of Rome – San Pietro in Vaticano (St Peter's); San Giovanni in Laterano; Santa Maria Maggiore; San Paolo fuori le Mura; and San Lorenzo fuori le Mura (the original five churches) and two accorded special veneration later – Santa Croce in Gerusalemme and San Sebastiano (Figure 5.23). This realized the ambition of Boniface VIII, adumbrated some 300 years earlier, to make Rome a worthy capital of Christendom.

Sixtus V was not, however, concerned only to facilitate religious ceremonial. He was well aware of the role the new streets could play in generating growth in the largely uninhabited, although climatically propitious, eastern and south-eastern districts. In this respect his contribution to the reconstruction of Rome has been misrepresented by a number of urban historians. Sibyl Moholy-Nagy over-emphasizes the religious function with her criticism that his nine processional routes through the sparsely built-up eastern section of town were designed for a blatantly anti-urban processional ritual of piety and penitence;[76] Lewis Mumford uncharacteristically gets his facts wrong in stating that 'the three great avenues that radiate from the Piazzo del Popolo – the conception of Sixtus V – were designed to make it easy for the pilgrim to find his way to the various churches and holy spots'.[77] (All three avenues were in existence before Sixtus's accession.) Sigfried Giedion, on the other hand, would seem to present a correct balance between religious and secular functions,[78] as also does Edmund Bacon in a beautifully illustrated section of his *Design of Cities* in which he argues the powerful proposition that Sixtus V saw clearly the need to establish a basic overall design structure in the form of a movement system as an idea.

Before describing the basis of Rome's late Renaissance street system, it is appropriate to relate the achievements of Sixtus V to those of Napoleon III in Paris during the two decades between 1850 and 1870. Each has suffered historical distortion. Both were effectively absolute rulers. Sixtus had Fontana and Napoleon III his Haussmann (although Haussmann's role as Prefect of the Seine made him more the Emperor's agent than executive planner). Both rulers were responsible for giving their ancient cities incomparable main route structures, and both assumed power with already formulated proposals. Napoleon's role has been slighted in this latter respect; usually Haussmann alone has been credited with restructuring Paris. This record has been set right by David Pinkney: 'When Georges Haussmann laboured as a little-known provincial prefect in the department of the Yonne, Louis Napoleon already had on paper his ideas for rebuilding the city and had been urging them on a reluctant prefect and municipal council. On the day when Haussmann took the oath of office as the Prefect of the Seine, Napoleon handed to him a map of Paris on which he had drawn in four contrasting colours (the colours indicating the relative urgency he attached to each project) the streets that he proposed to build.'[79] The reality of the Parisian situation is that emperor and prefect, and their professional advisers, together formed a powerfully effective team.[80]

The record of achievement is, however, still further distorted. In the same way that the Parisian boulevards have been criticized as having

Only members of the nobility and the ruling houses of Italy were usually elected to the papel throne. There were, however, exceptions, even in a period such as the end of the sixteenth century, when the steadily increasing privileges of the nobility had usurped the medieval rights of the people. So it was possible for Sixtus V, a man from the lowest strata of society to be invested with the highest dignity of spiritual and temporal power to which a mortal could aspire. It says much for the inner strength, vitality and instinct of the Catholic Restoration that it had the courage – at this very dangerous moment – to elevate a man such as Sixtus V to this office; a man who, regardless of his ancestry, was clearly born for action.

Sixtus V was the papal title chosen by the Franciscan mendicant friar, Felix Peretti, who had entered the order at the age of 12. His father, a small tenant farmer and gardener of Dalmatian stock, filled with visions of the future destiny of his son, had given him the name of Felix. This name Sixtus V – in contrast to other Popes – never laid entirely aside. He bestowed it upon the two projects that lay nearest his heart, and also nearest the place of his residence as a Cardinal, the Villa Montalto, north of Sta Maria Maggiore: the Strada Felice – Rome's grandiose, north-west south-east highway – and the Aqua Felice which brought life to the hills of the south-east.

(S Giedion, 'Sixtus V and the Planning of Baroque Rome', Architectural Review, April 1952)

76. S-M Nagy, *Matrix of Man*, 1969.

77. Mumford, *The City in History*; not only was the true purpose of the Piazza del Popolo that of providing a suitably impressive entrance to the city from the north, but also the three radiating streets date from 1516, well before Sixtus V's time (see p. 179).

78. Giedion, *Space, Time and Architecture*.

79. D Pinkney, *Napoleon III and the Rebuilding of Paris*, 1958.

80. The 'partnerships' of Sixtus V and Fontana, and Napoleon III and Haussmann were formally open and straightforward, whereas that in early nineteenth-century London between the Prince Regent (the future George IV) and John Nash, which produced Regent Street, operated in secret behind the scene (see Chapter 8).

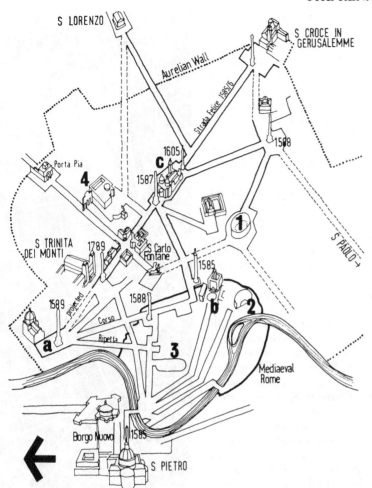

Sixtus V is the patron saint of American city planners who credit him with the first urban renewal miracle. It is doubtful whether any saint has ever been canonized for less valid reasons. There is no evidence that Sixtus intended to change Rome from a maze of medieval slums into a planned orthogonal city by opening up linear communications and designing effective 'reciprocal vistas'.

From the available documents, it is abundantly clear that he hated the worldly Romans, who were disenchanted by the stern structures of the Counter Reformation, and that his nine processional routes through the sparsely built-up eastern section of town were designed for a blatantly anti-urban processional ritual of piety and penitence.
(S Moholy-Nagy, Matrix of Man)

But:
The general result of the energy of Sixtus V. on the appearance of Rome is shown by the letter of a contemporary. 'I am in Rome,' writes Padre Don Angelo Grillo, 'after an absence of ten years, and do not recognise it, so new does all appear to me to be: monuments, streets, piazzas, fountains, aqueducts, obelisks, and other wonders, all the work of Sixtus V. If I were a poet I would say that, to the imperious sound of the trumpet of that great-hearted Pope, the wakened limbs of that half-buried and gigantic body which spreads over the Latin Campagna have replied – that, thanks to the power of that fervent and exuberant spirit, a new Rome has risen from its ashes.'
(N Young, The Story of Rome, 1926)

only a strategic military role, directed at discouraging and suppressing civil disturbances, Rome's new streets have been seen only as religious routes. The Paris boulevards certainly were well suited to artillery and cavalry – generally rendered impotent in the medieval winding streets – but this was only one of their functions, and Rome's improvements also had quite complex purposes. (It is of additional interest to note that London, which has never 'enjoyed' the attentions of a despotic ruler, has also never been given a comparable main route structure.)

It is important to distinguish between the work of Sixtus V and that of his predecessors. Figure 5.23 serves as a general map of Renaissance Rome; it shows his routes in heavy line (broken for those which were to remain only projects). Earlier streets are credited individually. This map also locates the obelisks which were erected at key intersections and other important locations. Sixtus V based Rome's new movement system on his Strada Felice – the main route from south-east to north-west which was constructed in the first year of his papacy. This street was to have linked the isolated Santa Croce in Gerusalemme directly to the Piazza del Popolo, some 2½ miles distant, by way of Santa Maria Maggiore. But although Fontana noted that 'so great a prince [Sixtus] has extended streets from one end of the city to the other, without concern

Figure 5.23 – Rome: showing the city's main street system related to the major churches, important ancient monuments and the extent of medieval Rome (as heavy outline). Key: 1, the Colosseum; 2, Theatre of Marcellus; 3, Piazza Navona; 4, ruins of the Baths of Diocletian; a, Piazza del Popolo; b, Capitol Piazza; c, S Maria Maggiore.

Sixtus V's 'landmark' obelisks are located with dates of erection. The streets laid out to his instructions are shown in heavy line: his unrealized proposals are shown dotted.

for either the hills or the valleys which they crossed, but causing the former to be levelled and the latter filled, has reduced them to gentle inclines',[81] he was defeated by the steep slopes of the Aventine and was forced to terminate the Strada Felice in front of Santa Trinità dei Monti. From here a magnificent flight of steps – the Scala de Spagna – was built between 1721 and 1725 down the hillside to the Piazza di Spagna, on the Strada del Babuino (Figure 5.30). The obelisk in front of-the church, terminating the Strada Felice, was erected in 1789.

Aware of Rome's immense planning problems, and his own limited time, Sixtus V devised a unique method of ensuring that his successors would be obliged to continue to implement his programme. 'Like a man with a divining rod [he] placed his obelisks at points where, during the coming centuries, the most important squares would develop.'[82] Sixtus V located four such obelisks: in the future Piazza del Popolo, at the intersection of the three routes; on the Strada Felice, immediately north-west of Santa Maria Maggiore; in front of San Giovanni in Laterano; and, most significantly in terms of its subsequent effect, in front of the still unfinished St Peter's. More obelisks were added by later popes.

During Sixtus's lifetime Santa Maria Maggiore was a typical asymmetrical Early Christian and medieval basilica. It occupied the key central position on the Strada Felice, and was connected to the San Giovanni in Laterano by the existing Via Gregoriana. Sixtus corrected the line of this street and also erected an obelisk at its southern terminal, in front of San Giovanni. North-west of Santa Maria Maggiore Sixtus's obelisk was to form the focal point of a new linear piazza when the church was remodelled by Rainaldi for Clement X between 1670 and 1676. Connecting this space to the eastern edge of the old city, marked by Trajan's Column, Sixtus V laid out the Via Panisperna. San Lorenzo fuori le Mura was integrated into the processional system by a street running east from the Strada Felice, on the far side of Santa Maria. North from Santa Maria Maggiore the Strada Felice cut across Pope Pius IV's Strada Pia – linking the Piazza Quirinal to the Porta Pia and constituting a main route leading out to the south-east. This intersection was effected almost at right-angles, a situation which invariably creates architectural problems but which was resolved, in this instance, by Fontana's successful use of four fountains, fed by the Aqua Felice, at the corners. Views along the Strada Pia were also improved by correcting its level. Other street systems were projected by Sixtus but were never actually constructed (see Figure 5.23). His death intervened to prevent completion of his building programme.

ENVIRONMENTAL CONDITIONS

Sixtus V was also concerned to improve health conditions in the city. In addition to increasing the supply of pure water he introduced dust carts for the regular removal of household refuse, improved the drainage system, and constructed public wash-houses. Although his programme of public investment provided work for thousands of men it failed to solve Rome's chronic unemployment problem; accordingly in the last year of his reign he embarked on an ambitious scheme to convert the Colosseum into a wool-spinning mill, with manufacturing space on the ground floor and dwelling apartments in the upper storeys. At his death he had already begun to excavate the earth and to level the street, working with seventy wagons and a hundred

Figure 5.24 – Rome: a section of Nolli's plan of 1748, between the Tiber and the Piazza Navona (north at the top). This area of low-lying ground was part of the ancient Campus Martius, which is believed to have consisted of a number of formally planned, if unrelated, parts and to have included many of the city's latter day architectural monuments.

Haphazard conversion and rebuilding resulted in a typical medieval form; many ancient remains being incorporated into later buildings. The Piazza Navona, top centre, follows exactly the shape of the central area of the Emperor Domitian's stadium (AD 81–96) with seats and corridors incorporated into the foundations of the surrounding buildings (see page 186).

81, 82. Giedion, 'Sixtus V and the Planning of Baroque Rome'.

labourers.[83] Had he lived only one more year, Giedion considers, the Colosseum would have become the first workers' settlement and large-scale unit of manufacture.

The work of rebuilding Rome did not end with Sixtus V; many important examples of urbanism in the city were carried out in later years. Moreover it should be noted that only in recent times has the city regained its ancient population level. Througout the Renaissance centuries it was only a fraction of its former size, with approximate populations of 35,000 in 1458; 55,000 in 1526; 80,000 in 1580; and 124,000 in 1656.[84]

It must be remembered that the examples of urbanism in Rome described below are essentially unrelated to each other, or, in several instances, to the main street structure. An important exception is the Piazza del Popolo which extends the structure through the north-western district. The illustrated sections of Nolli's 1748 map[85] clearly indicate the unplanned network of narrow winding streets and lanes still separating the city's Renaissance spaces, and the many ancient architectural remains restricting comprehensive redevelopment. Furthermore the examples described have to be seen in their correct historical perspective since, in several instances, they took centuries to complete.

THE CAPITOL PIAZZA

The Capitol Hill, the best known of the seven hills of Rome, was the seat of the Senate, the ancient Roman governing body, and the city's original religious sanctuary. Following the destruction of the ancient buildings between the eighth and twelfth centuries the market, and with it the seat of the prefect of the city, was transferred from the Forum Holiforium to the Capitol, so that the Capitol became the political centre of medieval Rome.[86] The Palace of the Capitol (later the Palazzo del Senatore) was known to exist on its present site in 1145 and was being rebuilt in 1299 and subsequent years in the form of 'a mediaeval town hall, a robber-baron stronghold, with corner turrets and parapets and a high central tower'.[87] In the early years of the fifteenth century the Capitoline Palace was so dilapidated that the municipal authorities were forced to sit in the church of Sta Maria in Aracoeli, which had been built in 1290 some distance away to the north-east. In 1429 Nicholas V, as part of his programme of improvement, converted an existing building into the Palazzo dei Conservatori – an arcaded medieval design – immediately to the north of the Palazzo del Senatore. These two buildings formed a sharp angle at their nearest corners (Figure 5.26). The Capitol, however, was just as disorderly as so many other places in Rome; the famous hill had been ploughed up by horsemen, and bushes grew at random over the uneven terrain.

From 1471 onwards there started, under Sixtus IV, a policy of collecting the surviving antique marble statues in the city and displaying them on the Capitol. With the growing interest in classical sculpture it was realized that this would give an added attraction to the city but there was neither any consideration of their arrangement nor any plan for comprehensive redevelopment of the Capitol Hill. It was during the papacy of Paul III (1534–49) that plans were finally made for the creation of a monumental square on the Capitoline Hill and Michelangelo was commissioned in 1537.

Michelangelo is arguably the most gifted of the many versatile artists

Hence we went towards Mons Capitolinus, at the foot of which stands the arch of Septimus Severus, full and entire, save where the pedestal and some of the lower members are choked up with ruins and earth. This arch is exceedingly enriched with sculpture and trophies, with a large inscription. In the terrestrial and naval battles here graven, is seen the Roman Aries (the battering-ram); and this was the first triumphal arch set up in Rome. The Capitol, to which we climbed by very broad steps, is built about a square court, at the right hand of which, going up from Campo Vaccino, gushes a plentiful stream from the statue of Tiber, in porphyry, very antique, and another representing Rome; but, above all, is the admirable figure of Marforius, casting water into a most ample concha. The front of this court is crowned with an excellent fabric containing the Courts of Justice, and where the Criminal Notary sits, and others. In one of the halls they show the statues of Gregory XIII and Paul III, with several others. To this joins a handsome tower, the whole 'facciata' adorned with noble statues, both on the outside and on the battlements, ascended by a double pair of stairs, and a stately Posario.

In the centre of the court stands that incomparable horse bearing the Emperor Marcus Aurelius, as big as the life, of Corinthian metal, placed on a pedestal of marble, esteemed one of the noblest pieces of work now extant, antique and very rare.
(J Evelyn, Diary)

Figure 5.25 – Rome: location of the Capitol Piazza, related to the ancient Forum Romanum, showing how this outstanding example of Renaissance urbanism was orientated towards the medieval city area, turning its back on the centre of ancient Rome. Key: 1, modern monument of Victor Emmanuel; 2, north-eastern hemicycle of Trajan's Forum; 3, Via Sacra through the Forum Romanum; 4, Arch of Titus; 5, the Colosseum. (See Figures 3.10 and 3.12 for details of the centre of ancient Rome.)

83. Giedion, 'Sixtus V and the Planning of Baroque Rome'.
84. Lanciani, The Ruins and Excavations of Ancient Rome.
85. G B Nolli, Nuova Pianta di Roma, 1748.
86. T Ashby, 'The Capitol, Rome', Town Planning Review, vol XII, 1927.
87. S E Rasmussen, Towns and Buildings.

of the Italian Renaissance.[88] Painter, sculptor, poet, architect and urbanist, his achievements are unparalleled and much of his work, if equalled, never surpassed except, possibly by Leonardo Da Vinci. In 1538, as the focal point of his plan for the Capitol, the bronze statue of Marcus Aurelius (the only surviving equestrian statue of ancient Rome) was erected on a pedestal designed by Michelangelo. In 1550 his proposals for the new piazza were published but in his last years he was preoccupied with designing the dome of St Peter's and little progress had been made on the Capitol project before his death in 1564. Final completion, following the original scheme, was delayed until 1664.

Thomas Ashby, describing the Capitol, lists six basic objectives in Michelangelo's plan: (1) to refine and simplify the existing Palazzo del Senatore, eliminating the medieval angle towers and battlemented parapets, substituting in their place a new organized elevation; (2) to clear the whole area of shops, houses, other unsuitable uses and the many ruins; (3) to reconstruct the Palazzo dei Conservatori, eliminating all medieval character, creating an elevational design compatible with that of the Palazzo del Senatore; (4) to build a new palace balancing the Palazzo dei Conservatori about the axis through the centre and tower of the Palazzo del Senatore and the statue of Marcus Aurelius; (5) to construct new access stairs up to the piazza on the main axis; and (6) to use the statue of Marcus Aurelius as the focal point of the piazza.[89]

In his use of a statue as the free-standing central focus of the piazza Michelangelo was breaking new ground. Previously 'sculpture was treated as part of a building, and if it stood alone it was placed as though under the protection of the building, close to its walls'.[90] As such the Capitoline Piazza is the first of the many 'monumental squares' which were built in all the countries of the Renaissance. The piazza is not a completely enclosed space. The three buildings form a trapezoid with the fourth side open along the edge of the hill, up which the monumental approach flight of steps has been cut, slightly wider at the top than at the bottom. It is a small space, 181 feet across at its widest and 133 feet at its narrowest, between the flanking buildings. This effect of false perspective, forced on Michelangelo by the existing alignments, accentuates the importance of the Palazzo del Senatore. For Edmund Bacon, 'one of the greatest attributes of the Campidoglio [the Capitol Piazza] is the modulation of the land. Without the shape of the oval, and its two-dimensional star-shaped pattern, as well as its three-dimensional projection in the subtly designed steps that surround it, the unity and coherence of the design would not have been achieved'.[91]

PIAZZA DEL POPOLO

The Piazza del Popolo is situated on the northern side of the city between the Tiber on the west and the steep slope of Monte Pincio on the east (Figure 5.28). The main approach road from the north to ancient Rome ran across this narrow area between the river and the hill, passing through the Porta del Popolo, built in the Aurelian Wall of AD 272, and continuing straight on through the city as the Via Flaminia, up to the northern slopes of the Capitol Hill. The Piazza del Popolo, immediately inside the gateway, was the main entrance place to the city but little is known of its form during the ancient Roman and medieval periods other than that the Via Flaminia ran across it and the

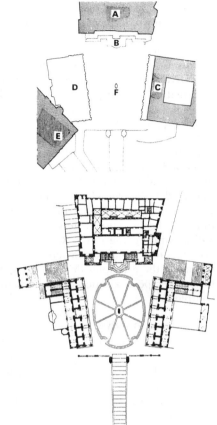

Figure 5.26 – Rome: the Capitol Piazza, before re-development of the area, showing the pre-existing buildings tinted (top); and after, with the detailed floor plans of the three buildings forming the new space (below). Key: A, the medieval Palazzo del Senatore built in part on the ruined Tabularium of the ancient Forum Romanum; B, the Renaissance extension and new north-west elevation; C, the Palazzo dei Conservatori, refaced as part of Michelangelo's design; D, the new building of the Capitoline Museum; E, the ancient church of Santa Maria in Aracoeli; F, statue of Marcus Aurelius.

88. For Michelangelo see: H Hibbard, *Michelangelo*, rev. 2nd ed., 1985; J S Ackerman, *The Architecture of Michelangelo*, 1961; and W E Wallace, 'Michelangelo's Drawings for the Fortification of Florence', *Journal of the Society of Architectural Historians*, vol XLVI, 1987.
89. T Ashby, 'The Capitol, Rome', *Town Planning Review*, vol XII, 1927.
90. S E Rasmussen, *Towns and Buildings*.
91. Third Edition Note: E Bacon, Design of Cities; with superb before and after vertical perspective drawings.

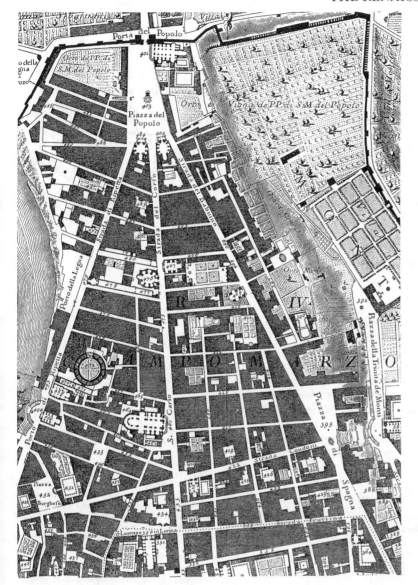

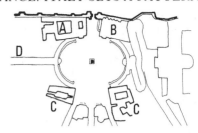

Figure 5.27 – Rome: the Piazza del Popolo as finally completed in the nineteenth century to a design by Guiseppe Valadier (1816–20). A new church (A) repeating the design of Santa Maria del Popolo (B), was built to the west of the Porta del Popolo; new buildings (C) were added on either side of the twin churches; a sweeping staircase and ramp up to the Pincio Gardens integrated them into the total design. At a still later date a street through from the Tiber (D) was added to complete the east–west cross axis.

Figure 5.28 – Rome: the Piazza del Popolo and the three main routes south into the centre of the city, from the Nolli plan of 1748 (north at the top and left, the Tiber). This plan shows the piazza at an interim stage in its development, before the remodelling of 1816–20 under Pius VIII.

The south-western corner of the plan shows part of medieval Rome; the narrow formless streets are in marked contrast to the formally laid-out, gridiron-based, streets of the Renaissance district between the Pincio ridge and the Tiber.

The Spanish Steps (Figure 5.30) are located to the bottom right, at the end of the Strada Condotti.

Via Ripetta started from it, aligned at an angle which just skirted the Tiber. A less important third road ran along the foot of Monte Pincio.[92]

The redevelopment of the Piazza del Popolo area, to make it a more impressive entrance to the city, started in 1516 under Leo X with the construction of a new third road, the Via Babuino, but the piazza was not completed in its present form until 300 years later (Figure 5.27). The alignment of the new Via Babuino is such that it intersects with the line of the Via Flaminia at the same angle, and at the same point, as the existing Via Ripetta. In 1589 Fontana, for Sixtus V, erected a red granite obelisk at this focal point, in the centre of the piazza. (Many references to this incorrectly state that the obelisk was positioned *before* the Via Babuino was planned.) Earlier in 1586 Sixtus V had hoped to bring his Strada Felice into the piazza, to the east again of the Via Babuino, but this was prevented by the slopes of Monte Pincio.[93] The

92. T Ashby and S R Pierce, 'The Piazza del Popolo', *Town Planning Review*, vol XI, 1924. See also G Ciucci, *La Piazza del Popolo*, 1974.
93. J A F Orbaan, 'How Pope Sixtus V lost a Road', *Town Planning Review*, vol XIII, 1928.

two twin-domed churches on the southern side of the piazza, in the angles formed by the three streets, were started in 1662 to the designs of Rainaldo and finished later by Bernini and Carlo Fontana. The other three sides of the piazza, completing its present form, were designed by the French architect-planner Valadier and date from 1816–20.[94]

PIAZZA NAVONA

The houses, palaces and churches of the Piazza follow precisely the layout of the stadium built by the Emperor Domitian (AD 81–96); indeed the well-preserved ruins of the seats and corridors are incorporated into the piazza's foundations (Figure 5.24). The final spatial organization of the piazza was carried out by Bernini between 1647 and 1651, although one of his three sculptured fountains was not added until the nineteenth century. The long and narrow form of the space meant that all views had to be designed as oblique perspectives.[95] The piazza contains three richly modelled fountains whose cascading waters are enhanced by the neutral backcloth of the surrounding houses and the two churches of San Giacomo degli Spagnuoli (1450) and Sant'Agnese (1652–77).

PIAZZA OF ST PETER'S

The great church of St Peter's was built between 1506 and 1626 but it lacked an appropriate entrance forecourt until 1655–67 when Bernini carried out the two major sections of a three-part piazza complex (Figure 5.29). These spaces are the *piazza retta*, directly in front of the church, and the vast *piazza obliqua* enclosed by the semi-circular colonnades.[96] The third section, the Piazza Rusticucci, has never been finally completed and is represented only in part by Mussolini's avenue linking St Peter's with the River Tiber. In preparing his layout, which was successful in competition with his leading contemporaries, Bernini had to incorporate the central obelisk erected in 1586 by Sixtus V, and fountains built by Maderna in 1613.

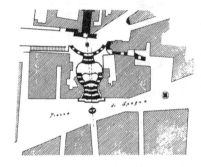

SPANISH STEPS

The 137 steps leading up from the Piazza di Spagna to the church of Santa Trinità dei Monti, were built by Alessandro Specchi and Francesco de' Santis between 1721 and 1725. Paul Zucker considers that they represent the climax of stage effects in Roman city planning on a larger scale; here nature lent a helpful hand to the spatial vision of the planner with the staircase, the link between two topographically different levels becoming the square. He adds that the Scala di Spagna is the only example in the history of city planning where a staircase does not merely lead to a square in front of a monumental structure, but where the stairs themselves become the visual and spatial centre.[97] The

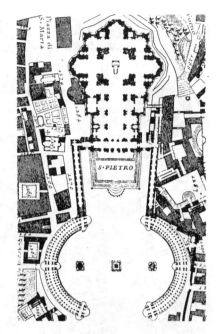

Figure 5.29 – Rome: the Piazza of St Peter's (north to the right). The Vatican City encloses St Peter's and extends east to the centre line of the oval *piazza obliqua*. The River Tiber is some 655 yards further east from the bottom of this plan extract, reached along the Via delle Conciliazione – the avenue which performs in part the internal functions of Bernini's Piazza Rusticucci. The scale of both the church and the spaces is vast: the *piazza retta* in front of the eastern elevation is 410 feet wide narrowing to 300 feet where it adjoins the *piazza obliqua*, and 320 feet deep; the *piazza obliqua* itself is not a true ellipse but consists of two semi-circles of radius of approximately 260 feet, with a rectangle in between giving a total width of 650 feet. From that which has been written earlier concerning the papal aggrandizement policy, the reader could well assume that there was a formal axial approach leading into the Piazza from the east (the bottom of the plan), thence past the obelisk to the grand front of St Peter's. See, however, the reality of the 1830 SDUK Map of Rome (Figure 5.32) which shows the Piazza as a spatial island, isolated from the Tiber and the city at large.

Figure 5.30 – Rome: the Spanish Steps, leading up the Pincio hillside from the Piazza di Spagna to the Church of S Trinita dei Monti. The plan cannot show the steeply sloping hillside third dimension of this extremely clever example of Renaissance urban design. Axial differences apparent on the plan do not register in reality; as a further distracting human dimension, there are the people, tourists and locals, who flock to this part of the city. (See B Rudofsky, *Streets for People*, 1969.)

94. See also E Bacon, *Design of Cities*.
95. See also P Zucker, *Town and Square*.
96. See Footnote 44 for Alberti's abandoned mid-15th-century proposal for a new district linking St Peter's to the River Tiber.
97. Zucker, *Town and Square*.

steps lead up in curved flights from the piazza, a triangular space formed by the oblique-angled intersections of five streets. They are finely adjusted to the slope, with their axial directions subtly varied to incorporate into the design the obelisk set at the top in front of the church façade.

ROME: A POSTSCRIPT

The city of Rome is potentially confusing, both for the student of its lengthy, multi-faceted history, and also for the visitor trying first to locate historic monuments and then to relate them to their respective periods in history. Profound personal dissatisfaction with the published maps of historic Rome led to the preparation of this book which, it is hoped, goes some way towards setting the buildings and monuments of Rome in their urban context. This account of the development of historic Rome is therefore concluded with W B Clarke's 'Plan of Modern Rome', published in 1830, related to Figure 5.31 identifying the individual buildings and places of interest. (See also Figures 3.6, 3.8, 3.10 and 3.12 for details of Imperial Rome.) It is important to note how much of the area of Imperial Rome within the Aurelian Wall still remained to be reoccupied.

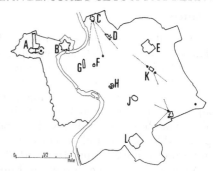

Figure 5.31 – Locational key to the eighteenth-century map of 'Modern Rome': A, St Peter's and the Vatican; B, Castle of St Angelo (Mausoleum of Hadrian); C, Piazza del Popolo; D, Piazza di Spagna; E, Baths of Diocletian; F, Pantheon; G, Piazza Navona; H, Capitol Piazza; J, Colosseum; K, Santa Maria Maggiore; L, Baths of Caracalla. (Sixtus V's obelisks are shown as solid circles.)

Figure 5.32 – Rome: the 'modern' map published in 1830 by the Society for the Diffusion of Useful Knowledge (this map is included in Branch, *Comparative Urban Design*).

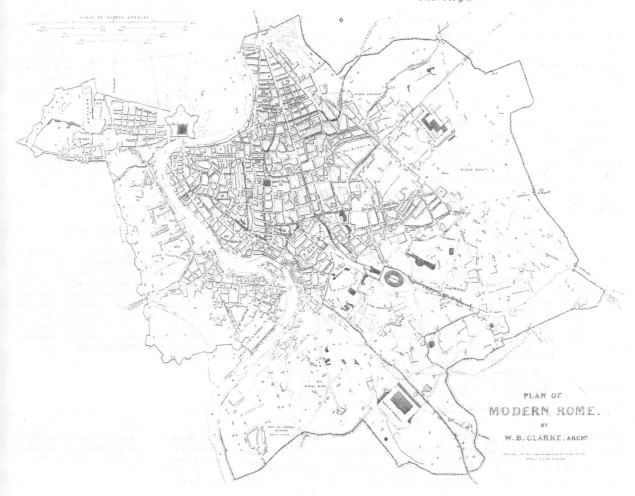

Venice

Venice, like the ancient city of Rome, has a mythical foundation date. The site – an easily defended group of islands in the north-west corner of the Adriatic – must, however, have attracted village settlement long before 'the 25th March, 421, at midday exactly'.[98] These islands on a lagoon were formed from sediment brought down by three ancient alpine rivers. Dominated at first by Ravenna and Byzantium, the Venetian Republic gradually developed into a major Mediterranean power able to exact full benefit from its strategic location on the most important trading routes. In his *Mediaeval Cities*, Henri Pirenne writes of Venice's debt to the Byzantium Empire: 'to it she not only owed the prosperity of her commerce, but from it she learned those higher forms of civilisation, that perfected technique, that business enterprise, and that political and administrative organisation which gave her a place in the Europe of the Middle Ages'.

Seeming deficiencies in their archipelago location were turned to advantage by the Venetians in controlling their city's growth: individual islands were zoned for specific functions. Most of the islands gradually coalesced into a tight-knit group, traversed by the labyrinthine canal system, but a rigid oligarchic government continued to differentiate between land-uses.

In the twentieth century Venice is no longer an island. It is linked to the mainland by the rail causeway, 3,000 yards long and completed in 1846, and a motorway built in 1931. The former terminates in a modern railway station, while all vehicle traffic is allowed no further than the vast parking garage at the end of the motorway. As James Morris notes: 'the lifestream of Venice arives on wheels but must proceed by water or by foot'.[99] To facilitate movement of people and goods there is a total of 177 canals amounting in all to some 28 miles of public waterway, complemented by the intricately woven network of pedestrian streets and lanes. If these indirect, and frequently congested routes have become quite impractical for modern city life, yet together they form an incomparable medieval townscape that has remained virtually unchanged and which attracts the year-round tourists upon whom the Venetian economy has come to rely.

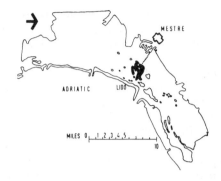

Figure 5.33 – Venice: the location of the city islands in the Lagoon; the access causeway from mainland Mestre is shown dotted.

*One lingers about the cathedral a good deal, in Venice. There is a strong fascination about it – partly because it is so old, and partly because it is so ugly. Too many of the world's famous buildings fail of one chief virtue – harmony; they are made up of a methodless mixture of the ugly and the beautiful; this is bad; it is confusing, it is unrestful. One has a sense of uneasiness, of distress, without knowing why. But one is calm before St Mark's, one is calm within it, one would be calm on top of it, calm in the cellar; for its details are masterfully ugly, no misplaced and impertinent beauties are intruded anywhere; and the consequent result is a grand harmonious whole, of soothing, entrancing, tranquillizing, soul-satisfying ugliness. One's admiration of a perfect thing always grows, never declines; and this is the surest evidence to me that it is perfect. St Mark's is perfect. To me it soon grew to be so noble, so augustly ugly, that it was difficult to stay away from it, even for a little while. Every time its squat domes disappeared from my view, I had a despondent feeling; whenever they reappeared, I felt an honest rapture – I have not known any happier hours than those I daily spent in front of Florian's, looking across the Great Square at it. Propped on its long row of low thick-legged columns, its back knobbed with domes, it seemed like a vast warty bug taking a meditative walk.
(M Twain, A Tramp Abroad, 1897)*

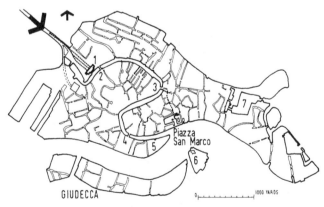

Figure 5.34 – Venice: general plan, showing the causeway from Mestre at top left. Key: 1, railway station; 2, Station Bridge over the Grand Canal; 3, Rialto Bridge; 4, Accademia Bridge; 5, Santa Maria della Salute; 6, San Giorgio Maggiore; 7, the Arsenal. (The Grand Canal, the city's main traffic artery, takes the form of a reversed 'S' between the railway station and the Lagoon, just west of the Piazza of St Mark.) This small-scale diagram can give merely an impression of the intricately entwined pattern of lanes and canals – history's extraordinarily different urban route system. See Branch, *Comparative Urban Design*, for the 1838 SDUK Map of Venice.

THE GRAND CANAL

The Grand Canal is the unique waterway high street of Venice. It follows the line of the ancient River Alto, its two-mile length sweeping through the city in three abrupt but majestic curves. The maximum

98. J Morris, *Venice*, 1974; see also G Rossi and F Masiero, *Venice from the Air*, 1988; C Hibbert, *Venice: The Biography of a City*, 1989; T Okey, *The Story of Venice* (Mediaeval Towns seies), 1914.
99. Morris, *Venice*.

width is 76 yards and the minimum is 40 yards between building frontages which, with few exceptions, rise directly from the water's edge. Other than the three bridges which cross the Grand Canal – the Station, Rialto and Accademia Bridges as Figure 5.34 – and limited quayside access, its waterfront is inaccessible to the general public and is best seen from onboard boat, preferably in the downstream direction leading up to the visual climax of the Doge's Palace and the Piazza linking through to the Piazza San Marco. On both sides of the Grand Canal, and on the separate Guidecca island, Venice is divided into six ancient segments – originally distinct islands in the archipelago but now unified into one continuous urban form. In addition to the main central Piazza San Marco, each district has its own local civic piazza.

THE PIAZZA SAN MARCO

One of urban history's most memorable spatial ensembles (and the author's personal favourite), the Piazza San Marco in reality consists of two linked piazzas – the piazza proper in front of the Basilica of St Mark and the piazzetta which connects it to the edge of the lagoon. The detached Campanile of St Mark's is located in the relatively narrow space between the two piazzas and acts as a perfectly positioned hinge (Figure 5.36). There is in addition a third, smaller piazza along the northern side of the basilica.[100] The site of this overall design was originally occupied by a market place lying outside the walls of the

Figures 5.35 and 5.36 – Venice: aerial photograph of the Piazza of St Mark from the south and key plan at the same orientation. Key: A, St Mark's Basilica; B, Doge's Palace; C, Procuratie Vecchie; D, Library; E, Procuratie Nuovo; F, Fabrica Nuova. The campanile and the small tempietto at its foot are shown as solid black, all other buildings are shown with their ground floor areas in heavy line.

The aerial view shows a marked contrast between the carefully organized massing and achitectural conformity of the piazza spaces, and the organic growth pattern of the surrounding buildings.

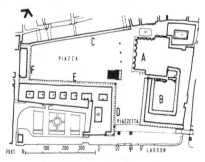

100. For the Piazza San Marco see M Marquese (ed) *Venice: An Illustrated Anthology*, 1988; see also the valuable section in Bacon, *Design of Cities*.

embryonic Venetian settlement. It began to be the central focus of the city from AD 827 when the Chapel of St Mark (originally a private chapel of the Doge) was built as a sepulchre containing the body of St Mark (Figure 5.37). The Doge's Palace, first constructed at the end of the eighth century as a fortress outside the walls, was rebuilt here between 1309 and 1424. The palace, together with the Basilica of St Mark, now forms the eastern side of the main piazza and the piazzetta.

During the early fifteenth century the piazza was still quite small and its surrounds consisted of uneven brick façade houses. Its present-day character dates from 1480–1517 when the Procuratie Vecchie was built on the north side. At this time the campanile – originally a timber construction dating from 888, but rebuilt in brick between 1329 and 1415 – was connected to the buildings on the southern side of the piazza. In redeveloping this side of the piazza, and the western side of the piazzetta, the width of the spaces was increased, isolating the campanile as a free-standing vertical element. The western side of the piazzetta is formed by the library building, designed in 1536 by Sansovino and completed by Scamozzi in 1584, after his death. The southern side of the main piazza was then formed by the Procuratie Nuova, designed in 1584 by Scamozzi and completed in 1640 by Longhena.

The final completion of the main piazza was not achieved until 1810 when the Fabrica Nuova was built across its western end. The three flag-poles in front of St Mark's, which play important secondary design roles, were erected in 1505. The paving treatment, which constitutes a vital unifying element in the design, was carried out between 1722 and 1735. The Campanile collapsed in 1902, destroying the small tempietto at its foot, but both have been successfuly restored.

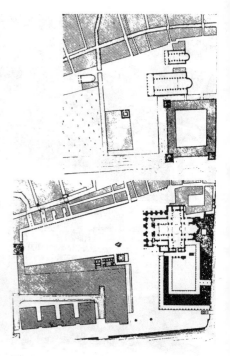

Figure 5.37 – Venice: two intermediate stages in the development of the Piazza of St Mark; its final, present-day form is shown in Figures 5.35 and 5.36.

Top, the late twelfth-century plan: a multi-purpose area in front of the old basilica served both as the main market and the city's public meeting space. The campanile, which dates from 888, was attached to the building in the south-west corner; the Doge's Palace was essentially a medieval castle in character, surrounded by canals in the form of a moat.

Below, the piazza in the early sixteenth century before reconstruction of the southern side, as the Procuratie Nuove, to a new building line, thereby giving the companile its free-standing location. By this date the piazza occupied its final area, but the western end, the Fabrica Nuova, was completed only in 1810.

Figure 5.38, Left – Venice, several of the minor squares and pedestrian route-spaces between the Piazza of St Mark (to the right) and the Accademia Bridge on the northern side of the Grand Canal (see Figure 5.34 for their location).

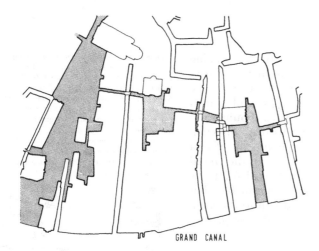

GRAND CANAL

6 – France: Sixteenth to Eighteenth Centuries

The arrival of the Renaissance in France is marked by the work of Italian artists at the courts of Charles VIII (1483–98) and Louis XII (1498–1515). Because of the advent of printing, the first books on architectural design and urbanism served as an early means of taking the new ideas from Italy to France and other European countries; reading about these revolutionary theories, however, had little impact compared to actually experiencing examples of their application.[1]

Many individuals had made visits to Italy during the latter part of the fifteenth century, including both architects and their potential clients, but the first contact of lasting significance was made in 1494, when the French army under Charles VIII marched south through the country as far as Naples. The king and his court were faced, as Rasmussen puts it, with a style that made all they had known appear confused, crabbed and petty. At the same time these bold square palaces, these columns, balusters and round-headed arches, these garlands and laurel wreaths were so disconcertingly novel that it took even the most progressive many years to digest them.[2]

The assimilation process took longest with architecture and urbanism but painters and sculptors could be commissioned immediately and a number of them returned to France with Charles VIII. Others followed, including Leonardo da Vinci, who died at Amboise in the Loire Valley in 1519. Early Renaissance architectural activity was centred in this part of France, with the sequence of châteaux of the Loire constructed or rebuilt for the king and his courtiers, before Paris emerged as the undisputed national capital. At Blois, Renaissance details were introduced for Francis I; Chambord, in the heart of the forest, was started in 1519 with a vast symmetrical plan. The first phase of Chenonceaux (1515–22) and Azay-le-Rideau (1518–19) were other important early designs.[3]

Under Francis I (1515–47) Paris became the capital of an effectively united French nation and scene of its brilliant royal court. The boundary of the city at that time was the defensive wall of 1367–83 and within this confined area the city displayed most of the characteristics of organic growth medieval urban form. The clarity of Lutetia's street pattern had been lost during the centuries after the Romans, when settlement contracted back on to the Ile de la Cité, and it was not re-established during the *laissez-faire* growth of the medieval city.

Pre-Renaissance Paris

As the Roman city of Lutetia, Paris was essentially a two-part

The movement was more closely linked with the cultural programme of the rulers than in any other country. It was not the individual struggle of artists and scholars alone, as in Germany and the Netherlands, which brought about the discussion of the new achievements of the Italian Renaissance. It was more the planned activity of the sovereigns which opened new possibilities to artists and writers. French culture in the sixteenth century was not the culture of burghers as in Germany and the Netherlands, but a court culture – and thus far a continuation of the mediaeval order. King Francis I's patronage of the fine arts gave a decisive turn to the whole development. His sister Marguerite of Navarre fostered humanism and literature. Henry II reaped the fruits of what his predecessors had planted ...

Francis I considered himself a connoisseur of painting, and he had the enthusiasm of a great collector. He saw his artistic ideal in the Italian Renaissance. When he was a young man, his expeditions and travels brought him into close contact with its sources. His agents in Italy tried to procure as many works of art as they could in order to transfer them to France. Francis even tried to transfer the masters themselves. Leonardo da Vinci spent the last years of his life in Cloux. Andreo del Sarto was in the service of the monarch for one year.

A systematic colonization of France by Italian artists began in 1528, when the king rebuilt the old castle in the idyllic surroundings of the forests and ponds of Fontainebleau ... They settled in the country, permeated it with a new artistic gospel, and at the same time assimilated themselves to its tradition. Thus an artistic culture of quite unique flavour arose, based on Italian Renaissance in form, on Latinism and Graecism in literary content, and on the French heritage in spirit.
(O Benesch, The Art of the Renaissance in Northern Europe)

---

1. The French Renaissance period, and Paris in particular, are well provided with general urban histories; see G Duby (dir.) and R Chartier, *La Ville Classique, Volume 3, Histoire de la France Urbaine*, 1981; P Lavedan, *Les Villes Françaises*, 1960; P Lavedan, *Histoire de l'Urbanisme: Renaissance et Temps Modernes*, 1959.
2. S E Rasmussen, *Towns and Buildings*, 1969.
3. For an introduction to French Renaissance architecture the reader is referred to N Pevsner, *An Outline of European Architecture*; and to P Lavedan, *French Architecture*, 1979.

settlement; the original Gallic inhabitants occupied the present-day Ile de la Cité – strategically a key crossing of the Seine – while the Romans themselves preferred the well-drained higher ground of the slopes and summit of mount Ste Geneviève, to the south (Figure 6.1). Roman Lutetia was divided approximately in two by the line of the Rue St Jacques, which continued across the twin bridges of the arms of the Seine.[4] (There was only a small, unimportant north bank bridgehead settlement.) Little is known of the detailed layout of Lutetia – a comparatively blank period in the city's history which it shares with London. The forum has been discovered between the Rue St Jacques and the Boulevard Saint Michel; the theatre and at least three bathing establishments have also been located on the right bank. Pierre Couperie, in his *Paris au fil du temps*, a beautifully meticulous sequence of maps showing the historical evolution of Paris, records that Lutetia occupied 480 acres, and had a population of around 10,000. The right-bank aqueduct delivered 2,000 cubic metres per day (compared with 75,000 cubic metres from its counterpart at the comparatively more important Roman city of Lyons). There is also the remarkably well-preserved arena, entered today through an inauspicious doorway in the Rue Monge, east of the Rue St Jacques, and incorporated into the little-known, attractively laid out Place Capitan. On the Island, the palace of the governor occupied part of the site of the Palais de Justice, and the Temple of Jupiter had already established a religious focal point, later confirmed by the building of Notre Dame.

Lutetia at most could have been only lightly fortified and during barbarian invasions of Gaul between 253 and 280 the left-bank settlement was destroyed. When re-established the city, which has been known as Paris since about AD 360, was moved on to the Ile de la Cité where an encircling wall could augment the river defences. From this second island nucleus Paris continued to expand until the twentieth century on the basis of clearly defined concentric rings, each delineated by the successive defensive systems as shown on Figure 6.13.

First Renaissance towns in France

Renaissance town planning did not arrive in Paris itself until the early years of the seventeenth century, by which time several planned new towns had been constructed elsewhere in the country embodying Renaissance principles. Thus Vitry-le-François (Figure 6.2) was built alongside the Marne for Francis I, from 1545, with a gridiron street structure set within symmetrical fortifications. It was notable for the fact that the four main cross-streets entered the central square at the mid-points of its sides: they did not run alongside the space, in the typical medieval bastide fashion.[5] Navarrenx, in the south of France, followed in 1548, and Philippeville two years later; this latter example was in fact built by the Spanish King, Philip II, to help guard his Dutch frontier with France. Philippeville's central square within a five-pointed fortification system is a straightforward application of Italian 'ideal city' theories discussed on pages 168–72.[6]

Nancy, which in the second half of the eighteenth century was to acquire a unique sequence of Renaissance spaces, was extended from 1588 by adding to the original medieval nucleus a regularly laid-out *ville-neuve*, designed for Duke Charles III by an Italian architect-planner, Jerome Citoni. Several gridiron blocks were left open and half

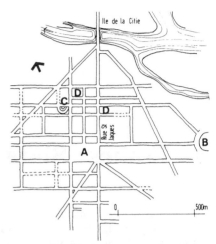

Figure 6.1 – Lutetia, the largely conjectural street pattern of the left-bank Roman Paris. Key: A, Forum; B, the arena; C, theatre; D, baths. As shown by Figure 6.5, the direct line of the main Roman north–south route across the Seine (the present-day Rue St Jacques) was lost during the Dark Ages, but subsequently restored. Although the routes shown radiating away from the Seine bridgehead are to have been expected as the city's fortunes revived after the departure of the Romans, and the original gridiron layout was either lost or altered, such a number of non-conforming diagonals would have been most unusual for a Roman plan. (See Chapter 3 for a comparable circumstance at London.) A probable reason is that this (conjectural) plan includes post-Roman radials.

Figure 6.2 – Vitry-le-François: the general plan.

4. For Paris see L Bergeron (ed.) *Paris: Genese d'un paysage*, 1989; P Lavedan, *Nouvelle Histoire de Paris* (Histoire de l'Urbanisme), 1975; P Couperie, *Paris through the Ages*, 1968; M C Branch, *Comparative Urban Design*, 1978, for the SDUK Map of Paris in 1834.
5. The positioning of the streets into a built urban space has much to tell us of the original city planning intentions. As at Vitry-le-François, with the cross-roads entries at mid-sides, the primary purpose was to accommodate traffic, with additional opportunity for aggrandizement of a central statue or monument.
Where the streets enter at the corners – as with the medieval bastides – through traffic could be kept out of the central space which was then free for commercial

of one of them utilized for the Place d'Alliance, Paul Zucker notes of this square that the absolute regularity of its layout is still emphasized today by a quadrangle of regularly trimmed trees; the houses show identical façades and the parallel horizontals of their ledges, eaves and roof tops tie the area of the square, the streets at its four corners running into the square unconcealed.[7] The form of Nancy in 1645, with the area of the *ville-neuve* fully developed, is given in the view reproduced as Figure 6.20. Citoni's regular gridiron can be seen to be in direct contrast with the organic growth form of the old city. Details of the Renaissance defences of both parts of the city are also clearly illustrated.

The last of the early provincial examples is Charleville, constructed between 1608 and 1620 (by which time the first Parisian Renaissance urbanism was under way). Charleville is based on a gridiron incorporating a main central space – the Place Ducale – and, originally, six secondary squares (Figure 6.4). The Place Ducale is a direct predecessor of the Place Royale (Place des Vosges) in Paris, the prototype European residential enclave and as such one of the earliest, if not the first, instances of the social grouping urban form determinant. Aspects of the designs in common include the arcaded two-storeyed houses, with their individual roofs, although the Place Ducale is basically a forecourt to the Palais Ducal (now the Hôtel de Ville) and the four streets entering it make it essentially a 'traffic-place' as opposed to the closed, balanced layout of the Place Royale. (As noted later, the Place du Champ à Seille, at Metz, probably had a greater influence on the design of the Place Royale.)[8]

Figure 6.3 – Philippeville: the general plan of this uncharacteristically planned Spanish frontier town (see Note 6).

Figure 6.4 – Charleville: founded in 1608 on the Meuse in north-eastern France, by Charles of Gonzaga, Duke of Nevers and Mantua, where, according to John Reps, 'the principles of Alberti and particularly those of Palladio found expression in new town planning' (*The Making of Urban America*). The Palace Ducale, one of the influences on the layout of the later Place Royale in Paris, is in the centre of the town. In addition to the traffic function, the three streets entering the Place Ducale at mid-sides all focus on the central statue, and that from the right also leads to the Palais Ducale. See P Zucker, *Town and Square*, 1959, for the oblique aerial photograph (Plate 59A) which shows the strongly individual steeply sloping house roofs.

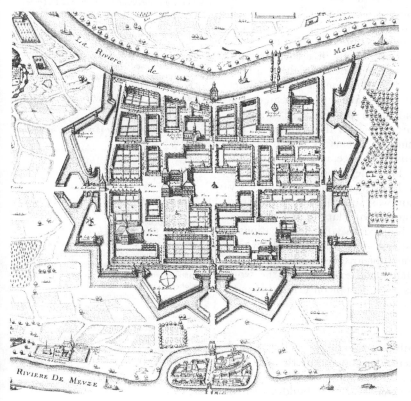

market-place activity; or as the Georgian residential squares of London and elsewhere, the centre was used for carriage parking before usual transformation as landscaped private gardens (see Chapter 8).

6. The centralized plan of 1550 for Philippeville is closely similar to that of Palma Nova of 1593. Both are aberrant plans, reasons for which can only be surmised. Philippeville was constructed for the Spanish Crown at a time when orthogonal geometry was the rule for the sixteenth-century new towns in Spain, of which Santa Fe, outside Granada, is the most important example (see Chapter 9). Similarly, Palma Nova (see Figures 5.17 and 5.18) differs fundamentally from the form of Scamozzi's own, theoretical, gridiron based ideal city plan. Aggrandizement of interests unknown is the probable answer.

7. P Zucker, *Town and Square*, 1959.

8. For the Place du Champ à Seille, at Metz, see P Lavedan, *Histoire de l'Urbanisme: Renaissance et Temps Modernes*, 1959.

Paris: Renaissance urbanism

In the Paris of Francis I there were no public urban spaces of major significance. Notre Dame was approached across a *parvis* far smaller in area than the present-day Place du Parvis Notre Dame. The ill-defined Place de Grève, now part of the Place de l'Hôtel de Ville, was the only other public space of any size. Views along the Seine from the three bridges – the Petit Pont from the southern (left) bank to the Ile de la Cité continued across to the northern bank as the Pont Notre Dame, and the Pont au Change across the northern branch only – were prevented by their being as continuously lined with buildings as the general street pattern. The banks of the Seine were also completely built-up.[9]

This does not mean that the city must be thought of as a human ant-hill, quite deprived of light and air. If public open spaces were rare, private ones were many, particularly convent gardens. The town was still half rural. Pierre Lavedan wrote that 'poultry yards, rabbit hutches, stables and fields were close to the houses, and the agricultural calendars carved on the façades of the cathedrals did not represent an escape for the citizen, but an everyday reality'.[10] These open spaces and the often extensive private gardens gradually disappeared during the seventeenth and eighteenth centuries, on occasion as the result of complex property speculation.

Renaissance urbanism in Paris covers the period between Francis I and the end of the eighteenth century. During these 250 years little was done to restructure the medieval core and except for establishing the Champs Elysées axis westward and laying out the Grands Boulevards, little of note was done which would predetermine the future form of the city. The work of Renaissance urbanists was essentially limited to creating isolated parts of the city, either on undeveloped land within and adjoining the city or by carving developments out of the existing fabric. The most important examples of this work are the five royal 'statue' squares described individually later – so called because either a condition for their being constructed or the intention of the promoter required the creation of an appropriately dignified setting for an equestrian statue of the king. In Paris and major provincial cities, these squares glorifying the monarchy are pre-eminently the grandest examples of this kind of urban aggrandizement.[11] Henri IV was honoured once, outside the Place Dauphine; Louis XIII once, within the contemporary Place Royale (renamed later Place des Vosges); Louis XIV twice, the Place des Victoires and the Place Vendôme; and Louis XV once, inauspiciously, for his successor was subsequently guillotined there in the Place Louis XV, now known as the Place de la Concorde.

The present-day pattern within the limits of the 1845 fortifications is the result of nineteenth-century restructuring. Napoleon I, whose ambition was to make his capital not only the loveliest town which ever had existed, but also the loveliest that could exist, began the work which culminated in the achievements of the Napoleon III and Haussmann collaboration of the 1850s and 1860s.[12]

PONT NEUF AND PLACE DAUPHINE

From its earliest history Paris had only one crossing over the Seine – the two separate bridges linking the central section of the Ile de la Cité to the north and south banks. During the medieval period both these bridges (and the later Pont au Change) were lined with buildings, in an

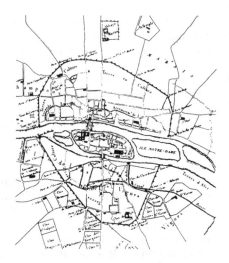

Figure 6.5 – Paris 1180–1223: the line of the Roman crossing of the Seine has been broken, with a new bridge from the Ile de la Cité to the northern bank further downstream from the old bridge. This map shows a considerable area of land under cultivation within the wall of 1180 (which accords with Lavedan's view: Note 10) but early urban mapping was frequently pictorial to an extent that distorted the reality of the time.

By 1609 Parisians were beginning to see the results of Henri IV's urbanism. The pont Neuf and Grande Galerie of the Louvre were completed, and construction of the Place Royale, Place Dauphine, and Hôpital St. Louis was well advanced. Two city maps publicizing the king's buildings and offering a global image of Henri IV's Paris were published that year. These accomplishments encouraged the king to pursue his building program. A northern gallery between the Louvre and the Tuileries was under discussion, and in the last months of 1609, the crown plotted two more projects: a square in north-east Paris and a royal college in the University district, on the Left Bank. Construction of the square and college was about to begin when the king was assassinated on 14 May 1610. Without Henry IV and his administration, the royal commitment that was needed to realize these projects was lost.
(H Ballon, The Paris of Henri IV: Architecture and Urbanism, 1991)

9. Construction of the quays alongside the Seine, as shown in Figure 6.6, was mainly carried out during the Renaissance period.
10. Lavedan, *French Architecture*.
11. While one or more statue squares are a feature in most major European cities, it was the French who entered most fully into this particular kind of autocratic aggrandizement. Elsewhere in Europe, civic squares were often originally developed without such a central focus, which was installed later, whereas in France the statue was ordinarily an original spatial determinant. In England the central spaces of the Georgian residential squares were open at first and only latterly soft-landscaped. In Spain, the Plaza Mayor was characteristically at first an open space; then a landscaped garden; and lastly, as today, an open hard-landscaped space (see Chapter 9).
12. See D Pinkney, *Napoleon III and the Rebuilding of Paris*, 1958.

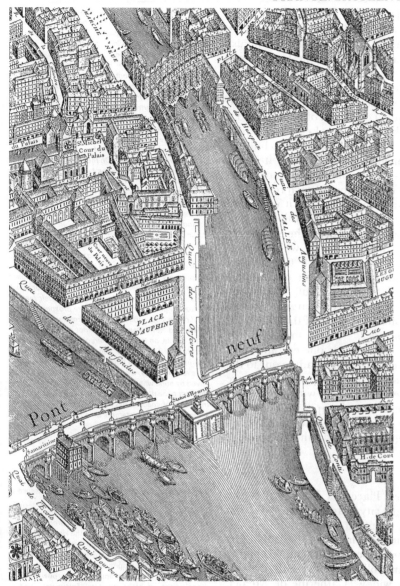

Figure 6.6 – The Place Dauphine and the Pont Neuf from the Turgot plan (1734–9). The old medieval bridge lined with buildings, at the top of the view, is in marked contrast to the open character of the Renaissance Pont Neuf. The Rue Dauphine, cut through the gardens of the Couvent des Grands Augustine, continues the line of the bridge to the right. (See L Berger (ed.) *Paris: Genese d'un paysage*, for an early view across the bridge; see also John Evelyn's eye-witness account below.)

While the Place Dauphine fulfilled the king's urban and economic goals, embellishing the city while serving the predominantly merchant and artisan community, in the years thereafter architects and agents of the crown found little to admire at the square. They did not share Henri IV's commitment to mercantile and artisanal activity and considered the small-scaled brick buildings inappropriate for so prominent a site. Despite repeated calls for the demolition of the Place Dauphine that began in the late 1600s and continued over the next two centuries, it survived intact until the mid 1800s, when two ranges were destroyed. But over time, the condition of the square deteriorated sharply, partly as a consequence of its intensive use, and partly because the crown saw little benefit in improving buildings that failed to meet emerging standards of urban design.

The development methods tested at the Palace Royale were adjusted at the Place Dauphine to enable merchants and artisans to build and occupy the shops and houses. The royal designers also demonstrated a growing interest in tying together different parts of the city, both visually and physically. Rather than viewing the Place Dauphine as self-contained, the designers joined the square into a network of interlocking elements with the Pont Neuf and the rue Dauphine, the Seine and the quais.
(H. Ballon, The Paris of Henri IV: Architecture and Urbanism, 1991.)

Figure 6.7 – The 'before-development' counterpart of Figure 6.6, showing the undeveloped marshy islands at the downstream end of the Ile de la Cité which, following the construction of the Pont Neuf, were developed as the Place Dauphine and extensions to the Palais Royal. (From the Legrand map of 1380.)

identical manner to London Bridge and the present-day Ponte Vecchio over the Arno in Florence. This crossing of the Seine had a national importance as a north–south route in addition to carrying city traffic. As Paris prospered, congestion on the bridges increased so that by the middle of the sixteenth century they were unable to cope. Under Henri III work started in 1578 on a second crossing – the two arms of the Pont Neuf – to connect the downstream end of the Ile de la Cité to the north and south banks. The Pont Neuf was constructed to the designs of Androuet du Cerceau and now belies its name as the oldest bridge in Paris, with the original structure still intact beneath restored surfaces. Progress was delayed between 1584 and 1598; Henri IV finalized the design in 1602 to exclude any buildings on the bridge, apparently as much to provide views of the river as to facilitate traffic. Access to the new bridge from the north was provided by the Rue de la Monnaie, which had been widened under Henri III, but a new road was required on the southern bank. This was constructed as the Rue Dauphine, 11 yards in width across·the gardens of the Couvent des Grands Augustins. Objections on the clerics' part were silenced by the king's comment that the money from rents would buy plenty of cabbages.[13]

The bridge was finally completed in 1604, opening up the end of the island for development. Before then this area west of the royal palace (now part of the expanded Palais de Justice) consisted of gardens and a number of small islets (Figure 6.7, taken from the Legrand map of 1380.) This western end was given by Henri IV in 1607 to Achille de Harlay, the first president of the Parlement, who laid out on it a new residential *place* in honour of the 6-year-old Dauphin, the future Louis XIII. The Place Dauphine was the first of a number of such residential precincts to be built in Paris during the Renaissance, but it was planned along unique lines. The essential difference between this *place* and later ones is not its shape, which tapers with the banks of the island, but the fact that the statue of the king – the erection of which was a condition of the development consent – is located outside the enclosed space. It is on its central axis but stands on the far side of the short length of road linking the two arms of the Pont Neuf. The statue undoubtably belonged to the *place* and not the city at large, however; although occupying a most prominent site on the end of the island the king was facing the square and not the river.[14] In the middle of the *place*'s wider end, facing the royal palace, there was one entrance into the Place Dauphine with the other at the narrow end leading out to the statue. The wider eastern end was demolished in 1874 to create space for an extension to the Palais de Justice, providing it with a new west front. Only the two skilfully restored houses at the apex, opposite the statue, show us the original pattern, but of the square as a whole we can have no idea except through engravings, particularly the fine plan of Paris attributed to Turgot (1734–9).[15]

PLACE DES VOSGES

The site of the Place Royale (renamed Place des Vosges after the Revolution) in the north-east corner of the city near the Charles V wall and close to the Bastille, was originally occupied by the buildings and gardens of the Hôtel des Tournelles, the town house of the Duc d'Orléans. After his assassination in 1407 the property was acquired by the Crown; Louis XII died there in 1515, and following the death of Henri II in a tournament in its grounds his widow, Catherine de Medici, caused the hotel to be demolished. The area remained derelict until Henri IV ceded half of it, temporarily, to Sully in 1594 as a horse market. In 1599 a M. Delisle put forward a proposal to build a *manufacture de velours* (velvet), which was eventually accepted in 1604. The king, however, had second thoughts and decided that the site was both too big and too important to be used merely as a factory. In the following year therefore, a new civic square was formed in front of the completed factory to the south, by the addition of three matching residential terraces, possibly intended to house its workers.[16]

The factory closed (or was closed) in 1606 and on its site an identical fourth, northern, side to the square was built. As finally completed in 1612 the Place Royale is of great significance in European urban history as the prototype of the residential square.[17] In contrast to the previously generally accepted method of fronting houses on to multi-

24th December 1643 – I went with some company to see some remarkable places without the city; as the Isle, and how it is encompassed by the river Seine and the Ouse. The city is divided into three parts, whereof the town is greatest. The city lies between it and the University in form of an island. Over the Seine is a stately bridge called Pont Neuf, begun by Henry III, in 1578, finished by Henry IV his successor. It is all of hewn freestone found under the streets, but more plentifully at Montmartre, and consists of twelve arches, in the midst of which ends the point of an island, on which are built handsome artificers' houses. There is one large passage for coaches, and two for foot-passengers three or four feet higher, and of convenient breadth for eight or ten to go a-breast. On the middle of this stately bridge, on one side stands the famous statue of Henry the Great on horseback, exceeding the natural proportion by much, and, on the four faces of a stately pedestal (which is composed of various sorts of polished marbles and rich mouldings), inscriptions of his victories and most signal actions are engraven in brass. The statue and horse are of copper, the work of the great John di Bologna, and sent from Florence by Ferdinand the First, and Cosmo the Second, uncle and cousin to Mary de Medicis, the wife of King Henry, whose statue it represents. The place where it is erected is inclosed with a strong and beautiful grate of iron, about which there are always mountebanks showing their feats to idle passengers. From hence is a rare prospect towards the Louvre and suburbs of St Germains, the Isle du Palais, and Notre Dame. At the foot of this bridge is a water-house, on the front whereof, at a great height, is the story of Our Saviour and the woman of Samaria pouring water out of a bucket. Above, is a very rare dial of several motions, with a chime, &c. The water is conveyed by huge wheels, pumps, and other engines, from the river beneath. The confluence of the people and the multitude of coaches passing every moment over the bridge, to a new spectator is an agreeable diversion.
(J Evelyn, Diary)

---

13. Lavedan, *Histoire de l'urbanisme; Renaissance et temps modernes.*
14. As shown by Figure 6.6, the downstream end of the Ile de la Cité is visually one of the most prominent locations in Paris, such that a statue of the king facing away from the Place Dauphine would have impressed his presence on the entire city; however, the need to be a very large statue for long-distance visual impact, meant that it could but have appeared overpowering, or meaningless, at close quarters. Fortunately neither Paris nor Moscow – where a vast 1930s statue of Stalin was proposed – suffered the reality of that kind of personal aggrandizement.
15. Lavedan, *French Architecture.*
16. Lavedan, *Histoire de l'Urbanisme: Renaissance et Temps Modernes.*
17. For this most influential early Renaissance square, see M Webb, *The City Square*, 1990, for an international context; L Bergeron (ed.) *Paris: Genese d'un paysage*, 1989; L Mumford, *The City in History*, 1966. Zucker in *Town and Square* (still generally a valuably perceptive reference) writes 'in terms of town planning, the closed square represents the purest and most immediate expression of man's fight against being lost in a gelatinous world, in a disorderly mass of urban dwellings ... the perfect realisation of this form, in its Platonic purity so to speak [is] the Place des Vosges'. While I agree with the concluding statement (and the sense of 'being lost in a gelatinous world'), I believe that the primary

purpose traffic streets, the built-form is used in a residential square to enclose a space from which extraneous traffic is excluded, or at least discouraged. Not content with this revolutionary plan form, as an important instance of the aesthetic determinant, the royal urbanist also decreed that all the buildings facing into the space were to be of the same elevational design. Jean-Pierre Babelon, in his contribution to *L'Urbanisme de Paris et de l'Europe 1600–1680*, considers it probable that the ordered, arcaded design of the medieval Place du Champ à Seille at Metz, which Henri IV visited in March 1603, influenced him in approving the design of the Place Royale.[18]

The *place* was formed of thirty-eight houses behind uniformly designed elevations. In the centre of the north and south sides, before the later roadway was brought in across the northern side, taller, dominant bays contained the only two arched entrances into the *place*. The entrance building on the southern side, the Pavillon du Roi, was intended by Henri IV for his own use but he died in 1610, two years before completion of the *place*. Above an arcaded ground floor, which gave sheltered entrance to the houses and provided a continuous undercover connection between them, were two upper floors capped by steep slate-finished individual roofs containing a row of dormer windows.

Construction of the Place Royale may also have been of great social significance. Previously the nobility had lived in country châteaux or in *hôtels* in various parts of the city. The Place des Vosges, as Rasmussen says, can be regarded as a visible effort to bring the aristocracy under the control of an integrating idea, that of forming a background for the monarchy; instead of a galaxy of petty princes opposing each other and the king, they were now to become part of the pageantry of the court.[19]

The centre of the *place*, between the buildings, was originally kept as a clear, gravelled space. Here the traditional use of the area as a tournament ground was continued for some years, with the upper floor windows of the houses providing direct views of the festivities. In 1639 Cardinal Richelieu, who lived at No. 21 for a period, presented an equestrian statue of Louis XIII for the central position and thereby superimposed on the residential function the character of a statue-square. With the addition to the central space of formal landscaping, based on fenced-in areas of lawn and symmetrically planted tree groupings, the character of the *place* was again modified, and the statue lost its dominant role. After an initial period when it was the focus of Parisian society, the fortunes of the Place des Vosges slowly declined; more modern residential areas arose, until it became no more than a feature in a slum housing district[20] – always, however, with the quality of the housing tempered by the attractively planted central open space, which has in recent years proved a main reason for the steady regeneration of the Place des Vosges. The current long-term renewal programme for the whole of the Marais district is furthering this upturn in its fortunes and it is once more a fashionable residential district.

PLACE DES VICTOIRES

This circular *place* originated in a proposal by the Maréchal de la Feuillade to develop a site near the north-east corner of the Palais Royal gardens. The centre-piece of his scheme was a statue of Louis XIV, surrounded by four groups of lanterns, illuminated at night as one of the earliest examples of street lighting. In 1687, when the king formally inaugurated the Place des Victoires, only a few of the houses

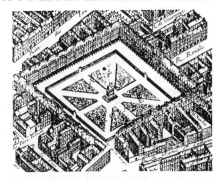

Figure 6.8 – The Place Royale (1605–12), renamed Place des Vosges after the Revolution, the prototype of the residential square from which extraneous traffic is excluded. The drawing shows the central space with the first stage of the landscaping work completed (the formal tree-planting is of a later date) and a new street access formed into the square at the bottom corner.

Within three years of Henri IV's death the treasury funds which had accumulated, amounting to nearly twenty million livres, were totally exhausted; during the same period expenditure rose by four million and revenues declined by five. The country was bankrupt. The collapse was disguised by exhibitions of pomp and circumstance which mesmerized contemporaries.

One such display was organised to celebrate the engagement of Louis XIII in 1612. The Place Royale was just completed and its noble residents were able to sit at their windows and for three days at Easter enjoy the celebrations ... the court, the nobility and fifty thousand people sat on scaffolds in the square to watch the gyrations of the nobly comparisoned houses, the fantastic floats drawn by reindeer and lions, the illuminations, fireworks and volleys of musketry and their ears were stunned by the salvoes overhead of one hundred cannon from the Bastille.
(D P O'Connell, Richelieu)

reason for the Place Royale was autocratic aggrandizement, rather than representing an early 'victory' in humankind's (continuing) campaign against a disorderly urban world. The Place, after all, was not for the city's benefit, as Zucker can be taken to mean. It was no more or less a privileged enclave totally separate from ordinary Parisians.

18. J-P Babelon, 'L'Urbanisme d'Henri IV et de Sully à Paris', in P Francastel (ed.) *L'Urbanisme de Paris et de l'Europe 1600–1680*, Editions Klincksieck, Paris, 1969.
19. Rasmussen, *Towns and Buildings*.
20. Sensitive restoration of the Place des Vosges is now well advanced, together with the adjoining historic district. In its changing fortunes the Place des Vosges is to be compared with the Covent Garden Piazza in London, for which it was the immediate influence (see Chapter 8). However, whereas the Place des Vosges is being restored to its original appearance, the enclosing buildings of Covent Garden Piazza are all replacements, with only a small part giving an impression of the original design.

were completed and the empty fronts of the remainder had to be filled in with canvas screens.

Jules-Hardouin Mansart was responsible for the layout and design of the uniform façade around the circumference of the space. Originally there were six minor streets entering the *place*, only two of which constituted an axis across it, but, as part of extensive alterations it has undergone, the Rue Etienne Marcel was added in 1883 as a major traffic route. Only one sector of the building circumference has survived in anything like its original form.

In spite of its identical façades, Zucker observes that 'the Place des Victoires does not achieve the impression of a closed square. Its openings to the surrounding streets (an anticipation of the typical *place percée* of the eighteenth century) are too numerous. It was in all likelihood just this condition between the square and its neighbourhood that made the Place des Victoires so much appreciated, especially by the theoreticians of the second part of the eighteenth century'.[21]

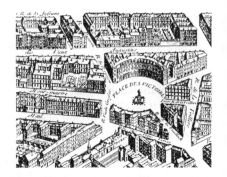

Figure 6.9 – The Place des Victoires (1687) designed by J-H Mansart. Although a carefully organized design in itself, there is clearly no considered relationship between the Renaissance space and the surrounding urban structure. Zucker was correct in recognizing the connection between the square and its neighbourhood as a form determinant. As such it was a traffic space, but with the added primary purpose of focusing attention on the status of Louis XIV in the centre.

PLACE VENDÔME

The early history of this residential *place* north of the Tuileries Gardens beyond the Rue de Rivoli and the Rue St Honoré is a fascinating record of the problems which can face speculative property developers and their architects.[22] As with many later schemes the initial idea was to realize the potential development value of land in an expanding urban area. The site was that of the impoverished Duc de Vendôme's town house and garden. After his death in 1670 these had to be set against his debts. Jules-Hardouin Mansart, a shrewd businessman as well as one of the leading French architects of the period, undertook, together with a number of financiers, to redevelop the site with leasable houses.

His first scheme soon ran into financial difficulties. Mansart subsequently put the project into the hands of Louvois, a wealthy courtier. But the Superintendent of Buildings meanwhile decided that instead of the proposed residential square the site should be used for a monumental statue-square dedicated to the glory of Louis XIV. Surrounding buildings were to include, among other prestige uses, various academies, a new national library, the Royal Mint and residences for foreign ambassadors. Louvois acquiesced in this; he was himself fired with an ambition to surpass the aspirations of the Duc de Feuillade, a rival courtier, to build his Place des Victoires. The site of the Hôtel Vendôme and its grounds, as well as the adjoining property of the Couvent des Capucines, was finally acquired in 1685. Shortly afterwards, however, the untimely death of Louvois removed the second of Mansart's clients and the scheme again fell into abeyance.

After further attempts to raise the necessary finances the land was ceded to the municipality of Paris in 1698. Mansart was able to maintain his professional involvement and finally in that year drew up outline plans of what was to become the present-day *place*. This was named the Place Louis XIV in honour of the king in 1699 and an immense equestrian statue was unveiled to his glory in the centre of the future *place*. There were, however, still no buildings and great difficulty in interesting potential tenants. Eventually, in desperation, the elevations surrounding the *place* were constructed in 1702 and the plots behind leased off over a period of years, in units of elevational bays as required by the different tenants. This unprecedented development procedure lasted until 1720.

Figure 6.10 – The Place Vendôme (1670–1720) designed by J-H Mansart. The contrast between the ordered central space and the uncoordinated buildings behind the elevations is typical of the way in which Renaissance examples of urban design exist as spatial 'islands' in otherwise uncontrolled growth contexts. This view shows the equestrian statue of Louis XIV. The originally limited through routes were subsequently extended north to the Place de l'Opera and south to the Tuileries Gardens.

21. Zucker, *Town and Square*.
22. For comparison with a different kind of historic speculative development project, implementation of which proved remarkably smooth, see Chapter 8 for Queen Square in Bath.

The plan of the Place Vendôme constitutes a rectangle with its corners cut off, thus creating an octagonal effect. The enclosed buildings are of three storeys with an additional row of dormer windows in the uniform roofs. The total height was specifically limited so as to be lower than the 54 feet-high statue. In the centres of its shorter side two relatively narrow streets lead from the *place*. Originally these outlets were only for minor traffic access and each terminated in vista-closing buildings short distances from the *place*. Later, however, they were both extended through to the main city network – as the Rue Castiglione and the Rue de la Paix. The centres of the two long sides of the *place* are emphasized by projecting pediments and columns and similar treatment is given to the four corners. The equestrian statue, perfectly scaled to suit the proportions of the *place*, was destroyed during the Revolution. Its replacement, the 144 feet-high Colonne d'Austerlitz, erected in 1810 by Napoleon, is obtrusive and out of scale.[23]

THE ILE ST LOUIS

At the beginning of the seventeenth century the area of the Ile St Louis was undeveloped and divided into two islets by a ditch which marked where the city wall continued across it. In 1609 a property group composed of a public works contractor, Christophe Marie, and two financiers, Le Regrattier and Poulletier, proposed to develop the islands. In return for constructing quays around the perimeter of the reunited halves of the Ile St Louis, and two new bridges, they were given the right to lease the building plots and to receive the rents for a period of sixty years. Royal assent to the scheme was given by Louis XIII in 1627 and by 1664 the work of preparing the island for housing purposes had been completed. It quickly became a desirable residential quarter with a distinctive character, which it still largely retains today.[24]

CHAMPS ELYSÉES

Catherine de Medici, widow of Henri II, tired of living in the cramped, enclosed courtyards of the Louvre, commissioned Philibert Delorme in 1563 to build a new palace outside the walls, with a spacious Italian style garden extending to the west. This was the Tuileries Palace, unfinished when Catherine died in 1568, and completed by Henri IV and Louis XIII. Henri linked the new palace to the Louvre by the Galerie du Bord de l'Eau. In the nineteenth century, under Napoleon I, the Grande Galerie du Nord was started along the south side of the Rue de Rivoli, thus creating an immense courtyard. This northern side was completed under Napoleon III between 1850 and 1857 but in 1871 the revolutionary Communards fired the Tuileries, leaving it a smouldering ruin. The Louvre itself was fortunate to escape; the damaged western ends of its north and south wings were restored in 1873–8, giving the present-day open-ended courtyard form.

At the end of the sixteenth century the Tuileries Palace and Gardens, though self-contained, still had an essentially directionless layout; there was no hint of the dramatic westward thrust to follow.[25] For the first time Parisian society could take attractive open-air recreation, enjoying the fountains, labyrinth and grotto.

Greater developments westward were to follow. Their first signs are seen in the Cours-la-Reine, created in 1616 for Marie de Medici. This did not extend the axis of the Tuileries Gardens west from the

Place Vendôme, Paris (previously Louis-le-Grand Square). Here the streams of architecture and town planning have joined to form a lake of repose in the bristling town compressed within its military walls; an architectural fashion owing much to the interior decorator and the scenic designer flowered in the salons and the anterooms. Salons to the glory of kings and princes. A fashion that soon flourished in the provinces abroad, wherever courts were held and courtiers dwelt.
(Le Corbusier, Concerning Town Planning)

The Ile St Louis is what it always was: a piece of unified town-planning which happened to get built at one of the noblest moments in the evolution of French architecture; and as it has been relatively little altered it still presents, on every hand, a look of grand and simple amenity. If much of it is a little shabby, we are reminded of how Baudelaire, one of its admirer-citizens, used to rub his suits with emery-paper in order to remove that look of newness 'si cher au philistin', as Gautier says, 'et si desagréable pour le vrai gentleman'.

Even ten years ago life on the Ile St Louis had still a note of discretion and retirement. The telephone alone seemed to link the island to the city proper: omnibus and Metro were excluded, dogs went to sleep in the middle of the main street, and if you chose to dream the day away by the water's edge nothing and nobody would disturb you. Today this is not quite the case. The island has been 'discovered'. Tenants of long standing dread the day when easy money will outwit the law and send them packing from their apartments. The insides are being ripped out of old houses for the sake of an 'amusing' interior ... Landlords cannot be blamed if they want to get something more than a peppercorn rent for some of the most delectable properties in Europe. Much of the Ile St Louis is in very bad condition and new money may prevent it from becoming, as it were, the Chioggia of Paris. And the island will in any case resist the attempt to 'hot it up'; certain places simply cannot be vulgarised, and one of them is the Ile St Louis.
(J Russell, Paris)

23. The positioning of a properly scaled king's statue, on the cross-axis of the Place Vendôme, is a straightforward instance of autocratic aggrandizement which, in its own terms, could but have been visually impressive; by comparison the present-day over-sized Colonne d'Austerlitz does little for the memory of Napoleon Bonaparte, or his victory.
24. For a detailed historic plan of the Ile St Louis see P Lavedan, *Nouvelle Histoire de Paris* (Histoire de l'Urbanisme), 1975.
25. See E Bacon, *Design of Cities*, 1974.

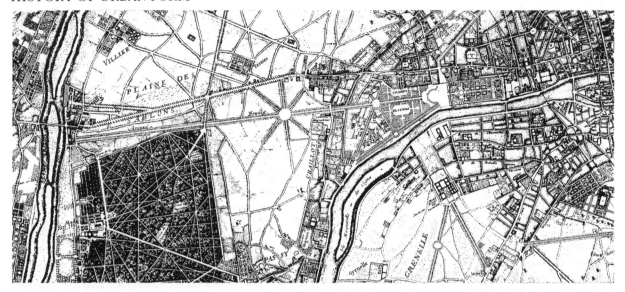

palace, however, but skirted the Seine as far as the site of the later Place d'Alma, at an angle to this axis. The Champs Elysées, with its dramatic perspective up the gentle slope of the Butte de Chaillot to the Place de l'Etoile, and the equally magnificent continuation down the far side to the bridge at Neuilly, and beyond, was begun by Le Nôtre, for Colbert in 1667 (Figures 6.11 and 6.12).[26]

The results of Le Nôtre's replanning are summarized by Edmund Bacon; the aim was 'to transform the whole nature of the Tuileries garden design from static to dynamic, with the thrust of the axis generated within the garden extended outward by the Avenue des Tuileries, now the Champs Elysées. Not yet present was the Place de la Concorde, which was later to occupy the open ground between the Tuileries garden wall and the planted area of the Champs Elysées and which connected the central Tuileries axis both with the area to the west and across the river Seine'.[27] Earlier in his perceptive analysis of the development of Paris, Bacon observes that with this scheme 'an entirely new breadth and freedom have been introduced in the art of civic design. The outward thrust of the movement systems, generated from firm building masses, penetrates further and further into countryside. It simulates similar axial thrusts originating in the châteaux and palaces about Paris, which also extend and intertwine, creating in the late eighteenth and early nineteenth century a form of regional development unique in the history of city building'.[28]

By 1709 extensive tree planting had matured sufficiently for the general area to become known as the 'Champs Elysées'. In 1724 the main axial avenue was extended as far as the top of the Butte de Chaillot, by the Duc d'Antin, director of the royal gardens. The further extension west down to the bridge at Neuilly by 1772 was the work of his successor, the Marquis de Marigny. The top of the Butte de Chaillot is shown on the 1740 map as a *rond-point* of eight avenues, set in the middle of an area of open fields (Figure 6.11). In 1774 Soufflot lowered the height of the hill by 5 yards, to ease the gradient. The Arc de Triomphe, set in the centre of the Place de l'Etoile, was originally proposed by Napoleon I in 1806 and was completed in 1836 by Louis-

Figure 6.11 – Paris, a section of the beautifully engraved map of 1740 showing the extent of the urban area and the line of the westward axis of the Champs Elysées. (Refer to Figure 6.12 for identification of parts of the city.) The Bois de Boulogne (bottom left) is shown with the formal landscaping layout, which was replaced by the present-day naturalistic design, under Haussmann, in the 1850s and 1860s. The front cover colour photograph shows the eastern end of the Champs Elysées as seen from the top of the Arc de Triomphe.

26. See H M Fox, *André Le Nôtre: Garden Architect to Kings*, 1962.
27. Bacon, *Design of Cities*.
28. In the second half of the twentieth century the historic main western axis has been taken way out beyond the bridge over the Seine to the new commercial skyscraper centre at La Defense, and beyond. Underground it is also the route of the new super Metro line.

Philippe. The present-day form of the Place de l'Etoile, with its twelve radiating streets, was created by Haussmann for Napoleon III from 1854 onwards: Hittorff was architect for the surrounding buildings.[29]

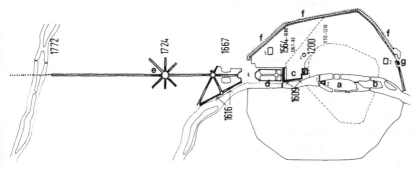

GRANDS BOULEVARDS

By 1660 the fortifications around most of northern Paris, between the Bastille in the east and the present-day site of the Madeleine in the west, were in poor repair. Internal peace in France and military successes, culminating in the campaigns of Louis XIV, which established strong frontier defences, had reduced the need for complex city fortifications. This enabled those of Paris to be demolished.

Between the Bastille and the Porte St Denis there was a section of the wall first built by Charles V (1363–83). The western portion had been constructed under Louis XIII (1610–43) for an extension of the city which included the Tuileries Gardens. The ground occupied by the walls and earthworks was cleared and levelled to form a linear open space of some 3½ miles in length. Within this a central driveway, wide enough for four lanes of carriages, was flanked by pedestrian *contre-allées*, each planted with double rows of trees. This work was completed in 1705 with a number of *arcs de triomphe*, including a new Porte St Denis (1672), marking the old entrances to the city. The term boulevard gradually became applied to this linear space as a corruption of, says Rasmussen, the Nordic *bulvirke* (bulwark) which means a palisade – a medieval form of defence used before the employment of real walls and ramparts; the boulevard is the line of the fortifications itself but when these were converted into wide tree-lined streets the designation boulevard was kept and today the word means simply a broad tree-lined avenue.[30]

Although not forming a physical barrier, the line of the boulevards continued to mark the limits of Paris; beyond them building was forbidden. The boulevards were put to no regular use until around 1750 when, probably as a result of the maturing of the landscaping, they became fashionable open-air recreation areas – the scene of the *promenade à la mode*. The main carriageway was paved in 1778 and asphalt all-weather pavements were provided in 1838.

When the boulevards were built up, the western part became a fashionable housing district, in contrast to the eastern sections which retained their leisure functions, now embodied in innumerable cafés, restaurants and theatres. Under Haussmann many changes were made to the form of the boulevards. The Place de l'Opéra and the Place de la République were created and the eastern section completely reconstructed. Today the Grands Boulevards are divided into three distinct

Figure 6.12 – A diagrammatic plan showing successive stages in the western expansion of Paris along the axis of the Champs Elysées to the bridge at Neuilly and beyond.

Key: a, Ile de la Cité; b, Ile St Louis; c, the Louvre; d, the Tuileries Gardens; e, the Arc de Triomphe location on the Butte de Chaillot; f, the line of the Grands Boulevards; g, the Bastille; 1, Pont Neuf; 2, Place Dauphine; 3, Place des Vosges; 4, Place des Victoires; 5, Place Vendôme; 6, site of the Place de la Concorde. The distance between the eastern side of the city (right-hand edge of diagram) and the Seine at Neuilly is 12.5 km. The complex of commercial skyscrapers at La Defense is 1.5 km to the west and the Seine at Nanterre (after its major loop to the north-east) is a further 5 km distant. (The Parisian planning authorities, favoured with their particular legislative context, have been able to concentrate the Paris skyscrapers at La Defense, on the historic axis, away from the historic nucleus.)

Paris, at the end of the eighteenth century, was still in some respects a homogeneous city. Though there were faubourgs like Saint-Antoine and Saint-Marceau, inhabited mainly by small masters and their journeymen, and new wealthy quarters like the Faubourg Saint-Germain, in much of older Paris, the homes of the well-to-do, of the middling people and of the poor existed under the same roof. It might also be said that class stratification was vertical, the rich and the poor entering by the same door, the former to mount by a short and broad staircase to the impressive apartments of the first floor, the latter to climb high up by ever narrowing stairs till they reached the attics in the mansard roof. (A Cobban, A History of Modern France)

29. See D H Pinkney, *Napoleon III and the Rebuilding of Paris*, for a description of the detail planning of this perhaps best known Parisian *place* where, 'to assure the symmetry of the Place de l'Etoile the Prefect (Haussmann) located the radiating streets so that uniformly shaped building lots remained between every pair of streets. Eight of the lots were identical in size, and the four on the opposite sides of the Avenue Kléber and the Avenue de Wagram were double size. An imperial decree required that the buildings on these lots should have uniform stone façades set off by lawns and that the lawns be separated from the street by decorative iron fences, identical before all the buildings. To provide access to these houses, whose monumental fronts must not be marred by entrances, Haussmann built a circular street around the back of the lots. The Prefect was proud of his development of the Place de l'Etoile and considered it one of the most successful undertakings of his administration.'
30. Rasmussen, *Towns and Buildings*.

sections: to the west the Boulevards de la Madeleine, des Capucines, des Italiens and de Montmartre are lined with high-class shops, boutiques, cinemas, restaurants and cafés, and many prestige offices; east to the Place de la République all the uses are of a significantly lower quality with diminished tourist interest; while further east again towards the Place de la Bastille there are mainly residential and commercial uses.

Many other towns and cities on the continent of Europe, following the early example of Paris, have benefited from this reversion to peaceful purposes of their encircling fortification zones. Such resulting urban structures incorporate inner ring boulevards still able to provide invaluable traffic routes around the city centre, and which serve in many instances as attractive open spaces. Vienna and Cologne are particularly clear examples of this radial-concentric growth which characterizes continental European cities. Such a pattern is in marked contrast to the radial-ribbon – with infilling – tendencies that are the general rule in Britain, where urban defences were not required after the early Middle Ages. Where medieval town walls were retained in Britain to define city limits for one purpose or another the strip of land they occupied was normally too narrow to enable new roads and linear open spaces to take their place. Medieval walls in Britain dated from a period before the age of effective offensive artillery; they were not supplemented by the extensive earthworks needed in fortifications after the fifteenth century. Such medieval open fire-zones as had been maintained were comparatively small and were encroached upon to an extent not permitted on the continent.[31]

Two further encircling walls were set around Paris. The first was the Wall of the Farmers General, about 14 miles in length, built under Louis XVI not for defensive purposes but rather to facilitate the collection of taxes on goods entering and leaving the city. The second was the last of the defensive systems constructed for the city: that by Thiers between 1840 and 1845. In turn both these rings were replaced by boulevards; those replacing the Farmers General wall were mainly to the north of the Seine; those of the 1840–5 fortifications form a complete ring of *boulevards extérieurs* at an average distance of 3 miles from the city centre.[32]

PLACE DE LA CONCORDE

Originally named the Place Louis XV, this vast square at the eastern end of the Champs Elysées was built to the designs of Jacques-Ange Gabriel (1698–1782) between 1755 and 1775. A competition had been held to obtain plans for a suitable setting for an equestrian statue of the king that had been commissioned by the city of Paris from the sculptor Bouchardon. Over sixty architects submitted their proposals for the location and treatment of the proposed statue-square. Their collected designs, published in 1765 in a volume of engravings, *Monuments érigés à la gloire de Louis XV*, present a fascinating record of contemporary urban thought. The extract from this book shows a number of the proposals plotted on to the map of the city (Figure 6.14).

Gabriel was the winner of the competition. He chose as his location for the statue the extensive, abandoned terrace which lay between the Tuileries Gardens and the Champs Elysées. On the Turgot map of 1734–9 (the section reproduced as Figure 6.16), this area appears as an informally used transition-space. It had a typically organic path pat-

Under Louis XV, Paris, pushing outwards, had swallowed up the limits traced by Charles V in the fourteenth century and their western extension of 1631 under Louis XII. The hated wall of the Farmers General, built in 1785 as a customs barrier with imposing monumental gates, took in a vast new area, stretching round the western and northern heights of Passy, Chaillot, Belleville, and Menilmontant, including – south of the river – the Faubourgs Saint-Victor, Saint-Marceau, Saint-Jacques, Saint-Germain, and curving back round the vast Champ de Mars. At the end of the eighteenth century much of this new territory was not yet built up; within the barrier, beyond the Bastille and the Temple in the East, were fields and scattered houses with their gardens. In the west the Champs Elysées were woodlands crossed by roads and wandering paths, and the Champ de Mars a huge open space. Paris proper still huddled together within the boulevards that marked the site of its former fortifications, a solid agglomeration of high, closely-packed, terraced houses separated by winding, narrow streets and alleys, noisy with street cries, busy with passers-by, crowded and dangerous with carriages and wagons of all kinds, strewn with rubbish and filth lying about in heaps or carried along in the torrents of water pouring after a storm down the wide gutters, across which pedestrians could only pass dry-shod by little plank bridges.

(A Cobban, A History of Modern France)

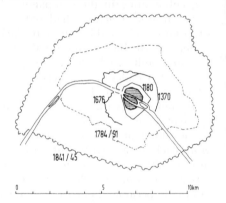

Figure 6.13 – Paris: the successive rings of fortifications which determined the city's growth pattern from the twelfth to the nineteenth centuries. The wall of 1784–91 is the customs barrier of the Fermiers Généraux. (Compare with diagrams of London's growth pattern drawn to the same scale as Figures 8.1 and 8.54.) The 1841–5 ring is now the route of the Boulevard Périphérique ring motorway, at approximately 5 km distant from the downstream end of the Ile de la Cité; the outer ring of forts is approximately 1–5.5 km further out.

31. See Chapter 8 for John Stow's contemporary account of the extreme state of disrepair of London Wall in 1603.

32. Whereas the circumferential zone of land occupied by the Thiers fortification was uncontroversially used for the Boulevard Périphérique ring motorway around central Paris; the proposed route for a comparable 'motorway box' in inner London constantly ran up against the urban grain, and aroused vehement local political opposition, such that only a short length was ever completed. (Although still little known, Paris had in fact been anticipated by Moscow in taking advantage of its long redundant defensive zone – see Figure 7.42.)

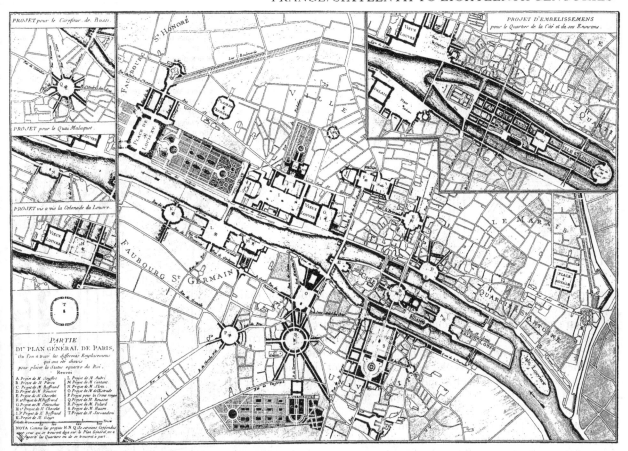

tern unaffected, surprisingly, by continuation of either the main Champs Elysées axis or the secondary Cours-la-Reine.

The royal squares of the seventeenth century were essentially closed spaces: landscape elements, infrequently used, and then only in minor roles, were dominated by the architectural built-form. During the eighteenth century, says Zucker, open space, in contrast to closed space, was the new ideal and its realization was assured by the inclusion of nature; out of these sentiments the king presented the city of Paris with the open land adjacent to the Jardin des Tuileries, the River Seine and the Champs Elysées and the quarter around the projected church of the Madeleine. Thus the new square could expand on three sides into open space.[33]

Gabriel's first problem was to define the *place* without separating it from the existing landscaped spaces to the west. His second problem was to organize a planned relationship of the four existing axes which crossed the site – those of the Tuileries Gardens/Champs Elysées and the Rue Royal (set out in 1732), and those of the Cours-la-Reine and its counterpart to the north of the Champs Elysées. In 1766 work started when the statue of Louis XV was erected at the intersection of the first two of these axes.

In the original plan landscape elements had played a more important space-defining role than buildings, which formed only the northern side of the *place*. The basis of Gabriel's layout was a 15 feet deep

Figure 6.14 – Paris: competition entries for the design of the 'statue-square' in honour of Louis XV plotted on a map of the city of 1765. The key at the bottom left identifies the locations proposed by individual architects; the inset top right shows a proposal to unify the Ile de la Cité and the Ile St Louis on the basis of a regularized street pattern with a symmetrical replica of the Louvre built on the left bank of the Seine. Neither this grandiose project, nor the proposal for the downstream end of the Ile de la Cité in the general map, shows any respect for the existing Place Dauphine. (Refer to Figure 6.12 for identification of parts of the city.) The Place Louis XV (subsequently the Place de la République and now the Place de la Concorde) is towards the top left of the map proper, with the Church of the Madeleine located on its central axis (see Figure 6.15).

33. Zucker, *Town and Square.*

203

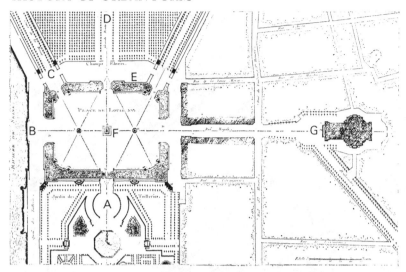

Figure 6.15 – The Place de la Concorde, originally the Place Louis XV, designed by Jacques-Ange Gabriel on the space between the Tuileries Gardens and the axis of the Champs Elysées. This plan is the 'after-development' counterpart of the land as shown in Figure 6.16. Key: A, Tuileries Gardens; B, the future Pont de la Concorde; C, Cours-la-Reine; D, Champs Elysées axis; E, the sections of *fosse* with bridges over; F, statue of Louis XV (replaced by the present-day obelisk); G, the Madeleine Church. As finally realized, the plan of the Madeleine is that of a classical temple, and it is now flanked by symmetrical boulevards.

Figure 6.16 – Section of the Turgot plan (1734–39). At the top can be seen the Louvre; and the Tuileries Palace at right-angles to the river at the end of the Pont Royal. Below the Tuileries, destroyed in 1871 (to the west), the central section of the drawing shows the Tuileries Gardens, with the main axis leading further west to the Champs Elysées, across the derelict area later to become the Place de la Concorde (see Figure 6.15, which is orientated through 180 degrees).

fosse, or ditch, with a surrounding balustrade: this defined the main central space. The corners were cut off, creating an octagonal effect, similar in shape to the Place Vendôme. This 'angle' treatment was to provide entries into the *place* for the Cours-la-Reine and its counterpart in the north-west corner. With axial entrances in the centres of the four sides, these were taken across the ditch by a total of six bridges. The ditch was some 60 feet in width and grassed at the bottom. In 1854 it was filled in and the Place de la Concorde acquired its present-day appearance.

Along the northern side Gabriel built the Garde-Meubles between 1760 and 1765 – two identical buildings with elevations respecting those of Perrault's Louvre. These buildings are symmetrically planned on either side of the Rue Royale axis. To the east the *place* is separated from the Tuileries Gardens by an imposing terrace and balustrade. As at the Place Vendôme only the façades of the Gards-Meubles were built at first: completion of the buildings behind took many more years. This speculative process was repeated in 1805, when the Rue de Rivoli was formed to the east; the architects of Napoleon I, Percier and Fontaine, designed only the architectural part of the façades, whose acceptance and construction was binding upon the purchasers.[34] The statue of Louis XV was demolished in 1792 during the Revolution. Louis XVI went to the guillotine on 21 January 1793 in the north-west corner of his father's *place*, then renamed the Place de la Révolution.[35]

The Madeleine Church was started in 1764. After a chequered early history, during which it was considered for a number of religious and secular functions, it survived proposals in 1837 to replace it by the first railway station in Paris and was consecrated in 1842. The imposing Pont de la Concorde across the Seine, on the Madeleine/Rue Royale/Place de la Concorde axis, was completed in 1790. The axial view is terminated by the dome of the Palais Bourbon (the Chamber of Deputies) on the left bank. The massive obelisk from Luxor, acquired from Egypt by Charles X, was erected in 1836 on what was earlier the site of the central statue. In 1854 Hittorff added two fountains, as originally had been proposed by Gabriel, on either side of the obelisk, at points where the axes of the Cours-la-Reine and its matching route in

34. Lavedan, *French Architecture*.

35. Chapter 8 relates how the English king Charles I walked out on to his scaffold in 1649 through a window of the Banqueting Hall, the only completed part of a grandiose, yet never to be realized scheme for a Whitehall Palace in London. Louis XVI, had he been so minded, might have consoled himself with a last look at the urban heritage which his family had created for Paris.

the north-west corner intersects with the Madeleine/Pont de la Concorde axis (see Figure 6.15).

Commencing with the palace and gardens for Catherine de Medici in 1563 the creation of this part of Paris took over 250 years. It has been shown that during this time successive clients, both royal and civic, employing a sequence of professional advisers, respected their inheritance from the past and contributed towards the eventual realization of one of history's most magnificently organized urban achievements.

Provincial Renaissance urbanism

Statue squares were built in many of the leading provincial cities of France, and were proposed for others. Those examples in Rheims, Rouen, Nancy (together with the related sequence of eighteenth-century spaces), Bordeaux and Rennes are described below.

During the Renaissance, France had effectively no need for new, ordinary commercial-function towns, but a number of new towns were planned and constructed for other particular purposes. Included among these towns, and also described in this chapter, are two service-towns: Richelieu and Versailles (both of which serve as a monument to the men who were the inspirational force behind their building – Richelieu and Louis XIV) and Neuf Brisach, one of the late seventeenth and early eighteenth-century frontier fortresses built by Vauban, the outstanding engineer-urbanist of Renaissance Europe.

RHEIMS AND ROUEN

The Place-Royale at Rheims (1756–60) originally contained a statue of Louis XV, and was designed as part of a plan to reconstruct a large area of the city centre. A new street, the Rue Royale (now the Rue Colbert) was created as the axial link between the existing square in front of the Hôtel de Ville and the Place Royale in front of the new Hôtel des Fermes. On either side of this axis are two smaller market squares (Figure 6.17).

Patte's *Monuments* includes a project by Lecarpentier for a large statue-square to Louis XV in front of the Hôtel de Ville in Rouen, but this skilful proposal to regularize the entry points of streets in from the rest of the city was never realized (Figure 6.18).[36]

NANCY

Nancy, the ancient capital of Lorraine, some 220 miles east of Paris on the main route to Strasbourg, was captured by the French in 1633. At that time the city was divided into two distinct parts, each with its own formidable defensive system. The old medieval town to the north, its street pattern typically formless and organic, constituted in effect the citadel; the ordered, gridiron-based Renaissance *ville-neuve* was to the south (Figure 6.20). In 1697, before returning the city to the Dukes of Lorraine, Louis XIV ordered the defences to be completely demolished. Both parts thus became surrounded, and separated, by areas of derelict fortifications and open fire-zones.

In 1736, having again taken possession of the city, Louis XV gave Nancy to his son-in-law, Stanislaus Leczinski, ex-king of Poland. Stanislaus was to prove an enlightened ruler, aware of planning deficiencies of his city in the context of urban progress during the three preceding centuries. The problem was essentially that of uniting the

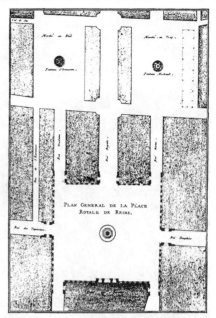

Figure 6.17 – Rheims: the Place Royale and the related streets and spaces in the centre of the city.

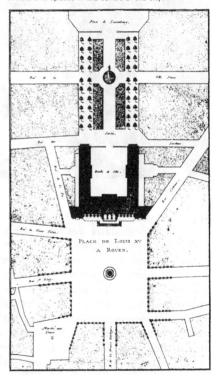

Figure 6.18 – Rouen: the Place Royale, a skilful theoretical exercise in planning an ordered relationship between existing streets and the new space.

36. For both Rheims and Rouen see G Duby (dir.) and A Chedeville, *La Ville Mediaeval, Volume 2, Histoire de la France Urbaine*, 1980.

two parts, medieval and Renaissance, into one integrated urban structure. Here Stanislaus was fortunate to find a gifted architect-planner, Héré de Corny (1705–63), as his executive collaborator. De Corny was commissioned in 1753 and developed a scheme in several parts (its outline is superimposed on the existing city in Figure 6.21). A new statue-square to Louis XV – now the Place Stanislaus – is the meeting point of two axial developments.[37] The first of these axes provided a new main east–west route across Nancy, just within the line of the fortifications in the *ville-neuve*. Running north into the old medieval city, at right-angles to this first axis, a second development took the form of three interrelated spaces – the Place de la Carrière; the short length of the Rue Héré, which links it to the Place Royale; and the Hemicycle in front of the Provincial Government Palace, which forms the northern termination of this second axis. Between the Rue Héré and the Place de la Carrière there is an imposing triumphal arch.

The whole scheme, as illustrated in Patte's *Monuments*, shows how the plan has been skilfully organized to take account of the remains of the

Figure 6.19 – Nancy: an aerial view of the sequence of eighteenth-century spaces seen from the west across the medieval city; the cathedral is in the foreground. The Place Stanislaus is to the top right. The viewpoint is marked with an open arrow. The Place Stanislaus (formerly the Place Louis XV/Place Royale) is at the right-hand side of the photograph.

37. For Nancy see Bacon, *Design of Cities*; G Duby (dir.) and R Chartier, *La Ville Classique, Volume 3, Histoire de la France Urbaine.*

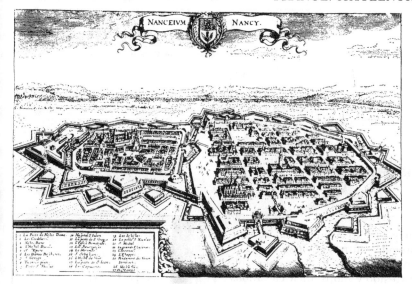

Figure 6.20 – Nancy in 1645, seen from the west, with the medieval city to the left. (This view has the reverse orientation from the general plan in Figure 6.21.)

Commercial prosperity and the growth of the state machinery inflated the size of the towns. Paris, in the middle of the century, had a population of half a million and was steadily increasing. Lyon had perhaps 160,000. Marseilles and Bordeaux were approaching 100,000. On the other hand, provincial capitals like Rennes, Dijon or Grenoble were towns of little more than 20,000. All told, by the end of the 'ancien régime' the urban population of France can hardly have exceeded two and a half million at most, which left a rural population of perhaps some 22 to 24 million. It is difficult to avoid the conclusion that France must have been suffering from intense and increasing rural overpopulation.
(A Cobban, A History of Modern France)

fortifications. In addition to these earthworks, de Corny's design had to incorporate an important existing building on the eastern side of the Place de la Carrière. This was the Hôtel de Beauvau Craon, the present-day Palais de Justice, designed in 1715 by Boffrand. Its elevation was repeated on the other side of this *place*, and its proportions became the basis of buildings framing the Place Royale. Houses on either side of the remainder of the Place de la Carrière were also given new unified elevations to accord with the overall design.

Nancy's Place Royale is notable for the magnificent curved wrought-iron screens which define its four corners. Those flanking the Hôtel de

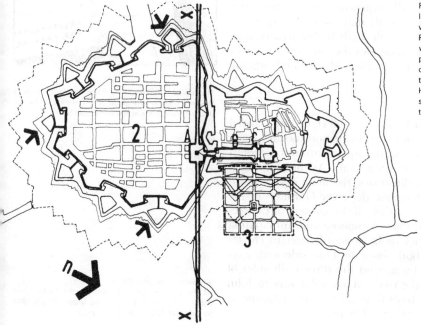

Figure 6.21 – General plan of Nancy showing the relationship of the eighteenth-century redevelopment with the other parts of the city. The late sixteenth-century Renaisance *ville-neuve* is to the left of the new east–west axis; the medieval city is to the right. This general plan and the detail arrangement of the eighteenth-century spaces as Figure 6.22, have the same orientation. Key: A, Place Royale; B, Place de la Carrière; C, the Hemicycle; 1, the medieval city; 2, the *ville-neuve* of the seventeenth century; 3, the later public park; X–X, the new main cross-axis.

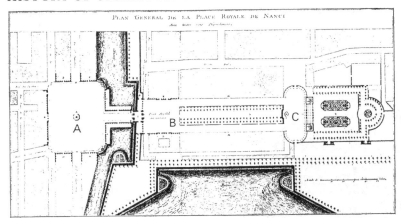

Figure 6.22 – Nancy: the detail plan of the Place Royale (Place Stanislaus), and the related space planned by Héré de Corny. A part of the defensive moat is shown at the bottom of the plan. Key: A, Place Royale; B, Place de la Carrière; C, the Hemicycle (Patte).

Ville, on the southern side, unfortunately have to admit traffic into the *place*, but those on the other corners provide controlled views out. The Rue Héré, leading from the Place Royale, was at first an open colonnade; later this was converted into two-storey shops, their elevations carefully related to these in the main *place*. The Place de la Carrière has two parts – the open paved space between the main buildings at its southern end and the long narrow directional space leading to the Hemicycle. This directional character is emphasized by two rows of formally clipped trees on each side of the central axis. Finally in this marvellous sequence of spaces, there is the Hemicycle itself, in front of the Government Palace. It has semi-circular open colonnades around its narrow ends, through which access is gained into the medieval town to the west, and into a large park laid out during the nineteenth century to the east. This was made possible by taking advantage of land which had once formed part of the old fortified zone.

Edmund Bacon uses this scheme to demonstrate that great work can be done without destroying what is already there and says that 'here a new element is added, that of symbolically expressing in new structures the spirit of what has been associated with that particular space in previous history. Thus the Arch of Triumph, built by Stanislaus, conveys the spirit of the fortified wall which divides the old mediaeval city from the new and recreates the feeling of the old bi-celled organic form'.[38]

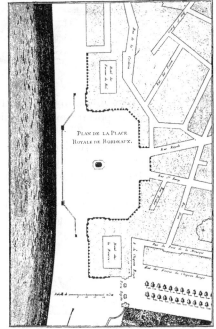

Figure 6.23 – Bordeaux: the Place Royale (Place de la Bourse) created alongside the Garonne River by Jacques-Jules Gabriel and completed by his son.

BORDEAUX

Bordeaux's Place Royale (Place de la Bourse) was designed by Gabriel the Elder (1667–1742) in 1729, and completed by his son Jacques-Ange Gabriel in 1743. It was designed as a statue-square around a monument to Louis XV (replaced in 1869 by a fountain). As the plan taken from Patte's *Monuments* (Figure 6.23) illustrates, the main design problem was to relate the new space alongside the river Garonne to the city's unplanned street pattern.[39] By cutting off the corners of the rectangle (following the example of the Place Vendôme and anticipating Gabriel's use of this motif in his Place de la Concorde design) entries into the *place* were limited to two, both from the land side with axes centred on the statue. Unified façades around the three built sides of the *place* frame the view out across the river, in a similar way to John Wood the Younger's integration of built-form and open landscape at the Royal Crescent in Bath (1767–75). (See Chapter 8.)

38. Bacon, *Design of Cities*.
39. For Bordeaux see Duby and Chartier, *La Ville Classique*; M C Branch, *Comparative Urban Design*, 1978, for the 1832 SDUK Map of Bordeaux.

RENNES

Considered by Pierre Lavedan as the finest of the provincial squares dedicated to Louis XIV, the Place du Palais at Rennes was designed by Gabriel the Elder as part of the rebuilding of the town after the fire of 1721 (Figure 6.24). This square provides a setting for the existing Palais de Justice (originally the Parliament of Brittany), designed by Salomon de Brosse, with the other three sides planned by Gabriel, with a great Ionic order over a high base; though it is almost unknown, it counts among the most perfect works of French architecture.[40] The statue of Louis XIV has been replaced by a fountain.

The Place du Palais is linked diagonally with the Place de la Maire (originally the Place Louis XV) through a narrow passageway. As both of these squares contained statues, pains were taken to vary the position of the monument. Louis XIV stood, as always, in the centre of the square. Louis XV was set right against the façade of the Hôtel de Ville. The Place de la Mairie is divided in two by a cross street, with five parallel rows of trees balancing the volume of the town hall, one of the first examples of planting employed as an architectural element within the town.[41]

RICHELIEU

In 1620 the Cardinal Richelieu decided to convert a small medieval chateau he had inherited into a vast new palace where he could entertain the king and his court retinue in the manner to which they were accustomed. He proposed to solve the problem of accommodating all the guests and their servants, in addition to his own enormous staff, by building a new town specially for the purpose. The palace has long since disappeared but the small town of Richelieu still exists, very much as it was in the cardinal's day, some 10 miles south of Chinon in the Loire Valley region.[42]

The palace and town at Richelieu pre-date, by some 40 years, the similar and suitably more grandiose work for Louis XIV at Versailles. It is a tradition that Richelieu's project was intended to show Louis XIII how he might also live in fitting style if he could only bring himself to move from his cramped, obsolete accommodation in the Louvre.

Richelieu's palace was started in 1620 to the designs of Jacques Lemercier, the leading French architect of the day. In 1631 the town was laid out by Lemercier, who was also responsible for designing the public buildings and the most important houses. In 1631 the cardinal obtained from the king permission to hold a twice-weekly market in the town, in addition to four annual fairs. These valuable privileges should have guaranteed the commercial success of the town, but Richelieu was in a relatively isolated position, away from the main roads. As a further incentive the first citizens were excused all taxes until there were 100 houses in occupation. In addition to these persuasive measures, the cardinal, determined to make his property the premier centre of the district, proceeded to buy up adjoining estates so that he could dispossess their inhabitants.[43] Within a short time of starting, over 2,000 men were at work on the project.

By 1635 the palace had been completed. After three years the town was well advanced: the walls, church, college, law-court and most of the houses were finished, but there were still few inhabitants. Some years later, after the cardinal's death in 1642, the English diarist John

Rennes, devastated by fire in 1720, was largely rebuilt in the eighteenth century; in other towns, even without this adventitious advantage, the work of demolition and rebuilding went on apace. Who would have guessed, even before the Revolution, that in 1700 Bordeaux had still been a mediaeval city? While private individuals built their eternal mansions on earth, elegant, grand, but in their frequent repetition of the same themes ultimately a little boring, town planning and the creation of imposing set-pieces on a larger scale was the work of the royal intendants, who rivalled one another in the task of beautifying the seats of their authority.
(A Cobban, A History of Modern France)

Figure 6.24 – Rennes: the Place de Louis-le-Grand (Place du Palais) in front of the Palais de Justice and the adjoining Place Louis XV (Place de la Mairie).

Two main routes traverse Poitou, one from Tours and the other from Saumur and they converge in the provincial capital, Poitiers. Between them on a tributary of the Vienne called the Mable is the little town of Richelieu. A regular quadrilateral with houses and two little squares of uniform style and elevation, it is the embodiment in stone of the spirit of Poitou, the province that almost simultaneously produced the master rationalist in philosophy, Descartes, and the master rationalist in politics, Armand-Jean de Plessis de Richelieu. It is, in fact, a village called into being by the Age of Reason, proclaiming its sense of order and authority ... of the château itself nothing remains. It was confiscated at the time of the Revolution and sold early in the nineteenth century on condition that it would be demolished, and its only relics are a pavilion and the moats of its mediaeval ancestor in the midst of a great park. The Cardinal, ill and over-worked, at a time of war and national crisis, did not see his creation completed or even more than half-finished, but he gave no indication of any regret; for the château and the village existed, not to accommodate him, but to symbolise the entry of the family of Richelieu into the ranks of the great houses of France ...
(D P O'Connell, Richelieu)

40. Lavedan, *French Architecture*. See also Duby and Chartier, *La Ville Classique*.
41. Whereas the use of trees in historic garden and park landscape design is well documented (see Note 47) the formal incorporation in urban design of trees and other natural landscape elements exemplified by the 1720s use at Rennes, and that at Nancy (Figure 6.22), is a comparatively obscure, un-researched subject. (See also Chapter 10 for the use of trees along Pennsylvania Avenue in Washington DC.)
42. See P Boudon, *Richelieu: Ville Nouvelle* (Collection Aspects de l'Urbanisme), Paris, 1978.
43. M S Briggs, 'Richelieu', *The Builder*, 12 January 1940.

Evelyn visited the town and observed that 'it consists of only one considerable street, the houses on both sides, as indeed throughout the town, built exactly uniform, after modern handsome design but it is thinly inhabited, standing so much out of the way and in place not well situated for health or pleasure'.[44]

Richelieu is a small gridiron town of rectangular shape, some 600 yards in length and 400 yards in width. The main street running the length of the town, exactly from north to south, originally constituted an axial link with the palace to the south. In this relationship the town was clearly subordinate to the palace: their respective axes intersected at right angles in the centre of the palace forecourt.

There are two *places* in the town, one at either end of the main street. Each is about 100 yards square. The southern one – the Place du Marché – has a distinctively designed church on its western side, facing the market hall. The Hôtel de Ville, originally the Palais de Justice, is also in this square. The Place des Religieuses at the northern end is less important now, lacking the theological college after which it was originally named.

The north–south Grande Rue is some 37 feet wide and comprises the twenty-eight large houses allotted to principal members of the cardinal's retinue. These houses have a street frontage of about 70 feet each; an entrance gateway leads to a central courtyard in which there is the house door. The elevations of these houses were originally completely identical, with a cornice height of about 30 feet beneath steeply sloping individual slate roofs. Generally, within the town, house sites were granted free of charge, provided that Lemercier's standard elevations were completed within a stipulated period.

There is a marked contrast between the faded Renaissance elegance of these Grande Rue houses – unfortunately now converted, in many instances, into shop fronts – and the poverty of the back land 'cottages' which could well be part of some obscure declining agricultural village. Other than the buildings forming the square and the main street there is every reason to suppose that the remainder of the dwellings were erected after the cardinal's interest in the town had ended. Richelieu has retained the complete circuit of its wall and gateways which probably had more of an aesthetic purpose than real military significance. The moat, originally about 70 feet in width and 10 feet deep, has been partially filled in.[45]

With the death of its founder and the subsequent demolition of the palace, the *raison d'être* of Richelieu was removed. From an estimated maximum population of around 5,000 to 6,000 the town steadily declined in size and importance. Today it exists as a clear illustration of Renaissance planning principles, with only relatively small suburban accretions of recent years. Richelieu can thus be compared to medieval Winchelsea in England (see Chapter 4), as the Renaissance counterpart of that bastide port which failed to survive the silting up of its harbour.

The work of Le Nôtre

The pre-eminent names in the history of landscape architecture are André Le Nôtre (1613–1700), whose greatest achievement was the layout of the park and town at Versailles, and Lancelot 'Capability' Brown (1716–83), perhaps best known for his work at Blenheim Palace.[46] Of the two Le Nôtre had by far the greater impact on the

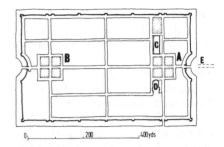

Figure 6.25 – Richelieu: general plan of the town as laid out by Jacques Lemercier in 1631. Key: A, Place du Marché; B, Place des Religieuses; C, the covered market; D, the church; E, the road leading, originally, to the château.

14 September 1644 – We took post for Richelieu, passing by l'Isle Bouchard, a village on the way. The next day, we arrived and went to see the Cardinal's palace, near it. The town is built in a low, marshy ground, having a narrow river cut by hand, very even and straight, capable of bringing up a small vessel. It consists of only one considerable street, the houses on both sides (as indeed throughout the town) built exactly in uniform, after a modern handsome design. It has a large goodly market-house and place, opposite to which is the church built of freestone, having two pyramids of stone, which stand hollow from the towers. The church is well-built and of a well-ordered architecture, within handsomely paved and adorned. To this place belongs an academy, where, besides, the exercise of the horse, arms, dancing, &c., all the sciences are taught in the vulgar French by professors stipendiated by the great Cardinal, who by this, the cheap living there, and divers privileges, not only designed the improvement of the vulgar language, but to draw people and strangers to the town; but since the Cardinal's death, it is thinly inhabited, standing so much out of the way and in a place not well situated for health, or pleasure. He was allured to build by the name of the place, and an old house there belonging to his ancestors. This pretty town is handsomely walled about and moated, with a kind of slight fortification, two fair gates and drawbridges. Before the gate, towards the palace, is a spacious circle, where the fair is annually kept. About a flight-shot from the town is the Cardinal's house, a princely pile, though on an old design, not altogether Gothic, but mixed, and environed by a clear moat. (J Evelyn, Diary)

44. J Evelyn, *Diary*, 1644.
45. See Lavedan, *Histoire de l'Urbanisme: Renaissance et Temps Modernes*, for the map relating the town and the palace, which was alongside the axial avenue ('E' in Figure 6.25) and not, as might have been expected, at its end. The town, therefore, enjoyed a certain symbolic independence; rather than being completely subservient to the palace, had it been located on the same axis.
46. See D Clifford, *History of Garden Design*, 1966, as a general history of landscape architecture; see also D Stroud, *Capability Brown*, 1975; and H M Fox, *André Le Nôtre: Garden Architect to Kings*, 1962, for detailed individual studies.

related field of urbanism (although his output of work was very much less). Le Nôtre, however, had the unique advantage of working in seventeenth century France where profligate expenditure of national resources on personal aggrandizement finds few parallels in all history. Two all-powerful clients gave Le Nôtre the opportunities to design on an enormous scale, dwarfing the landscaping achievements of all his predecessors and hardly equalled by subsequent work. These two commissions were at Vaux-le-Vicomte, for Nicholas Fouquet, and his continuous involvement with Versailles for Louis XIV.

Le Nôtre belonged to a family of landscape gardeners and designers. His grandfather had been an under-gardener, and his father chief-gardener, at the Tuileries and in 1637 at the age of 24, he was promised the succession to his father's post. Eight years earlier, in 1629, his father had been employed laying out the park for Cardinal Richelieu's chateau adjoining his new town, and it was here that young André probably gained his first experience of large-scale landscape designing. In 1649 he is known to have been a salaried designer in the royal gardens before being appointed Controller-General of Royal Buildings in 1657.

Le Nôtre's first chance to create a landscape park was given by Fouquet at his ill-fated chateau, Vaux-le-Vicomte.[47] Le Brun, the interior designer of the chateau itself, had been a painting student with Le Nôtre and recommended his friend for the work of creating a landscape setting for the building, and its related open-air social activities. The function of the landscape was 'not to be a spot in which a cultured man might take pleasure with his friends – the need which inspired the gardens of Medicis; still less was its principal purpose to supply seclusion for repose or love; it was primarily to be a stupendous theatre for fêtes.[48] The scale of operation, though less than at Versailles, presaged it. Three villages were demolished because they were in the way, and at times as many as 18,000 labourers were put to work.

It was at Vaux-le-Vicomte that Le Nôtre evolved his landscaping method. As summarized by Derek Clifford in his *History of Garden Design*, 'he seized on one great principle – that the whole extent of the enormous garden should be visible at a glance; accordingly whatever variety there might be within the parts, the parts themselves were to be subordinated to the whole. If the garden was to be seen at a glance it must be relatively narrow, but as it must be impressive by sheer size it had also to be long; the eye of a man on the uppermost terrace can look on and on, into the distance, but must not be asked to move from side to side'. The main vista, contained within densely planted flanking woods, is therefore arranged to be seen from the highest of a sequence of descending terraces. In addition to framing the vista these woods embrace secondary smaller gardens. Here groups of statues, pools and fountains provide alternative and more intimately scaled visual experiences to those of the dominant directional space.

VERSAILLES

In 1624 Louis XIII purchased an area of land near the small village of Versailles, some miles south-west of Paris. With plans attributed to Salomon de Brosse he built a royal hunting-box; this was a modest building of brick, with stone trim, and with a great slate roof such as could be seen at the time in Paris in the Place des Vosges.[49] The plan followed the standard French pattern: a main central block was flanked

Long before Le Nôtre, the religion, philosophy and way of life of a people had been expressed in the form of a garden in China and Japan, and again in Persia and India. This had occurred in Italy, too, and was to happen again in England. In each of these countries many artists designed the gardens. But in France it was only one artist, André Le Nôtre, who expressed the civilisation of his time through the medium of his work, and did it so vividly, aptly and brilliantly that his gardens are the perfect symbol of his era.
(H M Fox, André Le Nôtre: Garden Architect to Kings)

The rigid etiquette that Louis imposed on his court should be judged as the expression not of pettiness of mind but rather of political calculation. The object was to provide the necessary setting for a monarch who was to be the centre of the nation's life with all eyes turned on him. The court was a permanent spectacle for the people: the life of the king passed from birth to death in public. Louis XIV would as soon have neglected his council as the 'grand couvert', at which he dined in the presence of his subjects. Ill and able to do no more than sip a little moisture, he forced himself to the ceremony for the last time on 24 August 1715. He died on 1 September. A rigorous etiquette was needed if the impression was to be one of majesty and not of confusion. Given this, and a king like Louis XIV to play the principal part, and the court became the scene of a perpetual ballet performed before an audience of twenty million. A more classic background was needed for this than the ramshackle royal quarters in Paris. Although other reasons have been given why Louis left Paris for Versailles – the aversion from his capital induced by the troubles of the Fronde, his love of walking and hunting, the desire to relieve the royal mistresses of the embarrassment of life in a large city – not the least was the need to provide a glorious setting for the court. Versailles was described later by Voltaire as a great caravanserai filled with human discomfort and misery – he himself had been honoured for a time with a little room in the palace just above the privies – but the misery was not on view and the grandeur was.
(A Cobban, A History of Modern France)

47. See M Mosser and G Teysot (eds) *The History of Garden Design: The Western Tradition from the Renaissance to the Present Day*, 1991.
48. Clifford, *History of Garden Design*.
49. As a definitive general history of Versailles see J Custex, P Celeste and P Panerai, *Lecture d'une ville: Versailles*, 1979, which includes a sequence of maps showing the growth of the town in 1654, 1746, 1870 and 1907.

on either side by lower wings and these were joined at their ends by a low arcade to form an entrance courtyard on the Paris side. On the other side, a fraction of the extent of the present-day park, were small, formally laid-out gardens. Louis XIII frequently stayed at Versailles, both to hunt and to pursue his chaste love affairs. The village gradually expanded to accommodate visitors' servants and all the increasing ancillary services and industries. But only in the France of the seventeenth and early eighteenth centuries could the subsequent developments have taken place. Over the years the modest hunting-box became inflated into history's grandest palace – more than one quarter of a mile in length – with decoratively planted gardens contained by sweeping landscaped perspectives of a suitably magnificent park. The village expanded into a sizeable town, dependent on the palace for its livelihood, its three main streets constituting grand urban approaches and complementing the naturally formed spaces over the hill (Figure 6.26). All this was possible only because of the absolute power enjoyed by the monarchy during the 72-year reign of Louis XIV (1643–1715). Versailles was 'the place he designed for his magnificence, in order to show by its adornment what a great king can do when he spares nothing to satisfy his wishes'.[50]

Louis XIV had little altered his father's hunting-box when in 1661 he first set eyes on the magnificent combination of palace and landscaped park built by his finance minister, Nicholas Fouquet, at Vaux-le-Vicomte. Envious of this achievement, and suspicious as to how it had been paid for, Louis had Fouquet imprisoned and immediately requisitioned the services of his design team to create an even more magnificent scheme at Versailles. These designers were the architect, Le Vau, Le Brun, the interior designer, and André Le Nôtre. Le Nôtre was the first to see designs completed, replacing the earlier gardens with the vast park, the basic pattern of which, in spite of modification in detail, has never been altered. Lavedan describes it as 'a great vista flanked by two shrubberies and following the central axis of the palace, with symmetrical arrangements of lawns and flowers on either side of this line; statues everywhere, the orangery built by Le Vau, and a menagerie for rare animals. This was the Versailles of Louis XIV's youth, where Molière came to act his plays, the Versailles of the king's love affair with Mlle de la Vallière'.[51] Vaux-le-Vicomte served as a tree nursery and it was Fouquet's plantings which graced the new orangery. Two grand fêtes in honour of the king's mistresses were staged in the gardens, in 1664 and 1668, when over 3,000 guests were entertained on a lavish scale.[52]

The scope of the original house was by now inappropriate for such a setting, and completely inadequate to meet the demands of court festivities. It was now the turn of Le Vau and Le Brun to create an architectural masterpiece complementing Le Nôtre's park. The king, however, would not tolerate demolition of his father's original building and eventually it was decided to extend around it on three sides leaving intact only the original elevation on to the eastern entrance court. Accordingly between 1668 and 1671 Le Vau added extensions to the wings flanking the court – the Cour de Marbre – and constructed the immense central block of the present-day palace, facing the park. At the same time Le Brun completely remodelled the interiors.

Louis XIV had never felt at ease living in the Louvre in Paris; as an impressionable child he had experienced at first hand the latent power

The sedan chairs which carried people from one part of the château to another belonged to a company, like hackney cabs; none but the royal family were allowed to have their own. They were not allowed to go further into the King's part of the house than the guard-rooms and never allowed in the Cour de Marbre. They made tremendous traffic blocks in the Nobles' wing. One of the corridors there was called the Rue de Noailles, as its whole length gave onto flats occupied by that powerful but unpopular family. Such as they lived in splendour, but more humble folk could not be said to be well or comfortably lodged – in many cases the rooms they lived in had been chopped into tiny units with no regard for the façade – some had no windows at all, or gave on to dismal little interior wells. All the same, a lodging, however squalid, in the château, came to be more sought after than almost anything, as it was a sign of having succeeded in life. Those who could afford to also had houses or flats in the towns of Versailles; and the very rich began building themselves seats in the surrounding country.
(N Mitford, The Sun King)

The King treated the presence of the greater nobles at Versailles and Fontainebleau as a parade, and if he found any were missing he enquired sharply about the reasons for their absence. But the function of the nobility was almost totally ornamental. They were strictly excluded from all ministerial duties and from the royal counsels, and they were not allowed to do honest work of any kind. The only thing they were permitted to do was to fight, and consequently the more energetic among them pressed the King to go to war at frequent intervals. At the same time, although they were exempt from taxation, the French nobles had expensive estates to keep up and many onerous ceremonial functions to fulfil. The King seems deliberately to have encouraged them to indulge in the most extravagant luxuries. Thus, unless they were lucky enough to make a rich marriage, many of them were in due course ruined.
(M Ashley, Louis XIV and the Greatness of France)

Through building an all-embracing palace and an artificial city on a wasteland and living and working in it, surrounded chiefly by servile ministers and idle courtiers, Louis XIV cut himself off from his people, and, like Philip II of Spain, was unable to compensate by hours of unremitting toil at his desk for lack of contact with the realities of the everyday life of his subjects. As that profound and witty French historian, M. Lavisse, observed: 'The great events of the reign are not always those which at once spring to the mind. The establishment at Versailles was more important and had graver consequences than any of Louis XIV's wars or all of his wars put together.'
(M Ashley, Louis XIV and the Greatness of France)

50. M Ashley, Louis XIV and the Greatness of France, 1946; see also N Mitford, The Sun King, 1966, for a fascinating insight into everyday life at the court of Louis XIV.
51. Lavedan, French Architecture.
52. See Mitford, The Sun King, for an account of the other side of lavish court entertainment. Although there were public toilets available in the palace, it was a predominantly chamber-pot society.

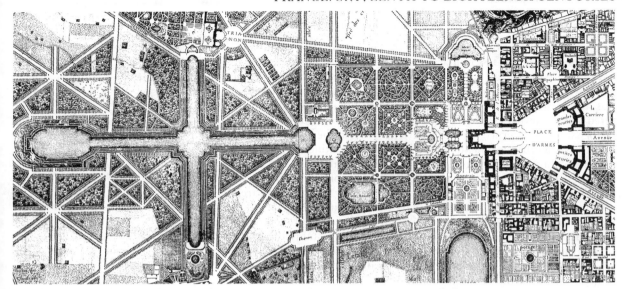

of the mob, which during the uprising known as the Fronde had proved so threatening that the boy king and his mother had been obliged to flee from the city. He therefore preferred to spend increasingly lengthy periods at Versailles and, around 1678, took the decision to expand still further the palace and town of Versailles, this time to a size where they could accommodate the entire court and government departments on a permanent basis. J-H. Mansart was the architect. He further extended the length of the palace by adding the south, or prince's, wing (1679–82) and the balancing north wing (after 1684). In 1682 this work was sufficiently far advanced for the Court to move out from Paris.

The work of laying out the town of Versailles had been proceeding concurrently with the palace expansions. The basis of the town plan is a gridiron, on which is superimposed three broad avenues which are centred on the king's bedroom in the middle of what remained of the original Louis XIII block. Between the palace and the town there is the vast expanse of the Place d'Armes – a regular trapezium some 1,300 feet across by 600 feet deep. Facing the palace across this space in the angles formed by the three avenues, J-H Mansart added the Royal Mews, barracks and other court offices. The horizontal scale is so vast as to make the whole composition meaningless, except when seen from the king's bedroom windows. Versailles received its charter in 1671 and had grown to about 30,000 population when Louis XIV died in 1715.

It is not known for certain who was responsible for the structure of the town but it is likely that Le Nôtre was significantly involved at some early stage. There is a direct connection between the use of the radiating routes motif in the town and much of the detail layout of the park; moreover, earlier in his career Le Nôtre had also been involved with similar radial axis planning of the Champs Elysées and the Cours-la-Reine, west of the Tuileries Gardens in Paris. This radial routes motif, which can be directly traced back to the plan of the Piazza del Popolo in Rome, in combination with the gridiron, also plays a key role in the layout of Washington DC, in the USA (see Chapter 10). It is highly significant that Pierre L'Enfant, the planner of Washington, spent his childhood in the park and town of Versailles.[53]

Figure 6.26 – Versailles: detail plan of palace, park and town in 1746 (north at the top). The horizontal scale is vast indeed: from the west front of the palace to the near end of the cruciform-plan lake is 430 *toises* (985 yd); from the palace to the central crossing of the lake is 750 *toises* (820 yd); the cross-arms of the lake are 525 *toises* (1,095 yd) across. However, the visitor was not intended to experience the park on foot; the intended time and motion scale was 10–15 miles per hour – the leisurely speed of a horse-drawn carriage. To *walk* from the palace to the near end of the lake can be tiring, even eventually boring. To walk on from there, round the lake to the distant horizon through dusty all but deserted *allées* is a further 1¾ ml by the shortest route.

As Derek Clifford describes it, 'the main plan of the garden was as it still is. The area occupied by it was roughly rectangular, but the impression to the eye was of an enormous vista stretching from the façade for three-quarters of a mile to the beginning of the Grand Canal, which in turn diminished to the horizon, suppressing any desire to explore its remote distances. In the immediate neighbourhood of the palace the home terrace gave laterally on to *parterres*, so that to right and to left impressive pictures appear, one downward to the famous fountain of Neptune, the other across the enormous orangery to the lake that was dug by the Swiss Guards. To ensure that the effect of distance was not dissipated Le Nôtre repeated the device of flanking the central axis with groves as at Richelieu and Vaux, although here it was done on an even vaster scale with even more triumph effect' (*A History of Garden Design*).

Compared with the palace and the park, the plan of the town of Versailles is characteristically neglected in urban histories as a poor third. Yet, in its combination of approach radials, to the glory of the monarch, and an expediency surveyor's grid, it is of paramount importance generally and more so in respect of L'Enfant's plan for Washington, DC (see Note 53 and Chapter 10).

53. The probable formative influence of the Versailles town plan on Pierre L'Enfant is described in Chapter 10.

The work of Sébastien Le Prestre de Vauban (1633–1707)

As a result of his lifetime's work in the service of Louis XIV, Vauban is recognized as the greatest military engineer in history. He is reputed to have taken part in nearly 150 battles, directed 53 sieges, fortified some 300 towns and built more than 30 new ones. His skill was such that it was acknowledged that 'a town besieged by Vauban was a town captured and a town defended by Vauban was impregnable'.[54] In this history we are concerned primarily with his work in the development of the science of urban fortification, and its effect on continental European urbanism during his lifetime and in the eighteenth and nineteenth centuries. His interests, however, extended far beyond this work, leading Voltaire to regard him as 'a genius of the age'. Vauban's career involved him in many other aspects of national life, including establishment of the French internal waterways sytem, advocacy of methods of overseas territorial expansion[55] and army reforms – notably the invention of the bayonet, which greatly increased the effectiveness of infantry. At the end of his life he also published a book concerned with radical reforms of the taxation system.

In 1651 the 18-year-old Vauban entered army service as a cadet, having a fair smattering of mathematics and fortification and being at the same time a passable draughtsman.[56] Four years later he received his commission as an engineer in ordinary to Louis XIV. The king had two main advisers – Louvois, responsible for war and the defence of the French frontiers, and Colbert, responsible for peace and the coasts and provinces. Vauban worked for Louvois, beginning that ceaseless tour of the frontiers which lasted the whole of his life, and which attained an epic grandeur. 'From Antibes to Dunkirk he journeyed unceasingly, studying the lie of the land, working on every kind of terrain, the rocky shores of the Mediterranean, the sand dunes of the north, the Alpine escarpments and the inundated tracts of Flanders.'[57]

At first he was mainly concerned with offensive military tactics for besieging fortress towns, but was also soon involved in the creation of new, improved defences to replace those he had successfully destroyed. In 1667 he was with Louis XIV in The Netherlands organizing successful assaults on Tournai, Douai and Lille, among several others. The subsequent reconstruction of the town and citadel at Lille in 1668–9 is considered by Sir Reginald Blomfield in his *Sébastien le Prestre de Vauban* as one of Vauban's most successful combinations of fortifications and urban planning.

Vauban himself regarded Maubeuge as one of his major achievements. His last important project was the building of Neuf Brisach; this small fortress town on the Rhine frontier is the clearest example of his methods as applied to a new site and is described separately later in this chapter. With Vauban, military requirements dominated all others in the layout of towns and their defences and he was 'much more interested in giving shape to the varying polygonal outlines of his fortifications than he was in the inner structure of the town. When Vauban remodelled existing settlements, he did not change the old street system of the inner town but confined his work to the construction of its fortifications. In his newly built fortress towns he took over the simplified gridiron scheme of the French mediaeval bastides with a quadrangular square in the centre, combining it with basic ideas of the Italian sixteenth-century theoreticians'.[58]

In 1687 Vauban invented a socket by which a bayonet could be attached to the musket without interfering with its firing. By this means pikemen could be abolished and the effectiveness of the infantry doubled ... at the same time the artillery and engineers began to come into their own largely through the exertions of Vauban, who made use of them in siege warfare, which became the principal feature of the many Flanders campaigns. In conducting a siege Vauban's method was first to surround the fortress which was being attacked with parallel lines of entrenchments and then to launch from them mortar bombs, the range of which was calculated with mathematical accuracy, upon the enemy forces. The whole plan of a siege, with the prescribed entrenchments, sapping, and mortaring, was usually worked out in such precise detail that the date of the final assault and capitulation could be exactly estimated in advance. Ladies would be invited as witnesses of the last stages of a siege, and the final assault would take place to the accompaniment of violins. Louis XIV loved a good siege – the bigger the better – and would graciously accept the credit for all Vauban's hard work.
(M Ashley, Louis XIV and the Greatness of France)

Despite his reputation as a fortress engineer, Vauban's major talent is revealed in his mastery of siegecraft. His influence on this type of operation persisted long after his death; the reduction of Antwerp, then held by the Dutch, by a French army in 1832 was virtually a textbook example of a Vauban siege operation.

As a designer of fortifications, Vauban is credited with three so-called 'systems', though the classification is not his; he once wrote that 'the art of fortifying does not consist in rules and systems, but solely in common sense and experience.'

His 'First System' differed little from the designs emanating from contemporary French and Italian engineers and has much in common with the elegant simplicities of Pagan's style. It consisted of a polygonal trace with bastions at the corners; there was a set standard dimension for the length of one front of fortification, that is, the distance between the salient angles of two adjacent bastions, so that it could be adjusted to conform with any given area. The front of fortification was fixed at 330 m (360 yd), and the dimensions of other features – the flanks and faces of the bastions, the width of the ditch, dimensions of ravelins and so forth – were specified as fractions of this basic measure. Where it was necessary to plan a larger or smaller work, the front of fortification was first measured and the various parts were then calculated on the same fractional relationship, so that every part was correctly proportioned. The First System also saw the return of the orillon or 'ear' on the shoulder of the bastion, formed by recessing the flank. Invented by Italian engineers, the orillon had fallen into disuse, but Vauban advocated it as a means of 'taking a breach in reverse', that is of firing on

54. R Blomfield, *Sébastien le Prestre de Vauban*. See also C Duffy, *The Fortress in the Age of Vauban and Frederick the Great 1660–1789*, 1985.
55. Vauban's proposals for the development of North American colonies on an agricultural basis, receiving a continuous flow of emigrants, could well have seen France much more firmly established, and better able to withstand British pre-emption than was to be the case (see Chapter 10).
56, 57. Blomfield, *Sébastien le Preste de Vauban*.
58. Zucker, *Town and Square*.

VAUBAN AT VERSAILLES

One of Vauban's least successful tasks, in the sense that he failed to convince Louis XIV of the merits of his counter-proposal, involved him in the ill-fated attempt to provide an adequate water supply for the gardens at Versailles. The idea of others had been to divert the River Eure to Versailles from a point some 50 miles to the south. By 1685 canals had been dug on both sides as far as the valley of the River Maintenon; the army, otherwise unoccupied during a period of peace, providing the massive labour force. The major problem was how to take the water across the valley and Vauban was assigned to this task in 1686. His was an original yet perfectly possible solution, based on the creation of an enormous siphon using cast-iron pipes. This, however, would have been far too unassuming visually for the king, who preferred a rival scheme for a gigantic aqueduct across the valley, surpassing the Pont du Gard and anything the Romans had built.[59] In 1686 work started on this aqueduct, which was to have been 5,000 yards in length. A division of 20,000 infantry encamped in the marshes to build it, but in these unhealthy conditions so many of the soldiers died that by the king's order nobody was allowed to speak of it.[60] Eventually a renewal of hostilities in 1688 brought an end to the project: only the lowest arcade had been completed, with 47 arches varying between 42 and 48 feet in height giving a length of 1,062 yards. The ruins of the aqueduct still stand in the valley of the Maintenon, and serve as a grim monument to the disastrous vanity and futility of Louis XIV.[61]

NEUF BRISACH

In 1967 the Treaty of Ryswick finally forced France to withdraw from the line of frontier fortresses which she had established on the eastern bank of the Rhine during the half-century after the end of the Thirty Years War (1618–48). One of the most important of these fortresses was Alt Brisach, which controlled the principal river crossing on the long stretch of river between Strasbourg and the frontier with the Swiss Cantons above Basle. A replacement west-bank fortress was required as quickly as possible.

Vauban selected a site above the seasonal flood level, just out of range of the guns in Alt Brisach on higher ground, some four miles away to the north-east, across the river. The location was the meeting point of important roads to Strasbourg, Colmar and Basle, as well as the Rhine crossing leading to Freiburg in Germany. In 1700 the Rhine here was a multi-branched river; the crossing consisted of several short lengths of bridge linking together a number of islands. On the largest of these the French had built a regularly planned *ville-neuve* during their period of occupation. A condition of the treaty required this town to be demolished, and it subsequently served as a source of building materials for Neuf Brisach. The first and second plans prepared by Vauban were rejected by Louis XIV; permission was given to proceed with a third in 1698. The military functions of Neuf Brisach were twofold: it was to serve as a strongly fortified garrison town, from which forces could be deployed as required and its guns controlled the river crossing. In recognizing the potential weaknesses of a civilian population in a military stronghold, Vauban omitted it from his first plan but this was rejected as impractical.[62] The intricate system of fortifications was constructed by 1708, but the town within was not finished until

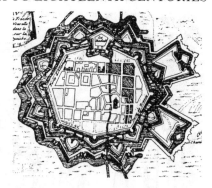

Figure 6.27 – Landau: an existing town, somewhat 'squared-up', within a typical Vauban fortification system. (See Duffy, *The Fortress in the Age of Vauban and Frederick the Great 1660–1789*, for a detailed plan of Landau and its fortifications.)

the rear of a storming party attacking a breach in the ramparts. Other features of the system included the use of the tenaille, a low outwork in front of the rampart between bastions, bonettes, covering the salient of the ravelin, and lunettes to guard the ravelin's flanks.
(I Hogg, *The History of Fortification, 1981*)

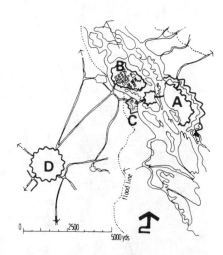

Figure 6.28 – Neuf Brisach: location of the Vauban fortress town related to the Rhine, which in the late seventeenth century had a considerably wider flood-plain than at the present day. Key: A, Alt Brisach; B, the French *ville-neuve*; C, left-bank bridgehead fortress; D, Neuf Brisach.

59, 60, 61. E Halevy, *Vauban*, 1924.
62. C Cooke, C Hennessey and D Wardlaw, 'Neuf Brisac', *RIBA Journal*, February 1965. See also Duffy, *The Fortress in the Age of Vauban and Frederick the Great 1660–1789*.

Figure 6.29 – Neuf Brisach and the surrounding countryside from the air (north at the top); the Rhine on the far right. This view also shows the strikingly contrasted form of several organic growth villages and the fascinating strip-farming patterns which, with the rural routes, comprise a typical pre-urban cadastre in the event of village expansions.

1772 when the church was completed. The quality of workmanship for the retaining wall section of the earthbanks was of a very high standard, and, with minimal restoration in the past, the completed ring has survived through to the present day as probably the best remaining example of eighteenth-century military engineering. Although densely overgrown in parts the original form is clearly visible.[63]

Within the defences the town was laid out to a typical engineer's grid of expediency. Nine streets in each direction, including the ring road inside the ramparts, form regular building blocks, around the central four-block Place d'Armes. This central space was primarily intended as the assembly area for the garrison: aesthetic considerations were dis-

63. When visited at Easter 1969, soldiers from the military garrison were clearing away the dense undergrowth from the fortifications, which have survived in good condition.

counted in its layout and the surrounding buildings are far too small in scale to be able to do more than define the extent of a bleak, somewhat anonymous, gravelled parade ground. The town church is in the north-east corner of the square, and the market place and town hall form a separate, much more happily scaled public space off the south-east corner. The four main roads into the town are brought in through the fortifications to the centre of the Place d'Armes.

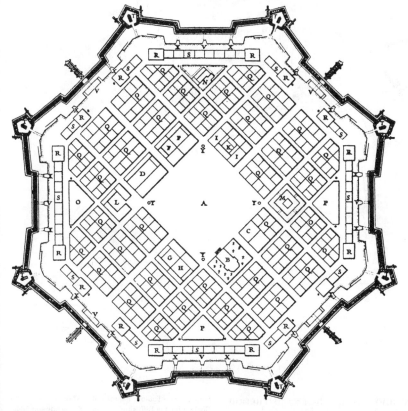

Neuf-Brisach, on the French frontier with Switzerland, is an almost perfectly preserved example of Vauban's Third System and was his last major work of fortification. Under the terms of the Treaty of Ryswick, Louis XIV had to surrender his existing fortress of Alt-Brisach, and he decided to replace it with a new work to cover the crossing over the Rhine. Vauban submitted three plans in 1698, and the fortress to the chosen plan was completed in 1708. The estimated cost was 4,048,875 livres. It was originally designed simply as a fortress; the town was built later.

Starting from the interior of the town, there is a 20-m (66-ft) wide rampart, a 20-m ditch, the 20-m wide tenaille, another 20-m ditch, a 22-m (72-ft) wide redoubt, a 9-m (30-ft) moat, a 35-m (114-ft) demi-lune, a 22-m ditch, a 9-m covered way, and finally a 36.5-m (120-ft) glacis, a set of dimensions which gives a good idea of the scale of Vauban's planning.

Neuf-Brisach was the only work in which Vauban's Third System was used. The key to the System is the use of 'tower bastions' which can be seen extending from the salient angles of the main bastions in this drawing. These towers are 4-m (13-ft) thick at the base, appearing to 2.5-m (8-ft) at the top, and have two storeys topped by a flat roof with parapet and gun embrasures. Inside each tower is a central magazine surrounded by a vaulted corridor, staircases to the roof and flues to carry off the gunsmoke from the casemates.

The fortress was never troubled by war until 1870, when it was besieged by the Prussian Army. after 34 days a more modern fort nearby surrendered, leaving Neuf-Brisach exposed to artillery fire. After two more days the garrison surrendered.

(I Hogg, The History of Fortification)

Figure 6.30 – General plan of Neuf Brisach, within the wall-proper of the defensive system. See the facing aerial view for the extent of the enveloping masonry and earthwork fortification; see also Duffy, The Fortress in the Age of Vauban and Frederick the Great 1660–1789, for ground-level photographs.

TOULON

Both Toulon, on the Mediterranean coast, and Le Havre, at the mouth of the Seine, came to prominence as French naval bases during the seventeenth century when Cardinal Richelieu established a regular French royal navy.[64] In their expanded nineteenth-century forms illustrated on these pages, both clearly exhibit the importance of their defensive systems. Construction of the harbour at Toulon commenced in 1589 and the two moles forming the Vieille Darse (or Port Marchand) were completed in 1610. Between 1680 and 1700 Vauban was responsible for the addition of the Darse Neuve (or Port Neuve) as a new royal dockyard and the Vieille Darse became the commercial harbour. After 1837 – the date of the map illustrated in Figure 6.31 – two other naval docks were constructed further west; the Darse de Castigneau under Louis Napoleon (from 1852), and the Darse de Missiessy under Napoleon III (from 1862).

LE HAVRE

Le Havre is of exceptional importance in the history of French urbanism for two main reasons; first, its foundation as a new planned port

64. For Toulon see Branch, Comparative Urban Design.

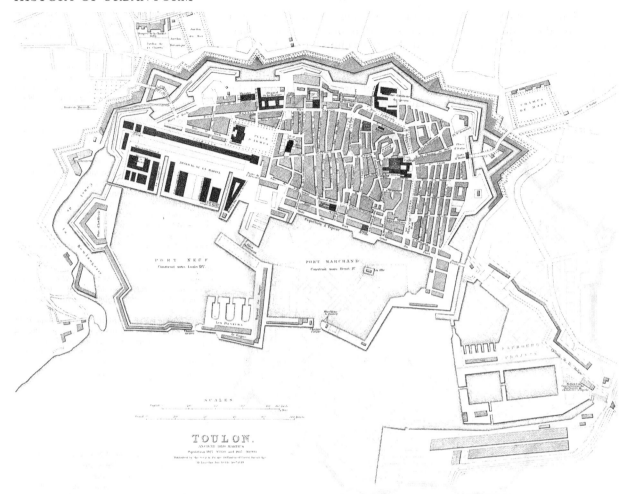

TOULON.

and town in the early sixteenth century; and second, the reconstruction of the old central part of the town after the almost complete devastation of the Second World War, on the basis of the plan drawn up by Auguste Perret. This present account takes development up to 1838, the date of the plan illustrated in Figure 6.32. This area approximates to that of the Perret plan – a minor masterpiece of modern town planning, which has been largely overlooked.[65]

Le Havre was a replacement harbour, like Winchelsea and Kingston upon Hull (described in Chapter 4) and resulted from the silting-up of the harbour of Harfleur. That ancient port at the mouth of the Seine in turn owed its medieval prosperity to the similar fate suffered by Lillebonne (Juliobona), the Roman predecessor further up the estuary. Harfleur did not readily surrender its commercial life and expensive canal works were carried out in the 1290s, and also from 1469, in order to provide for sea-going ships. Silting-up continued, however, and when in 1509 the French crown would not assist the with the costs of a third canal project, Harfleur was doomed.[66]

A survey of possible locations for a new harbour was undertaken, on the basis of which it was decided to develop a lagoon, protected by a sandbank, a mile and a half nearer the mouth of the estuary. On the

Figure 6.31 – Toulon: the map of 1837 published by the Society for the Diffusion of Useful Knowledge in 1840. The Arsenal Maritime of Toulon expanded to the west, with the addition of the two new dock basins; the Port Marchand developed to the south in the area of the projected *faubourg*, beyond which the Arsenal du Mourillon was developed from 1836. An extensive Arsenal d'Artillerie was also added to the north of the 1837 fortifications.

The Arsenal d'Artillerie retains its external fortifications; the only other surviving bastions are the two on the eastern side of the town flanking the present-day Porte d'Italie. With few exceptions, the street pattern within the fortifications remains.

65. For the Perret Plan for the Reconstruction of Central Le Havre, see P Collins, *Concrete: The Vision of a New Architecture*, 1959; P Dalloz, *Auguste Perret and the Reconstruction of Le Havre*, Casabella (Milan) April–May 1957.
66. The strategic importance of Le Havre was lost to Cherbourg, which was given a ring of outer forts comparable to those constructed around Portsmouth on the English coast (see Figure 8.39).

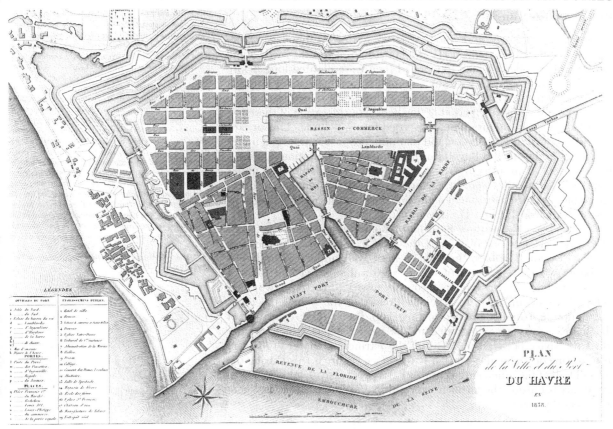

Figure 6.32 – Le Havre: the map of 1838 which relates to the diagrammatic growth map of Figure 6.33. The extent and complexity of the fortifications, embodying three lines of moats along the northern side, is remarkable and is proof of the importance attached by the French government to the port, in face of potential British assault. This fear prompted their retention through into the 1850s and 1860s with consequent constraining effect on the growth and prosperity of the commercial port.

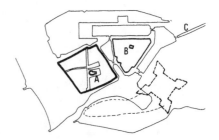

Figure 6.33 – Le Havre: the successive growth stages through to the map of 1838. Key: A, Church of Notre-Dame; B, Church of St-François; C, the Vauban Canal to Harfleur (now the location of the Basin Vauban). The citadel site is occupied by the modern Basin de la Citadelle. The extent of the original sandbank forming the lagoon is conjectural.

sandbank a small chapel had been built in 1189, known as Notre-Dame-de-Grâce, after which the new town took its name as Le Havre de Grâce. In 1517 Francis I commissioned Guillaume Gouffier, Admiral of France, to commence work on the port and the fortifications of the new town. A new entrance into the lagoon was cut; two quays were constructed on either side of the new harbour, that on the northern, town side subsequently becoming the present-day Quai de Southampton (where the cross-channel ferries dock); and both quays were extended seawards with twin towers commanding the entrance. The extent of the first town of Le Havre is shown in Figure 6.33, and the street layout of this original Quartier Notre-Dame was essentially that of the 1838 map.

A first extension of the harbour and its fortified urban area was carried out during the 1540s to plans prepared by Girolamo Bellamarti Sienne, who introduced into France from Italy the new science of fortification design based on projecting angle bastions, and which was used for the first time at Le Havre (see section on fortifications in Chapter 5). Bellamarti laid out the New Quartier Saint-François, added new harbour basins and refortified the town. A first citadel was commenced in 1571 and completed to a more modern design under Cardinal Richelieu. The second extension, which completed the form of 1838, was carried out to plans of François-Laurent Lamandé from 1787. The new fortifications of that period were retained until 1854, along the northern front, and until 1864 to the east.

Parisian postscript

'The history of Paris as a modern city,' writes Anthony Sutcliffe, 'clearly begins with Napoleon III and Haussmann. But the seeds of the improvement policies which burgeoned then were planted long before in the reign of Louis XVI'. Thus although several aspects of Parisian urbanism of the late eighteenth and early nineteenth centuries could be seen as falling within the period of this book, consideration of these has been held over for inclusion in a proposed future volume. This urbanism, which includes legislation and planning proposals prepared under Louis XVI, during the Revolution (including the ambitious plan for Paris produced in 1793 by the Commission of Artists), and under Napoleon Bonaparte, would therefore constitute the immediate background for detailed assessment of the mid-nineteenth-century programmes of Napoleon III and Haussmann.[67]

By the turn of the century Paris had a population of some 550,000 – probably well over one-quarter of the total French urban population which, to quote Cobban, 'must have been well under two millions. Probably some 95 per cent of France's 26 millions lived in isolated farms, hamlets, villages and small country towns'.[68] In 1851, two years before Haussmann took up office as Prefect of the Seine, the city's population had almost doubled to 1,050,000. By far the greater part of the increase was accommodated in the new districts which rapidly filled the space between the Wall of the Farmers General (1784–91) and the line of the fortifications of 1840–5 (see Figure 6.13). During the second half of the nineteenth century, Paris progressively annexed suburban areas outside the fortifications and had attained a population of 2,715,000 by 1900.

The provincial cities were not important enough to receive much imperial attention, and there the eighteenth century largely survived, but considerable steps were taken towards the spoiling of Paris. Napoleon required a grandiose setting for the capital of his Empire, and of course a classical one. Triumphal arches – such as the Etoile and the Carrousel – were 'de rigueur'. For public buildings the correct thing was temples; so we have the temple of finance – the Bourse with its sixty-four Corinthian columns, the temple of religion – the Madeleine, the temple of the laws – the palais Bourbon, all heavy pastiches and all equally unsuited to the purposes for which they were intended. The proportions of the place Vendôme were ruined by sticking in its middle, in place of the destroyed royal statue, a monstrous column, in imitation of that of Trajan. To enable the Napoleonic monuments to be seen, the process of driving long straight roads through Paris, which was to be carried much further under the Second Empire, was begun. Unlike the lath and plaster erections of the Revolution, the buildings of the Empire were made to last – unfortunately, for they embody too well the Emperor's chief aesthetic rule: 'Ce qui est grand est toujours beau'.
(A Cobban, A History of Modern France: Volume Two 1799–1945).

67. For Paris, and France generally, during the later nineteenth and twentieth centuries, see A Sutcliffe, *The Autumn of Planning in Central Paris*, 1970; A Sutcliffe, *Towards the Planned City 1780–1914*, 1981; A Cobban, *A History of Modern France: Volume Two 1799–1945*, 1961; H Malet, *Le Baron Haussmann et la renovation de Paris*, 1973; N Evenson, *Paris: A Century of Change 1878–1978*, 1979; P Lavedan, *Histoire de l'Urbanisme: Renaissance et Temps Modernes*, 1959; G Duby (dir.) and M Agulhon, *La Ville de l'Age Industriel, Volume 4, Histoire de la France Urbaine*, 1983; G Duby (dir.) and J Brun et al, *La Ville Aujourd'hui, Volume 5, Histoire de la France Urbaine*, 1985.
68. A Cobban, *A History of Modern France, Volume 2, 1799–1945*.

7 – A European Survey

In European countries other than Italy, France, Spain and Britain, which are given individual chapter coverage, significant urban developments of the fifteenth to eighteenth centuries (and later in a few exceptional instances) either were concentrated in the capital cities – which characteristically were undergoing extremely disproportionate growth relative to provincial urban centres – and economically favoured provincial centres, or, in a limited number of instances, took the form of new towns and cities.[1] Rather than spread attention too thinly over a large number of examples, with inevitable undue repetition, the approach followed in this chapter is to describe major city case studies in some detail in text and in captions to illustrations, whilst giving others necessarily more limited space in the form of captions only. The arrangement groups together cities in comparable European countries.

Amsterdam

In Chapter 4 Amsterdam was introduced as by far the most important example of that characteristic Dutch dike-and-dam form of urban settlement. The city dates from 1240, when the River Amstel was dammed, thereby creating the Damrack and the Rodin as the outer and inner harbours respectively. The dikes to the east and west, raised above flood level by the excavated soil from the new river channels, determined the lines of the two main streets of the original fishing village settlement – the Warmoesstraat (east) and the Nieuwendijk (west). Thirteenth-century Amsterdam was only a small fishing village, while Leyden, Delft and Haarlem were comparatively much larger and more important Dutch towns.[2]

During the fourteenth and fifteenth centuries the growth of Amsterdam as a trading centre was reflected in extensions of 1367, 1380 and 1450 which added a total of 350 acres to the original area of around 100 acres. This early expansion took a unique form, later also to be the basis of seventeenth-century development. New canals were excavated, in advance of building, at a more or less constant distance out from the existing urban perimeter. Eventually the ring of canals was to be complete to the west and south of the original nucleus; to the east it was modified by construction of larger basins and docks. The canals were used for direct barge access to the merchants' houses and warehouses – frequently combined in the same building – and at the time of their construction served as a defensive 'moat'. A first formal defensive

It was on a Sunday morning that I went to the Bourse, or Exchange, after their sermons were ended, to see the Dogmarket, which lasts till two in the afternoon, in this place of convention of merchants from all parts of the world. The building is not comparable to that of London, built by that worthy citizen, Sir Thomas Gresham, yet in one respect exceeding it, that vessels of considerable burthern ride at the very quay contiguous to it; and indeed it is by extra-ordinary industry that as well this city, as generally all the towns of Holland, are so accommodated with graffs, cuts, sluices, moles, and rivers, made by hand, that nothing is more frequent than to see whole navy, belonging to this mercantile people, riding at anchor before their very doors: and yet their streets even, straight, and well paved, the houses so uniform and planted with lime trees, as nothing can be more beautiful.

The Keizer's or Emperor's Graft, which is an ample and long street, appearing like a city in a forest; the lime trees planted just before each house, and at the margin of that goodly aqueduct so curiously wharfed with Klincard brick, which likewise paves the streets, than which nothing can be more useful and neat. This part of Amsterdam is built and gained upon the main sea, supported by piles at an immense charge, and fitted for the most busy concourse of traffickers and people of commerce beyond any place, or mart, in the world. Nor must I forget the port of entrance into an issue of this town, composed of very magnificent pieces of architecture, some of the ancient and best manner; as are divers churches.

(John Evelyn, Diary, 1644)

1. For general chapter references see E A Gutkind, *International History of City Development: Urban Development in Central Europe*, Volume 1, 1964; *Urban Development in the Alpine and Scandinavian Countries*, Volume 2, 1965; *Urban Development in East Central Europe: Poland, Czechoslovakia, and Hungary*, Volume 7, 1972; *Urban Development in Eastern Europe: Bulgaria, Romania and the USSR*, Volume 8, 1972. See also L Benevolo (trans. G. Culverwell) *The History of the City*, 1980.
2. For general Amsterdam references see S E Rasmussen, *Towns and Buildings*, 1969; G Burke, *The Making of Dutch Towns*, 1956; G Cotterell, *Amsterdam: The Life of a City*, 1973; A E J Morris, 'Historical Roots of City Planning', in A K Dutt and F J Costa (eds) *Public Planning in the Netherlands*, 1985.

system was begun in 1482 and marked the urban limits until the seventeenth century.[3]

Extensive fires in 1451 and 1452 destroyed or damaged nearly all the buildings. In 1521 a law was introduced which required brick and tile construction in place of timber and thatch. In 1533 another ordinance sought to improve public health in the city.[4] Rapid increase of population had resulted in the occupation of single houses by two or more families, and there being no sanitary provisions on the upper floors, slops were disposed of by emptying buckets out of windows into the canal or street; the ordinance therefore obliged house owners to install sinks emptied by lead soil pipes and it also forbade the building of covered drains or sewers unless they were fitted at suitable intervals with detachable inspection covers.[5] Further legislation of 1565, which in part remained in force until the early nineteenth century, required approval of piled foundations by municipal inspectors, a privy for each plot, and the levying of charges for making up roads, pavements and canal embankments. Moreover, this public health and building legislation was firmly enforced. In so doing the city created a respect for collective decisions – invaluable during implementation of its seventeenth-century master plan.

In 1570 the Spanish destroyed Antwerp, the leading port in the Netherlands. This led to an immediate increase in the trade handled by Amsterdam and by 1600 it had taken over the dominant role of Antwerp. The rapid growth in trade and wealth during the last quarter of the sixteenth century could have resulted in urban expansion pressures beyond the control of the city authorities – as occurred, for example, in countless instances during the Industrial Revolution, when individual short-term commercial gain was considered before community interest. Amsterdam, however, was able to control its future development. Four main factors were responsible for this unique historical achievement – three of these were form-determinants derived from the function of the city, its site and the need for a defensive system; the fourth decisive factor was that Amsterdam was like a great and flourishing corporation in which each citizen owned shares.[6] Inevitably, as will be noted later, there were those who saw the possibilities of personal gain, but they were few and the collective spirit prevailed.

PLAN OF THE THREE CANALS

Amsterdam developed from its late sixteenth-century area of around 450 acres to an early nineteenth-century total of nearly 1,800 acres, in accordance with the 'Plan of the Three Canals' (Figure 7.1). Hendrik Jz Staets was responsible for this plan, adopted by the city council in 1607 but not without objections from certain speculators among the councillors who had bought up land in anticipation of development.[7] The layout of the new parts of the city was determined by the first two of the four factors referred to above. Amsterdam's trading function required direct water access to individual merchant's houses and warehouses. This was only possible through communal action in the construction of new canals, the organized excavation of which required a plan, with the considerable investment needed for a defensive system imposing a limit on its extent.[8]

The three canals of the 1607 plan are the Herengracht, the Keizersgracht and the Prinsengracht, named in sequence out from the centre. The Herengracht, 80 feet wide, had been excavated in 1585 and the

In *The Making of Modern Holland*, A J Barnouw writes: 'The Dutchman is by nature a rugged individualist. But rugged individualism does not make for civil liberty.' He explains that 'the Dutch, who for various reasons, physical, economic, or political, joined their individual lives into communal units, learned by bitter experience that their strength lay in co-operation, and that co-operation was feasible only if all agreed to limit their personal liberties by common obedience to self-made laws.

By the end of the Middle Ages the majority of the Dutch people were living in urban centres. In 1500 Holland and Belgium number no fewer than 208 fortified towns and 150 large villages which, but for the lack of walls, might pass for towns'. Many of Holland's towns were built on the reclaimed floor of the sea, the so-called polders. To secure building land at all it was first necessary to dig canals, using the earth thus obtained to erect dikes around areas that had been filled up with sand brought there from long distances. And the erection of a house was no less toilsome. Piles had to be driven into the earth below ground-water, to procure a firm foundation.

Then a constant water-level had to be provided, to secure the piles from rotting and all the houses from collapsing. The first communal institution these individualists had to agree upon was a water-level office, whose duty it was to keep the water of canals and sluices at a permanent level.
(S E Rasmussen *Towns and Buildings*)

Control of development was strictest in the areas between the three great concentric canals. Here the plot sizes averaged 26 feet frontage and 180 feet depth, and the prescription of a minimum distance of 160 feet between the backs of buildings resulted in a minimum garden length of 80 feet for each plot. A maximum site coverage of 56 per cent was thus secured. A typical set of conditions of sale, published in 1663, stipulated that the following trades could not be carried on within the area: blacksmith, brewer, cooper, dyemaker, glassblower, gluemaker, soapboiler, stonecutter, sugar manufacturer and similar noxious or noisy occupations. It is remarkable that these restrictions have remained in force there ever since. The 'Building Order' for the same district required, inter alia, that outside walls be constructed only in Lekse, Leytse, Vechtse or Rijnse bricks, and that only blue bricks or Bremen stone could be used for drains.

So rarely are buildings of less than three stories to be seen in this area that the existence of some form of municipal control to secure a minimum, as well as a maximum, height might be presumed. Investigation has so far yielded no evidence of such control, and it is unfortunate that most of the city's older records perished in three major fires; but land values along such important frontages doubtless induced developers to obtain the maximum return in lettable space.
(G Burke, *The Making of Dutch Towns*)

---

3. See C Duffy, *The Fortress in the Age of Vauban and Frederick the Great 1660–1789*, 1985, for the peculiarly Dutch design of urban fortification – the 'Old' and 'New' Netherlands Method as they were termed; Morris, 'Historical Roots of City Planning'.
4. City planning and building control legislation requires a supervisory enacting bureaucracy; as described in Chapter 4 this was originally created in response to need for collective community action in the continual Dutch war against the sea. By comparison, one of the reasons why London could not be comprehensively redeveloped after the Fire of 1666 is that there was

Figure 7.1, Far left – Amsterdam in about 1400: the original settlement was on the east bank dike of the Amstel with the old church on lower ground. Key: a, Amstel; o, Oudezijds Voorburgwal; n, Nieuwezijds Voorburgwal. Amsterdam at that time exemplified the peculiarly Dutch dike-and-dam form of urban settlement.

Left – Amsterdam at the beginning of the nineteenth century with the area laid out under the 'Plan of the Three Canals' fully developed. The extent of the city at the end of the fifteenth century, before the plan was adopted, is shown in solid black. Key (where different from above): s, Singel Canal; h, Herengracht Canal; k, Keizersgracht Canal; p, Prinsengracht Canal; J, the Jordaan District; D, the main dock area. North at the top for both plans.

Figure 7.2 – Amsterdam: the central section of the map published by the Society for the Diffusion of Useful Knowledge in 1832, showing in greater detail the extent of the early fifteenth-century city shown in Figure 7.1 left.

Keizersgracht, 88 feet wide, in 1593. Both therefore predate the plan itself and had been located in the city according to the uniquely controlled organic growth pattern of concentric canal systems. The third canal, the Prinsengracht, 80 feet wide, was excavated later, in 1622. The three main ring canals were linked by radial waterways and all had spacious quays and roadways on both sides. Outside the ring, to the west, there was the Jordaan, an area zoned for industry. Around the entire area and encompassing the harbour district in the south-east was constructed a five-mile-long defensive system with twenty-six bastions and seven gates.

It is easier to draw up a physical plan than to create the legislative framework to put it into practice. However, in this latter respect Amsterdam obtained for itself the vital prerequisite of compulsory land purchase powers in 1609 and, as Gerald Burke tells us, exercised them as it became necessary for the various parts of the scheme.[9] Having acquired an area, the council prepared the land for building, divided it into plots of convenient size and shape and sold these in the open market subject to special conditions. Purchasers had to enter into covenants which bound their successors in title also, to the effect that the land would not be put to other than stipulated uses, that the plot coverage would be kept within prescribed limits, that the plot would not be subdivided by lanes or alleyways, that the party wall

not even the barest nucleus of an administrating bureaucracy.

5. Burke, *The Making of Dutch Towns*.
6. Rasmussen, *Town and Buildings* (with a particularly perceptive section on Amsterdam).
7. Also by comparison, it is doubtful if in 1666 the City of London mercantile interests viewed their burnt-out city in the same way (see Chapter 8).
8. For an account of the origins of Dutch urban bureaucracy see the Netherlands section in Chapter 4.
9. Burke, *The Making of Dutch Towns*; see also (pages 227–9) a comparable process used for the Vienna Ringstrasse development.

connections would be afforded to developers of adjacent plots and that only certain types of brick would be used for external walls. The civic administrator most responsible for implementing the scheme was surveyor-general Daniel Stalpaert (1615–76).[10]

The planned growth of seventeenth-century Amsterdam is a clear example of the rule that societies get the kind of cities they deserve. It is proof, if any is still required, that theoretical planning expertise is of little significance in the absence of community resolution. Without political direction, expressed in viable legislation, plans are just so much paper.[11]

Antwerp

Antwerp (Anvers) originated on the higher, drier eastern bank of the River Schelde, some 45 miles in a direct line from the open sea, but

Figure 7.3 – Amsterdam: aerial photograph looking towards Den Dam – the open space at the top – across the Singel and the Herengracht (the latter at the bottom of the picture). The high density of land development has not changed significantly since the seventeenth and eighteenth centuries, similarly the remarkably uniform building heights.

The richly close-grained urban form of Amsterdam is broadly characteristic of the historic nuclei of other Dutch *grachtenstadt* (see Chapter 4).

10. The vital role of Daniel Stalpaert, one of urban history's most gifted administrators, has been neglected in the specialist city planning studies.
11. Alternatively expressed, as I impress on students: 'town planning is the art of the politically possible'.
12. For Antwerp see *Plans en relief de villes Belges*, 1965; P Sica, *Storia dell'urbanistica L'Ottocentro*, 1991; M C Branch, *Comparative Urban Design*, 1978.

ANTWERP.
ANTWERPEN
ANVERS

considerably more by way of the winding estuary.[12] The width and depth of the Schelde at Antwerp – to which the city owes its fortune as an inland port – combined with the marshy left-bank (western) land, determined a one-sided, semi-circular growth pattern which has generally been followed through to the present day. A first waterside settlement dates from about AD 150, occupying part of the site of the *burcht* reconstructed by the Emperor Otto I from about 950. By then Antwerp was already well known as a leading commercial centre: a position consolidated from 1124 when the Church of Notre Dame became the regional religious centre.

Status and topographical vulnerability inevitably created a need for fortifications. From the end of the twelfth century they enclosed the cathedral quarter and the *burcht*, and were followed by incremental extensions. Subsequently a system of fortifications was constructed under Charles V between 1543 and 1545, which line – enclosed in turn by later works – still formed the limit of the city in the 1830s. In 1570, however, Antwerp was sacked by the Spaniards and lost its commercial supremacy to Amsterdam, recovering only slowly until the latter part of the nineteenth century. Before the Spanish occupation the city had a population estimated in excess of 100,000; afterwards it was reduced to less than 40,000.

Figure 7.4 – Antwerp: the map published by the Society for the Diffusion of Useful Knowledge (SDUK) in 1832; north to the right; the Schelde flows from left to right. The rampart and bastions of the complex system of fortifications were constructed under Charles V in 1543; the link through to the citadel, and the citadel itself, were constructed in the 1560s; and the outer works, with the second moat, were completed under Philip V, King of Spain, in 1701. The system was not demolished until the 1860s, by which time a ring of new independent forts had been constructed further out.

Figure 7.5 – Antwerp: the extent of the *burcht*, located immediately inland from the *Werf*, just downstream from the centre of the 1832 map.

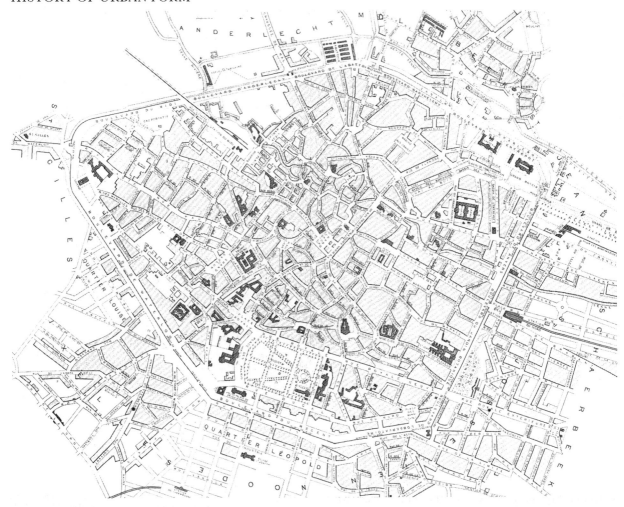

Brussels

Brussels (Bruxelles) was founded on an island in the River Senne, a tributary of the River Schelde, from where during the eleventh century it extended over on to the drier land of the right bank.[13] The original strongly defended *bourg* of the Dukes of Lower Lorraine was on the Île Saint-Géry, where it controlled the river crossing on the strategic main route between Ghent and Bruges, through to Cologne and the Rhine. Eleventh-century expansion followed the line of the trading route up the eastern valley slopes, at the top of which Duke Henry I of Brabant built a new castle from about 1200 as the nucleus of the 'upper town'. Between the Île Sainte-Géry crossing and the castle a flourishing market became established on the location of what was to become the Grand' Place – one of Europe's most impressive later medieval urban spaces. The prosperous lower town was fortified during the early thirteenth century and continued growth required a new *enceinte* of 1357–79 about 4½ miles in length, set imaginatively far out to cater for future expansion. Two centuries later, J van Deventer's map shows ample open land within the wall, which was not finally demolished until the 1830s although the gateways had been removed from 1782.

Figure 7.6 – Brussels: the Tallis Map of 1850, showing the extent of the historic inner city within the line of the mid-fourteenth century fortifications clearly revealed by the ring boulevard system laid out in the 1830s. (Note: north is to the right of the map.) The original Île Sainte-Géry has the Place Géry in its centre (towards the western side of the city) and the Coudenberg heights, where the Dukes of Brabant had their castle, are shown with the nineteenth-century Royal Palace and State Buildings separated by the formal park, to the south and north respectively. The clear contrast between the organic growth form of the historic nucleus, and that of the gridiron-planned early nineteenth-century suburban districts is a European characteristic. This map of Brussels illustrates the narrow width of ring-boulevard zone that was usually available to major European cities for redevelopment when their fortifications became finally redundant, in contrast to the comparatively broad swathe of land at the disposal of Viennese planners, as shown in the facing Figure 7.7.

13. For Brussels see M Culot, R Schoonbrodt and L Krier, *La Reconstruction de Bruxelles*, 1982; T Hall, *Planung Europaischer Hauptstadte*, 1986; Branch, *Comparative Urban Design*, for the 1837 SDUK Map of the city.

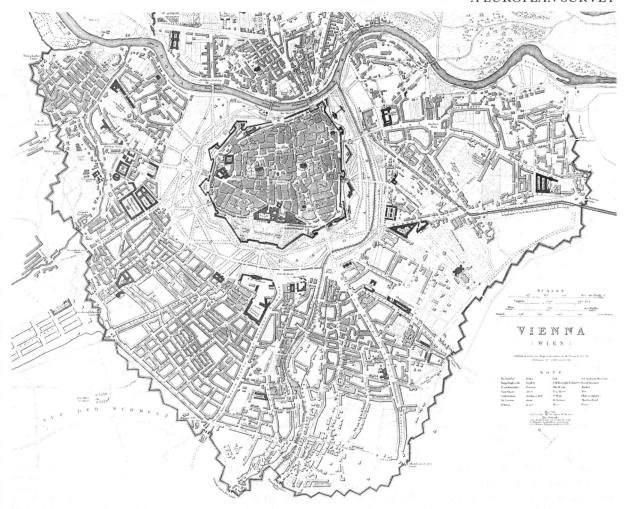

Vienna

From the beginnings of European history the site of Vienna has been an important meeting place of trading routes.[14] The River Danube was relatively passable at this point, using ferries to an island in the stream. The Romans recognized the strategic importance of the site, and in the first century AD they established a fortified camp on the southern bank of the river as a key part of their empire frontier defences. During the Crusades, Vienna was an important military centre, with trading interests following close behind. The first permanent bridge over the Danube was constructed between 1435 and 1440.

City defences were constantly improved during the fifteenth and sixteenth centuries and as a result Vienna was able to resist Turkish sieges in 1529 and 1683. The fortification system around the Altstadt, the old city of Vienna, followed the general pattern of Renaissance defensive works, combining formidable bastions and earthworks with a wide, cleared fire zone, as probably the most impressive – and expensive – system of its kind. Unlike many cities, however, the cost-effectiveness of the Viennese defences was proved by defeating actual assault, as opposed to untested but presumed gain from having

Figure 7.7 – Vienna: the map published in 1852 by the Society for the Diffusion of Useful Knowledge, showing the extent of the suburbs, beyond the open space zone (das Glacis), enclosed within the Linienwall of 1704. Even at this comparatively late date there is still ample open space remaining in the suburban zone. The Danube Canal passes across the northern side of the Altstadt; the main stream of the Danube is further to the north.

The map gives the following information: the city has 12 gates and 1,217 houses; the suburbs are divided into 34 *Vorstädte* with 10 barriers, containing 6,202 houses. Total population 255,000. This map is one of the handful of key illustrations referred to in my Introduction: not only does it exemplify the effect of being long-term fortified, and providing further contrast of organic growth and planned forms; the suburbs, outside the open fire-zone are also one of the best examples of the effect of the pre-urban cadastre.

14. For Vienna see C Sitte, *City planning According to Artistic Principles*, 1889, English translation 1965; E Lichtenberger, *Die Wiener Altstadt*, 1971; Hall, *Planung Europaischer Hauptstadte*; G Fabbri, *Vienna: Citta Capitale del XIX Secondo*, 1986; Sica, *Storia dell'urbanistica L'Ottocentro*.

deterred an attack. An outer ring of suburbs grew up during the sixteenth and seventeenth centuries to provide for the 'overspill' population from the ever more densely developed Altstadt and for houses with gardens for wealthier citizens. These suburbs, which were developed from the earlier village settlements around Vienna, were largely destroyed during the sieges, with their inhabitants sheltering within the Altstadt, but they were subsequently renewed.

After the 1683 Turkish invasion of central Europe, which was largely halted by Vienna's stubborn resistance, the cleared fire zone surrounding the defences was increased to a width of 1,700 feet involving the demolition of some 900 suburban houses, the Altstadt fortifications at that time consisted of twelve bastions, at first connected by earthen and later by stone walls, and of eleven outworks as an additional protection, the whole being surrounded by a wide moat.[15]

After the 1683 siege the rebuilt suburbs represented sufficient capital investment to justify their own subsidiary outer defensive system. The legal boundary of the city was extended to include the suburbs and a Commission of Defences was set up, with Prince Eugene of Savoy as chief consultant, to advise methods of protecting the new Greater Vienna area against attack. The outcome of this study was construction of the Linienwall of 1704 with its own 570 feet wide external zone supplemented by a 100 feet wide space along the inside of the walls. The Linienwall subsequently formed a customs barrier and was not completely demolished until 1893.

Throughout the eighteenth century Vienna developed on the basis of two distinct parts. Innermost was the congested Altstadt with narrow streets, tall houses, great old churches, the palaces of the aristocracy, and the Hofburg, the emperor's residence. Separated from it by a broad belt of defence works and open land lay outer Vienna in the form of extensive suburbs with gardens and spreading trees.[16] Napoleon captured Vienna in 1809 and rendered the city defenceless by demolishing the bastions and associated works.

From 1809 to the early 1850s the ruins of the fortifications remained untouched, with the broad space of the fire zone constituting an informally established park between the Altstadt and the suburbs. The ruins prevented any expansion of the Altstadt and the open space ring represented both a barrier separating the two parts of the city and an uneconomic use of land. In 1857, under the Emperor Franz Joseph, a competition was held for designs for the development of the derelict land. The competition conditions required that a large part of the area was to form residential sites, from the sale of which a building fund was to be created for financing the rest of the plan and erecting a number of large public buildings. In this respect the city was making maximum use of the fact that it owned all the previously fortified zone. Also required of competitors was the location of the given number of civic buildings, including the parliament building, *Rathaus*, university, museums, theatre and opera house. The winning design by Ludwig von Förster was approved in 1858 and carried out over the next decade. Its basis, as Figure 7.8 shows, is a broad ring boulevard, the Ringstrasse, the grandest of its kind, which is taken through the centre of the new area, circumscribing the Altstadt, with a linking quay along the bank of the Danube canal. The Ringstrasse is about 2 miles long and over 200 feet wide. It was laid out in the form of five straight sections, each of which determined the alignment of five gridiron planned districts.

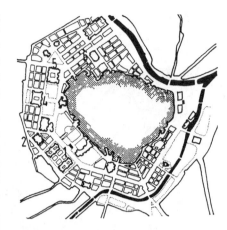

Figure 7.8 – The Ringstrasse development with the building blocks on either side of the new ring boulevard; the Altstadt is shown in outline only. Key: 1, Museum; 2, Palace of Justice; 3, Parliament House; 4, the new City Hall; 5, the University; 6, the Bourse.

The Ringstrasse project elicited much professional comment as it progressed, some of which relates to Sitte's own critique of it. In 1858 Eitelberger, who was to be his teacher, delivered a public lecture in Vienna on city planning and city building that was obviously promoted by these current events. Humanist that he was, Eitelberger took the occasion to give a sketch of the way in which the form of the city had served as an expression of culture and social systems from classical times to the present. It is clear that for Eitelberger city planning was a very serious and important matter. He emphasized the role of the architect as planner and the function of public buildings as the spiritual essence of the city – drawing copious examples from Greece, Rome, and the Middle Ages. Vienna in 1858 did not measure up to the status of a metropolis in these terms, he observed, but the project before them could produce the great artistic buildings necessary to fill the need. His lecture was a solid historical presentation and one that would do credit to a speaker a century later. It was not composed in Sitte's abstract analytical terms, but it does demonstrate the wealth of erudition that was available (for instance, Sitte's citations of Aristotle and Pausanias on his p. 8 are right out of Eitelberger). Eitelberger's nostalgic description of Pompeii recalls the opening of Sitte's book, and, as Sitte would do later, Eitelberger already bemoaned the indiscriminate use of the ruler in laying out the new Vienna.
(G R Collins and C C Collins, Camillo Sitte and the Birth of Modern City Planning, 1965)

15. See Duffy, *The Fortress in the Age of Vauban and Frederick the Great 1660–1789.*
16. Rasmussen, *Towns and Buildings.*

Thus, says Rasmussen, the new districts were made up entirely of uniform building blocks except at the corner which would not come out right; the streets had no face, they were simply voids, empty spaces between the cubic blocks, not pleasant outdoor rooms as in old Vienna.[17]

Vienna presents as clear a contrast as possible between the organic growth pattern of the medieval Altstadt and the formally planned Renaissance sections of the Ringstrasse. Much criticism of the visual aspects of the scheme – specially that of Camillo Sitte, who makes constant reference to his native city in his *The Art of Building Cities* – is derived from this fact. Functionally, although the Ringstrasse project added the necessary public buildings, conveniently located to the old core, the new districts in turn formed a barrier between the centre and the suburbs with the problems of even the existing metropolitan traffic not recognized.[18]

Prague

Prague developed at a strategically important location in the centre of the Bohemian plateau, where the River Vltava cuts deeply into the plateau and forms a dramatic escarpment on its left bank.[19] At this point there was a ford across the river which constituted a focal point for east–west trading caravans. Two castles, with related civil towns, were built on high ground commanding the river banks and crossing. The more important was the Hradcany, built from the ninth century onwards on the left bank escarpment above the sweeping bend in the river. The second castle, Vysehrad, was erected on the opposite bank, about 2 miles upstream on a 160 foot high hill. Its northern approaches were protected by the steeply sloping valley of a tributary stream.

Town expansion in the second part of the thirteenth century added the Mala Strana civil town, south of the Hradcany on the left bank. The first permanent bridge across the Vltava was constructed in 1153, with the existing Charles Bridge dating from 1357. The Stare Mesto

Figure 7.9 – Vienna: aerial photograph from the south (the same orientation as Figure 7.8, with Key to major buildings) showing the 'drawing-board' planning of the Ringstrasse buildings and open spaces contrasted with the organic growth Altstadt.

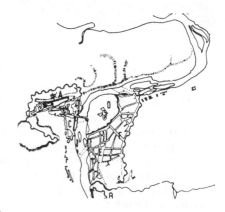

Figure 7.10 – Outline map of Prague in 1831 (north at top). Key: A, Hradcany Castle (the present-day national government centre); B, Vysehrad Castle; C, Mala Strana town, which grew up at the foot of Hradcany Castle hill; D, Stare Mesto (Old Town); E, Charles Square; F, Wenceslas Square. Both these squares form key parts of the mid-fourteenth-century expansion of the city under Charles IV.

17. Rasmussen, *Towns and Buildings*.
18. Although more than 100 years old, Sitte's perceptive if biased observations are revealing of late nineteenth-century anti-planning attitudes.
19. For Prague see V Hlavsa, *Praha Ocima Staleti*, 1967, with excellent maps, notably two of 1824, and general illustrations; Gutkind, *Urban Development in East Central Europe: Poland, Czechoslovakia, and Hungary*, Volume 7.

Figure 7.11 – Prague: the detail plan of the Stare Mesto (north at the top). The Old Town Square is in the centre, with the Ungeld Hotel project behind the Tyn Church to the east (solid black on the plan). The River Vitava forms the western and northern sides of the Old Town; the defensive wall of 1235, with its moat, was demolished in 1760 to form the line of the present-day east-bank inner-ring boulevard. The age-of-buildings key shows clearly the extent of the oldest areas (shown in heavy outline) with their typical organic-growth structure, in contrast to the nineteenth-century redevelopment of the northern section (shown hatched) organized around the new Parizska Avenue.

Key: A, Old Town Square; B, Town Hall (northern wing destroyed in 1945); C, Tyn Church; D, northern end of Wenceslas Square, which is neither square nor even rectangular; it is a wide street, in which the city's popular gatherings (including momentously those of the early 1990s) take place.

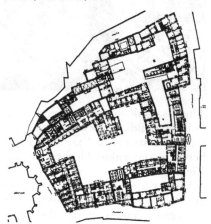

Figure 7.12 – The upper-floor plan of the Ungeld block to the east of the Old Town Square (shown in solid black in Figure 7.11). From the eleventh century onwards this part of the Old Town had served as a customs and taxation district for merchants arriving in the city. Here they were assessed for duty, from which the name Ungeld was derived. Individual buildings surrounding the main courtyard and the subsidiary spaces immediately to the north are all of basically thirteenth- and fourteenth-century construction, but elevations have been remodelled since then.

Prague has an embarrassment of such historic buildings. There is a limit to the number that can be used for museums, or specialist offices, and rehabilitation for residential purposes is extremely expensive. In the case of the Ungeld block the proposal was to convert the historic buildings for use as an hotel and cultural meeting centre.

settlement on the right bank had grown up around the approaches to the ford and, after completion of the bridge, increased in size and importance as the stopping place of the caravans. As a legal entity the Stare Mesto dates from 1232 and the line of walls and moat, followed by the present-day inner ring street, was built from 1235. Within this ring, the Stare Mesto forms a typically organic, medieval plan-form centred on the beautiful Old Town Square and crossed from east to west by the winding Royal Road which leads across the bridge to the Hradcany (Figure 7.11).

The northern part of the square and the entire north-western part of the Stare Mesto are, however, the unfortunate result of nineteenth-century 'improvements'. Parizska Avenue, which leads directly north out of the square, and as a vista street is in total contrast to the otherwise closed exits, is the worst aspect of this work.[20]

The great period of urbanism in Prague in the reign of Charles IV (1346–78) during which time the city, as the capital of the Holy Roman Empire, was the most important urban centre in Europe. Charles IV founded Europe's oldest university in Prague in 1348 and laid out the area to the south and east of the Stare Mesto as a controlled extension to the city. This development was planned around three squares, two of which remain: Charles Square, intended as the centre of the new district and until recent times serving as cattle and general markets, and Wenceslas Square, in effect a widened street nearly half a mile in length and 200 feet wide, and now the modern business centre of the city.[21]

20. The urban reality, when experienced, is more of an insensitive intrusion on the Square than the map might communicate.

21. Recent history has especially favoured Prague. Not only has its historic form and individual buildings been spared twentieth-century economic devastation; miraculously the city also escaped significant damage during the Second World War. (It was bombed, so legend holds – and then but lightly – only by US aircraft mistaking it for Dresden.) The wealth of architectural heritage, in the context of a depressed 'Eastern bloc' economy, has resulted, perforce, in a but shabbily maintained city. The momentous political events of the early 1990s, with their prospects for economic revival, could herald

Berlin

Berlin had its origins in two separate urban settlements: Colln, the oldest, located on an island in the River Spree, and Berlin proper, on the north bank of the river.[22] Colln is first mentioned in 1237 and Berlin in 1244. The two towns concluded a treaty in 1307 which effectively united them as Berlin-Colln, although each retained its own council. During the fourteenth century Berlin-Colln became one of the most important towns in the German Mark with, effectively, the status of a free imperial town. This assumed independence brought Berlin-Colln into conflict with the interests of the all-powerful Hohenzollern family, who became Masters of the Mark in 1415. Unsuccessful opposition to Elector Frederick II (1440–70) resulted in the town twice being deprived of all its privileges, in 1442 and 1448, and a fortified castle was erected on the Spree between 1443 and 1451 to keep the inhabitants under control. In 1448 each part of Berlin-Colln had about 6,000 population; throughout the sixteenth century and the first half of the seventeenth, however, the growth of Berlin-Colln was stunted, and the early suburbs which had grown up around the Altstadt of Berlin (Figure 7.13) were destroyed during the Thirty Years War (1618–48). Although Berlin-Colln itself was spared, the economic situation was so bad that the citizens considered emigrating en masse and abandoning the town.[23]

The city had been strongly fortified between 1658 and 1683, incorporating the new district of Friedrichswerder within the defences. (These were designed by a Dutch engineer and involved the construction of thirteen massive bastions.)[24] Outside this zone two new housing districts were laid out to Renaissance gridiron pattern – Dorotheenstadt from 1674 (named after the elector's second wife) and Friedrichstadt from 1688. Both of these districts were linked back to the Altstadt by the line of the Unter den Linden. This was originally an avenue of lime trees giving access to the open country and forests on the west; it had gradually been built up on both sides as an impressive 198 feet wide street, about one mile in length between the Royal Palace and the Brandenburg Gate. It thus has striking parallels with the earlier establishment of the Champs Elysées in Paris – also originally a direct link between the royal palace and country districts beyond the city limits (page 200). Their existence determined that the western sides of both Berlin and Paris would become the most fashionable, expensive residential areas.

By the early eighteenth century Berlin had a population of around 60,000 and covered an area of 2 square miles. In 1737 a customs wall 9 miles in length was built further out to protect the growing commercial and industrial interests of the city. This increased the city area to 5¼ square miles. At about this time the Dorotheenstadt and the Friedrichstadt were enlarged, the demolition of the defensive system was started, and new suburbs were added to the north and east. During the nineteenth century industrial expansion proceeded at a great pace, as evidenced by the population figures: 200,000 in 1820; 329,000 in 1840; 496,000 in 1860; 1 million in 1870; and 1.5 million in 1888. The customs wall was not demolished until 1850, when it was replaced by a ring-road system. The greater part of the population growth of the second half of the nineteenth century was accommodated in tenement housing districts between this line and the *Ringbahn* railway.

The division of Berlin into two parts – West and East – after the

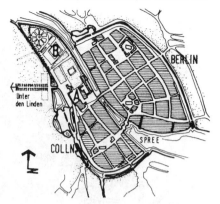

Figure 7.13 – Plan of Berlin-Colln around 1650, before the re-fortification of 1658–85. Key: A, Royal Palace, with the Unter den Linden leading west from the entrance gateway into the open country; B, the formal gardens of the palace. That which was not generally known of the divided post-1945 city is that the historic nucleus was entirely in the Eastern Zone.

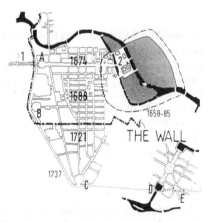

Figure 7.14 – Inner Berlin of the eighteenth century. The original Berlin-Colln area of Figure 7.13 is shown tinted, with the lines of the 1658–85 fortifications and the 1737 Customs Wall shown dotted. Key: 1, the Unter den Linden continuing west from the Brandenburger Tor (A) across the Tiergarten to the open country; 2, the Royal Palace facing on to Lustgarten Square. Gateways in the Customs Wall – A, Brandenburger Tor; B, Potsdamer Platz; C, Belle Alliance Platz; D, Thorbecken. (Note: this diagram is the same orientation as the preceding two Figures 7.13 and 7.15).

Prague's resurgence. It, and the South Bohemian towns of Chapter 4, are the exceptional survivors of central Europe. Hopefully they will survive the inevitable tourist industry assaults.

22. For Berlin see Hall, *Planung Europaischer Hauptstadte*; Sica, *Storia dell'urbanistica L'Ottocentro*; L O Larsson and A Speer, *Le Plan de Berlin 1937–1943*, 1983.

23. E A Gutkind, *Urban Development in Central Europe*, 1972.

24. See Duffy, *The Fortress in the Age of Vauban and Frederick the Great 1660–1789*.

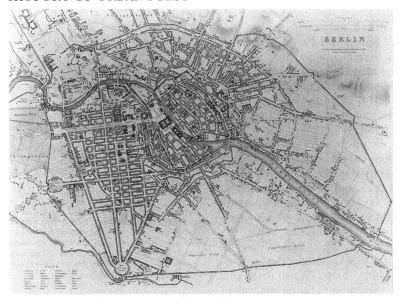

Figure 7.15 – Berlin, the SDUK Atlas map of 1833.

Second World War has no historical precedent; if the extent of devastation caused by Allied aerial bombardment and the final Soviet assault was no more, in proportion, to that suffered by other European and Japanese cities,[25] in combination the disruptive effect has been unequalled in all urban history. The historic nucleus was in the Soviet sector, east of The Wall (the line of which has been added to Figure 7.14) and a new Western centre had to be created. The long-term effects of the reunification in 1990 on the form of the city are beyond present conjecture.[26]

Budapest

The three original urban nuclei which were to be consolidated by subsequent growth into the modern city of Budapest were still separate well into the seventeenth century.[27] Obuda on the western (right) bank of the Danube occupied part of the site of Roman Aquincum, one of the most important legionary fortresses on the vulnerable eastern frontier, whose *canabae* were recognized as a *municipium* in the second century AD. Buda was centred on a commanding hill downstream alongside the river some 1¾ miles to the south, and Pest, the youngest of the three nuclei, had developed on the extensive river plain stretching to the east of the river, downstream again from Buda.

The Royal Palace in Buda was built in 1247 and from then until 1526 this city was the capital of Hungary. Between 1526 and 1686 all three urban centres were held by the Turks.[28] Pest was almost completely demolished and by 1710 had regained only a nominal population of 1,000. Maria Therese founded the university in Buda in 1777 (it was transferred to Pest in 1784); this marks the beginnings of the re-emergence of these urban nuclei and their transformation into an important European centre, in which Pest achieved dominance as the commercial and cultural core. The single municipality of Budapest was created in 1872. The defences around Pest were demolished in 1808 to be replaced by the present-day inner ring boulevard.

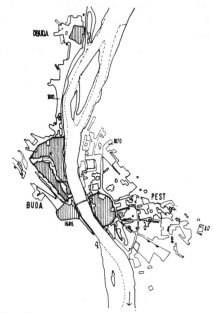

Figure 7.16 – Budapest, diagrammatic growth map (north at the top). The vertical hatched areas show the extent in 1685 of the twin cities of Buda and Pest, and the upstream settlement of Obuda (on the site of Roman Aquincum).

25. As 1980s background see P M Zimolo, *Berlino Ouest tra Continuita e Rifondazione*, 1987.
26. At the completion of third edition preparation there is much activity but as yet no Master Plan for the future of the re-united Berlin.
27. For Budapest see P Gabor, *Budapest Varosepitesenek Tortenete*, Volumes 1 and 2, with excellent maps and general illustrations; Hall, *Planung Europaischer Hauptstadte*.
28. At Budapest, in notable contrast to Athens, there is practically nothing to remind of its 150 years of Islamic occupation.

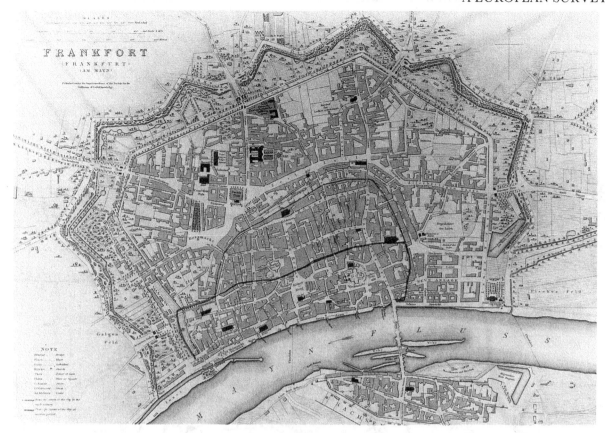

Frankfurt am Main

Located at a key crossing of the River Main, where ancient trading routes converged, Frankfurt has been for many centuries one of the most important commercial cities in Germany.[29] The site is first mentioned in 793 as the royal residence (*pfalz*) of 'Franconofurd' (ford of the Franks) where, in the next year, Charlemagne held an imperial convocation. From 876 it had become the *de facto* capital of the East Frankish Empire, and from the twelfth century most of the German kings were chosen there. Economic regional pre-eminence was assured from the sixteenth century when 'free imperial' market status was granted, with the two fairs sanctioned in 1240 and 1330 retaining importance through to the twentieth century.

The Pfalz was in effect a strongly fortified burg with a commanding high ground location on the northern right bank of the Main. The planned extension which was laid out around the Pfalz, during the later twelfth century, enclosed within its wall the city's historic Altstadt (which suffered extensive devastation in the Second World War). The second major historic extension, the Neustadt, was incorporated in 1333, and provided ample space within its fortifications well into the nineteenth century, as evidenced by the SDUK Atlas map of 1837. The encircling zone of land required when the enlarged city was strongly refortified between 1632 and 1650 provided for the characteristically attractively landscaped 'Anlagen' boulevard system when removed in 1806.

Figure 7.17 – Frankfurt am Main: the SDUK Atlas map published in 1837; and below, Figure 7.18, an analytical map identifying the areas of the Pfalz; the Altstadt; and the Neustadt.

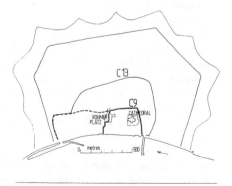

29. For Frankfurt see Branch, *Comparative Urban Design*; K Baedeker, *The Rhine*, 1906.

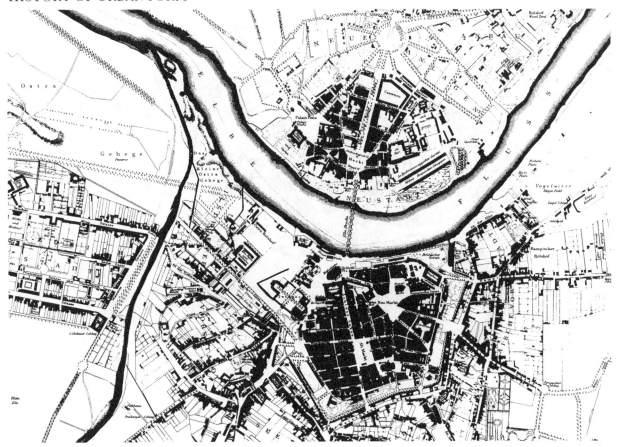

Dresden

Arguably the saddest, most pointless episode in the Second World War's record of European urban devastation took place in February 1945 when Dresden's incomparable Renaissance and Baroque urbanism was obliterated by Allied bombing: mainly, it would seem, because that late in the war it was the most important surviving German city.[30] Dresden's known history dates back to the twelfth century when a settlement is recorded at an important crossing of the River Elbe. Initially established on the northern right bank, it was the later left-bank settlement which was to provide the nucleus for early commercial prosperity under the protection of a burg; before, from the fourteenth century, the city became consolidated on both banks.

In 1685 much of right-bank Dresden was destroyed by fire and reconstruction of the 'Neustadt', as it was then called, was in accordance with Klengel's plan of that year, and a second plan of 1731. Royal patronage was responsible for the city's eighteenth- and nineteenth-century architecture and urbanism, notably that of Augustus the Strong, King of Poland and Elector of Saxony, of whom it was claimed that he 'found the city small and wooden, but left it of stone and splendid'. The major historic buildings were alongside the Elbe, in particular the group around the Schloss, including the Zwinger Museum.

Figure 7.19 – Dresden: the SDUK Atlas map published in 1833.

Figure 7.20 – Dresden: an analytical map identifying the several stages in the city's historic development.

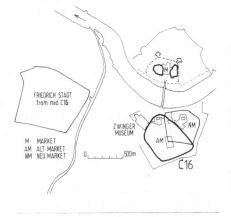

30. For Dresden see Branch, *Comparative Urban Design*; K Baedeker, *Northern Germany*, 1910; and D Irving, *The Destruction of Dresden*, 1963.

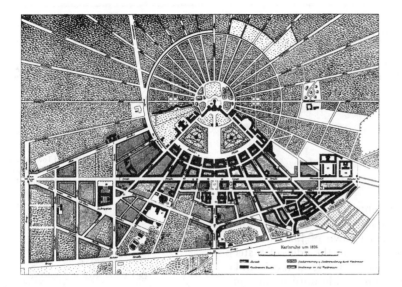

Figure 7.21 – Karlsruhe. The focal point of Karlsruhe – a palace, park and town example of autocratic planning second only to Versailles in Western Europe – was a forest hunting tower built in 1715 by the Margrave Karl Wilhelm, ruler of Baden Durlach. Although intended at first only as a simple retreat, the tower soon became the centre of a total of thirty-two radiating routes, twenty-three of which served as forest rides and the remaining nine as the structure of a town which was laid out to the south. The tower was replaced by a grandiose palace between 1752 and 1781 with flanking wings enclosing a formal forecourt. At the same time, the Langestrasse was laid out as a main cross axis through the town at right angles to that of the palace.

Early in the nineteenth century a formal western 'entrance' to the town was created by adding two streets at angles to the Langestrasse. Such drawing-board planning, while rigid and ingenious, nevertheless served as a perfect symbol of the 'l'État c'est moi' spirit of absolute rulers whose total command of resources and power was both their strength and their eventual weakness. (See E A Gutkind, *International History of City Development*, Volume I, *Urban Development in Central Europe*, 1961).

Mannheim

Although first recorded in AD 766, by the first decade of the seventeenth century Mannheim was still only a large fortified village settlement at the junction of the Rhine and Neckar rivers.[31] For almost 1,000 years the strategic advantages of the site, alongside one of Europe's major natural waterways, had been offset by seasonal flood dangers. However, by the seventeenth century the line of the Rhine had assumed renewed international boundary significance and in 1606 the Elector of the Palatinate, Frederick IV, started building a strongly defended citadel and town on the site of the village.

The star-shaped citadel of Friedrichsburg was about half as large as the town itself. This latter was laid out strictly according to the gridiron, within its own formidable defensive system. The citadel was destroyed for the first time in 1622. Vauban demolished the rebuilt defences in 1689, at which time Mannheim, as renamed, had a population of about 12,000. In the next reconstruction, which followed the Treaty of Ryswick (1697), both the citadel and the town were enclosed by one continuous defensive system. The last phase in the development of the plan of Renaissance Mannheim came with the final demolition of the citadel in 1720 to make way for the first stage of a monumental Baroque palace. This palace was extended between 1749 and 1760.

Mannheim by 1799 represented a straightforward exercise in unimaginative drawing board geometry. The general gridiron street pattern was modified only by the slightly wider main axis-streets in front of the palace and by the broad cross avenue. Two of the street blocks were left open as, at most, only incidental public open space. The defensive system had so diminished in importance as to have only an aesthetic significance as marking the boundary of the urban area.

Camillo Sitte is highly critical of both Mannheim and Renaissance grid planning in general. This much might be expected from the nineteenth century's strongest advocate of a return to the use of 'romantic' medieval urban form principles. Writing of the grid plan Camillo Sitte observes 'it was carried out already with an unrelenting thoroughness at Mannheim, whose plan looks like a chequerboard;

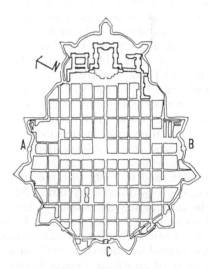

Figure 7.22 – Mannheim in 1799. The palace across the north-eastern end of the town occupies part of the former citadel area. The three gates into the town are marked as A, B, C. Paul Zucker has noted that 'this gridiron scheme, unique in Europe for such expansion, has often been mentioned as the prototype for the grids of eighteenth-century towns in the United States. Nor is it by chance that Mannheim is the only town in Europe where the streets are identified by numbers and letters instead of having individual names' (*Town and Square*).

31. For Mannheim see Baedeker, *The Rhine*.

there exists not a single exception to the arid rule that all streets intersect perpendicularly and that each one runs straight in both directions until it reaches the countryside beyond the town. The rectangular city block prevailed here to such a degree that even street names were considered superfluous, the city blocks being designated merely by numbers on one direction and by letters in the other. Thus the last vestiges of ancient tradition were eliminated and nothing remained for the play of imagination or fantasy'.[32]

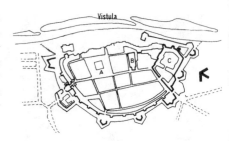

Figure 7.23 – Warsaw: the plan of the Old Town (Stare Mesto) in 1655, within its defensive perimeter. Key: A, the Old Town Square; B, St John's Cathedral; C, the Castle.

Warsaw

At the end of the thirteenth century town rights were granted to a small settlement which occupied the site of Warsaw's present Old Town.[33] This was probably a rebuilding of an earlier town of Jazdow, burnt down in 1262. The site of the Old Town was on the edge of a steep embankment, some 70 feet high, on the western side of the Vistula, which flowed in a flood plain about 7½ miles wide until it was contained from the eighteenth century onwards. The Old Town was strongly fortified by 1339, with a double wall on the land side and a single wall on the edge of the river embankment. Tributary streams formed natural moats to the north and south. Fourteenth-century Warsaw was a small town covering only 40 acres; other Polish towns were much larger, notably Cracow and Poznan. Within roughly oval walls the Old Town comprised some 150 building plots, each about 30 feet wide and 115 to 130 feet long. The street pattern was based on the gridiron and the market place in the centre was around 310 by 230 feet in area. A ducal castle (rebuilt later as the Royal Palace) was constructed from 1289, immediately to the south of the Old Town.

Warsaw's favourable location at the meeting place of the main east–west, north–south central European trading routes resulted in steady growth, and by 1408, with the area within the walls fully developed, expansion was necessary and a 'New Town' was built immediately to the north. After the disastrous fire of 1431 the town council decreed that no more wooden dwellings be permitted and Warsaw became a city of brick. The first permanent bridge dates from 1549 and in 1596 the capital was moved from Cracow to Warsaw, thereby ensuring its continued growth as a major European city.

The 1830 Warsaw was occupied by troops of the Russian Tsar and a citadel was built immediately north of the Old Town. It was not finally removed until after 1945 and its presence determined that the city expanded in an unbalanced fashion, mainly south along the Vistula, and to the west.[34]

The appalling record of the systematic destruction of Warsaw by the Nazi army in the last months of 1944, and the subsequent reconstruction achievements, is outside the period of this volume; other than to establish that the planned demolition of an historic city had few parallels in urban history, with only victorious Rome's treatment of Carthage coming to mind. Hitler's policy was to erase Poland from the map of Europe, and to reduce Warsaw to the status of a provincial market town. The post-war act of faith whereby an impoverished Poland commenced reconstruction of its historic capital, in order to recreate its past, is also scarcely precedented.[35]

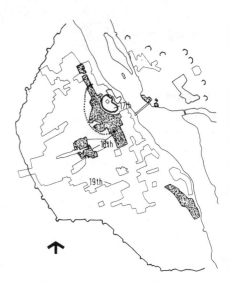

Figure 7.24 – Warsaw: a diagrammatic plan showing three growth stages of the seventeenth, eighteenth and nineteenth centuries. Nineteenth-century expansion northwards along the western bank of the Vistula was prevented by the existence of the Tsarist Russian citadel.

32. C Sitte, *City Planning According to Artistic Principles*, 1889/1965.

33. For Warsaw see Branch, *Comparative Urban Design* (the SDUK Map of 1831); K Baedeker, *Baedeker's Russia*, 1914.

34. See Figure 8.36 for Hull, a comparably one-sided English growth example.

35. See B Beirut, *The Six-Year Plan for the Reconstruction of Warsaw*, 1949.

Copenhagen

The first recorded reference to Copenhagen is as an insignificant fishing village, in 1043.[36] But the existence here of one of the best natural harbours in northern Europe, conveniently located for both the Baltic–North Sea and European–Scandinavian medieval trading routes, was soon to lead to its growth as a major trading centre. At first the port area was situated between the original village settlement, on the mainland of Zeeland, and the island of Amager. In 1167 a castle was built on Strandholmen, a small island in the harbour. Copenhagen was acquired by the Danish Crown in 1417, and by the middle of the century had become a chartered royal borough. (The university was founded in 1479.) By the end of the fifteenth century the city had a population of around 5,000. In 1535 its plan was that of a medieval walled and moated mainland centre, with the castle on Strandholmen controlling the harbour (Figure 7.27).

The introduction of Renaissance planning principles into Danish urbanism dates from the reign of Christian IV (1588–1648). During this time a number of small fortified towns were founded with regular gridiron plans incorporating numbers of squares. Christianstad (1614), Christiansand (1641) and Fredericia (completed by Christian IV's son) are typical examples.[37] Christian was also responsible in 1621 for the reconstruction of Christiana (present-day Oslo) in Norway.

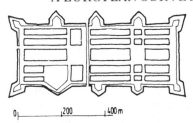

Figure 7.25 – Christianstad, founded in 1614; a two-part plan with a canal as the dividing element. Like Fredericia (Figure 7.26) this fortress-city was planned in the same era as Christiana (Oslo) and Gothenburg (Figure 5.5).

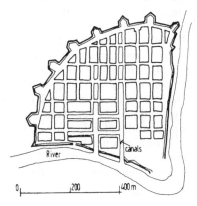

Figure 7.26 – Fredericia, founded in 1650: a plan notable for the penetration of canals into the defended area.

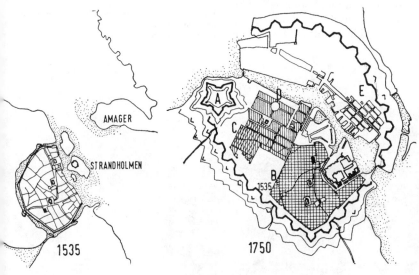

Figure 7.27 – Far left – Copenhagen in 1535; the mainland defended town with the castle on Strandholmen.
Left – The city in 1750. The medieval nucleus is shown cross-hatched and the mainland Renaissance addition with vertical hatching. Christianshavn on the island of Amager is shown in outline. The castle, and the other small islands in the harbour, have been joined to the mainland, and the island of Amager is in progress of being extensively enlarged. Key: A, Citadel; B, medieval nucleus; C, Renaissance district; D, the Amalienborg Square development; E, Christianshavn.

Christian converted Copenhagen's defensive system to a typical Renaissance pattern between 1606 and 1624. Within its perimeter was included a new district to the north of the medieval core. Christianshavn, on the island of Amager, was constructed as a new separate fortified centre from 1617 to strengthen the harbour defences. Further works intended to safeguard the city's trading interests included the fortification of the harbour entrance begun in 1627 and the construction of a citadel on the mainland side. Strandholmen island, with its old castle, was joined to the mainland. Under Christian V the defensive system around the central part of the city was extended to its furthermost extent, and in 1661, demarcation of the open fire zone was

36. For Copenhagen see S E Rasmussen, *København*, 1969.
37. See plans of these towns in Figures 7.25 and 7.26.

agreed. In 1662 a master plan for development of the city was produced but the citizens refused to accept the necessary land-use and architectural control measures. (A ring of outer fortifications around the suburbs was also proposed and never built.) The later seventeenth-century population of the city was around 60,000.

Architecturally the most important example of Renaissance urbanism in the city is the Amalienborg Square development, started in 1749 under Frederick V (1746–66). This scheme was designed by a Danish architect, Nicolai Eigtved (1701–54), on the site of a royal park and an adjoining drill ground. The year 1749 was celebrated as the 300th anniversary of the House of Oldenberg; in recognition of this the king originally intended to create on this land a new commercial centre with a harbour frontage. The project was advertised as 'a very advantageous neighbourhood, due to the proximity of the harbour and the Customs House, which will greatly contribute to the facilitation and advancement of commerce'.[38]

Rasmussen observes that from the first something more than an ordinary commercial centre must have been contemplated. On the original plan, which no longer exists, a great central space had been sketched in and the king expressly said that he would reserve the ground on which the four palaces were to be located which would form the central place. Four members of the Danish nobility were to be given the palace sites and their identical two-storeyed designs, with attics and high roofs became the basic structure of an octagonal square.[39] The other four sides were formed by two-storeyed pavilions, flanking the four streets entering the Amalienborg Square. The result is 'the unique

Figure 7.28 – Copenhagen: a section of a drawing of the Amalienborg Square with the harbour to the right and the Marble Church to the top left.

The freehold of building plots was to be given to applicants wishing to build houses: of these, timber merchants, with yards in the vicinity, were given priority in the choice of locations. Successful applicants were obliged to take up ownership immediately and were required 'to build within five years and, in every detail, adhere to plans approved by the King'.

Figure 7.29 – Copenhagen: the SDUK Atlas map published in 1837; see Figure 7.27 for identification of the main parts of the city.

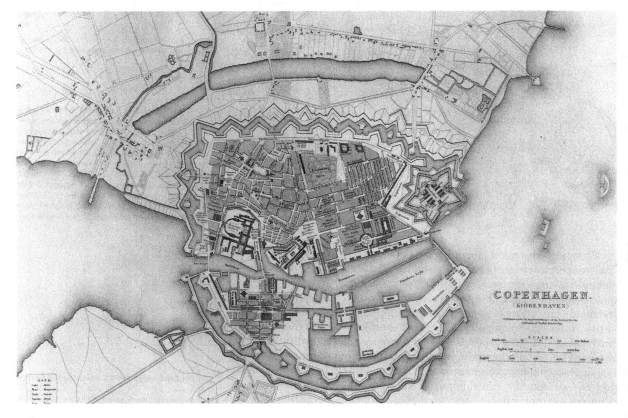

COPENHAGEN.
(KIÖBENHAVEN)

motif of four palaces, each flanked by two pavilions, which together form the large unit of the octagonal place'.[40] The two streets which link the space to the surrounding district are the Amaliengade, running parallel to the waterfront, and the Fredericksgade, which connects the waterfront to the monumental Frederickskirke. This church was completed towards the end of the nineteenth century as a much less impressive version of Eigtved's original design. At the intersection of these axes, in the centre of the square, an equestrian statue was erected to Frederick V in 1771.[41]

Paul Zucker considers that it is certainly no exaggeration to rank Amalienborg Square, although it is on a smaller scale, with the imperial fora, St Mark's Square and Piazzetta in Venice, St Peter's Square in Rome, and Nancy, as one of the most perfect realizations of grouped squares.[42] Nicolai Eigtved was undoubtedly influenced by contemporary work in Paris. The competition for the statue-square for Louis XV which had been held in 1748, the year before the Amalienborg Square was started, would have been known to him and there is a close similarity between his plan and that of the Place de la Concorde – the winning scheme in the competition with a similar waterfront/statue/vista-closing church (the Madeleine) sequence.[43] Zucker considers that the regular alternation between the height of the four monumental palaces on the one hand, and of the smaller entrance pavilions on the other hand, creates a fluctuating rhythm which lends a certain visual freedom to the square; this freedom corresponds exactly to the French ideals of the eighteenth century when the compactness of the previous seventeenth-century squares was criticized.[44]

Development of the old fortified zone, largely for housing purposes, early in the second half of the nineteenth century, is similar in many ways to the Ringstrasse development in Vienna (see pages 228–9).

We cannot fail to admire the ability of the eighteenth century to build up an entire new quarter at one stroke. The plans of the Amalienborg district were first laid in 1749 and when Eigtved died in 1754 the whole of the extensive quarter with its beautiful palaces, a handsome hospital, and its many fine houses, was entirely laid out and many of the buildings completed. This was possible only because Eigtved fully mastered his art. In all his undertakings he worked with dimensions and shapes with which he was entirely familiar. He was able to combine them to form extraordinary compositions like the really quite extravagant church project. But he could also use them for purely functional buildings, such as Frederik's Hospital, in which the dimensions of the long wards are based on the size of hospital beds, and the wards located to obtain the best light. And there is the warehouse of the Asiatic Company, down by the harbour, in which the hoisting lofts are the dominating motif. In all his works, small as well as great, Eigtved demonstrated that it is the architect's problem to join the various elements of a building together in a clear and convincing manner, and to group them into coherent units.
(S E Rasmussen, Towns and Buildings)

Figure 7.30 – Helsinki (Helsingfors): Finland gained national independence in 1917, and Helsinki was confirmed as the new capital, continuing its status since 1812 during the period of Russian rule. Earlier, from the twelfth century until 1809, the country was under Swedish dominance and the capital was at Turku further to the west.

The characteristic early history of Finnish cities is one of rebuilding after devastation by fire. In 1545 Turku suffered that fate and five years later Gustavus Vasa founded Helsinki as an alternative port at the mouth of a small river, situated some distance to the north-east of the later city which was planned in 1639. The new site was a rocky, indented peninsular with the port around Södra Hamnen and with gridiron districts conforming to the irregular topography. Helsinki was completely destroyed by fire in 1808 and the plan of 1914, as published in Baedeker's Russia of that year, shows the two main gridiron alignments which were subsequently consolidated as the basis of the modern city centre form. In 1812 the population was approximately 10,000; by 1900 it had risen to 60,000. (See K Baedeker, Baedeker's Russia 1914; H Lilius, Suomalainen puukaupunki (The Finnish Wooden Town – English summary), 1985.

38, 39. Rasmussen, Town and Buildings.
40. Zucker, Town and Square.
41. See Chapter 6 for the introduction to this form of aggrandizement.
42. Zucker, Town and Square.
43. See Chapter 6, Figure 6.15.
44. Zucker, Town and Square.

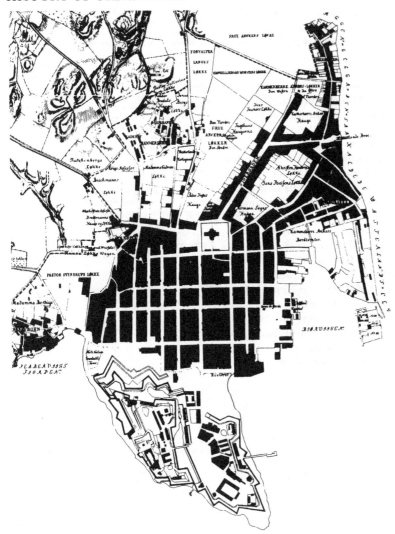

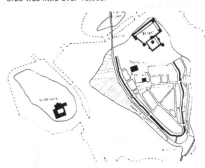

Figure 7.31 – Oslo: the city had its origins in a small fishing and trading settlement alongside the River Alna (modern Loevla) where it enters Oslo fjord some 80 miles from the Skagerack. Early in the fourteenth century Haakon V (1299–1319) made Oslo the capital of Norway ruling from the fortress of Akershus which was built on a rocky promontory. During the Middle Ages Akershus also served as a refuge for the civilian population; the town itself was not fortified. Apart from the main castle buildings, Oslo consisted of closely packed timber houses; the inevitable fires caused serious damage in 1352, 1523 and 1611 and were responsible for Oslo's slow medieval growth, until finally in 1624, a three-day fire completely destroyed the town. As rebuilt by Christian IV on a new site alongside Akershus to the north-east, the town was renamed Christiana; a title which it kept until reverting to Oslo in 1924. This new town had a simple gridiron plan and its spacious streets, combined with the compulsory use of stone, overcame the danger of future conflagrations. On this new basis Christiana grew steadily; all the original grid blocks were occupied by 1661, and with suburban additions the number of inhabitants had exceeded 5,000. In 1704 a plan was drawn up controlling growth on the basis of new gridiron districts. Economic incentives for rapid growth were lacking however, and by the end of the eighteenth century the population of city and suburbs was little over 10,000.

Figure 7.32 – Stockholm: the plan of Stadsholmen at the end of the thirteenth century. The present-day waterfront is shown dotted. The royal castle was located on the highest part of the island in the north-east corner.

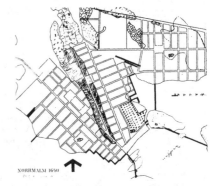

Figure 7.33 – Stockholm: the plan of lower Norrmalm in 1640; the old city of Stadsholmen is at the bottom. Stockholm's continuing central area regeneration pro-

Stockholm

Stockholm was founded in 1255 on Stadsholmen – an island in the narrow channel linking Sweden's extensive natural waterway system to the Baltic, by way of Lake Malaren.[45] Thus it controlled shipping along key trading routes and the north–south land crossing by way of its two bridges, Norrbro and Soderbro.

Early expansion on Stadsholmen required made-ground, gained by filling in between harbour jetties, and consequently took the form of long narrow blocks between the old wall and the new quayside. By the late fifteenth century Stockholm had become established as Sweden's leading town; continued growth resulted in sizeable bridgehead settlements and during the period 1620 to 1650, when the number of inhabitants increased from fewer than 10,000 to approximately 40,000, both Norrmalm and Sodermalm became extensively developed. Following the destruction by fire in 1697 of the old medieval palace, Tre Kronor, Nicodemus Tessin the Younger (1654–1728), architect to both the city and the royal court, designed the new palace of 1713 as part of a

45. For Stockholm see G Sidenbladh, 'Stockholm: A Planned City', *Scientific American*, 1965; Hall, *Planung Europaischer Haupstadte*.

grandiose attempt to unite Stadsholmen visually and symbolically with Norrmalm. However, as Kell Astrom relates, 'coming at a time of political and economic diversity, the whole project had to be kept

gramme – one of the 'most ambitious in the world – covers most of this part of lower Norrmalm and its implementation has been considerably aided by the existence of a regular grid structure.

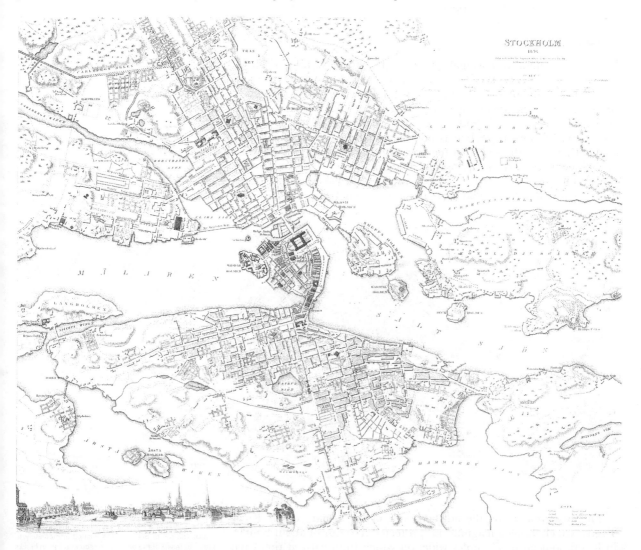

secret, known only to the king and his closest associates'.[46] Some of Tessin's proposals were evidently implemented in the 1770s under Gustavus III. Development of the mainland districts, especially Norrmalm and Kungsholmen further to the north-west, was such that by the beginning of the eighteenth century Stadsholmen had become an unfashionable backwater. The centre of Stockholm gradually moved across to Norrmalm leaving the island as the 'Old Town', with its shipping activities, warehouses and lower-class housing blocks.

The continuing story of Stockholm's mid-twentieth-century ambitious, seemingly successful, central area conservation and renewal programme is outside our present period. Although history has favoured Stockholm's modern planners – providing a grid structure which facilitated renewal of the modern commercial centre of

Figure 7.34 – Stockholm: the map published in 1835 by the Society for the Diffusion of Useful Knowledge. The old city of Stadsholmen is on the island in the main waterway (see the later thirteenth-century plan as Figure 7.32) with Norrmalm to the north, and Sodermalm to the south, which was in effect a second island, separated from the mainland by a partly artificial waterway. (The mid-seventeenth-century extent of Norrmalm, where the city centre renewal programme of the 1960s and 1970s has been concentrated, is given by Figure 7.33.)

46. K Astrom, *City Planning in Sweden*, 1967 (Swedish Institute for Cultural Relations with Foreign Countries).

Norrmalm, and a uniquely preserved historic nucleus on Stadsholmen – it must be stressed once again that one of history's main lessons is that 'paper planning', however seductive its images, amounts to nothing without political support.[47]

Lisbon

Lisbon, the capital of Portugal, has a lengthy history which has been traced back to before the Romans.[48] It was the centre of Portuguese political and commercial life during the 'Golden Age' of the sixteenth century, when the great overseas empire was created, and it had attained a population of almost 100,000 by 1550. In the mid-eighteenth century the history of Lisbon is notable for the systematic, planned reconstruction of the centre of the city, after devastation by an earthquake, which entailed redistribution of land holdings of the kind which would have been required, a century earlier, if London were to have implemented the 'Wren Plan' (see Chapter 8 and Figure 8.11).

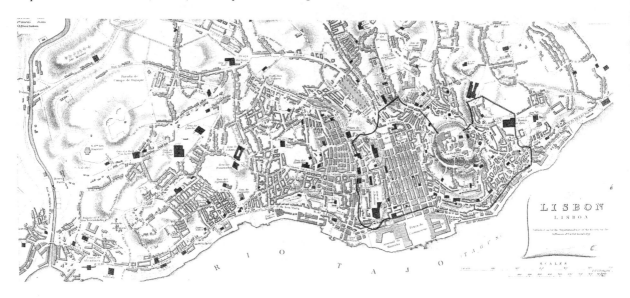

The city grew up on the northern side of the estuary of the River Tagus, which forms a magnificent, protected harbour of about 8 miles in length and up to 1¾ miles wide, on which from earliest times the commercial prosperity of Lisbon has been based. The first port was not on the main tidal stream of the Tagus, but was located in a small inlet known as the Baixa: a similar situation to that of Roman London, thirteenth-century Liverpool, and Philadelphia. And like them the first part of Lisbon was subsequently filled in and built over. An easily defended hill alongside the Baixa to the east formed the ancient nucleus of the city and was occupied by the Phoenicians and later the Carthaginians, before coming under Roman dominion in 207 BC. They knew it as Felicitas Julia, but although *municipium* status was gained at the beginning of the third century AD, it remained a comparatively unimportant town.[49]

After the end of the Roman Empire, Lisbon was taken in 419 by the Visigoths who fortified an area of some 15 hectares around and includ-

Figure 7.35 – Lisbon: the map published by the Society for the Diffusion of Useful Knowledge in 1833 (above right). The hill of Sao Jorge is located immediately to the right (east) of the distinctive regular gridiron blocks of the Baixa district, inland from the Praca do Commercio. The solid line enclosing the hill and the Baixa, but excluding the Praca (which was constructed later on reclaimed land) delineates the extent of the walls of 1373–5, the Cerca Fernandina.

---

47. It should be recognized that the Stockholm city planners for many years benefited greatly from the city's ownership of much of the land over which expansion took place.

48. For Lisbon see Branch, *Comparative Urban Design*; Gutkind, *International History of City Development: Urban Development in Southern Europe*, Volume 6.

49. See Chapter Three for description of the Roman division of Iberia into provinces, one of which comprised the future Portugal.

ing the original hilltop location. From 714 to 1147 the city was occupied by the Moors who retained those walls which were not extended until in 1373 the Castilians, under Fernando I, commenced work on a new extended circuit enclosing about 100 hectares. During the sixteenth-century decades of prosperity the city survived several earthquakes and, as progressively reconstructed and extended, a new waterside grid-structured centre was established over the Baixa.

Figure 7.36 – Lisbon: aerial photograph of the Plaza del Comercio, and the gridiron planning of the Baixa district, enclosed within the older organic-growth parts of the city.

Figure 7.37 – Oporto (Porto): the centre of the city taken from the SDUK Atlas map published in 1833.

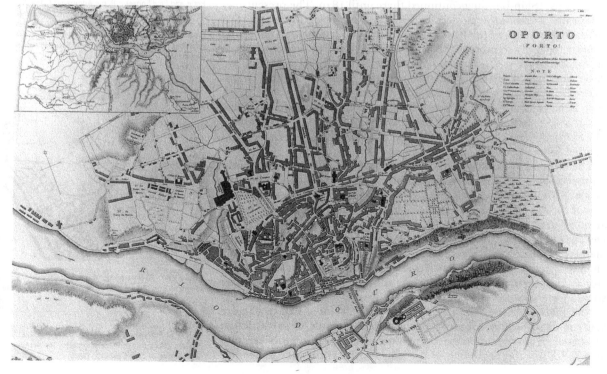

St Petersburg

St Petersburg (previously Leningrad) is the most important example of the application of fully developed Renaissance design principles to the large-scale planning of a complete new city.[50] In contrast to conditions under which the great majority of Renaissance urbanists had to work – either having to add new districts to medieval cores, or carving space for their projects out of the existing urban fabric – the planners of St Petersburg had two major advantages. These were first, an empty site regulated only by topography; and second, the fact that by the time St Petersburg was founded in 1712 there were the examples of almost three centuries of European Renaissance urbanism to draw on.

During the seventeenth century Russia had grown into a unified expansive nation. But she was still without an outlet either to the Black Sea, controlled by the Turks, or to the Baltic, where Sweden had conquered the territory around the Gulf of Finland. Peter the Great (1688–1725) gave great strategic priority to the re-establishment of the Baltic outlet and successfully regained control of the Gulf of Finland during the Northern War of 1700–21. To consolidate the position along the line of the River Neva, the Russians began construction in May 1703 of the Peter and Paul Fortress on Zayachy Island, at the broadest point of the Neva estuary. By 1712, the fortress and associated naval base of St Petersburg had emerged to rival Moscow as the capital of Russia.

The first, temporary, earthworks of the fortress were converted into powerful brick fortifications between 1700 and 1741. In 1712 the Cathedral of Peter and Paul was started as the first of a number of historic buildings within the fortress. Across the Neva the Admiralty Shipyard, also founded in 1703, was strongly fortified after 1705 and contained ten large covered slipways with ancillary storehouses and workshops. (Ship-building continued there until 1840. The present-day Admiralty Building, 1,340 feet in length, dates from 1806–23.) The increasing strength of the Russian navy and establishment of advanced frontier defences soon enabled St Petersburg itself to acquire a more peaceful character. On the left bank of the Neva, directly opposite the Peter and Paul Fortress, Peter laid out the Summer Garden (1704) where he dreamed of making a garden better than that of the French king at Versailles.[51] The Summer Palace was built between 1710 and 1714, to be followed by the Winter Palace (now the Hermitage Museum) of 1754–62. This latter was a rebuilding of an earlier and smaller fortified palace.

The most important element of the basic eighteenth- and nineteenth-century plan of St Petersburg is the centring of three main streets on the tower of the Admiralty. This gives to the immediate vicinity a combination of radial and gridiron streets which follows the example of the town area of Versailles (see Chapter 6), and presages the even more impressive appearance of the aggrandizement factor underlying L'Enfant's plan for Washington, DC (see Chapter 10). The road from the east was an existing route and is now, as Nevsky Prospect, the main street in the city linking the centre to the Moscow railway station. The two other streets were completed by 1800.

The plan of Palace Square, in front of the Winter Palace (Figure 7.40) is described by Edmund Bacon as an example of great design produced by accepting an existing plan and turning its problems into assets.[52] Until the beginning of the nineteenth century the space im-

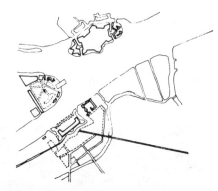

Figure 7.38 – St Petersburg in 1750, after nearly a first half-century's growth (see below for location of parts).

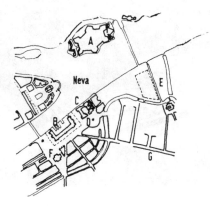

Figure 7.39 – Central St Petersburg in 1850. Key: A, the Peter and Paul Fortress; B, the Admiralty; C, the Winter Palace (now the Hermitage Museum and Art Gallery); D, the Alexander Column in the centre of the Palace Square, as Figure 7.40; E, Summer Garden; F, St Isaacs Cathedral; G, Nevsky Prospect (terminated at its distant end by the Moscow Railway Station).

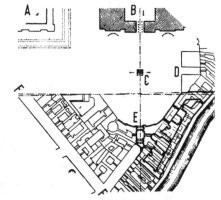

Figure 7.40 – St Petersburg: the Palace Square in front of the Winter Palace (north of the top). Key: A, the

50. For St Petersburg see Branch, *Comparative Urban Design*; Y A Egorov (trans. E Dluhosch), *The Architectural Planning of St Petersburg*, 1969.

mediately to the south of the palace had an unresolved form. The western side was open to the Admiralty; facing the palace and forming an uncomfortable trapezoidal shape, were unimportant buildings, with some façades parallel to and others almost at right angles to Nevsky Prospect. It was decided to reorganize the elevations forming the square, to make it an appropriate setting for the palace. The properties on the southern and eastern sides were bought up and in March 1819

Admiralty; B, The Winter Palace; C, Alexander Column; D, new buildings forming the eastern side of the square at right-angles to the palace elevation; E, the new arched entrance into the square on the central axis of the palace; F–F, Nevsky Prospect. The line of the pre-existing elevation across the south-eastern side of the square continued that of the linking street to Nevsky Prospect. The new building elevations of the General Headquarters (1819–29) are shown in heavy line.

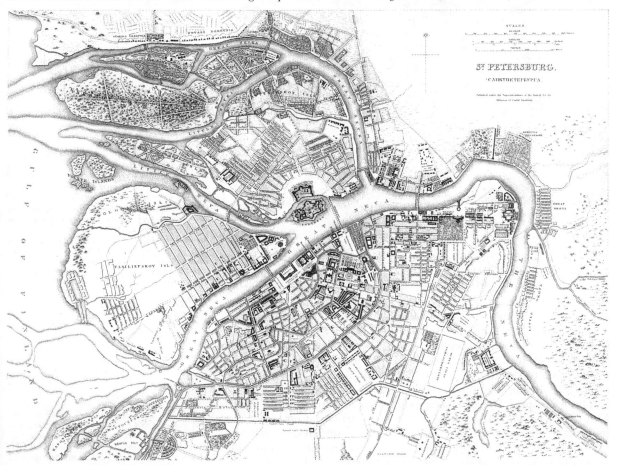

K I Rossie was commissioned to convert them into one building, known as the General Headquarters.

Work was finished in 1829 and the new elevation, some 640 yards in length, took the form shown by the heavy line on the plan. A magnificent triumphal arch on the central axis of the palace formed the new entrance into the square. An inspired touch linked this by curved sections to the retained south-western side and the new south-eastern side. New buildings closing the eastern side of the square, at right-angles to the palace, were added between 1840 and 1848, completing the design. Earlier in 1832, the 155 feet 6 inch high, 600-ton triumphal column was erected in the centre of the square.

Thus the design of the Palace Square in St Petersburg, in its respect of the existing city form, can be compared with Michelangelo's plan for the Capitoline Piazza in Rome, where two medieval buildings determined the form of Renaissance urbanism (see Chapter 5).

Figure 7.41 – St Petersburg: the map published by the Society for the Diffusion of Useful Knowledge in 1834 and which relates to Figure 7.39 for identification of parts of the inner city. The modern Nevsky Prospect terminates at the Moscow Railway Station at its eastern end. (Comparison with the map in the 1914 *Baedeker's Russia* shows the limited expansion of the city during the eighty-year interim.)

51. Egorov, *The Architectural Planning of St Petersburg.*
52. E Bacon, *Design of Cities*, 1974.

245

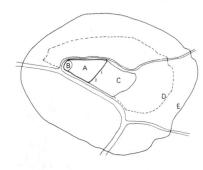

Figure 7.42 – Moscow in 1606, one of the clearest European examples of concentric-ring growth within successive defensive perimeters.

Figure 7.43 – Moscow: A, Kremlin; assumed extent of first settlement; C, Kitay Gorod; D and E, sixteenth-century perimeters; 1, site of Red Square; 2, St Basil's Cathedral.

At this period the city's growth was still contained with the sixteenth-century perimeters (D and E). The first of these rings was placed approximately 2 km out from the Kremlin, on the north bank only, and forms the line of a modern park boulevard system. The second perimeter, 15 km in circumference, was centred precisely on the belfry of the Cathedral of Ivan the Great in the Kremlin and today forms a continuous inner-ring motorway system including multi-level intersections with the main radial streets. Following the loss of the city's governmental functions to St Petersburg in the eighteenth century, Moscow devoted itself increasingly to trade and industry.

Moscow

The earliest known reference to Moscow is of 1147 although it seems most likely that the attraction of the site – a prominent ridge alongside and commanding the Moskva river – would have encouraged earlier occupations.[53] In 1147 Yuri Dolgoruki, Prince of Suzdal, constructed a fortress on the northern, left bank of the Moskva on high ground some 130 feet above the river, in the angle formed by the main stream and a tributory, the Neglinnaia. Within an oaken palisade of 1156–8 the first settlement occupied an area of about 10 acres at the western end of the ridge. After two centuries of steady expansion eastwards along the ridge – the area of the present-day Kremlin – approximately 75 acres had been occupied. (*Kreml*, meaning 'high town', is analogous to burg.) The first Kremlin wall of masonry was constructed from 1367; the immensely impressive, if longsince redundant, present-day wall was built between 1485 and 1508. This wall was crenellated, after the Italian fashion, and was strengthened by twenty towers.

From the fourteenth century the Kremlin consolidated its role as a city within a city, the political and administrative citadel of a ruling elite. Retention of the outer wall can but have contributed to its glorification, although that should perhaps be read as 'fear' during the decades of the worst Communist oppression. Hopefully the events of the early 1990s have seen the last of that symbolic political role.[54] Further to the east, beyond a moat cut across the ridge and the market place, later to become Red Square, a merchant's town, known as Kitay Gorod, became established from this time. (Although this name translates literally as 'China Town' its true derivation is from the Tartar word 'fortress'.) Kitay Gorod was walled between 1535 and 1538. The Kremlin and Kitay Gorod combined to form the historic *Gorodskaya Tchast* (the City Quarter) around which nucleus the further concentric rings of suburban development took place. The new wall of 1586–93 is marked by the modern *Bulvarnoe koltso* (the Boulevard Ring).[55]

Figure 7.44 – Samarkand: the plan published in the 1914 edition of Karl Baedeker's *Russia*. The city is in Uzbekistan, some 240 km south-west of Tashkent, situated in the fertile valley of the Zeravshan. Samarkand's ancient history dates from before the first recorded mention in 329 BC when, known as Maracanda, it was captured by Alexander the Great. Maracanda occupied

53. For Moscow see Baedeker, *Baedeker's Russia*, 1914; Branch, *Comparative Urban Design*, for the 1836 SDUK Map.

54. At the completion of third edition preparation, the Kremlin is still to be recognized as the symbolic centre of Russian political life, but with vastly diminished international and domestic significance following the demise of the unified USSR.

55. This defensive system was incomplete in 1591 when the city was last threatened by the Tartars, in face of whom a further outer ring of earthworks was hastily thrown up, to be replaced by a later stone wall ('E' on Figure 7.43). This line is followed by the city's present-day inner ring motorway. See Chapter 6 for the equivalent, but later, Boulevard Périphérique in Paris.

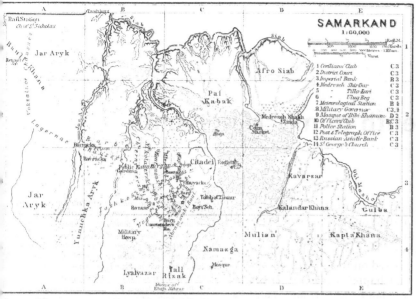

high ground to the north of later successor cities and is known today as the completely deserted and as yet only partially excavated area of Afro Siab.

The new city which grew up beneath Afro Siab was taken by the Arabs during the seventh century and then, in 1221, it fell to Jenghiz Khan. Samarkand's great period in history was under the rule of Timur (or 'Tamerlane') who gained possession in 1369 and who ruled from a magnificent fortified castle on the site of the 1914 citadel west of the old city. During his reign the city's several world-famous architectural monuments were constructed.

In 1868 Samarkand was taken by the Russians, in course of their Imperialist Eastern expansion, and from 1871 a new 'European' quarter was laid out to the west of the old native city. The contrast between the Islamic forms of the historic Uzbek nucleus and that of the planned European Russian quarter is comparable to that at Tunis where the French added their new district to the Islamic medina (see Chapter 11). (See J H Bater, *The Soviet City: Continuity and Change in Privilege and Place*, 1976; Baedeker, *Baedeker's Russia*, 1914; A Sheehy, 'Golden Samarkand', *Geographical Magazine*, vol XXXVI, no 8, 1963.)

Figure 7.45 – Kiev: the several stages in the growth of the historic nucleus of the modern Soviet city. Kiev is a further example of the multi-nucleus type of city which typically became established in linear form along the bank of a major river (see also Nijmegen, page 141, and Magdeburg, page 106). The location of the city is on the western, right bank of the Dnieper where a high-ground *gorodishche* ('burg') of the eighth century AD controlled a major river crossing. From the twelfth century the *podol*, or commercial town became established on the adjoining riverbank area; and the important Pecherski Monastery, with its *sloboda* ('civil town') grew up on a second ridge-top further downstream. Kiev's extremely important strategic location ensured growth and necessitated frequent re-fortification. By 1811 the population was 23,000; by 1863 it had reached 68,000. (See E A Gutkind, Ed. Gabriele Gutkind, *International History of City Development*, Volume 8, *Urban Development in Eastern Europe: Bulgaria, Romania and the USSR*, 1972.

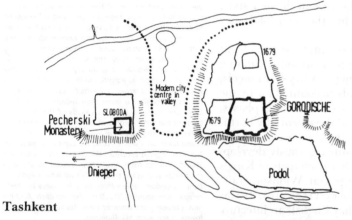

Tashkent

The city is the modern capital of the Uzbek Soviet Socialist Republic and is situated approximately 2,750 km south-east of Moscow.[56] Islam has been the religion of this part of the former Soviet Union for more than one thousand years, and its guideline determinants of the historic urban form of Tashkent, with that of Samarkand, are described in Chapter 11. The city is first mentioned in Chinese scripts of the second century BC, but the favourable oasis location on important ancient trading routes suggests earlier origins, most probably comparable to those of Samarkand. Historically Tashkent is over-shadowed by Uzbekistan's two other ancient cities of Samarkand and Bukhara. The attractive location left it open to attack and little remains of its undoubted historic architecture. This is also partly owing to the city's vulnerable seismic-zone position (the 7.5 Richter-scale earthquake of 26 April 1966 caused enormous devastation), and to the fact that Tashkent (unlike Samarkand) was not a royal capital during the formative Timurad cultural period. The native Uzbek city was taken by the Russian Empire in June 1865 when its roughly circular area of approximately 1.6 kilometre radius contained a population of some 75,000.

Figure 7.46 – Tashkent in 1865: the Islamic form of the native Uzbek city, to which a geometrically planned European quarter was added from the later 1880s.

56. See A E J Morris, 'City of Seven Earthquakes', *Geographical Magazine*, May 1974, pp. 409–16.

8 – Britain: Sixteenth to mid Nineteenth Centuries

In Britain the first example of Renaissance urbanism was Covent Garden Piazza in London, developed from 1630 onwards. As in Western Europe generally, the Renaissance in Britain in the plastic arts – painting, sculpture, architecture and urbanism – was preceded by the creation of a favourable intellectual climate, based on the growth of literacy and scientific humanism, as outlined in Chapter 5. The Renaissance had similar origins in both France and England; painters and sculptors were brought in from Italy to work at the royal courts in the new style; the dates are around 1500 in France and during the reign of Henry VIII (1509–47) in England.

The significant delay between the more or less simultaneous advent of Renaissance art in both countries, and the first urbanism (dating from 1545 in provincial France with the planning of Vitry-le-François, 1607 in Paris and the 1630s in London) is mainly accounted for by the firm hold retained by Gothic architecture in its concluding phases. Italy, on the other hand, had never fully accepted the essentially northern Gothic style and when the first Renaissance building was started in Florence in 1419 architecture in England was still vigorously developing its late Perpendicular masterpieces – notably the chapels of King's College, Cambridge (1146 onwards), St George's at Windsor (1481 onwards) and Henry VII's at Westminster Abbey (1503–19). It was only towards the end of the sixteenth century that English architecture saw the gradual introduction of Renaissance details on otherwise Gothic buildings, and it was not until 1621 that the first essentially Renaissance building was completed – the Banqueting Hall in Whitehall – to be followed within a decade by the first urbanism in London.[1]

London

PRE-SEVENTEENTH-CENTURY GROWTH

An account has been given in Chapter 3 of the devastation that followed the withdrawal of the Roman legions from Britain in the fifth century, with conjectural answers to questions posed by the difference between the Roman gridiron plan and the organic growth Saxon and early medieval street pattern.[2] Alfred the Great repaired the defences and consolidated London 'as a fortified garrison town which served the interests of the king by maintaining its own independence'.[3] Although not yet the national capital, London was by far the most important commercial centre.

It is probable that the population of England and Wales at the end of the Queen's reign (Elizabeth I) had passed four millions, about a tenth of its present size. More than four-fifths lived in the rural parts; but of these a fair proportion were engaged in industry, supplying nearly all the manufactures required by the village, or, like the clothiers, miners, and quarrymen, working for a more general market. The bulk of the population cultivated the land or tended sheep.

Of the minority who inhabited towns, many were engaged, at least for part of their time, in agriculture. A provincial town of average size contained 5,000 inhabitants. The towns were not overcrowded, and had many pleasant gardens, orchards, and farmsteads mingled with the rows of shops. Some smaller towns and ports were in process of decay. The recession of the sea, the silting up of rivers (which gradually put Chester on the Dee out of action as a port), the increase in the size of ships demanding larger harbours, the continued migration of the cloth and other manufactures in rural villages and hamlets, were all causes of the decline of some of the older centres of industry or commerce.

Yet the town population was on the increase in the island taken as a whole; York, the capital of the north; Norwich, a great centre of the cloth trade, welcoming skilled refugees from Alva's Netherlands; Bristol with mercantile and inland trade of its own wholly independent of London – these three were in a class by themselves, with perhaps 20,000 inhabitants each. And the new oceanic conditions of trade favoured other port towns in the west, like Bideford.

But, above all, London, absorbing more and more of the home and foreign commerce of the country at the expense of many smaller towns, was already a portent for size in England and even in Europe. When Mary Tudor died it may have had nearly 100,000 inhabitants; when Elizabeth died it may already have touched 200,000.

(G M Trevelyan, Illustrated English Social History)

---

1. For general references on Renaissance urban form in Britain see E A Gutkind, *International History of City Development*, Volume 6, *Urban Development in Western Europe: Great Britain and The Netherlands*, 1971; C W Chalklin, *The Provincial Towns of Georgian England: A Study of the Building Process 1740–1820*, 1974; for general architectural references see N Pevsner, *An Outline of European Architecture*, 1959; P Murray, *Renaissance Architecture*, 1971; J Hook, *The Baroque Age in England*, 1976.
2. For Saxon London see M Biddle, 'Towns', in D M

Early in the eleventh century the kings moved their residence from Winchester to Westminster – some miles upstream from the City of London – where a royal precinct was established on Thorney Island, consisting of church, monastery and palace. William the Conqueror confirmed the existence of the independent royal and commercial cities of Westminster and London. These twin centres were linked by Whitehall and the Strand, the land between this route and the river being mainly used for the town palaces of noble families. As early as 1175 the cities were joined by a ribbon of development and it was inevitable that the first stage in the expansion of the capital was their unification into one continuous urban area, with each, however, still zealously guarding its political independence.

By 1605 London's estimated population was 225,000, of which total '75,000 lived in the City, 115,000 in the liberties (which included such precincts as St Martin's-le-Grand, Blackfriars, Whitefriars, Duke's Place Aldgate, and the areas between the gates and outer bars, i.e. the suburbs), and 35,000 in the out-parishes'.[4] The population of the City in 1530 had been around 35,000 with the figure for the whole London area about 50,000. Only a small proportion of this increase was due to natural population growth; by far the greater part represented population movement into the capital, both from other parts of the country and from abroad. The internal migration was brought about by a number of factors, in particular the effect of the land enclosures. These had created more efficient sheep-farming conditions for wool production, but deprived many countrymen of their living and sent them to the towns in search of employment. Immigration from the continent was also influenced by the growing textile industry, with many weavers, among others, coming to settle around London.

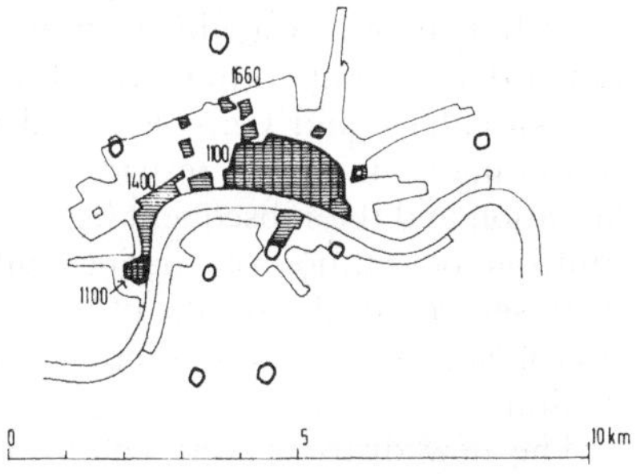

Parliament, the Crown, and the City of London authorities were all united in their resolve to arrest the drift to the capital. This policy, which proved to be markedly unsuccessful, was the result of a number of interrelated pressures, each involving the interests of the City merchants. First there was the political danger of revolutionary action by a large unemployed urban mob. This, by its existence, enabled wages to be kept at an extremely low level: a degree of unemployment was to the City's advantage. By the second half of the sixteenth century the situation was dangerously out of control. The problem was increased still further after the dissolution of the monasteries, with 'the friars who formerly had relieved the poor, now added to their numbers themselves'.[5] A second consideration was a dawning awareness of the connection between urban overcrowding and disease; the sequence of plagues, culminating in the visitation of 1665 which killed an estimated 90,000 people in the London area, represented a constant threat to the development of foreign trade. The third and most immediate challenge to the City's commercial interest, however, was the growth of suburban workshops operating outside the control of the long-established guilds, set up in many instances by unqualified apprentices in direct competition with their erstwhile masters.

Queen Elizabeth's famous proclamation of 1580 which 'doth charge and strictly command all manner of persons, of what quality soever they be, to desist and forbear from any new buildings of any house or tenement within three miles from any of the gates of the City of London',[6] was based on 'the good and deliberate advice of her Council, and the considerate opinions of the Lord Mayor, Aldermen, and other

The rising importance of parliament enhanced the capital's popularity and it was the home of busy tradesmen and merchants, of poets and painters, of beautiful women and witty and intelligent men. It was by far the largest port in the country and provided much employment; at the same time it contained poverty and disease and there were numerous slums. Though it was a gay city, above all when decorated by the full pageantry of the Court, it was also incredibly dirty and smelly. Hackney and Stepney were overcrowded and insanitary, while our modern suburbs such as Kensington and Hampstead were still country villages.
(M Ashley, England in the Seventeenth Century)

Figure 8.1 – Three stages in the early growth of London showing the unification of the Cities of London and Westminster along the line of the Strand, on the north bank of the Thames. Subsequent expansion east and north-east took the usual British form of radial-route ribbon-development, followed by infilling of the back land. West and north-west of the City of London development was partly planned – e.g. Covent Garden and Lincoln's Inn, and partly unplanned. (For the extent of the eighteenth- and nineteenth-century planned districts further west and north see Figure 8.2.) South of the river the original bridgehead settlement grew slowly by comparison, mainly as a waterfront ribbon along the Thames. The topographical determinant was a constraint on development on the south bank of the Thames. (See also Figures 8.2 and 8.3.)

Wilson (ed) The Archaeology of Anglo-Saxon England, 1976.

3. S E Rasmussen, London: The Unique City, rev. edn 1982.

4. N Pevsner, London: The Cities of London and Westminster, 1973.

5. 6. Rasmussen, London: The Unique City.

grave wise men in and about the City'.[7] This proclamation, which was embodied in an Act of Parliament in 1592, is a clear recognition of the fact that the government of the realm was economically dependent on London and London meant, to Elizabeth, the City.

It is essential to recognize the extent to which London dominated England's international trade. From 50 per cent of customs revenues in 1500, the proportion had increased to 66 per cent during Henry VIII's reign, and by 1581–2 it was over 86 per cent.[8] Rasmussen remarked that 'Elizabeth, who knew how to keep on the right side of the merchants, never called on the City in vain in time of need'.[9]

Retaining favour with the merchants inevitably resulted in one law for the rich and another for those who could not afford to buy 'planning' consents. An early law of 1588 prohibited building around London (and other towns) unless the house was to be built on four acres of ground, and the Crown was always prepared to waive the restrictions on receipt of suitable 'dispensation'. Legislation imposing the 3-mile limit was re-enacted in 1602; the distance was reduced to 2 miles in 1607 but increased again, to 7 miles, in 1615. Nikolaus Pevsner has noted that a 7-mile radius would have drawn this *cordon sanitaire* as far as Tottenham and Chiswick, and it must have been clear from the beginning that no such restriction could be enforced.[10] The 1580 proclamation and the subsequent legislation were also concerned to prevent multiple occupation of existing buildings. Penalties for contravention were severe but the results of bought permissions and lax administration meant a continuing increase in the extent and population of the capital.

The new districts were concentrated east and west of the City of London (Figure 8.1). To the north development was not taken far beyond the old medieval wall. Eastwards, Stow (writing in 1598) records 'half a mile of bad cottages in Whitechapel' and the existence of Wapping as 'a continual street or filthy passage'.[11] Expansion westwards at first covered the area between the Cities of London and West-

In England in the sixteenth and seventeenth centuries, for example, plague epidemics were common in the large towns like London, Norwich and Bristol, and each major outbreak brought a very severe toll of life. In London in 1603 at least 33,000 people died of the plague and 10,000 of other causes making a total of 43,000; in the plague of 1625 41,000 (63,000); in the Great Plague of 1665 (the last major English outbreak) 69,000 (97,000).

These figures are based on the Bills of Mortality and are minima; the true figures must have been higher and may have been very much higher. In between the big outbreaks there were many years in which plague deaths could be measured by the thousand. During the period 1600 to 1660 the population of London grew from about 200,000 to about 450,000. In plague years therefore between a sixth and a quarter of the population of a great city might die even in the seventeenth century.

(E A Wrigley, Population and History)

... sometimes the Bishoppe of Carliles Inne, which now belongeth to the Earl of Bedford, and is called Russell or Bedford House. It stretcheth from the Hospitall of Savoy, west to Ivie bridge, where sir Robert Cecill, principall Secretary to her Majestie, hath lately raysed a large and stately house of brick and timber as also leviled and paved the highway neare adjoining, to the great beautifying of that street and commoditie of passengers. Richard the 2, in the 8. of his raigne, granted license to pave with stone the highway called Strand street from Temple barre to the Savoy, and tole to be taken towards the charges and the like was granted in the 24. of H. the 6.

(J Stow, A Survey of London, 1603)

Figure 8.2 – Part of a map of the City of London before the Fire of 1666, showing the line of the medieval wall and ditch (already being built over to the west); Old St Paul's Cathedral; and Old London Bridge as lined with buildings. The wall all around the City, which followed the line of Alfred's defences, was long since militarily useless, compared with contemporary continental fortifications exemplified by Gothenburg (1630) and Cologne (1646) shown in Chapter 5, Figures 5.5 and 5.6 respectively. There were also London's completely 'open' suburbs, especially those of the wealthy to the west. When threatened by the Royalists during the Civil War of 1642–51, the Parliamentarians, who held the capital, hastily constructed an outer ring of earthwork forts and redoubts, etc, the dubious strength of which was never tested. For a map of these defences see C Duffy, *Siege Warfare, Volume 1, The Fortress in the Early Modern World 1494–1660*. Washington, DC was similarly fortified, and spared examination, during the American Civil War, in face of a particularly adventurous Confederate advance.

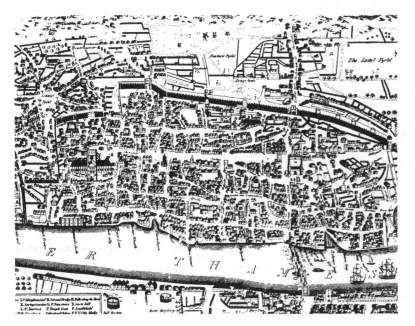

7. Rasmussen, *London: The Unique City*; J Summerson, *Georgian London*, 1978; D J Olsen, *Town Planning in London: The 18th and 19th Centuries*, 1982; Pevsner, *London: The Cities of London and Westminster*; M C Branch, *Comparative Urban Design*, 1978, for the 1843 SDUK Map of London; P Glanville, *London in Maps*, 1972.
8. G W Southgate, *English Economic History*, 1934.
9. Rasmussen, *London: The Unique City*.
10. Pevsner, *London: The Cities of London and Westminster*.
11. J Stow, *The Survey of London*, 2nd edn, 1603, Everyman's Library Edition, 1960.

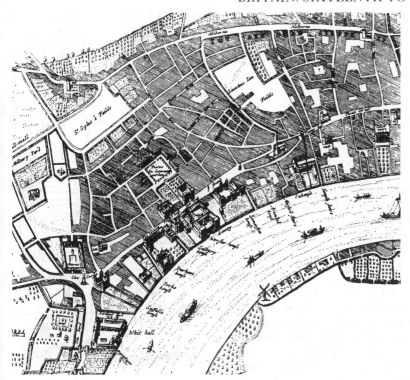

Figure 8.3 – A section of a map of London produced around 1667 showing the built-up area joining the City of London (to the right) and the City of Westminster (bottom left). A number of the mansions between the Strand and the Thames can be seen; Covent Garden Piazza is located left of centre, immediately north of the garden of the Duke of Bedford's mansion on the north side of the Strand. Lincoln's Inn Fields has yet to be formally developed (Kingsway, the present-day main north–south route to its west dates only from the early years of the twentieth century). Trafalgar Square at the south-western end of the Strand was the work of Nash in the 1820s.

The undeveloped southern (right) bank of the Thames is in total contrast to the already densely built-up northern district. The Thames served as a main highway for commercial and private movement; it was the most convenient way to travel between two cities and the numerous landing stages along the northern bank. (See Figure 8.4 for the contrast between the private, grand house waterfront, and the commercial redevelopment that followed.)

The south side of Covent Garden Square lieth open to Bedford Garden, where there is a small growth of trees most pleasant in the summer season; and on this side there is kept a market for fruits, herbs, roots and flowers every Tuesday, Thursday and Saturday, which is grown to a considerable account, and well served with choice goods, which make it much resorted to.
(J Strype, 1720 revision of Stow's A Survey of London)

minster and it is here, north of the Strand, that the Covent Garden development introduced Renaissance urbanism into the country.

COVENT GARDEN

This land, on 7 acres of which the original Covent Garden development was carried out,[12] was the first estate to be owned by the powerful Russell family. It was granted by Edward VI to the first Earl of Bedford in 1553, having been acquired by the Crown from the Abbey of Westminster in 1536. The name Covent Garden is derived from its previous use as part of the produce gardens of a convent. The Bedford Estate at first leased the land for grazing purposes – in 1559 the Earl is recorded as letting 28 acres on a 21-year lease for this purpose. Part of the Strand frontage was used for the Bedfords' town house in about 1560, with an enclosed and formally planted garden immediately behind, to the north. Unplanned, generally illegally constructed, groups of houses were built in the open fields north of this garden from 1603 onwards, as part of the uncontrolled growth of London. The Covent Garden district was clearly 'ripe for development' and the fourth Earl of Bedford, 'very intelligent in a practical business-like way'[13] as Sir John Summerson has described him, recognized its potential value. Summerson records that developments along Drury Lane and Long Acre were already bringing in £500 a year when the earl succeeded to the property.[14]

Obtaining the seventeenth-century equivalent of planning permission for an extensive upper-class housing development, as opposed to the tolerated process of gradual suburban accretion, proved difficult at first. Eventually, in return for a consideration of £2,000, a licence to

12. A C Downs jr., 'Inigo Jones's Covent Garden: the first 75 Years', *Journal of the Society of Architectural Historians*, vol XXVI, 1967: the comprehensive essential reference. In November 1974 the Covent Garden vegetable market (which had come to take the place of the original residential development) moved out to its new location at Nine Elms, Battersea. The first Greater London Council proposals for the extensive redevelopment of much of the area were on a grandiose scale, featuring high-rise commercial and office buildings and a new multi-level circulation system (see R Rookwood, 'A New Covent Garden', *Official Architecture and Planning*, May 1969). Concerted local community action forced the rejection of large-scale redevelopment in favour of rehabilitation and localized rebuilding, including new low-rental housing. The magnificent Central Market Hall occupying the main part of Inigo Jones' Covent Garden Piazza has been thoroughly restored and adapted for specialist small-scale shopping and related formal and informal tourist industry activities. The Covent Garden area has entered the 1990s as one of London's most popular attractions, and at time of writing (1992) it appears that the long-running saga of the Opera House's attempts to extend its facilities, and to develop adjoining land, may at last be coming to fruition.

13, 14. Summerson, *Georgian London*.

build was made out, on the king's own instructions, in 1630.[15] It was not, however, unconditional consent; the earl was required to carry out his speculative building on such a scale and in such a manner as to provide a distinguished ornament and not merely an extension to the capital'.[16] He also had to use Inigo Jones (1573–1652) as architect.

INIGO JONES

As Summerson has said, it is essential to think of Jones, from first to last, as a court architect imposing foreign formulae which were neither comprehensible nor particularly welcome to the ordinary Englishman.[17] In 1615, when he was 42 and recently returned from a second study tour in Italy, Inigo Jones was appointed Surveyor to the King (James I: 1603–25). He had previously been in great demand in court circles. His flair with the design of scenery for masques, in the now increasingly fashionable Italian Renaissance style, was incomparable. His post was one of great influence and personal opportunity. In London he was soon designing the Queen's House at Greenwich, built between 1618 and 1635 (now part of the National Maritime Museum), and the Whitehall Banqueting House of 1619–21, which was intended to form part of a grandiose Palace of Whitehall.[18] He was the first English architect thoroughly to understand the principles of Renaissance design, and the originator of Renaissance urbanism in England.

In 1625 Charles I came to the throne with definite and autocratic ideas about the improvement of his 'rather squalid and untidy metropolis'.[19] One of the king's first actions was to establish a commission for buildings as an attempt to control the development and form of London through enforcement of existing legislation, imposition of limits on new building and regulation of methods of construction. Inigo Jones was one of its members, but the powers of the commission were limited, with Parliament resolved that Charles was not to be allowed to play with his capital as if it was a royal pleasance.[20]

The Covent Garden Piazza took the form of a rectangle located immediately north of the Bedford House garden (Figure 8.3). Its northern and eastern sides were uniformly designed terraces; identical houses flanking the church of St Paul formed the western side. The southern side was kept open, maintaining the views northwards from the house and garden to the hills of Highgate and Hampstead, and reducing loss of privacy to a minimum. Two streets entered the piazza in the centres of the eastern and northern sides. In designing his piazza, Inigo Jones was undoubtedly aware of continental precedents, notably the first of the residential squares in Paris, the Place Royale, built by Henry IV in 1605–12.[21]

Covent Garden was at first an extremely fashionable residential district. But the demolition of Bedford House in 1703 and its replacement by poorer quality housing, together with the completion of new attractive squares further west, accelerated the decline in its fortunes. As early as 1671 the new Duke of Bedford was granted the right to have a daily vegetable market in Covent Garden.[22] At first this was an open-air market; later, in 1828–31,[23] informal permanent stalls were replaced by the earliest of the buildings still standing, though currently undergoing adaptation to other uses, with the fruit and vegetable market now housed at Nine Elms, Vauxhall (London).

Following the establishment of the market, and marking the decline, a number of the Covent Garden buildings came to be used for some of

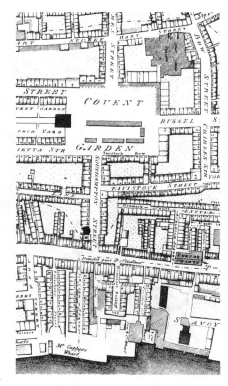

Figure 8.4 – Covent Garden: the area between the Piazza and the Thames in 1799, showing the first permanent market buildings, and the entrance to Covent Garden Theatre (the present-day Opera House, as rebuilt) through the arcade in the north-eastern corner of the Piazza. Bedford House and its gardens have been redeveloped as mean urban cottages, and similar fates have befallen Norfolk and Salisbury Houses between the Strand and the river (as shown on Figure 8.2). The surviving Savoy House at the right-hand (downstream) corner of the map illustrates the original private grand house waterfronts.

15. *Le Début de l'Urbanisme anglais, l'Urbanisme de Paris et l'Europe 1600–1680*, edited Pierre Francastel, Editions Klincksieck, Paris, 1969. J Summerson, *Covent Garden*.

16. Summerson, *Georgian London*.

17. J Summerson, *Inigo Jones*, 1964; see also J Harris (ed.) *The King's Arcadia: Inigo Jones and the Stuart Court*, Exhibition Catalogue, 1973.

18. The Palace of Whitehall, as proposed, would have rivalled Versailles for size but the English monarchy lacked the absolute authority to build on such a scale. Charles I was executed in front of the Banqueting Hall – the only completed part of his intended palace.

19, 20. Summerson, *Georgian London*. Since the mid-1980s Prince Charles – the future Charles III – has also revealed a personal interest in the appearance of central London, notably in the area providing the immediate setting for St Paul's Cathedral, where his not always behind the scenes encouragement of a another classical revival in architecture and urban design has occasioned much controversy.

21. See Chapter 6 for the Place Royale and its direct antecedent, the Place Ducale at Charleville.

London's earliest coffee houses, with the arcades around the northern and eastern sides making attractive and sheltered meeting places. One of the coffee houses became the venue of the theatrical fraternity and in 1731 John Rich, previously manager of the playhouse in Lincoln's Inn Fields, where *The Beggar's Opera* was first performed, leased the ground on which the present-day opera house stands. He also acquired the house of Dr Douglas on the north-east corner of the Piazza, to make the principal entrance from there; the frontage to Bow Street contained only the approaches to the stage door.[24] Covent Garden Playhouse opened in 1732. By then 'the neighbourhood had ceased to be a fashionable place of residence; apart from Lord Archer, all the aristocratic inhabitants had moved to the newly developed West End. Covent Garden became the centre of London's night life; here were taverns, bagnios and houses of ill-fame, patronized alike by noblemen and persons of the lowest class throughout the night'.[25] Part of the eastern side was demolished in 1769 and the remainder in 1880 and 1890. The western corner of James Street, on the northern side of the piazza, was reconstructed along the original lines and gives an impression of the character of the design. St Paul's Church was burnt in 1795 but subsequently restored.

LINCOLN'S INN FIELDS

A first attempt to develop the fields adjoining Lincoln's Inn to the west was made in 1613. The king was petitioned for a building licence but strong opposition from the Society of Lincoln's Inn, who wanted the area to remain a public recreation space, influenced the application's rejection. Five years later, following pressure from local inhabitants, the king instructed Inigo Jones to prepare a plan for landscaping the fields with tree-lined walks after the pattern of Moorfields, a long established and popular open space to the north of the City. In 1638 William Newton acquired the lease of the fields and was quick to point out to the king that in its undeveloped state the land was worth only £5 6s 8d to the Crown in annual rental. His application to build thirty-two houses was granted before the Society of Lincoln's Inn could object, but as Rasmussen notes, Newton 'made the agreement with them that the walks should remain an open square'.[26]

By August 1641 the houses forming the southern and most of the

Figure 8.5 – Covent Garden in 1751. The market in the central space has already assumed a permanent character, consolidating temporary stalls present since the 1670s. This view is looking north, as from across the garden of Bedford House. Modern James Street is the central street running north out of the piazza; the corner bays to the west (left in view) have been reconstructed as a replica of the original design, all the other residences have long since been demolished although the outline of the piazza can still be clearly traced. St Paul, Covent Garden (left), was rebuilt after fire in 1795. The entrance to the Playhouse, later rebuilt as Covent Garden Opera House, was through the arcade in the top right-hand corner of the piazza. The arrangement of the streets entering the Piazza in the centre of the eastern and northern sides is in accordance with French precedence whereby the church terminates a vista from the east, gaining thereby in status. However, although the approach from the north would have had Bedford House in view, it was only, most probably, not a very grand back side. (I now place a reduced emphasis on the value of the view northwards from Bedford House, towards the hills, as a determinant of the shape of Piazza.)

How Inigo Jones, the King's Surveyor, came to be employed on the work is not obvious from the documents but inferences can be drawn. The proclamation of 2 May 1625 was in the nature of an absolute prohibition of any building whatever, except on old foundations. The commissioners appointed to implement the Proclamation, however, were given some latitude. Nothing was said about building on old foundations and authority was given to any four commissioners, of whom the King's Surveyor of Works was to be one, to allot ground for the rebuilding of houses in such a way as to achieve 'Uniformitie and Decency'. This still did not allow for any increase in the number of houses but it did facilitate planned redistribution and, more significantly, established the principle that any new shaping of London's streets should come under the eye of the King's Surveyor – in other words Inigo Jones.
(J Summerson, Inigo Jones)

22. Olsen, *Town Planning in London: The Eighteenth and Nineteenth Centuries.*
23. Pevsner, *London: The Cities of London and Westminster.*
24, 25. H Phillips, *Mid-Georgian London,* 1964.
26. Rasmussen, *London: The Unique City.*

western sides had been completed. At this point the society, changing its tactics, petitioned Parliament for an order preventing further building. This was granted and despite Newton's lobbying he was unable to get it rescinded before his death in 1643. The next owners persuaded the society to agree that the new buildings on the north side should have the same proportions as to height and breadth as those already built on the south side, so that the whole formed a regular square.[27] Before it could be completed the society was to change its mind once more, obtaining an order from Cromwell, preventing building, in 1656. Finally, in 1657, the society and the developers reached agreement whereby Lincoln's Inn Fields was given its present-day form. There were to be continuous terraces on three sides – south, west and north – the garden wall of the inn itself making the east side. The whole was to be a large tree-planted square, which was opened to the public in 1894.

Preceding the developments of Lincoln's Inn Fields and forming a western extension of its northern side Great Queen Street had been built in about 1635–40. The elevational designs of both the square and the street were assiduously controlled by the Commissioners on Buildings, for in every case the houses were fronted 'in the Italian taste'.[28] Tradition associates Inigo Jones with their detailed design but as Summerson observes, 'his role in getting such things done was probably far more that of a civil servant than a professional architect'.[29] Great Queen Street is generally regarded as the first regular street in London and was to serve as the pattern for the integrated architectural design approach of the succeeding two centuries – an approach which was reinforced by the statutory legislation introduced after the Fire of 1666. Great Queen Street's buildings have long since been demolished and the street itself is now separated from Lincoln's Inn Fields by Kingsway, which was cut through a slum area between 1889 and 1906.

LONDON BEFORE THE FIRE

Between June and December 1665 an estimated total of 90,000 persons in the London area died in the outbreak of the plague. In four days at the beginning of September in the following year the Fire of London destroyed a total area of 437 acres – 373 acres within the city walls and 64 acres outside. Only about 75 acres remained undamaged within the city walls, the ruin being so complete that the Thames could be seen from Cheapside.[30] G M Trevelyan has observed that the fire was a unique event; the plague was merely the last, and not perhaps the worst, of a series of outbreaks covering three centuries.[31] This quotation must be qualified. The fire of 1666 was unique because of its extent, which far surpassed all previous outbreaks, but both disease and fire were ever-present hazards to life and property in an urban agglomeration the size of London, and far from unknown in towns in general.

London in 1666 was a typical medieval city with narrow, meandering streets and lanes. Most of the buildings were of timber construction, often with thatched roofs and corbelled out over the streets to obtain as much floor space as possible on their several upper floors. The monasteries and many of the merchants' mansions had been divided up into tenements and their gardens densely developed. The skyline was a romantic grouping of towers and spires: there were 109 churches in the City, dominated by the Gothic mass of St Paul's Cathedral. The fabric of the cathedral was, however, in an extreme state of disrepair – a

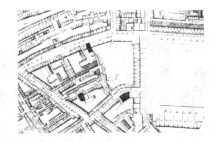

Figure 8.6 – Lincoln's Inn Fields (the western side) and the earlier Great Queen Street which had been built from 1635 as 'the first regular street in London'. (Kingsway, the modern main north–south street in this part of central London, runs parallel to the western side of Lincoln's Inn through what is shown – most probably misleadingly – as its open back gardens.)

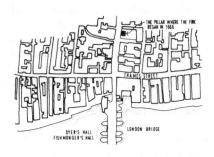

Figure 8.7 – London: the City streets and lanes immediately north of London Bridge redrawn from the John Ogilby and William Morgan map of 1676. The extremely 'close-grained' form of the City, as reconstructed after the Great Fire, adhered closely to the medieval pattern of land ownership and that of the related route system. With the exception of a few main streets (such as Thames Street) which were significantly widened after the fire, and otherwise minor improvements to the lanes, the map of the City of London was much the same as it had been before; but see below for the radical changes enforced in the construction of its new buildings. (Pre-fire 'maps' of the City of London are of the pictorial type and do not accurately show the detailed form of the City.)

27. Rasmussen, *London: The Unique City.*

28, 29. Summerson, *Georgian London.*

30. John Evelyn noted in *The Diary,* 2 September 1966: 'This fatal night, about ten, began the deplorable fire, near Fish Street, in London;' Next day he recorded: 'I had public prayers at home. The fire continuing, after dinner, I took coach with my wife and son, and went to the Bankside in Southwark where we beheld that dismal spectacle, the whole city in dreadful flames near the waterside ... and so returned, exceeding astonished what would become of the rest.'

31. G M Trevelyan, *England under the Stuarts,* 1944. See also D Keene, 'A New Study of London before the Great Fire', *Urban History Yearbook,* 1984.

condition reflecting that of the City in general. The street system was as obsolete as the buildings, which had gradually encroached on to the thoroughfares. It was totally unsuited to London's newly assumed role as a leader of world trade. The narrow, tortuous lanes leading down to the wharfs had already failed to accommodate packhorse traffic and could not possibly deal with the growth of cart and dray transport. As early as 1598 John Stow 'had been moved to describe carts and drays as one of the two great plagues of London, and in the intervening years their numbers had grown fast'.[32] Attempts were made to control this traffic by a system of licensing, broad wheels were made compulsory, and the number of horses was limited.

John Evelyn, the social commentator whose diaries give such a vivid picture of city life, had strongly condemned a number of its failings, notably the general use of soft coal, which kept a pall of smoke continually obstructing the sun. Evelyn's general observation that the buildings and surroundings in London were as deformed as the minds and confusions of the people aptly summarizes the situation in the London of 1666.

THE FIRE OF LONDON

The fire started in a baker's shop in Pudding Lane near London Bridge in the early hours of Sunday 2 September 1666. From being just another blaze of local interest this outbreak soon gathered force and, fanned by a strong north-east wind, spread from its origin just north of the bridge down to the warehouses and wharfs on the river. Before it was checked four days later the fire had destroyed 13,200 houses, the Royal Exchange, the Custom House and the halls of 44 of the city companies, Guildhall and nearly all the City buildings, St Paul's itself and 87 of the parish churches, besides furniture and commodities valued at over £3.5 million. In all the bill was reckoned at more than £10 million.[33]

About 80,000 people were made homeless by the fire. Tents were provided by the king, and the city authorities granted permission for temporary buildings on available open spaces. Royal proclamations ordered neighbouring parishes to provide lodgings, appointed new markets and instructed magistrates in the Home Counties to organize supplies. More important still Charles II broke down for the refugees the privileges of the corporate towns, commanding that all cities and towns whatsoever should without any contradiction receive the distressed persons and permit them the free exercise of their manual trades.[34] Of the 13,200 houses destroyed a large proportion represented both the homes and the work-places, warehouses and shops of the city merchants and craftsmen. Already suffering from the counter-attractions of rapidly developing suburban centres, the City's commercial interests had to start rebuilding as soon as possible. In addition, the City's own property was largely destroyed and its rental income effectively cut off.

Charles II was not a Londoner and 'its patent disadvantages were not dimmed for him by a kindly veil of familiarity'.[35] He had already endorsed Evelyn's 1661 criticisms of its failings with 'encouragement to press on to find remedies'.[36] He had been aware of fire hazards and the need for wide streets to accommodate growth of trade with its traffic, and he must have known of the proposals to improve the Paris of Louis XIV. The king was therefore sympathetic to the concept of a completely replanned City of London, as embodied in the several designs

16th July 1665 – There died of the plague in London this week 1100; and in the week following, above 2000. Two houses were shut up in our parish.

2nd August. A solemn fast through England to deprecate God's displeasure against the land by pestilence and war; our Doctor preaching on 26 Levit. v. 41, 42, that the means to obtain remission of punishment was not to repine at it; but humbly to submit to it.

8th I walked on the Duke of Albemarle, who was resolved to stay at the Cock-pit, in St James's Park. Died this week in London, 4000.

15th There perished this week 5000.

28th The contagion still increasing, and growing now all about us, I sent my wife and whole family (two or three necessary servants excepted) to my brother's at Wootton, being resolved to stay at my house myself, and to look after my charge, trusting in the providence and goodness of God.

7th Sept. Home. there perishing near 10,000 poor creatures weekly; however, I went all along the city and suburbs from Kent Street to St James's, a dismal passage, and dangerous to see so many coffins exposed in the streets, now thin of people; the shops shut up, and all in mournful silence, not knowing whose turn it might be next.

(J Evelyn, Diary)

John Evelyn, the diarist, was walking one day in the well kept grounds of Whitehall Palace when a cloud of smoke came up in the direction of London, which so invaded the Court that all the rooms, galleries and places about it were filled and infected. Men could hardly discern one another for the cloud and none could suffer it without choking. Indignant he wrote his 'Fumifugium; or the Inconvenience of the Aer and Smoak of London Dissipated', proposing remedies.

'It is horrid smoke', writes Evelyn, 'which obscures our churches and makes our palaces look old, which fouls our clothes and corrupts the waters, so that the very rain and refreshing dews which fall in the several seasons precipitate this impure vapour, which with its black and tenacious quality spots and contaminates whatever is exposed to it'. He bitterly complains that the gardens around London no longer bare fruit, instancing especially Lord Bridgwater's orchard at Barbican and the Marquis of Hertford's in the Strand.

(W G Bell, The Great Fire of London in 1666)

32. Stow, *The Survey of London*; while epidemic disease has been long since eradicated, nearly four centuries after Stow's stricture many Londoners would regard vehicle traffic as the city's most serious late-twentieth-century plague.

33. T F Reddaway. *The Rebuilding of London after the Great Fire*, 1951; see also W G Bell, *The Great Fire of London in 1666*, 1920. Forty years on, Reddaway is still the authority on the aftermath of the Fire, in particular the sections where he demonstrates that comprehensive replanning was out of the question and his account of that which was subsequently carried out.

34. See Evelyn, *The Diary*, 7 September 1666: 'subjoined is the [King's] Ordinance ... reprinted from the original half-sheet in black letter'.

35, 36. Reddaway, *The Rebuilding of London after the Great Fire*.

prepared for his consideration. But he recognized, and acted upon, the need to minimize delay in starting the reconstruction process. 'In this he was at one with the City'.[37]

On 13 September the king published a proclamation, in which he set out 'the decisions for the immediate present and his intentions for the future; the first step had been taken, and men could feel that the rebuilding was under control'.[38] Although only an interim step, the proclamation contained several of the measures later embodied in re-building legislation. Fire-resisting external building materials had to be used and important streets widened to constitute fire-breaks. The net-work of inconvenient and unhygienic lanes was to be replaced by wide streets; and, as a specific recommendation, there was to be a new quay along the bank of the Thames, maintaining the continuous contact between City and river revealed by the fire. The king undertook to rebuild the Custom House as soon as possible and to relinquish Crown property where it would be of common benefit. As the basis of financial provision for the rebuilding, the Crown promised to turn over all the coal tax revenue for seven years. The proclamation also instructed the City to prepare a survey of the devastated area, showing the existing land ownerships 'that provision may be made, that though every man must not be suffered to erect what buildings and where he pleased, he shall not in any degree be debarred from receiving the reasonable benefit of what ought to accrue to him'.[39] The time needed for the survey was to be utilized to agree a general plan for the rebuilding. In the interim a strict embargo was placed on unauthorized buildings. Those proven owners who wished to make an earlier start could do so if they conformed to the general plan.

By the end of September, realistic assessment of the complex obsta-cles in the way of complete replanning had led to general agreement that the existing urban cadastre street lines and property boundaries must be accepted. The property boundary survey, later to be aban-doned, was already proving extremely difficult;[40] ordinarily available revenue could not begin to pay for the land acquisition involved and the compulsory purchase legislation for street widening could not have been extended to comprehensive acquisition and redistribution of land, even if the political necessity to rebuild had been less urgent.

Because the fire occurred in September, it was possible to organize large-scale reconstruction. If the City had been devastated earlier in the year, the pressures to rebuild before the winter could have been irresistible. By the time the ruins had been cleared, a particularly severe winter had set in and had prevented any further work until spring. At the beginning of October six commissioners were appointed, nominally to supervise the survey but in effect to control all technical aspects of the rebuilding work. The king nominated three commission-ers – Wren, Hugh May and Roger Pratt. The City's representatives – styled surveyors – were Robert Hooke, Edward Jerman and Peter Mills. 'A better equipped body could hardly have been found ... it had to be improvised and as an improvisation the six nominated could not have been bettered.'[41] Wren, Pratt and May were architects though Wren was not much more than a student. Hooke was a mathematician and scientist, as Wren also was by training. Jerman and Mills, the City Surveyor, had practical building background and an extensive under-standing of the City's requirements.

Dr Christopher Wren – he was not knighted until 1674 – submitted

But tho' by the new buildings after the fire, much ground was given up, and left unbuilt, to inlarge the streets, yet 'tis to be observed, that the older houses stood severally upon more ground, were much larger upon the flat, and in many places, gardens and large yards about them, all which, in the new buildings, are, at least, contracted, and the ground generally built up into other houses so that notwithstanding all the ground given up for beautifying the streets, yet there are many more houses built than stood before upon the same ground; so that taking the whole city together, there are more inhabitants in the same compass, than there was before. To explain this more fully, I shall give some particular instances, to which I refer, which are living witnesses able to confirm.

For example, Swithen's Alleys by the Royal Ex-change, were all, before the Fire, taken up with one single merchant's house, and inhabited by one Mr Swithin; whereas, upon the same ground where the house stood, stands now about twenty-two or twenty-four houses, which belong to his posterity to this day.

Copt-Hall-Court in Throckmorton-street, was, before the Fire, also a single house, inhabited by a Dutch mer-chant; also three more courts in the same streets, were

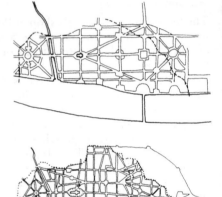

Figures 8.8 and 8.9 – Two of the three plans produced by John Evelyn, 'a typically wealthy, well-educated nobleman, who could live entirely for his hobbies and who divided his time equally between aesthetic and practical subjects' (S E Rasmussen, London: The Uni-que City). Although both plans are similar to Wren's, notably in the section west of the Fleet, his setting of St Paul's Cathedral is much less grand, and much more nearly that of the present day.

37, 38. Reddaway, The Rebuilding of London after the Great Fire.

39. W de G Birch, The Historical Charters and Constitu-tional Documents of the City of London, 1897.

40. For Leake's Map of 1666, which recorded the out-lines of the City after the Fire, see J Hanson, 'Order and Structure in Urban Design: the Plans for Rebuilding of London after the Great Fire of 1666', Ekistics, vol 56, 1989.

41. Reddaway, The Rebuilding of London after the Great Fire.

the first rebuilding plan to the king on 11 September. John Evelyn followed with his proposals two days later. Robert Hooke's plan was shown to the Royal Society on 19 September; 'the court of the Lord Mayor and aldermen had approved of it and, greatly preferring it to that of the city surveyor Peter Mills, desired that it might be shown to his Majesty'.[42] The surveyor's plan has not survived but three others have – two versions by Richard Newcourt (see page 339 for influence on the plan of Philadelphia) and one by a Captain Valentine Knight. With the exception of Wren's plan, which must be assessed at greater length, notes on the designers and their proposals are given as captions to the plans.

SIR CHRISTOPHER WREN (1632–1723)

In 1666 Wren, at the age of 34, was Professor of Astronomy at Oxford and an outstandingly gifted member of the Royal Society. He had put forward numerous scientific inventions and theories, many of which 'aimed right at the central problems of astronomy, physics and engineering'.[43] He had only two buildings to his credit – the Sheldonian Theatre at Oxford, designed in 1664, and Pembroke College Chapel, Cambridge, 1663–6. Sir Reginald Blomfield, 'while describing himself as not the least of Wren's admirers has to admit that between 1660 and 1670 he was the merest amateur in architecture and that his appointment had in it more of influence than of justice'.[44] Nikolaus Pevsner dismisses the two first buildings as 'evidently the work of a man with little designing experience'.[45]

Wren, however, had excellent connections. As early as 1661 he had been offered an appointment surveying and directing harbour works at Tangier,[46] and he was involved with the work of the royal commission (set up in 1663) on the state of St Paul's. From July 1665 to March 1666 he was in France studying architecture. No doubt he had the repair of the cathedral very much in mind, because a few weeks after his return he submitted a report with suggestions for its restoration. Wren was closely connected with John Evelyn, and Eduard Sekler is no doubt right in assuming that both men had pretty much the same views on the ideals to be pursued in town planning.[47] Certainly their respective plans for the City have much in common.

The history of urban form has its mythology, mainly centred around personalities ancient and modern. The claims that had been made for Hippodamus as 'the father of town planning' are probably fictional, and even his planning of Miletus could have involved only the early reconstruction areas. The Emperor Augustus may well have found Rome a city of brick, but he could have left it only partially a city of marble. In recent years more than one planner has achieved a reputation that will be undermined, if not demolished, sooner or later.

Wren has been accorded mythical significance on several interrelated counts. He has been described as an exceptionally gifted town planner, whose brilliant proposals for rebuilding the City were 'unhappily defeated by faction'.[48] These were the words of his apologist, the architect Gwynn, written in 1749, an observation restated by Lewis Mumford, who described the proposals as being foiled by tenacious mercantile habits and jealous property rights.[49] It has also been said that the rejection of his proposals and the rebuilding of the City as before, without improvements, was the greatest missed opportunity in urban history.

single houses, two on the same side of the way, and one on the other.

The several alleys behind St Christopher's Church, which are now vulgarly, but erroneously, call'd St Christopher's-Churchyard, were, before the Fire one great house, or, at least, a house and warehouses belonging to it, in which the famous Mr Kendrick lived, whose monument now stands in St Christopher's Church, and whose dwelling, also, took up almost all the ground, on which now a street of houses is erected, called Prince's-street, going through into Lothbury, no such street being known before the Fire.

Kings-Arms-Yard in Coleman-street, now built into fine large houses, and inhabited by principal merchants, was, before the fire, a stable-yard for horses and an inn, at the sign of the King's Arms.

I might fill up my account with many such instances, but 'tis enough to explain the thing, viz. That so many great houses were converted into streets and courts, alleys and buildings, that there are, by estimation, almost 4,000 houses now standing on the ground which the Fire left desolate, more than stood on the same ground before.
(Defoe, A Tour through the whole Island of Great Britain)

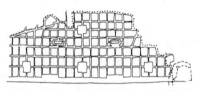

Figure 8.10 – 'Brilliant, cantankerous, secretive, always in ill-health, Robert Hooke was Curator of Experiments to the Royal Society, Professor of Geometry at Gresham College and already well known for the experiments which were later to bring him enduring fame' (T F Reddaway, *The Rebuilding of London after the Great Fire*). Hooke's plan has been lost, but a small drawing of a plan included in Doornick's *View of the Fire* may be of his proposals. The regular gridiron structure contains a range of civic spaces and a spacious quay along the Thames.

42. Reddaway, *The Rebuilding of London after the Great Fire*; see also Evelyn, *The Diary*, 13 September 1666: 'I presented his Majesty with a survey of the ruins and a plot for a new City, with a discourse on it'. (However, as a footnote by Ernest Rhys in the 1907 edition, Evelyn's 'survey of the ruins' would appear to have amounted only to an approximate outline of the devastated area.)

43. Pevsner, *An Outline of European Architecture*; for Wren generally see also G Beard, *The Work of Christopher Wren*, 1982 (with excellent illustrations); C Amery, *Wren's London*, 1988; M Whinney, *Wren*, 1971.

44. R Blomfield, *English Architecture in the Seventeenth and Eighteenth Centuries*.

45. Pevsner, *An Outline of European Architecture*.

46, 47. E Sekler, *Wren and his place in European Architecture*, 1956.

48. J Gwynn, *London and Westminster Improved, 1766* (republished 1969). Appendix A of Reddaway's *The Rebuilding of London after the Great Fire* deals most thoroughly with sources of the myth that Wren's plan was first accepted only to be discarded as a result of commercial pressures.

49. L Mumford, *The City in History*, 1961, page 442: Mumford, unfortunately, got it all wrong.

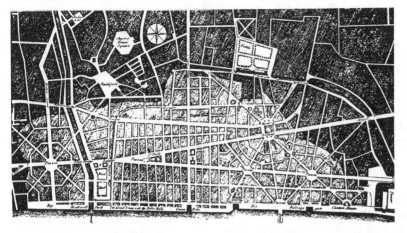

Figure 8.11 – Wren's plan for the rebuilding of London, as redrawn by John Gwynn in 1749 and presented for the 'consideration of the Lord Mayor, Aldermen, and Common Council of the City of London' under the totally misleading heading, 'Designed by that great architect Sir Christopher Wren and approved of by King and Parliament, but unhappily defeated by faction'. This claim by Gwynn (who was using Wren's plan as an example of what could be done, in his campaign for planning improvements in London) is a main source of the myth of Wren's plan. Later, in 1766, Gwynn wrote of 'the unaccountable treatment the noble plan of Sir Christopher Wren met with from the interested views of ignorant, obstinate, designing men, (notwithstanding it had the sanction of the King and Parliament) who by rejecting it did an irreparable injury to the city of London' (John Gwynn, *London and Westminster Improved*, London 1766, republished in 1969 by Gregg International Publishers Limited.)

THE WREN PLAN FOR LONDON

That Wren's plan was totally irrelevant to the needs of the City has already been established. As such it was neither more nor less invalid than all the other plans. Criticism on this count is therefore less of Wren himself than of his advocates. Also we do not know how seriously Wren regarded his proposals. It is recorded that he was greatly concerned to get in first with his plan, but perhaps this was the policy of an ambitious embryo architect seeking to establish his claim to a major share of the rebuilding commissions. Such pre-empting activity has not been unknown in more recent times, and it can be argued that his plan was the means to the end of being appointed Surveyor-General in 1669 and acquiring the commission to design the new St Paul's and sixty-six city churches, in addition to almost all the other worthwhile architectural work of the period.

On the other hand, his proposals could have had serious intent, and the plan therefore needs to be examined in greater detail. Steen Rasmussen regards them as the work of 'a mathematician who, starting from certain definite postules, has solved . . . an interesting geometrical problem. To him [Wren] it was a given thing (1) that the entrances to the town were its gates and its bridges; (2) that a town is composed of rectangular houses; (3) that all street corners should preferably be rectangular; (4) that the entrances should give easy access to the different parts of the town; and (5) that the centre of commerce, the Stock Exchange, and the religious centre, St Paul's, should have a dominating position. The problem was the plexus of streets. This he solved by means of the common form ideals of those days'.[50]

The Wren plan, in common with the others, would certainly have improved the street system, particularly the east–west routes, and it is competently related to the streets giving access to the City. But it would have suffered from the usual defects of a plan that imposes radial streets on a basic grid pattern, as in Washington, DC. Moreover there are also two inherent major counts on which it can be seriously faulted, the first being that it is not related to the topography of the site. This is not flat, but undulating, with the two hills which rise on either side of the valley of the Walbrook, more steeply sloping in 1666 than in modern times (Figure 8.12). The same conditions existed on both sides of the Fleet and the slopes down to the Thames. Superimposition of a grid on undulating ground results in visual effects contrary to the basic objec-

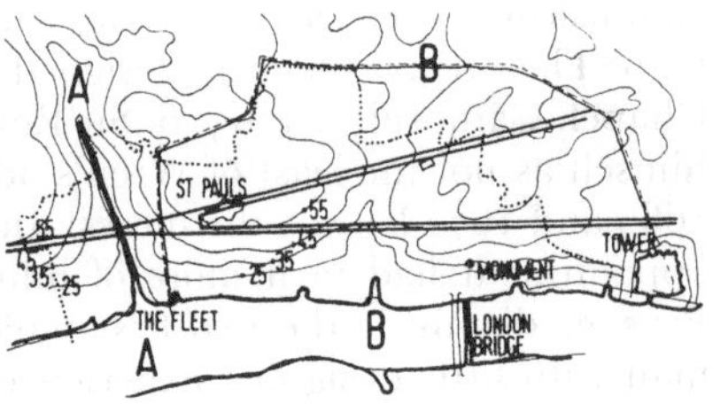

Figure 8.12 – The topography of the City of London today. The site is undulating, with the two hills on either side of the valley of the Walbrook (B–B) more steeply sloping in 1666 than in modern times. The same condition existed on both sides of the Fleet (A–A) and the slopes down to the Thames. The rise and fall along the lengths of Wren's main east–west streets are clearly shown.

In 1650 London was already a city of about 350,000. In spite of the immense loss of life in the plague of 1665 and the disruption brought about by the Great Fire in the following year, London was the biggest city in Europe by 1700 with a population of about 550,000. By 1800 the figure had reached 900,000 and London was about twice the size of Paris, her nearest rival. Already in 1650 about 7 per cent of England's population lived in London; a hundred years later 11 per cent (Paris contained about 2 per cent of the total French population at the later date). Some of the cities of the classical world, though much smaller, had a malign effect upon local economies and could with reason be termed parasitic. In the case of London, however, its growth brought great benefit to the economy and provided much of the impetus to the general transformation of English society.
(E A Wrigley, Population and History)

50. Rasmussen, *London: The Unique City*.

tives of Renaissance urbanism – that of creating a unified, ordered architectural entity. Because of the nature of the land Wren's main avenues, notably those flanking his new St Paul's, would have had vertically staggered cornice lines to accommodate the slope; the magnificent perspective vistas that seem to be created on Wren's two-dimensional diagram could never have existed in reality on the ground. The second main criticism is of the arbitrary divisions of his city into three parts, each of which is given a basically different structure. From east to west there is first an unresolved radial system; then a central, relatively straightforward grid section crossed by the two main avenues; finally, mainly west of the Fleet, there is a section arranged around a large traffic area.

It is surely not possible to see Wren's plan as more than an overnight exercise based on a superficial use of continental Renaissance plan-motifs. On balance, allowing for all that has happened since 1666, London as rebuilt to an improved version of its old plan is to be preferred. If a new plan had been adopted it would appear from evidence from the USA, where the gridiron seems flexible enough to accommodate radical change,[51] that a simple rectilinear network would have had most to offer – if adjusted to the topography.

The rest of the myth – that his plan was ignored and the City reconstructed without improvements to the existing form – is best refuted by describing the rebuilding process.

RECONSTRUCTION OF THE CITY

Statutory provision for the reconstruction was given by the first and second Rebuilding Acts of February 1667 and April 1670. The first Act generally followed the lines of the royal proclamation and incorporated the work of the commissioners. The three most significant improvements – the streets, the buildings and the Fleet Canal project – are described later in this chapter. Proposals for the Thames Quay were omitted from the first Act but included in the second.

The first Act implemented the king's proclamation by granting a duty of one shilling per ton on coal brought into the port of London for ten years – a sum which Professor Reddaway describes as so niggardly that the City might almost have appealed in vain. Reddaway shows that over the ten years it would have raised around £150,000 and that 'the land required for the quay ... alone would have taken approximately that figure'.[52] In addition to paying for riverside land purchases, the duty was intended for road widening acquisitions and the rebuilding of the city prisons.

The second Act recognized the financial shortcomings of the first and increased the coal duty to three shillings per ton, effective until 1687 – ten and a quarter years after the end of the original period. The City was granted one-quarter of the increase, giving it one shilling and sixpence per ton to pay for land purchases, work on the Fleet Canal and improvements to several public buildings. The remainder of the increased duty was for the rebuilding of the city churches – one-quarter of it for St Paul's. This income from coal tax and its ability to raise loans on security enabled the City to finance the rebuilding as fast as the availability of labour and materials allowed.

At a meeting on 11 October 1666 the commissioners determined the street widths of the new city as: 'Key, 100 feet; high streets, 75 feet; some other streets 50 feet and others 42 feet; the least streets 30 feet or

The familiar story of Wren's rapid production of a plan for rebuilding the City after the Fire of 1666, and the fate that befell it, needs no recital here. The merits of the plan were considerable, substituting method, broad streets, and generous spaces for cramped building congestion and tortuous lanes that had long given encouragement to pestilence and fire. Adoption of the plan would have vastly affected for good the future of the city, so needlessly stifled within its wall, and the greater metropolis that was soon to stretch out indefinitely beyond it. Not to take advantage of the exodus of population, that followed plague and fire, to carry out radical reforms must now be regarded as lack of foresight carried to the point of folly. At the same time it must be acknowledged that the position created was a very difficult one, and the impatience of the displaced citizens to return to their homes quite understandable ... Wren's reasoned scheme of replanning was, accordingly, set aside in favour of the status quo, or largely so, and the opportunity for generous roads and spaces, and much hygienic gain, was neglected.

What was even worse was that this major misfortune revealed a complete lack of any effective centralised control over town development, which was to continue as a drag on urban reform and civic administration throughout the whole country.
(F R Hiorns, Town-Building in History)

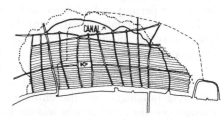

Figure 8.13 – Captain Valentine Knight, an army officer, devised a plan that laid greater emphasis on the regular reorganization of building plots, in contrast with the plans of Wren and Evelyn, who put the street system first. Main streets 60 feet wide form a 'super-grid', which is broken down by 30 feet wide secondary streets, parallel to the river, into building plots of around 500 feet by 70 feet, which would allow two rows of buildings separated by a central back alley. Pavements were to be as arcades beneath the buildings (as at Covent Garden) and a 30 feet canal was to run around the central section of the new City from the Fleet, at Holborn Bridge, to Billingsgate – an interesting idea but topographically impossible at other than great expense.

51. See Chapter 10; there is also Stockholm, where the centre city renewal programme has been greatly facilitated by the historic grid structure.
52, 53. Reddaway, The Rebuilding of London after the Great Fire. During the early 1990s, royal-inspired criticism of existing and proposed replacement buildings in the St Paul's Churchyard vicinity characteristically bemoan the 'missed opportunity' presented by the Great Fire; moreover, all too frequently critics make the mistake of believing that because Wren was an architect of genius, then he was ipso facto a great urbanist.

25 feet; alleys, if any, 16 feet'.[53] Although the Thames Quay was never started and the street widths were subsequently modified to range from 50 feet down to 14 feet, the new provisions nevertheless were an enormous improvement on previous dimensions.[54] The post-1666 street pattern was essentially a reinstatement of the old system, with its capacity greatly increased; of the very few new routes created the most important was 'the formation of King Street and the transformation of an ancient lane into Queen Street, the combined thoroughfares giving direct access from Guildhall to the Thames'.[55]

In addition to standardizing street widths the commissioners also insisted on having standard house types. Sir John Summerson has described how the whole of the houses were divided into four classes, for better regulation, uniformity and gracefulness. In the high and principal streets (six only were classified as such) houses were to be neither more nor less than four storeys in height; in the streets and lanes of note three storeys was the rule; while in by-lanes two storeys were prescribed. A fourth class was reserved for houses of the greatest size, which did not front the street but which lay behind, with their courtyards and gardens. Their height was limited to four storeys. (Figure 8.14).[56] Building construction was standardized; thicknesses of the walls at various heights and the sizes of floor and roof timbers were rigidly controlled. To ensure conformity with regulations 'knowing and intelligent persons in buildings' were appointed.

The only materials permitted for the elevations were brick and stone. There was not, in effect, a free choice. Stone had high carriage surcharges on it and was in relatively short supply, whereas London was surrounded by large areas of readily accessible brick-earth. The new city was therefore predominantly of brickwork, with stone reserved for the civic buildings and the churches and St Paul's. An invaluable encouragement to rebuilding was the introduction of party-wall legislation; this required common boundary walls to be set out equally on both sites, with the first owner erecting the entire wall and the second owner paying half the cost, plus 6 per cent interest for the intervening period.

THE FLEET CANAL AND THAMES QUAY

By the middle of the seventeenth century the Fleet river, which had been navigable from the Thames up to Holborn Bridge, had gradually deteriorated into a shallow and evil-smelling sewer.[57] The second Act included proposals for the redevelopment of the valley of the Fleet. Reddaway says that this, if successful, would both have eliminated the troubles within the area of the city and have made a valuable addition to the wharves and storage accommodation of the port; a dangerous nuisance would have been disposed of, much needed facilities provided, and the streets serving the older wharves relieved of their overgreat burden of traffic.[58] The scheme was to canalize the Fleet, making it 40 feet in width for its nearly half-mile course below Holborn Bridge. Both banks were to provide 30 feet wide wharves, built over underground warehouses, with second-category houses fronting on to them. The City aldermen approved the plan drawn up by Wren and others on 9 March 1670. Early in April statutory provision for the work was given with the passing of the second Rebuilding Act and on 23 April the lines of the canal and the wharves were set out on site.

After a slow start to the actual work, largely resulting from the need

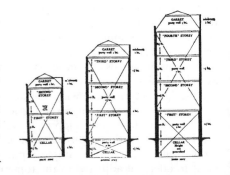

Figure 8.14 – Sections through the three main types of building authorized by the Rebuilding Act of 1667. The four-storey type was to front the 'high and principal streets', the three-storey type was for the 'streets and lanes of note and the River Thames', and the two-storey version was for 'by-streets and lanes'.

54. C H Holden and W Holford, *The City of London – A Record of Destruction and Survival*, 1951.

55, 56. Summerson, *Georgian London*.

57. Stow, *A Survey of London*, 1603, notes that, 'in the yeare 1502, the 17. of Henrie the 7. the whole course of Fleete dike, then so called, was scowred (I say) downe to the Thames, so that boats with fish and fewel were rowed to Fleete bridge and to Oldburne bridge, as they of olde time had beene accustomed, which was a great commoditie to all the inhabitants in that part of the citie. In the year of 1589 was granted a fifteene, by a common Council of the citie, for the cleansing of this Brooke or dike: the money amounting to a thousands marks was collected and it was undertaken, that by drawing diverse springs about Hampstead heath, into one head and course, both the citie should be served of fresh water in all places of want and also that by such a follower as men call it, the chanell of this brooke should be scowred into the river of Thames, but much money being spent, ye effect fayled, so that the Brooke by meanes of continuall in crochments upon the banks getting over the water, and casting of soylage into the streame, is now becoming woorse cloyed and choken than ever it was before.'

58. Reddaway, *The Rebuilding of London after the Great Fire*.

to accumulate capital and to acquire the land, and in the face of considerable difficulties, the Fleet Canal was completed in 1674 for the sum of £51,307 6s 2d.[59] It was not a commercial success. Few tenants were found for the warehouses and the canal dues did not pay the necessary dredging and maintenance costs. The open wharves slowly degenerated into general storage areas and, reflecting London's change from water to land transport, became increasingly used by carts and carriages as a direct north–south route. In 1733 the canal was arched over between Holborn and Fleet bridges, with the southern section similarly treated in 1766. Blackfriars Bridge was opened to cross-river traffic in 1769, at the southern end of the still existent culverted Fleet.

In contrast to the Fleet Canal project, proposals for a new quay along the north bank of the Thames failed so completely that today no trace of it remains, and even its history is confused. For decades, however, the need for improvements was discussed: there was general agreement that the city's waterfront was in a disgraceful condition. Public landing places serving London's predominantly water-orientated east–west transport routes shared the river front with many contrasting commercial users and refuse dumps. Reddaway describes how the whole area, landing places, lay-stalls and wharves, was approached by an inconvenient network of lanes so narrow that the drays in their passage endangered houses and pedestrians and so steep that the drays themselves were endangered every time a horse stumbled.[60]

All the rebuilding proposals, official and unofficial, took advantage of the opportunity to create a new riverfront with a quay extending between the Tower and the Temple. There was complete accord between the 'city beautiful' party and the commercial interests, who anticipated improved port facilities at minimum cost. The proclamation of 13 September included the construction of a 'fair key or wharf' and the commissioners gave it a width of 100 feet at their 11 October meeting, later reducing it to 80 feet.

The subsequent history of the Thames Quay is that of a universally agreed improvement which failed, for a combination of reasons, to get built. It was omitted from the first Rebuilding Act, although redevelopment was prohibited within 40 feet of the riverside in order not to compromise its future construction. The second Act made statutory provision for the quay but although several attempts were made to get it started it has remained only a planners' dream.

THE LONDON SQUARES

Before describing a number of the more important late seventeenth-century and eighteenth-century squares the general background to their development must be established.[61] The plague of 1665 and the fire of 1666 gave impetus to a trend on the part of the nobility and other wealthy families to leave the City of London and move to new houses in the country immediately to the west, within easy carriage distance of both cities. In addition to this local population movement, large general migration to the capital from other parts of the country included a considerable proportion of noble and wealthy families attracted by the flourishing social life. At first the pattern in the western fields was a scattering of mansions and large houses and ancillary service buildings, within their own spacious parks and gardens. As Rasmussen describes it, 'the arrival of each noble family increased the population not only by

The canal or river, called Fleet-ditch, was a work of great magnificence and expense; but not answering the design and being now very much neglected, and out of repair, is not much spoken of, yet it has three fine bridges over it, and a fourth not so fine, yet useful as the rest, and the tide flowing up to the last; the canal is very useful for bringing of coals and timber, and other heavy goods; but the warehouses intended under the streets, on either side, to lay up such goods in, are not made use of, and the wharfs in many places are decay'd and fallen in, which make it all look ruinous.
(D Defoe, A Tour through the whole Island of Great Britain)

The typical English residential square – and the vast majority of all English squares are residential squares – may be defined as a green framed by architecture, just as the French park and formal garden have been characterised as architecture built of greenery. But it had not always been like that; when the first squares were established in London they were not yet planted. Only in the eighteenth century did the adulation of nature by the English become so strong that they felt almost a moral obligation to plant every free area.
(P Zucker, Town and Square)

59, 60. Reddaway, The Rebuilding of London after the Great Fire. John Evelyn noted in The Diary, 6 March 1667: 'I proposed to my Lord Chancellor, Monsieur Kiviet's undertaking to wharf the whole river of Thames, or quay, from the Temple to the Tower, as far as the fire destroyed, with brick, without piles, both lasting and ornamental'.

(A footnote to the 1907 Diary edition suggests that 'Monsieur Kiviet was probably the same person described by Pepys as "Kevet, Burgomaster of Amsterdam", who would have known about waterside construction and brickwork; yet there are also mentions in The Diary of a certain Sir John Kiviet who seemed to have been in business of making bricks – IE, September 26th, Sir John Kiviet dined with me. We went to search for brick-earth, in order to a great undertaking.' If not the same person, it's likely that they were related; 1667 was a great time of opportunity in the brick fields.) Also on 6 March, Evelyn recorded: 'Great frosts, snow and winds, prodigious at the vernal equinox; indeed it had been a year of prodigies in this nation; plague, war, fire, rain, tempest and comet.'

61. For general references on the London Squares, which I believe to have been the most significant British contribution to European Renaissance urbanism, see M Webb, The City Square, 1990 (an international history which sets a context for the London squares); H Phillips, Mid-Georgian London, 1964. (E Beresford, History of the Squares of London, 1907, is the most comprehensive account.)

For the social implications, see P J Atkins, 'The Spatial Configuration of Class Solidarity in London's West End 1792–1939', Urban History Yearbook, 1990; and as a counterpart A Palmer, The East End: Four Centuries of London Life, 1989.

Figure 8.15 – The western section of John Roque's 1769 map of London (north at the top), showing the completed development of the estates south of Oxford Street through to Park Lane (the eastern boundary of Hyde Park) and the work in progress north of Oxford Street. The line of the New Road, constructed in 1756 as a direct link between Paddington in the west round to the City of London, bypassing the congested existing streets especially Oxford Street, is at the top of the map. Forty years later the fields north of the New Road at Marylebone were being developed to John Nash's plan as the Crown Estates Regent's Park scheme. (See Figure 8.19.) John Nash's early-nineteenth-century Regent Street, linking the new park to the St James's area, follows approximately the line of Swallow Street north to Oxford Street.

Lincoln's Inn is on the right-hand edge of the map: St James's Park, in front of the Queen's Palace (Buckingham Palace), is shown with the formal layout which was changed early in the nineteenth century to the present day 'naturalistic' design; south of the river, growth by 1769 was still comparatively very slow, but, stimulated by the construction of Westminster Bridge in 1738–49, the fields shown by Roque were soon being built over.

Any town planner would envy the power the great landowner of London had within the boundaries of their own estates ... What has given the ground landlords of London and other cities in the British Isles their particular power and responsibility has been the enormous size of their buildings. The freeholder of two or three lots, or even two or three streets, may have some discretion as to the sort of houses his lessees erect; but he can hardly aspire to anything worthy of the name of planning.

The Duke of Westminster, in contrast, had the whole of northern Mayfair together with what is now Belgravia and Pimlico to work with. Lord Portman and the Duke of Portland had nearly the whole of Marylebone between them. The Duke of Bedford had Bloomsbury and Covent Garden. The Marquis of Northampton had Clerkenwell and Islington. The whole character – social, architectural and economic – of a neighbourhood could be determined by the kind of street plan the landlord chose to impose, the kind of leases he chose to grant and the kind of control he chose to exercise over his tenants.

(D J Olsen, Town Planning in London – the Eighteenth and Nineteenth Centuries)

the family itself and its many servants and their relatives, but also by merchants, artisans and others who lived on the aristocracy. Besides London, the town of producers, the capital of world-trade and industry, there arose another London, the town of consumers, the town of the court, of the nobility, of the retired capitalists. Where a little room was left between the big mansions the middle classes settled in groups of smaller houses, which sprang up as best they could'.[62]

London expanded during the seventeenth and eighteenth centuries on the basis of clearly defined social-class districts. Upper-class and middle-class families established a respectability for the West End, conveniently situated for the relatively short journeys east to the commercial City of London and south to the Court and increasing central administration activities of the City of Westminster. As new squares and streets were created further out, pockets of working-class housing developed around, and eventually in a number of the original squares. Principal working-class districts were to the north and east of the City of London, where the commercial and industrial activities of the capital were rapidly taking hold. In marked contrast to the spacious character of the West End, the East End was densely built up in all stages of its expansion.

Rasmussen observes that 'when an earl or a duke did turn his property to account, he wanted to determine what neighbours he got. The great landlord and the speculative builder found each other, and

62. Several determinants in combination were responsible for the location of the moneyed-class new residential districts: first, the convenience of location not only to both the City of London and the City of Westminster, but also the regional and national routes north, south and west, avoiding the former as it expanded to the north; second, the favourable prevailing wind micro-climatic conditions – blowing the soft coal-fire smoke away to the north-east; third, the readily developed brick-earth flat land, literally on top of its own local vernacular material; fourth, the large-unit, pre-urban cadastre of estates owned by development conscious landed nobility; and fifth the lower lying, less well-drained land to the east, downstream of the City of London.

together they created the London square with its character of unity, surrounded as it is by dignified houses, all alike'.[63] Sir John Summerson establishes three clear principles of these squares' development. First, the principle of an aristocratic lead – the presence of the landowner's own house in his square. Second, the principle of a complete unit of development, comprising square, secondary streets, markets and perhaps church. Third, the principle of the speculative builder, operating as a middleman and building the houses.[64]

The great estates to the west of London were generally developed by granting building leases, a system peculiar to England. The first building leases were granted in 1661 for the properties fronting Bloomsbury Square and set the pattern whereby the great landowners retained both ownership of the land and control of buildings erected on it. Under such leases the tenant paid a low ground rent on the understanding that the lessee built at his own expense a house (or houses) of substantial character, which house (or houses) at the end of the lease, became the property of the ground landlord. Summerson observes that this system represents a convenient device by which land can be rendered profitable over and over again.[65] This was an essential consideration for such estates, which were generally entailed in a family, or which were held in trust, and which required an Act of Parliament before they could be sold.

The squares and their surrounding streets were given extremely simple layouts, invariably based on the gridiron principle. A number of squares followed the example set by Covent Garden, with the development of the gardens on the northern side of the landowner's house. The southern side was generally left open, until the house itself was redeveloped. Frequently a wide street out of the centre of the northern side was stipulated in order to preserve the highly esteemed views of Hampstead and Highgate Hills.

Bloomsbury Square

Following the example set by Lord Bedford, Lord Southampton in about 1636 applied for permission to build on his Bloomsbury Manor estate. This was refused. By the late 1650s he had negotiated consent, however, for the erection of a mansion for himself and an associated residential square planned to front it on the south. The square originally consisted of two rows of houses – Allington Row and Seymour Row – west and east respectively of the Southampton House (later Bedford House) axis. The house was pulled down in about 1800 and terraces were added along the north side to complete the square. At about the same time the gardens in the square were laid out by Humphry Repton, who also laid out Russell Square on the site of the Southampton House gardens and thus gave London the largest square it had had down to that time.

St James's Square

The first proposals for St James's Square date from 1662 when Lord St Albans obtained as a favour from the king a sixty-years' lease of an area of land near St James's Palace,[66] which he proposed to develop as 'a square consisting entirely of really large mansions – only three or four on each side – to be built and occupied by the very best families, including his own'.[67] Market research revealed a reluctance to build such expensive houses on leasehold sites. St Albans then successfully

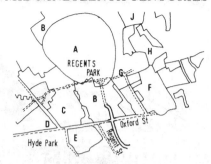

Figure 8.16 – London showing the extent of the major north-western estates in the eighteenth and early nineteenth centuries. Key: A, Crown (Regent's Park as developed in the early nineteenth century; see Figures 8.20 and 8.21); B, Portland; C, Portman; D, Bishop of London; E, Grosvenor; F, Bedford; G, Southampton; H, Somers (developed as Somerstown); J, Camden (developed as Camden Town).

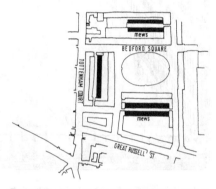

Figure 8.17 – London: Bloomsbury, a detail section of the map of 1799, showing Bedford Square (constructed from 1776) and streets east of Tottenham Court Road. The combination of houses fronting the prestigious streets and 'mews' at the rear for stables and coachmen's accommodation was the property developer's response to the needs of a horse-powered transit age. As such, it is an interesting and little remarked precedent for the rear garaging of motor vehicles which is characteristic of much modern international housing estate planning. The combination of house and mews was not limited to British eighteenth- and nineteenth-century middle- and upper-class urban housing (see also Craig's New Town at Edinburgh, Figure 3.31 for a Scottish example) but it was more widely used in Britain than on the continent, above all in London where numerous variations on the theme have survived through to the present day; the old house more often than not now used for offices, or divided into flats, while the mews cottages and stables, expensively converted, represent probably the most sought after top-of-the-market housing in central London. The location of the streets at the corners of Bedford Square, and not entering at the centres of the sides, is a characteristic of British residential square development (see Note 89).

63, 64. Rasmussen, London: The Unique City.

65, 66. Summerson, Georgian London.

67. St James's Palace was the royal residence until the end of the eighteenth century.

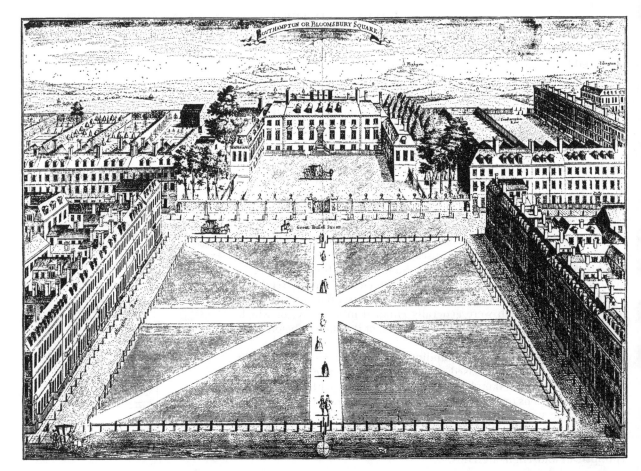

asked, in 1665, for the freehold of the land to be given to him and the king's wish to have suitable near-neighbours undoubtedly prompted him to acquiesce. The concept of an exclusive enclave was dropped in favour of a scheme which closely paralleled Bloomsbury Square, with a total of twenty-two plots let to wealthy individuals for their own houses and to professional speculators. Also following the lead of Bloomsbury Square, St Albans built a market and reserved a site for a church on the central axis of his square – St James's built by Wren between 1776 and 1784 alongside Piccadilly.

Soho Square

This was laid out as a speculative development in 1681 by an architect, Gregory King, after whom it was originally named. It was built immediately to the north of Monmouth House, which, set back behind its forecourt, provided an impressive central building on the southern side. Soho Square was initially a highly favoured address – the rate books record four dukes and a total of at least twenty-four earls and barons among its residents.[68] The diarist Evelyn wrote in 1689 that he went to London with the family to winter at Soho in the great square.[69] Extensive sub-letting was a feature of Soho Square. Thus Hugh Phillips (in his detailed study *Mid-Georgian London*) quotes an advertisement in the General Advertiser of 10 May 1740: 'To be lett in Soho Square, a convenient house, ready furnished with clean Furniture, free from Bugs. With Coach House and stables for a middling family'.

264

Figure 8.18 – Bloomsbury (Southampton) Square, from the south, in 1731, looking towards Lord Southampton's London mansion, and beyond to the twin hills of Hampstead and Highgate – a highly thought-of view. The square is shown with the usual unlandscaped, open character of the seventeenth- and eighteenth-century London squares – Bloomsbury Square's magnificent plane trees (still surviving) are the result of later planting. The eastern (right-hand) terrace has been replaced by a monstrous commercial development: the western side has retained its original scale, as also has the northern side with the later terrace on either side of the street which replaced the mansion in about 1800. Russell Square, of which Southampton Row (right, top) formed the southern side, was not laid out until the 1830s.

68. Summerson, *Georgian London.*
69. Evelyn, *Diary*, 27 November 1689: 'I went to London, with my family, to winter at Soho, in the great square.'

Red Lion Square

Nicholas Barbon, 'speculative builder and bogus doctor',[70] carried out property developments between about 1670 and 1698, the year of his death, on an unequalled scale. Sir John Summerson notes that he was active all over London, building here a square, here a market, here a few streets or chambers for lawyers; he completely grasped the advantages accruing from standardization and mass-production, in housing; it was not worth his while to deal little, he said – that a bricklayer could do.[71] In 1684 he laid out Red Lion Square on fields to the west of Gray's Inn Walk. His workmen had to resist physical assaults by the Gentlemen of Gray's Inn endeavouring to protect the open space; Wren protested against the project and the Middlesex Justices issued warrants. But Barbon was not to be thwarted, and the square was completed.

The Mayfair squares

John Roque's great plan of London of 1769 is reproduced in part as Figure 8.15. The great estates south of Oxford Street had been completely developed through to Tyburn Lane (Park Lane) by that date, the eastern boundary of the Royal Hyde Park followed about 1800. Hanover Square and adjoining streets were built between 1717 and 1721. Grosvenor Square was started in 1725, with thirty-two houses on the northern side and fifteen on the southern side completed by the following year. With an area of 6 acres Grosvenor is the largest Mayfair square. Berkeley Square was laid out on the gardens to the north of Berkeley House (built in 1664 and demolished in 1733 to make way for Devonshire House) from 1739. The streets around the gardens had been constructed earlier, however, in 1675.

Of the Mayfair streets, Old Bond Street, running north from Piccadilly, was started in 1686. Its northern extension through to Oxford Street, New Bond Street, was delayed until 1721. Swallow Street, later to be widened into Regent Street by John Nash, was the main north–south link between Piccadilly and Oxford Street. Until 1719 residents on the west side of Swallow Street had been able to look out from their back windows across meadows as far as Park Lane, where a magnificent avenue of walnut trees stretched from Oxford Street to Piccadilly on its western side.[72]

DEVELOPMENT NORTH OF OXFORD STREET

The year 1769 saw development of estates as far as the New Road well under way. The New Road had been built in 1756 as a direct bypass route – possibly the first of its kind in history – into the City of London from Paddington in the west, initially through fields still clear of development to the north of the new residential districts, before turning south into the City of Moorgate. Later renamed Marylebone, Euston, Pentonville and City Roads, in sequence from the west, this route formed a northern limit of development until the early nineteenth century and subsequently marked the southern termini of the main railway lines from the north.

The first of the squares to be started north of Oxford Street was Cavendish Square. This was an initial phase of an ambitious plan to develop the Cavendish-Harley estate. From 1717 the square, the surrounding streets, a market (still known as Market Place) and a church (St Peter, Vere Street) were under construction. In spite of publicity

Georgian London was a city made up almost entirely of these long narrow plots with their tall narrow houses and long narrow gardens or courts. Practically the whole population lived in one version or another of such houses. A handful of aristocrats had their isolated palaces; and the unemployable and criminal classes had their centuries-old rookeries; but the remainder, from earls to artisans, had their narrow slices of building, now called for no very good reason, 'terrace houses'.
(J Summerson, Georgian London)

As for the plan of the house itself, nothing could be simpler. There is one room at the back and one at the front on each floor, with a passage and staircase at one side. On a site as narrow as 24 feet hardly any other arrangement is possible; in broader sites it is still a perfectly satisfactory and economical arrangement. There is no escape from it. Mariners' humble cottages in the East End have this plan; and so have the great houses in Carlton House Terrace.
(J Summerson, Georgian London)

In other words, the concept of these squares represents the greatest imaginable contrast to contemporaneous French squares and continental squares under French influence. There, representative display, and, if possible, monumentality were the artistic aim, based on integration of the area into the total structure of the city, as exemplified by Patte's plan of Paris. In London the aim was privacy, the privacy of the pedestrian, residential comfort, and seclusion from the life of the surrounding neighbourhood. There can be no doubt that architecturally, and also emotionally, the basic concept of these squares was rooted in the collegiate of the Middle Ages.
(P Zucker, Town and Square)

70. Pevsner, London: The Cities of London and Westminster, 3rd edn, revised B Cherry, 1973 (in the Buildings of England series); although primarily an architectural study of great breadth and depth, this volume (as others) also provides excellent descriptions of main parts of London.
71. Summerson, Georgian London.
72. Phillips, Mid-Georgian London.

stressing its relative nearness to Westminster, Cavendish Square was not a great success, and its completion had to await the building boom of the 1770s. Portman Square was laid out around 1761 and partly completed by 1769. The extensive Queen Anne Square projected for north of Foley House and shown in outline on Roque's plan was not built. Other squares in this district included Manchester Square, 1776, and Bryanston and Montagu Squares, both 1811.

The most important street architecture of this period is Portland Place, planned in 1774 by the Adam Brothers as the widest street in London to preserve the view north from Foley House. Portland Place was later incorporated into John Nash's *via triumphalis* between Regent's Park and Carlton House. Fitzroy Square was another Adam project, commenced in 1793–8 but not completed until 1827–35. East of Tottenham Court Road, the map of 1769 shows no development north of the buildings fronting Great Russell Street, the two most important of which were Montagu House, which had contained the British Museum since 1759, and Bedford House, in front of which was Bloomsbury Square. By 1769 plans for Bedford Square had been discussed, and construction was started in 1776. Nikolaus Pevsner considers that it remains without any doubt the most handsome of the London squares, partly because it is preserved completely on all sides.[73] Gower Street, which runs north–south across the eastern end of Bedford Square, dates from 1790 but the general development of the Bloomsbury squares is a nineteenth-century enterprise.

Duke's New Road and Queen Square

Southampton Row and its northern extension, Woburn Place, are shown on Roque's map as the Duke of Bedford's New Road, leading out across Lamb's Conduit Fields to the New Road proper. East of the Duke's New Road much of the land between it and Gray's Inn Lane was owned by the Foundling Hospital, which in 1790 instructed its architect, Samuel Pepys Cockerell, to report on its development potential. Cockerell recommended the creation of two new squares, with related streets, one to the west and one to the east of the Hospital buildings. These were later named, respectively Brunswick and Mecklenburg Squares. This project was not completed until about 1810, by which time development further north, reaching to the New Road, was under way.

Queen Square, south-west of the Foundling Hospital, when completed by 1720, 'was unique in having three sides and a view – the beautiful prospect of the hills, ever verdant, ever smiling, of Hampstead and Highgate'.[74] The northern side of Queen Square remained open until finally built over as part of the Foundling Hospital development. This took place, despite vigorous protests and legal action by the residents to retain their visual amenity. Within twenty years Queen Square had lost its fashionable appeal. Hugh Phillips quotes a resident saying, around 1820: 'When I came to the Square I was the only lady who did not keep a carriage. Before I left I was the only one who did.'[75]

JOHN NASH (1752–1835)

John Nash is undoubtedly one of the most important and fascinating personalities in the history of urbanism. Coming from obscure origins he acquired a knowledge of architecture in Robert Taylor's office, was bankrupt at 31 (losing a legacy in his first property venture) and subse-

Farther west, in the same line, is Southampton great square, called Bloomsbury, with King-street on the east side of it, and all the numberless streets west of the square, to the market place, and through Great-Russel-street by Montagu House, quite into the Hampstead road, all which buildings, except the old building of Southampton House and some of the square, has been formed from the open fields, since the time above-mentioned, and must contain several thousand of houses; here is also a market, and a very handsome church new built.

From hence, let us view the two great parishes of St Giles's and St Martin's in the Fields, the last so increased, as to be above thirty years ago, formed into three parishes, and the other about now to be divided also.

The increase of the buildings here, is really a kind of prodigy; all the buildings north of Long Acre, up to the Seven Dials, all the streets, from Leicester-Fields and St Martin's-Lane, both north and west to the Hay-Market and Soho, and from the Hay-Market to St James's-Street inclusive, and to the park wall; then all the buildings on the north side of the street, called Piccadilly, and the road to Knight's-Bridge, and between that and the south side of Tyburn Road, including Soho-Square, Golden-Square, and now Hanover-Square, and that new city on the north side of Tyburn Road, called Cavendish-Square, and all the streets about it.
(D Defoe, A Tour through the whole Island of Great Britain)

... It is a little surprising that among the great number of squares in London, not one is to be found that is regularly built, on the contrary it is hardly possible to conceive any thing more confused and irregular than the generality of them are.
(John Gwynn, London and Westminster Improved, 1766)

Bedford Square is unique. It has four sides of uniform palace-fronted terraced houses which form a 'perfect' symmetrical square, with stucco-faced pedimented centres surrounding a leafy garden. Built between 1775 and 1783, its chief importance lies in the fact that it was the first example in London of a square with such consistent uniformity: the builders of the Square managed to achieve what hitherto had been squabbish attempts at some overall coherence. These earlier essays usually comprised four unequal sides: two sides of arcaded terraces, a wall, and a church (Covent Garden, 1630s); three sides of houses, not uniform, with the fourth side taken up by a large house (Bloomsbury Square, 1660s) or with a palace-fronted terrace on one side only (Queen Square, Bath, 1720s). Moreover, the achievement at Bedford Square was never again exactly repeated. Fitzroy Square, for example, had only the south and east sides built to Robert Adam's elevations in the 1790s; by the time building could continue after the French Revolution, taste had changed and the latter blocks, influenced by Greek Revival ideas, are, according to Bolton, 'confessedly inferior'. Other squares suffered from being built in fits and starts, and none display the four palace-fronted terraces with pedimented centres and end pavilions that make up Bedford Square.

73. Pevsner, London: The Cities of London and Westminster.

74, 75. Phillips, Mid-Georgian London.

quently self-exiled to Wales before returning to London. A prosperous period in partnership with Humphry Repton, the leading landscape architect of the day, established Nash in high society and led to him becoming a protégé of the Prince of Wales at the turn of the century. In 1798 Nash exhibited at the Royal Academy a drawing of a conservatory he had designed for the prince. At the end of that year he married, and if one believes the rumours that his wife had a relationship with the prince which was still continuing, the mystery concerning Nash's sudden rise to fame and favour is explicable. The key to his future success was to be his appointment in 1806 as architect to the Department of Woods and Forests. This was a comparatively obscure post but Sir John Summerson is in no doubt that Nash 'must have guessed, or was perhaps persuaded, that his humble appointment would lead somewhere'.[76]

'In the eighteenth century,' says Summerson, 'the Crown lands were loosely and uneconomically managed, with assets neglected and liabilities nursed, and a commission, set up on George III's suggestion in 1786, found itself faced with much difficult research.' As a result of the commission's investigation the post of Surveyor-General of His Majesty's Land Revenue was established, and John Fordyce was appointed in 1793. Summerson recognizes that 'London owes much to Fordyce. It was he who guided the development of Marylebone Park on to farsighted lines, who first propounded as an urgent necessity the need for a great street from Marylebone to Charing Cross, and who forcibly stated to the Government of his day the superiority of comprehensive planning over piecemeal alteration.' In September 1793 Fordyce raised the question of Marylebone Park, an extensive area of Crown land 'over which it is probable that on the return of peace the town may be extended'.[77] Fordyce was authorized by the Treasury to hold a competition for proposals for developing the park. At the time of his death in 1809 little interest had been shown in the competition, and there was still no plan.

No successor to Fordyce was appointed, and the Office of Land Revenue was combined with that of Woods and Forests, under the direction of three commissioners. With less than two years remaining before the reversion of the lease the competition was abandoned, and the official architects of each of the two departments – Nash and Leverton, his Land Revenue counterpart – were instructed in October 1810 to prepare their proposals of the area ostensibly in competition. The two reports were submitted in July 1811. It is unlikely that the result of the competition was other than predetermined for Nash, with his rivals very likely used simply as a check on his proposals and estimates. Very little is known of the role that Nash's patron, the Prince Regent, played in all this, but he is quoted in October 1811 as being 'so pleased with this magnificent plan (which) will quite eclipse Napoleon'.[78] Nash's proposals were duly recommended to the Treasury. His Land Revenue rivals, Leverton (designer in 1785 of Bedford Square) and his partner Chawner, avoided the problem of the new street and based their uninspired scheme on a straightforward extension northwards of the existing gridiron structure south of the New Road, punctuated by 'bigger and better Bedford Squares'.[79] The Nash report is unquestionably the work of a very special kind of genius, that of great creative ability wedded to the political expertise required to realize ideas in practice. In his report he dealt thoroughly with all aspects of the problems of

Socially he was successful. We hear that he 'associated with all the best men in the country', and he was not the sort of man who would let lack of means or deficient education stand in his way. Nobody was ever more completely free from a sense of inferiority. His unassailable self-possession, irritating at times to his acquaintances, carried him everywhere. Eighteenth-century society was broadminded enough to tolerate a bounder, provided that he was an amusing bounder, and Nash's irrepressible good humour, mischievous wit, and quite extraordinary knowledge of all sorts of subjects easily made up for a tendency to pushfulness and a slight deficiency in breeding.
(J Summerson, John Nash, Architect to King George IV)

Nash's achievement as a town-planner is his paramount claim to the attention of posterity. He grasped the essentials of town-planning as nobody else had done; he grasped, without effort or conscious search, the social, as well as the economic and aesthetic aspects. The real cause of his success lay in this: that he was not only an architect but a man of the world, a Londoner who had known his town in poverty and wealth, failure and success, as a carpenter and a courtier. He was a man who lived his life adventurously and to whom human relationships were more important for their multiplicity and their variety than for their intrinsic value. Everything essential to the pattern of contemporary life found some counterpart in his experience; he was a Londoner first, a contriver second, an artist third ... Nash stands on the threshold of modern town-planning not as a pioneer, but as a personality emerging naturally from a concurrence of historical phenomena. His point of view was not that of a man who saw further than his generation; he was most essentially of his generation, and stands at the end of a tradition rather than the beginning. Modern town-planning is a process of analysis leading to a corresponding synthesis: Nash, remote as he was from the scientific departmentalism of today, had something of the objectivity which a modern town-planner must have if his work is to be real and effective.
(J Summerson, John Nash, Architect to King George IV)

76. J Summerson, *John Nash: Architect to King George IV*, 1949, is the main reference; see also T Davis, *John Nash: The Prince Regent's Architect*, 1973.
77, 78, 79. Summerson, *John Nash*.

Figure 8.19 – John Nash's 1811 proposals for Regent's Park; compare with Figure 8.20 which shows the layout as actually constructed. In most respects change was for the better; the proposed circus at the crossing of the New Road (Marylebone Road), for example, would have provided a far less satisfactory extension of Portland Place, as the main entrance into the park, than the combination, as built, of semi-circular Park Crescent and Park Square, further extended as an avenue of trees. In addition it would seem that Nash's real-estate instincts had at first led him to provide for too many residences within the park – both as detached villas and in the grandiose Great Circus. (The site for the Life Guards and Artillery Barracks, across the northern side of the park, is that occupied in part by London Zoo.)

The Regent's Park, above all, is a scene of enchantment, where we might fancy ourselves surrounded by the quiet charms of a smiling landscape, or in the delightful gardens of a magnificent country house, if we did not see on every side a countless number of mansions adorned with colonnades, porticoes, pediments, and statues, which transport us back to London; but London is not here, as it is on the banks of the Thames, the gloomy commercial city. Its appearance has entirely changed; purified from its smoke and dirt, and decked with costly splendour, it has become the preferred abode of the aristocracy. No artisans' dwellings are to be seen here; nothing less than the habitations of princes.
(C d'Arlincourt, The Three Kingdoms, 1844)

developing the park and providing the new residences with a suitably imposing connection with the St James's and Westminster district.

Regent Street, as this new route south from the park was to be known, has been erroneously attributed by several town-planning historians to the Prince Regent's desire to link his Carlton House (which occupied a site at the St James's Park end of Lower Regent Street) to the new Royal Park. This was, it is true, a secondary benefit resulting from the line of the street, but its main function was determined by real-estate commonsense.[80] Nash also redesigned the landscaping of St James's Park and proposed for it a peripheral development which would repeat the terrace motif adopted for Regent's Park. Both the park and the street must therefore be seen in this wider context.

REGENT'S PARK

The 1811 report contained proposals for the layout of the park which, although subsequently modified in some respects, were to form the basis of the final plan. Nikolaus Pevsner has aptly summarized the basic concept as 'the combination of palatial façades with a landscaped park. The terraces are urban but their setting is countrified. Tenants paid the price for a three-windowed terraced house and obtained the illusion of living in a vast mansion in its own grounds. The scheme catered for the genuine love of the English for a life away from the town, and also clearly for snobbery'.[81] Humphry Repton's influence is clearly apparent in most aspects of the design, from the basic Reptonian strategy of 'apparent extent' down to the detailed use of trees and water.[82]

The point of entry into the park from the south was from the top end of the existing Portland Place, which had been built by the Adam

80. I maintain that real-estate expediency was the primary determinant of the line of Regent Street; nevertheless, behind the scenes, as it were, royal aggrandizement was also a main consideration. It is a measure of Nash's genius that he was able to satisfy the requirements of both the royal estate agents and the Prince himself.
81. Pevsner, London: The Cities of London and Westminster.
82. For Humphry Repton see D Stroud, Humphry Repton, 1962; as applied for the landscape setting of the Regent's Park residential terraces, the Reptonian strategy of 'apparent extent' is an example of a peculiarly English approach to moneyed-class domestic aggrandizement.

Figure 8.20 – Part of North London, mapped in 1832 (north at the top), showing Regent's Park as originally completed with Portland Place integrated into John Nash's new route south to St James's and Westminster. (See Figure 8.23 for the detail treatment of the junction with Oxford Street.) Immediately to the east of the park, across Albany Street, there are the three squares intended as markets for the new development – York Square, Clarence Gardens and Cumberland Market, Cumberland Basin, on a branch of the Regent's Canal, has now been filled in (1942–3). Further north, beyond the Horse Barracks, on both sides of the canal, Nash laid out his two Park Villages – East and West – as detached and semi-detached villa developments. In these estates embryonic twentieth-century suburbia can be discerned. This map also shows the western Bloomsbury squares and streets, including Bedford Square: the first London University building in Gower Street (present-day University College); Euston Square, and the mainly working-class housing streets to the north through which the railway was brought to the station in 1836.

Brothers from 1774, and which was taken by Nash as the determining line of the northern section of his street proposals. The New Road was crossed through a large circus, with the stipulated church in its centre. A second double-circus was located on high ground near the centre of the park, with the required 'Valhalla' in the middle. The long lines of the residential terraces are stiffly laid out, surrounding the open space in which Nash proposed the location of between forty and fifty villas, on sites ranging from 4 to 20 acres in area. This original Nash plan for the park (Figure 8.19) was approved in August 1811 and in October, moving with commendable speed, the Treasury authorized preliminary works, including a driveway around the park and the planting of young trees to ensure an attractive landscape setting by the time houses were completed.

The southern half of the entrance circus was started in 1812, but the original lessee, Charles Mayor, was bankrupt before the end of the year; work was suspended until 1815 and completion of the south-eastern quadrant was delayed until 1822. In that year the two quadrants north of the New Road were deleted from the design and replaced by the eastern and western sides of Park Square (1823–5). The canal was removed from the interior of the park to a less conspicuous perimeter route in 1812. The number of villas was drastically reduced to eight and the double-circus with its 'Valhalla' was omitted in order to enhance the illusion of a rural park.

The whole unappropriated area of the Regent's Park is now thrown open to the public. The first object appears to have been to make the whole of its disposable area available as early as possible in the season. In addition to the five entrances already made, a sixth will be formed, to afford admission into the park from what is termed the Inner Circle. The ornamental water will be crossed by a suspension bridge of nearly 150 feet span, and the line of the path of which it is to form the connection will extend, with scarcely any deviation, from the entrance of York Gate to the summit of Primrose Hill. To secure the privacy of the villas, the Commissioners of Woods and Forests have granted the Marquess of Hertford four additional acres to be inclosed in his plantations, two to Mr. Goldsmid (now Sir Isaac Lyon), and three to Mr. Holford, fronting North Lodge Gate. Plans for connecting the property recently acquired near Primrose Hill with the Regent's Park, from which it is now divided by a public road, are under consideration.
(The Gentleman's Magazine, Part II., p. 418, 1841)

Mr Nash is a better layer-out of grounds than architect, and the public have reason to thank him for what he has done for Regent's Park. Our gratitude on that point induces us to say as little as we can of the houses there, with their toppling statues, and other ornamental efforts to escape from the barrack style. One or two rows of the buildings are really not without handsome proportions, those with the statues among them; and so thankful are we for any diversity in this land of insipid building, where it does not absolutely mortify the taste, that we accept even the bumpkins of Sussex Place as a refreshment ... we have reason to be thankful that the Regent's Park has saved us from worse places in the same quarter; for it is at all events a park and has trees and grass and is a breathing space between town and country. It has prevented Harley and Wimpole-streets from going further; has checked, in that quarter at least, the monstrous brick cancer that was extending its arms in every direction.
(L Hunt, The Townsman)

The section of an 1832 map of London (reproduced as Figure 8.20) shows the arrangement of terraces as built. The first to be started was Cornwall Terrace in 1821 and all the others were under way by 1826. York Terrace is about 1,080 feet long, in two equal sections on either side of York Gate; the parish church of Marylebone is axially located at the end, on the southern side of the New Road. Individual residences comprising the various parts of York Terrace were all entered via porches in the service road at the rear. Avoidance of a succession of front doors in the park elevations meant that the palace illusion was preserved on one side at least. In general Nash controlled only the basic elevational designs of the terraces, working drawings and supervision of construction was delegated to others, who were by no means as careful

Figure 8.21 – Regent's Park seen beyond the line of the New Road of 1756 (as Figure 8.15), 'A–A' from right to left (east to west). Portland Place is the wide street running north from the British Broadcasting Corporation's building at the bottom of the photograph (the spire of All Souls, Langham Place is just visible in the centre of the bottom edge); the crossing of Portland Place and the New Road (Marylebone Road) was effected with Park Crescent and Park Square (as Figure 8.20); and the line of Portland Place is shown continued through into the park by the avenue of trees. The regular gridiron structure of this part of northern Mayfair is clearly shown; Harley Street and Wimpole Street are the two routes parallel to and west of Portland Place.

as they should have been. Many of the sections of terrace spared by Second World War bombing have required reconstruction in recent years, care being taken to preserve or, in the cases where this was no longer possible, to reconstruct the original elevations.

REGENT STREET

The Roque 1769 map of London (reproduced in part as Figure 8.15) shows most of the existing urban fabric through which John Fordyce stipulated the driving of the new main street south from the park. Incomplete for Roque's survey, Portland Place, the 'grandest street in London', ended just short of the line of the New Road and seemed to Nash to form the obvious northern section of the new route. The southern end was determined by the position of Carlton House, on the south side of Pall Mall, and the new street would both connect the Prince Regent's house to the park and provide it with a suitably impressive approach. The problem now was to create an axial route north from Carlton House to Portland Place.

Nash's flair for separating the essential problem from all the conflicting overlying issues revealed that along a certain line the town abruptly changed character; the mean streets of Soho stopped and the spacious criss-cross West End began. In Nash's own words from his 1811 report the new street should constitute 'a boundary and complete separation between the streets and squares occupied by the nobility and gentry, and the narrow streets and meaner houses occupied by mechanics and the trading part of the community.[83] The central section of the street was therefore pencilled in along a line drawn north–south across the eastern ends of the Mayfair streets, roughly following the existing north and south part of Swallow Street.

The next problem was how to relate this central section to the southern and northern ends. Here the plan contained in the report differs from what was built and, to a great extent, from the present-day layout of the intersection with Piccadilly (construction of Shaftesbury Avenue from 1870 required demolition of the north-eastern corner of the junction). In building Piccadilly Circus Nash abandoned his initial thoughts for a grand square crossed diagonally by the line of the street, with a secondary circus at the Piccadilly intersection, and evolved his famous quadrant, swinging west and north from a small square adjoining the circus as one unbroken curved section of street.

From Oxford Street to Portland Place the line was determined by the need to keep a respectful distance from the Cavendish Square houses, and the 1811 plan shows a grand circus at Oxford Street enabling a simple change of direction to be made. Objections on the part of the Cavendish Square residents forced the line still further east. Additional complications posed by the presence of Foley House, on the axis of Portland Place, provided Nash with an opportunity to achieve a second brilliantly contrived change of direction at the circular-spired vestibule of All Souls Church which 'is made to do its work in the larger scheme of things so well that it still conducts the movement around the difficult turn in Regent Street with power and grace'.[84] In 1813 Parliament passed, by a substantial majority, the Act 'for making a more convenient communication from Marylebone Park and the northern parts of the metropolis ... and for making a more convenient sewage for the same'. In addition to its primary purpose the Act also provided for the widening and eastern extension of Pall Mall, and creation of what is

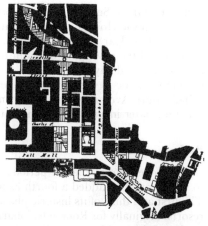

Figure 8.22 – Detail plan at the southern end (Lower Regent Street) of Nash's *via triumphalis*, with Carlton House at its terminal point, showing the arrangement of Piccadilly Circus where Regent Street curves away to the west and north. Nash's proposals for Trafalgar Square are also shown, to the bottom right.

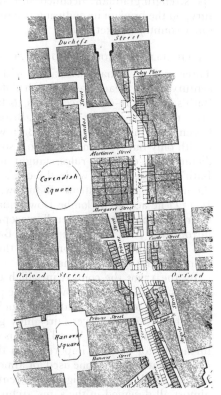

Figure 8.23 – Regent Street at the crossing of Oxford Street, showing the displacement to the east required to avoid disturbance to Cavendish Square properties, and the return curve needed to join with the existing line of Portland Place (at the top of the plan).

83. Quoted by Summerson, *John Nash*.
84. E Bacon, *Design of Cities*, 1974, a major section with excellent maps and illustrations.

today Trafalgar Square, and the provision of a 'more commodious access from the Houses of Parliament ... to the British Museum'. The shape of Trafalgar Square was regularized by Nash, but his proposals to surround the new space by an imposing sequence of buildings, with the Royal Academy in its centre, were not carried out. Neither was his plan for the new street from the square to the British Museum. (Shaftesbury Avenue, which today only indirectly serves this purpose, was built later in the nineteenth century.)

Bath

Bath enjoyed three separate periods of historical fame and prosperity, to which must be added a fourth as a present-day tourist attraction.[85] The first and third of its historic phases were both as health and leisure resorts: originally for Roman legionaries who came to Aquae Sulis from as far afield as Metz and Trier in distant Gaul; and latterly made famous by the eighteenth-century aristocracy for whom its incomparable architectural heritage was created. In between times it was a flourishing centre of medieval cloth manufacture. In turn this commercial strength gradually declined and by the end of the seventeenth century, on the brink of its enduring rise to fame, Bath was no more than a quiet country market town.

RALPH ALLEN AND JOHN WOOD THE ELDER

Three men are usually associated with the rise to fame of eighteenth-century Bath: Beau Nash, Ralph Allen and John Wood the Elder. Bath's social pre-eminence seems to have dated from Queen Anne's visits to the city in 1702 and 1703. Beau Nash was appointed Master of Ceremonies in 1704: before that Bath was a place visited for reasons of health only but Nash made it into a social centre and, with the help of dictatorially enforced rules, taught it elegance.[86] Ralph Allen was Bath's richest citizen; while the city's postmaster he had reorganized the national postal system – to his own considerable financial benefit. In 1727 he acquired Combe Down quarries and the chance to exploit their stone products is generally accepted as one of Allen's reasons for introducing John Wood the Elder to Bath. The Woods, father and son, are by far the most important of the eighteenth-century architect/planners who worked in the city.

By the 1720s the context in which the Woods were to operate had been established. The city was on the point of rapid expansion: money was starting to flow into Bath and much more was on the way. The city itself had to expand over fairly hilly ground – a factor brilliantly exploited by the younger Wood with his design of the Royal Crescent. A further factor of great importance is that Bath is essentially a single-material city – local stone used for almost all the buildings giving a rare architectural unity; the eighteenth-century additions to Bath were also almost all designed in the same architectural style.

John Wood the Elder (c. 1700–54) moved to Bath in 1727, the year in which his son and successor, John Wood the Younger, was born. Little is known of the father's early life or where and in what form he received his architectural training. It is generally assumed that while he was engaged in Yorkshire as a road surveyor he met Ralph Allen and was either persuaded to move to Bath or, having heard of the prospects, 'was independently attracted'.[87] Having first checked on the city's de-

Looking at old prints of Regent Street we are apt to think of it as a series of quasi-classical blocks making up a fairly balanced 'Regency' thoroughfare with the Quadrant as its most spectacular element. In fact it was nothing of the sort; the blocks, ionic, corinthian, astylar, elaborate or plain, provided a progress of great variety and were sometimes interspersed by a church, an assembly hall, a hotel or other non-terraced buildings. (T David, John Nash: The Prince Regent's Architect)

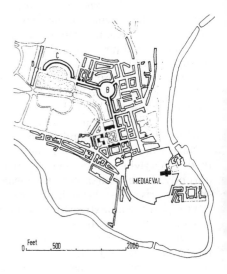

Figure 8.24 – Bath (north at the top), showing the relationship of the Renaissance spaces and streets to the north and west of the medieval nucleus. Key: A, Queen Square; B, the King's Circus; C, the Royal Crescent; D, Marlborough Buildings.

There is one thing very observable here, which tho' it brings abundance of company to the Bath, more than ever us'd to be there before; yet it seems to have quite inverted the use and virtue of the waters (viz.) that whereas for seventeen hundred or two thousand years, if you believe King Bladud, the medicinal virtue of these waters had been useful to the diseased people by bathing in them, now they are found to be useful also, taken into the body; and there are many more come to drink the waters, than to bathe in them; nor are the cures they perform this way, less valuable than the outward application; especially in colicks, ill digestion, and scorbutick distempers.

This discovery they say, is not yet about fifty years old, and is said to be owing to the famous Dr Radcliff, but I think it must be older, for I have myself drank the waters of the Bath above fifty years ago: But be it so, 'tis certain, 'tis a modern discovery, compar'd to the former use of these waters. (Defoe A Tour through the whole Island of Great Britain)

85. For Bath see J Elliot, The City in Maps: Urban Mapping to 1900, 1987, which includes 'A New and Correct Plan of the City of Bath', H Godwin, 1810; Bacon, Design of Cities; N Pevsner, North Somerset and Bristol (Buildings of England series), 1958.
86. Pevsner, North Somerset and Bristol.

velopment potential Wood notes, in his own account, 'I procured a plan of the town, which was sent me into Yorkshire, in the summer of the year 1725, where I, at my leisure hours, formed one design for the ground at the north-west corner of the city, and another for the land on the north-side of the town and river.'[88] Wood returned to London with his designs, which he discussed with the owners of the land involved – first with a Mr Gray and second, in March 1726, with the Earl of Essex. Gray's land north-west of the existing medieval city limits was considered most suitable for development, both by reason of its altitude and immediate proximity, and this was acquired.

A ninety-nine-year lease of land was obtained from Gray sufficient for the east side of what was to become Queen Square, the first of the sequence of spaces which have gained for Bath and the Woods a unique place in the history of urbanism. With Queen Square, John Wood was in control of all aspects of the development process: he was at one and the same time architect, contractor and estate agent. Between November 1728 and October 1734 Wood took up further leases as building progress required and in seven years Queen Square was built.

QUEEN SQUARE

As it was originally planned, Wood intended Queen Square to be dominated by the palatial composition of the northern elevation – a single unified architectural ensemble, with the eastern and western sides of the square forming a 'palace' forecourt. The eastern side was started in January 1729 and was completed according to plan, with Wood subletting parcels of land 'to such persons as were willing to build in direct or near conformity with his designs'.[89] There are six houses on the east side, with those at the ends having their entrances from side streets. The houses are stepped down in accordance with the sloping ground. Wood had intended to level the site, but he saved some £4,000 by omitting this work. Queen Square is dominated by the magnificent northern elevation, comprising seven large houses organized with great skill into a symmetrical composition. The southern side is noted by Walter Ison as being merely a pale echo of its splendid opposite number.[90] The western side shows the main departure from the plan: the repeat of the eastern elevation is abandoned in favour of a design based on a large central mansion set back from the street line and flanked by buildings forming an entrance court. The central garden was enclosed within a low balustraded wall, with imposing entrance gates in the centre of each side. This space was crossed by gravel walks, with four formally planted parterres at its corners. In the centre a water-basin of a diameter of 44 feet provided the setting for the obelisk (69 feet in height) which was erected in 1738 in honour of Frederick, Prince of Wales.

Queen Square was built as a speculative development. Wood's great achievement was to create an ordered architectural composition out of the varying requirements of his tenants. Walter Ison describes his working method as follows: 'Wood devised houses of different size and degree conforming to six definite standards, which he classed as first-rate to sixth-rate, in that ascending order of magnitude. . . . Having first designed the elevation he sub-leased sites for individual houses to builders or building tradesmen, giving them full liberty to plan the interiors to suit their prospective tenants, but demanding strict adherence to his exterior design'.[91]

Here, in three spinal extensions, Gay Street, the Circus and the Royal Crescent, with the adjacent Queen Square at the south end of Queen Street, one had, in miniature, the new order of planning at its captivating best. Even now, after a century and a half of change, the heart of Bath has qualities of design that even the best examples in Paris, Nancy, London, or Edinburgh do not surpass. The excellence of Bath shows the advantage of a strict discipline, when it is supple enough to adapt itself to challenging realities, geographic and historic. The placing of the Royal Crescent on a height that commands the whole valley, protected by the part that spreads below, shows that it was no mere application of an arbitrary geometric figure; and while nothing in the rest of the eighteenth century reaches this level of planning, the further building of Bath, right through the Regency, never fell too short of its standard. Not less notable than the preservation of the park-like environs was the generous allotment for gardens in the rear: gardens visible through their iron gates, spacious and richly textured, as shown in the plan of 1786, and still often handsomely kept up today.

This is a superior example of open planning, combined with a close urban relationship of the buildings, which are treated as elements in a continuous composition. In short, Bath's eighteenth-century town planning was as stimulating and as restorative as the waters, and the money invested has brought far higher returns in life, health, and even income than similar amounts sunk into more sordid quarters.

(L Mumford, The City in History)

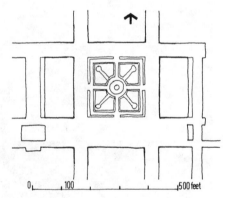

Figure 8.25 – Queen Square, Bath: the original plan of John Wood the Elder (north at the top) in accordance with the orientation of Figure 8.26.

87. W Ison, The Georgian Buildings of Bath, 1969.
88. John Wood the Elder, An Essay towards the Description of Bath, 1742.
89. Ison, The Georgian Buildings of Bath; while from the inside Queen Square is dominated by the grand, if deceptively divided, northern side, that status it not enhanced by location at the end of a main street entering the square in the centre of the facing side, as would have been the likely case in French and other continental cities. Woods's new streets in this part of Bath are clearly there for the sole purpose of urban movement, without any thought of domestic aggrandizement.
90, 91. Ison, The Georgian Buildings of Bath.

THE KING'S CIRCUS AND ROYAL CRESCENT

The continuation northwards of the southern side of Queen Square –
Gay Street, started by John Wood the Elder and finished by his son in
1760 – leads into the Circus (originally the King's Circus) as one of its
three radial connections. The building of the Circus started in February 1754. In the following May, John Wood died, and his design was
completed by his son. The Circus is 315 feet in diameter, with a total of
thirty-three houses forming three equal-length segmental elevations of
eleven, twelve and ten houses respectively. Although the plans vary
considerably, each house consists of three principal floors with a base-
ment and an attic, giving a uniform height of 42½ feet. Pevsner, in the
Bristol and Somerset volume of *Buildings of England*, describes it as the
most monumental of the elder Wood's works, even more so if one
remembers that the old plane trees which are now so much more

Figure 8.26 – An aerial view of the sequence of spaces
in Renaissance Bath, looking north. (This view and the
general layout, Figure 8.24, have the same orientation.)
The ground rises fairly steeply from bottom to top of
this view. Key: A, Queen Square; B, the King's Circus;
C, the Royal Crescent; D, Lansdown Crescent. The
grand Georgian architecture of the northern side of
Queen Square is shown, as also the absence of a vista
street entering the square in the centre of the southern
side.

splendid than the buildings did not exist and were not projected. The centre was paved-stone and had no greenery.[92]

John Wood the Elder's plan for the Circus provided for short lengths of street leading out to the north-west and north-east, each terminated by a suitably imposing building. Continued demand for houses resulted in his son changing the plan for the street radiating to the north-west (Brock Street). This he made a link to yet another residential development, the Royal Crescent, built between 1767 and 1774, and a work of true genius. With its integration of built form and natural landscape it is in complete contrast to his father's equally outstanding approach. Brock Street (1767–8) leads directly out of the circus, with continuous building frontages, and was designed as an architecturally subdued approach to the magnificent curve of the Royal Cresent. The thirty houses that make up the crescent follow the practice of the earlier developments. They have a strictly controlled elevational design, within which there are different plans. The basic form is that of half an ellipse, with a major axis of 170 yards, built around a sloping lawn on the side of the hill; the southern side was open to the magnificent view of then unspoilt natural landscape. The ground floor of the houses is plainly designed and serves as a base for the range of giant Ionic columns, 22 feet high, spaced generally at 8 feet centres, framing one window on each of the two upper floors. Set back behind the cornice line there is an attic floor. The view out to the north-west across the Royal Crescent is stopped by Marlborough Buildings (c. 1790) where, as Pevsner observes, urban Bath comes as abruptly to an end now as it did 150 years ago.[93]

Edinburgh

Edinburgh, one of the world's most dramatically situated cities, is also of great interest in urban history for three main reasons.[94] First, it is one of the oldest known British urban settlements dating back to some 500 years BC. Second, its early medieval ridge-top beginnings with twin castle and monastery nuclei are a classic of their kind. Third, it offers a direct contrast between the organic growth form of the medieval 'old' town and the planned Renaissance layout of its late eighteenth-century counterpart, known after its planner as 'Craig's New Town'. Many cities show similar contrasts in form between medieval cores and Renaissance additions: Berlin (Figures 7.13 and 7.15) and Nancy (Figure 6.21) are notable examples. Neither, however, has such a clear-cut distinction between its two parts as Edinburgh.

ORIGINS AND MEDIEVAL BACKGROUND

Edinburgh's history begins with a Bronze Age fortress constructed on the great volcanic rock which dominates the surrounding Lothian countryside some two miles south of the Firth of Forth. Rising to 440 feet above sea level and approachable only along a ridge sloping gradually upwards from the east, Castle Rock, the magnificent Acropolis of the future city, properly claims for Edinburgh the title 'Athens of the North'.

Subsequently the story is unclear until in 1018, under Malcolm II, Castle Rock became the location of a new Scottish capital, with the castle itself occupied as a royal residence under Malcolm III. By the later eleventh century Edinburgh had assumed the form of a strongly

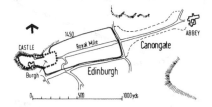

Figure 8.27 – Edinburgh: diagrammatic plan of the original ridge-top city, relating the castle and its burgh to the abbey, and showing the extent of walled Edinburgh in 1450 with undefended Canongate as it had grown up on either side of the route (now the 'Royal Mile') linking castle and abbey.

I am now at the gates of Edinburgh; but before I come to describe the particulars of that city, give me leave to take it in perspective, and speak something of its situation, which will be very necessary with respect to some disadvantage which the city lies under on that account.

When you stand at a small distance, and take a view of it from the east, you have really but a confused idea of the city, because the situation being in length from east to west, and the breadth but ill proportioned to its length, you view under the greatest disadvantage possible; whereas if you turn a little to the right hand towards Leith, and so come towards the city, from the north you see a very handsome prospect of the whole city, and from the south you have yet a better view of one part, because the city is increased on that side with new streets, which, on the north side, cannot be.

The particular situation then of the whole is thus. At the extremity of the east end of the city stands the palace or court, called Haly-Rood House; and you must fetch a little sweep to the right hand to leave the palace on the left, and come at the entrance, which is called the Water Port, and which you come at through a short suburb, then bearing to the left again, south, you come to the gate of the palace which faces the great street.

From the palace, west, the street goes on in almost a straight line, and for near a mile and a half in length, some say full two measured miles, through the whole city to the castle, including the going up the castle in the inside; this is, perhaps, the largest, longest, and finest street for buildings and number of inhabitants, not in Britain only, but in the world.
(D Defoe, A Tour through the Whole Island of Great Britain, 1724–26. (Letter 11.))

Edinburgh indeed was an extreme example of the French type of town, kept within its ancient limits for reasons of safety and defence, and therefore forced to find room for growth by pushing its tenement flats high in air – in contrast to the ground plan of the easy-going peaceful towns of England, that sprawled out in suburbs ever expanding, to give each family its own house and if possible its own garden. French influence and the disturbed condition of Scotland in the past had confined the capital within its walls and pushed its growth up aloft.
(G M Trevelyan, English Social History)

92, 93. Pevsner, North Somerset and Bristol.
94. For general Edinburgh references see A J Youngson, The Making of Classical Edinburgh; T A Markus (ed) Order in Space and Society: Architectural Form and its Context in the Scottish Enlightenment, 1982; J Gifford, C McWilliam and D Walker, The Buildings of Scotland, revised 1988.

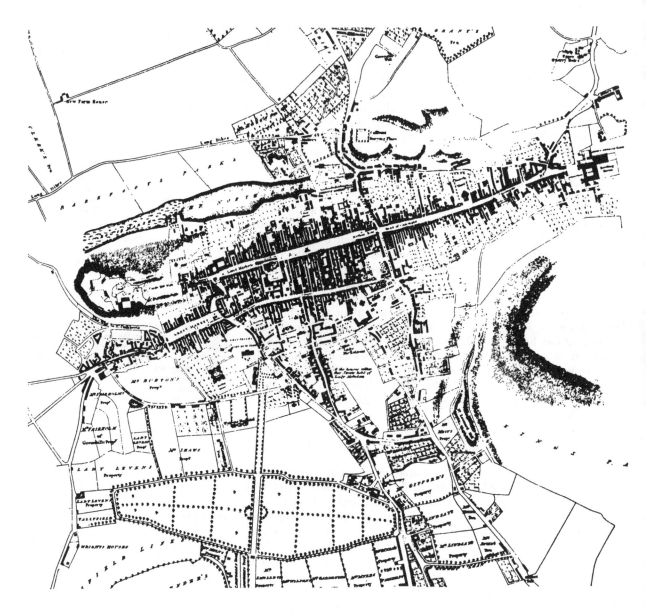

fortified burgh, as shown in Figure 8.27. Early in the twelfth century David I (1084–1153) granted land about one mile east of his burgh for the establishment of the Abbey of Holy Rood, thereby creating the second of the future city's medieval nuclei. In a charter of 1128 he also gave the monks a tax income from land in 'my burgh of Edinburgh' – a first recorded mention of the name. The township that grew around the abbey gate was given separate burghal status, as Canongate, with a dividing line set approximately halfway along the ridge.

The later medieval growth of the old ridgetop town is summarized in the caption to Figure 8.27, related to the early eighteenth-century map as Figure 8.28. Prior to the expansion from 1765 through the addition of Craig's New Town, Edinburgh had consolidated its characteristic medieval form with only tentative ribbons of development leading away from the axial 'Royal Mile', across the less steeply sloping southern fields.

Figure 8.28 – Edinburgh in the first half of the eighteenth century before the addition of Craig's New Town on the fields to the north of the North Loch. The castle (at left) and Holyroodhouse (at right) terminate, respectively, the High Street and Canongate. The pattern of fields and roads to the south of the old city is another example of the pre-urban cadastre urban form determinant; effects of which can be clearly seen in the 1850 Tallis map on the facing page.

CRAIG'S NEW TOWN

A pamphlet entitled *Proposals for carrying on certain Public Works in the City of Edinburgh* was published in 1752 as a first move in a sustained campaign to improve the city. In very favourably comparing London to Edinburgh, the author noted as the latter's failings that 'Placed upon a ridge of a hill, it admits but of one good street and even this is tolerably accessible only from one quarter. The narrow lanes leading to the north and south, by reasons of their steepness, narrowness and dirtiness, can only be considered as so many unavoidable nuisances. Confined by the small compass of the walls, and the narrow limits of the royalty, which scarcely extends beyond the walls, the houses stand more crowded than in any other town in Europe, and are built to a height that is almost incredible.... The principal street is encumbered with the herb market, the fruit-market, and several others; the shambles are placed along

Figure 8.29 – Edinburgh: the central section of the Tallis map of 1850 relating the extent of Craig's New Town to the old ridge-top city (with the contrasting organic growth suburb to the south) and to the later planned extensions further north and to the west. Craig's New Town comprises the eight grid blocks (as two rows of four) between, and including, St Andrew's Square to the east, and Charlotte Square to the west. The three main east–west streets are Princes Street (facing the old city). George Street and Queen Street. By this date haphazard 'planned' estates had been added to the New Town, and the city's subsequent economic prosperity was to engulf the historic nucleus within extensive organic growth working-class housing districts.

277

what was the side of the North Loch, rendering what was originally an ornament of the town a most unsufferable nuisance.'[95] The pamphlet stressed that there were powerful motives prompting the improvement of Edinburgh and that the time was right, with several of the principal parts of the town now lying in ruins. Many of the old houses were decayed, several had already been pulled down, and probably more would soon be in the same condition. The most important of the several detailed proposals put forward advocated an Act of Parliament for extending the royalty to enlarge and beautify the town by opening new streets to the north and south, removing the markets and shambles, and making the North Loch a canal, with walks and terraces on each side.

In March 1766 it was announced that the land to the north of the city had been surveyed, and in April notice was given inviting architects and others to submit plans of a new town, making out streets of proper breadth, and by-lanes, and the best situation for a reservoir, and any other public buildings which might be thought necessary.[96] Six plans were submitted. In August 1766 James Craig's proposals were selected, and after discussion and revision his design was finally adopted in July 1767.

Craig's plan is extremely simple – three long east–west streets, run the length of the ridge and seven shorter streets cross them at right angles; the central long street links two squares, one at each end of the ridge (Figure 8.29). A total of eight large building blocks is formed by this gridiron structure, each divided into two parts by a service road giving access to mews. Four smaller blocks, also served by mews, form the remaining sides of the two squares.

Neither Craig nor the Town Council seems to have seriously con-

Figure 8.30 – Edinburgh: aerial view looking past the castle, at the western end of the old ridge, across the North Loch towards Craig's New Town. Beyond the line of trees forming its northern boundary is the nineteenth-century sequence of circuses and squares planned by Reid and others. Princes Street forms the southern, near side of the New Town. The layout of the buildings on the Mound, linking the old and new parts of the city, across the North Loch, dates from the early 1830s; this is seen to the right of the photograph.

On the north side of the city, as is said above, is a spacious, rich, and pleasant plain, extending from the lough, which as above joins the city, to the river of Leith, at the mouth of which is the town of Leith, at the distance of a long Scots mile from the city. And even here, were not the north side of the hill, which the city stands on, so exceeding steep, as hardly, (at least to the westward of their flesh-market) to be clambered up on foot, much less to be made passable for carriages. But, I say, were it not so steep, and were the lough filled up, as it might easily be, the city might have been extended upon the plain below, and fine beautiful streets would, no doubt, have been built there.
(D Defoe, A Tour through the Whole Island of Great Britain, 1724–26. (Letter 11.))

95. *Proposals for carrying out certain Public Works in the City of Edinburgh.*
96. A J Youngson, *The Making of Classical Edinburgh,* 1975.

278

sidered the possibility or desirability of designing 'standard' unified elevations to which individual buildings would have to conform, as was the rule for example in London and Bath. It was accepted that, with a limited housing market, this might risk alienating possible tenants. The council did, however, take special steps to control building, initially with an Act of July 1767 and later with stricter legislation of 1782 and 1785. The first controls laid down continuity of building lines, established pavements as 10 feet wide and made provision for a sewer in George Street. Youngson records that no one in Edinburgh had adequate knowledge of this subject and that Craig himself was paid thirty guineas to carry out the necessary research in London. By 1782 St Andrew's Square was completed and work was under way as far west as Hanover Street. The new legislation of that year was much more specific in content and included the following provisions:

(1) That no feus shall be granted in the principal streets of the extended royalty for houses above three storeys high, exclusive of a garret and sunk storeys, and that the whole height of side-walls from floor of sunk storey shall not exceed 48 feet.

(2) That the Meus Lane shall be solely appropriated for purposes of building stables, coach houses or other offices, and these shall in no case be built on any of the other streets of the extended Royalty.

(3) That the casing of roofs shall run along the side-walls immediately above the windows of the upper-storey, and no storm or other windows to be allowed in the front of the roof, except skylights, and that the pitch of the roof shall not be more than one-third of the breadth or span over the walls.

A J Youngson in his detailed study of *The Making of Classical Edinburgh* has this to say regarding the qualities of the plan: 'The New Town owes its superiority partly to situation, partly to the whole being built to conform to a regular and beautiful plan. But in the Dictionary of National Biography the same plan is damned as utterly destitute of any inventive ingenuity or any regard for the natural features of the ground. The truth is that the plan is entirely sensible and almost painfully orthodox.' Sensible yes, but certainly not painfully orthodox. Working at almost exactly the same time that John Wood the Younger was creating his Royal Crescent, Craig also adopted the revolutionary, but right and obvious, response to his own ridge-top site. His two bounding side streets, Princes Street (on the old city side) and Queen Street were designed with houses forming the inner side only, and aspects outwards across the street, in the one case over the low ground towards the castle and High Street, in the other down the slope towards the Firth of Forth and the distant hills of Fife.[97]

Urban fortification

It is generally true to state that internal peace, within the defensive 'moat' of the Channel, made urban fortifications unnecessary in England after the end of the fourteenth century, and that subsequently walls which were retained or constructed served merely as trading or 'political' barriers. But there are exceptions to those rules which have hitherto been overlooked in the urban histories (including the first edition of this volume).[98]

Five instances of English urban defence systems from the sixteenth century and later are described: two were provided for frontier towns

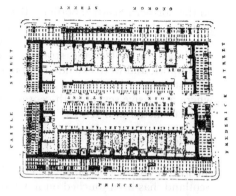

Figure 8.31 – Edinburgh: a typical gridiron block on the southern side of Craig's New Town between Princes Street, George Street and Castle Street, to the right and Frederick Street to the left. Also shown is the arrangement of the rear service mews and the minor internal terraces fronting the secondary cross-street. Building elevations generally conformed to parapet-line uniformity so that, although there were minor variations of detail, the overall effect would have been one of pleasant unity. (It is not clear why the line of Frederick Street is not parallel to Hanover and Castle Streets and is thus a non-conforming element in an otherwise rigidly rectilinear plan.)

97. The open one-sidedness of Princes Street may seem to have been the obvious way of relating the old and the new parts of Edinburgh, yet I doubt it was so apparent at the time. In the belief not only that it requires genius to recognize the obvious, but also and more so to be able to put it into practice, James Craig must be linked with John Wood the Younger as a city planning genius.

98. For references on military engineering aspects of urban fortification in general (bearing in mind from Chapter 5 that there is as yet no definitive work dealing with urban fortification from the standpoint of city planning), see C Duffy, *Siege Warfare, Volume 1, The Fortress in the Early Modern World 1494–1660; Volume 2, The Fortress in the Age of Vauban and Frederick the Great 1660–1789.*

on the northern Scottish border – Berwick-on-Tweed on the east coast, and Carlisle, its western counterpart (as described in Chapter 4). Two other systems, more typical of the complex, continental scale, were developed to protect the strategic naval bases of Portsmouth and Devonport, and were extended through to the late nineteenth century. The other case is Kingston upon Hull, or Hull as it is now known, where a massive seventeenth-century citadel was added to replace the obsolete medieval defences.

BERWICK-ON-TWEED

The history of Berwick-on-Tweed, as the vital frontier town controlling the lowest crossing of the River Tweed on the main east coast route to Scotland, has been touched on in Chapter 4, with its capture from the Scots by Edward I in 1296. Notwithstanding the immediate repair and strengthening of the castle and replacement of the wooden palisade by a stone wall of 1297–8, the town was retaken by Robert Bruce in 1318 and successfully defended against Edward II's counter-attack of the following year. Between then and 1482, when the English finally gained permanent control, Berwick was frequently fought over, most notably in 1405 when Henry IV's army made one of the earliest known uses of artillery.

Chapter 5 has established the extent to which artillery revolutionized the science of military assault, thereby requiring comparable advances in defensive methods and technology, the most important component part of which was the bastion, further illustrated in Figure 8.32. At Berwick the earliest artillery fortifications date from 1522–3 when, under Henry VIII, Windmill Bulwark was constructed as a massive detached earthwork in front of the eastern wall. Also at that time the vulnerable north-eastern corner was strengthened by the addition of Lords Mount in 1539–42, a circular masonry structure of more than 100 feet radius, with casemented walls over 19 feet thick. Iain MacIvor regards this to be 'of outstanding interest as the final development in the transitional stage of artillery fortification, now quite transformed from mediaeval concepts'.[99]

Under Edward VI a citadel was designed with corner bastions, but it had not been completed by 1557, in which year England was again at war with France. The Scots were being persuaded to march south into England, and with the English mindful of the fate of Calais which had fallen, poorly defended, to the French, modernization of Berwick's fortifications became a vital necessity. Sir Richard Lee (the eminent military engineer who had already refortified Portsmouth) designed the new system, work on which commenced in 1558. His plan involved abandoning the northern part of the town and constructing a ring of bastions and connecting ramparts within the temporary protection of the old walls (Figure 8.33). Work started on the northern side consisting of the massive central Cumberland Bastion, and Brass Bastion and Meg's Mount at the two corners. By 1560 these were sufficiently advanced to enable construction to commence on the eastern side. Here an Italian consultant, Giovanni Portinari, controversially advised modifying Lee's plan to extend the northern front across the peninsular. His recommendations were disregarded and the eastern side was completed as intended, comprising a modified Brass Bastion at the angle, and Windmill Bastion and King's Mount. (Only an earthwork traverse was excavated across the peninsular.)

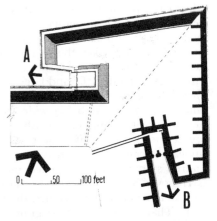

Figure 8.32 – Berwick: the plan of Brass Bastion as proposed by Lee, with a symmetrical shape and with small, two-storey 'flankers' equipped with guns to fire along the northern rampart (A), and the eastern rampart (B). The top half-plan shows the earthworks, and the bottom half-plan at ground level shows the masonry structure and the access tunnel into the flanker.

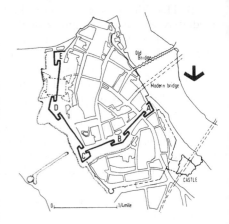

Figure 8.33 – Berwick: the plan of the Elizabethan Ramparts, drawn with north at the bottom, in accordance with the orientation of the aerial photograph. Key: A, Meg's Mount; B, Cumberland Bastion; C, Brass Bastion; D, Windmill Bastion; E, King's Mount.
The Edwardian citadel is shown dotted.

From hence lay a road into Scotland, but at present not willing to omit seeing Berwick upon Tweed, we turned to the east, and visited that old frontier, where indeed there is one thing very fine, and that is, the bridge over the Tweed, built by Queen Elizabeth, a noble, stately work, consisting of sixteen arches, and joining, as may be said, the two kingdoms. As for the town it self, it is old, decayed, and neither populous nor rich; the chief trade I found here was in corn and salmon.
(D Defoe, A Tour through the Whole Island of Great Britain, 1724–26. (Letter 11.))

99. I MacIvor, *The Fortifications of Berwick-upon-Tweed*; see also Duffy, *Siege Warfare, Volume 1, The Fortress in the Early Modern World 1494–1660*; F Graham, *Berwick: A Short History and Guide*, 1987.

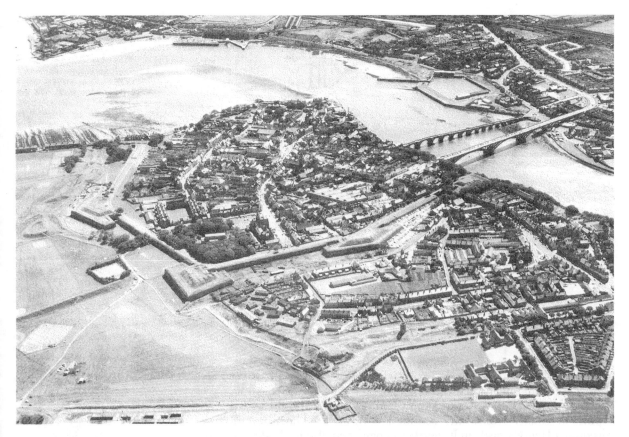

By 1563 the military emergency had diminished and Lee's projected ramparts and bastions between Meg's Mount and King's Mount were abandoned, leaving the medieval wall and the river as the only defences on those sides. Neither Meg's nor King's Mounts were completed as planned. The ditch was only partially excavated and the counterscarp wall never begun. Furthermore the upper earthworks on top of the ramparts were not added until 1639–53, before and during the English Civil War, and then not fully in accordance with Lee's proposals. Nevertheless, in their shortened and incomplete form, the 'Elizabethan Ramparts' at Berwick still stand as a superb example of mid-seventeenth-century urban artillery fortifications, which show what might have been a characteristic British necessity and consequent general determinant: but for the English Channel.

HULL

The familiar modern name of the town is used for this summary description of the sixteenth- and seventeenth-century fortifications which were added to supplement and then to replace the old medieval defences of Kingston upon Hull.[100] After an apparently brief period of strategic importance as Edward I's main east coast base, the town's role in this respect was taken over by Berwick, leaving it to consolidate a prosperous commercial and manufacturing economy. Nevertheless it remained important enough for Henry VIII to inspect the defences in 1541 and to order repairs to the wall and moat, together with the addition of the system east of the River Hull depicted in Hollar's view of 1665 (Figure 4.48).

Figure 8.34 – Berwick: aerial photograph of the Elizabethan Ramparts from the north; see Figure 8.33 for identification of the bastions. The clear-cut distinction between the town and the open fields immediately outside the western (left-hand) side, exemplifies the 'fire-zone' originally surrounding all such systems of latterday urban fortification.

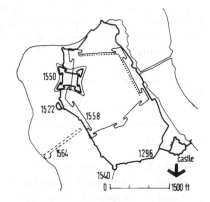

Figure 8.35 – Berwick: growth stages of the town's fortifications showing the full extent of Lee's bastioned 'inner' town proposal, the abandoned sides as dotted lines.

100. For Hull see E Gillett and K A MacMahon, *A History of Hull*, 1980; A E J Morris, 'Cities Built by a King', *Geographical Magazine*, November 1975.

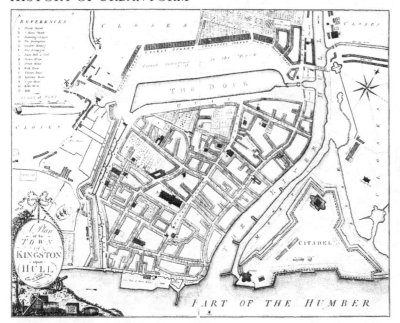

Figure 8.36 – Hull as recorded by Thew's map of 1784. 'The Dock' across the north-western side of the town was constructed in 1775–8 when it was known as the 'New Dock'; subsequently it was successively renamed the 'Old Dock' (when a new dock was added in 1803–9, itself to be renamed the Humber Dock), and then Queen's Dock in 1854, before being closed in the 1930s ultimately to be filled in and landscaped as present-day Queen's Gardens. (See Figure 8.37 for the dates of the entire system of enclosed docks constructed outside the line of the old wall and representing, in effect, a conversion of the medieval moat.) There was as yet only embryonic suburban expansion to the north and west, and the River Humber frontage was still that of the Edwardian bastide, before the reclamation of land in the early nineteenth century which added two streets beyond the original Humber Street.

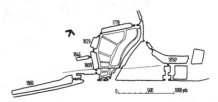

Figure 8.37 – Hull: a diagrammatic plan giving the dates of the system of four non-tidal, enclosed docks on which the port's prosperity of the first half of the nineteenth century was based. In response to the related problems of congestion and needs of larger ships, the harbour activity gradually moved away from these docks, and the comparably obsolete River Hull staithes, to immediately beyond the citadel on the River Humber, and latterly further downstream where extensive modern port facilities are located.

The new mid-sixteenth-century defences effectively precluded establishment of any major trading activity on the east bank of the River Hull: a constraint which was consolidated after 1680 by the construction of an extensive new citadel almost half the area of the old walled town. The citadel, which later became known as 'the Garrison', incorporated Henry's castle in its northern angle, and also his bastion at the entrance to the River Hull. Thew's map of 1784 relates the citadel to the old town and locates the three Henry VIII works – the second of the bastions is alongside the bridge over the River Hull. This map also shows the first of Hull's eventual system of four non-tidal enclosed docks.

The determining effect of the location of the citadel on the future form of Hull is without parallel in English urban history. It is logical to presume that, in the absence of its constraining presence through until the mid-nineteenth century, Hull would have developed, at least initially, on both sides of its river: a process exemplified by Philadelphia, around its Dock Creek, and by Roman London with the Walbrook. Instead, nineteenth-century expansion was essentially uni-directional, away from the old town, leaving it to decay within the ring of docks and staithes when a new twentieth-century centre became established to the north-west.[101]

PLYMOUTH, DEVONPORT AND STONEHOUSE

Plymouth, which today embodies the once separate towns of Devonport and Stonehouse, is of exceptional interest not only for the complex eighteenth- and nineteenth-century urban fortifications of Devonport, but also on account of its unusual origins in the form of three closely related, yet distinct urban entities.[102] Furthermore, there is the importance of Plymouth – by far the oldest of the trio – as the Mayflower's eventual port of departure with the Pilgrim Fathers in 1620.

Plymouth itself grew out of small twelfth-century fishing settlements grouped around Sutton Pool, and consolidated its location along the

Plymouth is indeed a town of consideration, and of great importance to the public. The situation of it between two very large inlets of the sea, and in the bottom of a large bay, which is very remarkable for the advantage of navigation. The Sound, or bay is compassed on every side with hills, and the shore generally steep and rocky, though the anchorage is good, and it is pretty safe riding. In the entrance to this bay, lies a large and most dangerous rock, which at high-water is covered, but at low-tide lies bare, where many a good ship has been lost, even in the view of safety, and many a ship's crew drowned in the night, before help could be had for them.

Upon this rock, which was called the Eddystone, from its situation, the famous Mr Winstanley undertook to build a light-house for the direction of sailors, and with great art, and expedition finished it; which work considering its height, the magnitude of its building, and the little hold there was, by which it was possible to fasten it to the rock, stood to admiration, and bore out many a bitter storm.

... the town of Plymouth is, as will always be a very considerable town, while the excellent harbour makes it such a general port for the receiving all the fleets of merchants' ships from the southward, as from Spain, Italy, the West-Indies, &c.

(D Defoe, A Tour through the Whole Island of Great Britain, *1724–26. (Letter 3.))*

101. The exceptional one-sided growth of Hull, determined by the downstream presence of the citadel, has a close central European equivalent at the city of Warsaw (see Figure 7.24).

102. For Plymouth see B Cherry and N Pevsner, *Devon*, 2nd edn (The Buildings of England series), 1989; P MacDougall, *Royal Dockyards*, 1982.

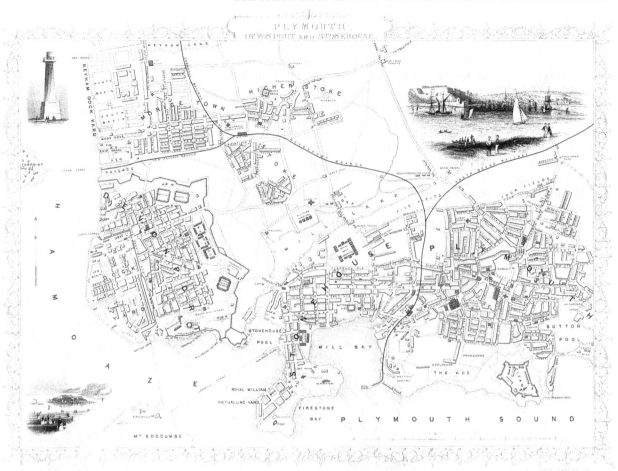

Figure 8.38 – Plymouth, Devonport and Stonehouse: recorded by the Tallis map of about 1850. Plymouth originated around Sutton Pool on the extreme right-hand (eastern) side of the map, and Devonport was founded as 'the Port in Hamaoze', some two miles away to the west. (Hamaoze is a local alternative name for the Tamar.) Stonehouse, in between, would appear to have arisen as a service-township on either side of the direct connecting route which was consolidated after 1810 as a prematurely named Union Street.

Devonport in 1850 had in effect two walls: an inner enclosure to the dockyard itself, separating it from the town, and the rather more complex, late eighteenth-century outer defensive system with its open fire-zone. There is a clear contrast between Plymouth's typically organic growth form, and Devonport's more or less regular gridiron planning.

In the second half of the nineteenth century and through until after the Second World War, Plymouth – or more precisely Devonport – continued to expand as a naval base. In addition, after the arrival of the Great Western Railway in 1876, Mill Bay was developed as a terminal for ocean liners, thereby saving several hours on the extended voyage further up the Channel. This latter reason for commercial prosperity was comparatively short-lived, fading away with the advent of larger, faster liners (which could not berth in Mill Bay) and provision of facilities at Southampton and Liverpool.

protected western beach behind the sheltering slopes of the Hoe. By 1254 there was mention of the Bailiffs of the Port of Plymouth and the Poll Tax of 1377 shows that, with a population of perhaps 7,000, the town could claim to rank as high as the fourth largest in England after London, York and Bristol. A connection with the Navy was long-established and the superb location between the Rivers Tamar and Plym, which opens out into the vast Sound, assured the port's future prosperity. Wealth and strategic importance entailed construction of town walls, to which Henry VIII added blockhouses in 1537–9, including one which was replaced by the magnificent citadel of 1666–72.

The Navy's needs had greatly increased by the late seventeenth century, requiring improved, special dockyard facilities and in 1691 under William III, construction work started on 'the Dock in Hamaoze'. Originally some 5 acres in extent, located at the south-western corner of the Dockyard of 1851, Plymouth Dock acquired its own civilian township from 1700. By 1821 it had a population of 33,578, compared with Plymouth's total of 21,591, and two years later it took the new name of Devonport to mark its independence. The origins of Stonehouse, which became an urban district in 1874, are unclear. The unification of the 'three towns' took place as late as 1914.

PORTSMOUTH

Portsmouth is traditionally the home of the British Navy, an association which dates back to the twelfth century when the future extensive naval base was first established as a small port at the south-western corner of Portsea Island, by the entrance to Portsmouth Harbour. This extensive area of sheltered tidal water had also served the Romans, but their base had been a fort at Portchester, further inland alongside the main east–west coast road. The eighteenth- and early nineteenth-century fortifications of Portsmouth comprised in effect the three separate systems around Portsmouth and Portsea – with 'The Common' (Southsea), the third nucleus of later-nineteenth-century Portsmouth, and Gosport described separately in the caption to Figure 8.43. Portsmouth Harbour was again refortified during the 1850s and 1860s, with the outer ring of forts known familiarly (but perhaps unreasonably) as 'Palmerstone's Follies'. This further intriguing episode in the history of English urban fortifications is beyond our present period.[103]

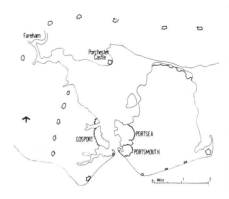

Figure 8.39 – Portsmouth Harbour: the general location map for Portsmouth, Portsea and The Common (the three nuclei of the modern city); Gosport, the separate town across the entrance to the harbour; and also the distant ring of forts constructed during the 1850s and 1860s and known familiarly as 'Palmerstone's Follies'.

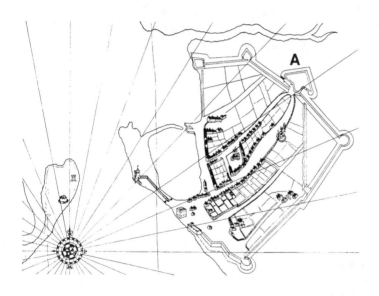

Figure 8.40 – Portsmouth: the 'old' town as refortified under Henry VIII; indicating an as yet considerable area of open space within the ramparts. This is Portsmouth at an interim stage in the development of artillery fortifications with its new entrance bastion (A) but still with obsolete, simple rounded bastions at the eastern and southern corners.

Twelfth-century Portsmouth developed as a result of increased trade and traffic with France following the Norman Conquest, and to meet the need for a permanent south coast naval base. Southampton, further up Southampton Water, had the better land communications and gradually assumed local commercial supremacy. Portsmouth on the other hand was nearer open water, which facilitated naval readiness, and it also had the requisite defensive potential within natural waterway moats on its comparatively isolated island location. The port was founded by Richard I in 1179 as a royal new town on the eastern side of the Camber, a natural inlet off the entrance to Portsmouth Harbour which provided a readily defended tidal anchorage. The area occupied by the town is that of present day 'Old' Portsmouth where the original gridiron street layout still largely survives. A first charter of 1194 granted a weekly market and an annual fair, and the Guild Merchant was chartered in the 1230s.

Figure 8.41 – Portsmouth: the mid-eighteenth-century extent of de Gomme's system, modified in some minor respects by later engineers. The ramparts and King's Bastion at the south-eastern angle, both of which still stand, are drawn in heavy outline (indeed these were still in use for gun emplacements during the Second World War).

Key: A, King's Bastion; B, Long Curtain; C, Round Tower; D, Pembroke Bastion; E, East Bastion; F, Town Mount Bastion; G, Guy's Mount Bastion; H, Beeston's Bastion; J, Legg's Demi-Bastion; K, Camber Bastion; L, Amherst's Redoubt.

103. A T Patterson, *Palmerstone's Folly: The Portsdown and Spithead Forts*, 1967.

Commercially, however, Portsmouth remained second to South-ampton, from where the local customs were administered; a condition resulting partly from naval and military priorities in the port, and also from frequent raids by the French. The early defences were ineffective and even after investigation by a commission of 1386 it seems that they amounted to no more than earthworks within a moat. Henry VIII instigated major improvements from 1538 in face of threatened Franco-Spanish invasion, adding the large new entrance bastion and extending the ramparts, as shown in Figure 8.40. This work marked the commencement of a continuing programme of repair, renewal and replacement of the Portsmouth fortifications which lasted, in part, through until the Second World War.

Sir Richard Lee, already encountered through his work at Berwick, was mainly responsible for the comprehensive reconstruction of the 1550s and 1560s which consolidated the inner ring of ramparts and bastions much as they were to remain for the next three centuries or more. This inner system is shown in Figure 8.41 related to the far more extensive outer works commenced under Charles II, after the restoration of the monarchy in 1660 and not finally completed until about 1750. The military engineer responsible was Sir Bernard de Gomme, a Dutch contemporary of Vauban (see Chapter 6) who believed in the same defence-in-depth principles involving the construction of 'ravelins' in an extended moat, with a counterscarp and a covered way within an extensive 'glacis'. De Gomme was also responsible for the contemporary main works at Gosport.

The area of Portsmouth contained within its fortifications (or for convenience that of the 'old' town) sufficed until 1495 when Henry VII ordered the establishment of a separate royal dockyard further into the Harbour, beyond the Mill Pond, thereby creating the second of Portsmouth's historic nuclei. At first the dockyard workers 'commuted' from the old town, and perhaps from Gosport, until new housing land was needed from the 1660s to provide for greatly increased dockyard activities. Old Portsmouth could not expand and the answer – similar to that adopted at Devonport – was to develop a 'new' town for the dockyard, soon known as Portsea.

By the 1770s it also warranted separate complex fortification and when in turn the enclosed area was fully developed, additional land for housing and related service industries, etc. could be provided only beyond the statutory open fire-zone outside the defensive glacis. Already there was well-established organic growth suburban development alongside the main approach roads, known generally as 'The Common', which Defoe described in 1730 as 'a kind of suburb, or rather a new town built on the healthy ground adjoining the town which is so well built, and seems to increase so fast, that in time it threatens to outdo for numbers the town itself'.[104] The 1801 census shows this to have long since happened, recording 7,839 inhabitants for Old Portsmouth compared with The Common's total of 24,327. The Common, of which artisan Croxton Town and Landport formed main parts, developed rapidly during and after the Napoleonic Wars, and Southsea grew up as a middle-class suburb between it and the seafront. The map of 1833 shows the extent of the suburban growth which comprises the comparatively amorphous third nucleus of the city of Portsmouth. Figure 8.42 also locates the Hilsea Lines across the top of Portsea Island and the several forts along the southern sea-front.

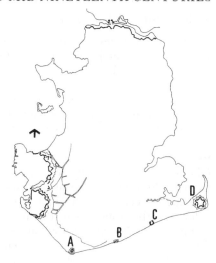

Figure 8.42 – Portsmouth: the Island of Portsea, as redrawn from the Lewis Map of 1833. Full details of the defences of both Portsmouth and Portsea are given by this map, which also provides the layout of the Portsea Dockyard. Intriguingly by the 1840s, for security reasons, these details are omitted from maps of the island, which simply show white areas of land. Hilsea Lines are shown across the northern end of the island with the main entrance bastion on the main London road. Key to the southern seafront forts: A, Southsea Castle, famous and well-restored Henry VIII fort; B, Lumps Fort; C, Eastney Fort; and D, Cumberland Fort, a good example of its kind.

Figure 8.43 – Gosport: the Lines in 1742. (See Figure 8.39 for the relationship to the contemporary Portsmouth Lines, and also to the future mid-nineteenth-century ring of outer forts.) Although the Gosport Lines were an integral part of the fortifications of Portsmouth Harbour, the town of Gosport itself was completely separate from Portsmouth, both physically across the Harbour entrance, and also administratively – a distinction it has enjoyed to the present day.

104. D Defoe, *A Tour through the whole Island of Great Britain*, 1724/1971 English edition.

Major British cities: origins and pre-railway growth

Chapter 8 concludes with summary description of the origins and pre-1850 histories of eight British cities of particular significance. (Dublin, the capital of the Republic of Ireland, was in effect a British city during its formative period through until 1922.) An 'epilogue' at the close of the essentially historic period in British urban history provides a link with the extensive mass urbanization of the second half of the nineteenth and of the twentieth centuries.

GLASGOW

Glasgow originated at the lowest permanent fording point of the river Clyde, a location which was used by the Romans for a military road crossing.[105] The Antonine Wall passed just to the north, and although a fort had been located in the area, little else is known of possible Roman, or earlier site occupation. In the fifth century a village by the ford on the northern bank was chosen by St Ninian for a 'cathedral' of the Celtic Church, a religious function consolidated in the mid-sixth century when St Mungo (more correctly St Kentigern) was made 'Bishop' of Glasgow by the King of Strathclyde. The diocese was strengthened by King David I in 1124 and a Papal Bull of 1172 recognized Glasgow as a city. Civil powers granted in 1176 gave burghal status to which an annual fair was later added.

Twelfth- and thirteenth-century Glasgow was tightly confined to the Cathedral Hill, some two-thirds of a mile north-east of the ford, where a stone bridge was constructed from 1345. Subsequent growth took place along the road between the Cathedral of St Mungo and the bridge, with the city centre moving down to the cross at the intersection with the main east–west route following the river. By the mid-eighteenth century Glasgow had consolidated around the cross as the organic-growth nucleus from which the nineteenth-century gridiron districts were extended.

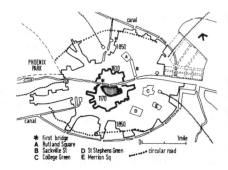

* first bridge
A Rutland Square
B Sackville St
C College Green
D St Stephens Green
E Merrion Sq
----- circular road

Figure 8.44 – Dublin: the extent of the city in 1850 (as the Tallis map) showing the main phases in the growth of the city. The estuary of the River Liffey was recorded as a safe anchorage by Ptolemy, who called it Eblana, but it was not until AD 841 that a permanent settlement was founded by Norse Vikings, known as Dyfflin. Trade was facilitated by a ford across the Liffey – the Atha Cliath in Gaelic, hence the ancient name Baile Atha Cliath – and under Olaf the White a fortified township was constructed around 852 on the steeply sloping ridge where Dublin Castle and Christ Church were later to be located.

In 1014, on Good Friday, the Irish under Brian Boru gained control and Dublin entered a first prosperous phase. Decline followed capture by the Anglo-Normans in 1169, after which Dublin was for seven and a half centuries the centre of English rule in Ireland. In 1312 the walls were extended to the riverbanks and prosperity slowly returned, only to be lost again following Cromwell's mid-seventeenth-century depredations. Recovery, nevertheless, commenced once more and by 1700 Dublin was becoming one of Western Europe's major cultural centres with its old medieval buildings and cramped civic spaces being replaced by new architecture and spacious squares and streets. In 1850 Dublin was a compact city of some 250,000 population, reaching out to, but not completely covering the area within the Circular Roads. (See M C Branch, *Comparative Urban Design*, 1978 (the 1836 SDUK Map with accompanying historical text); J Malton, *A Picturesque and Descriptive View of the City of Dublin*, 1799 (reproduced 1978); D A Chart, *The Story of Dublin*, 1907 (reprinted 1932 in the Medieval Towns series).)

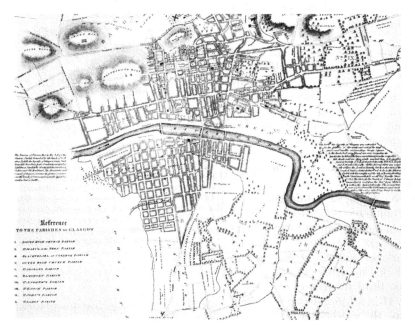

Reference
TO THE PARISHES OF GLASGOW

Figure 8.45 – Glasgow: the map of 1822 (by James Cleland and David Smith) showing the old nucleus with the High Church in the north-east corner of this extract, and the late eighteenth- and early nineteenth-century gridiron districts which formed the basis of the city's subsequent rapid expansion. The modern city centre was in process of consolidation in the central north-bank district. (Population statistics: 1650 – approximately 4,500; 1750 – 23,000; the census of 1801 gave a total of 77,000 and that of 1851 recorded 255,000.)

105. For general Glasgow references see T A Markus (ed) *Order in Space and Society: Architectural Form and its Context in the Scottish Enlightenment*; 1982; M D Lobel (ed) *The British Atlas of Historical Towns, Volume 2*, 1975.

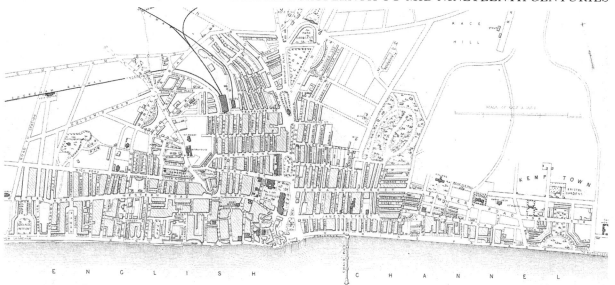

BRIGHTON

Brighton's exposed English Channel location lacked sheltered harbour facilities and it was no more than a small fishing township until the advent of sea-bathing as a fashionable pursuit in the 1750s laid the basis of future prosperity – and mild notoriety.[106] Dr Russell provided the initial impetus with his *Dissertation concerning the Use of Sea Water in Diseases of the Glands*, published in English in 1753. He followed this up by 'conveniently' discovering a nearby chalybeate spring thereby ensuring both his own and the town's fortunes. A first ballroom opened in 1766 and from 1783 the Prince of Wales' patronage guaranteed a future social success made available to all when the railway from London was opened in 1841.

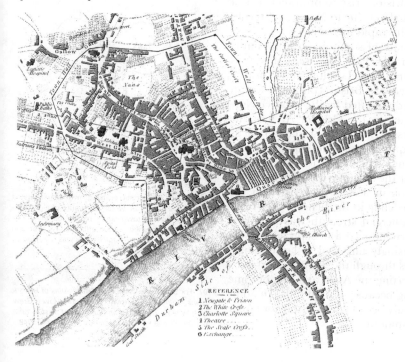

REFERENCE
1 Newgate & Prison
2 The White Cross
3 Charlotte Square
4 Theatre
5 The Scale Cross
6 Exchange

Figure 8.46 – Brighton: the waterfront section of the Tallis map of about 1850. The old fishing town grew up on level ground, beyond reach of the sea immediately to the west of the Steyne – a dry valley running down to the sea (in the centre of the map). It was bounded by North, East and West Streets and by the beach. (The eastern corner of the old town is the popular 'Lanes' district of modern Brighton.)

The Marine Pavilion, first built for the Prince of Wales in 1786–7 to Henry Holland's plans, was rebuilt by John Nash between 1815 and 1822 as the present-day Royal Pavilion. The Brighton seafront of 1850 was largely the creation of the 1820s and 1830s, taking the population from 7,339 in 1801, to 65,569 in 1851 (and to 102,320 by 1901).

Figure 8.47 – Newcastle upon Tyne, shown as part of the *British Atlas* map of 1808; togeher with Gateshead, the latter in the form of a main road 'ribbon' development leading south from the bridge over the River Tyne (north at the top; the Tyne flows to the east).

Hadrian's Wall as first planned terminated in a bridge across the tidal River Tyne, known as Pons Aelius and when extended to Wallsend (Segedunum) on the coast, the north-bank bridgehead was provided with a fort, named after the bridge. The fort covered the area of the later Norman castle and there would have been a civilian settlement between it and the bridge. After the withdrawal of the legions Pons Aelius appears to have rapidly disappeared and there is no record of any significant Anglo-Saxon reoccupation of the site. However, the river crossing must have retained its importance and as early as 1080 the Normans built a major castle to ensure its control.

A first defensive wall was constructed during the 1260s and the enclosed area reached the full medieval extent when the riverside Pandon suburb was incorporated within the extended wall in 1307. The form of the early nineteenth-century city is still essentially medieval, notably with the long and narrow burghal plots fronting on to the spacious quayside.
(See also L Wilkes and G Dodds, *Tyneside Classical: The Newcastle of Grainger, Dobson and Clayton*, 1964.)

106. For Brighton see E M. Gilbert, *Brighton: An Ocean's Bauble*, 1975; A Dale, *Fashionable Brighton 1820–60*, 1967.

BRISTOL

The original settlement was located on a narrow neck of land between the main stream of the River Avon and a minor tributary, the River Frome, approximately 6 miles inland from the estuary of the River Severn and 8 miles downstream from Bath.[107] The Avon was navigable to beyond Bath and it is somewhat surprising that such a readily fortified site, controlling the river-trade, was overlooked by both the Romans and the early Anglo-Saxons. Bristol's rise to prominence as a Domesday Book town of 1086, rated equal in value to Norwich, Lincoln and York, accordingly denotes unusually rapid growth.

J C Russell's 1086 population estimate of some 2,300 inhabitants placed Bristol sixth in size; the comparable 'Poll Tax' total of 1377 is given as about 9,500 and third ranking after London and York.[108] Figure 8.49 shows the extent of early medieval Bristol contained within either natural or man-made moats and dominated by the massive Norman castle – a sure sign of importance to the Crown. At first the adjoining banks of the Avon, and to a lesser extent the Frome, served as the port, but increasing prosperity required quayside expansion and from 1239 the Frome was diverted through a new channel half a mile in length and about 40 yards wide to enter the Avon further downstream.

Major late medieval expansions took in a new area between the rivers, and also the entire suburb of Redcliffe in the bend of the Avon. The basis of seventeenth-century commercial strength was the West India trade, but the abolition of slavery induced a decline which was accelerated as the port became increasingly unsuited to larger ships. Heroic efforts to modernize the facilities through the 'floating harbour', created between 1803 and 1809 by diverting the Avon at a cost of some £1 million, was ultimately self-defeating because of high tolls. In addition passage of the Avon Gorge eventually became impossible for larger ships and the port moved out to Avonmouth.

Figure 8.48 – Bristol from the Tallis map of around 1850, north at the top. The 'Floating Harbour', formed between 1803 and 1809 by diverting the course of the River Avon to the new line, was entered through the Cumberland Basin, with a secondary Bathurst Basin further upstream. The northern half of the River Frome Basin was filled during the 1920s. The 1851 census population was 100,000 (68,000 in 1801) with suburban expansion of Bristol and neighbouring Clifton well under way to give a total of 337,000 at the end of the nineteenth century.

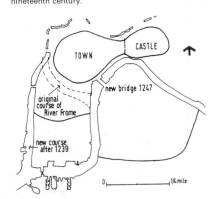

Figure 8.49 – Bristol: successive growth stages related to the Tallis map, from which the 'figure-of-eight' outline of the Norman town can be readily identified. The castle was demolished in 1654, after the Civil War, and its considerable area of some 6 acres was occupied by densely built-up housing.

107. For Bristol see Lobel, *The British Atlas of Historic Towns, Volume 2*; D Brown, *Bristol and How it Grew*, 1978; W Ison, *The Georgian Buildings of Bristol*, 1952.
108. J C Russell, *British Mediaeval Population*, 1948.

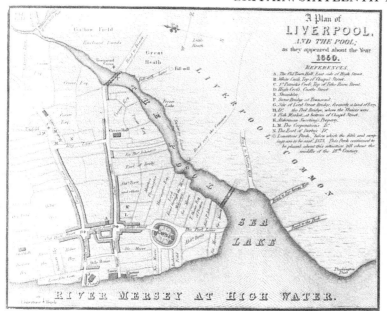

Figure 8.50 – Liverpool: the plan of the city in 1650, 'copied from the original drawing ... published by Thos Kaye, 1829'. The town as refounded by King John was located on a broad sandstone ridge between the Pool and the eastern, right bank of the estuary of the River Mersey, where it is some three-quarters of a mile wide. The streets were in the form of a double cross: the main route along the ridge from north to south, leading up to the castle, comprising three separate streets. Old Hall, Joggler and Castle Streets; and the two intersecting routes leading away from the Mersey comprising Water and Dale Streets (nearest the castle) and Chapel and Tithe Barn Streets. There were two market crosses, one at each of the street intersections.

With the exception of Joggler Street, which lies beneath the present-day Exchange and Town Hall Buildings, the other streets have retained their identities in the modern city centre. The castle site is occupied by St George's Circus.

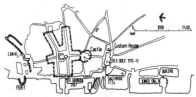

Figure 8.51 – Liverpool: a diagrammatic growth plan based on the *British Atlas* map of 1807. The Old Dock was constructed between 1710 and 1715 to provide for 80–100 ships. At the same time the upper part of the Pool was filled in and Paradise Street and Whitechapel were created. (The Old Dock later suffered the same fate and is now Canning Place.) The names and dates of construction are given for the early docks on the Mersey riverfront.

LIVERPOOL

At its mid-seventeenth-century stage of development (shown by Figure 8.50) Liverpool clearly exemplifies the use as a harbour of a comparatively sheltered minor stream where it enters a major tidal waterway.[109] The combination of the Pool and the estuary of the River Mersey, as a determining locational factor, can be compared with those of Roman London – the Walbrook and the River Thames (Figure 3.49); Kingston upon Hull – the River Hull and the estuary of the River Humber (Figure 4.48); and Philadelphia – the Dock Creek and the River Delaware (Figure 10.22). Again like London and Philadelphia, Liverpool was to develop by filling in and building over its original harbour.

The favourable location was occupied during the Bronze Age but it was ignored by the Romans whose military road from Chester to Lancaster crossed the Mersey some 16 miles upstream. Neither was it important enough to warrant separate mention in Domesday Book although there was a village settlement by the Pool, from which grew the small town first recorded as Liverpool in a charter of 1192.

The foundations of Liverpool's future prosperity were laid when King John, in 1207, took possession of the location for a new royal port through which to ensure control of the increasing trade and military traffic with Ireland. Chester, the obvious existing alternative, enjoyed 'independent' Palatine status and was unsuited to the purpose, but remained Liverpool's greatest rival until the eighteenth century when the berthing requirements of larger ships could be met only along Liverpool's open water frontage (see Chapter 4). It is reasonable to assume that, as refounded by King John, Liverpool was 'planned' on the basis of the more or less regular gridiron street layout of the map of 1650, which remained the extent of the town until the expansion phases of the eighteenth century. These, and the subsequent modern growth of the city, are summarized in the caption to Figure 8.51, which shows the 'Old Dock', created out of the lower Pool, and the beginnings of the Mersey docks.

Dr. Formby, a respectable physician of Liverpool, whom I accidentally met in the promenade-room at Harrogate, assured me that as many visiters now proceed to Liverpool for the benefit of sea-bathing, as are known to attend Harrogate for the purpose of drinking those mineral waters, or bathing in them.

That people should select a place which in reality does not enjoy the advantage of genuine sea-water, with an intention of taking sea-baths, somewhat puzzled me, until the learned Doctor explained to me that of late years a new sea-bathing place had been created, exclusively for the accommodation of the wealthier classes in and about Liverpool, who, having now nearly deserted the once fashionable Park Gate, on the Cheshire coast, gladly availed themselves of the new establishment to which the emphatic title of New Brighton had been given.

One hails, therefore, with satisfaction any manifestation on the part of a large community like that of Liverpool, spread over a quasi-maritime region, to secure to themselves the benefit of sea-bathing; and, in this respect, the people of Liverpool and the country around – ill-favoured as they are with regard to position in reference to genuine sea-water – have shown their own conviction of the efficacy of sea-baths, by forming new sea-bathing establishments, and supporting them when formed, although placed at a somewhat inconvenient distance from the town.

(A B Granville, Spas of England and Principal Bathing Places, Volume 2: the Midlands and the South, 1841)

109. For Liverpool see M C Branch, *Comparative Urban Design*, 1978 (the 1836 SDUK Map with accompanying historical text).

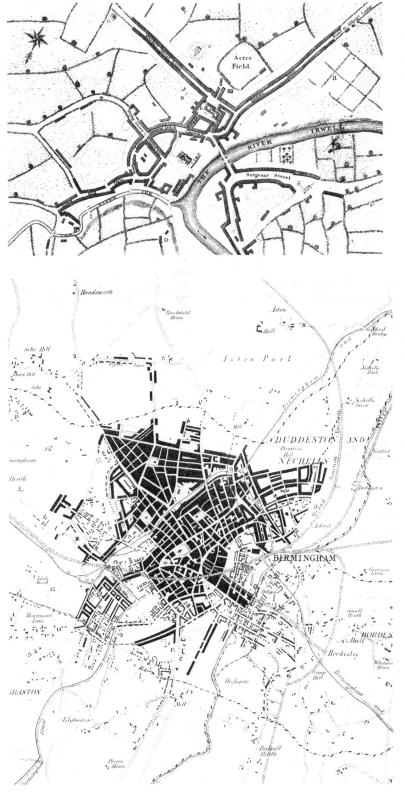

Figure 8.52 – Manchester: the extent of the town in 1650, affording direct comparison with the comparable date map of Liverpool, its great Lancastrian rival. The small Roman fort of the first century AD known as Mancunium (or Mamucium) was not of any civil significance; neither, apparently, was the settlement mentioned in the Anglo-Saxon Chronicle as 'Mameceastre', from which the city's name derived in the late Middle Ages.

Manchester grew from trading village origins situated at the confluence of the River Irwell with a minor stream, the River Irk. On the map of 1650, Manchester is located in the angle of the Irwell and the Irk, on the right-hand eastern bank of the Irwell; Salford is contained within the bend of the Irwell. (Note: north is diagonally to the bottom left.) A royal charter of 1227 confirmed an annual fair, and in 1301 privileges accorded 'my burgesses in Manchester' were endorsed.

In about 1375 Edward III settled a colony of Flemish weavers in the town, initially working with wool and linen, and turning to specialize in cotton from the mid-seventeenth century. Subsequently the history of the city is that of the Lancashire textile industry, closely related to the late eighteenth- and nineteenth-century innovations in canal and rail transport. For Manchester see J Tait, *Mediaeval Manchester and the Beginnings of Lancashire*, 1904.

Figure 8.53 – Birmingham: the extent of the city in 1871. Two main reasons are usually given for Birmingham's extraordinarily rapid rate of growth during the early decades of the Industrial Revolution, and the comparatively slower changes experienced by other West Midland towns, of which Coventry is the main example (see Chapter 4). First, Birmingham's unincorporated status enabled new mass-production technology to be readily accepted, with comparatively free influx of non-guild 'certified' craft labour; and second, it was free from the religious restrictions imposed by the Clarendon Code against non-conformists (see Conrad Gill, *History of Birmingham* Volume I). These advantages stemmed from Birmingham's relatively late rise to 'urban' status during the sixteenth and seventeenth centuries, and they enabled the local potential of iron and coal supplies to be rapidly exploited. (Birmingham, with Manchester, did not receive urban recognition until their charters of 1838, remaining in effect villages, in local government terms, during the early phases of the Industrial Revolution.)

Coventry on the other hand was exceptionally well organized on a self-protection guild basis within a 'customs' wall constructed as late as 1355–1400. With hindsight it may be argued that long-term benefits were gained from avoiding the early industrial and commercial prosperity, and from being spared the many negative side-effects of nineteenth-century industrialization.

The map of 1731 shows Birmingham as a classic instance of the organic growth processes, already well advanced in its eighteenth-century change from a loosely formed metal-working centre of some 12,000 or 15,000 inhabitants, into a great manufacturing town with a population of more than 70,000. The two medieval focal points were the moated manor house (bottom of map) and the church, north from which grew the main market street. In 1671 there were 69 forges working in the immediate area; by 1683 this total had increased to 202. The advent of canal and turnpike road transport at the end of the eighteenth century consolidated Birmingham's supremacy in this field and during the Napoleonic Wars over two-thirds of the firearms supplied to the army and navy were manufactured in the town. For Birmingham see M C Branch, *Comparative Urban Design*, 1978 (the 1839 SDUK Map with accompanying historical text).

Epilogue

In addition to London, Bath and Edinburgh, the three principal centres of Renaissance urbanism in Britain, the majority of cities and towns of appreciable standing included in their eighteenth- and early nineteenth-century expansion at least one carefully laid-out residential district or estate, frequently embodying one or more unpretentious, landscaped Georgian squares. However, these attractive developments catered exclusively for the rising ranks of the moneyed middle and upper classes; less fortunate urban families lived at rapidly increasing densities and under steadily deteriorating standards, either in older parts of town or in new working-class areas which spread around factories in the early industrial centres.

Britain's Industrial Revolution has been exhaustively chronicled, perhaps best through the eyes of a Frenchman, Paul Mantoux. Readers are referred to his admirable book, *The Industrial Revolution in the Eighteenth Century*, for a background to nineteenth-century urban development in Britain.

Industrialization goes hand in hand with urbanization. From its origins in England around the middle of the eighteenth century, the Industrial Revolution effected a gradual but inexorable reverse of the proportion of people living in rural and urban areas. In 1750 probably fewer than 20 per cent of England's population lived in towns; by 1900 fewer than 20 per cent remained in the country.

The last major works of the Renaissance in Britain must therefore be seen against the sombre background of the uncontrolled expansion of the industrial towns. Figure 8.55 shows the nineteenth-century rates of growth of eight major industrial centres from the first census of 1801. Industrial urban growth usually progressed through distinct phases, commencing with a thickening-up of the original organic-growth, medieval nucleus. This was followed by small, expediency-grid, residential developments interspersed with factories, as the tempo increased around the turn of the century. But, contrary to a general impression, these did not develop into the seemingly limitless expanse of two-storey by-law housing – Britain's particular urban nadir – until legislation in the second half of the century imposed some minimum standards for hygiene and for the provision of light and air. M W Flinn in his introduction to the 1965 edition of Edwin Chadwick's famous *Report on the Sanitary Condition of the Labouring Population of Great Britain* (first published in 1842) aptly observes that 'some towns expanded during the early nineteenth century at rates that would bring cold sweat to the brows of twentieth century housing committees', and that as a result of the appalling living conditions 'the history of British towns in the first half of the nineteenth century is, to a considerable degree, the history of typhus and consumption'. For further reading on this subject reference should be made to Chadwick's *Report* and to Friedrich Engels' *The Condition of the Working Class in England*, which is much more coldly factual than might be imagined.

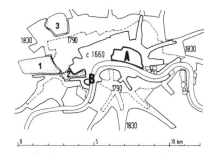

Figure 8.54 – The growth of London showing three stages (1660, 1790 and 1820) related to the City of London (A) and the Whitehall area of the City of Westminster (B). Three royal parks are shown: 1, Hyde Park; 2, Green Park and St James's Park; 3, Regent's Park.

This map can be compared with Figure 8.1, showing the extent of London at the ealier dates of 1100, 1400 and 1650. This map can also be compared with Figure 6.13, which shows Parisian growth stages within successive rings of fortifications: a pattern which is in total contrast to London's militarily unconstrained, main-route ribbon and infilling development. In 1830 the major constraint determining London's physical expansion was the lack of public transport, a factor which was overcome from the early years of the railway age. (See also the 1843 SDUK Map of London, with accompanying historical text, in Branch, *Comparative Urban Design*.)

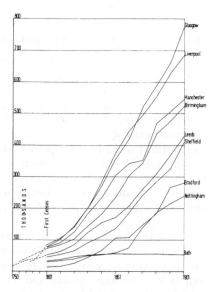

Figure 8.55 – Nineteenth-century urban growth represented by example cities in England and Scotland. (Bath has been included for comparison as a non-industrial example.)

9 – Spain and her Empire

The chapter is in four parts: first, a summary of political, social and economic changes in Spain from the commencement of the *reconquista* (as introduced in Chapter 4) to the end of the eighteenth century; second, a general account of urbanism in Spain of the period, with illustrated descriptions of major cities; third, the Spanish Latin American empire, introduced separately (pages 302–4); and fourth, a summary of settlement in the Spanish Philippines, and the Portuguese colonization of Brazil.[1] The settlement of those originally Spanish parts of the USA is described in Chapter 10. Special acknowledgement is made to J H Parry's *The Spanish Seaborne Empire* as a main reference source for this chapter.[2]

Spain: the *reconquista* and Spanish unification

The eighth-century conquest of Spain by Islam has been described in Chapter 4. Two prerequisites for the Christian *reconquista* of the country were first, need for separate northern kingdoms to gain strength and then to unite against the Muslim south, and second, for the Moors to be seen to be vulnerable.[3] The latter encouragement existed from the early eleventh century; nevertheless, as Fisher cautions, 'so inveterate was the localism of Christian Spain that only the most powerful motive would have sufficed to overcome it ... and such a motive was wanting'. It was thus no accident, he explains, 'that the great period of the Spanish reconquista, which begins with the taking of Toledo in 1085 and ends with the recapture of Murcia in 1266, corresponds with the epoch of the crusades'.[4]

The three leading northern kingdoms were Leon, the oldest but which became only a secondary realm when united in 1230 with Castile, and Aragon, which had been created in 1035. Navarre was a fourth, less important realm, and there was also the County of Barcelona. Castile emerged as the dominant kingdom from origins as 'Old Castile' – in the region around Burgos. Its leading families were to provide the majority of the ruling class of the unified Spain of the later fifteenth century, and it was their life-style and values which very largely determined the nature of the Latin American Empire. Immediately upon their unification, Castile and Leon commenced a new sustained offensive against the Moors, taking Cordoba in 1236 and then Seville in 1248. Aragon contributed with the regain of Valencia in 1238. The remaining Moors retreated into a mountain fastness centred on Granada where they were tolerated, in return for annual tribute,

1. See C R Boxer, *The Portuguese Seaborne Empire*, 1969/1977. See also C Gibson, *Spain in America*, 1966: 'the Portuguese marine attack of 1415 upon the fortified Moslem city of Cueta was ... the first act of state-directed imperialism of modern European history'.

2. The main general chapter reference is J H Parry, *The Spanish Seaborne Empire*, 1977, a volume in the History of Human Society series; the other most useful historical background references are J H Elliott, *Imperial Spain 1469–1716*, 1963/1983; G Pendle, *A History of Latin America*, 1976/1990; J B Lockhart and S B Schwartz, *Early Latin America: A History of Colonial Spanish America and Brazil*, 1983/1989. The general urban histories which have been found most useful are J E Hardoy, 'Two Thousand Years of Latin American Urbanisation', in J E Hardoy and C Tobar (eds) *Urbanisation in Latin America*, trans. F M Trueblood, 1975 (as an updated and expanded version of the article of the same name in *La Urbanizacion en America Latina*, 1969); various of the chapters, as individually referenced, included in R P Schaedel, J E Hardoy and N S Kinzer (eds) *Urbanisation in the Americas from its Beginnings to the Present*, 1978; and also chapters in G H Beyer (ed.) *The Urban Explosion in Latin America*, 1967.

3. See G Goodwin, *Islamic Spain*, 1991; R Collins, *The Arab Conquest of Spain*, 1989; see also Chapter 11 and Figure 4.92 which shows the main successive phases of the Christian reconquest.

The Islamic conquest of Spain commenced in the early eighth century AD and by mid-century the Moors (the Spanish Muslims) held virtually the entire Iberian peninsular, before gradually withdrawing to the south. As an example, Toledo was an Islamic city for more than three and a half centuries, and Granada for nearly seven. As related in Chapter 4, their historic morphology is that of Islamic tradition, in marked contrast to Western European Christian counterparts. See L T Balbas, *Ciudades Hispano Musulmanas*, 1988; T F Glick, *Islam and Christian Spain in the Early Middle Ages*, 1979.

4. H A L Fisher, *A History of Europe: From the Earliest Times to 1713*, 1935.

until 1492. In this auspiciously coincidental year, Castile and Aragon, united under Ferdinand and Isabella, completed the *reconquista* with their assault on Granada, and Columbus set sail west, out across the unknown Atlantic.[5]

Far from creating Pan-Iberian unity, however, the *reconquista* served to consolidate differences between the only nominally united Spanish kingdoms of Castile, Aragon and Navarre, with Portugal maintaining its western independence. The interests of Aragon were directed eastwards into the Mediterranean and Navarre looked northwards across the Pyrenees, leaving Castile in control of the main central and southwestern regions, from whence came the great majority of the eventual 'new world' *conquistadores*.

'From its first emergence,' writes Parry, 'Castile was a frontier kingdom, whose people, prompted by pugnacity, land hunger and religious zeal, looked to expansive war and conquest for the satisfaction of their ambitions.'[6] Shortage of readily farmed land created the *meseta*, a pastoral economy, involving constant movement of vast numbers of sheep and cattle to fresh grazing areas. As a result, settled peasant life in fields came to be despised in favour of mobile independent herding on horseback. An absolute distinction came about between 'town' and 'country'. Families of importance were based in towns from where they conducted their affairs, and 'a gentleman was a man who owned a horse and was prepared to ride it into battle in his lord's support': a social mode destined to be of determining military significance in the one-sided defeat of the natives of Latin America. Castillian kings made Seville their favoured residence-city, preparing it for a future role controlling the Indies. They also created numerous planned 'new town' settlements[7] for loyal subjects throughout Andalusia (as the hitherto Moorish south became known) thereby acquainting Castillian leaders of Latin American colonizing expeditions with the processes of new urban settlement.

Religious intolerance, a further primary characteristic of the *conquistadores*, derived from the Crusade against the Moors and culminated with the Inquisition set up in 1478. One result was the expulsion from Spain in 1492 of some 170,000 Jews, an act of extreme intolerance that deprived Spain of many of its economically most active inhabitants just when they could have been of great assistance in their financial organization of complex 'new world' ventures. Instead, the Crown had to turn to Italian and German financiers, mortgaging in advance an unnecessarily high proportion of the returns from Empire. Later, in the early seventeenth century, the cause of Catholic uniformity in face of repeated international reverses, also required expulsion of the Moors, 'numbering half a million of the most skilled agriculturalists and artificers of the country, whereby Spain was rendered so much the less able to sustain the burden of her far-reaching empire'.[8] The implications of finance, and population, are further developed in the adjoining column.

Castile had also acquired maritime expertise embodying skills from various seafaring parts of Iberia and the western Mediterranean.[9] This meant that in 1492, with Granada retaken and Castilian knights looking for fresh fields to conquer, Spain was ready and above all able to accept the challenge, brought back by Columbus, of a 'new world' for the taking.

Figure 9.1 – Spain: the map locating the major cities included in this chapter and two smaller maps of about AD 1100 and AD 1250 showing the emergence of the unified Kingdom of Castile and Leon, whose leading families were to provide the majority of the ruling class of the united Spain of the later fifteenth century. See also the map in the Spanish section of Chapter 4 chartring the progress of the Christian *reconquista* of the Islamic central and southern parts of the country.

The history of Spain in the seventeenth century reveals a decline in wealth, influence and power, verging, in the later part of the century, upon collapse. The degree and speed of decline varied greatly, as might be expected, from one field of activity to another. In the arts and in the related fields of manners, dress and social behaviour Spanish influence and leadership survived well beyond the middle of the century. In literature and drama, in painting and allied arts, Spain continued to produce men of commanding and creative genius. Royal, noble and ecclesiastical support was maintained on a lavish scale, which increasingly exceeded what the country could afford.

(J H Parry, The Spanish Seaborne Empire, 1966)

5. For Spain on the eve of Empire, see J H Elliott, *Imperial Spain 1469–1716*, 1963/1983; J W Reps, *The Making of Urban America*, 1965.

6. Parry, *The Spanish Seaborne Empire*; see also D E Vassberg, 'Land and Society in Golden Age Castille', *Journal of Historical Geography*, 1984.

7. For the Spanish medieval 'New Towns' see J I Linazasoro, *Permanencias y Arquitectura Urbana*, 1978 (Spanish); A B Correa, *Morfologia y Ciudad*, 1978; J E Hardoy, 'European Urban Forms in the 15th to 17th Centuries and their Utilisation in Colonial America', in R P Schaedel, J E Hardoy and N S Kinzer (eds) *Urbanisation in the Americas from its Beginnings to the Present*, 1978; see also Chapter 4, page 147.

8. H G Koenigsberger and G L Mosse, *Europe in the Sixteenth Century*, 1968/1972.

9. The coincidental coming together of the several factors underlying Spain's imperial venture is of special note. It is only of diverting interest to conjecture where that crusading energy and nautical expertise might have been directed, had not a 'new world' appeared on the horizon. Suffice that it would be a different world today.

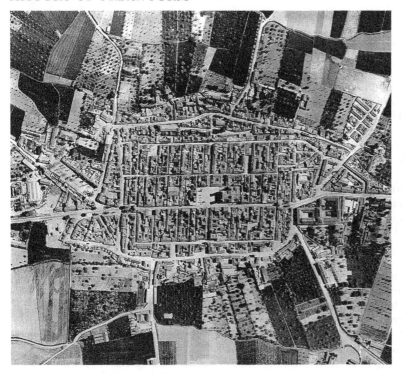

Figure 9.2 – Santa Fe: 5 miles (8 km) west of Granada, the plan of the new town constructed, by repute, in eighty days in 1491 as the military base for the concluding stages of the campaign to regain Granada from the Moors (the Spanish Muslims). Their capitulation was signed there in November 1491, and in April 1492 it was where Columbus received his 'new world' commission from Queen Isabel.

Figure 9.3 – Santa Fe: an aerial photograph showing the clearly retained original gridiron street layout within a ring of organic growth suburbs (and with the roads and field pattern of the surrounding pre-urban cadastre).

As one of urban history's intriguing coincidences, Nicolas de Ovando, who was present at the siege of Granada, later became the Spanish New World Governor of Hispaniola, with responsibility for the foundation of its capital Santo Domingo on the basis of a gridiron plan (Figure 9.21). If Ovando had also been present when Santa Fe was laid out and constructed within eighty days, he could not but have been impressed with the value of its expediency plan. This would have been a formative urban planning experience directly comparable with that of Pierre L'Enfant, whose home town of Versailles, France, was to be a primary influence when he was drawing up his plan for Washington, DC (Chapter 10).

However, it should be noted that while exemplifying Spanish (Castilian) gridiron new town planning on the eve of empire, Santa Fe did not serve as a model or preferred imperial city plan (see pages 305–6, which deal with the Laws of the Indies).

In the same period, from the late sixteenth century, Spain itself entered upon an economic and demographic decline which was not reversed until the beginning of the eighteenth century. Demographic data, curiously, are more scanty and more difficult of interpretation for Spain than for New Spain; but the evidence is clear for the decline in the population of most big towns and for a corresponding decline in economic activity, in sharp contrast with the steady growth of the early and middle years of the sixteenth century. As for the country-side, there is evidence that not only Spain but most countries bordering the western Mediterranean suffered severely from over-grazing, soil exhaustion, erosion and consequent depopulation in the seventeenth century. Pestilence also played its deadly part. Epidemics of unusual severity occurred in 1599–1600 – accompanied by widespread crop failure – and in 1649–51. The second of these affected chiefly Andalusia, which for many months was commercially cut off from the rest of Spain. In Seville it caused 60,000 deaths, about half the total population of the city.
(J H Parry, The Spanish Seaborne Empire)

Spanish urbanism: fifteenth to eighteenth centuries

Had there been continuing great wealth from Latin America at the disposal of the Crown, the leading families and the Church, or had Spain's population increased from the sixteenth century in accordance with the European average rate, resulting, perhaps, in large-scale preindustrial urban growth, then Spanish cities could but have been markedly different in 1800 as compared with three centuries earlier.[10] Neither of these prerequisites for urban change occurred, and with few, but significant exceptions, Spanish cities came through to the nineteenth century largely unaltered. Accordingly there are few examples of extensive grid-planned new residential districts, such as characterize the eighteenth- and nineteenth-century growth of major cities elsewhere in Europe. Cordoba is an example of a leading Spanish city which failed to regain its historic Islamic extent until a western gridiron suburb was planned at the beginning of the twentieth century; similarly, historic Seville embodies a regularly laid out north-western district, replacing a derelict Moorish suburb (Figure 9.14). Exceptionally medieval (Christian) San Sebastian was redeveloped to a new grid plan after devastation by fire in 1813. La Barceloneta is a grid-planned suburb of Barcelona dating from 1755 (Figure 9.12); Las Palmas, on Grand Canary island, was a planned foundation in 1478 by Juan Rejon.

The outstanding Spanish contribution of the period to European urbanism was the creation in a number of so-favoured cities of planned *plazas mayores* (main central squares). In providing for uniquely Spanish urban activities, which determined their forms and architectural

10. For Spanish cities generally see E A Gutkind, *International History of City Development*, Volume 3, *Urban Development in Southern Europe: Spain and Portugal*, 1967; R L Kagan (ed.) *Spanish Cities of the Golden Age: The Views of Anton van den Wyngaerde*, 1989; R Ford, *Murray's Handbook: Spain*, Parts I, II and III, 1845 (9th edn 1898 as the reference).

Figure 9.4 – Cordoba: a sketch plan (redrawn from the modern map), showing the regular street layout of the late nineteenth- and early twentieth-century district developed to the west of the old Islamic nucleus, on the site of an abandoned Islamic suburb (see Figure 4.84). Cordoba affords one of the clearest contrasts between Islamic urban form and modern Western European gridiron planning. To walk at leisure through the surviving Islamic area around the extraordinary cathedral is one of urban Europe's uniquely rewarding experiences. (Key C, cathedral.)

If I had been disappointed by the mosque, I was enchanted by the little courtyards of the Jewish quarter, for I suppose they represent one of the perfect sights that a stranger can come upon in Europe, like the Spanish Steps in Rome or the island museums of Oslo. They are a series of small, informal patios strung out by accident along unimportant back streets, and they number perhaps a hundred. You see them casually through doorways as you walk along, and occasionally one acquires a greater importance than the others because it stands behind formal arches and has had professional attention. Mostly, however, they are family gardens, but unlike any you have seen before.
(J A Michener, Iberia, 1968)

Figure 9.5 – San Sebastian: the plan of the city towards the end of the nineteenth century. The old town is at the foot of Monte Urgull, which commanded the eastern end of the Bahia de la Concha, with the estuary of the River Urumea on its eastern side. The strongly defended harbour provided a base for the Castillian fleet, and after a fire in 1489, the town was rebuilt of stone. In 1813, when taken by the British army commanded by the future Duke of Wellington, the town was devastated, after which it was rebuilt to a more regular gridiron layout. The fortifications were removed in 1863. The extensive new gridiron district subsequently developed over the neck of the peninsular and along the bay (as illustrated) has the same 45 degree corners to the grid blocks as those of the better-known Plan Cerda of 1859 for Barcelona. (See also J I Linazasoro, *Permanencias y Arquitectura Urbana*, 1978 (Spanish), for plans of alternative proposals for the rebuilding of the city.)

Figure 9.6 – Salamanca: the magnificent Plaza Mayor, one of the largest in Spain, was the scene of bullfights as recently as 1863, attended by 16,000 to 20,000 spectators. *Murray's Handbook for Travellers* (9th edn, 1898) notes that 'the centre is occupied by a pleasant garden', happily now replaced by paving.

... the Plaza Mayor is unique in that its spacious area is bordered on all four sides by what amounts to one continuous building, four stories high and graced with an unending arcade of great architectural beauty. It is the most harmonious plaza extant, with its repetitious balconies and windows providing just enough accent and its blending colors creating a vision of amber loveliness ... the outstanding feature is the endless arcade; in heat of day café chairs are moved off the plaza and under the arcade, but at other times it forms a graceful promenade lined with fine stores.
(J A Michener, Iberia)

295

designs, they are the distinctive national equivalent of the eighteenth-century royal statue-squares and grand boulevards of France, the Georgian residential squares of London, and elsewhere.[11] In Spain, even villages have their formal public gathering places; large cities may well have several important plazas, although only a minority have a *plaza mayore* as such. With the exception of Toledo, they are missing from the long-term southern Islamic cities, for reasons which are explained in Chapter 11.

Plazas and *plazas mayores* alike were where public spectacles were staged. Bullfighting on horseback was one traditional activity, from which evolved the special-purpose *plaza de toros*. The Inquisition, however, introduced new, peculiarly Spanish public spectacles when, on 6 February 1481, six heretics were burnt as the first *auto de fe*, staged in the main plaza at Seville;[12] the *plaza mayore* at Vivarrambla was used in 1500 for the ceremonial burning of more than 1 million proscribed books, including unique works of Moorish culture.[13] During its first eight years alone, the Seville tribunal burnt more than 700 heretic, and the *auto de fe* became a commonplace throughout Spain and her empire. The enclosing buildings served as 'grandstands' for these events, with the first purpose-planned *plaza mayore* constructed at Vallaloid from 1561, with accommodation for 24,000 spectators (replacing, ironically, the old plaza which had itself just burnt down, along with much of the old town). Salamanca's *plaza mayore* (1729–33) is the author's personal favourite; that at Madrid is described below (Figure 9.9).[14] In addition to the *plazas mayores*, many cathedrals and major churches were provided with appropriate urban settings for their dominant roles in Spanish society.[15] Characteristically, larger urban churches were part of a complex of college, monastery, convent and hospital buildings, with private gardens which came to comprise extensive religious districts. The most important example is at Santiago de Compostela (Figure 9.7).[16]

In addition to Madrid, Barcelona and Seville, which are described in both main text and captions, other cities are included as plans and captions.

MADRID

The modern capital of Spain probably originated as a Moorish castle-town on a small left-bank hilltop partially encircled by the Manzanares river to the south and west.[17] Although it was not until the middle of the fifteenth century that the small but strongly fortified town came to prominence with the building of a nearby, popular royal hunting lodge; its location near to the geometric centre of the Peninsula was to determine the future choice as national capital. Charles V favoured Madrid as a semi-permanent residence, departing from his predecessors' practice of moving from city to city, and by the mid-sixteenth century its area had increased considerably to house an estimated population of 20,000–25,000.

On 8 May 1561 Philip II decreed that Madrid was to be given the formal status of *unica corte*, with the royal chancellery transferred from Toledo. In addition to the geographical consideration, a second reason for choosing a comparatively insignificant town was the political need to consolidate the unification of Spain by removing the seat of national government from otherwise jealously competing historic cities such as Toledo, Burgos, Leon, and others.[18] Subsequent growth was both

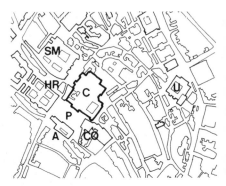

Figure 9.7 – Santiago de Compostela: the outline plan of this most important pilgrimage centre at the far north-western corner of Spain, where during the Middle Ages as many as 2 million people a year visited the reputed burial place of St James the Great (Santiago). The first chapel was enlarged into a cathedral in 874–99. The Saint's tomb survived destruction by the Moors under Almansor in 997, following which it was rebuilt during the twelfth century in much of its present-day form, as the focus of what was soon to become a prosperous centre for the crowds of pilgrims coming from all over Europe.

The cathedral occupies the eastern side of the spacious Plaza de Espana (formerly the Plaza Mayor); the Colegio de San Geronimo – one of the city's numerous religious foundations – forms the southern side; the Ayuntamiento (City Hall) on the west was originally a seminary; and the northern side was the large Hospital Real built in 1501–11 by command of the Catholic monarchs as a pilgrim-hostelry. (It was converted in 1954 into a luxury hotel.) Key: P, Plaza Mayor (Plaza de Espana); C, Cathedral; A, Ayuntamiento; HR, Hospital Real.

11. For the *plaza mayore* in general see M Defourneaux, *Daily Life in Spain in the Golden Age*, trans. N Branch, 1966/1979.

12. Henry Kamen, *The Spanish Inquisition*, 1992.

13. For an account of the importance of Islamic culture in Spain, when the rest of Europe was still slowly emerging from the Dark Ages, see J Mathe, trans. D Macrae, *The Civilisation of Islam*, 1980; N Daniel, *The Arabs and Mediaeval Spain*, 1979.

14. For Salamanca see Ford, *Murray's Handbook: Spain*, Part I; see also Chapter 4.

15. Writing of Valladolid, Defourneaux quotes Pinheiro: 'the monasteries are big enough to be towns in themselves and I am surprised Valladolid can support so many convents and churches. The one Franciscan monastery with its 200 monks occupies half the city' (*Daily Life in Spain in the Golden Age*).

16. For Santiago de Compostela see Ford, *Murray's Handbook: Spain*, Part I; A Castro, *The Spaniards*, 1971; C Gasquoine, *Santiago de Compostela*, 1912 (Mediaeval Towns Series).

17. For Madrid generally see D R Ringrose, 'Madrid and the Spanish Economy 1560–1850', *Journal of Historical Geography*, 1983; Ford, *Murray's Handbook: Spain*, Part I, 1898; Defourneaux, *Daily Life in Spain in the Golden Age*; M C Branch, *Comparative Urban Design*, 1978, for the SDUK Map of the city in 1831.

18. Madrid is the earliest of modern national capitals founded for reasons whereby entrenched local political interests were not to be allowed to influence national considerations. Other such new capitals include Washington, DC, 1792; Ottawa, 1858; Canberra, 1908; Brasilia, 1960.

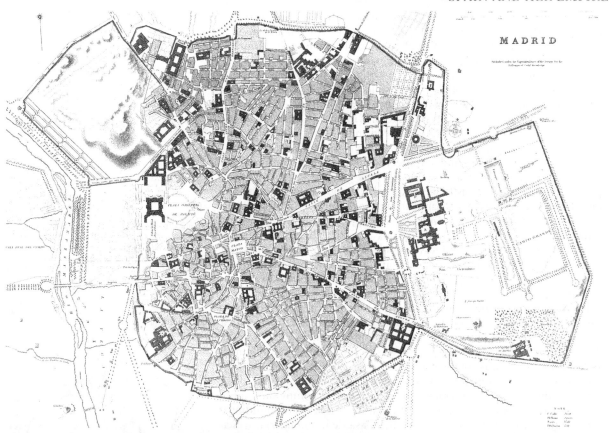

MADRID

Published under the Superintendence of the Society for the
Diffusion of Useful knowledge

Figure 9.8 – Madrid: the map of the city in 1835, as published by the Society for the Diffusion of Useful Knowledge (north at the top). The organic-growth form has pronounced radial streets within the seventeenth-century Customs Wall – an administrative boundary of negligible strategic significance – which enclosed an area of 787.5 hectares. Madrid was never seriously fortified and could grow horizontally, free of need to build vertically within a constraining girdle, thereby making possible the city's unusual if not unique single-storey housing development. (In effect, a result of the absence of the fortification determinant, compared with most other major Spanish cities.)

The superb Plaza Mayor is situated to the south-east of the Palace, which occupies much of the area of the early settlement, in the centre of the western side of the mid-nineteenth-century city. The city's monasteries were suppressed in 1836 and many of the sites were redeveloped for apartment houses. The old Customs Wall remained the city boundary until 1868, beyond which new residential districts were constructed later in the nineteenth century. (See M C Branch, *Comparative Urban Design*, 1965, for this map and descriptive text.)

Figure 9.9 – Madrid: aerial photograph of the Plaza Mayore showing the way that the space for the city's heart had to be carved out of the medieval urban fabric. In 1898, as described in *Murray's Handbook for Travellers in Spain*, the Plaza 'is now converted into a beautiful garden'; today, a century later, it is an appropriately paved pedestrian concourse. (See also A B Correa, *Morfologia y Ciudad*, 1978 (Spanish), for numerous other illustrations of the Plaza Mayore.)

rapid and uncontrolled. By 1600 the urban area had quadrupled and its population had passed 100,000, a rate of growth which was not maintained, and the Customs Wall of 1625 was to serve as the administrative boundary until the renewed large-scale expansion of the latter part of the nineteenth century.

Following the removal of the royal court from Toledo, residential accommodation in Madrid was so scarce that Philip II ordered all owners of two or more storeyed houses to provide rooms for members of the court and government officials. Accordingly, from then on, most new housing construction in the city was of only one storey, known as *casas de malicia* and by the mid-seventeenth century they outnumbered the taller buildings – the *casas de aposento* – by 5,436 to 1,470.[19] Consequently the urban scale of the new capital was modest in the extreme and occasioned frequent criticism through into the twentieth century. John Murray's *Handbook for Travellers* of 1845 was totally dismissive: 'unlike the many time honoured capitals of Spain, this is an upstart favourite without merit. ... Madrid is built on several mangy hills that hang over the Manzanares, which being often dry in summer can scarcely be called a river. ... [it] is not even a city or a Ciudad; it is only the chief of villas'.[20]

The city's *plaza mayore* dates from 1581 when additional land was purchased in order to enlarge the extent of an old provision market area which had grown up outside the wall of the medieval town.[21] Juan de Herrera prepared a first design but it was to the plans of Juan Gomex de Mora that work was commenced in 1617. In 1623 it was the scene of a tournament in honour of the future Charles I of England, for which type of public spectacle the plaza was reputed to accommodate more than 50,000 spectators, some 4,000 of them in the enclosing buildings. Architecturally the chief building was the Casa de la Panederia, which served as a bread shop on the ground floor and housed the royal box on the first floor (Figure 9.9).

Redevelopment of the areas of insignificant housing and the sites of the monasteries, which were suppressed in 1836, gradually brought about a more impressive urban character which further twentieth-century rebuilding and replanning had continued to enhance. Nevertheless, Madrid can still appear an oddly unsatisfactory national capital compared with other historic claimants to that role.[22]

BARCELONA

As Roman Barcino, the city had enjoyed importance as a major naval base and the prosperity of a well-situated trading centre.[23] It was held by the Moors only briefly from 713 to 801 before being regained by the Franks, who made it the capital of the County of Barcelona. Devastated as the Moors marched north in 986, recovery afterwards was initially slow until the general revival of Mediterranean trade from the twelfth century. The city spread down from its historic nucleus on Monte Taber, consolidating its commercial waterfront and growing to an area of some 250 acres by the beginning of the fourteenth century. At that time the defensive system comprised the conventional medieval wall with approximately 100 towers (see Figure 9.11).

Continued economic growth created an increasingly serious land shortage within, and in about 1350 the decision was taken to add a new fortified district – the *arrabal* – almost equal in area to the existing city. Development of the *arrabal*, however, proceeded only very slowly and as

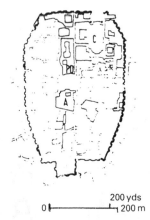

200 yds
0 ⊢————⊣ 200 m

Figure 9.10 – Barcelona: the Roman nucleus, known as the Barrio Gotico, contained within the line of the fourth-century walls, parts of which are visible with others in process of restoration. Narrow, indirect lanes are reminiscent of Islamic urban form although Barcelona was occupied for only ninety years. Key: C – the Cathedral, known to the Catalans as La Seu, mainly constructed between 1298 and 1329; A – the Ayuntamiento (City Hall) facing the Palacio de la Disputacion (PD) across the Plaza de San Jaine.

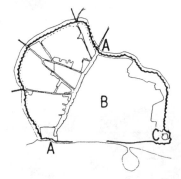

Figure 9.11 – Barcelona: the two-part city as drawn from a map of 1492, showing the *ciudad* (B) and the area of the *arrabal* suburb separated from it by the linear open space on the site of the earlier wall (A–A) which has become the modern city's famous ramblas boulevard. (The original nucleus of Roman Barcino is centred on the key letter 'B' within the *ciudad*.) 'C' denotes the medieval castle which was to provide a basis for Phillip V's citadel constructed in the early part of the eighteenth century.

19. See Defourneaux, *Daily Life in Spain in the Golden Age*; this 'social legislation' was an extremely unusual, if not unique urban form determinant.

20. Ford, *Murray's Handbook: Spain*, Part I.

21. For the Plaza Mayore at Madrid see Ford, *Murray's Handbook: Spain*, Part I: 'on this square the autos de fe and the royal bull-fights were celebrated. Here our [the English] Charles I beheld one given in his honour by Philip IV. The locality, 434 ft. long by 334 wide, was well adapted for spectacles. By a clause in their leases the inmates of houses were bound on these occasions to give up their front rooms and balconies which were then fitted up as boxes'.

22. 'The gross mistake of a most faulty position was

Figure 9.12 – Barcelona during the latter part of the eighteenth century, within its improved fortified enceinte, and with the massive citadel. The fortress of the Montjuich (left) on its dominant hilltop is now a military museum and pleasure park, from whose esplanades (and redundant bastions) magnificent views are afforded to the city centre. The planned suburb of La Barceloneta was built from 1755 on reclaimed land at the head of the harbour bay.

late as 1492 amounted to little more than organic-growth street frontage buildings along the main routes across it from the *ciudad*.[24] Meanwhile, within the old city, building densities and heights had continued to increase.

After being besieged by the French in 1697, the city invested in the modern defensive system shown by the map (Figure 9.12). After the end of the War of Succession, Philip V ordered the construction of the massive citadel for which the Ribera, a waterside suburb of some 2,000 dwellings, was demolished. Many of the displaced families moved across to the spit of land which had formed along the eastern side of the harbour, where they constructed, as best they could, a shantytown (*barrida*). This, in turn, was removed in 1752 to be replaced by the new suburb of La Barceloneta, laid out from 1755 to an elementary gridiron plan produced by the military engineer Pedro Cermeno. From 1778 Barcelona was permitted to commence trading with the Latin American Empire and further growth ensued.[25]

Authority to demolish the constricting and obsolete fortifications was given in 1854, on condition that a general master plan was drawn up for the city's future growth. This was prepared by the government surveyor Ildefonso Cerda y Suner on a much criticized, combination gridiron and radial street basis, implementation of which commenced in 1860 with the removal of the fortifications.[26]

soon felt and Philip III, in 1601, endeavoured to remove the court back again to Valladolid, which, however, was then found to be impracticable. ... Charles III, a wise prince, contemplated a removal to Seville; so also did the intrusive Joseph, but the thing was impossible' (Ford, *Murray's Handbook: Spain*, Part I).

23. For Barcelona see M de Sola-Morales, *Barcelona: Remodelacion Capitalista o Desarrollo Urbano en el Sector de la Ribera Oriental*, 1974 (Spanish); Ajuntament de Barcelona, *Urbanisme a Barcelona: Plans Cap al 92*, 1987 (Spanish, with excellent maps and illustrations); T Hall, *Planung Europaisker Hauptstadte*, 1986 (German).

24. The routes through to the gates, and the related landownership pattern, are an unusual example of a pre-urban cadastre within the walls of a city.

25. See Parry, *The Spanish Seaborne Empire*, for Barcelona and trade with the Indies.

26. For the Cerda Plan for Barcelona see Ajuntament de Barcelona, *Urbanisme a Barcelona: Plans Cap al 92*, 1987.

SEVILLE

Situated some 60 miles up the winding, muddy and in part hazardous Guadalquivir River, the city of Seville had from the beginnings of its sixteenth-century imperial prosperity, those inherent locational disadvantages which were to bring about its eventual late seventeenth-century decline.[27] Nevertheless, geographical deficiencies were over-ruled by political and commercial interests, and Seville, as the capital of New Castile, was favoured by the Spanish Crown to be the sole port of entry for the immensely valuable trade with the New World, or the Carrera de Indias as it became known. The Casa de la Contratacion de las Indias was established in Seville in 1503 as the first organ of Span-ish colonial administration, 'to promote and regulate trade and naviga-tion to the New World'. Although the Seville monopoly 'has often been cited as an example of Spanish indifference to commercial interest, of a preference for bureaucratic regulation over economic enterprise', Parry explains, 'in fact, the Crown was merely giving official approval to a choice already made by most commanders experienced in the Indies navigation. The ships then employed in the trade were neither large enough to run much risk of grounding on the San Luca bar (at the mouth of the Guadalquivir), nor numerous enough to strain unduly the harbour facilities of Seville'.[28]

Figure 9.13 – Seville: the city's location on the River Guadalquivir, and other ports. Key: 1, San Juan de Aznalfarache; 2, Los Horcades; 3, Borrego; 4, Puerto de Santa Maria.

The port of the Indies was not merely Seville, but the whole stretch of swift, winding, muddy river from Seville to San Lucar, crowded with ocean-going ships and hurrying small craft. Just as a minimum limit was set to exclude ships too small to defend themselves at sea, so a maximum limit was set to exclude vessels too big to use the river safely.

Figure 9.14 – Seville: a map of the city of the latter part of the nineteenth century, still contained largely within the line of the sixteenth- and seventeenth-century walls. The River Guadalquivir flows from left to right; the bridgehead settlement of Triana is at the bottom right-hand corner. The regularity of street layout of the north-western quarter is in direct contrast with the Islamic organic-growth of the main part of the city, and would indicate that this was a 'planned' rebuilding of a previously abandoned area.

Sevilla was also a major capital of the Moors, having been occupied by them in one capacity or another for 536 years, yet today one finds in the city even fewer of the Muslim memories that make Córdoba and Granada such noble testaments to the Moorish influence in Spanish history. Even the graceful Giralda has had its Moorish origin submerged in Christian additions, while in the nave of the massive cathedral the Moorish pillars are lost in heavy Gothic shadows. Whenever a con-queror departed, Sevilla quickly reestablished itself as a Spanish city, jealous of its prerogatives and marvel-ously insular in its attitudes. If one seeks a whole body

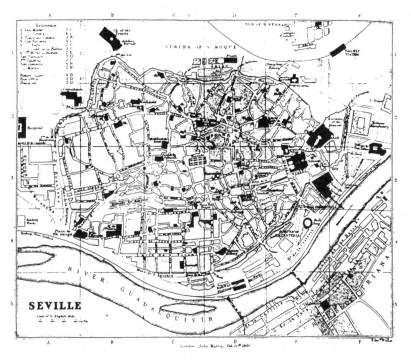

SEVILLE

Cadiz was in some respects the natural first choice but it was isolated at the end of a rocky peninsular and its otherwise excellent harbour was vulnerable to enemy attack; its turn was to come later.[29] Other Andalu-sian Atlantic harbours were too small; the Castilian Mediterranean ports of Malaga and Cartagena were too distant, and preoccupied with that trade; Galician possibilities such as Bilbao and Corunna were also remote.

27. For Seville see W M Gallichan, *Seville* (Mediaeval Towns Series), 1903; J Goss, *Braun and Hogenberg's The City Maps of Europe: A Selection of 16th Century Town Plans and Views*, 1991; H Kamen, *Spain 1469–1714*, 1983, who writes 'as merchants of all nations realised the potential of direct investment in American trade, Andalucia became a hub of international activity: from over 95,000 people in 1561, Seville increased to about 122,000 in 1588 and 150,000 in 1640 making it the largest city in the peninsula'.
28. Parry, *The Spanish Seaborne Empire*.
29. For Cadiz see Ford, *Murray's Handbook: Spain*, Part II, 1898.

GRANADA

Scale of English Mile

of people who have refused to acknowledge the advent of change, few can compare with the citizens of Sevilla. (J A Michener, Iberia)

Figure 9.15 – Granada: the city of 1869, from *Murray's Handbook for Travellers*, with a diagram identifying the historic parts and showing the main growth phases. The map is deceptive not only because the site is strongly three-dimensional, but also because the streets and lanes – notably in the historic Albaicin (Islamic) district – are both more numerous and less wide.

Granada originated as Elibyrge, a Celtic settlement, occupying the high ground of the later Albaicin, and with the steeply rocky ridge of the later Alhambra as its fastness, across the valley of the river Darro. The Romans and Visigoths knew it as Illiberis, an obscure unimportant settlement. Under the Moors, however, Granada came to unequalled fame in the Europe of the thirteenth and fourteenth centuries; its material prosperity enabling Moorish art to flourish, notably in the architecture, courtyards and interiors of the Royal Palace in the Alhambra.

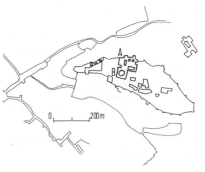

Figure 9.16 – Granada: the plan of the Alhambra, a ridge about 800 metres in length and 200 metres at its widest. Key locating the main buildings as A, the Alcazaba; B, the incomparable Moorish Royal Palace, C, the Palace of Charles V, an incongruous sixteenth-century intrusion: architectural merit, but on the wrong site; D, the Generalife, the summer palace of the sultans set in extensive landscaped hillside grounds.

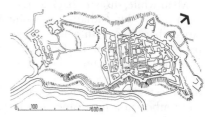

Figure 9.17 – Tarragona: redrawn from the map of 1802, showing the city still contained within the sixteenth- and seventeenth-century fortifications, on the site of the higher-ground main part of the Roman city, with the separate port established to the south-west. The cathedral at the northern end was constructed on the site of the Roman Temple of Jupiter. The fortifications along the south-western side of the city were demolished early in the second half of the nineteenth century, and the city and port rapidly grew together. The line of the fortifications is that of the Ramblas Nova.

Figure 9.18 – Granada: a view of the Alhambra from the south.

301

The Spanish Latin American empire

Description of the settlement of the Spanish Latin American empire is on the basis of separate phases of colonial expansion, grouping the countries concerned in the following sequence: first, the Caribbean Islands, the Gulf and the Isthmus; second, mainland Mexico, the Aztecs and the Mayas; third, Peru and the Incas; and fourth, the South American Agricultural Colonies. The sequence is that of historic origins, but there is considerable overlapping, such was the speed of Spanish conquest.[30] Emphasis is accorded to early settlement activities, with a concluding section summarizing the empire's latterday history and its ultimate disintegration in the face of early nineteenth-century national independence movements.

As the prelude to Empire, Christopher Columbus's epic voyage of 1492 needs no retelling.[31] It followed some fifty years of progressive exploration down the African coastline as far as the Cape of Good Hope, which was rounded by the Portuguese explorer Bartholomew Diaz in 1486. Columbus (c. 1451–1506), with three ships, had hoped to discover a new, direct western sea-route to the Indies – as China and India were known – in competition with the Portuguese, who were committed to the eastern way around Africa. His discovery of Cuba, which he imagined to be Asia, and Haiti (thought to be Japan) is well known; less so the fact, as Reps records, that 'when in December 1492 he built a crude fortress from the timber of the wrecked *Santa Maria* on the northern coast of the island of Espanola (as Haiti was first named), he began an era of city planning in the Americas'.[32]

Forced to leave thirty-nine of his company at La Navidad, as the 'colony' was named, he returned to Spain where Charles V immediately petitioned the Pope for sole rights to the Indies. The Portuguese promptly registered their objection, whereupon papal compromise established a boundary between the two national spheres of influence, drawn 100 leagues west of the Azores. In 1494 this was extended by agreement to 370 leagues west of the Cape Verde Islands, which explains how it was that Portugal was entitled to Brazil.[33]

Meanwhile, disaster back at La Navidad, whose inhabitants had all been killed by Indians, as Columbus discovered on returning in November 1493 with a fleet of seventeen ships carrying some 1,200 people. With them he founded the intended permanent colony of Isabella, some 30 miles further east. This was replaced in 1496 by Santo Domingo, located more safely, it was believed, on the east bank of the estuary of the River Ozama. Within six years, however, it had succumbed in a hurricane, and it was the new Governor, Nicolas de Ovando, who authorized its replacement by the west bank Santo Domingo, the first permanent European settlement in the Western Hemisphere (described below).[34]

THE CARIBBEAN ISLANDS, THE GULF AND THE ISTHMUS

Nicolas de Ovando had arrived on Hispaniola in 1502 taking over as Governor and bringing with him a fleet of thirty ships carrying some 2,500 new settlers, including seventy-three families. He imposed a disciplined order throughout the island, establishing it as a secure base for further exploration in search of worthwhile plunder and, still, the mythical direct route to the Indies. Santo Domingo was refounded in 1502 on the west bank of the Ozama as the capital city of Hispaniola;

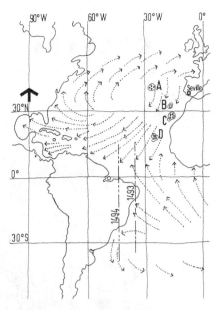

Figure 9.19 – The North and South Atlantic oceans, showing the favourable prevailing winds westwards from Seville, and the four first-visited island groups. Key: A, the Azores; B, Madeira (both to Portugal); C, Canary Islands; D, Cape Verde Islands (both to Spain). The dividing line between the spheres of interest of Portugal and Spain, set in 1494 at 370 leagues west of the Cape Verde Islands, gave Portugal her fortuitous dominion over the area of future Brazil.

30. See J E Hardoy, *Pre-Columbian Cities*, 1964 (1973 English edn) as the comprehensive account of the urbanism of the ancient civilizations (from which was abstracted, with key illustrations, Hardoy, *Urban Planning in Pre-Columbian America*, 1968). All of the pre-Columbian cities of this chapter are described in Hardoy, 1964/1973 and references here are only in respect of specific significance. See also F C Goitia and L T Balbas, *Planos de Cuidades Iberoamericanas y Filipinas*, 1951, for extensive coverage of Spanish (post-Columbian) urban maps held in the Archives de Indias, at Seville in Spain, as individual references. For general historical background, see J Lockhart and S B Schwartz, *Early Latin America: A History of Colonial Spanish America and Brazil*, 1983/1989; G Pendle, *A History of Latin America*, 1976/1990; W W Sweet, *A History of Latin America*, 1919, included not least because it is the eminently readable book I took with me down into Mexico in 1961.

31. The epic voyage of Columbus needs no re-telling, that is in respect of the Spanish discovery of the Americas; however, at the time of writing, the 500th anniversary, literary (and television) industries are promising new interpretations in their 1992 programmes.

32. J W Reps, *The Making of Urban America*, 1965; although primarily concerned with the history of urban settlement in the USA, Professor Reps's definitive study also establishes a background for the southern and south-western Spanish colonies (see Chapter 10).

33. For the papal agreements see Parry, *The Spanish Seaborne Empire*.

34. For Santo Domingo see Reps, *The Making of Urban America*; Goitia and Balbas, *Planos de Cuidades Iberoamericanas y Filipinas*; J M Houston, 'The Foundation of Towns in Hispanic America', in R P Beckinsale and J M Houston (eds) *Urbanisation and its Problems*, 1968.

its plan is described with Figure 9.21. As the capital of the Dominican Republic – the eastern two-thirds of the island – it is now the oldest of the European cities founded in the Americas. Haiti – its capital city Port au Prince – occupies the western one-third of the island, and was ceded by Spain to France in 1697.

Before leaving Seville, Ovando had been given a set of instructions (or 'standing orders') prepared in 1501, one section of which dealt with the location and planning of new colonial settlements. 'Because it is necessary to establish some towns in the Island of Espanola,' it was stated, 'and it is not possible to render specific instructions from here; examine the locations and situations of said Isle, and in conformity with the quality of the Land and the people now resident in those towns which now exist, establish towns in those places which seem proper to you.'[35] As a planning brief, this could hardly have been more non-committal, in contrast to the formal, precise directions later to be embodied in the all-embracing 'Laws of the Indies'. Nevertheless, the unknown 'city planner' on Ovando's staff established from the outset at Santo Domingo the dominant imperial urban characteristic of regular gridiron street blocks, centred on the one main plaza around which the several major civic buildings were grouped.[36]

Figure 9.20 – The Caribbean, the Gulf of Mexico, and the Isthmus, locating the first Spanish settlements.

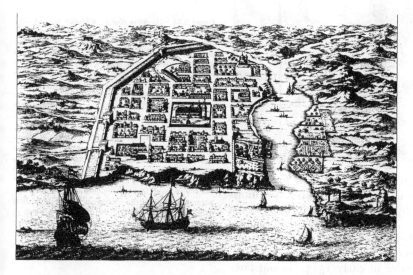

Figure 9.21 – Santo Domingo: an artist's impression of 1671, which had changed but little from an earlier view of 1586. In 1526 a resident avowed it to be 'superior in general to any town of Spain ... the many streets were more level and wide and incomparably straighter ... it was laid out with a ruler and a compass, with all the streets being carefully measured' (J W Reps, *Cities of the American West*). After the lightly defended town of that time had been sacked by the English fleet under Sir Francis Drake, the plans for artillery fortifications drawn up by Antoneli (see Figure 9.22) were not implemented and Santo Domingo became only a locally important Spanish city.

The three other major islands colonized from Santo Domingo were Puerto Rico (from 1508), Jamaica (from 1509) and lastly and most importantly Cuba (from 1511). The island of Puerto Rico had been claimed by Christopher Columbus on 19 November 1493, early in his second voyage, only to be neglected, initially, while Hispaniola was colonized. In 1508 Juan Ponce de Leon was authorized to explore Puerto Rico and founded a first settlement on the northern coast at Caparra, located alongside an excellent natural harbour which he named Puerto Rico. The first urban site proved unhealthy and in 1521 it was replaced by the present-day location, where San Juan del Puerto Rico was founded (see Figure 9.22).[37] (The island gradually acquired its name Puerto Rico through common usage.)

Jamaica, settled by Juan de Esquival from 1509, is not important in this account of the Spanish Latin American Empire. It played no great part in the Spanish conquest and from 1655 the island was a British

35. Reyes Catolicos to Nicolas de Ovando, Granada, 16 September 1501 (Instruccion ... al Fray Nicolas de Ovar.do ...), as translated and quoted in D Garr, 'Hispanic Colonial Settlement in California: Planning and Urban Development on the Frontier', unpublished PhD dissertation, Cornell University, Ithaca, New York, 1972, prepared under the direction of Professor J W Reps, and quoted in his *Cities of the American West*.

36. The primary determinants of the typical Spanish Latin American city plan were the gridiron, topography, religious and less so political aggrandizement, construction materials, and social segregation, with the Indians relegated to outer grid blocks, if allowed to live within the city at all. See also example plans below in the description of effects of the Laws of the Indies.

37. For San Juan see K R Andrews, *The Spanish Caribbean: Trade and Plunder 1530–1630*, 1978.

colony.[38] Cuba, on the other hand, not only was one of the first of Columbus's discoveries of 1492, but also remained longest in Spanish hands, until 1899. It was conquered during two brutal years between 1511 and 1513 by a force of about 300 men, and already by 1515 there were six major colonies in existence, including Santiago del Cuba and the original site of Havanna (possibly present-day Barabano), for which see the caption to Figure 9.23.[39] Cuba in turn became the base for explorations of mainland North America, and through Havanna, its principal port, the main supply and communications base for the east-bound shipping trade with Spain.

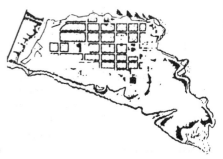

Figure 9.22 – San Juan del Puerto Rico: the plan of the city at the end of the sixteenth century, within the fortifications designed by an Italian engineer, Juan Bautista Antonelli, who was employed by Spain for more than twenty years on this and other Caribbean urban and port fortifications. Investment at San Juan proved its worth when the city held out against Drake in 1595. See D A Iniguez, *Bautista Antonelli: Las Fortificaciones Americanas del Siglio XVI*, 1942, quoted in K R Andrews, *The Spanish Caribbean: Trade and Plunder 1530–1630*, 1978. Further late eighteenth-century strengthening added the massive Vauban-style forts of San Felipe del Morro and San Cristobal.

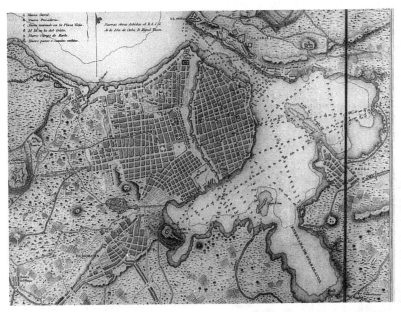

Figure 9.23 – Havanna (Cuba): a French map of 1762 showing the extent of the magnificent deep-water harbour, approached through a narrow heavily defended channel, controlled by the Fort du Maure to the north with fifty-seven cannon, and the Fort de la Pointe with twenty-seven cannon. The Vieux Chateaux mounted twenty-two cannon and there were many more in the total of ten bastions strengthening the land wall.

Havanna was founded in 1514 as San Cristobel de la Habana and moved to its present site shortly afterwards. At first only lightly defended, it was sacked by the English fleet in 1537 and 1556, before being refortified by Antoneli (see Figure 9.22). In 1592 Havanna was raised to city status with the Governor of Cuba moving across from Santiago del Cuba. By 1608 the population was about 10,000; this total was doubled by 1655 when the city received Spanish refugees from Jamaica. Impregnable for two centuries, the city fell to a British fleet in 1762, thereby ensuring the international fame of 'Havanna' as the finest cigar tobacco when it was taken back to Europe by British servicemen.

The history of Spanish settlement in the Gulf and their first moves south from the Isthmus commences in 1509 when two expeditions sailed from Santo Domingo: one, led by Alonso de Ojeco, went to the northern coast of South America; the other, under Diego de Nicuesa, was sent to the western coastline. Their most important foundations were Nombre de Dios and Santa Maria del Darien, from where Vasco Nunez Balbao, appointed as commander in early 1513, set out later that year to investigate Indian reports of a 'great sea and a golden kingdom to the south'.[40] He reached the Pacific on 29 September, claiming it for Spain, and was planning to explore to the south when Pedrarias Davila took over command. By 1519 the city of Panama had been founded, linked back to Darien across the Isthmus by a rough track cut through the forest. After preliminary excursions, Francisco Pizarro set sail from Panama in December 1530 for what was to be the conquest of Peru. Meanwhile, Hernando Cortés, setting out from Cuba in 1519, had conquered Mexico as the first of the Spanish mainland Latin American territories.

By 1519, although firmly established in the Caribbean and poised for undreamt-of continental expansion, 'urban life in America was precarious', writes Jorge E Hardoy. 'Almost all the Spanish lived in isolated and precarious cities and towns; the pompous title of city was a legal figment, not a physical reality. The coats of arms given to certain

38. For Jamaica and its principal city, Kingstown, see M M Carley, *Jamaica: The Old and the New*, 1963; J H Parry and P M Sherlock, *A Short History of the West Indies*, 1963.

39. For Havanna see N Sapieha, *Old Havanna*, 1990; Goitia and Balbas, *Planos de Cuidades Iberoamericanas y Filipinas*; Andrews, *The Spanish Caribbean: Trade and Plunder 1530–1630*.

40. See Hardoy, *Two Thousand Years of Latin American Urbanisation*, who emphasizes the Spanish quest for precious metals; see also D McCullough, *The Path between the Seas* (which is primarily an account of the creation of the Panama Canal 1870–1914) for the foundation of Panama City at its southern, Pacific coast end. Confusingly, the isthmus linking North and South America runs from west to east. Old Panama City, founded in 1519, was destroyed by the pirate Morgan in 1671, and the present Panama, a walled city, was begun some three years later, at the head of the bay approximately 5 miles to the south-west.

Hispanic cities had no significance other than that of a royal gift, easily bestowed'.[41]

The *conquistadores*, even those of immediate peasant origins, had no intention of engaging in personal manual labour and the *repartimiento* system was quickly introduced whereby leading settlers were allocated native communities for that purpose.[42] *Repartimiento* was also intended to facilitate evangelization of the natives as required by Papal Bulls of 1493, but as Hardoy cautions, 'we should not confuse the vision and generosity of certain laws of those first decades with the possibilities and the true interests of royal representatives, conquerors, and the new landed proprietors in applying them'.[43]

The inevitable gridiron basis of the 1502 plan for Santo Domingo has been discussed above, related to the presumed absence of any formal city planning instructions issued to Ovando. It was not long, however, before leaders of expeditions were specifically directed in this respect; notably Davila, who in 1513 was instructed by Ferdinand V: 'let the city lots be regular from the start, so that once they are marked out the town will appear well ordered as to the place which is left for the plaza, the site for the church and the sequence of the streets; for in places newly established, proper order can be given from the start, and thus they remain ordered with no extra cost: otherwise order will never be introduced'.[44] These instructions were consolidated and extended in 1521 when Charles V issued the code of 'city planning practice' which was subsequently incorporated into the 'Laws of the Indies'. They were the uniquely widespread, most important instance in our period of the legislative determinant of urban form; the city planning sections are described and assessed below.

THE LAWS OF THE INDIES

Although the *Recopilacion de Leyes de los Reynos de las Indias* was not formally published by the Spanish Crown until 1681, the town planning sections had been separately in force for over a century from 1573, when Philip II had issued a royal ordinance governing the founding and physical planning of new towns throughout the Empire.[45] (In turn, the 1573 ordinance had embodied existing imperial 'standing orders'.) The city planning legislation remained in effect throughout the period of the Latin American empire (as also in the Spanish Philippines) and it also applied to those Spanish towns founded in the future United States (see Chapter 10). However, while rigidly framed, the legislation was not strictly enforced, such that the innumerable applications represent variations on a basic theme, rather than mere monotonous pattern-book repetition.[46]

In their formulation, the regulations were based on the experience of the first settlers, as reported back to Seville. By 1526, Reps concludes, some kind of a prototype town plan had been prepared by the Council of the Indies (set up in Seville in 1524), citing as evidence the report on 26 June 1524 of the founding of the Villa de la Frontera de Caceres in Honduras, by Bartolone Celada, who had instructed his surveyors: 'according to the custom of these monarchs and lords, a plan be made conforming to that of Seville, immediately marking off on it lots for the church and for the plaza, for the hospital, for the Governor, and for myself'.[47] Although seemingly not specified as such in the Laws — probably because it was taken for granted, a gridiron street layout was the determining morphological rule. Within the framework of streets,

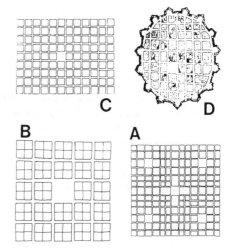

Figure 9.24 – The Laws of the Indies: four examples of minimalist city gridiron 'plans'. Key: A, Guatemala (1776); B, Concepcion (1765); C, Mendoza (1561); D, Trujillo (1760), which is an extremely rare (perhaps the only instance) of a Spanish plan following the Italian 'ideal city' pattern (see Chapter 5).

41. Hardoy, *Two Thousand Years of Latin American Urbanisation.*

42. See B W Diffie, *Latin American Civilisation*, 1945, on this aspect of Spanish colonial settlement: 'it was only natural that the repartimiento and encomienda, given their long historical development in Spain, should migrate to America with the first settlers'.

43. Hardoy, *Two Thousand Years of Latin American Urbanisation.*

44. For Davila's instructions see J W Reps, *Cities of the American West*, 1979. (This wording has the same intent, but differs from that given by Reps in *The Making of Urban America*.)

45. The main references for the Laws of the Indies are L Hanke, 'Spanish Ordinances for the Layout of New Towns, 1573', in L Hanke (ed.) *The History of Latin American Civilisation, Volume 1, The Colonial Experience*, 1967; D Stanislawski, 'Early Spanish Town Planning in the New World', *Geographical Review*, who is over-preoccupied with Vitruvius as a precedent (see Chapter 5); Reps, *Cities of the American West*.

46. As variations on a basic gridiron planning theme, Spanish cities of Latin America are later and different equivalents of the medieval European bastide towns (see Chapter 4); although the morphology of even the most basic bastide was a work of art compared with the banality of the simplest Spanish layouts (see Figure 9.24).

47. Reps, *The Making of Urban America*; it is accepted by Reps that the requirement for a plan to conform with 'that of Seville', is in respect of a (lost) standard city plan drawn up in Seville, rather than an instruction to repeat, or follow, the actual layout of that city which was nearly all a response to the centuries of Islamic occupation, rather than 'planned' *per se* (see Figure 9.14).

the regulations, in theory, determined the detail arrangements.

There were more than three dozen specifications and advisory clauses: one near the beginning had the needs of future growth in mind, requiring 'the plan of the place, with its squares, streets and building lots is to be outlined by means of measuring by cord and ruler, beginning with the main square from which streets are to run to the gates and principal roads, and leaving sufficient open space so that even if the town grows it can always spread in a symmetrical manner'.[48]

The central, distinctive component part of imperial Spanish city planning was the plaza, which was given appropriate symbolic emphasis. For the waterfront towns a location was prescribed appropriately near the sea or the river; for inland towns 'the central square should be in the centre of the city, of oblong shape, with the length at least one and a half times the width, since this is the best proportion for festivals in which horses are used, and for other celebrations. The size of the square shall be proportionate to the numbers of inhabitants, bearing in mind that the cities of the Indies, being new are subject to growth; and indeed it is our intention that they shall grow. For this reason the square shall be planned in relation to the possible growth of the city. It should not be less than 200 feet wide and 300 feet long, nor wider than 500 feet or longer than 800 feet. A well-proportioned square of medium size shall be 600 feet long and 400 feet wide'.[49]

In addition to the bureaucratic legislative determinant, religious aggrandizement played a major role, whereby, the church was to be the dominant building on the plaza, set back from it in order to accentuate its symbolic pre-eminence: 'in the cities of the interior the church should not stand on the perimeter of the square, but at such a distance as to appear free, separate from the other buildings so that it may be seen on every side, in this way it will appear more handsome and more imposing. It should be built above the level of the ground, so that the people must climb a series of steps to gain entrance to it'.[50] Churches in waterside cities were to face directly on to the plaza and were to be strongly constructed to serve as emergency fortifications.[51]

THE MAINLAND: MEXICO, THE AZTECS AND THE MAYAS

In direct contrast to their findings in the islands, which were occupied by only isolated groups of primitive tribesmen, the *conquistadores* encountered on the American mainland three distinct higher civilizations.[52] The map (Figure 9.25) locates them as the Aztecs of Mexico, the Maya of Yucatan, and the Inca of Peru. The sequence of conquest was from north to south, commencing with the Aztecs, followed by the Incas, and then the other Central and Southern American colonies. The achievements of the Maya, who had been in progressive decline for several centuries, are summarized at the end of this Mexican section.

'The Aztecs,' writes G C Vaillant, 'were a concentrated population of independent groups living in the Valley of Mexico and later welded into an empire, whose authority reached out to dominate much of southern and central Mexico.'[53] They were the last of a series of pre-Columbian civilizations in Mexico, and flourished from the fifteenth century, preceded in turn by other cultures with significant urban activities, including the Zapotec capital of Monte Alban and the Olmec ceremonial centre of Teotihuacan (Figure 9.26). The earliest permanent village settlements in the valley of Mexico date from about

Figure 9.25 – Latin America; a map locating the three major pre-Columbian Central and South American civilizations. (See also Figure 1.2 showing these civilizations related to the other first civilizations of the Middle East and Asia.)

48. Although by 1573 there was ample evidence that irregular, unbalanced expansion of Spanish cities in Latin America was an inevitable response to unpredictable socio-economic market forces, as well as changed and new urban form determinants, the planning theoreticians and legal draughtsmen back in Seville persisted in imagining balanced, neat and tidy growth. (See the critique of similar optimism at Philadelphia, USA, in Chapter 10.)
49. This proportion 3:2 is coincidentally similar, but of no greater significance, to that advocated by Vitruvius. See Stanislawski, 'Early Spanish Town Planning in the New World', for a detailed, divertingly pointless comparison of the Laws of the Indies and Vitruvian 'theory'; see also the general critique of the relevance of Vitruvius in Chapter 5.
50. To my knowledge this requirement is unique in history as direct expression of the religious aggrandizement determinant in a code of city planning legislation.
51. This doubling of roles as church and expedient citadel is also a characteristic of European medieval bastide planning (described in Chapter 4). Where building materials, technical expertise – and time – were in short supply, this represents a logical emergency use of the most strongly constructed building in a city.
52. See G Daniel, *The First Civilisations*, 1968/1971: Chapter 1's international introduction to the first higher civilizations.
53, 54. G C Vaillant, *Aztecs of Mexico*, 2nd edn, 1962.

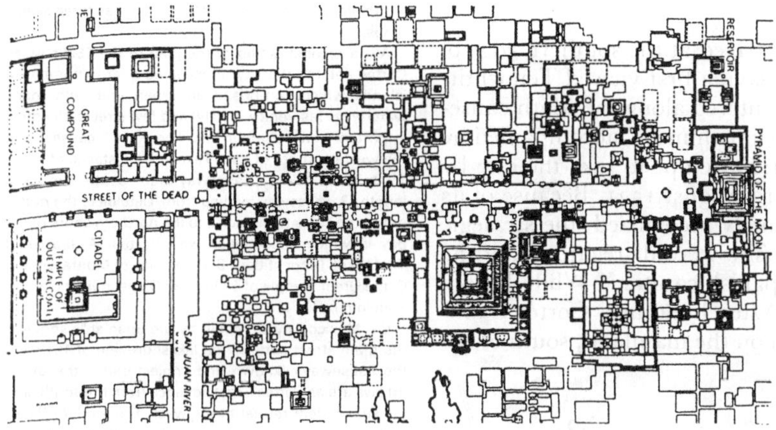

[Map image at top left with labels: GREAT COMPOUND, RESERVOIR, STREET OF THE DEAD, CITADEL, TEMPLE OF QUETZALCOATL, SAN JUAN RIVER, PYRAMID OF THE MOON, PYRAMID OF THE SUN]

Figure 9.26 – Teotihuacan: detail map of the central part of the city. This immensely important archaeological site is located some 25 miles north-east of Mexico City (Tenochtitlan). In the 1520s, when the *conquistadores* arrived in the Valley of Mexico, it was about 8 miles from the shore of Lake Texcoco, across from Tenochtitlan. Hardoy notes that 'from the very last centuries of the first millennium B.C. Teotihuacan was the residence of a large and permanent population occupied in government duties, services, trade and handicrafts without the necessity to leave the city to earn their living'. While the structure of Teotihuacan includes parts that are recognizably gridiron-based, it is preferable to regard this plan in general as consisting of a large number of various sized square and rectangular precincts and buildings, organized according to the same orientation. The general layout dates from the first three or four centuries AD and takes the form of groups of buildings arranged on either side of the main north–south axis. This main street, called the Street of the Dead, was some 145 feet in width and led up to the impressive Pyramid of the Moon running from the south for about 3¼ miles.

1500 BC. It was a propitious environment for eventual synoecism: '7,000 feet above sea level, high mountain chains walled in a fertile valley in which lay a great salt lake, Texcoco, fed at the south by two sweet water lagoons, at the north-west by two more, and the north-east by a sluggish stream which drained the fertile Valley of Teotihuacan. The lakes were shallow and their marshy shores, thick with reeds, attracted a teeming abundance of wild fowl. On the wooded mountain slopes deer abounded. During the rainy season thick alluvial deposits, ideal for primitive agriculture, were washed down along the lake shore'.[54]

Tenochtitlan

The Tenochas were a small but warlike group living on islands in Lake Texcoco who became the most successful of a number of competing predatory tribes. Their main settlement was Tenochtitlan, established around AD 1325 on an island located 3½ miles from the south-western shore and about 1 mile south of a second small island, Tlateoco, which is believed to have been occupied from about AD 1200.[55] The two islands gradually coalesced into the fragmented complex of islands and canals which, through the domination of the Tenochas, came to be known as Tenochtitlan. Two processes determined this topographical change: first, the gradual natural drying out of the lake; and second, the formation of intensively cultivated *chinampas* (floating islands) using the extremely fertile soil dredged up from the lake bed (a technique still to be seen south of the modern city, at the 'floating gardens' of Xochimilco). When absorbed into the one extensive urban area, Tlateloco became the main sub-centre and the district for the *pochteca*, a class of merchant travellers.

During the reign of Moctezuma I (1440–68) the Aztec state greatly extended its boundaries and Tenochtitlan rapidly developed as the political, military, religious and cultural capital. An aqueduct was built to bring in pure water and a dike 10 miles in length was constructed to prevent flooding and to minimize effects of increasing lake-water salinity on the cultivated *chinampas*. The cruciform basis of the city's quasi-gridiron plan was probably determined by the preceding Aztec king, Itzcoatl (1498–1540), who constructed the three causeways linking the islands to the mainland.[56]

Figure 9.27 – Tenochtitlan (Mexico City): the extent of the Aztec city in 1519, on the island in Lake Texcoco approached by causeways, with the route followed by Cortés shown by a heavy dotted line. (Teotihuacan was located some 12 miles (19 km) north-east of the northern end of the Lake.)

55. For Tenochtitlan, and Mexico City, which replaced it, see J Kandell, *La Capital: The Biography of Mexico City*, 1988; B Trueblood, *Mexico City Architecture*, 1978; Goitia and Balbas, *Planos de Cuidades Ibero-americanas y Filipinas*.
56. See R C Smith, 'Colonial Towns of Spanish and Portuguese America', *Journal of the Society of Architectural Historians*, vol. XIX, 1955, who discusses Aztec uses of the 'gridiron'; see also Chapter 1 where that urban form determinant is introduced. My view is that the Aztec planning of straight causeways and main routes at right-angles should not be regarded as 'gridiron planning', *per se*.

Mexico City

The Spaniards under Hernando Cortés arrived at the southern edge of Lake Texcoco in early November 1519, having first viewed Tenochtitlan from the enclosing mountains. They entered along the south causeway on 8 November, to be met by Moctezuma himself, accompanied by a thousand nobles. Cortés failed in a first attempt to take the city by ruse and it eventually fell after siege on 13 August 1521. Because of its 'renown and importance', as Cortés astutely recognized, the strategic and topographical disadvantages of the site were disregarded and he decided to rebuild the city as the new Spanish capital.[57] While Indian labourers completed razing the major Aztec buildings, Cortés established the first government of New Spain on the mainland, south of the city, at Coyoacan.

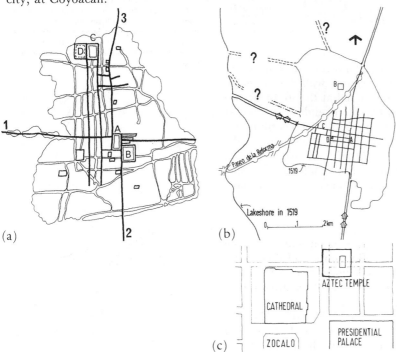

(a)   (b)   (c)

Figure 9.28 – Tenochtitlan (Mexico City) as three sketch maps.

A – a reconstruction of the plan of the 1520s when Cortés entered the city. It occupied a large number of islands which comprised an urban unit with many waterways (as Venice in Italy) and few streets, the most important of which are in white. Key: 1, causeway to the western shore of Lake Texcoco, a distance of some 1½ miles in 1520; 2, causeway to the south, 4 miles in length; 3, causeway leading from Tlatelolco, the northern of the city's twin nuclei, to the northern shore of the bay in which Tenochtitlan was located; A, the Great Temple of Tenochtitlan; B, the palace of Montezuma II; C, the Great Temple of Tlatelolco; D, the market place of Tlatelolco.

B – Mexico City as rebuilt by the Spaniards showing the main street pattern, the lakeshore line of 1519 and the causeways (known and conjectural) – the broad arrows are as Cortés's first advance from the south, and his subsequent retreat to the west. The modern Pasea de la Reforma is located across the north-western side of the Spanish city. Key: A, the Zocalo (the modern Plaza de la Constitucion), the city's main square; with the cathedral on the northern side; B, the location of the Tlatelolco market centre; C, the modern Almeda Park occupying the site of a Spanish market place where heretics were burned during the Inquisition; D, the modern Latin American skyscraper.

C – the detail plan of the city centre relating the location of the Spanish Cathedral to the site of the Aztec Temple, and other buildings around the Zocalo (The Plaza).

Then Montezuma took Cortes by the hand and told him to look at his great city and all the other cities that were standing in the water and the many other towns and the land around the lake ... So we stood looking about us, for that huge and cursed temple stood so high that from it one could see over everything very well, and we saw the three causeways which led into Mexico ... and we saw the (aqueduct of) fresh water that comes from Chapultepec, which supplies the city, and we saw the bridges of the three causeways which were built at certain distances apart ... and we beheld on the lake a great multitude of canoes, some coming with supplies of food, others returning loaded with cargoes of merchandise, and we saw from every house of that great city and of all the other cities that were built in the water it was impossible to pass from house to house except by drawbridges, which were made of wood, or in canoes; and we saw in those cities Cues (temples) and oratories like towers and fortresses and all gleaming white, and it was a wonderful thing to behold!
(B D del Castillo, True History of the Conquest of New Spain, c 1560, various English editions and widely quoted)

The plan of the new city was a more or less regular gridiron following the original orientation and incorporating major Aztec streets, notably that of the main causeway to the south. Ground level generally was raised on rubble fill, and Aztec walls provided substantial foundations for Spanish buildings, in particular those around the Zocalo, or central square, where the cathedral on the northern side was constructed above the sacred precinct, and the presidential palace to the east which took the place of Moctezuma's royal palace. Lake Texcoco has also long since virtually disappeared, the result both of continued infilling and natural drying out. Of all the world's historic cities, it is Mexico City that bears least resemblance to its original location and form.[58]

NEW SPAIN

In the northern regions of Mexico the Spaniards came up against first the Chichimecas ('wild people') and then the Apaches and Comanches who learnt to fight on horseback and who proved unconquerable. Their stubborn resistance and the arid terrain was unsuited to the *encomienda*

57. Cortés also recognized that in order to facilitate conversion of the Indians to Christianity it was necessary to erase the Aztec religious symbols, notably the Great Temple, for which see E M Moctezuma, *The Great Temple of the Aztecs*, trans. Doris Heyden, 1988.

In Tlatelolco, the sister city of Tenochtitlan, the lower levels of the temple pyramid remain as a platform for the Church of Santiago de Tlatelolco, for which see M P Weaver, *The Aztecs, Maya and Predecessors*, 1981.

58. Boston, Massachusetts, and, arguably the more so, Bombay, are other waterside cities the extents of which have changed greatly as results of landfill construction activity. (But see G Pendle, *A History of Latin America*, 1976/1990, who notes that the Anahuac basin is not naturally drained to the sea and that a ditch and a tunnel were dug for that purpose in 1607–8.)

system, and those factors also combined to limit production of the silver mines discovered in New Galicia; that at San Luis Potosi, for example, 'being left unworked for many years because of danger of native attack'.[59] The mines at Zacatecas and Guanajuato, discovered in 1548, however, could be fully exploited and numerous variously short-lived shanty towns were established. The cities of New Spain were mainly founded near centres of peaceful indigenous population with gridiron plans which were characteristically modified in the course of time. Spanish mining towns were typically unplanned, as also were the international ports of Vera Cruz and Acapulco, which really came to life only when the respective Atlantic or Pacific fleets were in port.[60]

The viceroyalty of New Spain was created in 1535 with its capital at Mexico City; the university there was founded in 1551. By 1591 the city's population was 50,000; it had reached 112,926 in 1790. New Spain was attractive to Spanish emigrants and the white (or reputedly white) population steadily increased from, possibly, 1,000 in 1520, to 125,000 by 1650, and 1 million in 1800. During the same period 1520–1650, however, the native Indian population was reduced from approximately 25 million to a little more than 1 million.[61]

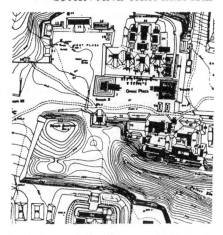

Figure 9.29 – Tikal: the enlarged plan of the western part of the central area. Coe notes: 'within a little over six square miles there were about 3,000 structures, ranging from lofty temple pyramids and massive palaces, to tiny household units of thatch-roofed huts' (M D Coe, The Maya, 1966/71).

The Mayas

The *conquistadores* found the Maya living in decline in the tropical forests and mountainous areas of Guatemala, Honduras and Yucatan in what is present-day southern Mexico. Daniel notes that 'they have been called the Greeks of the New World, and in their achievements they certainly outstripped other American pre-Columbian high cultures'.[62] However, although they had considerable knowledge of mathematics and astronomy (their calendar had leap year corrections almost as accurate as those of the present day), they had no metal technology except that which produced a few gold and copper ornaments. Their magnificent stone buildings were fashioned with stone implements, and their agricultural methods were comparably rudimentary, being based on the 'slash and burn' clearance of small fields in the jungle (the *milpas*) which had a fertile life of only a few years.

The generally agreed periods in Maya history comprise the Formative (1500 BC to AD 150) during which time village settlements were established in all Maya areas; a short Proto-Classic (AD 150–300); the Classic (300–900); and the Post-Classic phases which lasted until the Spanish Conquest of the 1520s.[63] Maya civilization reached its peak of achievement during the Classic period, notably in the lowlands. Subsequently it went through a series of retrogressive phases and was effectively a dead culture by the time of the Spanish invasion. Michael Coe believes that the 'great culture of the Maya lowlands during the Late Classic period is one of the "lost" civilisations of the world, its hundreds of ceremonial centres buried under an almost unbroken canopy of tropical forest'.[64] Opinions differ as to whether or not these centres amounted to 'cities'. Coe is in no doubt: 'none of the great sites . . . were anything of the sort'; Hardoy is uncertain, however, observing 'in every Mayan settlement a clear differentiation has to be made between (1) the central complex; (2) an intermediate sector; and (3) the agricultural countryside. Because of their high concentration of buildings and superior architecture the central complex and the intermediate sector may be qualified as urban by comparison with the countryside, although

The Maya are hardly a vanished people, for they number an estimated two million souls, the largest single block of American Indians north of Peru. Most of them have resisted with remarkable tenacity the encroachments of Spanish American civilization. Besides their numbers and cultural integrity, they are remarkable for an extraordinary cohesion. Unlike other more scattered tribes within Mexico and Central America, the Maya are confined with one exception to a single, unbroken area that includes all of the Yucatán Peninsula, Guatemala, British Honduras, parts of the Mexican states of Tabasco and Chiapas, and the western portions of Honduras and El Salvador. Such homogeneity in the midst of a miscellany of tongues and peoples testifies on the one hand to a lack of interest on the part of the ancient Maya in military expansion, and on the other to their relative security from invasions by other native groups.
(M D Coe, The Maya, 1971)

59. Elliott, *Imperial Spain 1469–1716*.

60. Reasons for the organic growth exceptions included the difficult hilly topography and their characteristically spontaneous occurrence on a basis of scattered mine-holdings.

61. For population loss see Sanchez-Albornoz, *The Population of Latin America*, 1974; and N D Cook, 'Demographic Collapse: Indian Peru 1520–1620', *Journal of Historic Geography*, 1954.

62. Daniel, *The First Civilisations*.

63. For the Maya see M D Coe, *The Maya*, 1966; G F Andrews, *Maya Cities: Placemaking and Urbanisation*, 1975/1977; S G Morley, G W Brainerd and R J Shearer, *The Ancient Maya*, 4th edn, 1983.

64. Coe, *The Maya*.

they had neither the density, nor the layout, nor visual characteristics of what would be considered as urban today, or even then, among other Middle American cultures'.[65]

Densities in the *milpas* areas of Mayan settlements were comparatively extremely low, perhaps two or three persons per hectare, living in scattered, unplanned groups of generally fewer than five houses. In the intermediate zones the density of houses and *milpas* was markedly higher and temples and palaces also appear. A description of a typical central complex is given by Coe as consisting of a series of stepped platforms topped by masonry superstructures, arranged around broad plazas or courtyards.[66] In the really large sites, such as Tikal, there may be a number of building complexes interconnected by causeways. 'Towering above all are the mighty temple pyramids built from limestone blocks over a rubble core ... the bulk of the construction however is taken up by the palaces, single-storied structures containing plastered rooms, sometimes up to several dozen in the same building.'[67] The central area of the ruins at Tikal, a meticulously mapped site of around AD 900, is illustrated in Figure 9.29.

PERU AND THE INCAS

The empire of the Incas ultimately extended along the Pacific coast of South America for some 2,500 miles, from Equador in the north to midway down the coast of Chile in the south. It included the narrow seaboard plains but was essentially a highland civilization centred on the valleys and plateaux of the Andean range. The Incas had no form of writing but could keep arithmetic records by means of knotted strings called quipus. Daniel tells us of their technology that 'the Incas were in a full Bronze Age when discovered by the Spaniards. Their craftsmen made knives, chisels, axes of a mixture of copper and tin, but these tools were (often) not hard enough to cut rock'.[68]

Although it is believed that indigenous plant domestication had reached its maturity by the third millennium BC, and that agriculture production along the Peruvian coastline was rapidly increased following the introduction of maize corn from Mexico around 1500 BC, little is known of pre-Inca urban cultures during the first millennium AD. The period 1000–1400 is known as the 'urbanist period' in Peruvian history, during which many 'city states', the results of protracted synoecism, struggled for local mastery before gradually being drawn into the one Inca empire. Most valleys supported their own urban nucleus, of which Chan-Chan (alongside the Pacific near present-day Trujillo) grew to be the capital of a sizeable kingdom (Figure 9.30).[69]

The urbanist period was followed by the 'imperialist period' from 1440 to 1532. Inca history was traditional and was based on royal reigns, commencing in about AD 1200 with eight undated and generally insignificant emperors, followed by five rulers of the Inca empire proper. The first of these was Pachacuti (the ninth in succession) whose reign lasted from 1438 to 1471. Pachacuti was both a great military leader and apparently a skilled urbanist. From their capital Cuzo, he and his son, Topa inca Yupanqui (1471–93), extended imperial power for 2,500 miles along the coast and over an area of about 350,000 square miles containing, possibly some 6 million inhabitants. Inca history credits Pachacuti as the 'planner' of Cuzco. Betanzos writes: 'thus it was done; Inca Yupanqui (Pachacuti) planned the city and ordered clay figures made just as he thought to build and make it'.[70] He is

Figure 9.30 – Chan-Chan: the general arrangement of the citadels.

Within the urban area of Chan Chan are crowded dwellings, small compounds, isolated burial platforms, a large area of melted earth, wells and, on the seaward side of the site, a vast cemetery. Along the eastern and northern edges of the site are three large mounds that were originally truncated pyramids. In the central part of the site are ten large compounds called ciudadelas. The best preserved architecture and most sophisticated planning at Chan Chan is found within the ciudadelas. A

65. Hardoy, *Two Thousand Years of Latin American Urbanisation.*

66. Coe, *The Maya.*

67. Coe, *The Maya.*

68. Daniel, *The First Civilisations.* For the Inca see also J A Mason, *The Ancient Civilisations of Peru,* 1957/ 1969. For the Spanish cities in Peru, see R A Gakenheimer, 'The Peruvian City of the Sixteenth Century', in G H Beyer (ed.) *The Urban Explosion in Latin America,* 1967.

69. For Chan Chan see K C Day, 'Urban Planning at Chan Chan', in P J Ucko, R Tringham and G W Dimbleby (eds), *Man, Settlement and Urbanism,* 1972.

70. H de Betanzos, 'Suma y Narracion de los Incas', 1551, quoted in G Gasparini and L Margolies, *Inca Architecture,* trans. P J Lyon, 1980.

reputed to have greatly increased the size of the city, and to have demolished all the villages for 6 miles around to increase its agricultural land. The old village of Cuzco itself was evacuated to lay out the plan of the new city.

After leaving Panama for Peru in December 1530 (as above), Pizarro's expeditionary force landed at Tumbez in mid-January 1531. He marched south down the coast conquering the native towns and founding San Miguel as a temporary base in 1532, before turning inland to the important Inca city of Caxamarca which he entered on 15 November 1532. The next year his augmented force set out in September for Cuzco, the Inca capital, making use of the excellent Inca system of main roads.

complex system of passages inter-connects the ciudadelas and a system of roads leads from the centre of the site to other parts of the valley and beyond.
(K C Day, 'Urban Planning at Chan Chan, Peru', in P J Ucko, R Tringham and G W Dimbleby (eds), Man, Settlement and Urbanism, 1972)

The organized nature of urban colonization in the Viceroyalty was fundamental to the incipient urban pattern. The conquerors did not consolidate fronts close to the shores against the alien environment, pushing cautiously inland and stabilizing their achievements behind – with the knowledge that much more land lay beyond than behind. Instead they advanced immediately to the loci of indigenous habitation in the central mountains and expanded from these points back to the coast and in other promising directions, gaining a fairly full perspective on the potentialities of the entire region in a short time. Thus, mineral wealth and arable land were soon made scarce – scarce in the sense that they were finite, that the quantities of them were known to be limited. There was no frontier.
(R A Gakenheimer, The Peruvian City of the Sixteenth Century, in G H Bayer (ed.), The Urban Explosion in Latin America, 1967)

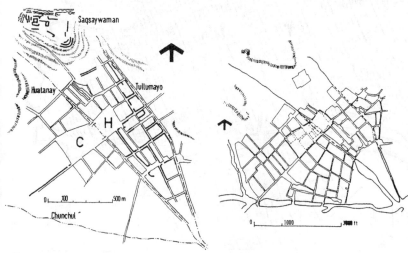

Figure 9.31 – Cuzco (far left): the general form of the pre-Columbian city in 1533; 'H' denotes the Plaza of Haucaypata, and 'C' the Plaza of Cusipata.

Figure 9.32 – Cuzco (centre): redrawn plan of the city at the end of the nineteenth century, showing the combined Inca plazas in thick outline, with the new infilling Spanish house blocks shown in dotted line.

Cuzco

The city is located on a ridge between the two rivers Huatanay and Tullumayo with the higher ground fortress of Saqsaywaman at the north-western end of its elongated rectangular form.[71] Figure 9.31 shows the extent of the Inca city as it is believed to have existed when the Spaniards arrived in 1533. It is possible that only a first phase of a larger city plan had been completed by then and that provision for expansion south-west of the Huatanay as far as the Chunchuyl River had been made by Pachacuti. Garcilaso supports that theory: 'to the west of the Huatanay River the Inca Kings had not built; there were only the surrounding outlying settlements ... that site was set aside so that the succeeding Kings might build their houses'.[72] Cuzco was taken by Pizarro in a single afternoon in November 1533, the inhabitants, lacking effective leadership, accepted passively and had not even occupied the Saqsaywaman citadel.

Figure 9.33 – Callao: redrawn from the 'Plan of the Town of Callao, as it was before the Earthquake in 1746'; the road to Lima leading away to the north-east (north is to the bottom left). The substantial fortifications, strengthened by five seaward and eight landside bastions, defined the irregular extent of the gridiron plan of the seventeenth-century town; outside which there became established the two native suburban settlements: the 'Old Petipiti' (OP) and the 'New Petipiti' (NP). Key: M, the Mole; A, the Administration; WP, the watering-place. (The Parish Church is located one block directly in from the Mole.) The waterfront length of the town was approximately 1,200 yards (1,100 metres).

Lima

In contrast to Cortés's reinstatement of Aztec Tenochtitlan as Mexico City, capital of New Spain, Cuzco was not destined to become the capital of Spanish Peru. Pizarro regarded it as being too isolated from Pacific seaboard supply bases, and instead, on 18 January 1535, he founded Lima as a new capital city on the south bank of the River Rimac, conveniently near the ocean.[73] Pizarro is reputed to have personally traced out the plan on the ground, in the form of a central plaza

71. For Cuzco see Gasparini and Margolies, Inca Architecture; F Katz, 'A Comparison of Some Aspects of the Evolution of Cuzco and Tenochtitlan', in R P Schaedel, J E Hardoy and N S Kinzer (eds) Urbanisation in the Americas from its Beginnings to the Present, 1978; M Webb, The City Square, 1990.
72. G de la Vega, Royal Commentaries of the Incas (1609 and 1617), trans H V Livermore, 1966.

311

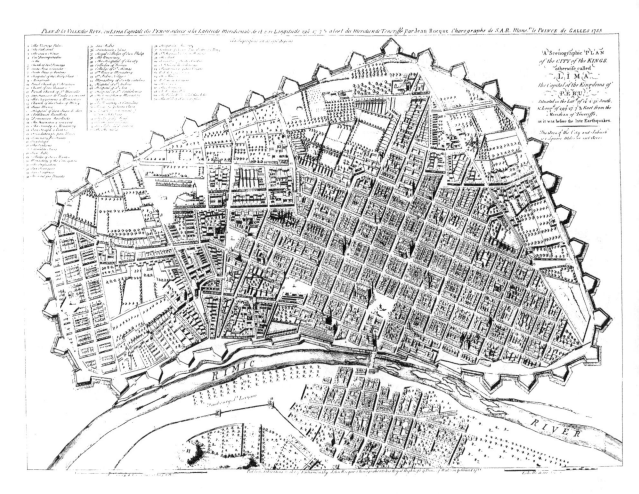

'A Scenographic 'PLAN of the CITY of the KINGS, otherwise called LIMA the Capital of the Kingdoms of PERU

(the Plaza de Armas) and a total of 117 gridiron blocks, the plaza located one block in from the riverbank and towards the eastern end. As first Viceroy of Peru he took up residence in 1544 and an archbishop arrived the next year. The University of San Marcos – the first in South America – was founded by Charles V in 1551. Nevertheless, during its early years Lima was an unimpressive city, lacking the architectural monuments of Mexico City, until, from the 1560s it had control of the vast wealth of the silver mines at Potosi.

Huanuco Pampa, Potosi and La Paz

In addition to Cuzco and Lima, three other Spanish settlements in Peru are of particular interest. The first is Huanuco Pampa within whose vast open plaza was laid out a short-lived Spanish gridiron plan, as illustrated and described in the caption to Figure 9.35.[74] The second is Potosi (in modern Bolivia) where a prodigiously rich silver mountain was discovered in 1545. Its full exploitation from the 1560s created the shanty-town at Cerro de Potosi as the largest concentration of 'urban' population in the Spanish empire, totalling some 140,000 inhabitants by 1630. Potosi's fame was also comparatively transient and its population had declined to about 30,000 by the early eighteenth century.[75] The third is La Paz, which had its origin in 1561 as a small way-station at the southern end of Lake Titicaca, on the long and difficult route north along the Andes from Potosi to Cuzco.[76]

Figure 9.34 – Lima: 'A Scenographic Plan of the City of the Kings, otherwise called Lima – as it was before the late earthquakes,' published by John Rocque in 1755. (The major earthquake was that of October 1746.) North is at the bottom, the Rimic flows from east to west (left to right). The overall east–west dimension is 2¼ miles; the north–south dimension from the bridge to the furthest bastion is 1¼ miles. The area enclosed within the fortifications of the 1680s is given as 1,306 acres (52 ha). The considerable proportion of open garden space within the fortifications, some three-quarters of a century after their construction, reveals that growth during that period had been slower than anticipated.

Lima in 1755 had a population of 54,000; other figures are as follows: 1561 – 2,500 in the city proper and approximately 100,000 in the city region (the 2,500 can be safely assumed to have been the Spanish settler population); 1599 – 14,262; 1614 – 25,434; 1700 – 37,234; and 1812 – 64,000.

Mid-eighteenth-century Lima exemplifies the type of city which having been founded on the basis of a plan, subsequently outgrew that controlled origin through

73. For Lima see J Lockhart, Spanish Peru 1532–1560, 1968; J Descola, Daily Life in Colonial Peru 1710–1820, 1968.
74. For Huanuco Pampa see C Morris and D E Thompson, Huanuco Pampa, 1985. Huanuco Pampa was taken over in 1539 as The Very Noble and Royal City of the

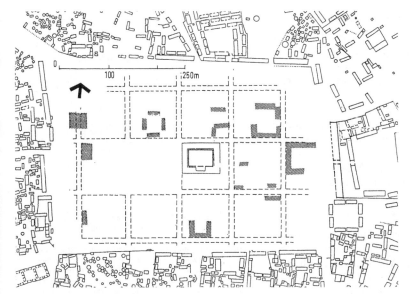

CHILE

Chile was settled directly from Peru by Valdivia who arrived in late 1540 in command of 200 Spaniards and a large force of Peruvians. In the face of determined native resistance he advanced inland to found the city of Santiago de Chile on 12 February 1541, which soon became

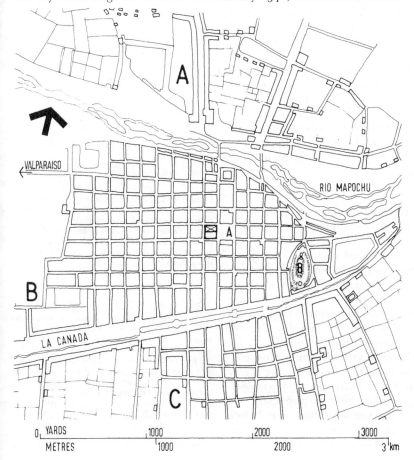

addition of spontaneous organic-growth 'suburbs'. Pizarro's plan provided for nine ranks each of thirteen grid blocks, inland from the river. Of the total of 117 blocks, only 67 can be regarded as having been regularly developed before the plan lost its controlling status. The Viceroy's Palace on the northern side of the Plaza would have controlled the river bridge in its secondary role as citadel; the Cathedral occupies the eastern side and the Town House the western side; only to the south is there the central street stipulated by the Laws of the Indies. Lima's foundation in 1535 predates the codified issue of those city planning regulations, but Pizarro can be expected to have had available an earlier version. (Map by courtesy Historic Urban Plans, Ithaca, New York.)

Figure 9.35 – Huanuco Pampa: the vast 'plaza' at the centre of the Inca city, measuring 540 × 370 metres on plan, showing the grid blocks laid out by the Spaniards within its space, as the basis of their short-lived occupation of the site. The new Spanish plaza was intended to be formed of the two grid blocks west of the Inca usnu (temple) on which dominant platform the Spanish Cathedral was to have been constructed.

Figure 9.36 – Santiago de Chile: redrawn from the 'Plano de Santiago – 1831' by Cl. Gay, and printed in Paris. The city of 1541 was located between the major Rio Mapocho and the minor stream La Canada, at their confluence. The regular gridiron plan was orientated to suit the shape of the triangular site, and is not in accordance with the Laws of the Indies. The main square – the Plaza de Armas (previously the Plaza de la Independencia) – is centrally located (small 'A') with the cathedral at the western corner. The three suburbs of 1831 are: A, La Chimba – the bridgehead settlement across the Rio Mapochu; B, Chuchunco; C, La Canadilla. The Alemeda de la Canada, along the southern side of the city, is the renamed present-day Alameda Bernado O'Higgins.

Knights of Leon of Huanuco with the projected Spanish buildings contained within the vast Inca plaza. However, the Lima authorities jealously guarded their interests and forced the abandonment of Huanuco Pampa, its settlers moving to Huanuco, located about 150 km distance by road.

75. Of Potosi, Gakenheimer has written that it was 'prevented from ever achieving the rank of a city – even when its mines had made it the largest place on the continent – by pressures from the city of Sucre, in whose jurisdiction it was located' ('The Peruvian City of the Sixteenth Century', in G H Beyer (ed.) *The Urban Explosion in Latin America*, 1968). See also L Hanke, 'The Imperial City of Potosi: Boomtown Supreme', in L Hanke (ed.) *The History of Latin American Civilisation, Volume 1, The Colonial Experience*, 1967; L Hanke, *The Imperial City of Potosi*, 1956.
76. For La Paz see F Violich, *Cities of Latin America*, 1944.

313

established as 'a modest but soundly based farming community in one of the loveliest and most fertile valleys in the world'.[77] Santiago is described with the plan as Figure 9.36. Valpariso was founded in 1544 as the main supply port, and La Serena (1544) and Concepcion (1550) were the respective northern and southern frontier cities of the early Spanish colony. Valdivia was killed by natives in 1553 and recurrent uprisings through to independence in 1817 required a permanent Spanish military presence, the cost of which from 1600 made Chile a 'deficit area' within the empire, requiring Crown subsidies. From Chile a first expedition across the Andes into present-day Argentina resulted in the founding of Mendoza (1561) and Tucuman (1565).

South American agricultural colonies

Mexico and Peru were colonies with primarily mineral-based economies and as such they were accorded appropriate strategic status and favour. By comparison, the vast South American agricultural lands of Venezuela, New Granada, and along the Rio de la Plata, were the neglected colonies of the Latin American Empire.[78]

VENEZUELA

The coastline had been sighted by Columbus in 1498 and in the following year Ojeda, accompanied by Amerigo Vespucci, carried out a detailed seaboard exploration, naming the country 'little Venice' after the huts built on piles above the swampy ground, but finding nothing to warrant immediate settlement. Only in 1527 was the coastal city of Coro founded as the first seat of government. In 1550 Venezuela became a captaincy-general, the capital of which was moved in 1576 under the name of Santiago de Leon de Caracus, for which see Figure 9.37.[79] Barcelona was founded in 1617 and became a major centre of agricultural produce, including cocoa which had been illegally introduced. Spanish government restrictions generally on agriculture and prohibition of exports, constrained the Venezuelan economy until more liberal trading policies were introduced in the eighteenth century. Other early settlements of note were Maracaibo – one of the main ports for the Carrera – and inland Trujillo.

NEW GRANADA

Santa Marta, founded in 1525 as the first permanent settlement, amounted initially to little more than a slave-catching post and New Granada effectively dates from the creation of Cartagena in 1533 (see Figure 9.38). Although not a major commercial harbour, Cartagena was a naval and military base of great strategic importance and from the mid-sixteenth century it was strongly fortified.[80] By the latter part of the century the victualling requirements of the Carrera convoys, and its own needs, had opened up the fertile and temperate farmlands of Antioquia, which were settled largely by northern Spaniards as 'one of the few genuine colonies of agrarian settlement in the Indies'.[81]

Expansion into the hinterland took place virtually simultaneously from north to south, with Gonzalo Jimenez de Quesada forcing a way up the valley of the Magdalena from Santa Marta in 1537 with 800 followers and 100 horses, into the populous, high savannahs of the Chibchas with their capital at Bogota, arriving some two years before Belalcazar who had marched north from Quito by way of Popayan. Quesada was confirmed governor in Santa Fe de Bogota, which he had

Figure 9.37 – Caracas: the original twenty-five gridiron blocks of the averagely mundane 1567 plan, which were extended to about twenty-five in each direction by the end of the city's historical period. It was located at the western end of an extremely isolated valley, which although only about 8 miles inland from the Caribbean, was separated from it by the 7,000 feet mountain wall of the Cordillera de la Costa. The remainder of the valley was occupied by twenty-five to thirty extensive sugar farms which developed in time as suburban village centres on either side of the main central road alongside the river.

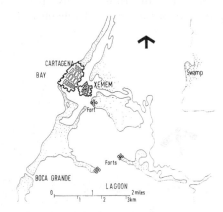

Figure 9.38 – Cartagena: the town and the related naval base on the island of Xemem, within the seventeenth-century fortifications protecting the anchorage in the Lagoon, which was approached from the open sea through the Boca Grande and a second entrance further south. Cartagena was also part of the Antoneli fortification programme (see Figure 9.22).

77. For Chile see G Guarda, 'Military Influence in the Cities of the Kingdom of Chile', in R P Schaedel, J E Hardoy and N S Kinzer (eds) Urbanisation in the Americas from its Beginnings to the Present, 1978.
78. For the South American agricultural colonies see J M Houston, 'The Foundation of Towns in Hispanic America', in R P Beckinsale and J M Houston (eds) Urbanisation and its Problems, 1968.
79. For Caracas see T J Sanabria, 'Urbanization on an Ad Hoc Basis: A Case Study of Caracas', in G H Beyer (ed.) The Urban Explosion in Latin America, 1967; F Violich, 'Caracas: the Focus of a New Venezuela', in H Wentworth Eldredge (ed.) World Capitals; J V

founded in 1538. The city was made capital of the vice-royalty of New Granada established in 1739 when the cumbersome viceroy of Peru was sub-divided.[82]

THE RIO DE LA PLATA COLONIES

Until Charles V, fearing French pre-emption, made a contract in 1534 empowering the wealthy Don Pedro de Mendoza to conquer and colonize the River Plate, Spain had expressed little interest in the eastern seaboard. Mendoza's fleet of eleven ships carrying more than 1,200 emigrants sailed up the muddy estuary of the River Plate in January 1536. 'After a considerable search,' writes James R Scobie, 'the explorers chose a site near the mouth of a meandering stream, later named the Riachuelo, and traced the outline of a plaza'.[83] The settlement was named Nuestra Senora Santa Maria del Buen Aire, after an Italian patron saint of Mediterranean navigators. The first Buenos Aires, as it became known, was badly located and the remnants of the expedition were forced to move up-river under the leadership of Mendoza's lieutenants, Juan de Ayolas and Domingo Martinez de Irala. Irala decided to stay on the Paraguay River where he founded the colony of Asuncion. The local Indians proved peaceful and the settlement grew into a quietly prosperous centre for agricultural production, becoming the capital of independent Paraguay on 14 May 1811.[84]

Buenos Aires

By 1580 Asuncion had grown strong enough to refound Buenos Aires on a site a few miles north of the abandoned settlement. At first it was controlled from Asuncion, until in 1618 Buenos Aires was made the seat of a royal governorship with authority over an extensive area of coastal Argentina and modern Uruguay.[85] Such political status was scarcely warranted by a mud-brick 'city' of only about 1,000 inhabitants, but, as ever, the Spanish Crown was anxious to be seen to be firmly established on the vital Rio de la Plata. However, after being created Spain's fourth viceregal capital in 1776, imperial bureaucracy resulted in rapid population growth to 25,000 in 1780 and more than 40,000 by the end of the century. 'Merchants built splendid fortunes, of a size previously known only in Lima or Mexico City. The number of hides exported increased from an annual 150,000 in the 1750s to 700,000 in the 1790s. Royal revenues, largely collected at the custom-house in Buenos Aires, jumped tenfold, from 10,000 pesos in 1774 to one million pesos in 1780. Money poured into the construction of houses, which while still built of sun-dried bricks were now plastered and whitewashed, enclosing spacious interior patios'.[86] The city also began to acquire some of the intellectual and social trappings that befitted vice-regal status: 'a printing plant provided a modest supply of pamphlets and broadsides, while theatrical performances regularly entertained audiences of increasing size and sophistication'.[87]

The foundation of Buenos Aires marked the close of the continental phase in the history of the Spanish Latin American empire, following which, emphasis was accorded gradual consolidation of the vast areas gained, with only the slow expansion northwards into Texas and California (described in Chapter 10). The organization of the Latin American empire in 1784 is given by the map (Figure 9.43) which also shows the boundaries of the republics into which it was divided following the independence movements of the early nineteenth century.

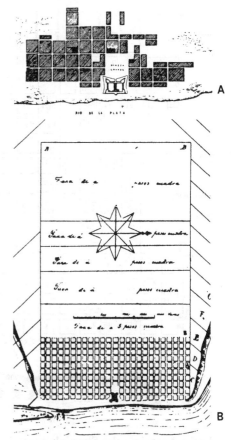

Figure 9.39 – Buenos Aires: two early plans. A – the first grid blocks around the fort and the central plaza; B – the full extent of the initial grid plan of the city, and related land allocations.

Lombardi, *Venezuela*, 1982; 'Venezuela', a special edition of *Architectural Design*, August 1969.

80. For Cartagena and the Carrera convoys, see J M Houston, 'The Foundation of Towns in Hispanic America', in R P Beckinsale and J M Houston (eds) *Urbanisation and its Problems*, 1968; D E Worcester and W G Schaffer, *The Growth and Culture of Latin America, Volume 1, From Conquest to Independence*, 2nd edn, 1970; J Goss, *Braun and Hogenberg's The City Maps of Europe: A Selection of 16th Century Town Plans and Views*, 1991.

81. For Bogota see C L Narvaez, *Santa Fe de Bogota*, 1960; Worcester and Schaffer, *From Conquest to Independence*.

82. J R Scobie, *Buenos Aires: Plaza to Suburb, 1870–1910*, 1974.

83. For Asuncion see H S Quell and J Rubiani, *Asuncion de los Recuerdos*, 1984 (early photographs and a city map).

84. Scobie, *Buenos Aires: Plaza to Suburb, 1870–1910*; S R Ross and T F McGann (eds) *Buenos Aires: 400 Years*, 1982.

85, 86, 87. Scobie, *Buenos Aires*.

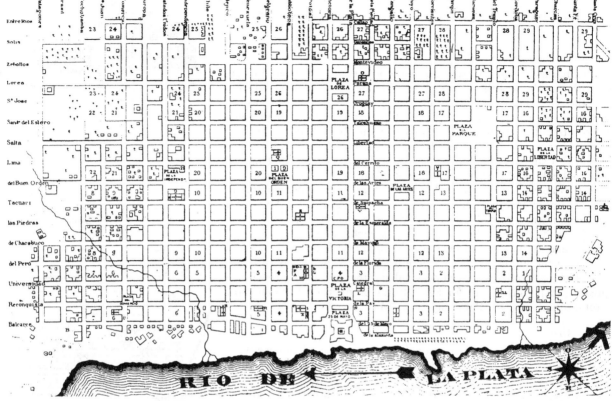

Figure 9.40 – Buenos Aires: the extent of the city as it had occupied the grid squares by 1822.

Latin America: the Spanish contribution

The wars of independence, which created the sovereign states of Latin America out of the Spanish empire, fall into two distinct phases, writes Parry, 'from 1808 to 1814 and from 1816 to 1825'; subsequently 'there was no commercial rapprochement between Spain and the Indies comparable with that between England and North America in the late 18th century. Impoverished Spain had little to offer, in goods or markets, to its former colonies; and the Creoles were in no mood to offer reconciliation, help or sympathy'.[88] The different conclusions to empire were rooted in their vastly different origins. Whereas British settlement in New England was a gradualistic process of slow expansion inland by self-supporting agricultural colonies, the Spanish 'policy' was one of earliest possible military control of centres of indigenous population; their dominion then consolidated through 'cities established in unknown and often hostile territories as centres of conquest and political control'.[89] In direct contrast to the early New England towns which emerged *to serve* (as also almost all others of the USA); Spanish Latin American cities were characteristically intended *to subdue*. From them, 'the Spaniards moved out into a hostile environment to conquer, control and to indoctrinate the surrounding populations'. As power-bases the Spanish cities were planned in response to 'the strategic requirement of concentrating scarce human resources in a restricted and therefore militarily defensible perimeter'.[90] (Numerically, the Indian missions were comparatively insignificant.)

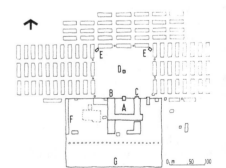

Figure 9.41 – San Jose, Chiquitos, Bolivia: the general arrangement of the buildings of this characteristic Jesuit Mission. Key: A, the college; B, the residence of the Jesuits; C, the church; D, the Great Cross in the centre of the square; E, chapels; F, workshops; G, gardens. The Indian houses comprising the main part of the Mission are outside the walled Jesuit enclosure. (See Chapter 10 for examples of Praesidios established by the Spaniards in Texas to protect their northern frontier.)

88. Parry, *The Spanish Seaborne Empire*.
89. Hardoy, *Two Thousand Years of Latin American Urbanization*.

Spanish and Portuguese colonial settlement provided republican Latin America with comprehensive, ready-made urban systems, with the great majority of present-day major cities having developed as enlarged and often distorted versions of those founded at the start of the imperial period. Of the twenty most populous Latin American cities in 1970, fifteen were founded during the years 1520–80, and the first eight were all of that period. (The comparable data for the USA reveal that only ten major cities were founded before the War of Independence: see Chapter 10.)

As negative aspects of imperial inheritance, with the exceptions of Mexico City, Bogota, La Paz and Asuncion, the coastal locations of leading republican cities has created peripheral patterns of urbanization which their governments have found hard to change. Communications focused on the capitals and provincial centres were isolated from each other, thereby encouraging centralized militaristic government, and rural drift to the capitals. Parry notes that 'another characteristic of Spanish-American government with roots reaching far back into colonial times, is the phenomenon which may best be described as "passive administration" whereby transaction of any kind of legal or public business involves dealing with numerous petty officials who are not expected to be actively helpful to the public; they are there to insist that the forms are observed'.[91]

Socially there were two worlds in Latin America; the new cities, planned exclusively for those of European descent (the *ciudades des espanoles*), and the Indian *barrios*. 'After initial confusion in the early days of colonisation in Hispaniola,' notes Houston, 'a form of apartheid was practised . . . in New Spain, no Spaniard, negro, mestizo or mulatto was permitted to live in an Indian village.'[92] In Mexico City, the separate areas for Indians were laid out in 1570, and their quarters at Lima have been described as 'no better than concentration camps'.[93] The Indian *barrios* and separate townships near Spanish cities provided the necessary supply of menial labour; further away, in the unsettled hinterlands, the Indian village communities were also urbanized, ostensibly for evangelical purposes, but also to provide pools of labour. Crowded together under strange, insanitary urban conditions and exposed to foreign diseases for which they had no natural immunity, the Indian populations of Latin America rapidly declined in numbers.[94]

At the end of the Empire, the average Spanish city was not only small, but also of characteristically modest appearance. Hardoy observes that 'during the colonial period there does not appear to have been great interest in embellishing the cities, which for the most part were only simple villages, precariously built and served . . . urban complexes of significant architectural value did not exist in colonial cities, except for those formed around the Plaza de Armas in Mexico City, or certain public gardens, such as the Alameda of Lima or Mexico City'. Of Caracas in 1804: 'the city possessed no public buildings other than those dedicated to religion. The captaincy-general, the royal *audiencia*, the Intendancy, and all the courts of justice occupied rented houses. The Army hospital was located in a private house. The accounting office or treasury was the only building belonging to the King and its construction was far from mirroring the majesty of its owner'.[95]

Figure 9.42 – The Spanish Latin American empire locating the sequential centres of conquest.

Figure 9.43 – Latin America: as it had become consolidated by 1784, on the basis of the Vice-Royalties of New Spain, New Granada, Peru and La Plata (Brazil to Portugal).

90. A Portes, *The Economy and Ecology of Urban Poverty*.
91. Parry, *The Spanish Seaborne Empire*.
92, 93. Houston, 'The Foundation of Towns in Hispanic America'; see also Schaedel, Hardoy and Kinzer (eds) *Urbanisation in the Americas from its Beginnings to the Present*, 1978.
94. Houston, 'The Foundation of Towns in Hispanic America', writes: 'the urban agglomeration of Indians in creating a pseudo-Hispanic culture ... subjected the natives to frightful mortality, in exposing them to European diseases, against which they lacked immunity. As the towns became decimated more natives were brought into them, to add to the mortality rates. Thus, whenever urbanism was promoted among the Indians, the decline of the indigenous population was catastrophic, notably in Mexico; the population there declined from about 11 million in 1520 to approximately 2.5 million in 1650'. See also Sanchez-Albornoz, *The Population of Latin America*, 1974; N D Cook, 'Demographic Collapse: Indian Peru 1520–1620', *Journal of Historic Geography*, 1954.
95. Hardoy, *Two Thousand Years of Latin American Urbanization*.

THE SPANISH PHILIPPINES

The Philippines comprise an archipelago of some 7,100 islands, of which Luzon, to the north, and Mindanao, to the south, are by far the largest at 40,420 and 36,537 square miles respectively.[96] The Spaniards had been preceded by Chinese traders, Arab Muslim missionaries and the Portuguese. By 1565, when Miguel Lopez de Legazpi arrived from Puerto de Navidad, Muslim communities in Sulu and Mindanao were sufficiently integrated to be able to defend themselves to some effect. (The Muslims were referred to as *Moros*, from Spanish homeland experience, a term which has come down to the present day.)

Cebu was founded as the first Spanish settlement in the Philippines, which were named after King Philip II of Spain. Encountering local opposition from the Portuguese, Legazpi moved his base first to Panay Island in 1569, and then further north to Luzon where Manila was founded in 1571. Following their colonial model, by then well established in Latin America, the Spaniards located the primary urban settlements where there were existing centres of native population. 'In 1565 Cebu was a minor port of call,' writes Doeppers, while 'Manila (Maynilad) was apparently the point of importation for the large Tagalog population living around Lagunda de Bay.'[97] Although Manila had a native population of the order of 2,000, including traders, a culverin founder and possibly goldsmiths, Doeppers concludes 'there is no evidence that an important fulltime artisan class had developed, or that many of these specialists were not essentially fishermen, hunters or agriculturalists. ... Manila (and certainly Cebu) was probably a relatively large agricultural and fishing village with an unusually strong secondary trade function ... lacking both the size and power of a city'.

Figure 9.44 – The main islands of the Philippines archipelago, locating the Spanish settlements at Manila, Paray and Cebu, showing Spanish dominated regions as 'S' and those of the Muslims (*Moros*) as 'M'.

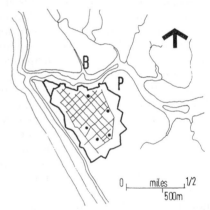

Figure 9.45 – Manila: a sketch map of the city, its gridiron street plan contained within a strongly fortified perimeter (the area known as the *Intramuros*). The Chinese trading quarter was consolidated from 1581 in the Parian (market) area alongside the *ciudad* – located as 'P'; subsequently it was removed across to the Binondo district on the other side of the Pasig River – located as 'B'.

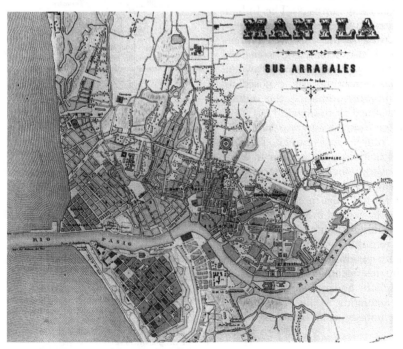

Figure 9.46 – Manila: the map of the city in 1848.

96. For background history to the Philippines see K Lightfoot, *The Philippines*, 1973; Parry, *The Spanish Seaborne Empire*.
97. For Manila see D F Doeppers, *Manila 1900–1941: Social Change in a Late Colonial Metropolis*, 1984; Goitia and Balbas, *Planos de Cuidades Iberoamericanas y Filipinas*; D H Burnham and E H Bennett, 'Plan for the Development of Manila', in *The Plan of Chicago*, 1906.

Early in their occupation, Legazpi realized that the Philippines offered neither gold nor spices in large readily exploited quantities on which to base a colonial economy. Accordingly, he opted to develop trading activities around the South China Sea, and perhaps further abroad, for which purpose Manila was best situated (see Figure 9.45 and caption description). In addition, the local agricultural potential was greater than elsewhere in the islands. By 1600 the Spaniards had established a systematic settlement pattern over the islands north of predominantly Muslim Mindanao, comprising the cities of Santismo Nombre de Jesus (Cebu – Figure 9.47), Nueva Caceres (Naga), Nueva Segovia (Lallo), and the towns of Arvelo (on Panay) and Fernandia (Vigan). Other major settlements were founded later, related to internal trading routes.

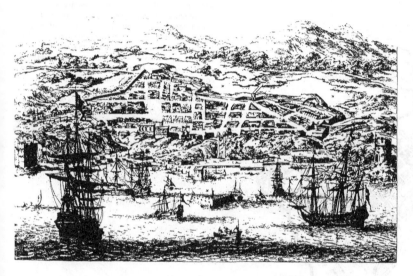

The Portuguese colonization of Brazil

The first Portuguese landfall in 1500 is believed to have been accidental: a fleet commanded by Pedro Alvares Cabral was probably blown off course around Africa, but it sufficed to claim that unknown land on the basis of the papal boundary with Spain (described above).[98] Following up, the Portuguese found little of interest. When they eventually decided to colonize Brazil (which was named after its valuable 'brazil-wood'), they did so more to deter the Spanish and the French, than with expectations of economic return. The Portuguese approach to colonization was to divide up the coastline between the Amazon and Sao Vicente in 1534 into twelve hereditary *capitanias* (captaincies) of widths varying between 30 and 100 leagues but with indefinite extension into the interior. Of the twelve *donatarios*, only two were successful in establishing relatively important centres of population and economic growth, at Sao Vicente and Pernambuco. By 1548 when the *capitanias* were abolished and replaced by a captaincy-general for the entire country, a number of ports were conducting prosperous trade with Lisbon, notably Olinda (1537), Porto Seguro (1535), Espiritu Santo and Igaracu (1536), Santa Cruz Cabralia and Ilheus (1536), and an expanded Sao Vicente. When Tome de Souza was appointed as the

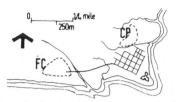

Figure 9.47 – Cebu, relating the extent of the Spanish gridiron *ciudad*, with its protecting fort on the headland, to the Chinese Parian to the north ('CP'), and the Filipino Cabacera to the west (FC).

Figure 9.48 – Bahia (Salvador): a seventeenth-century artist's impression, from which the essentially unplanned nature of the city can be seen. Topography played the major part in determining the form; 'a triangularly-shaped platform 25 metres above sea level, limited by ravines at either end' was enclosed by fortifications following the most readily defended line. Within the walls, 'the Rua Direita was the principal axis and united the two gates to the city. The other streets were also rectilinear, ending at the city walls. Bahia's ordinate plan was adjusted, therefore, writes Hardoy, 'to topographical characteristics of a site selected for defensive and strategic reasons – a radical change from Olinda, Vila Velha and other first settlements'.

... *the similarities between Spanish colonial settlements in Mexico and the Philippines are particularly striking. The distinctive features of Philippine communities are the result of the presence of large numbers of Chinese performing highly specified roles. The Chinese came initially to trade in Chinese export goods. Soon they were performing many of the services required by the colonial elite – services which the Tagalog villagers around Manila were initially unable or unwilling to provide. Thus the Chinese formed a vital economic link between the subsistence-oriented Filipinos and the external trade-oriented Spaniards. These circumstances supported the rapid growth of the Chinese community in and about Manila to an extent that the Spaniards were rapidly outnumbered.*

The capital now assumed a new social-spatial order. At first the Chinese pattern of residence was reasonably dispersed – some even resided in the walled city. But in 1580 those Chinese who were not married to Filipinas were required to live and trade in a specified area. This was ordered because the Spaniards recognized their declining defensive position and because they wanted to concentrate the merchants and craftsmen in one area in order to regularize the exchange of goods. Furthermore, the religious community was convinced that contact with 'infidels' was a 'hindrance to the growth of the faith and morals of the natives'. The initial Chinese-market district proved inadequate, and in 1581 a new site was selected in a marsh literally under the guns of Intramuros. This area was called the Parian or market. Chinese along with some Japanese were required to live and maintain their shops within its confines.
(D F Doeppers, *The Development of Philippine Cities before 1900, Journal of Asian Studies, Vol. 31, 1972*)

98. For Brazil generally see R M Morse, 'Brazil's Urban Development', *Journal of Urban History*, vol 1, 1974; D Alden (ed.) *Colonial Roots of Modern Brazil*, 1973; C R Boxer, *The Portuguese Seaborne Empire*, 1969; J Dickenson, *Brazil*, 1982; R M Morse, 'Cities and Society in 19th Century Latin America: see the Illustrative Case of Brazil', in R P Schaedel, J E Hardoy and N S Kinzer (eds) *Urbanisation in the Americas from its Beginnings to the Present*, 1978.

first captain-general in 1548, he was charged (*inter alia*) with the choice of a capital city, for which purpose he selected Bahia (later renamed Salvador).[99] Sao Paulo (1554), Rio de Janiero (1555) and Paraiba (1585) were founded in the following decades, with Rio de Janiero (Figure 9.49) taking over as the capital in 1763.[100]

Portugal was far less systematic than Spain in the foundation of colonies and no codified urban planning method was developed comparable with the 'Laws of the Indies'. Portuguese colonial administrators were acquainted with the Iberian 'new town' planning procedures familiar to their Spanish counterparts, and until the earthquake of 1755 which devastated much of Lisbon, requiring its planned renewal, their national capital had retained its medieval form and appearance. Nevertheless, expediency grid plans were a usual basis of initial land distribu-

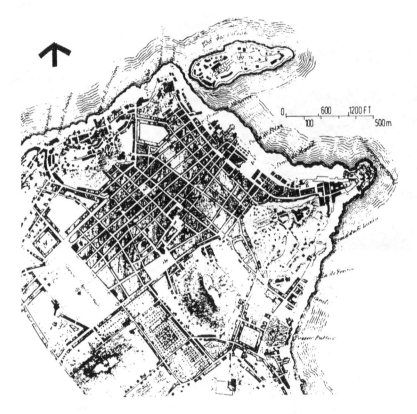

However satisfactory the Amerindian male might prove as a hunter, fisher, fighter, or slave-raider in the service of the white man in some areas, and however willing the Amerindian female might be to serve as his wife, concubine or handmaiden, it was the African Negro slave who formed the mainstay of the plantation economy in the three (relatively) populous coastal regions of Pernambuco, Bahia and Rio de Janeiro. These slaves were obtained mainly from various West African regions north of the equator before about 1550, and chiefly from the Congo and Angola during the second half of the sixteenth century.

Conditions in parts of Portugal were often such that many people had no alternative but to emigrate. Brazil, with all its drawbacks, gave them opportunities for a better life than they could hope to find at home. Portugal, no less than Brazil, suffered from capricious and ill-distributed rainfall, and from organic poverty of the soil in many regions. The mother country was severely ravaged by bouts of plague in the sixteenth and seventeenth centuries, something which did not occur in Brazil until the yellow fever visitations of the 1680s. Overpopulation and pressure on the land in certain fertile regions (Minho) in northern Portugal, and in the Atlantic islands of Madeira and the Azores, provided a constant stream of emigrants; and from about 1570 onwards increasing numbers of these people sailed for Brazil rather than for 'Golden' Goa and the East. Degredados or exiled convicts had formed 400 out of the original 1,000 settlers of Bahia in 1549; but thenceforward the voluntary emigrants heavily outnumbered those who were deported from their country for their country's good.

(C R Boxer, The Portuguese Seaborne Empire, 1969)

Figure 9.49 – Rio de Janiero: the early nineteenth-century city, showing the regular grid blocks extending inland from the linear waterfront origins.

tion, as exemplified by Rio de Janiero. In 1822, when Brazil gained its independence, there were twelve cities and sixty-six towns in the country, the largest being Rio de Janiero with a population of about 30,000. The national total of approximately 4 million, comprised 1 million free whites, about 2.4 million negro slaves, 400,000 Amerindians (reduced from the original estimated 1 million), and some 200,000 mulattos and free blacks.

99. For Bahia see Boxer, *The Portuguese Seaborne Empire*; R M Delson, *New Towns for Colonial Brazil*, 1979.

100. For Rio de Janiero see Delson, *New Towns for Colonial Brazil*; J Lockhart and S B Schwartz, *Early Latin America: A History of Colonial Spanish America and Brazil*, 1983/1989.

10 – Urban USA

This chapter covers the first three centuries of urban settlement in the United States of America, from its commencement by the Spaniards in 1565, through to the Civil War of 1861–5, which is taken as the watershed dividing the period of this volume and that of the modern industrial age in US urban history.[1] An introductory description of the Spanish, French and English settlement processes, together with general geographical, economic and social factors, is followed by accounts of individual city origins and growth during the period.

Spain, moving northwards from Mexico (see Chapter 9) was first to colonize systematically the territory of the future USA, but only on a characteristically scattered and small-scale basis, dictated, in the main, by strategic considerations. The French, who came next, initially entered down from the Canadian north but with essentially limited trading ambitions. The British – ultimately the dominant colonizers – arrived on the scene later still, in company with a number of less important nations, of whom the Dutch were the most significant. That the future USA was to become an English-speaking nation was due to their establishment from the outset of permanent agricultural settlements. Portugal was the only maritime nation capable of challenging Spain in the 'New World', during the half-century or so following its discovery by Columbus in 1492; but Portugal respected her part of the agreement of 1492 whereby Portuguese global interest ended at the line of latitude 370 leagues west of the Cape Verde Islands.[2]

Spanish settlements

The present concern is with urban origins in Florida, New Mexico, Texas and California. San Francisco and Los Angeles are introduced here, and both are further described below, at their mid-nineteenth-century stages of development. The three types of formal Spanish settlement were, first, *presidios*, the fortified military bases, garrisoned by regular soldiers with families and servants; second, *pueblos*, centres for trade and agricultural production, often based on existing Indian villages; and third, missions, the religious foundations for conversion of the Indians.

'While in theory *pueblos*, missions and *presidios* were established for different purposes, and took different forms,' writes the eminent American urban historian Professor J W Reps, 'in practice the distinction between them was not always so clear cut.'[3] Spanish colonial policy intended that missions became self-governing civil towns when their

1. This chapter accepts the view that there were no pre-European (pre-Columbian) settlements in the territory of the future USA that could be considered of 'urban' status, against one or more of the criteria discussed in Chapter 1. If current archaeological interpretation points the contrary, then such settlements could only have been exceptionally few and far between.
2. For Portugal see C R Boxer, *The Portuguese Seaborne Empire 1415–1825*, 1969/1977. For general historical chapter background see R B Nye and J E Morpurgo, *A History of the United States, Volume 1, The Birth of the USA*; *Volume 2, The Growth of the USA*, 1955 (this collaborative history by an Englishman and an American presents an eminently readable interpretation of people, trends and events which is different from that to be found in other brief histories of the United States); H Brogan, *The Penguin History of the United States of America*, 1985/1990 (as a comparatively recent publication, Brogan's Note on Further Reading is a valuable further reference); C P Hill, *A History of the United States*, rev. 3rd edn, 1973/1991; S E Morrison, *Oxford History of the American People*, 1965.
3. J W Reps, *Cities of the American West: A History of Frontier Urban Planning*, Princeton University Press, 1979: Professor Reps' definitive work which, with his earlier book, *The Making of Urban America: A History of City Planning in the United States*, Princeton University Press, 1965, has provided reference to otherwise unavailable original source material and served as the main basis of this chapter. *Cities of the American West* is essential further reading for detailed information and a wealth of illustrations of the Spanish period in American urban history, as a main part of its wide-ranging content. *The Making of Urban America* is the definitive general work on the subject through to the First World War. It is also profusely illustrated with original map reproductions and contemporary early city views.

The cities of this chapter are all included in these two books; they are acknowledged only where there are no additional references available.

Most of the maps and views contained in these books are available as high-quality cartographic reproductions through *Historic Urban Plans Inc.*, Ithaca, New York, USA. A number of these prints have been made available as illustrations in this chapter and also Chapter 9 as individually credited.

Figure 10.1 – The southern and south-western USA, related to Cuba and Mexico, and showing the broad directions of Spanish advances with their key settlements. French and English colonies are also located, with the boundaries of the future States of the USA.

Indian inhabitants had been converted and raised to acceptable levels of cultural and practical education. Their concentrations of population created local economic demands, and missions were typically quasi-urban settlements comparable with monasteries in medieval Europe. Similarly, *presidios* also generated trade and Royal regulations of 1772 specifically required their commanders to encourage such activity and also the civilian settlement of nearby lands. The *pueblos* came within the 'Laws of the Indies'. At San Antonio, Texas, the *presidio* and the *pueblo* occupied closely adjoining sites; the mission of San Antonio de Valera was only a short distance away across the river.

FLORIDA

For the half-century or so following their conquest of New Spain (Mexico) between 1519 and 1521, the Spaniards concentrated on exploiting that extensive territory (as also their vast South American lands). They had made several tentatively unsuccessful attempts to move across from Cuba to the south-eastern continental mainland of the future Florida, Georgia and the Carolinas; it was not until 1564, when a party of French Huguenots set up Fort Caroline, near the mouth of St John's River, that they were provoked into action to protect their interests. These, by then, were twofold: first, to deny any other nation the right of settlement, particularly if Huguenots; and second, the need to ensure naval mastery of the vital Florida Straits through which increasingly rich homeward-bound merchantmen convoys were sailing. With Havanna controlling the southern side, a comparable base was needed on the south-eastern coast of Florida.[4]

In 1565, Philip II ordered Menendez, Governor of Cuba, to oust the French and to plant in their place a Spanish colony numbering 500

The term 'Spanish borderlands' refers to northern Mexico and those areas of the southern United States, from Florida to California, that were once colonized by Spain. In the United States Southwest, the Spanish language is widely spoken, and Hispanic toponyms continue to be current as far as Montana and Oregon. In parts of the United States borderlands, land titles and water rights derive from original Spanish grants and surveys. Remnant mission buildings dot the landscape, and the Spanish heritage is still taken seriously in the twentieth century. The tradition embraces Old St. Augustine, Santa Fe, Capistrano, the Alamo, and other appealing elements of our folklore. It is obvious that borderlands culture has influenced our conception of all Spanish America. But there were important differences between the northern frontier and the more developed parts of the Spanish colony, and it is for this reason, as well as because sections of the United States itself are involved, that we devote our final chapter to this subject. (C Gibson, Spain in America, 1966)

4. For Florida see J H Parry, The Spanish Seaborne Empire, 1966/1977; C M Andrews, The Colonial Period in American History, 1964.

white male settlers and an equal number of slaves. The former were to comprise 100 soldiers, 100 sailors and 300 artisans, craftsmen and labourers, 200 of whom were to be married, plus four Jesuits and ten or twelve other monks. A fleet sailed from Cadiz in July 1565 and after calling in at Cuba for supplies, it arrived, writes Reps, 'at what is now the site of our earliest city on St Augustine's Day, August 28th'.[5]

On 6 September, Menendez formalized the foundation of St Augustine, on a site from which it was moved a short distance in the following year. The plan of St Augustine in about 1770 is illustrated by Figure 10.2 and described in the caption. Although the formally codified town planning section of the Laws of the Indies dates from 1573, it seems most likely that Menendez had been issued instructions defining its arrangement, perhaps along similar lines to those given by Ferdinand V in 1513 to Davila (which were probably applied at Panama City), or those issued more generally by Charles V in 1521. (See Chapter 9 for this key aspect of Spanish Latin American urbanism.)

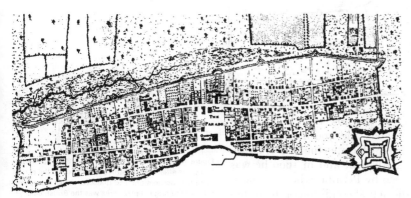

St Augustine was well located as a naval base, but it was never able to establish an adequately productive hinterland. 'Far from being self-sustaining,' writes Brown, 'Florida received direct financial aid from the Crown, and occasional delays in the receipt of annual installments of the subsidy caused distress and anxiety.'[6] At its peak the city probably numbered 3,000 inhabitants; the total population of the colony may have reached 6,000. Fear of the local Seminole Indians – 'a strong people, jealous of their rights' – restricted Spanish attempts to establish settlements in the hinterland. The inhabitants of those which had been founded by the middle of the eighteenth century were then 'sent reeling back to the safety of their coastal walls and forts'. In 1763 the Spanish government ceded Florida to the British (taking it beyond the scope of this section).

NEW MEXICO

The second oldest city of Spanish origin is Santa Fe, New Mexico, founded in 1609 by Don Pedro de Peralta, as La Villa Real de la Santa Fe de San Francisco. 'Except for a period of Indian occupation from 1680 to 1698,' Reps writes, 'Santa Fe has served continuously as capital, first of the Spanish and then Mexican province; then the Territory, and now the State of New Mexico.'[7] Settlement of New Mexico dates from 1595, when Juan de Onate was granted authority to 'settle and pacify' the valley of the Rio Grande.

In 1492 America was inhabited by a brown-skinned (usually light brown, certainly not black) people whom Columbus and his successors, because of a mistaken belief that they had found lands off or near the coasts of India, called Indians. These people were scattered fairly sparsely through the continent. In certain areas in Central and South America – the regions of Mexico and Peru to-day – they had developed what was in some respects a high level of civilisation. They had settled in fertile areas and established towns, built temples, houses and palaces, and developed an elaborate organisation of government, in addition to creating numerous objects of high artistic worth. Outside these areas, and especially in North America, the Indians lived in a far more primitive way. Permanent settlements were rarer and more scattered. The North American Indians – the Redskins – lived mainly by hunting, ranging over wide areas, each of which some Indian tribe or 'nation' claimed as its own property.
(C P Hill, A History of the United States, 1974)

Figure 10.2 – St Augustine, Florida, in about 1770. The city combined all three functions of Spanish settlement as *presidio, pueblo* and mission; the seemingly rudimentary perimeter defences of the urban area were augmented by the waterfront fort at the northern end, with the central plaza open to the water. The 1556 grid-iron street pattern is less formal than would be the case with later Spanish cities laid out in accordance with the 'Laws of the Indies' (see Chapter 9). The density of development is apparently low, with spacious garden areas at the rear of the houses.

Spanish colonisation was mainly the work of those tireless and courageous pioneers, the Jesuit missionaries. In spite of huge difficulties of climate and transport, and of Indian attacks, they gradually spread their mission stations farther northwards during the 17th and 18th centuries. A number of Spanish landowners also established themselves in California, building their haciendas, or farm-mansions, in the fertile lands of that region. There was little traffic by sea. The magnificent harbour of San Francisco Bay was not discovered until 1769, and then only by accident and overland! The whole process of Spanish colonisation was extremely slow, partly because of the decline of Spanish power in Europe after the end of the 16th century, and this early Spanish settlement of the far western and southwestern parts of the U.S.A. has left little permanent impression. Few traces of it remain, apart from many place-names and a romantic background for film scenarios: in this the Spanish legacy to the U.S.A. is not unlike that of the Red Indians.
(C P Hill, A History of the United States)

5. Reps, *The Making of Urban America*.
6. R H Brown, *Historical Geography of the United States*, 1948.
7. Reps, *The Making of Urban America*; see also R E Twitchell, *Old Santa Fe*, 1914.

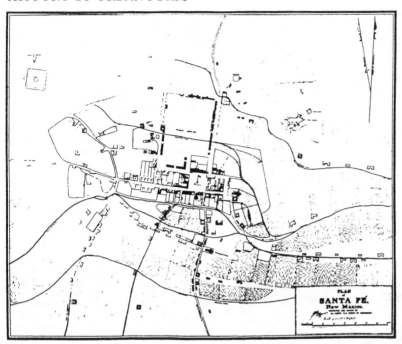

Figure 10.3 – Santa Fe, New Mexico, in about 1846: surveyed and drawn by 1st Lt J F Gilmer, US Corps of Engineers, during the Mexican War (north at the top). The eastern half of the plaza had been built over, thereby reducing the original rectangle to a square of sides 325 feet, with the Governor's House across the northern side. The parish church therefore remained outside the plaza, to the east, where it terminated the vista along the Calle de San Francisco, the town's main street across the southern side of the plaza. North of the town, the original *presidio* buildings had been enlarged to enclose an extensive parade ground some 1,200 by 900 feet in size. The American army had added Fort Marcy on a hilltop to the north-east, as a citadel commanding the town.

Travelling north from San Geronimo, in January 1598 his party of 129 soldiers (some with families), 83 wagons and 700 head of cattle, reached the river across from present-day El Paso at the end of April. Onate failed to establish a permanent settlement and in 1609 he was replaced as royal governor of New Mexico by Peralta, who was specifically instructed to 'inform himself of the actual condition of the settlements, endeavouring first of all to found and settle the villa that has been ordered built, so that they may begin to live with some order and decency'.[8] For the city itself, six districts were to be marked out, with a square block for government buildings and other public works.

Peralta selected a site on the northern bank of a minor river flowing into the Rio Grande, where he founded Santa Fe. The plan of the town in 1846 is given in Figure 10.3. It was abandoned during the Pueblo Indian revolt of 1680–93 and when recaptured, the Spaniards found 'the churches destroyed, council houses erected in the plaza and other adobe structures built by the Indians enclosing its east, south and west sides'.[9] The original plan was reinstated, but a visitor of 1776, Father Francisco Dominguez, was extremely critical: 'the appearance, design, arrangement and plan do not correspond to its status as a villa; nor to the very beautiful plain on which it lies. It is a rough stone set in fine metal'.[10]

Governor Juan Bautista de Anza, appointed in 1777, proposed raising Santa Fe and rebuilding it across the river. Thwarted in this ambition by concerted public reaction, he turned his attention towards reorganizing the settlement pattern along the Rio Grande valley, and the replanning of Santa Cruz, located 25 miles north-west of Santa Fe. These three foundations were the only attempts by the Spanish to found true cities in New Mexico, although they were augmented by numerous villages and small *pueblos* along the river valley by the end of the eighteenth century.

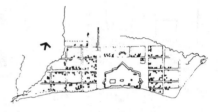

Figure 10.4 – Pensacola, Florida, in 1778, some twenty-five years after its reconstruction and during the period of British occupation. The massive citadel occupying most of the original plaza area was the British enlargement of the original Spanish fort.

8. 'Instructions to Don Pedro de Peralta ... March 30th 1609', from Louis de Velasco, Archivo General de Indias, Audienca de Mexico, legajo 27, as translated in Hammond and Rey, *Don Juan de Onate, Part II*, 1087–88 (quoted Reps, *Cities of the American West*).

9. Twitchell, *Old Santa Fe*; Reps, *Cities of the American West* (note 46 on page 47), adds from Governor Diego de Vargas's description of Santa Fe on the day of his reoccupation: 'it had only one gate; its entrance built and constructed for the defence of its ravelin, a redoubt entrenched above in the form of a half-tower with its trench ... with two plazas and its dwellings three storeys high and many of four, and in truth most perfectly planned in its capacity and amplitude', quoted from the *Spanish Archives of New Mexico*, II, 117.

Santa Fe is of exceptional interest in American urban history because it is the most important regularly planned colonial settlement to have been subsequently occupied by Indians for a significant period of time. In their evident disregard for European planning niceties, through superimposition of their organic-growth pattern, there is an intriguing American parallel with fates suffered by numerous planned settlements of the Roman Empire after its fall (see Chapter 3) and by Damascus when it became an Islamic city (see Chapter 11).

10. E B Adams and Fray Angelico Chavez (eds and trans) *The Missions of New Mexico*, 1956 (quoted Reps, *Cities of the American West*).

TEXAS

The record of settlement in the south-eastern part of the modern State of Texas, is that of indecision and belated, half-hearted attempts by the Spanish to secure their border lands. Herbert Bolton has observed of this period: 'fears of aggression largely inspired the activities of the Spanish government in Texas; when such fears slept, Texas was left pretty much to itself'.[11] As with Florida, it was the French who first threatened their interests when La Salle, in 1682, having journeyed

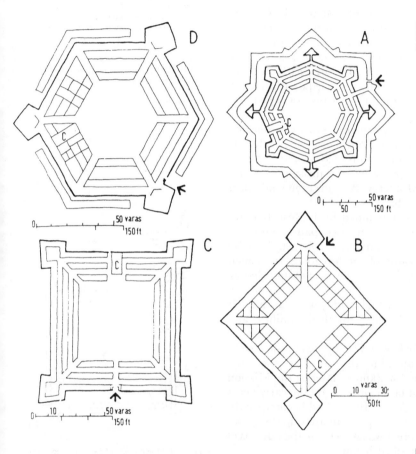

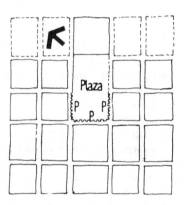

Figure 10.5 – Four Texan *presidios*: A, Los Dolores de los Tejas; B, Los Adaes; C, San Antonio de Bejar; D, Loreto en la Bahia del Esperitu Santo.

Figure 10.6 – San Antonio, Texas: the plan of 1730. John Reps writes 'it is evident that the Laws of the Indies governed most features of the town plan. The orientation of the streets so that the corners of the rectangular blocks faced the cardinal points of the compass were as specified; so too were the dimensions of the plaza which if the bordering streets were included measured exactly 400 by 600 Spanish feet as appropriate for "a well-proportioned medium size plaza". The plan also showed an arcade (*portales*) around the plaza as called for in the Laws'.

down the Mississippi, named the territory at its mouth Louisiana for the French King. (That was also the year of the British foundation of Philadelphia.) Although La Salle's attempted colonization of 1684 was a failure, further attempts were rumoured, prompting the leader of a Spanish exploration party, Alonso de Leon, to recommend the establishment of a line of *presidios* and related missions in eastern Texas. One mission, that of San Francisco de los Tejas (near modern Weches), was founded in 1690, some 300 miles from the nearest Spanish base, in the belief that the local Indians were peaceful. They were not, and after three years the mission had to be hastily abandoned.

In November 1698 renewed French threats resulted in the Spanish foundation of Pensacola, on an offshore island about halfway between Florida and the Mississippi. It was captured by the French in 1719, restored to Spain four years later, and then rebuilt on a new site after

11. H E Bolton, 'Spanish Activities on the Lower Trinity River, 1746–71', *Southeastern Quarterly*, vol 16, 1940; see also H E Bolton, *The Mission as a Frontier Institution in the Spanish American Colonies*; H E Bolton (ed.) *Spanish Exploration in the Southwest 1542–1706*).

hurricane devastation in 1754, it was held by the British between 1763 and 1781. The plan (redrawn as Figure 10.4) shows the rebuilt town in 1778.[12] Three years after Pensacola, the French constructed their Fort Louis, located on Mobile Bay, only some 55 miles to the west, with the related small gridiron township of Mobile. After flooding in 1710, this was also refounded in a safer position. Its form in 1770 is shown by Figure 10.13 below.

CALIFORNIA

As elsewhere, the Spaniards became seriously interested in the Californian seaboard only when they realized that the English or the Dutch – possibly even the Russians(!) – might settle there first. The Viceroy of New Spain, Francisco de Croix, and Jose de Galvez, a *visitador* (inspector-general), alerted the Crown to that possibility early in 1768, and before the end of the year they had organized, with Father Junipero Serra, the Father-President of the Franciscans, an expedition to colonize Alta California.

Already well familiar with the coastline, San Diego Bay was selected for their first settlement, where the mission of San Diego de Alcala was founded in July 1769, with a nearby small palisaded *presidio*.[13] Other Californian settlements (through to 1782 when the mission at Santa Barbara was the tenth) included Monterey Bay in 1770 and most auspiciously San Francisco Bay in 1776.

The *presidio* at San Francisco was commanded by Lt Jose Joaquin Moraga, who laid it out (see Figure 10.7). Its first occupants comprised '16 soldiers, all married and with large families; 7 colonists likewise married ... some workmen and servants of the foregoing, herdsmen and muleteers'.[14] In 1792, a visiting English naval captain, George Vancouver, recorded: 'only three of the four sides had been enclosed by a wall (of timber and turf) about 14 feet high and 5 in breadth'.[15] It was then garrisoned by thirty-five soldiers, with their families, and their Indian servants.

In 1794 the Spaniards built a fort below and to the east of the *presidio* commanding the 'Golden Gate' entrance into the bay. The associated mission was that of San Francisco de Asis (known later as the Mission de Dolores) located several miles east of the *presidio*. By the early years of the nineteenth century the *pueblo* of Yerba Buena had become established on the southern shore of the bay, approximately midway between the two earlier settlements, where it was to become the nucleus of the future city of San Francisco (described below).

Meanwhile, when in 1774 Felipe de Neve was appointed Governor of California, he inherited the major problem of ensuring adequate food supplies for the new colony. It was necessary to augment the inadequate production by *presidios* and missions, through foundation of agricultural *pueblos*, the first of which was San Jose de Guadalupe, settled in November 1777 with a total of sixty-eight persons. Lt Moraga led the party, 'marking out for them the plaza for the houses and distributing the houselots among them. He measured off for each one a piece of land for planting; ... they also proceeded to build a dam to take water from the Guadalupe River to irrigate the fields'.[16] San Jose was moved in 1798 to a new site about one mile south, where its population had reached 165 by the end of the century.

The legal basis of Californian settlement was consolidated by *reglamento* of January 1781 which incorporated the relevant town plan-

Characteristically the congregated Indian houses were laid out in an orderly pattern within the mission confines. Surrounding the central quadrangle were the smith's shop, workshop, granary, tannery, and stables. The church was the largest building, dominating all others. This standard physical form was developed by Jesuits for northern Mexico and it achieved its culmination in the Franciscan establishments of California. In New Mexico, on the other hand, the tendency was to retain existing Indian communities intact, to build the churches on the edge of the settlements, and to divide Indian life between a town-oriented secular aspect and a church-oriented religious aspect. These differences were less the result of differing preconceptions and philosophies among the religious bodies than of the nature of Indian societies. In New Mexico the Franciscans adapted their methods to sedentary Indian peoples. In California they dealt with natives who had no permanent communities, who knew neither agriculture nor pottery making, and who lived on acorns and whatever else they could gather.
(C Gibson, Spain in America)

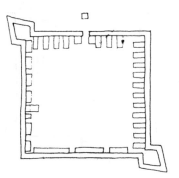

Figure 10.7 – The *presidio* of San Francisco, California. (See later in the chapter for the subsequent history of the city, with the map of 1852.)

12. For Pensacola see Reps, The Making of Urban America.
13. Reps, Cities of the American West.
14. G E Dane (trans) 'The Founding of the Praesidio and Mission of Our Father Saint Francis; being Chapter 45 of Fray Francisco Palou's Life of the Venerable Padre Fray Junipero Serra ... 1787' (quoted Reps, Cities of the American West).
15. G Vancouver, 'Voyage of Discovery, San Francisco in 1792', Californian Historical Society Quarterly, XIV, 1935.
16. F Palou, Historical Memoirs of New California (quoted Reps, Cities of the American West).

ning Laws of the Indies, with building lots and streets around a plaza, and farming tracts 200 varas (555 feet) square. In the spring, Neve travelled south from Monterey to Mission San Gabriel in order to plan the second *pueblo*, which was founded on 4 September 1781, for just eleven families, as El Pueblo de la Reina de los Angeles.

Los Angeles was given a plaza of extent 100 × 75 varas (277 × 208 feet) from which, Neve specified, 'four main streets shall extend, two on each side; and besides these, two other streets shall run by each corner'.[17] Each family was to receive four of the 200 varas square farm tracts, two of them capable of irrigation. The total of town lots to be marked out at 20 × 40 varas was to equal the number of irrigated tracts. 'The front of the plaza looking towards the East shall be reserved to erect at the proper time the Church and Government Buildings, and other public offices.' Figure 10.8 shows the plan of Los Angeles in 1793 and the caption explains how it differed from Neve's intentions. In 1818 the site of the church was moved further south, where it was in the centre of the north-western side of a new irregular shaped plaza, as shown below by the map of 1849.

In 1812 the Russians did indeed appear on the Californian coast when they constructed their Fort Ross, 60 miles north of the Golden Gate.[18] The Spanish counter-move was to place two missions on the northern side of San Francisco Bay, one in 1817 and the other in 1823, by which time the Russian intentions were seen to be peaceful. Nevertheless, just in case, Governor Figueroa in 1835 instructed the foundation of the *pueblo* of Sonoma, north of the Bay, in the river valley of that name, incorporating the earlier mission of San Francisco Solano.

Mariano Vallejo was its planner, who wrote later: 'I undertook the task of planning the new town, tracing first a great plaza of 212 varas square [688 feet] leaving the small building which had been constructed as a church on the east side of the plaza I built a barracks about 100 varas west of the church and then traced the streets and houselots as prescribed by the law.'[19] The caption to Figure 10.9 notes departures from the Laws of the Indies. Reps concludes that Sonoma was 'a notable achievement, marking the first time that in California a complete town, rather than a mere plaza surrounded by building lots, had been planned'.[20]

The USA: The Spanish contribution

Summary assessment of the urban achievements of the Spanish Latin American empire concluded Chapter 9, to which are added the following observations specific to their parts of the future USA. When in 1845 and 1848, respectively through the Texas Annexation and the Mexican Cession, the US Government gained from then independent Mexico the territory of the future States of Texas, New Mexico, Arizona, Utah, Nevada, California, and the larger part of Colorado, there were no urban settlements that even began to compare with those of the East and the rapidly emerging mid-west.[21]

As late as the 1890s, with the exception of uniquely favoured ('Gold Rush') San Francisco, which was nearing the 300,000 mark, other south-western urban populations were below the 50,395 total of Los Angeles (as evidenced by the accompanying table). In terms of urban population growth, the story of the south-west is that of waiting to be caught up in the westward expansion of the USA from its English-

Figure 10.8 – El Pueblo de la Reina de los Angles, California, as recorded by Jose Arguello in 1793. Neve's orthodox orientation was implemented (north on top left-hand corner); but the specified street layout was modified to locate the street across the northern end of the plaza some 28 feet away from the corners, and the street at the south-western corner was omitted. The space at the south-eastern corner was probably intended for the church.

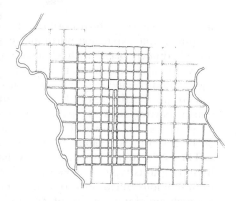

Figure 10.9 – Sonoma, California, as re-surveyed and extended by 1850. The plaza is a square, instead of the specified rectangle of the 'Laws of the Indies'; the streets entering the plaza were incorrect and the grid orientation as approximately north–south is also at fault. (Early in the twentieth century, the city hall and courthouse were rebuilt on sites in the centre of the plaza.)

17. D Garr, 'Hispanic Colonial Settlement in California: Planning and Urban Development on the Frontier, 1769–1850', unpublished PhD dissertation, Cornell University, Ithaca, New York, 1972.
18. A Russian threat was real enough: Alaska was a Russian colony until purchased by the USA in 1867 for $7.2 million. (See O W Miller, *The Frontier in Alaska and the Matanuska Colony*, 1975, who notes that in 1867 there were fewer than 700 Russians and about 2,000 Creole Indians in the Territory.)
19. Reps, *Cities of the American West*.
20. For Sonoma see Reps, *Cities of the American West*.
21. Hill, *A History of the United States*, summarizes Spanish and English colonial differences; see below for comparison with the English colonies in the east and Chapter 9 for Spanish settlement in Latin America.

Figure 10.10 – Tucson, Arizona: the plan of the city in 1862. Four years earlier a visitor had described it as giving 'the impression that it had originally been a hill, which, owing to the unexpected but just visitation of Providence, had been struck with lightning; and the dissipated mud walls ... were the residium left in the shape of mud deposits, for not a white wall nor a green tree was to be seen there' (quoted by Reps, *Cities of the American West*).

Cities, 1860

1. San Francisco, Calif.	56,802
2. Sacramento, Calif.	13,785
3. Salt Lake City, Utah	8,236
4. San Antonio, Tex.	8,235
5. Leavenworth, Kans.	7,429
6. Galveston, Tex.	7,307
7. Houston, Tex.	4,845
8. Denver, Colo.	4,749
9. Marysville, Calif.	4,740
10. Santa Fe, N. Mex.	4,635
11. San José, Calif.	4,579
12. Los Angeles, Calif.	4,385
13. Lewiston, Calif.	4,348
14. Grass Valley, Calif.	3,840
15. Nevada City, Calif.	3,679
16. Stockton, Calif.	3,679
17. Austin, Tex.	3,494
18. Portland, Ore.	2,874
19. Brownsville, Tex.	2,734
20. Bridgeport, Calif.	2,686

Cities, 1890

1. San Francisco, Calif.	298,997
2. Omaha, Nebr.	140,452
3. Denver, Colo.	106,713
4. Lincoln, Nebr.	55,154
5. Los Angeles, Calif.	50,395
6. Oakland, Calif.	48,682
7. Portland, Ore.	46,385
8. Salt Lake City, Utah	44,843
9. Seattle, Wash.	42,837
10. Kansas City, Kans.	38,316
11. San Antonio, Tex.	37,673
12. Dallas, Tex.	36,067
13. Tacoma, Wash.	36,006
14. Topeka, Kans.	31,007
15. Galveston, Tex.	29,084
16. Houston, Tex.	27,557
17. Sacramento, Calif.	26,386
18. Pueblo, Colo.	24,558
19. Wichita, Kans.	23,853
20. Fort Worth, Tex.	23,076

(Source: U.S. Census)

speaking eastern seaboard origins. There was not even the beginnings of comparable expansion north and east from Spanish colonial bases; there had never been any likelihood of its taking place. The map of the USA (Figure 10.11), showing the limits of the settled areas in 1800 and 1900, could be modified to include for the earlier date the valley of the Rio Grande north from El Paso, to Santa Fe, but only as an inconspicuously narrow strip of Spanish settlement. Yet their arrival in that fertile region had been just two years later than the foundation in 1607 of the first English colony of Jamestown, in Virginia.

The explanation, in main part, is that New England was populated by immigrants typically seeking land for farming by themselves, freed from religious constraints (or active persecution) back home. Potential Spanish emigrants, on the other hand, had no interest in farming by their own labour, and religious circumstances were equally prescriptive at home and in the Americas. The power of the Spanish Catholic church precluded immigration from other countries; whereas New England became increasingly open to northern Europeans generally.[22] Not only was there minimal Spanish interest in voluntary emigration, but also when the local Spanish imperial authorities tried to encourage settlers there was delay and lack of interest on the part of the Crown. When in 1722 Aguayo perceived need for settlers to support the East Texas presidios and missions, his urgent request for 400 families to be sent, produced (eight years later) no more than sixty persons! And yet, in 1630 alone, well over 1,000 settlers had arrived in the Massachusetts Bay colony. Between 1630 and 1643 60,000 people were conveyed to the New England and Virginian colonies.[23]

Before turning to concentrate on eastern and mid-western settlement by the subsequently dominant English-speaking emigrants, with brief mention of the French involvements, it remains to observe that although Spanish urban achievements were modest in the extreme: so too were their intentions. What might have come about had Spain not lost population during the fourteenth and first-half of the seventeenth centuries – thereby having a comparably high proportion of potential emigrants; or if they had admitted settlers of other nationalities into North America, is but an intriguingly academic question.

22. See Parry, *The Spanish Seaborne Empire*; C Gibson, *Spain in America*, 1966, for Spanish imperial policies; see also Chapter 9.

23. Brown, *Historical Geography of the United States*, provides the following estimated breakdown of the white population in 1790:

Stock	Percentage of Total
English	60.1
Scotch (incl. Ulster)	8.1
Irish Free State	3.6
German	8.6
Dutch	3.1
French	2.3
Swedish	0.7
Spanish	0.8
Unassigned	6.8

(from Report of the Committee of Linguistic Stocks, American Council of Learned Societies, 1931)

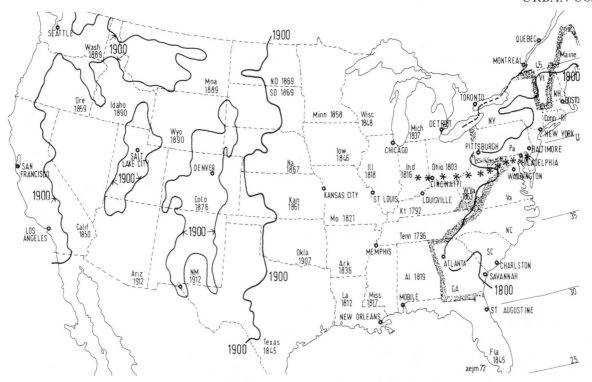

French settlements

France entered the New World over 100 years later than Spain, having failed to follow up the explorations of Jacques Cartier between 1534 and 1542. From 1604 onwards French settlements were established along the Canadian coastline and up the St Lawrence River, commencing with Sainte Croix on the island of Douchet in what is now Maine. Quebec dates from 1608 and Montreal from 1642. Louisburg, founded on Cape Breton Island in 1712 was originally intended as the main French fortress in Canada but after capture by the British in 1758 it was completely destroyed. With the St Lawrence as their baseline and water transport the only effective means of movement, the French were able to penetrate south and create a widely distributed network of trading posts at a time when the British colonists in New England were still struggling through the forests a few miles inland from their coastal origins. But as with Spain in the south and south-west, French policy did not include the establishment of permanent agricultural colonies, and their settlements were comparatively few and far between.[24] Detroit was founded in 1701 by Antoine de la Mothe Cadillac to control the key Detroit River link between the Erie and St Clair Lakes, and Fort Duquesne of 1724 (renamed Pittsburgh after capture by the British in 1758 had a similar strategic role where the Allegheny and Monongahela Rivers unite to form the Ohio, the main waterways to the west and south. St Louis on the Mississippi was founded in 1762 and marks the end of French urban settlement. At the southern extremity of their vast inland territory Mobile had been founded in 1711 and New Orleans in 1722.

Figure 10.11 – General map of the United States. The extent of the first colonies is indicated by the shaded boundary; the limits of the settled areas in 1800 and 1900 are marked by heavy lines. The asterisks across eastern states show the western movement of the centre of population at the end of each decade from 1790 to 1890.

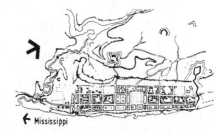

Figure 10.12 – St Louis (Missouri) in 1796. The city was founded in 1762 and named in honour of Louis XV and his canonized ancestor.

24. See Chapter 6 for the disregarded recommendations of S Le Prestre de Vauban, the leading French military engineer, in respect of the need for an agricultural settlement policy.

329

English settlements

Henry VII had missed the opportunity to sponsor Columbus but made amends by authorizing one of his countrymen, Giovanni Gaboto – known as John Cabot – to seek out new unknown lands and to claim them for the English Crown. This Cabot did during a short voyage in the summer of 1497, the details of which are lost to history except that he is known to have made a brief landfall on Cape Breton Island.[25] Cabot's landing was invaluable as the basis of the 1583 English claim to all of America north of Cape Florida, whereby (as S E Morrison put it) England set aside the Pope, Spain and Portugal and used convenient, if almost imaginary, history to give authority to the fulfilment of national desires. England had no intention of recognizing Spanish monopoly in the Americas.[26]

In 1584 Sir Walter Raleigh was commissioned to find a suitable site for a colony and despatched an expedition fleet for that purpose. On its return with optimistic reports, Queen Elizabeth permitted Raleigh to name the new lands Virginia. The first English settlement followed in 1585, when Raleigh sent out seven ships with 108 men on board – hardly a balanced emigrant party, which probably meant that a military base was the real objective. This party established a small fort on Roanoke Island, in North Carolina, but Ralph Lane, its commander, 'seems to have contrived to turn the Indians from friendly and curious neighbours into violent and treacherous enemies'.[27] Luckily Sir Francis Drake arrived in time to rescue them. The next move was to despatch three ships with 117 settlers, including this time seventeen women and nine children, to re-establish Roanoke. Unfortunately open war with Spain intervened, culminating in the 1588 defeat of the Armada, and it was not until 1589 that a fresh expedition was sent. This found Roanoke deserted, with no sign of its inhabitants.[28] Thus the sixteenth century ended with no permanent English foothold in America.

During the seventeenth century English colonization was reorganized on a new basis, passing from individual adventuring to corporate enterprise through the medium of royal-chartered joint-stock companies. The first of these date from 1606 with the Virginia Company of London and the Virginia Company of Plymouth authorized to settle lands roughly between the 34th and 45th parallels, from Cape Fear River in North Carolina to the coast of Maine. The London Company kept to the south and the Plymouth Company to the north.

Three ships with about 120 settlers were sent out by the London Company in December 1606 and after exploring the lower James River a site was finally agreed about 30 miles from the coast where they founded Jamestown in May 1607. This first English settlement was triangular in shape, with one side along the riverfront of 400 feet and the others each of 300 feet. The area within the fortified palisade was about one acre and contained the church, storehouses and probably a double row of houses flanking a single street. The original settlement was later destroyed by flooding of the James River. John Reps observes that there was little here reflecting the status of civic design in the Europe of that era. Not even the bastide towns of the late Middle Ages in France, Spain and England were so modest in scale and primitive in character. If the first settlers had ever believed that they could at once create an English village in the New World, this illusion was quickly dispelled.[29]

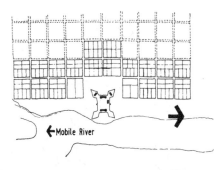

Figure 10.13 – Mobile (Alabama): founded by the French in 1711. For this exceptional instance of artillery fortification in the future USA, see W B Robinson, 'Military Architecture at Mobile Bay', *Journal of the Society of Architectural Historians*, vol XXX, March 1971.

25. Nye and Morpurgo, *History of the United States*.
26. Morrison, *Oxford History of the American People*. See also Brogan, *The Penguin History of the United States of America*; Parry, *The Spanish Seaborne Empire*.
27. Nye and Morpurgo, *History of the United States*.
28. Roanoke might have become permanently established if, like the later Jamestown, it had been supported from England. However, as Morrison observes, 'it was the wrong time to look for help from home. A Spanish armada was being prepared to invade England, where nobody could spare the effort to succor a tiny outpost in Virginia' (*Oxford History of the American People*).
29. Reps, *The Making of Urban America*. See also A B MacLear, *Early New England Towns*, 1967; D J Boorstin, *The Americas: The Colonial Experience*, 1958/ 1988.

The early history of the English colonies proved in reality to be ceaseless struggle to maintain a subsistence agricultural economy. The first great improvement in their fortunes came with the discovery that a cross between West Indian and local Virginia tobaccos produced a smooth smoke which captured the English market. Virginia went tobacco mad; it was even grown in the streets of Jamestown.[30] By 1619, with tobacco plantations extending some 20 miles along the James River, the colony had a population of about 1,000. Over 50,000 pounds of leaf were exported to England during the previous year. The early disadvantages of this single-crop economy – including the need to import food from England and the unsympathetic local Indians – were overcome: by the mid-1620s Virginia had entered a prosperous consolidation era.

Early urban growth

From well before 1600 fishermen had been familiar with the coastline of the future New England but a first attempt by the Northern Virginian Company to establish a trading post on the Kennebec River failed after precariously surviving only the winter of 1607–8. The history of permanent settlement in New England starts in the autumn of 1620 after the arrival of the Pilgrim Fathers in the *Mayflower*. With their first rudimentary huts they were to establish the settlement of Plymouth in the December of that year. Although no plan of Plymouth has survived, John Reps is able to quote a visitor's observations of the late 1620s, from which he infers that the neat regularity of this little village plan reflected the tight social and economic organization of the Pilgrim group.[31]

BOSTON

Plymouth did not attain urban status in time to offer a challenge as the dominant city in New England. The Massachusetts Bay Company was created in 1629 with Charlestown, on a peninsular between the Charles and Mystic rivers, intended as the colony's capital. The location was readily defended and afforded excellent harbour opportunities, but it lacked a reliable drinking water supply – that other vital prerequisite for waterside existence – and following an outbreak of fever in 1630, Governor Winthrop transferred the seat of government just across the Charles river to the south where another rocky peninsular offered all of Charlestown's advantages, plus a good water supply. In the autumn of 1630 work commenced on the new town and port of Boston, so named after the home town of many of the settlers in Lincolnshire, England.

The topography of central Boston around and inland from the 1630s' nucleus has altered greatly over the years and confusingly shows little resemblance to the original shore-lines. The 1775 map of 'Boston, its environs and harbour', gives the original condition, modified only in minor respects, which is shown in Figure 10.15 as part of the overall changes through to the end of the nineteenth century. (Figure 10.16 also locates Charlestown and Cambridge, which was the temporary seat of government until 1636, when Boston took over, and where in the same year the first buildings of Harvard University were commenced.) The peninsular site of Boston was in effect virtually a high-tide island, connected to the mainland by only a narrow neck of marshy land. In 1775, under siege by the 'rebels', the causeway was cut and access into Boston was across a strongly defended bridge. Subsequently extensive

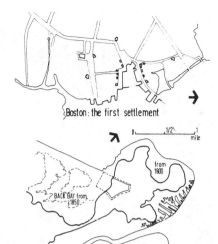

Figure 10.14 – Boston: the waterfront of the harbour and rivers in 1775 (as Figure 10.16) shown in heavy line, with subsequent major areas of land reclamation given with their dates. The Back Bay, which was laid out on a gridiron from 1850, is an extensive early example of topographical adjustment.

Figure 10.15 – Boston: a reconstructed general plan of the peninsular in about 1640, and the enlarged plan of the first port facilities which made use of the minor coves and inlets off the main eastern bay. Boston grew up around its first dock and adjoining Dock Square (A) and the continuation of the main land route across the causeway (B). Great Street (present-day State Street) leading away from the centre of the bay became the main street with a market square at its intersection with main highway. Long Wharf, which so dramatically changed the nature of the harbour, was in effect a direct continuation of the line of Great Street (C).

30. Morrison, *Oxford History of the American People*. The importance of tobacco cannot be over-estimated; see Nye and Morpurgo, *A History of the United States*, and Boorstin, *The Americas: The Colonial Experience*.
31. Reps, *The Making of Urban America*.
32. For Boston see Reps, *The Making of Urban America*; W M Whitehill, *Boston: A Topographical History*, exp. 2nd edn, 1968; for later nineteenth-century history there is S B Warner, jr, *Streetcar Suburbs: The Process of Growth in Boston 1870–1900*.

land reclamation from the harbour, occurring both naturally and as a result of planned programmes, has almost completely altered the original natural topography.[32]

Already by 1775 after a century and a half's growth to a population of some 17,000, the original small harbour cove off the main eastern bay had been filled in, to be replaced to a large extent by the 'Long Wharf' stretching out some 1,500 feet into the bay; on the northern side the 'Mill Pond' was built over in the early nineteenth century. Largest by far of the projects, however, was the 'Back Bay' reclamation of 1850 onwards which created the extensive regular gridiron district, centred on Commonwealth Avenue, which is in such direct contrast to the organic growth form of old downtown Boston.

Figure 10.16 – Boston: the 1775 map of its environs and harbour, with the Rebels Works raised against that town; from the observations of Lieutenant Page, of His Majesty's Corps of Engineers, and from the Plans of Captain Montresor. At this date the city with a population of about 17,000, had occupied almost all the land on the original 'island' with the exception of the northern, more hilly area north of the Common which was shortly to be developed. The Common itself has remained a jealously guarded open space, directly adjoining the crowded downtown central business district, where one of urban America's seemingly most successful renewal and redevelopment programmes of the 1960s and 1970s has been carried out. By comparing this map and that of some 135 years earlier (Figure 10.15), the effect of that pre-urban cadastre can be seen.

Boston prospered from the outset and within four years of its foundation, William Wood, a visitor from England could observe: 'This towne although it is neither the greatest, nor the richest, yet it is the most noted and frequented, being the Centre of the Plantations where the monthly Courts are kept.' Although in origin a 'planned' new town, Boston shows no evidence of any co-ordinated 'master plan' for its physical development, as was to be the case, for example, at Philadelphia. It is true that the first plots of land for residential and harbour-side purposes would have been disposed according to a simple gridiron-based ground scheme, thereby inferring a 'plan' of sorts. But from then on Boston's early growth was on the basis of street lines and property boundaries determined by unavoidable topographical variations in both extent and height of the originally available land. Hence Boston's characteristically unplanned, organic-growth downtown street pattern: fascinating for pedestrian tourists with its European medieval echoes, so different from urban America's otherwise monotonous grids, but which, to quote John Reps, 'has so persistently remained to plague the modern driver.'[33]

THE SEVENTEENTH CENTURY

The foundation and early growth of Boston coincided with the great migration of the Puritans. In 1630 alone well over 1,000 settlers arrived in the Massachusetts Bay Colony.[34] In the way these successive waves of Puritan immigrants established themselves in New England there is a fascinating instance of history repeating itself. The Anglo-Saxon and other settlers of the fourth to tenth centuries in England had also either observed the territorial rights of earlier arrivals and moved on, or created new settlements with overspill population drawn from the existing villages. In both the old and new countries the effect was the same – a steady expansion of the settled area, but whereas in England the process must have been based as much on the rule of strength as respect for the rights of others, in New England the rule of law prevailed from the start.

The original settlements had neither the resources nor, in most instances, the inclination to take in newcomers, most of whom, with sturdy independence, were determined to make their own way. As with the early seventeenth-century villages in England, which provided the majority of the settlers, agriculture was the basis of the New England communities who were only one crop removed from starvation.[35] Given the same economic base, the physical form of these communities also closely followed that of the contemporary English village. Each settler received a 'home lot', generally as one of the number which comprised a village, and a share of the surrounding common fields – frequently in the form of long and narrow strips.[36] The total extent of the built nucleus, its immediate home lots and the common fields, was called a town or township, a term which applied later, as one example, to the six-miles-square units of the 1786 sale of the first seven ranges of townships west of the Ohio River (see Figure 10.19).

The 1620s and 1630s along Massachusetts Bay and this first incursion into the mid-west after the War of Independence resulted, however, in totally different landownership patterns, which over the years have each in turn led to the development of contrasting urban forms. Whereas the New England pattern is best defined as controlled organic growth – with the planned field and street layouts determined by

The New England town of the seventeenth century was a village community settled for purposes of good neighbourhood and defence. Its most characteristic features resulted from the topography of the country, and from the ideas of the nature of a town which the colonists brought from England. Forced by the geographical features of New England and by the necessity of protection, the colonists, already acquainted, settled in groups, and at once began organizing their settlements in accordance with the type familiar to them – the old English manor. Between this and the New England town many analogies may be drawn, showing the Germanic origin, not only of the government with its democratic features, but of the form of settlement – a compact town with outlying fields – and of the land system, with 'the houses and home lots fenced in and owned in severalty, with common fields outside the town, and with a surrounding track of absolutely common and undivided land used for pasturage and woodland under communal regulations.
(A Bush Maclear, Early New England Towns)

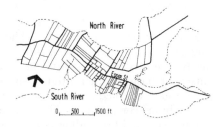

Figure 10.17 – Salem: the oldest of the Massachusetts Bay villages. This plan shows the home lots in 1670 with the line of present-day Essex Street along the high ground between the North and South rivers. With Boston (Figure 10.15) this land subdivision exemplifies the New England organic growth pre-urban cadastre.

33. Reps, *The Making of Urban America*. Unlike the later gridiron-based American cities, there was no preconceived idea of the City of Boston, other than that of a hopefully well-situated shelter for immediate survival and subsequent growth. Topography, waterside mobility and then the adjoining pre-urban cadastre were primary determinants of the unplanned organic growth nucleus of modern Boston.

34. G M Trevelyan has observed that, 'both patriotic and religious motives inspired many of those who supplied the fund, the ships and the equipment for the enterprise. Between 1630 and 1643 £200,000 was spent in conveying 20,000 men, women and children to New England in 200 ships; in the same period 40,000 more emigrants were conveyed to Virginia and other colonies' (*English Social History*), 1944. See also Boorstin, *The Americas: The Colonial Experience*; W Notestein, *The English People on the Eve of Colonisation*, 1954.

35. See MacLear, *Early New England Towns*; J Archer, 'Puritan Town Planning in New Haven', *Journal of the Society of Architectural Historians*, vol XXXIV, 1975.

36. For a description of the settlement of New England villages – notably Salem, Dorchester, Watertown, Roxbury and Cambridge – see MacLear, *Early New England Towns*. See also the sections in Chapter 4 describing village settlement patterns in England, in particular the aerial photograph of Appleton-le-Moors, Yorkshire, a street village with characteristic long and narrow village land-holdings (Figure 4.18).

natural, topographical boundaries – by 1786 in the mid-west the inflexible surveyors' gridiron, which essentially ignores topography, had long been accepted as the only practical method of controlling the growth of the nation. New England therefore contains almost all of the USA's historic non-grid urban cores, notably Boston and, further south, New York.[37]

Meanwhile, on the eastern sea-board, the Dutch had established New Netherland, at one time comprising the entire Hudson Valley and the shores of Delaware Bay and Long Island,[38] with Albany (Fort Orange) founded in 1624 and New York City (New Amsterdam) of two years later as their main trading posts. The French were active further north and by establishing Quebec in 1608 and Montreal in 1620 Champlain put his king in possession of the one great river valley that led from the heart of North America to the east coast.[39]

The end of this eventful seventeenth century nevertheless saw only a 50-mile wide fringe of country settled inland from navigable water. From Maine to North Carolina the development was fairly continuous, but from there to South Carolina there was a gap of 250 miles where the Indians were still undisturbed.[40] The future of North America as an English-speaking continent, however, was by now ensured. Although the British had come last into the New World they created from the outset the basis for their ultimate supremacy by sustaining the flow of emigrants – accepted regardless of nationality and capital into agriculturally based communities. This was in direct contrast to the exploitative mineral and fur trading posts of Spain and France.[41]

EIGHTEENTH-CENTURY EXPANSION

In 1763, with the War of Independence of 1776 already inevitable, Morrison notes that the settled area included about 1.5 million people, a total which was in excess of 2.25 million by 1775. 'The bulk of the population was involved in agriculture, but visiting Europeans regarded the country as a wilderness because over 90 per cent of it was still forested. Only near the Atlantic, in sections cultivated for over a century, could one have found anything resembling the farming areas of Iowa, Illinois or Nebraska today. Elsewhere and especially in the south, farms and plantations lay miles apart, separated by forest.'[42]

Of the 1.5 million people in 1763 almost one-third were slaves. This social situation had its origins in the earliest years of the European colonization and is today arguably the most demanding of urban America's problems.[43] Within twenty years of Columbus reaching the New World, African negroes transported by Spanish, Dutch and Portuguese traders were arriving in the Caribbean Islands. The first negroes were landed in colonial America at Jamestown in 1619. By the beginning of the Revolution there were 500,000 slaves – 20 per cent of the population with three-quarters of them in the south where they amounted to 40 per cent of the total. In Virginia in 1756 there were about 120,000 negroes out of a total of around 293,000; in the tidewater counties they outnumbered whites by at least two to one.

The New England colonies also had their slave populations: Massachusetts had the most – 5,250 out of 224,000 in 1746 – but Rhode Island with 3,000 out of 31,500 in 1749 had by far the largest proportion. At the outbreak of the Civil War in 1861, largely fought over the issue of slavery, the south had 4 million slaves out of a total of 12 million. Although of immense continuing social significance, slavery

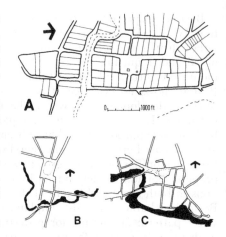

Figure 10.18 – Three organic-growth New England urban nuclei: A, Hartford, Connecticut; B, Fairhaven, Vermont; C, Ipswich, Massachusetts.

37. See *The Making of Urban America* for the plans of a number of other, smaller organic growth towns, notably Exeter, New Hampshire, in 1802; Woodstock, Vermont, in 1869; Lebanon, New Hampshire, in 1884; Fairhaven, Vermont, in 1869; Ipswich, Mass., in 1872.
38. In 1621 the Dutch East India Company received exclusive trading rights from the Dutch Government for New Netherland. Morrison records how in return for New Amsterdam (New York), the British renounced claims in parts of the East Indies, in favour of the Dutch. See also C R Boxer, *The Dutch Seaborne Empire*, 1965/1977.
39. For French settlement see W J Eccles, *France in America*, rev. edn, 1972; K McNaught, *The Penguin History of Canada*, 1988.
40. Morrison, *Oxford History of the American People*.
41. See Chapter 6, page 214, for reference to French policy in North America.
42. Morrison, *Oxford History of the American People*.
43. *Urban USA: Population Growth and Movement*, 'Official Architecture and Planning', December 1968.
44. The statement that the 'resultant minority ethnic group problem (of slavery) had little direct effect on the pattern of urban settlement in the emergent United

and the resultant minority ethnic group problem had little direct effect on the pattern of urban settlement in the emergent United States.[44]

Factors affecting development

Before describing the establishment, layout and early development of key cities before the early years of the nineteenth century, we must consider briefly several determining factors, with eighteenth-century origins, which influenced the expansion of urban USA.

The extent of the settled areas of the United States in 1800 is shown in Figure 10.11. The population was around 4 million, of whom only some 170,000 were settled west of the Alleghenies. The conclusion of the War of Independence removed restraints which had limited further western expansion and heralded the beginnings of the inexorable drive west which ended only in 1906 when the United States were finally constituted as today.

THE GREAT AMERICAN GRID

One of the most controversial issues which had to be resolved in framing the Articles of Confederation was the question of the undeveloped western lands. Seven states were claiming western projections of their boundaries and only in 1802 did Congress gain full control of the situation. A national land policy was required and the Land Ordinance of 1785 represented the results of compromise between the Government's desire to raise public funds from the sale of land and mounting pressure from the countless thousands who wanted enough land, preferably as a grant, for a farm.

The law required that the territory to be sold should be laid out in rectangular townships, 6 miles square. Each township was to be divided into 36 square sections of 1 square mile or 640 acres. Half the land was to be sold by townships, the other half by sections.[45] Section 16 in each township was reserved for the support of schools. This rectilinear survey basis obviously disregarded topography, with many anomalous results, but it was easy to locate a purchase in the wilderness and to avoid boundary disputes.[46]

The first seven ranges of townships were surveyed by 1786, immediately west of the Ohio River (Figure 10.19). Given the speed of western expansion, there was no alternative to general adoption of these regional grids-of-expediency, and their existence as property boundaries reinforced the preference for gridiron urban structuring. Section boundaries proved natural rural road lines and villages either established themselves at junctions, or were laid out and promoted by property speculators. In its effect the great American grid is the most far-reaching instance of a planned pre-urban cadastre.[47]

TRANSPORT

A formidable deterrent to prospective western settlers had been the lack of transport facilities to get farm produce from the Ohio region to the urban markets of the east. Originally the United States was settled on the basis of water transport. To be distant from the sea or navigable rivers meant effective isolation. Roads were non-existent. Goods haulage was by horse and cart or pack animals, both inefficient and costly. Personal movement was only possible by horseback for most of the year.

One of the earliest political incentives to provide improved

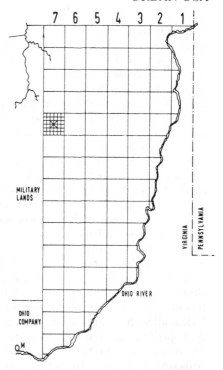

Figure 10.19 – The first seven ranges of townships, as surveyed by 1786, west of the Ohio (numbered 1 to 7 from east to west). An example township, in the seventh range, is shown as divided into its thirty-six one square mile sections each of 640 acres. Key: M, in south-western corner – the town of Marietta platted in 1788.

'Today,' writes John Reps, 'as one flies over the last mountain ridges from the east, one sees stretching ahead to the horizon a vast chequer board of fields and roads. With military precision, modified only on occasion by some severe topographic break, or some earlier system of land distribution, this rectangular grid persists to the shores of the Pacific' (*The Making of Urban America*). The 'Great American Grid' is by far the most extensive instance in rural and urban history of a predetermined pre-urban cadastre.

States' is retained, but it is stressed in this edition that this observation applies only to the emergent cities of the late seventeenth and eighteenth centuries. Even by then disadvantaged minority group districts were forming which, while they had little effect on two-dimensional plan-form, were to result in diminished land values with consequent effect on redevelopment potential and the urban scene. (See G Osofsky, 'Harlem Tragedy: An Emerging Slum', in A B Callow jr (ed.) *American Urban History: An Interpretive Reader with Commentaries*, 1973.

45. Reps, *The Making of Urban America*.
46. E C Kirkland, *A History of American Economic Life*, 1969.
47. By far the most extensive planned basis of a pre-urban cadastre, the Great American Grid has an ancient precedent in the Roman *centuriato* whereby arable land was subdivided prior to the establishment of a colony (see Chapter 3 for the *centuriato* and N J W Thrower, *Original Survey and Land Subdivision*, 1966, for the American Grid).

communications resulted from the admission of Ohio as a state of the Union in 1803. Because of its relative isolation behind the mountains there had been fears that the newly organizing territory might turn further west and seek allegiance with Spain, so as to gain use of the Mississippi trade routes. To ensure the entry of Ohio the Government promised a new road to the east across the Alleghenies. This 'National Road' was authorized in 1806 and opened in 1818 between Cumberland on the upper Potomac and Wheeling on the Ohio River – a distance of 130 miles. It served primarily to connect navigable waterways. Before its provision, to transport 100 pounds of freight from Philadelphia to Pittsburgh cost from 7 to 10 dollars; afterwards the cost was under 3 dollars.[48] British expertise in new road construction had earlier been used to build the Philadelphia–Lancaster road.

New York business interests were well aware that the National Road would take western trade to Baltimore. A long-discussed canal project to link the Hudson River with Lake Erie became an economic necessity and was started in 1817. By 1824 it was completed to Buffalo, 363 miles from the Hudson. At once the charge for transporting one ton of freight from Buffalo to New York dropped from 100 to 10 dollars.[49] In 1825 Buffalo, previously an obscure village, built over 3,400 houses and reached over 18,000 inhabitants by 1840. Rochester, Syracuse and Utica, all with canal-side locations grew rapidly into prosperous cities. As hoped, the greatest effect of all was on New York, which grew from 152,000 in 1820 to 391,000 in 1840, achieving unchallenged eastern seaboard dominance. In its first part-year of operation the Erie Canal recouped one-seventh of its cost in tolls and gave impetus to many other projects.[50]

The first steamship service in the USA was started in 1790 between Philadelphia and Trenton, 40 miles away up the Delaware. Robert Fulton's 'Clermont'[51] inaugurated a New York to Albany service in 1807 and in 1809 the Ohio and Mississippi Rivers were surveyed for steamboat possibilities. Both rivers had long been used for bulk transport, but in effect only in the downstream direction with freight floated on dispensable rafts. In 1811 the first steamboat built west of the Alleghenies was launched at Pittsburgh, with a Boulton & Watt engine imported from England. She was named the *New Orleans* and duly reached that port after a triumphant passage down river. This inaugurated the riverboat era of American internal transport, which in turn resulted in the rapid development of riverside settlements.

The Baltimore and Ohio, chartered in 1827, was the first commercial railway in the USA,[52] and had the intention of competing with the Erie Canal. By 1830 the economic advantages of rail transport were such that all established urban centres were engaged in railway promotion. The effect of the railway construction on the location, form and function of urban America cannot be over-emphasized. For an established settlement it meant certain growth. The existence of the lines themselves led to inevitable track-side development. To be bypassed meant stagnation, if not effective commercial death.

INDUSTRY AND URBANIZATION

Before the War of Independence the British government had attempted to minimize industrial activity in the colonies, hoping to maintain them as markets for the developing home industries. In 1765 legislation barred emigration of skilled operatives from Britain. This was reinforced in

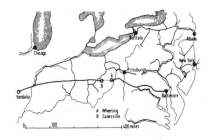

Figure 10.20 – Canals in the north-eastern states as completed by 1850, in heavy line; related rivers in lighter line. Also shown is the line of the 'National Road', the first section of which was opened between Cumberland, on the Potomac, and Wheeling, on the Ohio, in 1818. Later extensions linked Baltimore with Vandalia.

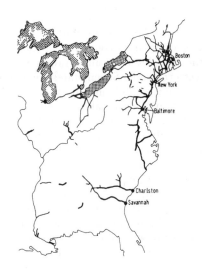

Figure 10.21 – The railway system in 1840 (heavy line) and 1850 (lighter line extensions). Completed mileages as:

	1830	1840	1850	1860
New England States	—	517	2,508	3,660
Middle States	30	1,566	3,105	6,354
Trans-Mississippi	—	40	80	2,906
Southeast	10	522	1,717	5,351
Old Southwest	—	74	336	3,392
Old Northwest	—	111	1,276	9,583
Total US	40	2,818	9,022	31,246

(From E C Kirkland, *A History of American Economic Life*)

48, 49. Kirkland, *A History of American Economic Life*. See also C Goodrich (ed.) *Canals and American Economic Development*, 1961.

50. 'In the first year of operation the Erie Canal carried 13,111 boats and 40,000 persons west' (Nye and Morpurgo, *A History of the United States*, Volume 2).

51, 52. F M Reck, *The Romance of American Transportation*, 1938. See also J F Stover, *American Railroads*, 1961.

1774 when the export of machinery was prohibited. These factors, however, played only a small part in the slow development of American engineering industry before 1860, until which year the emphasis was on food and clothing, with the most compulsive of the new demands coming from transportation – for steam engines, locomotives, and rails.[53]

The Industrial Revolution in the United States followed that in Britain. The years from 1815 to 1830 were sometimes remembered in New England as 'the period when attention shifted from the wharf to the waterfall; from foreign trade to local manufacturing'.[54] It took the next century, through the First World War, for the labour force in manufacturing to exceed that in agriculture, and not until the census of 1920 did the enumerators discover that slightly more than half the American people lived in urban communities, if those with as little as 2,500 could be esteemed as such.[55]

The established urban areas of the eastern states, with their excellent water transport and power facilities, were the natural locations of industry during the first half of the nineteenth century. From about 1820 the iron industry was concentrating around Pittsburgh. This western shift in industrial location was accelerated after 1870 with the triangle contained by Pittsburgh, Cleveland and Chicago forming the centre of heavy industry. Pig-iron production, an excellent index with which to relate urban growth, totalled 54,000 tons in 1820. Although this had increased to around 540,000 tons in 1850 the USA was importing about twice this amount. Subsequent growth was such that the USA surpassed British production in 1894 and the combined British and German totals in 1906. Coal production is a second valuable index; in 1860 the USA mined 14,610,000 tons as against nearly thirty times that figure in 1914. Only one person in ten had lived in a town of more than 1,000 inhabitants in 1800. In 1790 the first census had found no city exceeding 50,000 (although Philadelphia was almost there); by 1860 there were 16 cities with this total; by 1900 78; by 1920 144; by 1940 199. Moreover, by the beginning of the twentieth century there were 38 cities of over 100,000 population.

City examples

PHILADELPHIA

Until the early years of the seventeenth century the valley of the Delaware was heavily forested and sparsely populated, the domain only of the Leni Lenape tribe, a branch of the great Iroquois people. The year 1609 saw the first European contact with this region when Henry Hudson's 'Half Moon', chartered by the Dutch, entered Delaware Bay. The first settlement on the Delaware was under Swedish auspices, although led by Peter Minuit of earlier 'Manhattan Purchase' fame. This was in 1638 at the mouth of Christina Creek, now part of present-day Wilmington. The Dutch captured this settlement in 1655, before in turn giving way to the English in 1664. These activities set the historical context for the founding in 1682, by William Penn, of Philadelphia.

William Penn (1644–1718) inherited on his father's death in 1670 a substantial estate, including a 'sizeable but uncertain' asset of £16,000 in the form of a debt owed by Charles II.[56] Penn had been an active participant in developing the New England colonies in both East and West Jersey from 1670; in return for cancellation of the debt, Charles II granted him a charter in 1681 establishing his authority as governor and proprietor of Pennsylvania.

Population breakdown – 1700 and 1790

		1700	1790
New England			
New Hampshire		10,000	141,885
Massuchusetts Bay		80,000	378,000
Rhode Island		10,000	68,825
Connecticut		30,000	237,946
	Total	130,000	727,443
Middle Colonies			
New York		30,000	340,000
New Jersey		15,000	184,139
Pennsylvania and Delaware		20,000	434,373*
	Total	65,000	958,632
Chesapeke Colonies			
Maryland		32,258	319,728
Virginia		55,000	691,737
	Total	87,258	1,001,465
The Carolinas			
North Carolina		5,000	393,751
South Carolina		7,000	249,073
	Total	12,000	642,824

* This figure is for Pennsylvania only, not Delaware

Population breakdown: 1700 and 1790.

53. Kirkland, *A History of American Economic Life.*
54. J Gottmann, *Megalopolis,* 1964.
55. Kirkland, *A History of American Economic Life.*
56. Reps, *The Making of Urban America.* See also R S Dunn and M M Dunn (eds) *The World of William Penn,* 1986. For Philadelphia generally see S B Warner jr, *The Private City: Philadelphia in Three Periods of its Growth,* 1968; R S Wurman and J A Gallery, *Man-Made Philadelphia,* 1972; M C Branch, *Comparative Urban Design,* 1978, for the 1840 SDUK Map of the city; E Bacon, *Design of Cities,* 2nd edn, 1974, for descriptions of the city where Mr Bacon was a long-serving Executive Director of the City Planning Commission.
57. Extract from *Certain Conditions and Concessions Agreed Upon by William Penn,* 1681, as included in Samuel Hazard's *Annals of Pennsylvania,* 1850 (quoted by Reps).

In July of that year Penn published his general scheme of colonization including in the first paragraph the statement that 'so soon as it pleaseth God a certain quantity of land or ground plat shall be laid out for a large town or city and every purchaser and adventurer shall, by lot, have so much land therein as will answer to the proportion which he hath bought or taken up upon rent'.[57] Three commissioners were selected to lead the first group of settlers and Penn, who was not intending to leave London until the following year, drew up a detailed brief for them in September 1681, shortly before their departure. John Reps aptly sums up these instructions as specific, practical and to the point – the work of a man to whom planning was no longer a novelty.[58] For the location of the new city Penn directed that 'the rivers and creeks be sounded on my side of the Delaware River and be sure to make your choice where it is most navigable, high, dry and healthy, that is where most ships may best ride, of deepest draught of water, if possible to load or unload at the bank or key side, without boating or lightening of it. Such a place being found out, for navigation, healthy situation and good soil for provision, lay out ten thousand acres contiguous to it in the best manner you can, as the bounds and extent of the liberties of the said town'.[59]

The Surveyor-General for the new colony was Captain Thomas Holme, whose arrival on the Delaware in June 1682 was probably too late for him to be involved in the siting of Philadelphia. He did, however, lay out the first section of the city so that a drawing of lots for town plots could be held on 9 September. In his work Holme and the commissioners had received further precise instructions from Penn 'to settle the figure of the town so as that the streets hereafter may be uniform down to the water from the country bounds; let the place for the storehouse be on the middle of the key, which will yet serve for markets and statehouses too. This may be ordered when I come, only let the houses built be in a line, or upon a line'.[60] Penn himself arrived at Philadelphia in October 1682 to find the city located as he had directed, on the neck of the peninsula between the Delaware and Schuylkill. Here the Delaware has scoured a deep channel forming a steep bluff on its western bank, permitting inshore mooring of large ships. There was also a small creek, called the Dock, penetrating inland to provide shelter for small boats. The site met all Penn's requirements. It was high, free from flooding, well wooded, provided with fresh water and reasonably flat.

Holme's Philadelphia plan, of which there is no surviving copy, extended only about half way across to the Schuylkill. One of Penn's first actions was to extend the city over the peninsula to give it a frontage to both rivers. The combined Holme–Penn plan for Philadelphia was drawn up in 1683 and used as the basis for advertising the colony in London (Figure 10.22). In Holme's words 'the city consists of a large Front-Street to each River, and a High-Street (near the middle) from Front (or River) to Front, of one hundred foot broad, and a Broad-Street in the middle of the city from side to side of the like breadth. In the centre of the city is a square of ten acres; at each angle are to be houses for Public Affairs, as a Meeting House, Assembly or State House, Market-House, School-House, and several other buildings for Public Concerns. There are also in each quarter of the city of eight acres, to be for the like uses, as the Moor-fields in London; and eight Streets (besides the High-Street), that run from Front to Front, and twenty Streets (besides the Broad-Street) that run across the city,

Philadelphia, the Expectation of those that are concerned in this Province, is at last laid out to the great Content of those here, that are any wayes Interested therein; The situation is a Neck of Land, and lieth between two Navigable Rivers, Delaware and Skulkill, whereby it hath two Fronts upon the Waters, each a Mile, and two from River to River. Delaware is a glorious River, but the Skulkill being an hundred Mile Boatable above the Falls, and its Course North-East towards the Fountain of Susquahannah (that tends to the Heart of the Province, and both sides our town) it is like to be a great part of the Settlement of this Age ... It is advanced within less than a Year to about four Score Houses and Cottages, such as they are, where Merchants and Handicrafts, are following their Vocations as fast as they can, While the Countrymen are close at their Farms; Some of them got a little Winter-Corn in the Ground last Season, and the generality have had a handsome Summer-Crop, and are preparing for their Winter-Corn. They reaped their Barley this Year in the Moneth called May; the Wheat in the Moneth following; so that there is time in these parts for another Crop of divers Things before the Winter-Season. We are daily in hopes of Shipping to add to our Number; for blessed be God, here is both Room and Accommodation for them.

(From William Penn's letter to the Society of Traders, 16 August, 1683; quoted in B McKelvey's The City in American History)

---

58. Reps, *The Making of Urban America*.

59, 60. Hazard, *Annals of Pennsylvania*.

61. *A Short Advertisement upon the Situation and Extent of the City of Philadelphia and the ensuing Platform thereof*, by the Surveyor General, London, 1683.

62. The Philadelphia gridiron served the first purpose of quickly dividing up the urban land into equal-sized plots such that they could be allocated by lottery on 9 September 1682. However, unlike innumerable subsequent urban gridirons throughout the USA, which were no more than real estate distribution expedients, at Philadelphia there were planned intentions – however unrealistic – for which a grid was the inevitable bureaucratic framework.

Penn, in reaction to the unhealthy, inflammable organic growth conditions of London, not only intended a planned city, but also, above all, had the idea of a healthy, safe city. His requirement that each house should stand in its own garden open space was, however, the antithesis of traditional European urban form. Although Newcourt's intentions in that respect are not known, I believe that whereas his Plan could have made an enduringly great London, the Penn–Holmes Plan – had it been perpetrated – would have resulted in at best a transient 'suburban' character, to be lost when rising land values forced higher density redevelopment. In the event, Philadelphia early assumed the usual European form of continuous street front development.

There were no churches provided for in the Plan, because, Garvan explains, Quaker worship was with one another in their own homes, a version of the religious determinant that enabled low-density rural settlement, compared with the Puritans in New England. Neither is there other than passing reference to the need to accommodate trading activities along the riverfronts, or indeed recognition of the fact that the Delaware was by far the more important river in that respect. Eventually the city centre was in the right place, between the rivers, but in the short term there was no urban focus (see A N B Garvan, 'Proprietory Philadelphia as an Artefact', in O Handlin and J Burchard (eds) *The Historian and the City*, 1966).

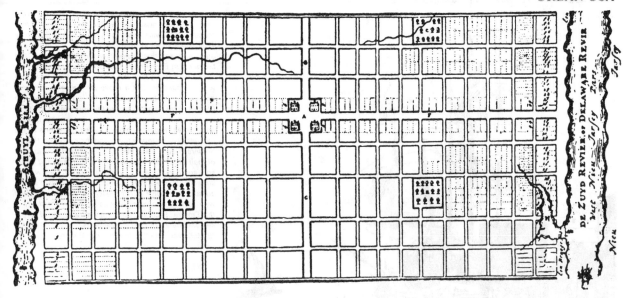

from side to side; all these streets are of Fifty-Foot breadth'.[61]

In the plan of Philadelphia in Figure 10.22 the gridiron is clearly being used as a means to an end.[62] Both the drawing and Holme's description refer to a closed urban programme which is in total contrast to the open-end uses of the gridiron as an end in itself, as generally employed across the American continent by later city 'planners' (including also subsequent expansion of Philadelphia itself).

Two distinct influences determining the Philadelphia plan can be traced back to English precedents. The first of these is a general one, and is embodied as such in Penn's original briefing; the second is specific and relates to post-1666 plans for the rebuilding of London after the fire. Penn himself had been in London during both the great plague and the fire. His detail instructions included the advice that 'every house be placed, if the person pleases, in the middle of its plot, as to the breadth way of it, so that there may be ground on each side for gardens, or orchards, or fields, that it may be a green country town, which will never be burnt, and always be wholesome'.[63]

In 1665 Penn had been living in Lincoln's Inn. As has been noted, this was then the centre of extensive building operations involving the creation of regular-façade, residential squares. Covent Garden was still almost the latest fashion, as yet unspoilt by market activities and the Bloomsbury squares were taking shape. Holme's direct reference to Moorfields in London is of great significance: this open space was available to all rather than only to those actually living in a square.

The second influence is not directly acknowledged by Penn or Holme, but it is probable that both were familiar with the plans produced for the city of London after the fire (see pages 255–8). The plan drawn up by Richard Newcourt, with its main central square and four smaller corner ones (Figure 10.23) bears a close resemblance to that of Philadelphia.[64]

After three years of development there were some 600 houses in the city: this had increased to more than 2,000 by the turn of the century, most of them stately and of brick, generally three storeys high, after the mode in London, and as many as several families in each.[65] The Dela-

Figure 10.22 – Philadelphia, Penn and Holmes's plan of 1683 (north at the top). The city was approximately two miles in length, from the Delaware in the east to the Schuylkill in the west, by one mile in width. The two main cross streets were 100 feet wide; the eight east–west and twenty north–south minor streets were 50 feet wide. The grid blocks were 425 feet by 675 feet and 425 feet by 500 feet. The main central square was of 10 acres, the four minor squares were each of 8 acres. The 'Dock', off the main stream of the Delaware, was the first harbour (bottom right of plan).

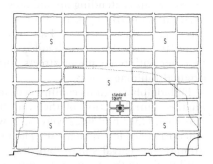

Figure 10.23 – Richard Newcourt's plan for rebuilding London after the fire of 1666. The grid blocks were 855 feet by 570 feet, each with a church in the centre. (See also page 348 for possible relationship between Newcourt's London plan and the layout of Savannah.)

63. Hazard, *Annals of Pennsylvania*, 1850, quoted in Reps, *The Making of Urban America*.
64. The similarity between Newcourt's London Plan and Holmes's for Philadelphia does not extend down in scale, however, to the arrangement of individual grid blocks. Newcourt's five city squares are supplemented by the open spaces containing the local church in each of the blocks. Holmes, in accordance with Penn's instructions, does no more than divide each block up into rectangular street-frontage building lots.
65. G Thomas, *An Historical and Geograhical Account of Pennsylvania and West New-Jersey*, 1698.

With continued growth in the eighteenth century, all the colonial ports assumed additional responsibilities. Muddy streets were improved and extended; Boston by 1720 had a street system that excelled any in the colonies and, as Bridenbaugh has shown, rivalled all but London's in England. Its drainage sewers and its new public market, a gift by Peter Faneuil in 1742, were likewise outstanding. But a move by the selectmen, endorsed by the town meeting, for a bridge over the Charles River failed to secure the approval of the Governor's council. In similar fashion, inadequate powers and limited resources often checked the efforts of the officials in other towns to expand their services. Householders in each place had to sweep the streets in front of their properties, dispose of their garbage and nightsoil, and hang lanterns over their doors if they wanted street lights. Philadelphia was the first in 1750 to secure authorization for a public system of street lamps and won acclaim two years later as the best lighted city in the Empire. Each such accomplishment added to the self-confidence of the colonial ports.
(B McKelvey, The City in American History)

Figure 10.24 – Philadelphia: eighteenth-century waterfront activity.

ware had great advantages over the Schuylkill for shipping and it was soon apparent that Penn's concept of equal river-front growth was unrealistic. Development formed a crescent, centred on High or Market Street and extending up and down the Delaware for about a mile in each direction. By 1755 an additional street between Front Street and the river – Water Street – had encouraged commerce to concentrate along the waterfront. A detailed map of 1762 shows this intrusion, together with other minor streets added to break down the size of the original building blocks.

Philadelphia with its commercial, maritime and civic activities based on the central section of the waterfront became the busiest port in America. Civic buildings were built where sites were available, as and when the need for them arose. The master plan had provided for them around the central square but despite the city's prosperity it was not until well into the nineteenth century that it had expanded so far. As a

Figure 10.25 – Philadelphia: an eighteenth-century waterfront view. The contrast between the rigid orthogonal geometry of the ground plan (Figure 10.22) and the totally uncontrolled three-dimensional form was a general characteristic of American townscape which is discussed further in the conclusion of this chapter.

result Philadelphia's historic civic buildings are scattered around the eastern quarter of the city, in an area which is currently undergoing comprehensive redevelopment.[66]

Thus only towards the end of the nineteenth century did Philadelphia reflect the original intentions to have its focus around Centre Square. A new city hall was eventually completed there in 1890. This new 'central business district' attracted legal, financial and commercial enterprises away from the waterfront, at a time when ocean shipping requirements were determining a move away from the old restricted wharf district. These factors, together with the growth of the railways, resulted in gradual decay of the original port area and the earliest residential districts immediately behind.[67]

CARLISLE

Cumberland County as first established comprised much of Western Pennsylvania, stretching between the Susquehanna and Ohio Rivers. By the middle of the eighteenth century the county had a total population of about 3,000 and Shippensburg, which had been laid out in 1742, became the temporary county seat in 1750, 'being,' as Charles Beetem notes in his *Colonial Carlisle*, 'the most centrally located place in the part of the county which had been settled'. In 1751, however, Governor James Hamilton under proprietary instructions from Thomas Penn located a permanent county town approximately 110 miles west of Philadelphia which was named Carlisle.

In laying out towns in Pennsylvania, the general practice of the Proprietors, led by the Penns, was to follow their own planning example at Philadelphia and there are numerous town plans comprising similarly simple street grids with central squares, entered by main cross routes at their mid-sides. Carlisle, most probably, would have been just another unexceptional grid-town in early American urban history, but for local departures made from the Penns' standard planning practice in the layout of its central square, which has come to represent an intriguing instance of organic growth modification of a plan.

In May 1750 Thomas Penn had sent over from England instructions for the new county town to be fully half-a-mile square, with wide grid streets and a large central square of north–south sides 600 feet in length, and an east–west dimension of 560 feet. The difference was the result of two back-alleys 20 feet wide entering from the eastern and western sides, in the corners. There was to have been a total of thirty-two lots of 60 feet frontage facing onto the square, which was to have contained civic buildings within a 100 feet wide driveway. Governor Hamilton reduced the dimensions to 480 feet and 440 feet respectively, and turned the lots forming the square through 90 degrees, thereby isolating the back-alleys from the square. By 1772 the alleys were still 'dead ends' but in response to real estate interests they were soon diagonally extended into the square.

NEW YORK

In contrast to Philadelphia, its great eastern seaboard rival, New York started without a plan and it was only after almost a century and a half of organic growth that the first continuous gridiron districts were laid out.[68] The foundation of the city by the Dutch in 1624 as New Amsterdam and the real estate 'bargain' Peter Minuit negotiated in 1626 needs no further recounting. The original settlement took the form of a

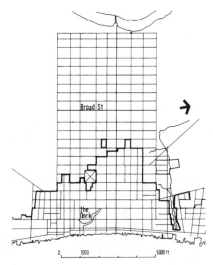

Figure 10.26 – Philadelphia: the central part of the city between the Delaware and Schuylkill Rivers, showing the extent of the built-up area in 1794. The general form of Philadelphia in 1794 was triangular: the base formed by the Delaware water front; the other two sides running inland, intersecting on High Street, somewhat less than half way across the city and still well short of Broad Street, the main cross axis. Only one of the four residential squares – that in the south-east corner – is shown on this 1794 plan. The main central square at the intersection of High Street and Broad Street is also not marked as such in any way.

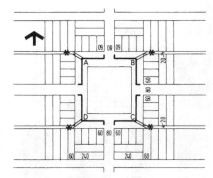

Figure 10.27 – Carlisle, Pennsylvania: the plan of the central square as actually laid out in 1751 shown in heavy line, superimposed on the plan proposed by Thomas Penn in his letter of May 1750.

The original proposal provided for the back alleys to conveniently enter the square at each corner; whereas they were to be 'dead ends' in accordance with Governor Hamilton's unexplained reduction in area of the square. 'Organic growth' pressures subsequently brought about the most unusual diagonal corner entries of the alleys into the square.

66. Bacon, *Design of Cities*.
67. *Urban USA: Redeveloping the Waterfront. Penn's Landing. 'Official Architecture and Planning'*, December 1969.
68. For New York City see S E Lyman, *The Story of New York*, 1975; J A Kouwenhoven, *The Columbia Historical Portrait of New York*, 1953. M C Branch, *Comparative Urban Design*, 1978, with the 1840 SDUK Map of the city.

small fortified village located at the extreme southern end of Manhattan Island, between the broad Hudson River to the west, and the East River.

When the British captured the city in 1664, renaming it New York, the population was around 1,500. Present-day Wall Street, the line of a 1633 defensive system, marked its extent. Chambers Street was reached by 1775, when the population was 23,000. By 1767 growing demand for housing land had already brought about a change from small-scale, sporadic, uncontrolled individual-street development to the laying out of extensive new districts, based inevitably on the gridiron. The Ratzen Plan shows the proposed development of Delancey property including the city's first planned public open space – the Great Square (see Figure 10.28). Unfortunately the War of Independence intervened; subsequently the land, which had become city property, was sold off piecemeal and the Great Square concept was abandoned. (British occupation during the war, and two disastrous fires, reduced the population to an estimated 5,000.) Around 1820, with the Erie Canal having its effect, 150,000 was passed; 300,000 in 1840; 515,000 in 1850 and 942,000 in 1870 show the rate of development. To cater for just

Figure 10.28 – New York City in 1767 (based on the Ratzen Plan). The line of Wall Street across Manhattan Island – marking the defensive wall of 1633 – is shown by asterisks at its Hudson and East River ends. Key: C, the Great Square.

Figure 10.29 – An early nineteenth-century view on Wall Street. (Reproduced courtesy of the Museum of the City of New York.)

Figure 10.30 (right) – New York City in 1850 as recorded by the map published by John Tallis. The southern start-line of the Commissioners' grid of 1811 can be readily identified against the earlier, random alignment grids.

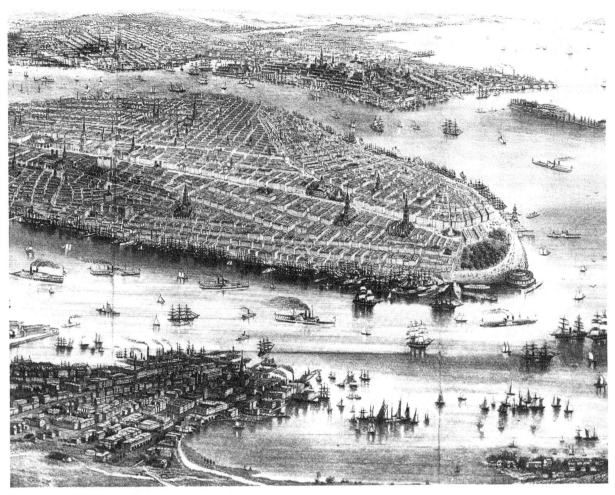

such a rapid growth, the city in 1807 was authorized by New York State to appoint commissioners to lay out the undeveloped main part of Manhattan Island, north of Washington Square.

By this date the gridiron had been used for some additions to the city, but only for unrelated sites. The commissioners' plan of 1811 imposed a completely uniform grid over the island, based on twelve 100 foot wide north–south avenues, and 155 east–west streets, 60 feet in width between the rivers. The rocky topography of Manhattan was ignored by the plan; it also allowed for inadequate open space and had other faults which have been revealed by time.[69] In its favour it can be said to have been far-sighted, by extending much further north than was originally considered necessary, and it did provide the means to control the extremely rapid nineteenth-century expansion. John Reps nevertheless regards the plan as an unequalled opportunity for speculators, with the commissioners motivated mainly by narrow considerations of economic gain.[70] In 1850 42nd Street was the northern limit and the entire island was covered by about 1890. In 1898 the present-day five-borough City of New York was constituted with Brooklyn, the Bronx, Queens and Staten Island added to Manhattan. The inexorable process of filling empty spaces in the commissioners' grid had reached

Figure 10.31 – New York City: an imaginary aerial view of 1869 seen from the western New Jersey bank of the Hudson River, and with Brooklyn in the distance, across the East River which was not bridged until 1883 when the famous Brooklyn suspension bridge was completed. The main concentration of shipping activity on Manhattan Island is along the Hudson River frontage which was extended further north as the port activities developed.

The Manhattan Island skyline is characterized at that time by the uniform three- and four-storey high buildings, punctuated by numerous church spires. (Figure 10.29 provides an impression of the streetscape on the island.) The southern end of Manhattan Island was Castle Garden, an entertainment park and opera house which occupied the old Fort Clinton site, and which was used between 1855 and 1890 to process the admission of more than seven million immigrants to the USA. (Reproduced courtesy of the Museum of the City of New York.)

69. The overriding concern of the New York City Commissioners in 1807 was to establish as quickly as possible a framework for the anticipated rapid urban expansion northwards up Manhattan Island. In so doing, it is clear that they did not allow themselves to be diverted by any thought of the niceties of city planning.

42nd Street by 1850 without any provision having been made for major public open spaces. A campaign for a park had been launched in 1844 by William Cullen Bryant, editor of the New York Evening Post, but as an issue it lacked political support until 1851 when it was taken up by newly elected Mayor Kingsland. Real-estate interests failed to block a project for a large park and by mid-1856 some 810 acres had been acquired, for nearly 8 million dollars, between 59th and 106th Streets and Fifth and Eighth Avenues. (A further land purchase extended the area north to 110th St.) In April 1858 Frederick Law Olmsted and

Figure 10.32 – New York City: the overall extent of Manhattan Island, locating the three early bridges across the East River from Manhattan to Brooklyn, and the George Washington Bridge across the Hudson River to New Jersey. The four motor vehicle tunnels linking the island to Brooklyn and to New Jersey are also shown. Made possible by nineteenth- and twentieth-century construction technology, these river crossings enabled New York City to expand far beyond its original island limits. (The small drawing to the right shows the extent of Manhattan Island compared with that of the city of Paris within the line of the 1845 fortifications – see Figure 6.13.)

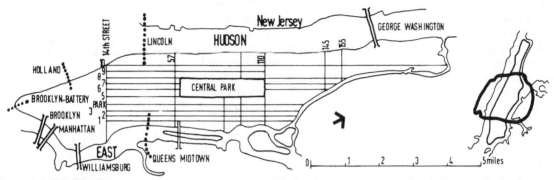

Calvert Vaux won the competition for the design of Central Park and construction work on America's first major public park began.[71]

During the past 100 or more years, innovations in construction technology have played major roles in determining both the two- and three-dimensional forms of cities. Although essentially outside the period in American urban history covered by this volume, the dramatic effects of technological changes on the later nineteenth-century form of New York City must be mentioned briefly here. In 1869 the city was still confined to Manhattan Island, occupying only about its southern half. There were no bridges across either the comparatively narrow East River, or the much more formidable Hudson River: neither were there any signs in the engraving reproduced as Figure 10.31 of the skyscrapers for which, literally above all else, Manhattan has become renowned.

But revolutionary change was in the offing, made possible by, and in response to, the technological advances in structural engineering. These made possible the bridges which facilitated the city's horizontal expansion, and the tall buildings of its incomparable vertical growth. Brooklyn Bridge – the first of the East River bridges – was in fact commissioned in 1869; its central suspended span of 1,595 feet, opened to traffic in 1883, was made possible by the availability of steel wire, of which the four innovatory 11-inch diameter cables were formed. Not only was it the steel frame method of building construction (in succession to cast and wrought-iron) that made skyscrapers structurally possible; it was the same steel wire which supported the 'safety' passenger elevators (or lifts) providing access to the upper floors.[72] The first 'elevator' building to use Elisha Graves Otis' invention was the ten-storey Western Union Building of 1873, from which comparatively small beginnings grew the downtown Manhattan towers, topped by 1914 with the Woolworth Building at 60 storeys and 792 feet high – the world record holder until the Empire State Building was completed in 1931 at 102 storeys and 1,472 feet high. (Currently the two 110 storey World Trade Center towers are Manhattan's tallest buildings.)

Of all history's grids of real-estate expediency, that for Manhattan Island is the most intellectually banal, not least because it has proved the most important.

How could the inevitable grid have been humanized? What provisions might the Commissioners have made for that which I choose to term the 'niceties' of urban life?

The most glaring omissions are parks and waterfront esplanades, but in fairness to the Commissioners – and as a repeat of the landownership circumstances in the post-Fire London (England) of 1666 – the City of New York had no effective legislative means whereby private land could be brought into public ownership for those purposes, even if the funding had been available. To have 'zoned' or otherwise earmarked specific grid blocks as open space would have immediately aroused landowner opposition. Similarly any proposals that would have prevented waterfront commercial development. In the event the Commissioners left it to civic interests and individuals to finance the few open spaces, notably Central Park, as described.

70. Reps, *The Making of Urban America*.

71. Consideration of the design of Central Park is outside the scope of this present volume. Readers are referred to Reps, *The Making of Urban America*, and T Kimball, *Frederick Law Olmsted*, 1928, for further details.

72. The skyscraper did not evolve on Manhattan Island: it originated in Chicago in the late 1860s before its appearance in New York: see C W Condit, *American Buildings*, 1968, as the recommended single volume reference. See also J M Fitch, *American Building: The Forces that Shaped it*, 2nd edn, 1966; M Whiffen and F Koeper, *American Architecture 1607–1976*, 1981.

As a little remarked contributory determinant of the transformation of the downtown 'Wall Street' skyline, that was the part of Manhattan where the direct current electricity relied on by the first elevators was available. (Land values in the Wall Street district, however, were already such that either alternating current elevator motors would have been that much more quickly devised, or a direct current supply brought in.) For the

CHARLESTON AND SAVANNAH

In 1660, when Charles II succeeded to the throne, there were no formally organized white settlements between Virginia and Spanish Florida. Three years later, Carolina territory was established as a proprietary colony and after a false start in April 1670, on an unhealthy site near the mouth of the River Ashley, Charleston was founded in 1672 further inland on the peninsula between the Ashley and the Cooper.[73] The original layout of Charleston consisted of eight irregular grid blocks within a fortified perimeter. By 1717 its population had reached 1,500 and the danger of Indian attack had receded enabling the town to expand beyond the wall. Figure 10.33 shows the extent of Charleston in 1739, with the fortifications of 1672 apparently still retained.

Planning for growth was limited to merely projecting outwards into the countryside the lines of the original gridiron streets, suitably straightened out in one or two instances. At the intersection of the two main streets there was an open square: this space and the attractive harbour promenades would have given the grid some distinction, but 'so little attention was given over the years to preserving the square as an open plaza that one corner was occupied by the market as early as 1739, a church was constructed in another corner in 1761, and in 1780 and 1788 the remaining two corners furnished the sites for an arsenal and the courthouse'.[74] Such 'colonial' neglect of inherited civic space should not give British readers any sense of superiority – a strikingly similar process had already despoiled London's Covent Garden Piazza by the mid-eighteenth century. And what is more, this was encouraged by the landlords, the Dukes of Bedford, who were supposed to know better. John Reps observes that 'when compared to New Haven or Philadelphia, the Charleston plan comes off distinctly second best. The Lords Proprietor were never known for particularly lavish expenditures on behalf of the well-being of the colony, and one is forced to conclude that they carried over this niggardly attitude when they decided on the plan of their capital city. In fact they might have copied the plan of Londonderry, doubled the scale, added an extra tier of blocks all around, and laid it off in their delta site in the Carolinas'.[75]

The southern part of Carolina territory was conveyed by George II to a new group of trustees, in 1732, with the object of founding the colony of Georgia. James Oglethorpe, Member of Parliament and a leading prison reformer, undertook to establish the first settlements and he sailed with the first group of 114 colonists in November 1732. The following February, after calling in at Charleston, Oglethorpe's pioneering party was at work among the pine trees on the southern bank of the River Savannah, clearing a site for the first houses of the future town of that name. An engraving of 1734 (Figure 10.35), 'dramatises the pioneer hardships and portrays the underlying ordering of the ground'.[76]

In two letters of February 1833, Oglethorpe informed the trustees back in London of his intentions; 'I fixed upon a healthy situation about ten miles from the sea. The river here forms a half-moon, along the south side of which the banks are about forty feet high, and on the top flat ... the plain high ground extends into the country five or six miles, and along the riverside about a mile. Ships that draw twelve foot of water can ride within ten yards of the bank. Upon the riverside in the centre of this plain I have laid out the town ... I chose the situation for the town upon an high ground, forty feet perpendicular above high-

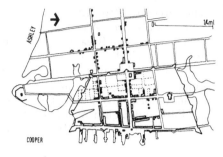

Figure 10.33 – Charleston, South Carolina, in 1739. The defensive perimeter of the original settlement is shown dotted and the beginnings of the extension of the grid outwards into the surrounding countryside can also be seen.

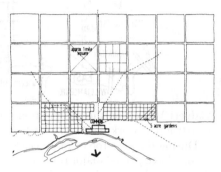

Figure 10.34 – Savannah, the city and land-subdivision in the surrounding countryside at the beginning of the nineteenth century. The original city blocks are in black; the garden and farm lots are in outline.

related structural engineering background to the East River bridges see Condit, *American Building*. Modern New York City is unequalled in its having overcome the topographical deficiencies of starting life on an island, thereby enabling its enormous commuter population to travel to work from both the mainland and Long Island (see Figure 10.32).

73. For Charleston see W J Fraser, *Charleston! Charleston! The History of a Southern City*, 1989.

74. F R Stevenson and C Feiss 'Charleston and Savannah', *Journal of the Society of Architectural Historians*, December 1951.

75. Reps, *The Making of Urban America*. See Figure 4.63 for the plan of Londonderry. Reps is properly scathing in his criticism of the unambitious, unimaginative nature of the Charleston city plan. Boorstin is also highly critical of 'the basic error of the Trustees, from which many other evils flowed, [which] was the rigidity of their rules for the ownership, use, sale, and inheritance of Georgia's primary resource – land (Boorstin, *The Americas: The Colonial Experience*).

76. Bacon, *Design of Cities*, with characteristically beautiful illustrations.

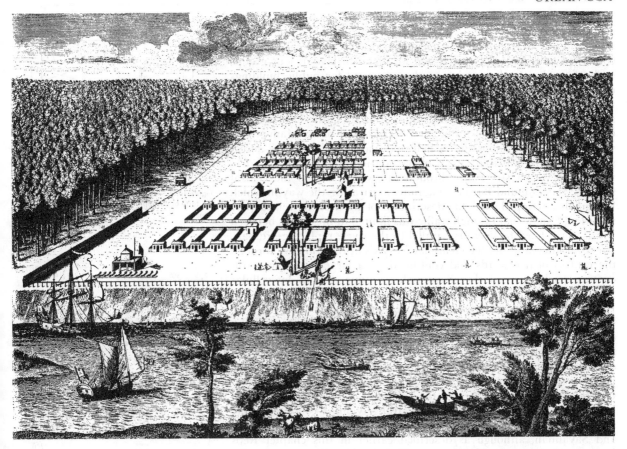

water mark; the soil dry and sandy, the water of the river fresh, springs coming out from the sides of the hill. I pitched upon this place not only for the pleasantness of the situation, but because the above mentioned and other signs I thought it healthy; for it is sheltered from the western and southern winds (the worst in the country), by vast woods of pine trees, many of which are an hundred and few under seventy feet high.'[77]

Oglethorpe's plan for Savannah extended far beyond the immediate limits of its urban nucleus. As shown in Figure 10.34, three kinds of land allocation were provided for: each settler received a house plot in the town, 60 feet wide by 90 feet long; a garden plot of 5 acres near the town; and farmland further out, making a total area outside the town of 50 acres. It was stipulated that a house must be completed within eighteen months; ten years was the time allowed to clear at least 10 acres of the farmland and to put it into production. These wholly reasonable requirements contributed greatly to the successful consolidation of the British colonies, and closely follow those accompanying medieval bastide land grants.[78]

Few American cities used the gridiron as more than an equitable expedient: Savannah is probably the most important exception and the orthogonal geometry of the urban mid-west might well have been less monotonously debasing under its influence had it not been isolated from the immigrant tide in a southern backwater. John Reps sees this

Figure 10.35 – Savannah, a view of 1734. In the most advanced, bottom left-hand ward, three of the central public building sites are shown occupied. The uniformity of house building (cabins more likely) was not just artistic economy, according to Mills Lane, who notes that 'at least the first forty, perhaps the first seventy timber framed houses were all to the same model and dimension, each one and a half storeys raised 2 ft above ground level and 24 ft wide by 16 ft deep (*Architecture of the Old South: Georgia*, 1986).

77. Oglethorpe's letters quoted by Reps, from Georgia Historical Society, *Collections 1 and 2*, Savannah 1840 and 1842.
78. But see Boorstin, *The Americas: The Colonial Experience*, for the further criticism (as Note 75) that 'fifty acres of Georgia pine-barren proved insufficient to support a family'.

as 'one of the great misfortunes of American town planning'.[79] Savannah was different because it was laid out on the basis of finite cellular units and not as an infinitely extensible grid. The units, called wards, contained forty house plots and had an identical layout: four groups each of ten house plots and four plots reserved for public buildings enclosed a public square (Figure 10.36). Four wards were laid out at first, providing for a total of 160 houses (Figure 10.35). The route structure which organized the cells into an urban entity clearly differentiated between types of traffic on a hierarchical basis. Main streets were 75 feet wide, minor streets half that width, and the back access service lanes (reminiscent again of bastide planning) were 22½ feet wide. The widest street ran through the wards, across the central squares, but Edmund Bacon's diagrams illustrating the way in which Savannah developed, and his description of its modern central city situation,[80] show that through traffic was limited to the streets between wards. Tree-lined boulevards parallel to the river, replacing ordinary streets at intervals, and the creation of an axis at right-angles, through the centres of five wards and continued by a large late nineteenth-century park, ensured that Savannah's unique growth by cellular repetition did not lose coherence with size.

American urban historians have been greatly concerned to trace back the origins of Oglethorpe's plan. Clearly, as in the case of Philadelphia, the London squares were an immediate source of ideas, more so for Oglethorpe than for Penn and Holme; by the 1730s Mayfair had demonstrated their use, combined with grid streets, over a large new part of the city (as noted on page 261). Reps establishes that Oglethorpe was fully aware of the planning of Londonderry as an English plantation in Ulster.[81] But he strangely omits pointing out the similarity between Newcourt's plan for London after the Fire of 1666 and Savannah, although as described earlier in this chapter Reps argues an influence by Newcourt on Philadelphia (Figure 10.23). By comparison Savannah owes much more to Newcourt than does Philadelphia. Turpin C. Bannister writing in the *Journal of the Society of Architectural Historians* puts forward the argument that Renaissance ideal city planning was the most probable source of the Savannah plan, in particular a design of 1598 by one Robert Barret.[82] Bacon takes a similar approach and cites a plan by Pietro di Giacomo Cataneo included in his *L'Architettura* published in Venice in 1567 (Figure 5.14).

Whatever its origins and whoever assisted him, Oglethorpe's name coupled with that of Savannah merits a far higher position in urban history than hitherto accorded. Great credit is also due to those anonymous city fathers who refused, long after their planner's death, to allow the design to be compromised.

LOUISVILLE AND JEFFERSONVILLE

As established earlier, points of trans-shipment from one means of transport to another were an almost certain location of urban settlement. A survey of 1824, on which Figure 10.38 is based, shows that five such settlements had been established alongside the Ohio, each of which claimed to be the future metropolis of the falls area and engaged in the typical antics of town promotion.[83] Clarksville and Louisville were first on the scene, followed by Jeffersonville, Shippingsport and Portland as river traffic increased and the possibilities of canals by-passing the Falls were discussed. Later a sixth city, New Albany, was

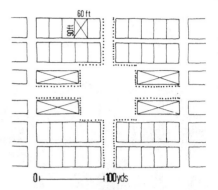

Figure 10.36 – Savannah: the basic ward unit, comprising forty house plots and four reserved public building sites.

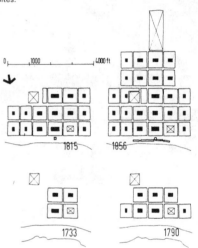

Figure 10.37 – Savannah: four growth stages of 1733, 1790, 1815, 1856. The standard ward unit (Figure 10.36) is in outline with the central open squares in black. Not all the ward units were the same size.

79. Reps, *The Making of Urban America*. However, when a gridiron city was well endowed with open space at the outset – eg Louisville (Figure 10.39) and Jeffersonville (Figure 10.40) – it was not long before the city fathers succumbed to development pressures and traded in that birthright. (Further proof, if necessary, that a city plan is only as good as its political trustees.)
80. Edmund Bacon observes that 'when one is within any of these squares one feels entirely removed from the rushing traffic of the surrounding streets, which crosses but does not parallel the lines of sight ... traffic has been allowed to park only on the square which was originally the market' (*Design of Cities*).
81. For Londonderry see Chapter 4: caption to Figure 4.63; see also J S Curl, *The Londonderry Plantation 1609–1914*, 1986.
82. T C Bannister, 'Oglethorpe's Sources for the Savannah Plan', *Journal of the Society of Architectural Historians*, May 1961. Bannister, and less so Bacon, can be seen to be aiming both too far, and too high in seeking respectable European intellectual precedents for American colonial plans in Renaissance 'Ideal city' exercises. Healthy expediency, perhaps tempered by a certain humanizing, was the extent of ambition, and a simplistic gridiron its inevitable vehicle.
83. Reps, *The Making of Urban America*.

founded on the north bank of the Ohio.

Clarksville was founded in 1783 on the northern Indiana bank, but benefited little from the developing trade which became monopolized by the southern Kentucky-side cities. Louisville, eventually the dominant city, was founded on 24 April 1779 when 'the intended citizens of the Town of Louisville' met to draw lots of half-acre sites in the township. The new owners undertook to clear off the undergowth and begin to cultivate part of the lot by 10 June and to build a good covered house 16 by 20 feet by December 25.[84] These first citizens formed a part of a small military force which had arrived at the Falls of the Ohio, many with their families, early the previous summer. Their first fortified encampment was on Corn Island. A previous attempt (1773) to found a town in the vicinity on 2,000 acres of land granted to a Doctor John Connolly for military services had been abandoned during the war.

Louisville was named in honour of the French king and the plan, attributed to George Rogers Clark, the military commander, comprised 'a number of lots, not exceeding 200 for the present, to be laid off, to contain half an acre each, 35 yards by 70, where the ground will admit of it, with some public lots and streets'.[85] This plan was notable for the reservation of the area between Main Street and the river, together with the row of plots along the southern boundary, as public land (Figure 10.39). Reps mentions a local tale which credits Clark with the intention of repeating this strip of common land every third street to the south. In reality, however, Louisville had to dispose of this potentially invaluable asset very early on in order to meet debts of Connolly's, for which, as the result of political manoeuvres, it was held responsible. On a city plan of 1836 there is evidence of an attempt to revive this feature which would have been as attractive and functional as it would have been unique[86] but such community amenities were disregarded in later growth.

By 1797 Louisville is recorded as having some 200 houses, while Clarksville was a village of about 20 houses. Shippingsport, downstream from Louisville at the foot of the Falls, was platted in 1803 and by 1806 had become the favoured port for upstream traffic, thus complementing Louisville. Further downstream Portland was platted in 1814, in anticipation of the construction of the canal around the Falls. This was completed on the southern side in 1830, after years of vicious wrangling between rival Kentucky and Indiana business interests, each determined to capture the future toll income.

Jeffersonville on the Indiana shore was first established in 1802; its novel layout is shown in Figure 10.40. The commercial intention was to hive off Louisville's trade. President Jefferson was notable for a keen interest in urbanism; he had been closely involved in the planning of Washington, supplying L'Enfant with details of contemporary European design, and he had been concerned with drafting the 1785 Land Ordinance. Jefferson had observed the ease with which yellow fever epidemics were spreading through American cities and recommended a more open form of planning than their built-up grids. Using the chequer board as an example he proposed that the black squares only be building squares, and the white ones be left open, in turf and trees; every square of houses would be surrounded by four open squares, and every house would front an open square.[87]

Unfortunately, the executive planner, John Gwathmey, 'fancied

Figure 10.38 – Towns at the Falls of the Ohio: Louisville, 1779; Clarksville, 1783; Jeffersonville, 1802; Shippingsport, 1803; Portland, 1814. (The Ohio flows from right to left.)

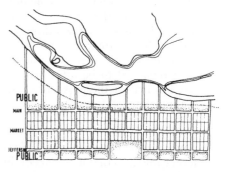

Figure 10.39 – Louisville, 1779, showing the open strips of common land on each side of the original two rows of grid blocks.

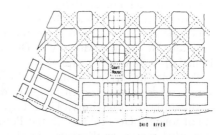

Figure 10.40 – Jeffersonville, 1802, showing the original chequer-board pattern of open and built squares and the diagonal street system.

84. Brown, *Historical Geography of the United States.*
85. Reps, *The Making of Urban America.*
86. John Reps writes that 'for 3,300 dollars the town disposed of its lands'.
87. Yellow fever is carried by a mosquito and it is doubtful whether the open spaces would have contributed to its control or eradication. In equating low-density urban form with health, Jefferson was following the examples of Charleston, Philadelphia and Savannah.

himself something of a L'Enfant' and combined a diagonal street pattern with the building grid. The town proprietors opposed the plan with its 'wasteful' use of space. In this they were prompted by the demand for lots in a likely candidate for the metropolis of the Falls region, and they petitioned Congress for 'an Act to change the Plan of the Town of Jeffersonville'. This was passed in 1816 and under the new plan only one of the open squares survived.

Louisville profited greatly from the increase in river traffic following the introduction of steam boats in 1840 and the completion of the Kentucky–Ohio Canal. There were the usual doubting businessmen, frightened as to the possible damaging effects of the canal on the city's trade, but they were silenced by Henry McMurtrie's question 'do the gentlemen really believe that Louisville draws her importance solely from the obstruction to the navigation of the river, or do they pretend to assert that a canal or river has ever deducted from the population, wealth or business of a town through which it has passed'.[88] Before the canal was built through-traffic in the city was very heavy, but as industries established themselves an increasing proportion of the river-borne freight remained to be processed in Louisville. Nevertheless the city today has neither the prosperity it desired nor the unique plan it did not appreciate.

WASHINGTON, DC

From the outset the new United States government intended to create a federal capital, so that national administration could be conducted free from local political pressures, and without favouring any one State with its presence.[89] The first president, George Washington, was authorized by the Residence Act of 1790 to select a site, not more than 10 miles square, on the Potomac River between the mouth of its Eastern Branch and the Connogocheague. He was to appoint three commissioners to supervise the survey, and have the new city ready for its national functions by the first Monday in December 1800. (During this period Philadelphia was to be capital.)

Possible sites were investigated during the autumn of 1790, and in January 1791 the President announced the choice of the southern end of the authorized territory. The selected area had the well-established small town of Georgetown on the Potomac to the west, and included two other platted settlements, Hamburgh and Carrollsburg, neither of which, however, contained many buildings. Two independent advisers were responsible for surveying the site: Andrew Ellicott, a professional surveyor, was appointed in February 1791 to determine the boundaries, followed one month later by Major Pierre Charles L'Enfant, later to be responsible for planning the new capital city but with an initial commission providing only for the internal site survey.

L'Enfant (1754–1825) was born in Paris and spent his childhood in and around the palace and park of Versailles. The son of a gifted painter he studied at the Académie Royale de Peinture et de Sculpture before emigrating to America in early 1777, to take up a commission in the revolutionary army.[90] He was first mentioned in the *Journal of Congress* as being present at the siege of Savannah in the autumn and during the later stages of the war he served under Washington as a military engineer. After the war, L'Enfant settled for a while in New York City where, in practice as an architect, his work included alterations to Federal Hall in January 1785 to accommodate Congress. On 11

Louisville, at the falls of the Ohio, is about the size of Lexington, or perhaps, at this time, more populous. From a commercial point of view, it is by far the most important town in the state. The main street is nearly a mile in length, and is as noble, as compact, and has as much the air of a maritime town, as any street in the western country. It is situated on an extensive sloping plain, below the mouth of Beargrass, about a quarter of a mile above the principal declivity of the falls. The three principal streets run parallel with the river, and command fine views of the villages and the beautiful country on the opposite shore. It has several churches, among them churches for the presbyterians, baptists and Roman catholics. The mouth of Beargrass affords an admirable harbour for the steam boats, and river craft. The public buildings are not numerous, but respectable; and the people are more noted for commercial enterprise, than for works of public utility.
(From T Flint, A Condensed Geography and History of the Western States, or the Mississippi Valley, Cincinnati, 1828; quoted in B McKelvey's The City in American History)

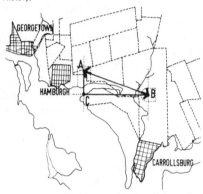

Figure 10.41 – Washington: the site of the national capital in 1791 showing the three existing platted towns – Georgetown, substantially developed and able to entertain the president when he visited the locality; and Hamburgh and Carrollsburg which were essentially 'paper towns' with few houses. This map also shows the property boundaries in dotted line, and superimposed in heavy line, the Federal Triangle (ABC), from which the plan of the city was developed.

Key: A, the President's House (White House); B, the Capitol (Jenkin's Hill); C, location for a memorial (the Washington Memorial).

88. Brown, Historical Geography of the United States.
89. Washington is therefore one of a number of federal capitals which have been created; others with significant urban form importance include Brasilia (the idea for which predates that for Washington) and Canberra.
90. For Major Charles Pierre L'Enfant, the most significant city planner in American history, if only because of his Washington (Master) Plan, see P Caemmerer, The Life of Pierre Charles L'Enfant, New York, 1970; J W Reps, Monumental Washington, 1967. L'Enfant's military service made him known to the right people in the right place at the right time; similarly well placed – as a modern parallel – Albert Mayer had served in India during the Second World War, meeting Jawaharlal Nehru as a fortunate introduction to his being appointed the original planner for Chandigarh, the new capital city of the East Punjab. (Le Corbusier inherited Mayer's Plan and modified it into his final version.) See A E J Morris, 'Chandigarh: the Plan Corb tore up?', Built Environment Quarterly, December 1975.

September 1789, Paul Caemmerer records that 'while yet the idea of having a capital city was still unsettled, L'Enfant wrote to President Washington asking to be employed to design the capital of this vast Empire'.[91]

By the end of March 1791, when delimitation of the federal district and compensation of landowners were agreed, L'Enfant was further commissioned to plan the city itself. In his site analysis report he identified Jenkin's Hill – now Capitol Hill – as the best location for public buildings, and he stressed the need for a plan which would '... render the place commodious and agreeable to the first settlers while it

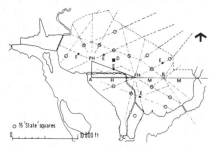

Figure 10.42 – Washington: the L'Enfant plan, a copy of the original manuscript drawing, made in 1887. (See Figure 10.45 for details of the central section of the city as given by Ellicott's version of 1792.) The manuscript contained numerous marginal notes, giving an indication of L'Enfant's detailed planning intentions. Unfortunately these notes cannot be reproduced in their original (as copied) handwriting form, but they include the following points, keyed on the related sketch plan below.

References (see sketch plan above):

A, The equestrian figure of George Washington, a monument voted in 1783 by the late Continental Congress.

B, An historic column ... from whose station (a mile from the Federal House) all distances of places through the Continent are to be calculated.

D, This Church is intended for national purposes ... and assigned to the special use of no particular sect or denomination ...

E, Five grand fountains intended with a constant spout of water NB: there are within the limits of the City above 25 good springs of excellent water abundantly supplied in the driest season of the year ...

Squares ... being fifteen in number are proposed to be divided among the various States in the Union, for each of them to improve, or subscribe a sum additional to the value of the land for that purpose, and the improvements around the Square to be completed in a limited time ...

Every house within the City will stand square upon the streets, and every lot, even those on the divergent Avenues, will run square with their fronts, which on the most acute angle will not measure less than 56 feet and many will be above 140 feet.

H, Grand Avenue, 400 feet in breadth and about a mile in length, bordered with gardens ending in a slope from the houses on each side ...

M, Avenue from the two bridges to the Federal House, the pavement on each side will pass under an arched way under whose cover shops will be most agreeably situated. This Street is 160 feet in breadth and a mile long.

may be capable of being enlarged by progressive improvement ... which should be foreseen in the first delineation in a grand plan of the whole city'.[92] Working with commendable speed L'Enfant presented his final plan to the president in August 1791. (The reproduction in Figure 10.42 is from a copy of the damaged original made in 1887.) L'Enfant did not enjoy a smooth relationship with the commissioners during the design phase, but his obvious planning expertise made him indispensable, added to which he had the president's whole-hearted support. As soon as the implementation phase was under way, L'Enfant's position quickly became much less secure. He had his own site development ideas, based on establishing the main streets and public buildings first, before selling to the public a proportion only of the 15,000 or so lots. He argued that the better situated land would then realize its potential, higher value. However, for neither the first nor last time in urban history pressing political considerations made essential as rapid a start as possible. An anti-federal-capital campaign was gathering momentum and Philadelphia's claims in particular made any delays inexpedient.[93] The commissioners therefore instructed L'Enfant to have 10,000 copies of his plan prepared for the public land auction of 17 October 1791.

91. Caemmerer, *The Life of Pierre Charles L'Enfant.*
92. L'Enfant's complete initial site report is included in E S Skite, *L'Enfant and Washington 1791–92,* 1929 (quoted by Reps).
93. Brasilia, the new Federal Capital of Brazil (from 1960) is a modern parallel in respect of the need to become quickly established during the four-year presidency of Dr Juscelino Kubitschek.

Not only did L'Enfant fail to produce the copies – blaming engraving difficulties – but he also refused to allow his original plan to be exhibited at the auction which was held, rather unsatisfactorily, on the basis of the plats of individual blocks. Earlier, in August, L'Enfant had directed the demolition of a new house not in conformity with his plan, without consulting the commissioners. In December there was further friction over instructions to contractors. Still L'Enfant enjoyed the president's confidence and survived, but his days were numbered. He continued to obstruct publication of the plan, so much so that early in 1792 the president was forced to direct Ellicott to carry out the necessary engraving and printing work. L'Enfant refused him access to his master copy and Ellicott was forced to reconstruct the design on the basis of his own survey information, incorporating, it would seem, modifications put forward by various people, including Jefferson. (The central section of the Ellicott plan is given in Figure 10.45.) The changes, although not of great consequence, provoked L'Enfant to write to the president alleging that his plan was 'most unmercifully spoiled and altered ... to a degree indeed evidently tending to disgrace me and ridicule the very undertaking'.[94] Whether by design or accident L'Enfant's name had also been omitted from the printed version. In late February, after he had spurned the final offer of continued involvement, made on the condition that he accepted the authority of the commissioners, L'Enfant's stubbornness gave the president no alternative but to end his appointment. Of L'Enfant's brief eleven-month career as capital city planner,[95] well under half the time – about 144 days in all – had sufficed to produce the fully developed design. This was remarkably fast work. Even allowing for the fact that L'Enfant was known to have been determining the basis of the plan while occupied with his earlier survey work – giving him six months at the outside – it would clearly have been impossible for anyone less a genius than L'Enfant, or less familiar with the design idioms that make up this vast essay in civic development.[96] Judged on his plan L'Enfant in all probability was a genius and apparently one of the most uncompromising kind. Inspired adaptation of existing ideas to a site, rather than significant innovation was his forte but, as discussed below, widely differing views are held as to the merits of his work.

Three particular influences can be traced in L'Enfant's plan: first, his French family background; second contemporary late-eighteenth-century European urbanism; third, proposals for Washington put forward by Jefferson. The significance of L'Enfant's childhood at Versailles has already been introduced; there he had first-hand experience, at a formative age, of urban history's pre-eminent example of autocratic aggrandizement, whereby the town's three main diagonal streets, superimposed on a gridiron, are focused on the king's bedroom in the centre of the palace. He was also a student in Paris where he would have appreciated the carefully arranged 'use of monuments or important buildings as terminal vistas to close street views'.[97] It is possible that the Place de la Concorde's fusion of architectural and landscape elements found an echo in his treatment of the Mall and the setting for the President's House. His proposal (unfortunately never realized) to line main streets in Washington with arcaded shops has a precedent in the shops under construction near the Palais Royal when L'Enfant revisited Paris in the winter of 1783–4.[98]

Thomas Jefferson, the second president, whose urban planning in-

... a territory not exceeding 10 miles square (or, I presume. 100 square miles in any form) to be located by metes and bounds.

3 commissioners to be appointed
I suppose them not entitled to any salary.
(if they live near the place they may, in some instances, be influenced by self interest, & partialities: but they will push the work with zeal, if they are from a distance, & northwardly, they will be more impartial, but may affect delays).
the Commissioners to purchase or accept 'such quantity of land on the E. side of the river as the President shall deem proper for the U.S.' viz for the federal Capitol, the offices, the President's house & gardens, the town house, Market house, publick walks, hospital, for the President's house, offices & gardens, I should think 2. squares should be consolidated, for the Capitol & offices one square. for the Market one square, for the Public walks 9. squares consolidated.
The expression 'such quantity of land as the President shall deem proper for the U.S.' is vague, it may therefore be extended to the acceptance or purchase of land enough for the town: and I have no doubt it is the wish, & perhaps expectation, in that case it will be laid out in lots & streets. I should propose these to be at right angles as in Philadelphia, & that no street be narrower than 100. feet. with foot-ways of 15. feet. where a street is long & level, it might be 120. feet wide. I should prefer squares of at least 200. yards every way, which will be of about 8. acres each.
The Commissioners should have some taste in architecture because they may have to decide between different plans ...
In locating the town, will it not be best to give it double the extent on the eastern branch of what it has on the river? the former will be for persons in commerce, the latter for those connected with the Government.
Will it not be best to lay out the long streets parallel with the creek, and the other crossing them at right angles, so as to leave no oblique angled lots but the single row which shall be on the river?
(T Jefferson, as quoted by S K Padover in Thomas Jefferson and the National Capital, Washington: United States Government Printing Office, 1946)

94. L'Enfant in a letter to Tobias Lear, Secretary to the President, dated Philadelphia, 17 February 1792. Beginning 'Dear Sir, Apprehending there may be some misconstruction of my late conduct and views' L'Enfant seeks to blame Ellicott for delays etc. (Caemmerer, The Life of Pierre Charles L'Enfant).
95. L'Enfant was appointed shortly before arriving on the site of future Washington on 9 March 1791; he was dismissed formally in a letter of 27 February 1792. Immediately after his Washington commission was terminated, L'Enfant was engaged in planning Patterson in New Jersey, as a city that would 'far surpass anything yet seen in this country' (L'Enfant quoted by Caemmerer). It is not clear exactly what he achieved but this second involvement was also short-lived. In 1794 L'Enfant was the architect for Robert Morris's proposed mansion in Philadelphia; on account of its cost, this became known as 'Morris's Folly'.
96, 97. Reps, The Making of Urban America.
98. Reps, Monumental Washington. L'Enfant's idea of recreating Parisian urban niceties in Washington ('M', Figure 10.42) exemplifies the boundless 'other world' optimism of early American city planners. In real life such an idea stood no chance whatsoever of implementation.

terests have been described earlier in this chapter, generously passed on to L'Enfant a collection of some twelve European city plans, including Paris, Karlsruhe, Milan and Amsterdam, after hearing that his own hopes of planning Washington were not to be realized. Jefferson in fact produced what were most probably the original sketch plans for the new city, first for the area covering the site of Carollsburg and second for the central section of the land subsequently taken for the city by L'Enfant (Figure 10.44). The latter's debt was essentially to this second sketch; this located the President's House in more or less its final position and related it eastwards to the Capitol by a riverside walk along the northern bank of Tyber Creek. The Capitol, however, was only about half-way between the President's House and Jenkin's Hill (its eventual location) giving Jefferson's plan, with its simple gridiron

Figure 10.43 – Washington: aerial view looking west from a point just north of the axis Capitol Hill (A) to the Washington Memorial (C). The White House (B) completes the Federal Triangle, connected to the Capitol by Pennsylvania Avenue. The Pentagon, on the Virginia bank of the Potomac, is keyed D. Georgetown is at the top right.

353

structure, only a modest uninspired scale, which would most probably have proved unsuited to Washington's future role. Jefferson was an advocate of the gridiron and looked no further than Philadelphia for his inspiration but his Washington, although given a certain individuality by its waterside location, must surely have been just another expediency-grid city had it expanded on that base without later, distinctive, restructuring elements.[99]

THE L'ENFANT PLAN

Because of its modest extent Jefferson's plan – if it had been implemented – could quickly have attained an urban character and might perhaps have grown by the controlled addition of new blocks along the lines of Savannah. L'Enfant's plan meant that there was to be about a century of sporadic urban development at Washington, but the plan was fulfilled in the long term. It proved able to cope with unforeseeable future requirements, functional and symbolic, rather than mere short-term practicalities. Indeed, as noted below, the unsatisfactory and sporadic implementation of his plan may well also have been avoided had L'Enfant's development proposals been adopted. His plan has detailed failings, both inherent and imposed, but above all else it has ensured a magnificently scaled setting for the government of a dominant world power.

The apparent complexity of the layout was developed from a simple base: that of the right-angled triangle formed by the Capitol, the Presi-

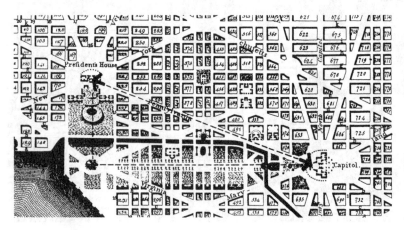

dent's House and, at the intersection of the respective east–west and north–south axes, the Washington Memorial (Figure 10.45). These axes were also those of a regular gridiron – the essential basis of land subdivision.[100] The third side of the triangle cut across the grid as a direct link between the Capitol and President's House. This is Pennsylvania Avenue, ostensibly the nation's most important street, and it symbolizes the distinct but mutually dependent relationship between legislature and executive. In contrast to the autocratic self-centring of Versailles, from where L'Enfant drew inspiration, it was his genius that created urban history's pre-eminent example of democratic aggrandizement. Based on the diagonal line of Pennsylvania Avenue, a system of symmetrically arranged main routes was superimposed on the grid to connect other key internal locations and to relate to important regional roads.

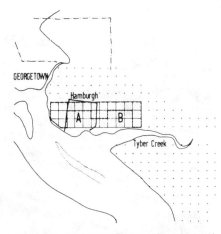

Figure 10.44 – Washington: Thomas Jefferson's plan of 1791 for the capital city, located between Georgetown and Tyber Creek and comprising a total of only thirty-three grid blocks, fourteen of which were earmarked for the President's House (A), the Capitol (B), and the connecting Public Walk along the northern bank of Tyber Creek.

Figure 10.45 – Washington: detail plan of the centre of the city, taken from Ellicott's version of the L'Enfant plan; the Federal Triangle has been added by the author. The approximate scale of this extract is 1:36,000; from the President's House to the Capitol is just over 1½ miles.

Reciprocity of views between the three most important of L'Enfant's objects – President's House, Capitol and the Washington Memorial (denoted by an asterisk) – has not been retained; the US Treasury building blocks the view along Pennsylvania Avenue (see Figure 10.62, a plan of Lafayette Square, 1857), and the Washington Memorial, eventually built from the mid-nineteenth century as a 580 feet high obelisk, was located some 300 feet east of its intended position at the intersection of grid axes through the White House and Capitol, because more expensive foundations would have been necessary for it to have been placed in the correct place.

Philadelphia April 10, 1791
To Major L'Enfant
Sir

I am favoured with your letter of the 4 instant, and in compliance with your request I have examined my papers and found the plans of Frankfort on the Mayne, Carlsruhe, Amsterdam, Strasburg, Paris, Orleans, Bordeaux, Montpelier, Marseilles, Turin, and Milan, which I send in a roll by this Post. They are on large and accurate scales, having been procured by me while in those respective cities myself. And they are connected with the notes I made in my travels and often necessary to

99. Jefferson's Washington could also have incorporated the symbolic linked distinction between the President's House and the Capitol along a grand waterside promenade. But if as intended the President had faced down the Potomac, and the legislature along the waterfront, then their comparatively close proximity would have revealed them to be looking in different directions. As it is, the greater separating distance of the L'Enfant Plan makes the different orientations less significant.
100. The main text adequately establishes L'Enfant's idea for the city, summarized here as symbolic diagonal linkages superimposed on an expediency real-estate gridiron.

Implementation of the plan starting slowly, Washington still deserved the various epithets that were coined, such as 'the city of magnificent distances' and 'the city of magnificent intentions'. Francis Baily, an English visitor of 1796, wrote: 'there are about twenty or thirty houses built near the point as well as a few in South Capitol Street and about twenty others scattered over in other places: in all I suppose about two hundred: and these constitute the great city of Washington. The truth is, that not much more than one half of the city is cleared: the rest is in woods; and most of the streets which are laid out are cut through these woods, and have a much more pleasing effect now than I think when they shall be built; for now they appear like broad avenues in a park, abounded on each side by thick woods; and there being so many of them, and proceeding in so many directions, they have a certain wild, yet uniform and regular appearance, which they will lose when confined on each side by brick walls.[101] Little did Bailey know how true his observation was to be: how beautiful those distant views must have seemed but how ugly they were to become when transposed into undisciplined bricks and mortar.

Although as assessed below, Pennsylvania Avenue's non-existent street architecture has deserved all the criticism it has received, Jefferson, immediately following L'Enfant's dismissal, did ensure that the physical and symbolic link between the President's House and the Capitol would at least be defined by planned tree planting. Accordingly he invested one-third of the limited funds that were available in a landscaping programme that subdivided L'Enfant's route into a roadway flanked by footpaths and reservations.[102]

However, the raising of money for the construction of federal buildings through sale of lots reserved for the purpose, and by lotteries, proceeded very slowly. As also the population which, from an estimated 3,000 in 1800, had only attained 23,000 by 1846 when Congress ceded back to Virginia that part of the Federal District lying across the Potomac. In the second half of the century, Washington grew much more rapidly: 61,000 in 1860 had become nearly 110,000 ten years later, under the stimulus of the Civil War. Only then were the extensive open spaces revealed on a detail map of 1854 even beginning to be occupied.[103] Washington's magnificent intentions had long since been abandoned; by mid-century it had resolved itself into 'a plan without a city'. The late nineteenth- and twentieth-century proposals to resolve the worst of the two-dimensional anomalies and to create, at least for the major avenues, a suitable street architecture are outside our period, other than to note that during the 1970s and 1980s (in effect the gestation of this book) renewed efforts to enhance the status of Pennsylvania Avenue are at last coming to fruition.[104]

This chapter, establishing significant aspects of the origins of urban America, is concluded with a brief mention of a further seven city examples, and a summary of reasons for the characteristic total contrast between the two-dimensional order of urban plans, and the consequent three-dimensional anarchy, embodied in visual poverty, which resulted when these plans were put into practice.

DETROIT

Detroit was established in 1701 by Antoine de la Mothe Cadillac to control the Detroit River for the French, between Lakes Erie and St Clair.[105] The original fortified settlement was about 600 by 400 feet in

explain them to myself, I will beg your care of them and to return them when no longer useful to you, leaving you absolutely free to keep them as long as useful.

I am happy that the President has left the planning of the Town in such good hands, and have no doubt it will be done to general satisfaction ... Whenever it is proposed to prepare plans for the Capitol, I should prefer the adoption of some one of the models of antiquity, which have had the approbation of thousands of years, and for the President's House I should prefer the celebrated fronts of modern buildings, which have already received the approbation of all good judges. Such are the Galerie du Louvre, the Gardens meubles, and two fronts of the Hotel de Salm. But of this it is yet time enough to consider, in the mean time I am with great esteem Sir &c

TH. Jefferson
(From H A Washington, The Writings of Thomas Jefferson, 1859; quoted in B McKelvey's The City in American History)

101. F Bailey, Journal of a Tour in Unsettled Parts of North America in 1796 and 1797, London, 1856. I have retained the phrase 'how ugly they were to become' in this edition in order to emphasize the difference between a Washington that might have been – with urban architecture in keeping with the Plan – and the uncontrolled reality of the Plan implementation.
102. The role of trees in urban design has been introduced in Chapter 6 related to the mid-eighteenth-century French Renaissance urban design at Nancy (Figure 6.22) and the early eighteenth-century Place Louis XV at Rennes (Figure 6.24). There were also earlier examples of tree planting as the basis of grand-scale landscape designing, notably by Andre Le Notre at Vaux le Vicomte and Versailles (Figure 6.26) of which the latter would have been familiar to Pierre L'Enfant.
103. For this magnificent map of 1854 (prepared by the US Coast Guard and Geodetic Survey) which plots every building in existence at that time, see Of Plans and People, a study of the plan of Washington prepared by the Washington-Metropolitan Chapter of the American Institute of Architects, May 1950.
104. For the later history of Washington, readers are referred to the following sources: Reps, The Making of Urban America; Of Plans and People; The Grand Design, an exhibition catalogue tracing the evolution of the L'Enfant plan and subsequent plans for developing Pennsylvania Avenue and the Mall area, organized jointly by the Library of Congress and the President's Temporary Commission on Pennsylvania Avenue (The Library of Congress, Washington, 1967); and a summary of proposals for the redevelopment of Pennsylvania Avenue and the creation of a new square at its White House end, based on the author's discussions in 1968 with planning staff of the president's commission (Official Architecture and Planning, January 1969). In the mid-1970s the proposals for a grand new square at the western end of Pennsylvania Avenue were quietly dropped in favour of a much smaller 'Western plaza' formed in part by rehabilitated existing buildings – see The Pennsylvania Avenue Plan 1974 (published by the Pennsylvania Avenue Development Corporation).
105. For Detroit see A Pound, Detroit: Dynamic City, 1940 (with evocative drawings of the contemporary city by E H Suydam).

extent and strongly reminiscent of a medieval French bastide. In June 1805, fire destroyed the fort and some 300 dwellings in the adjoining civil town, but earlier that year Detroit had been designated as the capital of the new Michigan Territory of the USA. One of the three judges appointed to administer the Territory was Augustus Woodward. Judge Woodward, only 31 years old, was the man 'shortly to focus his numerous talents on the exciting task of creating a metropolis in the west'.[106] He had already been an experienced property speculator in his native Washington, where he knew both L'Enfant and Jefferson.

Arriving in Detroit some three weeks after the fire, Woodward was in time to prevent the citizens from redeveloping their old sites, and managed to have himself appointed 'a committee of one to lay out the new town'. Under an Act of Congress of 1806, 'To Provide for the Adjustment of Title of Land', the city area was surveyed and its new layout approved. This design (shown in Figure 10.46), was based on a combination of rectilinear and diagonal streets, with a variety of open spaces, and was clearly influenced by the plan for Washington (see Figure 10.45).

SAN FRANCISCO

The Spanish foundation in March 1776 of the *presidio* of San Francisco, and the related mission of San Francisco de Asis (known as Mission de Dolores) has been described above. In 1834 the *presidio* was given the status of a municipality (*pueblo*). The commercial growth which ensued was centred on Yerba Buena cove, where an informal trading anchorage had become established, and commenced with a warehouse erected in the spring of 1835 by William Richardson. He delineated a street across the front of his plot as the Calle de Funacion, from which, it was assumed, a regular gridiron would be extended. Only one month later, however, the first mayor, Francisco de Haro, laid out a second street at an acute angle to Richardson's, on which inauspicious basis began the planning of San Francisco. In 1839 Governor Juan Alvarado ordered its reorganization by Jean Jacques Voiget – a Swiss surveyor, sailor and tavern-keeper – in which latter capacity, perhaps, was produced the even more odd-shaped layout shown in Figure 10.47 with his grid streets intersecting at 2½ degrees off a right angle?[107] Voiget's anomalous layout was rectified and extended in February 1847 by an Irish engineer, Jasper O'Farrell, commissioned by the mayor, Lt Washington A Bartlett, who in the January had officially changed its name to San Francisco. 'O'Farrell's Swing', as noted by Reps, straightened out the grid blocks, which were confirmed as containing six lots, each 50 varas (139 feet) square. The new plan was extended out into the shallow cove where the lots were smaller at 16½ by 50 varas. Diagonally across the southern side of the first section, O'Farrell planned a second grid area at an angle of 45 degrees, separated by Market Street, Justification of the misalignment, 'the Y- and T-shaped intersections of which have plagued the city from his day to ours', includes the direction of Market Street parallel to the plank road leading to Mission Dolores; the comparatively level ground on that line; and the creation of First Street parallel to the southern shore line of Yerba Buena cove (Figure 10.48).[108]

In addition to the two-dimensional conflict, the sharply hilly topography of San Francisco is also constantly at odds with the rigidly

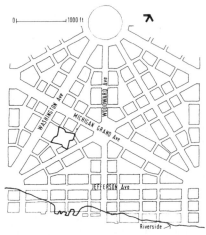

Figure 10.46 – Detroit: a part of the 1807 Woodward plan for the centre of the city, with the Grand Circus of 5½ acres at the top. The problems of implementing such an involved plan had direct parallels with the situation in London after 1666, and, perhaps inevitably, a similar outcome: within months it was largely rejected in favour of a conventional gridiron pattern. Half of the Grand Circus, was, however, constructed, and between it and the riverfront the downtown central business area has retained some of the main diagonals in combination with more or less rectilinear building blocks.

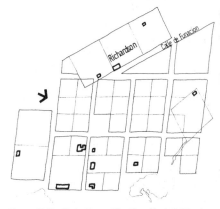

Figure 10.47 – Yerba Buena (San Francisco), California: the plan of 1839 showing the two different grid orientations, and also the off-square form of the Voiget section. (Richardson's lot, and his street, are across the south-western edge.)

106. Reps, *The Making of Urban America*.
107. J H Brown, *Reminiscences and Incidents of the Early Days of San Francisco 1845–1850* (quoted by Reps, *Cities of the American West*).
108. See G Clay, *Close-up: How to Read the American City*, 1973, where Clay's Figure 24 shows five cities, in addition to San Francisco, where there are differently orientated centre-city gridiron districts: New Orleans, Seattle, Las Vegas, Minneapolis and Denver (for which see Figure 10.61). Clay takes a more favourable 'motorist's-view' than Reps of the street intersection misalignments: 'such breaks are handy navigation zones for getting one's bearings in a strange city'.

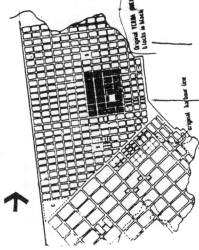

Figure 10.48 – San Francisco; the map published in 1852 by the US Coast Survey for which the survey by R D Cutts was completed in February 1852. North is at the top; the distance as scaled from North Point to Steamboat Point is 2.2 miles. The contours are at 20 feet intervals. Along the 'Plank Road', from the bottom left-hand corner to Mission Dolores, is a further 1 mile. (Market Street as the division between the two grid sections is to the north-west of and parallel to the Plank Road (as Mission Street through the southern section of the Bay). (Map by courtesy Historic Urban Plans, Ithaca, New York.)

Figure 10.49 – San Francisco: the outline grid plan of 1849, showing the original Yerba Buena blocks filled in solid black and the original shore line dotted.

rectilinear street pattern, provoking recurring criticisms of its unsuitability compared with imagined benefits of an organically derived form. Yet to criticize San Francisco's gridiron is to deny its success in controlling the city's extraodinary growth rates resulting from the discovery in January 1848 of gold in the Californian Sierras. From a population of perhaps 850 it had reached 5,000 by July 1849, and when the mines closed down for that winter the total was of the order of 20,000. During 1849 91,415 passenger arrivals were recorded, the great majority of whom headed straight for the diggings some 40 miles distance.[109]

San Francisco was their supply base and by the end of 1852 its population was recorded at 42,000. One measure of the city's booming economy is the price of $40,000 paid for a lot on Portsmouth Square in late 1849. This had been worth $500 in the summer of 1848, having been sold a year previously for $15.62½.[110] Modern San Francisco may require its own extra-special car-parking regulations for the more precipitous streets – and for the inexperienced a taxi ride is a never to be forgotten experience: likewise, though, the city's unique visual characteristics, which literally rise above the limitations of a gridiron.

109. Brown, *Historical Geography of the United States.*
110. E Gould Buffum, *Six Months in the Gold Mines* (quoted by Reps, *Cities of the American West*).

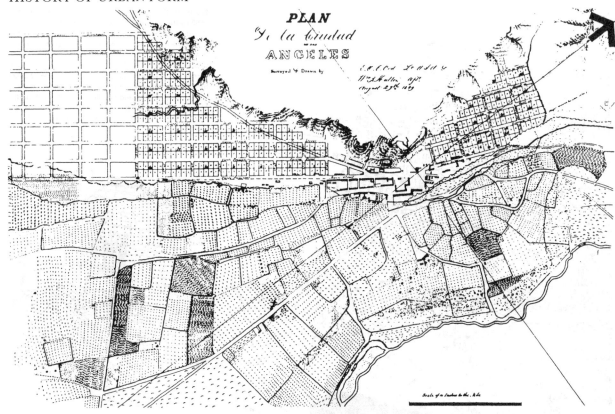

LOS ANGELES

The Spanish origin in 1781 of El Pueblo de los Angeles has been described above. Growth through to the American occupation of California in 1846 had been slow but steady and with about 1,400 inhabitants Los Angeles was the largest of the Californian settlements. In accordance with American practice, the new military Governor of California, General Bennett Riley, required an accurate survey of the town, for which purpose Lt Edward Ord was engaged in June 1849 by the Los Angeles council for a fee of $3,000 to record the developed area, and to 'lay out streets and blocks where there are no buildings'.[111] His work is shown by Figure 10.51, dated 29 August 1849. The buildings of old Los Angeles are mostly to the south-west of the church (at the intersection of the cardinal point lines) on either side of Main Street, with the Plaza in front. Ord's plan has two new gridiron sections of different orientations, based on Main Street on either side of the church. Compared to San Francisco, early growth was exceedingly slow, the initial sale of lots failing to raise enough for Ord's fee. By 1870 the population was still less than 6,000 but the stimulus for subsequent rapid development was provided in 1876 when the Southern Pacific Railroad reached the city from the north. When the rival Santa Fe Railroad also arrived in 1887, with fares from Kansas City falling as low as one dollar, the future of Los Angeles was assured.[112]

In his *Cities of the American West*, Reps includes a map of East Los Angeles, surveyed in December 1873, showing the lines of the South Pacific Railroad, one of which runs close by the southern corner of the plaza along the main route shown out to the west.

Figure 10.50 – Los Angeles, California: the 'Plan de la Ciudad de los Angeles' by Lt Edward Ord, dated 29 August 1849; an extract showing the recorded (few) existing buildings and the two new grid sections. By the end of 1873 there was still only a scatter of buildings on Ord's northern section. (Map by courtesy Historic Urban Plans, Ithaca, New York.)

Figure 10.51 – Los Angeles: diagrammatic modern street pattern for the area of the city's historic nucleus.

111. Reps, *Cities of the American West*.
112. See Reps, *Cities of the American West*, for further description of the early Los Angeles property booms.

CHICAGO

Fort Dearborn was built in 1803, on the site of the present-day city centre, to command the portage between the Ohio–Mississippi Basin and the Great Lakes. A canal had been proposed here as early as 1673 and eventually in 1826 Congress authorized federal aid for the project. This took the form of a land grant, to the State of Illinois, of alternate sections on each side of the canal for 5 miles. To pay expenses a town was laid out in 1830 in Section 9, Township 39, Range 14. This was the site of Chicago, hitherto consisting of the fort and a small village.[113]

Although the canal was not operating until 1848, the promise of the trade it would bring served to create the city's first land boom. By the summer of 1833 Chicago had a population of about 350 and its harbour entrance was being improved. In 1837 Chicago, with a population of 4,170 was chartered as a city, and surviving the 1837 crisis the population stood around 20,000 when the canal opened. The first two railways reached the city from the east in 1852–3 and within years these and eight other lines made Chicago the undisputed railroad centre of the middle west. Industry developed rapidly, stimulated by the Civil War, such that the population figure for 1865 was around 180,000. This figure increased fourfold over the following two decades.

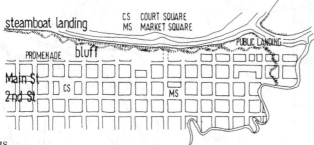

MEMPHIS

Memphis originated around 1820 as one of a large number of towns which were established along the Mississippi after the general introduction of steamboat transport on the river in the years following the first successful Pittsburgh–New Orleans journey of 1811. By 1827 its population had grown to about 500 and Memphis was incorporated as a town.[114] Competition for the river trade was intense with Memphis being rivalled by Raleigh and Randolph which lay some 25 miles upstream. Its eventual dominance as the regional centre was assured only when the railways chose Memphis as their western terminus, on the east bank of the river; at the end of the nineteenth century, in spite of recurring yellow fever epidemics, the population had exceeded 100,000.

ATLANTA

In March 1838 construction work started on the newly authorized Western and Atlantic Railway. The route was planned to run from the Tennessee state border to some point on the south-eastern bank of the Chattahoochee River. The site selected for this southern terminus and the future junction for lines to be built to other cities in Georgia was known simply as Land Lot 78 in what was then Dekalb County. A small village called Terminus grew up around the end of the line, but the question of a permanent base for the railway remained unresolved until 1842, when land in the vicinity was purchased for depot buildings. This location, later that of Union Station, determined the precise

Figure 10.52 – Chicago in 1834. The space between the eastern side of the platted area and Lake Michigan was reserved in federal ownership, subsequently much of it became railway land. 'S' denotes blocks reserved for schools, one of which, Reps records, was sold for 38,000 dollars in 1833 and valued at nearly $1.25 million three years later, during the city's short-lived land boom.

Figure 10.53 – Memphis, Tennessee: the rudimentary gridiron plan of the city in the late 1820s. It was laid out on 1 May 1819 by John Overton and General Winchester, with William Lawrence, their surveyor and attorney. Tennessee was settled from the 1750s and became the 16th State of the Union in 1796.

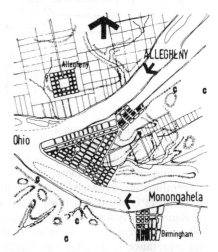

Figure 10.54 – Pittsburgh, Pennsylvania: the plan of about 1815. The original foundation was French and dated from the construction in 1724 of Fort Duquesne at the confluence of the Allegheny and Monongahela Rivers. After capturing it in 1758, the British renamed it Fort Pitt. Ten years later, along with an extensive territory in western Pennsylvania, the small but strategically and commercially well-located settlement became the property of heirs of William Penn (of Philadelphia fame). In 1784 the triangular area between the rivers was platted to give a waterfront grid alignment on both sides. The neat little settlement of Allegheny around its central square, across the river of that name, dated from 1784–8.

113. For Chicago see W Cronon, Nature's Metropolis: Chicago and the Great West, 1991; H M Mayer and R C Wade, Chicago: Growth of a Metropolis, 1969; J S Ackerman, The Architecture of Michigan, 2nd edn, 1986.
114. For Memphis see G M Capers jr, Memphis: Its Heroic Age, 1939.

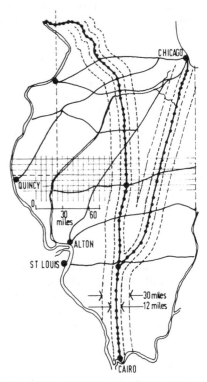

Figure 10.55 – Atlanta: plan of the central part of the city in 1847 within a circle of a radius of 1 mile. The plan shows how the early unrelated gridiron districts of the first railway settlement were subsequently resolved into a standard north–south, east–west orientated pattern (a procedure also followed some years later at Denver where there are three differently orientated downtown grids – Figure 10.61).

Its future assured, the town grew rapidly, achieving city status in 1847 with a charter giving it 'authority and jurisdiction extending one mile from the State (Railway) depot in all directions'. One of the council's first actions was to appoint a surveyor to plat the city within the one-mile radius circle. With a population of 6,000 by 1853, Atlanta successfully petitioned for the creation of a new county, Fulton County, with itself as the centre of government.

The Civil War which at first boosted Atlanta's importance as a key industrial and transportation centre, by December 1864 saw its burning by the Union forces under General Sherman, immortalized in the epic *Gone with the Wind*, after months of near siege. From 1865 the rebuilding process was under way, with an enlarged 3 mile diameter boundary of 1866 and a new, stronger charter of 1874.

Figure 10.56 – The Illinois Central Railroad lines and stations in 1860; with the other railroads, related to the western lands settlement grid. The dotted lines on either side of the railroads show the extent of land grants within the 6 mile and 15 mile limits.

centre of Atlanta, between Pryor and Central Streets.[115] Around the depot the new town, still known as Terminus, was laid out on the basis of seventeen land lots, its layout largely determined by the railway tracks and the existing footpaths.

Terminus was renamed Marthasville in 1843 and was chartered by the Georgia General Assembly, with a council composed of five commissioners. Early attempts on their part to restructure the town with a uniform, convenient, gridiron street pattern were frustrated by a general reluctance to assume this expense. It was felt that Marthasville would not survive the completion of the railway's construction programme; early growth therefore took place without an overall plan. By 1845, however, the Georgia Railway was brought to the Western and Atlantic's terminus and the first passenger train linked Marthasville to Augusta. Later in the same year the W & A Railway was in full operation as far as Marietta, and the commercial advantages of the key junction point were quickly recognized. In turn, Marthasville was renamed Atlanta in honour of its lifeline on 26 Decemeber 1845.

Railroads and mid-western gridirons

The establishment of urban settlements in the broad band of states between the Canadian border and Texas, east of the Rocky Mountains (Figure 10.11) not only was facilitated by the railroad companies, but also to a considerable extent was sponsored by them through disposal of federal land grants awarded to finance their western route

115. For Atlanta see J M Russell, *Atlanta 1847–1890: City Building in the Old South and New*, 1988; for an unusual Civil War episode in American urban history (when Atlanta was briefly fortified) see Samuel Carter III, *The Siege of Atlanta, 1864*, 1973.

116. See J F Stover, *American Railroads*, 1961.

extensions.[116] This process commenced in Illinois which by 1850 was faced with a rapidly increasing number of farmers who could not market their produce economically without railroad connections with the eastern urban centres. The railroads countered that the low-density settlement of Illinois, and of other western regions gave them no possibility for recouping their initial capital investments, without federal government aid.

The US Government owned about one-third of the unproductive land in Illinois and the formula eventually proposed was that a proportion would be given to the state, which in turn would donate it to the railroad for sale, in order to finance the new lines. Specifically the Illinois Railroad would receive alternate square-mile sections on either side of a 6 mile wide right of way. Congress passed such a law in September 1850 through which the Illinois Central received 2,595,133 acres of land, worth some 20 million dollars over the next two decades. Illinois got its transport and the railroad magnates were in the town development business. During the next twenty-two years the Government granted a total of between 150 and 160 million acres of western lands in many states, creating and extending a trans-continental system across scarcely inhabited regions.

As Reps observes, 'early western railroad promoters quickly realised that town development and railroad company profits went hand in hand'.[117] Congress prevented them from locating stations on their own land and developing towns themselves, but under other names unscrupulous company directors made vast profits from knowing in advance where the stations were planned. For the layout of the towns themselves the gridiron came into its own and there was even a standard plat used for a total of thirty-three towns developed by one group in Illinois (Figure 10.57).

Further west unsettled land was made available on specified days at given times, on a first-come, first-served basis, which precipitated the great Oklahoma land rush of 1889. The mid-west may have been won with the 'Colt 45': thereafter its use must have been inevitable in keeping the real-estate peace, were it not for indisputable gridiron boundaries defining the new land-holdings.[118] How else could Oklahoma City have been open prairie at noon on 22 April 1889, when cannon signalled the start of the rush from the territory boundary, and a 'settled' township by nightfall? This process was repeated throughout the territory and on numerous occasions elsewhere (Figure 10.58). A description of the kinds of towns created overnight on their anonymous gridirons is given in photographs of early Denver, Colorado.

DENVER

Denver, Colorado, by far the largest of the mid-western cities and centre of one of today's major regions of growth in the USA, was 'founded' in 1858 when a group of prospectors led by one William Green Russell believed that the Colorado gold fields extended to include Cherry Creek, near where it joins the River Platte. Events proved otherwise and Russell moved on, leaving behind the reluctant remaining residents of the two separately platted townships of Auraria and Denver City, which in 1860 joined together as Denver.[119]

Seemingly doomed to become yet another dead mining village, Denver still succeeded in making its future from the gold rush. 'Businesses came quickly to the little town that had no reason for being there,'

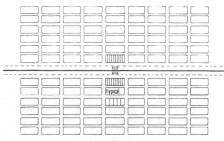

Figure 10.57 – The Illinois Central Associates standard gridiron town plat. (The 'Associates' comprised four one-time company directors and the engineer in charge; one of them, David Neal, had been vice-president of the company until 1855.)

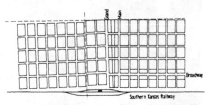

Figure 10.58 – Oklahoma City: the first 'plan' of 1889. The two disjointed, differently orientated grids were the result of claims advanced by two rival town companies. They were adjusted before site development commenced.

'Denver's founding was an accident of circumstances at a time when even the circumstances bid fair to be transitory, and when advantages and resources may be said to have been unknown. In her infancy, she was the temporary objective point for thousands who cared nothing for the country, nothing for the town; who had no intention of permanently identifying themselves with either and with but the one thought of getting what they came for before somebody else got it, and then getting away again. It was not until the rare beauty of the accidental location, the grandeur of the region, the charms of the climate, and the enormous permanent resources of the country became fixed in the minds of the people, that these alien feelings and purposes disappeared.'

The first few years of the 1860's were unstable and rather loose years in the history of Denver. Buildings were cheap, wooden, and temporary. Early inhabitants came and went with great abandon, and the city had a relatively large temporary population. The early houses of Denver were cabins, not built to last for long and ineffective at keeping out the rain and the cold.

Cottonwood and willow logs cut from trees along the banks of the river apparently provided materials for several hundred cabins in the late 1850's and early 1860's. The first cabins were about fifteen by twenty feet...
(J C Smiley, History of Denver, 1901)

117. Reps, The Making of Urban America.
118. It was usual for a new town gridiron to be in accordance with the National Land Grid, but there were interesting exceptions, for which see Denver (Figure 10.61); for grid anomalies generally, see Clay, Close-Up: How to Read the American City.
119. For Denver see S Dallas, Cherry Creek Gothic, 1971; C McL Green, American Cities in the Growth of the Nation, 1957.

Figure 10.59 – Denver: photograph of c. 1858–60 looking north along 14th Street (see viewpoint on plan as Figure 10.61). The Methodist Church on Lawrence Street to the right; Larimer and Old Market Streets beyond.

It is interesting to observe the simple one- and two-storey buildings set within their gridiron plots with the rivers and the empty prairie beyond. Only the main commercial streets could afford boarded side-walks above winter-time mud, and those were the only streets where there was continuity of building facades, often, however, of a deceptively grand nature. (Photograph courtesy of the State Historical Society of Colorado.)

Figure 10.60 – Denver, view on Capitol Hill of the early 1880s looking south between Sherman and Grant Streets (left and right) towards the site of the future State Capitol.

Contrasting with Figure 10.59, this shows Denver's wealthy quarter in process of development up on the drier, healthier elevation of Capitol Hill where irrigation made it possible to create within a few years densely planted gardens, framed by tree-lined streets. With the exception of the occasional full-panoplied classical example, these mansions are fully illustrated in Sandra Dallas's aptly titled book, *Cherry Creek Gothic*. (Photograph by W H Jackson, reproduced courtesy of Denver Public Library, Western History Department.)

writes Sandra Dallas. 'There was no gold in Cherry Creek, so Denver rationalized its existence by becoming a supply point.'[120] A promotion booklet, *Denver and Auraria: The Commercial Emporium of the Pike's Peak Gold Regions*, claimed 420 structures and some 3,000 permanent residents in 1859 but visitors generally reported otherwise and it was to be some time before Denver realized that ambitious claim. When it did, from the early 1870s, the factor that consolidated the future prosperity was the arrival of the railroad. From perhaps 5,000 inhabitants in 1870 (compared with nearly 1 million on New York's Manhattan Island

120. Dallas, *Cherry Creek Gothic*.

alone, and 674,000 in Philadelphia), Denver had grown to 46,000 in 1880 and 100,000 was passed in the next decade.

Figure 10.61 gives the grid structure of what is today the centre of the city of Denver, with the original extents of Auraria and Denver City. Each had its own grid orientation, to suit the river frontages, and both were departures from the National Land Grid which was adopted from the late 1860s with the setting out of Broadway and Colfax Avenue.

Reproductions of cowboy main streets have provided settings for numerable 'Western' movies – some it would seem many times over – and scarcely a reader can fail to summon up a mental image of their visual characteristics. In addition, places such as Denver were photographed almost as soon as they were born and western historians are gradually republishing fascinating archival reminders, including Figures 10.59 and 10.60 (further described in the caption to the early plan of the city).

Conclusion

Within a necessarily small compass this chapter has been concerned first with establishing the origins of urban settlement in the USA, and second to trace the early development of a number of significant city examples through to the second half of the nineteenth century. In effect this is a history of the urban USA before the advent of mass-production, factory-system industry which, although originating on a small scale before the Civil War (1861–65), subsequently came to awesome fruition. The extent to which cities of the USA became submerged beneath the rising tide of national urbanisation, and consideration of those underlying social, economic, aesthetic and political modern-age determinants that have continued, one hundred and more years later, to shape their forms awaits separate in-depth assessment. However, before leaving the urban USA at mid-nineteenth century, its ubiquitous gridirons energetically going places – without knowing where; it is necessary to explain, by way of interim conclusion, how it was that from seemingly auspicious planned origins, American cities came to embody many of the least desirable visual characteristics of laisser-faire, unplanned growth processes? Offset only by the quality of exceptional small, isolated residential enclaves, and a handful of historic centre-city architectural groups.

Although much of the blame for visual failings can be levelled at the gridiron, nevertheless it must be acknowledged that it frequently provides present day American city planners with a simple, logical geometric framework for structured incremental inner-city renewal which is the envy of European counterparts struggling to modernize, while conserving, their organic growth inheritance. Also on the debit side, though, because the gridiron gave a semblance of 'planning', its legal status characteristically pre-empted need for further controls, once growth pressures had extinguished any remaining civic ambitions.

Frontier towns – and in the main sense that includes just about all of the urban USA – had neither the political unity enjoyed by long-established European communities when rebuilding or expanding – as with Amsterdam's Plan of the Three Canals, or Edinburgh's New Town; nor the despotic, or equivalent prescriptive control, necessary to carry through extensive planned building programmes – as at Bath, Nancy and Karlsruhe. Neither the American West, nor for that matter

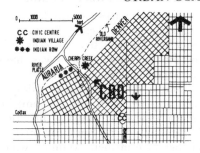

Figure 10.61 – Denver, Colorado: the modern centre-city plan with Auraria and Denver City as founded on opposite sides of Cherry Creek, immediately above the junction with the major River Platte. The National Land Grid was imposed with the alignment of Broadway and Colfax Avenue. (The viewpoints of the two photographs of old Denver are given as their respective figure numbers.)

Denver is an excellent place with which to exemplify the processes of early mid-western urbanization. Although its original mean cowboy and mining town buildings have been replaced, their forms have been recorded, and there are fortunately still numerous surviving photographs of that contrasted, but overlooked aspect of edge-of-prairie life whereby prosperous families built themselves imposing mansions and created 'garden suburbs' from the treeless desolation.

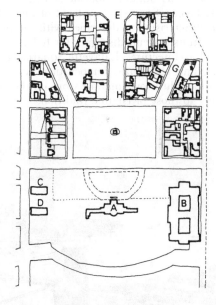

Figure 10.62 – Lafayette Square, Washington DC, at the middle of the nineteenth century. Key: A, the White House; B, US Treasury; C, War Department; D, Navy Department; E, 16th Street NW; F, Connecticut Avenue; G, Vermont Avenue; H, St John's Church.

the East, was settled by people who would tolerate more than minimal interference with their rights to do whatever they pleased with their property. The fact that patterns of land-use emerged early on in the post-frontier growth phases was characteristically the result of natural, freemarket economic motivation.

Compared with European cities the youthfulness of urban America must also be taken into account; but not just, it is stressed, at the superficial level of being less 'historic'. The essential historical reality is that compared with European predecessors, American cities missed out on the stage in their evolution which could have provided a unified historic centre-city urban architecture. Whereas in the old world the mediaeval first buildings were replaced by unified, more or less classical designs of a second phase, prior to the advent of modern construction technology; throughout the urban USA the evolutionary process was ordinarily that of direct replacement of the first domestic-scale buildings by those of the third wave; not only in a planning control vacuum but also during the continuing period when architectural fashion determines that each building is different from its neighbours.

To this day, Pennsylvania Avenue exemplifies, and highlights, the absence of street architecture in the European image. Much of the south side has variously suitable federal buildings whereas to the north, but changing recently, there has been a sequence of disparate domestic scale buildings interspersed with vacant lots. Not only is there no coherent historic urban third dimension but also there are hardly any public spaces as suitably formal settings for civic, and federal buildings. Lafayette Square ... in front of the White House ... exemplifies this characteristic, as described with the plan in the adjoining column.

In its extreme disregard for Jefferson's concept of a 'regulated classical architecture', Lafayette Square, just as the Downtown Manhattan skyline, epitomizes democratic every-man-for-himself aggrandizement: a thought provoking last word on the Western-world city, before turning to those of Islamic tradition.

In addition to the absence of street architecture, the United States also had hardly any historic urban spaces worthy of consideration and the few exceptions were mostly informally domestic in character, including Lafayette Square. It had been intended as a forecourt to L'Enfant's designated 'Palace', but by around 1850 all that had developed was an unrelated and above-all unplanned collection of individual private houses, admittedly those of important people, known as 'the lobby of the White House'. Jefferson's ideal of a regulated classical architecture, complementing the plan, was so irrelevant to the needs of society that it failed even to create a uniform square to provide an adequate setting for the President's House.

Figure 10.63 – New York City, the lower Manhattan Island (downtown Wall Street) skyline of the early 1950s, before the arrival of the twin World Trade Centre towers and the other more recent skyscrapers. The view is looking south-west down the East River (Brooklyn is to the left) with Brooklyn Bridge (1883) beyond Manhattan Bridge – see Figure 10.32. The distinctive stepped profile of some of buildings was determined by the sunlight angle of latterday city planning legislation aimed at preventing vertical 'canyon-side' street development. In response to irresistible economic motivation, this most famous of western-world skylines represents the triumph of construction technology over topographical constraints.

11 – Islamic Cities of the Middle East

The addition of this chapter enables the third edition of this book to conclude with a full-circle return to Iraq (ancient Mesopotamia), among other countries of the Middle East, where the history of urban form originated with the cities of the Sumerian civilization.[1] It also serves to bring our subject up to the 1960s and 1970s in respect of those Islamic States whose modern, industrial age urban transformations had awaited full availability of their national oil revenues.

Chapter 1 has described how favourable environmental circumstances for urban settlement first came about during the fourth and third millennia BC, in and around the valleys of the Tigris and the Euphrates in Mesopotamia. Coincidentally, that was where beneficent pre-historic geology had laid down oilfields which, with others in the Arabian Gulf and elsewhere, have provided economic means for revolutionary late-twentieth-century urban changes. A summary account of the discovery of oil in our subject states is given in the adjoining column; notes updating major cities to the late 1980s are included with the descriptions in the main body of text, or at the end of the chapter.

In addition, reference back is made to the aerial photograph of the Islamic city of Erbil (ancient Arbela) in north-east Iraq (Figure 1.10, which shows the historical nucleus of that city in the 1930s, much as it most probably would have appeared four to five thousand years earlier. With the ancient Sumerian housing area at Ur (Figure 1.30), where the arrangement of houses and access routes was virtually the same as that of the historic Islamic city,[2] Erbil has been taken as the basis of the introduction to Islamic urban form given in Chapter 1, in advance of the description in Chapter 4 of the Islamic cities of medieval Spain.

Chapter 1 introduced Professor Besim S Hakim's book, *Arabic Islamic Cities: Building and Planning Principles*, as the first of three main English-language reference works supporting this chapter.[3] Although based on detailed study of the historic city of Tunis, as it evolved within the Maliki School of Law, 'the uniform legislative guidelines and the almost identical socio-economic framework created by Islam', resulted, explains Hakim, 'in remarkable similarities in approach to the city building process'. After two introductory sections, the main part of this chapter addresses the question, more correctly a set of interrelated questions: what were the determinants of these 'remarkable similarities'?

The second pertinent reference is *The Islamic City* (edited by R B Serjeant), a compilation of selected papers by mainly Muslim authors, presented at the colloquium held in July 1976 at the Middle East

OIL AND THE MIDDLE EAST
The first modern drilled oil well is attributed to 'Colonel E L Drake, who on 27 August 1859 struck oil at a depth of 69½ ft at Titusville, Pennsylvania, USA. 'Oil was hardly unfamiliar to mankind,' though, writes Yergin: 'in various parts of the Middle East, a semisolid oozy substance called bitumen seeped to the surface through cracks and fissures, and such seepages had been tapped far back into antiquity – in Mesopotamia, back to 3000 BC. The most famous source was at Hit, on the Euphrates, not far from Babylon (and the site of modern Baghdad) ...' (D Yergin, *The Prize: the Epic Quest for Oil, Money and Power*, 1991.)

By 1870 there were commercial oilfields in the USA, Canada, Italy, Rumania, and Russia, and before the turn

1. For general references on the past and present of our subject area see T Mostyn and A Hourani (eds) *The Cambridge Encyclopaedia of the Middle East and North Africa*, 1987; P M Holt, K S Lambton and B Lewis (eds) *The Cambridge History of Islam Volumes 1 and 2*, 1970; P Mansfield, *The Arabs*, 1980; A T Welch, 'Islam,' in J R Hinnells (ed.) *A Handbook of Living Religions*, 1985.

For architectural references see G Michell (ed.) *Architecture of the Islamic World: Its History and Social Meaning*, 1984; D Hoag, *Islamic Architecture*, 1977.

2. On the close similarity between the ancient Mesopotamian houses discovered at Ur and those of traditional Islamic culture, see Lloyd, 'the houses themselves were so exactly like those of a small town in Iraq today' (*Foundations in the Dust*, 1947); see also Chapter 1.

3. B S Hakim, *Arabic-Islamic Cities: Building and Planning Principles*, 1986. From my earliest interest in the traditional Islamic city it had been clear that there were close morphological links with ancient Sumerian predecessors in Mesopotamia. That which remained unanswered, until I first read *Arabic-Islamic Cities* in late 1989, were the essential questions concerning the how' and the 'why' of extraordinary consistencies in the form of Muslim cities throughout the widely established Islamic world. I had assumed that Islam, in some manner that I did not comprehend, embodied mandatory 'city planning legislation' which was superimposed on the natural-world determinants as the cause of that uniformity. Although I can now see that I was heading in the right direction, it is entirely to Hakim that credit is due for the answer. Serious students of the Islamic city – especially those like me working outside their own urban culture – can but commence with Hakim.

Centre, Faculty of Oriental Studies, Cambridge, UK.[4] The third source, 'Traditional Houses of Makkah: the Influence of Socio-cultural Themes upon Arab-Muslim Dwellings', by Yousef Fadan,[5] is one of several contributions to the published papers of a Symposium on the subject of *Islamic Architecture and Urbanism*, held at King Faisal University, Damman, Saudi Arabia, in January 1980. Other main English language references are as annotated,[6] and several personal acknowledgements in respect of this chapter have been given at the beginning of the book.

The chapter is in seven parts:

1 Introductory time scale and definitions
2 Arabia: the seventh-century origins and spread of the Arab Islamic empire
3 Islam: original an later urban form determinants
4 Islam: The Prophet Muhammad and the Holy Law
5 Islamic urban form: Makkah and Medina and the urban guidelines
6 Islamic urban form: city types by origin, characteristics and building types
7 Islamic urban form: illustrated descriptions of major cities through to the end of their historical periods.

Within a necessarily limited compass, other key cities are incorporated as examples into the general text.

Introductory time scale and definitions

The chapter requires both geographical containment and historical definition. To write in general of 'Islamic urban form' would have required recognition of the wider Islamic world, of which the Middle East is but a part. Not only would this have greatly increased the size of the chapter, but also it would have broadened the Islamic world beyond the author's personal experience.[7] Neither would 'Arab cities' have been suitable as a title, given the historic urban significance of Iran – the Persia of the ancient world – and Professor Hakim has already written of 'Arabic Islamic Cities'. Figure 11.1 delineates the Middle East of the present consideration, within the wider Islamic world, and locates the cities which are described.

Furthermore, in writing of 'pre-industrial' cities of the Middle East it is necessary to establish various national dates separating the period of this volume from that of the succeeding modern industrial age. Broadly, these watersheds occurred when the countries concerned came within the 'Western' sphere of influence; initially as results of European imperial expansion or mandated authority, and latterly through Western need to secure their vital oilfields. Whereas Egypt came under French influence at the beginning of the nineteenth century (to be followed by the British when taking over control of the Suez Canal in 1882),[8] there are those countries of the Arabian Gulf and Peninsular that were of minimal interest until a century or so later and whose oil-financed, Western-inspired urban expansions commenced only in the 1950s and 1960s – even as recent as the 1970s, in the case of the Sultanate of Oman.

Elsewhere, after Egypt, the French annexed Algeria (1848), gained both Tunisia (1881) and Morocco (1912) as Protectorates, and they were the dominant influence in Syria and Lebanon;[9] while for her part,

of century they were joined by Germany, India, Indonesia, Japan, Peru and Poland. By 1908 oil had also been found in Argentina, Mexico, Trinidad and ... Persia (Iran): the first of the Middle East oil-producing countries. The first concession in the Arabian Gulf (then known as the 'Persian Gulf') had been granted by the Persian government in 1901 with the first oil discovered on 26 May 1908 at Masjid-i-Sulaiman in the south-west of the country. The Anglo-Persian Oil Company was formed in which Burma Oil was the major shareholder and the British Admiralty a prospective customer. A refinery was constructed at Abadan on an island in the Shatt-el-Arab, the extended estuary of the historic Tigris and Euphrates rivers.

In the years 1912–13, preceding WWI, the British Royal Navy was converted to oil; the merchant fleet following its example. Yergin credits Winston Churchill, then newly appointed First Lord of the Admiralty, with the decision in the summer of 1911 that Britain would have to base its naval supremacy upon oil. 'Two British adventurers, Julius De Reuter and William Knox D'Arcy,' writes Michael Tomkinson, 'take the credit for starting an industry that was to permeate, embroil and finally dominate the destinies of most Middle Eastern states, and to give the Arab shaikhdoms and republics their present economic physique.' (*The United Arab Emirates: an Insight and a Guide*, 1975.)

Key dates from then on include 1928 Iraq – the Kirkuk field; 1932 Bahrain; 1938 Kuwait – the Burgan field, first exports in 1946; 1935 – Saudi Arabia – the first Dhahran field; 1940 Qatar –first exports in 1949; 1956 – Algeria; 1959 Libya – the Zelten field; 1960 – Abu Dhabi – the Murban field; 1966 Dubai – offshore field; 1969 (British) North Sea – production from 1975.

4 R B Serjeant (ed.) *The Islamic City*, UNESCO, Paris, 1980. Acknowledgement to individual papers is given separately.
5. Y. Fadan, 'Traditional Houses of Makkah: the Influence of Socio-cultural Themes upon Arab-Muslim Dwellings', in the published papers on *Islamic Architecture and Urbanism*, a Symposium of 1983, King Faisal University, Damman, Saudi Arabia.
6. Other English language reference works by Muslim authors on the traditional Islamic cities of the Middle East include J Akbar, *Crisis in the Built Environment: The Case of the Muslim City*, Concept Media, Singapore, 1988. (Most useful in respect of the early Islamic 'planned' military cities); I Serageldin and S Sl-Badek (eds) *The Arab City*, 1982.

Works by Western-world authors include G H Blake and R I Lawless, *The Changing Middle East City*, 1980; I M Lapidus, *Muslim Cities in the Later Middle Ages*, 1967; I M Lapidus (ed.) *Middle Eastern Cities*, 1969; C L Brown (ed.) *From Madina to Metropolis: Heritage and Change in the Near Eastern City*, 1973;
7. From 1976 to its closure in 1987, I was Architectural Correspondent for the monthly magazine *Middle East Construction*, during which period I visited the Arabian Gulf countries and Egypt, Tunisia and Algeria, writing regularly on a wide range of architectural, planning and construction subjects.
8. For Egypt see P J Vatikiotis, *The History of Egypt*, 1985; A Moorhead, *The Blue Nile*, 1962; A Moorhead, *The White Nile*, 1960.
9. For French involvements generally see R Betts, *Tricouleur: The French Overseas Empire*, 1978 (with a detailed bibliography); for specific countries H Cobban, *The Making of Modern Lebanon*, 1985;
10. For the Italian involvement in North Africa see G L Fowler, 'Italian Colonisation in Tripolitania', *Annals of the Association of American Geographers*, vol 62, December 1972; J Wright, *Libya: A Modern History*, 1982.

Figure 11.1 – The Middle East: a map showing the political boundaries of the States included in this chapter, and named in capitals the cities described in the text and accompanying captions. (Other cities are in smaller lettering.)

The inset map gives the boundaries within the UAE which was established on 2 December 1971 as a union comprising seven Sheikhdoms of the former Trucial Coast: Abu Dhabi (see Figures 11.2 and 11.3); Ajman, Dubai (see Figures 11.27 and 11.28); Fujairah; Ras al-Khaimah; Sharjah and Umm-al-Qawain.

See also Figure 11.5 for maps relating the Middle East to the broader present-day world of Islam, which includes Indonesia, as summarized in Appendix C; and for the extent of the Turkish Ottoman Empire. For the Islamic occupation of parts of Spain, see Figure 4.80.

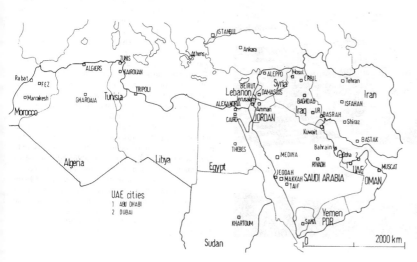

11. For the British presence in the Persian (Arabian) Gulf see R S Zahlan, *The Origins of the United Arab Emirates*, 1978, with an extensive bibliography; A O Tayan, *The Establishment of the UAE 1950–85*, 1987; T Sadik and W P Snavely, *Bahrain, Qatar and the United Arab Emirates*, 1972.

British law-keeping in the Persian Gulf was initially administered from Bombay, for which see H L Hoskins, *British Routes to India*, 1966; see also Sultan Muhammad Al-Qasimi, *The Myth of Arab Piracy in the Gulf*, 1986.

12. For Iraq see Z Saleh, *Britain and Mesopotamia*, 1966.

13. For Kuwait see A M Abu-Hakima, *The Modern History of Kuwait 1750–1965*, 1983.

14. For Persia (Iran) see R L Greaves, *Persia and the Defence of India 1884–1892*, 1959; F Kazemzadeh, *Russia and Britain in Persia: A Study in Imperialism*, 1968.

15. The involvement of the USA in Saudi Arabia is well covered by I H Anderson, *Aramco: The United States and Saudi Arabia*, 1981; see also A H Cordesman, *Western Strategic Interests in Saudi Arabia*, 1987; M Abir, *Saudi Arabia in the Oil Era*, 1988.

Saudi Arabia is by far the most important Islamic State not to have been 'colonized' or otherwise brought under direct Western-world control. The extent of US construction investment in a defence infrastructure was, however, revealed during the 'Desert Shield/ Desert Storm' Gulf War of 1991.

16. The time scale of the chapter means that the subject Islamic cities were variously in their historic periods when visited by modern Western-world travellers, many of whom were British. Numerous cities were recorded and photographed at what were effectively medieval stages in their evolution, most importantly Makkah (see Figure 11.10). When considering the truly revolutionary impact of modern change on traditional Islamic cities, it is essential to keep in mind that many were transposed overnight, as it were, from historic to late-twentieth-century urban forms. Compared to European cities, many have been taken directly from their small-scale, vernacular medieval periods without passing through an intermediate, larger-scale renaissance architectural phase.

17. See Chapter 1 for introductory description of Makkah and Medina in their Arabian Peninsular context.

18. Istanbul (then named Constantinople), one of the most important Islamic cities (and the only one in Europe), has been described in Chapter 3 to enable continuity within its Greek, Roman and Byzantine periods. Prior to 1453, when captured by the Turkish Muslim army, it is probable that climate response and an absence of wheeled traffic would have determined an urban morphology already effectively that of an Islamic

Italy colonized Libya (1912).[10] Along the Gulf, the British involvement had been of an essentially paternalistic law-keeping nature, 'safeguarding routes to India': until, that is, the Royal Navy's early-twentieth-century conversion from coal to oil and consequent British strategic interests.[11] Iraq had seen Turkish (Ottoman) dominion and then a British-mandated presence from 1920 to 1932.[12] The Ruler of Kuwait made a treaty with Britain in 1899 enabling it to become a self-governing protectorate.[13] Persia – renamed Iran in 1935 – had usually succeeded in maintaining her autonomy.[14] After early British interests, the USA has been the major late-twentieth-century influence in Saudi Arabia.[15]

The different local time scales are such that at the beginning of the twentieth century, while Cairo and Alexandria, and Algiers, Tunis and Beirut were major Mediterranean-world cities; Riyadh and Abu Dhabi, among others, were no more than tribal 'townships'. Other cities of this chapter, including Kuwait, Baghdad, Jeddah, Dubai, Bahrain, Damascus, Tripoli (Libya) and Muscat, were variously on their ways to becoming major urban centres.[16] Of central importance in this chapter, Makkah and Medina, the two Islamic holy cities of Saudi Arabia, were exceptional long-established centres of trade and annual pilgrimage; without, however, maintaining other than small resident populations on the early-twentieth-century Middle Eastern scale.[17]

Constantinople (renamed Istanbul in the 1920s) was – as it has remained – the exceptional Islamic city of Europe.[18] Characteristically, Spain's Islamic medieval cities described in Chapters 4 and 9, have kept historic forms while changing their religion. See also descriptions in Chapter 7 of Samarkand and Tashkent, two of Soviet Uzbekistan's historic Islamic cities which have extensive reminders of their urban past, and where Islam is tolerated, for reasons of political expediency.[19] Many countries originally annexed for colonial agricultural settlement and economic exploitation, or allocated for protection, have also subsequently proved to have oilfields.

Figures 11.2 and 11.3 – Abu Dhabi, UAE: the aerial photograph shows the Gulfside township of 1958, and the photograph of the waterfront was taken in 1961. Abu Dhabi originated at the seaward end of an approximately rectangular coastal island about 18 km in length and 6 km wide, linked to the mainland by a short causeway. From 1961 the city has expanded to the south-east (right as view) to cover the island on the basis of a 1.2 km supergrid, with a 1991 population of 798,000.

'The village of Abu Dhabi was first settled in 1761,' writes J E H Boustead, 'when some Beni Yas bedu discovered that there was drinking water there. Before then there had been no settlements anywhere on the coast between Doha in Qatar and Sharjah' ('Abu Dhabi, 1761–1963', *Journal of the Royal Geographical Society*, July/Oct 1963). The Beni Yas lived as semi-nomads in the interior, with gardens watered by the Liwa oases, until 1793, when the Ruling Family established themselves on Abu Dhabi Island. Fishing and pearling became established as alternatives to date-growing and herding, and Abu Dhabi emerged as a leading township on the Arabian Gulf. In 1939, as the largest in area of the Trucial Sheikhdoms, its population was about 10,500. (See also F Heard-Bey, *From Trucial States to United Arab Emirates*, 1982.)

Figure 11.4 – Mutrah, Oman: the beach on the southern side of the bay photographed in the mid-1960s, looking east towards Muscat (see Figure 11.40). In 1868, W B Palgrave noted: 'two miles to the west of Muscat lies al-Matrah, the gate of the interior. In the centre of the town facing the sea the Khoja community is found [whose] houses are built in a square mass with all the doors inward, thus forming a large fort which can only be entered by two gates – one in front on the sea shore' (*Narrative of a Year's Journey through Central and Eastern Arabia*, 1868). Until 1970 the town gates of Mutrah were closed at sunset, but as motor traffic went through a hole in the town wall it was not affected. The beach was replaced by the main Corniche Road in 1970.

city. The major evident change would have been construction of the series of superb mosques, for which see below.

19. In addition to the profoundly disquieting events of 1990–91 in the Arabian Gulf and Iraq, completion of this new chapter has been taking place in the context of momentous, continuing independence movements within the Soviet Union, culminating in December 1991 with the breaking away of five Islamic States from the Union. From visits to the Republic of Uzbekistan since 1965, I formed a high regard for the evident ways whereby the central Moscow authorities respected the Islamic traditions of a people who had been conquered by the Russians only in the 1860s.

DEFINITIONS

The glossary in the adjoining column gives the English translation and summary explanation where necessary, of the most important Arabic words used in this chapter. Introductory definition is required concerning the terms 'Islam', 'Islamic' and 'Muslim' (the terms 'Arabia' and 'Arab' were defined in Chapter 1). 'The word Islam has several meanings,' explains Lewis, 'in the traditional sense, as used by Muslims, it means the one true divine religion, taught to mankind by a series of prophets, each of whom brought a revealed book ... Muhammad was the last and greatest of the prophets and the book he brought, the Qur'an, completes and supercedes all previous revelations. In a still wider sense the word Islam is often used by historians and especially non-Muslim historians, as the equivalent not only of Christianity but also of Christendom, and denotes the whole rich civilisation which grew up under the aegis of the Muslim empires.'[20]

The word 'Muslim' connotes either a follower of the Islamic religion, or that which relates to the Muslims or their religion.[21] Although Arabic is the language of the Qur'an, the terms Arab and Muslim are not synonyms. The Arabs represent only about 15–20 per cent of the total Muslims, and the Soviet Union contains more Muslims (about 50 million) than Egypt (38 million). Indonesia, India, Pakistan, Bangladesh and the USSR comprise, sequentially, the largest number of Muslims.[22]

The Qur'an (the preferred spelling, replacing the obsolete 'Koran', from the Arabic, al-qur'an, which translates as 'the recitation') is the Islamic scripture, and with the multi-volume collection of accounts called hadiths (from the Arabic hadith, 'story, tradition'), forms the basis of the Sharia, the 'Islamic Holy Law'. However, it is stressed at the outset that although the traditional Islamic urban guidelines had quasi-legal status, there was no formal historic 'town and country planning legislation' as variously known in the Western world.[23]

Arabia: the seventh-century origins and spread of the Arab Islamic empire

The prelude to the Arabian Peninsular in Chapter 1 has established the climatic context for its early urban settlements, and their direct cultural descent from the Sumerian cities of ancient Mesopotamia. Subsequently, the Romans came to know of the Arabian peninsular in two parts; distinguishing between 'stony Arabia' (Arabia Petrea) as the inhospitable desert north, and 'happy Arabia' (Arabia Felix) as the comparatively settled, fertile south.[24] Although Rome's interest in the peninsular had extended to a serious, if unsuccessful attempt at conquest in 24 BC by Allius Gallus, Prefect of Egypt under Caesar Augusta; nevertheless, writes H A L Fisher, 'for the first six centuries of the Christian era no European statesman had occasion to remember the existence of Arabia. It was a land of mystery, doing a little trade with Syria and Egypt and contributing some mercenaries to the Persian and Byzantine armies, but otherwise as remote and as inhospitable as the frozen north. Nothing likely to be reported from this scorching wilderness would be calculated to disturb the bazaars of Damascus or Alexandria'.[25]

And yet, within the next hundred or so years the Arabs had become a major world power, controlling an empire extending from the Punjab

GLOSSARY

Although Arabic is the language of Islam there are local differences in terminology between the three broad regions of Iran (ancient Persia), the Arabian Peninsular, and North Africa.

Ab water
Ab-anbar water storage cistern
Bab city gate
Bagh garden
Band dam
Badgir windcatcher ventilator
Bat'ha small minor public space (Bahara/Saha/Rahba)
Bayt dwelling
Bazar complex of suqs and related buildings
Burj tower in defensive wall

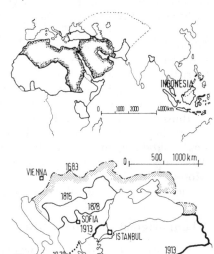

Figure 11.5 – Islam: an outline map relating the Middle East to the wider Islamic world, which includes Indonesia (see Appendix D) and showing (inset) the furthest extent of the Turkish Ottoman Empire into south-east Europe.

20. B Lewis, 'The Faith and the Faithful', in B Lewis (ed.) The World of Islam, 1976.
21. See M Rogers, The Spread of Islam, 1976; see also F Robinson, Atlas of the Islamic World since 1500, 1982, for comprehensively statistical map and illustrated histories of the respective periods.
22. For Indonesia see Appendix D.
23. See B S Hakim, 'Islamic Architecture and Urbanism', in J Wilkes (ed.) Encyclopedia of Architecture, Design, Engineering and Construction, 1989.
24. A Guillaume, Islam, 1956/1983, has observed:'so far as our knowledge goes the first use of the name Arab is on the inscription of the Assyrian king Shalmaneser III, who in the year 853 BC defeated a coalition of small western states in which Ahab, king of Israel, took a prominent part, being supported by a certain "Jundibu the Arab" ... the first known use of the term by the Arabs themselves is on an inscription in the Nabataean script which records the exploits of a certain "King of all the Arabs". His rule cannot have been recognised farther south than central Arabia'. See also P Marechaux, Arabia Felix: The Yemen and its People, 1980.
25. H A L Fisher, History of Europe, 1935.

in the east through into northern Spain in the west, from where they penetrated north of the Pyrenees before that tide of expansion was reversed. Later, it was only the exceptional fortifications of Constantinople that kept Islam out of south-eastern Europe until 1453, when the city fell to the Muslim Turkish army (see Chapter 3). Vienna took over as Europe's bulwark in face of militant Turkish Islam and survived within impregnable defences when besieged by them in 1529 and 1683 (see Chapter 7).[26]

How had this come about? What had aroused the Arabs and sustained their campaigns for power? Religion can be only part of an answer. That 'they rode, battled, and conquered to extend the faith' does not entirely accord with the fact, Fisher explains, that 'during the early years of Arab expansion, the conquerors were at no particular pains to make converts'.[27] Latterly, though, religious conversion to Islam by more or less peaceful persuasion did become a main consideration.[28] This, in direct contrast to the Spaniards, who imposed Christianity by force from the outset in Latin America.

By 636 the Arabs were strong enough to conquer Syria; a year later they had reduced Ctesiphon and gained Iraq, and in 642 they were established in Alexandria. From Egypt the Arabs moved westwards, taking in first Cyrenaica and Tripolitania (modern Libya), before finally subduing obdurate Berber tribesmen of the Maghreb (modern Tunisia – where Kairouan was founded in 670 – and Algeria and Morocco). And then, in the early eighth century, the Arab and Berber armies invaded a weak, disorganized Visigoth Spain (see Chapter 4).

Islam: original and later urban form determinants

Set against the preceding introductory background, a primary concern of the chapter is to identify the major determinants which shaped the historic Islamic cities of the Middle East, explaining their roles in general and in respect of a range of individual city examples. The essential characteristics of Islamic urban form have been introduced in Chapter 1. Summarily restated, there are two contrasted characteristic kinds of traditional Islamic urban morphology.

First the more usual, generally recognized form, with adjoining densely packed climate-response courtyard houses on one or two storeys, access to which was gained by way of narrow, indirect cul-de-sac alleys, entered off marginally wider through routes leading from the city gates to the centrally located main mosque. These thoroughfares – for which 'street' is a misleading European term – were mostly sun-shade roofed over and lined by the small individual shops of the *suq*, the city's markets. This form is shown diagrammatically (Figure 11.6) and is exemplified by the aerial photograph of Fez (Figure 11.8).

Caravanserai   fortified hostel on trade route
Dar   house (dwelling around courtyard)
Dar Al Islam   territory ruled by Law of Islam
Derb, Ghair, Nafidh   cul-de-sac residential access alley (also Bunbast, Sikka Munsadat al-Asfal, Sikka Ghair Nafidha, Zuqaq, Zanqa)
Diwan   courthouse, public audience hall
Driba   residential access space, entrance hall
Fina   domestic courtyard and exterior space adjoining house walls
Funduq   warehouse, inn
Hajj   pilgrimage to Makkah
Hammam   public baths (hammamat)
Hara   quarter/district of a city
Haram serai   private dwelling
Harim, Harem   women's domestic quarters
Jami   *see* Masjiid-I-Jami
Kadi   judge
Kasbah   castle, citadel, urban fortress
Khan   inn, hotel (caravanserai)
Khadaq   main sewer line (moat)
Khanih   house
Khutbah   Friday mid-day Sermon
Ksar   palace, fortified dwelling
Madina/Medina   city (or the old city)
Madrasa   theological school, Quaranic college
Mahalla   segregated religious or ethnic group neighbourhood
Maidan   extensive public open space
Maqbara   cemetery (Jabbana) Marfaq
Maristan   hospital, infirmary
Mashrabiya   screened balcony/window
Masjid   mosque
Masjid-i-Jami   Congregational ('Friday') mosque (Jami)
Massassa   public water fountain
Mida'at   public ablution facility
Minaret   symbolic mosque tower, platform for call to prayer
Minbar   pulpit in a mosque
Mihrab   recess in mosque wall indicating direction of Makkah
Misr   metropolis
Musalla   extensive open air public assembly space adjacent to medina
Mushrabiya   *see* Rawashin
Pankeh   ceiling fan
Qalah   fortress
Qanat   underground water conduit
Quibla   direction of prayer to Makkah
Rabad   suburb adjoining medina
Rawashin   wooden privacy ventilation screen
Sabat   room built over a street
Sahn   interior court of mosque
Shari   public through street (also Tarik Nafidh/Nahj)
Skifa   house entry room
Sufli   single-storey dwelling
Suq   marketplace
Sur   city wall, ramparts
Talar   domestic upper floor reception room
Ulvi   two-storey home
Wekala   travelling merchants building
Yakh-chal   ice-house

Figure 11.6 – Islamic urban form: diagrammatic 'low-rise' courtyard housing, exemplified by Erbil (Figure 1.10).

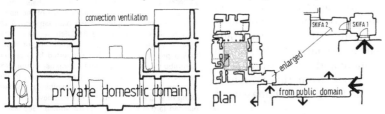

26. Islam, however, has been able to maintain a long-term presence in parts of south-east Europe; Albania today, for example, is about 70 per cent Muslim

The second form which occurs much less frequently and which is much less well known,[29] is that of comparatively 'high-rise' houses on

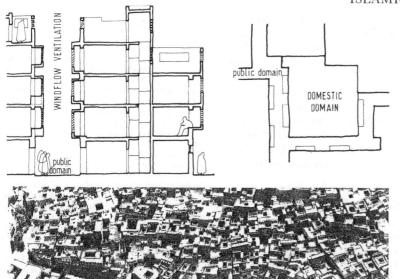

Figure 11.7 – Islamic urban form: diagrammatic 'high-rise' street frontage housing, exemplified by Makkah as the 1880s photograph of Figure 11.11 and accompanying house plans, and Jeddah as Figure 11.38 (see also the map of Sa'na in 1879, as Figure 11.39).

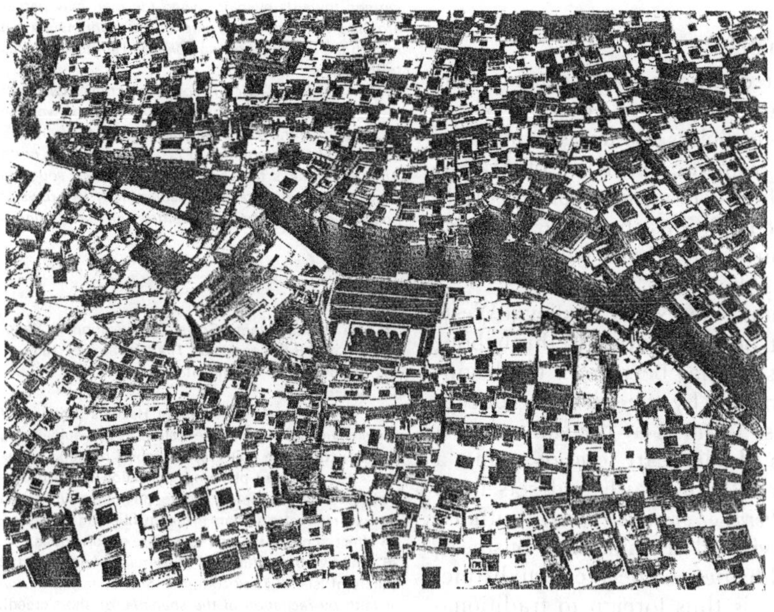

Figure 11.8 – Fez, Morocco: an aerial photograph exemplifying the contiguous cellular essence of Islamic urban form. Fez is a two-part historic city: Fes el Bali ('Old Fez') which was founded in 808, and Fez Djedid ('New Fez') which was added in 1276 as a political quarter. Since then the capital of present-day Morocco had alternated between Fez and Marrakesh, with a brief interval at Meknes, before being moved to Rabat following the French occupation of 1891.

Military defence was not a paramount locational consideration; rather it was availability of an ample supply of water in the valley of the Oued Fez. This was brought into the city at higher levels by potable water canals, used for the various purposes, and then taken out, 'as 'dead' water, by complementary lower ditches as an exceptional instance of an historic waterborne drainage system (see T Burckhardt, 'Fez', in R B Serjeant (ed.) The Islamic City, 1980).

Fes el Bali became established as separately fortified quarters on the valley slopes of the river, that to the south receiving Muslim refugees from Cordoba in 818, hence its name, Adwat al-Andalus, with the Mosque of al-Andalus founded in 861. The more important Mosque of Qarawiyin, which became a major centre of learning, had been founded in that quarter two years earlier. Fes Djedi was added on adjoining higher ground to the north-west, beyond which there is the (fortunately distant) planned French Ville Nouvelle of 1916. (See also 'Fez: The Ideal and the Reality of the Islamic City', in Architecture as Symbol and Self-Identity (Aga Khan Awards) 1980; S Bianca, 'Fez: Towards the Rehabilitation of a Great City', in Conservation as Cultural Survival (Aga Khan Awards) 1980.

three or four, exceptionally five, storeys which face on to the city streets, taking their light and ventilation by way of screened windows. The all-important example is Makkah – the Prophet's hometown. Other major examples are located in southern Arabia, notably San'a; in hot-humid waterside micro-climates, in particular Jeddah; and in densely developed, major city centre residential districts such as Cairo and Baghdad.[30] This form is shown diagrammatically by Figure 11.7.

In considering the Islamic urban form determinants we shall divide them into two groups: original and later, distinguishing, respectively, between those of ancient Mesopotamian and Arabian precedents, and those that became established in response to requirements of Islam.[31]

ORIGINAL DETERMINANTS

The original determinants inherited from the past were of both the 'natural world' and 'man-made' types (as described in Chapter 1). While the three determinants of 'natural world' origin – topography, climate, and construction materials – maintained their ancient roles in shaping Islamic urban morphology, several 'man-made' determinants are conspicuous by their absence from traditional Islamic urban culture. These include the gridiron, an essentially Western-world, Graeco-Roman concept; legislation, in the sense of formally codified civil law;

although officially a secular State. (See also Chapter 2 for Greece and Athens in this respect.)

27. Fisher, History of Europe.

28. See A Guillaume, Islam, 1956/1983. who observes that differential taxation was one means of persuasion.

29. Arguably the main reason for the emphasis accorded the low-rise courtyard house in the urban histories, is that this was the form of the house built for the Prophet on his arrival at Medina.

30. For Cairo central area housing see A Raymond, 'The Residential Districts of Cairo during the Ottoman Period', in I Serageldin and S El-Sadek (eds) The Arab City: Its Character and Islamic Cultural Inheritance, 1982; for Baghdad see J Warren and I Fethi, Traditional Houses in Baghdad, 1982.

31. I am further indebted to Glyn Daniel for his use of 'first' and 'later' as the means of avoiding 'primary' and 'secondary' when referring to the first civilizations, as quoted in Chapter 1. That terminology, modified here in respect of the 'original' and 'later' determinants of historic Islamic urban form, avoids possible pejorative inference that Islam was only a secondary influence (The First Civilisations).

('Original' means existing from the beginning, that is original and continuing.)

aggrandizement and considerations of civic aesthetics, with partial exception of the pre-eminence accorded the mosque; and, by and large, social segregation – although religious and ethnic grouping was to be of considerable significance.[32] On the other hand, the pre-urban cadastre and urban mobility continued to play leading roles. Within Islam, although strength of urban fortification was less important than in continental Western European countries, internal city security was a primary consideration. (Reasons why certain of the ancient man-made determinants were of little or no significance have been introduced in Chapter 1, with further consideration below.) The historic Islamic city was shaped by the urban guidelines, in combination with locally relevant original determinants.

LATER DETERMINANTS: THE URBAN GUIDELINES

Hakim makes no comparably clear-cut distinction between ancient, pre-Islamic determinants – our 'original' category – and the 'later' guidelines of Islamic urban culture; nevertheless, he lends support to this concept in stating 'Islam emerged from the heartland of Arabia, which had strong pre-Islamic Arab traditions in various spheres of life including building practice. Many of the traditions that were not compatible with Islamic values were prohibited but others, with modification became part of Islamic civilisation.'[33]

Unlike Christianity, the parallel religious context for medieval Western European urban history, the effect of Islam extends throughout Muslim everyday life. 'For Muslims, Islam has been from the beginning much more than what is usually meant by the Western concept "religion",' writes Welch, who explains 'Islam is at the same time a religious tradition, a civilisation and, as Muslims are fond of saying, a "total way of life". Islam proclaims a religious faith and sets forth certain rituals, but it also prescribes *patterns of order for society* (author's italics) in such matters as family life, civil and criminal law, business, etiquette, food, dress, and even personal hygiene. The Western distinction between the sacred and the secular is thus foreign to traditional Islam.'[34] (As will be seen, Welch could also have included houses and urban form in his examples.) In effect, the later Islamic determinants are of two types. First, the everyday urban guidelines – as we shall term them – by way of combining the wording of Welch's 'religious patterns of order' and that of 'legislative guidelines' which forms the basis of Hakim's comparable approach.[35] Their observance was an everyday requirement of even the most humble of the Faithful, and their cities were directly shaped by them, in combination with locally effective original determinants. Second there is the effect of the five essential Muslim duties, known as the 'Pillars of Islam', which are summarized separately.[36]

As a further contrast to the role of Christianity (for which no separate introduction was needed in Chapter 4), our investigation of the origins, nature and effects of the later determinants requires that we look next at the life and times of the Prophet Muhammad and his successors, during the early formative Islamic period.

Islam: the Prophet Muhammad and the Holy Law

The Prophet Muhammad was born around AD 570 in the city of Makkah. He was the posthumous son of Abd Allah of the Hashim clan of

The bedouin tent is a direct expression of the pastoral-nomadic way of life which required one to travel with one's shelter. It is also climatically suitable, as the covering of woven black goat's hair provides shade without which it would be difficult to camp in the desert. The nomadic way of living requires that the tent should be light, flexible, and easy to carry and erect. The bedouin tent of the Middle-Eastern desert nomads is designed to do all of the above. The herds of goats, the camels, and the black tent are the basic necessities of the bedouin way of life.

The tent awnings are woven from goats' hair by the bedouin women. In some cases camel hair, cotton, or wool may also be used. The tent widths, depending upon the desired size of the tent, are woven in lengths of 7 m to 45 m. The tents are usually erected by bedouin women towards evening, when the men may gather around a water-pipe for a smoke. The children may join in with the women erecting poles and stretching ropes to the stakes.

The woven goat-hair tent covering of the Arabian tent is supported on three longitudinal rows of wooden poles. The central row is kept higher than the front and the back rows. According to C. G. Feilberg, the north Arabian nomad's tent differs from those of southern Arabia because its frame is made of only two rows of poles placed in the front and centre. The rear awning is

32. See R B Serjeant, 'Social Stratification in Saudi Arabia,' in R B Serjeant (ed.) *The Islamic City*, 1980.
33. Hakim, *Arabic-Islamic Cities*.
34. Welch, 'Islam'.
35. I have chosen not to adopt Hakim's wording of 'legislative guidelines'; I use instead 'urban guidelines', in order to ensure that Western-world readers, accustomed to their city planning legislation being of formally codified mandatory nature, do not presume such status for the roles of Islam in the city building processes. (In effect, though, the two terms are synonymous.)
36. The five basic duties of Muslims are known as the 'Pillars of Islam'. They comprise *Shahadar* (profession of faith by recitation of the shahada, or short creed), *Salat* (prayer), *Zakat* (alms-giving), *Saum* (fasting) and *Hajj* (pilgrimage).

Michon explains that 'certain conditions which precede or accompany the performance of prayer have considerably influenced the design and function of the cities of Islam, namely (a) the stage of ritual purity, by means of the greater or lesser ablution; (b) respect for the time of prayer; (c) the facing towards Makkah; (d) the existence of a site large enough to accommodate all the faithful at the communal prayers on Friday, which are obligatory in all important centres of population.

The first requirement led to the provision of lavatories, pools, fountains and public baths (*hammam*), the domes of which are sufficient to acclaim the presence of a Muslim city. The second ... led to the construction of minarets for the muezzin who delivers the call to prayer; the last two requirements determined how mosques were to be built.... The communal prayer is at times performed outside the centre of the mosque, as is the case when the two great (annual) feasts are celebrated by Muslims. The faithful then make their way in a crowd to the masalla situated outside the walls of the mosque. ('Religious Institutions', in R B Sergeant (ed.) *The Islamic City*, 1980).

The numbers of pilgrims participating in the annual Hajj has had primary determining effect on the forms of Makkah and Medina, as described below; it has also affected other minor centres of pilgrimage such as Karbala and Najaf.

the Quraysh, the most powerful of the city's mercantile tribes.[37] His uncle, Abu Talib, an elder of the tribe, looked after him and took him on lengthy caravan expeditions to Syria. This experience gave Muhammad an early awareness not only of contemporary Arabian and eastern Mediterranean societies, but also of the Jewish and Christian religions with their established communities throughout Arabia, most notably that of the Jews (or, more properly, Judaized Arabs) at Yathrib (Medina). At 25 Muhammad married Khadija, a wealthy widow fifteen years older, who was later to become the first convert to Islam. Then about the year 610, during retreat in the mountains near Makkah, Muhammad began seeing visions and receiving revelations which he perceived as recitations by God.[38]

Some three years later, when Muhammad began preaching in Makkah, the city's leading families, 'fearing economic repercussions against the pagan deities worshipped at the shrine of the Ka'ba'[39] persecuted the Prophet and his followers and imposed an economic boycott against his clan'. When in 619 both Khadjia and his uncle Abu Talib died, Muhammad lost his clan protection and was forced to seek refuge elsewhere for his small Muslim community. He first investigated Ta'if, a pleasant oasis town in the mountains about 110 km to the south-east, but was rejected.[40] Back at Makkah, however, he met a group of pilgrims from Yathrib, visiting the pagan shrines, with whom he commenced negotiations that led to the migration there in 622 – the *Hijra* of Muhammad and his followers. At Medina, as Yathrib came to be called, from *madinat al-nabi* (the city of the Prophet), Muslim strength steadily increased, culminating, after a period of warfare, in the peaceful surrender of Makkah in 630. Aged 61, Muhammad died on 8 June 632, after a brief illness, and was buried in his mosque at Medina.

THE PROPHET'S SUCCESSORS

The Prophet had left no instructions for selecting a successor as leader of Islam. His companions improvised by creating the post of the *caliph*, which Arabic word combines the concepts of successor and delegate. The third Caliph was Uthman, of the Umayyad, another of the leading tribes in Makkah under whose firm leadership the Arab Muslims completed the conquest of Egypt, Iran, Iraq and Syria. Uthman was assassinated in 656 and succeeded by Ali, a cousin and son-in-law of the Prophet, who was also murdered, in 661. At this point the Muslim community divided, with the Umayyad founding a new caliphate and moving their capital from Medina to Damascus, in Syria, thereby ensuring that ancient city's leading role in Islamic history.

The followers of Ali – which translates as *shii* in Arabic – considered that only direct descendants of the Prophet should succeed to the caliphate. Henceforth known as the Shiites, in 680 they attempted to overthrow the Umayyad but failed; their movement then becoming the first of several religious sects within Islam.[41] Allied with other dissident groups, the Shiites were successful in overthrowing the Damascus caliphate in 750, replacing it with the Abbasid dynasty, who were descendants of al-Abbas, the Prophet's uncle. However, they were never able to rule the entire Muslim world as had their predecessors; their very accession in 756 resulted in the establishment of an independent amirate in Spain by a fleeing Umayyad prince. In 762 the Abbassids moved from Damascus to a new, planned capital – the 'Round City'– which they founded near ancient Csetiphon, on the River

stretched and supported by the anchor ropes attached to the stakes. Fewer supports may have been preferred because this would mean reduced weight to carry and reduced erection and dismantling time.

The roof of the Arabian tent is gabled, corresponding to each row of poles there is a stake to provide tensile strength against the desert wind. A typical central Arabian Najd tribe tent may have three rows of three poles each. Wooden pins and hemp ropes are used as fastenings. Sometimes thick woolen cloth my be used as an additional covering. The tents are generally divided into two sections – one for men and one for women – by a cloth made of the same material as the walls. Colourful strips with geometric designs may be included among the cloths which make enclosures (walls) as well as the partition. The awning (roof) is generally black.
(K Talib, Shelter in Saudi Arabia. 1984)

37. For English language references on the life and times of the Prophet Muhammad see K Armstrong, *Muhammad: A Western Attempt to Understand Islam*, 1991, a most important reference that became available as this chapter was in preparation; N Rodinson, *Muhammad*, trans. A Carter, 1971; Welch, 'Islam'; Guillaume, *Islam*.

38. See N J Dawood (trans) *The Koran*, 1974/1984, a translation in contemporary English with explanatory footnotes.

39. See Armstrong, *Muhammad*, for the importance of the shrine of the Ka'aba. 'At the time of Muhammad, the Ka'aba was officially dedicated to the god Hubal ... but the pre-eminence of the shrine as well as the common belief in Mecca seems to suggest that it may have been dedicated to al-Lah, the High God of the Arabs'. (In addition to the shrine of the Ka'aba, there is also at Makkah the nearby sacred spring of Zamzam, the mystical significance of which can be presumed to have predated, and also probably provided the reason for construction of the Ka'aba.

40. For a late nineteenth-century description of Taif, a pleasantly situated, ancient and present-day mountain summer resort town, see C Doughty, *Travels in Arabia Deserta*, 1880: 'from the next rising ground I saw el-Tayif! The aspect is gloomy for all their building is of slate-coloured stone ... the gate where we entered is called Bab es-Seyl and within is the open space before the white palace of the Sherif, of two stories; and in face of it a new and loftier building with latticed balconies and the roof full of chimneys, which is the palace of Abdullah Pasha, Hasseyn's brother. The streets are rudely built, the better uses are daubed with plaster and the aspect of the town, which is fully inhabited only in the summer months is ruinous. The ways are unpaved'.

Karen Armstrong notes that 'in the walled city of Taif was the shrine of al-Lat' (one of the three shrines close to Makkah that were dedicated to the three daughters of al-Llah (banat al-Lah); she describes how Muhammad 'walked through attractive gardens, orchards and cornfields' as he approached the walled city on the hill (*Muhammad: A Western Attempt to Understand Islam*, 1991).

41. See Robinson, *Atlas of the Islamic World since 1500*.

Tigris, thereby ensuring the continuing importance of its successor, Baghdad, on the international scene. Later, in 929 the Spanish Umayyads were to declare their own caliphate at Cordoba, increasing to three the number of separate caliphs in the Muslim world, of which the Abbasid was the most widely recognized.

From the ninth and tenth centuries a distinctive Persian identity began to emerge within Islam. Subsequently the history of Islam is fragmented and several cities gained enduring, or transient importance. Cairo, for example, became the capital of the Mamluke caliphate, re-establishing the old Arab Muslim culture in Egypt until conquered by the Turkish Ottomans in 1517. Samarkand, outside our Middle Eastern limits,[42] and Isfahan in Persia enjoyed passing pre-eminence. Istanbul (Constantinople) – the exceptional enduring Islamic city in Europe – and Athens, which experienced nearly four centuries of Islamic rule, both lost and then regained their international importance.[43]

ISLAMIC HOLY LAW

'Islamic orthodoxy and modern critical scholarship agree that the contents but not the final arrangement of the Qur'an go back to Muhammad', states Welch, adding 'it is also virtually certain that Muhammad began but did not complete the task of compiling a written text of the Qur'an'.[44] It was the third Caliph, Uthman, who ordered the preparation of a canonical text with a copy to be kept in each Friday mosque. 'Next to the Qur'an stand the Hadith works, multi-volume collections of accounts called Hadiths (from the Arabic, *hadith*, "saying(s)") that report on the sayings and deeds of the Prophet Muhammad. After the Qu'ran, the Hadiths provide a guide for all aspects of Muslim daily life, for which the Prophet stands as an exemplar par excellence.'[45]

Together the Qur'an and Hadiths form the basis of the Shari'a (the Holy Law) which 'covers all aspects of the public and private, communal and personal lives of the Muslims'.[46] The Shari'a is based on legal decision of early Muslims, as they confronted immediate social and political problems, to derive systematic codes of Muslim behaviour from the Qur'an and the Hadiths. In this way four main Schools of Law became established: the Hanifi School, which grew up in the Abbasid capital of Baghdad and believed to have been founded by Abu Hanifa (d 767); the Maliki School, based on the practice of the Medina judge Malik Bin Anas (d 795); the Shafii School, evolved by a disciple of Malik; and the Hanbalite School.[47]

While there are formal differences of emphasis and technique between the first three schools, they are agreed on important matters, including those affecting urban development. In principle, interpretation of the Shari'a was the responsibility of scholars appointed as Kadi, or judge, and its administration came to be in the hands of political rulers. Hakim includes numerous case-study examples of the law in practice relating to building and urban development matters, and he explains that 'although the tradition of documenting problems and providing solutions and opinions to each case is pre-Islamic; it was in the "Fiqh" that this tradition was fully developed'.[48]

The *Fiqh* is the Arabic term of jurisprudence, or the science of religious law in Islam. It concerns itself with two spheres of activity: *ibadat* – matters concerning ritual observances – and *mumalat* – legal questions that arise in social life, e.g. family law, law of inheritance, of property, of contracts, criminal law, etc. In essence, therefore, *Fiqh* is the science

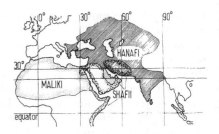

Figure 11.9 – The distribution of the Islamic Schools of Law.

A distinction must be made between the rapid political and military expansion of empires ruled by Arabs and other Muslims and the spread of Islam or religious conversion, which proceeded at a much slower pace. Those areas where the rulers and the majority of the people became Muslims came to be called Dar al-Islam, the House of Islam. Eventually the overwhelming majority of the people across northern Africa, the Fertile Crescent and Anatolia converted to Islam from the various forms of Christianity, and an even higher percentage of the people of Iran converted from the Zoroastrian faith; a very small percentage of the Jewish people, on the other hand, converted to Islam. Since the time of the 'Abbasids, Islam has continued to spread, mostly by peaceful, missionary means, eastwards through Asia to parts of China and South-East Asia, notably Malaysia and Indonesia where a large majority are now Muslim, and also across a wide band of northern sub-Saharan Africa, where Islam continues to gain new converts.
(A T Welch, 'Islam', in J R Hinnells (ed) A Handbook of Living Religions, 1985)

42. See Chapter 7 for description of Samarkand, with references.
43. See Chapter 2 for Athens, the forgotten sometime Islamic city of Europe.
44, 45. Welch, 'Islam'.
46. Lewis, 'The Faith and the Faithful'.
47, 48, 49. Hakim, *Arabic-Islamic Cities*, and in correspondence with the author.

of laws based on religion and is concerned with all aspects of public and private life and business. Most importantly for our purpose, 'problems arising from the building process were viewed by the "Fiqh" in the same light as other problems, resulting from activities and interaction between people'.[49]

The next question to be addressed concerns the pre-Islamic Arabian traditions which, with suitable modification, were to become part of Islamic urban culture. Without becoming deeply involved in religious matters, it is clear that the Prophet was aware of need to retain as much as possible from the staunchly traditional Arabian way of life.[50]

Hence the answer to this question requires that we consider the physical forms of Makkah and Medina – the Prophet's birthplace and refuge – and salient social characteristics of the pre-Islamic Arabian communities they housed.

Islamic urban form: Makkah and Medina and the urban guidelines

Makkah and Medina were introduced in Chapter 1 in their Arabian Peninsular climatic and historical context.[51] In the late sixth century it can be presumed that the two cities had taken markedly different forms in response to their locational and original determinants. Makkah was situated in a rocky valley with demands on limited building land, which, with stone masonry construction technology, created its exceptional three and four-storey house types. Its limited water supply gave minimal opportunity for agriculture. Medina, on the other hand, was on flatter land with horizontal growth possibility and adequate water for comparatively extensive oasis agriculture. The Makkan economy relied on regional and international trade, and the revenue from pilgrims visiting its shrines, whereas Medina had its agricultural base and local trade. Both had broadly similar hot-dry climates.[52] For the pre-Islamic traditions that were to become embodied in Islam, we shall look first at Makkah, the Prophet's birthplace and where he grew to maturity.

MAKKAH IN THE LATE SIXTH CENTURY

Through to the immediate pre-Islamic period Makkah had continued to flourish, stimulated latterly by need for safe overland caravan routes in face of Red Sea piracy.[53] Pilgrims visiting the historic pagan shrine of the *Ka'aba* provided a further major source of income. The *Ka'aba* (Arabic 'cube' or 'chamber') which was to become the most important of the historic Arabian symbolic monuments to be incorporated into Islam, is an approximately cubic building about 13 metres in height and off-square on plan with unequal length sides.[54] The Black Stone which the Prophet declared had been given to Abraham by Gabriel is set at about chest height in its external wall, where it is revered by Muslim pilgrims as the focus of the *Hajj*.[55] In 683 much of the *Ka'aba* was destroyed by fire and it was also badly damaged by flooding in 1627.[56] Today it forms the centrepiece of the courtyard of the Great Mosque and a summary of its role in the *Hajj* ceremonies is given in the adjoining column.

There is an excellent eye-witness description and a plan of historic Mecca published in 1829 by John Lewis Burckhardt in *Travels in Arabia* (Figure 11.10).[57] There are three photographs of the Great Mosque

THE ISLAMIC CALENDAR

The Prophet's migration in 622 – the Hijra – from Mekkah, his home-town, to Yathrib, subsequently to be known as Medina, from madinat al-nabi, 'the city of the Prophet', where he found shelter and support; marks the beginning of the Muslim era, and established at its outset the importance of the two cities as the pre-eminent holy places of Islam. AD 622 was subsequently designated the Muslim year AH1, but, as Roberts explains 'the difference between the Islamic and the Christian Gregorian calendar is not simply the date of its inception. Islam officially uses a lunar calendar of 12 lunar months giving the year only 354 days'. (H Roberts, *An Urban Profile of the Middle East*, 1979.) This means that all the months commence about 11 days earlier each year, thereby making it impracticable to attempt to give the calendar dates in this chapter on the twin basis 1/622 and so on. Accordingly, only the Gregorian calendar is used in this chapter.

50. For an uncomplicated assessment of the everyday political realities confronting Muhammad see J Mathe: 'As the prophet launched himself into his task of disseminating Islam, the Middle East was already being pulled in opposite directions by the other two great revealed religions, Judaism and Christianity. Muhammad knew both of them and respected them. The Islam which he preached was essentially the adaptation of the same monotheistic principle of his own race ... it was reform rather than competition that he had in mind' (*The Civilization of Islam*, 1980).

51. For general background see E Essin, *Mecca the Blessed: Medina the Radiant*, 1963; see also the special topic references below.

52. See K Talib, *Shelter in Saudi Arabia*, 1984. The palm groves and other vegetation and planting at Medina would have had a modifying effect on the climate, in addition to providing welcome open air shade. Not all of the oasis area at Medina would have been fertile land and it can be presumed that its conservation was a major factor determining location of buildings.

53. See P Crone, *Meccan Trade and the Rise of Islam*, 1987; M Bloch, The Social Influence of Salt, *Scientific American* vol 209, 1963.

54. The present-day *Ka'aba* plan dimensions are: the north-east wall 12.63 m, the eastern wall 11.22 m, the western wall 13.10 m, and the north-west wall is 11.03 m. A new *kiswah* (cover or cladding) is made for the *Ka'aba* each year; it is a black cloth with black calligraphic patterns woven into it, and a band of Koranic calligraphy in gold thread woven around the top portion.

55. Neither the Black Stone nor the *Ka'aba* are objects of worship, but they represent a sanctuary consecrated to God since time immemorial, and it is towards the *Ka'aba* that Muslims orientate themselves in prayer.

56. The *Ka'aba* has extremely ancient origins and has been rebuilt on several occasions, including that of 605 when Muhammad, then about 35 years old and yet to commence his prophetic mission, resolved differences between the clans of the Quraysh concerning who should have the honour of replacing the Black Stone. The most recent major repairs were carried out in 1957. (For the sacred mosque, al-Masjid al-Haram, which contains the *Ka'aba*, see Figures 11.30 and 11.50).

57. J L Burckhardt, *Travels in Arabia*, 1829.

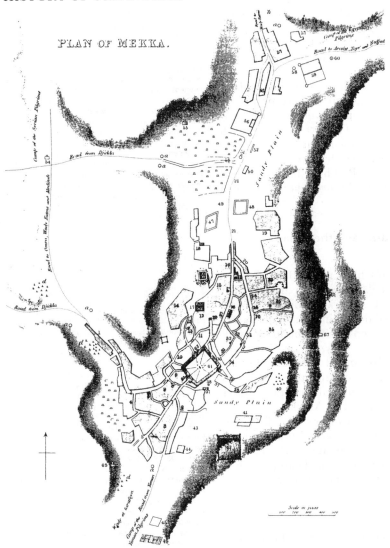

PLAN OF MEKKA.

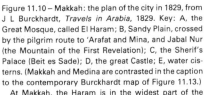

Figure 11.10 – Makkah: the plan of the city in 1829, from J L Burckhardt, *Travels in Arabia*, 1829. Key: A, the Great Mosque, called El Haram; B, Sandy Plain, crossed by the pilgrim route to 'Arafat and Mina, and Jabal Nur (the Mountain of the First Revelation); C, the Sherif's Palace (Beit es Sade); D, the great Castle; E, water cisterns. (Makkah and Medina are contrasted in the caption to the contemporary Burckhardt map of Figure 11.13.)

At Makkah, the Haram is in the widest part of the main south-flowing wadi and main streets follow other tributary valleys. In history the courtyard of the mosque was enclosed by what were called 'oratory houses' which had first-floor balconies on the roof of the courtyard colonnade. These were the most sought-after lodgings in the city because their occupants could pray at home while observing the Prophet's injunction that whoever lived near the mosque should pray in it. See Figure 11.11 for a photograph of 1884–5 and Figure 11.31 for the present-day Grand Mosque.

taken in 1884–5) by S Hurgronje, illustrated in *The Historical Mosques of Saudi Arabia* by G R D King, one of which is shown here as Figure 11.11.[58] From these eye-witness sources it is clear that the character of mid-nineteenth-century Makkah differed in several major respects from that of the traditional low-rise, courtyard-housing Islamic city. Indeed, as a result of the long-established pagan pilgrimage to the Ka'ba, it is most probable that Muhammad's birthplace was already an atypical Arabian city.[59]

The first major departure from the characteristic Mesopotamian climate-response morphology is in the arrangement of the houses, of which Burckhardt wrote: 'Mecca (like Djidda) contains many houses three storeys high ... [and] it is in the houses adapted for pilgrims and other sojourners, that the windows are so contrived as to command a view of the streets,' to which he adds 'the numerous windows that face the streets give them a more lively and European aspect than those of

58. G R D King, *The Historic Mosques of Saudi Arabia*, 1986.

59. Contrary to possible misbelief, Muhammad was not a tent-dweller of the desert. 'The old tribal ethic affected the Qu'ranic message,' writes Armstrong, 'but the new religion was first received by the Arabs of Mecca in an atmosphere of cut-throat capitalism and high-finance. Like all the great confessional religions and the philosophical rationalism of Greece, Islam is a product of the city' (*Muhammad*).

60. Burckhardt, *Travels in Arabia*.

Figure 11.11 – Makkah: the Grand Mosque in its multi-storey urban context, photographed in 1884–5 by C Snouck Hurgronje and illustrated in his *Mekka*, 1889 (this most important historic photograph, with two others, is included in G R D King, *The Historical Mosques of Saudi Arabia*, 1986.

Egypt or Syria, where the houses present but few windows towards the exterior'.[60] Supporting evidence in this respect is provided by Fadan whose research informs that in the tenth century, al-Muqaddasi, an Arab author, wrote 'the houses of Makkah are built of black, smooth stones and also of white stones but the upper parts are of brick. Many of them have large projecting windows of teak wood and are several stories high, white-washed and clear'.[61] Two centuries later, quotes Fadan, a well-known Arabian traveller, Ibn Jubayr, said of his stay in Makkah: 'we passed the nights on the roof of the house and sometimes the cold of the night air would fall on us and [we] would need a blanket to protect us from it'.[62] In 1978 Fadan studied fifteen houses in Makkah and could confirm: 'both descriptions are applicable to Makkan houses over 700 years later in the nineteenth and early twentieth centuries'.[63]

Not only were these multi-storey Makkan houses essentially different from the ancient one and two-storey courtyard precedent, but also as a second difference Burckhardt found that whereas 'in most towns of the Levant the narrowness of a street contributes to its coolness, and in countries where wheel-carriages are not used, a street width that allows two loaded camels to pass each other is deemed sufficient. At Mekka, however, it was necessary to leave the passages wide, for the innumerable visitors who here crowd together'.[64]

Although English language corroborative archaeological evidence is lacking, it is probable that the main streets of pre-Islamic Makkah were also wider than usual to accommodate the pagan pilgrims. As a further possible difference, late sixth-century Makkah can be assumed to have contained an open space centred on the *Ka'aba* and perhaps enclosed as a building courtyard, to provide space for the pagan worshippers; the corollary to which proposition is that ordinary pre-Islamic Arabian urban form did not contain any public spaces. (The 'ordinary' quali-

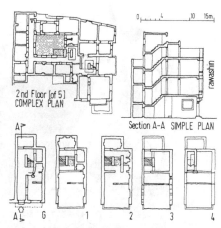

Figure 11.12 – Makkah: multi-storey house plans, after Y Fadan, *Traditional Houses of Makka*, 1983.

61, 62, 63. Quoted in Y Fadan 'Traditional Houses of Makka: the Influence of Socio-cultural Themes upon Arab-Muslim Dwellings'.

64. Burckhardt, *Travels in Arabia*.

65. See Armstrong, *Muhammad*, for description of other shrines including Najran in the Yemen, al-Abalat south of Makkah, and Taif, as above.

fication is because there were a few other Arabian urban settlements with pagan shrines that may also have required an appropriate spatial enclosure.)[65] Of Makkah, Burckhardt wrote 'the only public place in the body of the town is the ample square of the Great Mosque' – an observation confirmed by Hurgronje's fascinating photographs, and which applies to the historic Islamic city in general. (Further description of historic Makkah is the caption to Figure 11.10.)

MEDINA

From Burckhardt's description it is clear that Medina in 1829 was also, but less so, an aberrant Islamic city, although as explained below it is unlikely that its form was other than in the conventional southern Arabian mould through until it received the Prophet and the first community of Believers.[66] 'The precious jewel of Medina,' Burckhardt wrote, 'which sets the town almost upon a level with Mekka, and has even caused it to be preferred to the latter, by many Arabic writers, is the great mosque, containing the tomb of Mohammed. Like the mosque of Mekka, it bears the name of El Haram, on account of its inviolability; a name which is constantly given to it by the people of Medina....' The ground-plan will show that this mosque is situated towards the eastern extremity of the town, and not in the midst of it'.[67] (For further description of Medina see Figure 11.13).

Medina is well built, entirely of stone; its houses are generally two stories high, with flat roofs. As they are not white-washed, and the stone is of a dark colour, the streets have a rather gloomy aspect; and are, for the most part very narrow, often only two or three paces across: a few of the principal streets are paved with large blocks of stone; a comfort which a traveller little expects to find in Arabia. The principal street of Medina is also the broadest, and leads from the Cairo gate to the great mosque: in this street are most of the shops. Another considerable street, called El Belat, runs from the mosque to the Syrian gate: but many of its houses are in ruins: this contains also a few shops, but none are found in other parts of the town; thus differing from Mekka, which is one continued market. In general, the latter is much more like an Arab town than Medina, which resembles more a Syrian city.

Gardens and plantations surround the town of Medina, with its suburbs, on three sides, and to the eastward and southward extend to the distance of six or eight miles. They consist principally of date-groves and wheat and barley fields; the latter usually enclosed with mud walls, and containing small habitations for the cultivators.

Few of the date-groves, unless those dispersed over the fields, are at all enclosed; and most of them are irrigated only by the torrents and winter rains. The gardens themselves are very low, the earth being taken from the middle parts of them, and are heaped up round the walls, so as to leave the space destined for agriculture, like a pit, ten or twelve feet below the surface of the plain: this is done to get at a better soil, experience having shown that the upper stratum is much more impregnated with salt, and less fit for cultivation, than the lower.

(J L Burckhardt, Travels in Arabia, 1829)

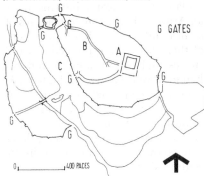

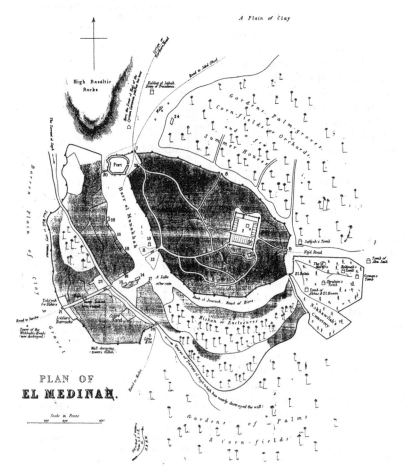

PLAN OF
EL MEDINAH.

Scale in Paces

Figure 11.13 – Medina: the plan of the city in 1829, from J L Burckhardt, *Travels in Arabia*, 1829. Key: A, the Haram or Prophet's Mosque (noted with unfinished porch to north); B, street leading to the Bab el Salam (it is the principal Bazaar, or *Suq*); C, low-lying land – 'a lake after rain'.

Medina can be presumed to have originated as several oasis agricultural villages, situated on higher ground above the seasonal flood level in the central wadi. The extensive courtyard house built for the Prophet on his arrival at Medina would have been on the edge of that part of the city, which explains its off-centre position in 1829. In direct contrast, the central position of the Great Mosque at Makkah can be explained by its evolution around the *Ka'aba*, which was located alongside the Spring of ZamZam, sacred from prehistoric times, as the central *raison d'être* for the city.

---

66. Burckhardt, *Travels in Arabia*; King, *The Historic Mosques of Saudi Arabia*.
67. Burckhardt, *Travels in Arabia*.

ISLAMIC URBAN GUIDELINES: THE INFLUENCE OF MAKKAH AND MEDINA

The marked differences between the presumed form of late sixth-century Makkah and that of the conventional Islamic city, requires that we look beyond the urban fabric and into the homes for the formative influences on the young Muhammad. Crucially, we find that it was not the physical form of Makkah that profoundly influenced Muhammad: rather it was the underlying social traditions of Arabian society, in particular that of domestic domain privacy, expressed in terms of avoidance of overlooking, both inter-dwelling between neighbouring families and also from the external public domain – ordinarily that of the street. Within, there was clear separation of male and female parts of the home. This requirement also provided for protection by the menfolk of the vulnerable members of the family, the women and children – a staunchly held ancient tradition upheld in the Qur'an and the Hadith which taught the virtues and importance of privacy: the right to it and respect of it, thereby establishing domestic privacy as the most important and far-reaching of the urban guidelines.

Thus it can be seen that the Islamic requirement was not that of the courtyard house *per se*, as might be imagined, but rather one of domestic domain privacy. In response to domestic privacy guidelines, the traditional Islamic house took a variety of forms, as illustrated by Figure 11.14. Street-frontage houses, exemplified at Makkah and also Jeddah (Figure 11.38), were cleverly designed to enable maximum advantage to be taken of the cooling benefit of such breezes as arose, blowing into the house rooms by way of rawashin (mashrabiya) that screened the interiors from public-domain view.[68]

The second pre-Islamic morphological characteristic to be accorded guideline status was that of the narrow indirect urban access routes that were possible because only pedestrian and pack-animal traffic needed to be accommodated.[69] Ancient tradition required that two laden camels should be able to pass in the thoroughfares, a width of the order of 7 cubits (3.23–3.50 m); and that the culs-de-sac should allow passage for one laden camel, which requires about 4 cubits (1.84–2.00 m). Hakim gives a Hadith as the source of this guideline, based on the Prophet's words: 'if you disagree about the width of a street, make it seven cubits'.[70]

THE URBAN GUIDELINES

The evolution of urban guidelines can be presumed to have commenced in AD 622 when the Prophet Muhammad settled in Medina, and by the end of the third century of Islam (about AD 900) a legal consensus had become established against which individual interpretations were made. The guidelines are primarily concerned with the nature of the individual family house and its access system, thereby extending, in effect, over by far the greater part of the Islamic city. On an essentially case-law basis, Hakim indicates that 'their development paralleled that of Islamic law, and soon became semi-legislative in nature'.[71]

The most important 'principles and behavioural guidelines' – our urban guidelines – given by Hakim with morphological impoications, include (with reference overpage to both Qur'anic verse 'V' and/or Hadiths 'H').[72]

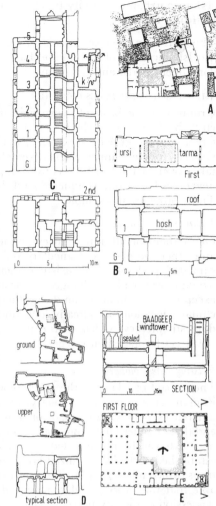

Figure 11.14 – Islamic urban form: a range of example house types.

A, Bukhara, two-storey courtyard house (after K Herdeg, *Formal Structure in Islamic Architecture of Iran and Turkistan*, 1990).

B, Baghdad – street frontage courtyard house (after J Warren and I Fethi, *Traditional Houses in Baghdad*, 1982).

C, San'a' – free-standing multi-storey house (after S Hirshi and M Hirshi, *L'Architecture au Yemen du Nord*, 1983).

D, Ghardaia – small courtyard house (after A Raverau, *Le M'Zab: une leçon d'architecture*, 1981).

E, Dubai – wind-tower house (after A Coles and P Jackson, *A Windtower House in Dubai*, AARP, 1975).

68. For photographic and line illustrations of traditional rawashin at Jeddah see T M Kamel Kurdi, 'Influence of

Islamic urban form: city types by origin, characteristics and building types

This part is in three sections: a summary description of the origins of the Islamic cities of the Middle East; the general characteristics and component parts; and the main Islamic urban building types.

THE ISLAMIC CITY: TYPES OF ORIGIN

Grouped on the basis of their origin there are three types of Islamic city: the first two of which were those existing urban settlements of either organic growth – exemplified by Erbil (ancient Arbela) – or Graeco-Roman planned origins exemplified by Damascus, which were taken over by the Muslims as their empire expanded; and third, those new cities founded in conquered lands by the Muslims armies of which Tunis provides an example.[73]

Existing cities: organic growth

Other than the all-important exception of construction of the Friday and other mosques, and ancillary religious buildings, it can be presumed that the forms of existing organic growth Middle East cities would have been little changed by the advent of Islam.[74] With possible exceptions of settlements on the northerly, colder fringes of Islam, their climatic location had already long since ensured conformity with the domestic privacy guidelines as embodied in the courtyard or comparable street-frontage house types.

Existing cities: planned

The impact of Islam on the morphology of cities of planned Graeco-Roman origins was revolutionary: probably more so in its way than differences between European organic growth medieval urban form and that of underlying Roman gridirons. Damascus provides a fascinating, extensively documented example of ways whereby a long-established, formally organized Western-world gridiron city was subtly transformed to meet the later Islamic determinants, although it is probable that this process had commenced before the Muslim army took possession of the city in 635.[75] Damascus is an extremely ancient city located in a rich agricultural valley fed by the river Barada. Syria was proclaimed a Roman province in 64 BC and the city was replanned in the 2nd century AD as a walled gridiron of sides 1,500 × 900 metres. The building blocks were 100 × 45 metres, each containing two back-to-back rows of four courtyard houses. While a proportion of the houses and grid blocks would have been deserted if not ruined when taken over by Muslim families, thereby facilitating the change to an Islamic urban route-system; overall it would have been a slowly incremental process, the morphological result of which is unique in urban history.

Damascus is also of first importance in respect of the change from the Byzantine Christian Basilica of St John the Baptist, in succession on the site to the Roman Temple of Jupiter Damascenius, into the Islamic Great Mosque. As illustrated in Figure 11.31, the Muslims at first shared the courtyard of the Basilica of St John with the Christians, placing their Mihrab in the southern wall which conformed to the Qibla direction of prayer. In 705, Caliph al-Wahid purchased the Basilica and demolished it in order to use the entire ancient courtyard for the magnificent new Congregational Mosque which was completed in 714/15. A conjectural way whereby arcaded streets of the Roman city

Arabian Tradition on the Old City of Jeddah: House Form and Culture' and *Jeddah: Old and New*, Stacey International, 1980.

For *shanashil*, the Iraqi equivalent, see J Warren and I Fethi, *Traditional Houses in Baghdad*, 1982 (*Mashrabiya* is another locally applicable Arabic word.)

69. See R W Bulliet, *The Camel and the Wheel*, 1975; H Fathy, *Constancy, Transposition and Change in the Arab City*, 1973.

70, 71. Hakim, *Arabic-Islamic Cities*.

72. Principles and behavioural guidelines given by Hakim; Qur'anic verse 'V' and/or Hadiths 'H'

1 Harm, which prevents actions that would result in harm to others, the most frequently quoted of which is Saying 34: 'no person or party to be harmed for another to benefit (V: 6, 18, 19; H: 34, 35).

2 Interdependence, of which verse 5 encourages self-policing behaviour (V: 1, 2, 3, 5; H: 7, 8, 12, 13, 14, 15, 33, 39, 40, 42, 44, 45).

3 Privacy, in particular the private domain of the home (V: 13, 14, 15, 16; H: 29 to 32).

4 Rights of building higher within one's air space (V: 17; H: 26, 44).

5 Respect for the property of others (V: 17; H: 10, 11, 16).

Those with references only to the Hadith include

1 Rights of original (or earlier) usage (H: 17 to 20).

2 Pre-emption, the right of neighbour or partner to purchase adjoining property when offered for sale (H: 46 to 51).

3 Minimum width of public thoroughfares (H: 4).

4 Excess of water should not be barred from others (H: 12 to 15).

73. Hakim gives as his 'three types of cities in the Arab and Islamic world: (a) the planned and designed city, e.g. the round city of Baghdad; (b) the renewed and/or remodelled pre-Islamic city, e.g. Aleppo; and (c) the "spontaneously" created and incrementally grown city'; (or organic growth city). 'Arab-Islamic Urban Structure', *Arabian Journal for Science and Engineering*, vol 7, no 2, 1982.

74. See B S Hakim, 'Islamic Architecture and Urbanism', in J A Wilkes (ed.) *Encyclopedia of Architecture, Design, Engineering and Construction*, 1989: 'the location of most territories of the Islamic world between latitudes 10 degrees and 40 degrees, and the resulting similarity in macroclimatic conditions, contributed towards certain unifying influences in building practice.'

75. For Damascus see M Ecochard, 'Damascus,' in I Serageldin and S El-Sadek (eds) *The Arab City: Its Character and Islamic Cultural Inheritance*, Proceedings of a Symposium, Medina, Kingdom of Saudi Arabia, 1981; Nikita Elisseff, 'Physical Layout', in R B Serjeant (ed.) *The Islamic City*, 1980; L Benevolo, *The History of the City*, 1980; *Murray's Handbook for Travellers in Syria and Palestine*, 1892.

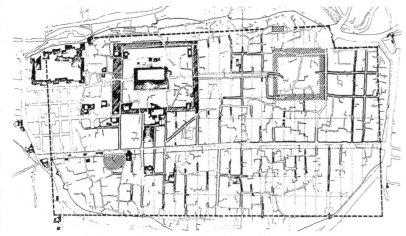

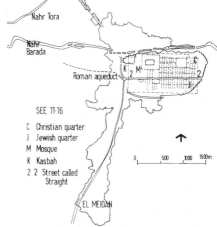

Figure 11.15 – Damascus: the historic city showing the organic growth streets and culs-de-sac which gradually modified the pre-existing gridiron plan, the residual influence of which can still be seen, especially in the retained line of the main east–west axis – 'The Street called Straight' (the Via Recta of the Acts of the Apostles).

were converted into the characteristic Islamic *suq* configuration is included in Figure 11.33.

Although there is little reliable archaeological evidence of their Graeco/Roman gridiron plans, other major cities to have experienced comparable Islamic change can be presumed to include Alexandria (see Figure 11.17) and Cordoba in Spain (see Figure 4.82).

Figure 11.16 – Damascus, Syria: the extent of the late nineteenth-century Islamic city (after *Murray's Handbook for Travellers in Syria and Palestine*, 1892), showing the historic Islamic nucleus (dotted) and the rectangular outline of the Roman and Byzantine predecessor.

Figure 11.17 – Alexandria, Egypt: the city in 1855, a diagrammatic map based on the plan by C Muller in J H G Lebon, 'The Islamic City in the Near East', *Town Planning Review*, April 1970. Key: A, French forts; B, walls and moat repaired and constructed by Mohammed Ali, 1808–48; C, Arab villages and hut settlements; D, two lengths of disused Hellenistic aqueduct.

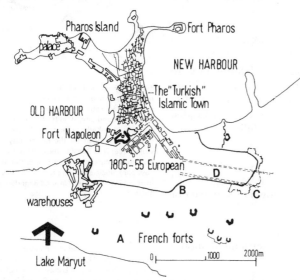

Hellenistic and Roman Alexandria has been introduced in Chapter 3 where Figure 3.30 shows the gridiron plan of this most important Mediterranean city, second only to Rome until the fourth century AD. When Alexandria was taken by the Muslim army in 642 it was still a major port and naval base but from then, through until the later nineteenth century, its history was that of progressive decline into the obscurity of an outpost of Rosetta, which was twice as populous. When Napoleon's French army engineers surveyed Alexandria in 1800, they found the original city area, within its circuit of ruined walls, deserted except for scattered agricultural settlements, although still known as the 'Arab Town'. Alexandria had by then become the 'Turkish Town' which had grown up along characteristic Islamic lines on the broad sandy isthmus which marine currents had formed against the ancient mole linking Pharos Island to the mainland. The plan of 1800 is essentially the same as that illustrated of 1855, the main difference being the embryonic gridiron European quarter which was then becoming established on the mainland alongside the 'new' or East Harbour. During the British period in the second half of the nineteenth century the Arab Town was reoccupied on a mainly gridiron plan, and later the Turkish Town was cleared and rebuilt on a similar basis.

New Islamic cities

There is not much archaeological evidence of the original forms of the settlements founded by the Muslims as their empire expanded. Moreover, with a main exception of Baghdad – an aberrant circular 'planned' Islamic city, English-language literature had little to contribute on this topic until Jamel Akbar's study on early Muslim towns became available.[76] It is stressed that the term 'planned' applies only in respect of a political decision to establish a permanent settlement on a given site. Akbar explains that 'the early pattern of growth which was characteristic of such military colonies as al-Basrah and al-Kufah was rapid and without any real awareness of the formal elements of city planning'.[77] Al-Basrah and al-Kufah, both founded in 638 in

76, 77. Akbar, *Crisis in the Built Environment: The Case of the Muslim City*, 1988.

present-day southern Iraq, needed to be settled immediately.[78] Another example, Al-fustat, founded by the army commander in 641 or 642, south of the subsequent medieval nucleus of Cairo, was settled over a long period of time.[79]

The most important of the 'planned' cities was Baghdad, founded in 762 as Madinat as-Salam, the presumed site of which has never been seriously excavated and whose descriptions rely on historic accounts that 'vary with regard to its dimensions'.[80] Akbar quotes Lassner's interpretation of the arrangement within the circumference: 'the city was divided into three zones. The central zone, ar-rahbah, was open space accommodating the Palace of al-Mansur, the congregational mosque and other buildings for the chief of police and the chief of guards. In the inner ring lived the younger sons of al-Mansur and their servants. The army chiefs and their supporters lived in the outer ring. The overall diameter was about 2,750 metres.

The quarters were allotted to important individuals who laid out their buildings to meet their own requirements, possibly in the same way that el-Amarna was developed in Ancient Egypt, as described in Chapter 1. The Muslim army's 'new town' of Fustat, a direct predecessor of Cairo described below, was established on a similar land allocation basis. The concept of the Round City of Baghdad is unique in Islamic urban history; however, it is probable that laisser-faire implementation on the ground would have meant that precisely circular imagery existed only in the mind.

ISLAMIC URBAN FORM: GENERAL CHARACTERISTICS AND COMPONENT PARTS

In establishing characteristic attributes of the historic Islamic city, to be followed by description of urban components and main building types, consideration must first be made of criteria whereby the Muslim 'city' is distinguished from a 'village'. Hakim refers to Al Maqdisi (c 380/990), the Jerusalem-born Arab geographer who defined a hierarchy of settlement types comprising

1 *Amsar* (sing. *Misr*) – metropolis
2 *Qasabat* (sing. *Qasabah* – provincial capitals
3 *Mudun* (sing. *Madina*) – provincial towns, a main town of a district or a market town
4 *Qura* (sing. *Qaryah*) – villages.[81]

To qualify for urban status, Hakim concludes that a settlement should have

1 a *Mesjid al-Jami* which is recognized as a Friday mosque where the Friday sermon is given and which should serve the residents of the city and its dependants living outside it
2 a governor and/or a *Kadi* (judge) who can execute his duties within the city's area of jurisdiction
3 a *Suq* (market) serving the needs of the people in the city and the surrounding countryside.[82]

In addition, some Arab authors mention the *Hamman* (public baths) as another essential attribute of urban status. It should be noted that while church and market are requirements in common with European (Christian) urban status, a need for a governor and/or a *Kadi* is special to Islam, and similarly the provision of *Hamman* (see below).

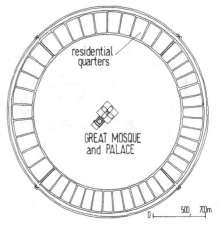

Figure 11.18 – Baghdad: the plan of the Round City, or Madinat al-Salam founded in 762 by al-Mansur near the west bank of the River Tigris. There are no evident remains of this exceptional circular planned Islamic city, and the site has not been excavated.

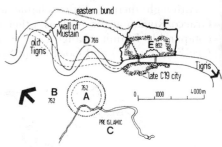

Figure 11.19 – Baghdad: the location of the Circular City related to that of the later right-bank nucleus of the modern Iraqi capital city. Key:

A. the Round City; B, al-Harbiya – the northern army suburb; C, al-Karkh – an area of existing agricultural villages developed for the Round City's construction work-force; D, al-Rusafa – the east bank palace complex built for al-Mansur's son in 769; E, Dar al-Khilafa – the east bank nucleus of the present-day city commenced in about 892; F, the wall around Baghdad demolished in 1860.

78. Both al-Basrah and al-Kufah were heavily involved in the 1991 Gulf War and the extent of damage to historical monuments is not known at time of writing.
79. For al-Fustat see S Lane Poole, *The Story of Cairo*, 1906.
80. Akbar, *Crisis in the Built Environment*, who quotes J Lassner, 'The Caliph's Personal Domain: The City Plan of Baghdad Re-Examined', in A H Hourani and S M Stern (eds), *The Islamic City*, 1970. See also A A Duri, 'Governmental Institutions', in R B Serjeant (ed.) *The Islamic City*, 1980, who writes: 'Baghdad was planned to be the capital and the seat of a new Islamic regime. Hence it had Caliphal institutions in addition to city institutions'; *Cook's Travellers' Guide to Palestine, Syria and Iraq*, 1934, informs us 'the Bazaars of Baghdad are extensive but are relatively modern'.
81. Hakim, *Arabic Islamic Cities*. Other authorities include first, the noted Andalusian geographer Abn Obeid al-Bakri (487/1094) who required only a *Mesjid al-Jami* and a *Suq*; second, Malik, the founder of the

382

The following are major elements comprising the main parts of a 'mature and relatively large Arabic-Islamic city', related to the general plan of Tunis (see Figure 11.21):

1 *Medina* (also *Madina*) – this term can be applied in two ways; first, in reference to an entire city, and second, to the older (historic) nucleus of a city, if it is physically distinguished from contiguous later additions, or suburbs, which are ordinarily referred to as *Rabad*.

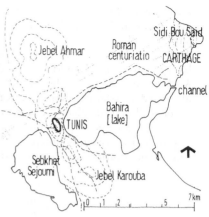

Figure 11.20 – Tunis: the Islamic city related to the ancient site of Carthage (and locating also the attractive, if over-restored historic fortified cliff-top village of Sidi Bou Said.)

When in 698/79 the Islamic leader Hassan b. al-Numan captured Carthage (see Chapter 3) the first decision taken was to site its replacement on more readily fortified high ground further inland. The second determined the location of the Zaytuna Mosque at the intersection of two established routes, possibly, writes Hakim, 'the Cardo and Decumanus of a pre-existing Roman network' (*Arabic-Islamic Cities*).

The coastal location that enabled Tunis to prosper economically as a centre of European/North African trade, also made it vulnerable to attack. In the first half of the twelfth century, as one of numerous such necessities, it was refortified by Ahmad of the ruling Banu Khurasan, who also constructed the first citadel (al-kasr). The northern and southern suburbs (Rabad) were established from that time.

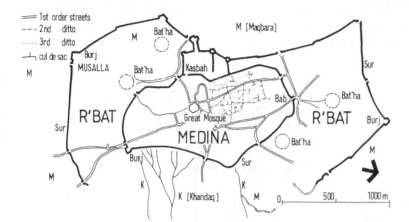

Figure 11.21 – Tunis: diagrammatic plan exemplifying the major elements of a mature, large Islamic city (after Hakim, *Arabic Islamic Cities*). See Figure 11.22 for the nineteenth-century French ville nouvelle.

Figure 11.22 – Tunis: an outline map showing the relationship of the historic Islamic medina and the late nineteenth-century European district laid out along main cross-axes of the gridiron drawing-board lines after the French annexation in 1881. The central axis, the Avenue Habib Bourguiba, leading from the medina to what remains of the Lake of Tunis, was the pre-existing determinant of the French plan. Within a ring boulevard, the medina has retained much of its traditional form and Tunis, as visited today, most clearly illustrates the diametrically opposed Islamic and Western European urban morphologies.

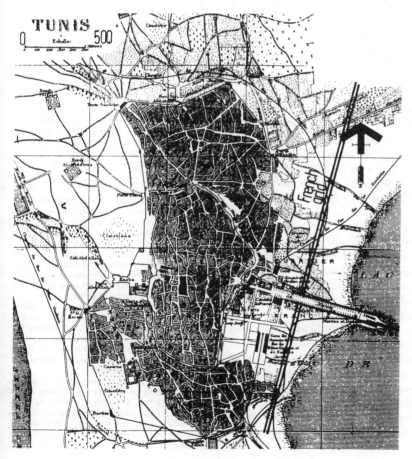

Maliki School of Law, who recognized a *Mesjid al-Jami* only in those settlements which had a *Suq*; third, Al-Shaf'i, the founder of the Shaf'i School, who wrote: 'when there are, all about an important place, secondary centres with inherited connections with this place, and only this place, and when this place has a Suq where the people from these secondary centres go to acquire their provisions, then I cannot allow that any inhabitants of these centres be dispensed from attending Friday prayer in the Mesjid al-Jami of the principle place'; and fourth, Abu Hanifa, founder of the Hanafi School, who required the residence of a governor and a

2 *Rabad* (also *R'bat*) – adjoining districts or quarters: a *Rabad* had its own name and it was characteristically enclosed within its own defensive wall.

3 *Mahalla* – the quarters of families of a common ethnic or religious background, which could be locked up for security purposes.

4 *Kasbah* (also *Kasaba*) – the citadel, a primarily North African term, also used in Islamic Spain: the *Kasbah* was usually attached to the outer wall of a fortified city, with independent external access.

5 *Musalla* – an open space large enough for the entire adult male population to assemble for prayer on special occasions: this spatial requirement ordinarily determined location outside the city wall.[83]

Figure 11.23 – Model One, entitled 'A Typical Traditional Islamic Town', from *Settlements in Dry Countries: A Design Approach*, by A G Sheppard Fidler and Associates / Derek Lovejoy and partners / Mauder Raikes, and Marshall, 1977. This model indicates the courtyard house type yet omits the access culs-de-sac, misplaces the main mosque and includes as typical a Public Garden although in reality they were conspicuous by their absence.

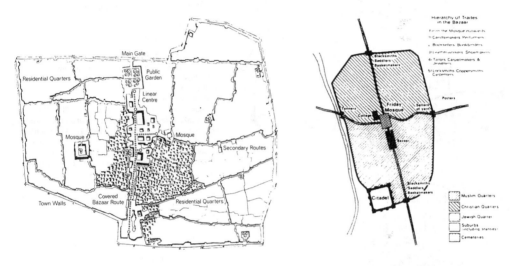

ISLAMIC URBAN FORM: THE RICH DIVERSITY

Other writers on the subject unfortunately give an impression that there can be an 'ideal' or geographer's 'model' historic Islamic city, whereby standard component parts are characteristically assembled. Such models (two of which are illustrated)[84] may well have validity for other purposes, but their present relevance is limited to demonstrating usual parts in theoretical combination valid only for cities of the Arabian hinterland and comparable hot and dry inland locations.

Such models ignore the great riverside cities of Cairo and Baghdad and the ports of Jeddah, Tripoli and Algiers. Topographical aberrants such as Alexandria and Muscat/Mutrah are denied and similarly Arabian Gulfside settlements exemplified by Dubai and Abu Dhabi. Islamic modification of numerous Roman imperial gridirons, notably Damascus and Cordoba, is excluded; moreover, the models have minimal, if any, relevance to Makkah and Medina.

Although it might be imagined that consistent application of urban guidelines would have had an effect of standardizing the appearance of Islamic cities along, possibly, monotonous repetitive lines; the reality of historic Islamic urban form is that of rich diversity.[85] The equal, if not surpassing, that of Christian medieval Western Europe in defying other than general description of characteristic parts, exemplified in combination as broadly as possible across a range; at the extremes of which there are two of the author's favourite historic Islamic cities: Ghardaia and Dubai.

Figure 11.24 – Model Two, from J. M Wagstaffe, 'The Origin and Evolution of Towns 4000 BC–AD 1900', in G H Blake and R I Lawless (eds) *The Changing Middle Eastern City*, 1980. Greatly over-simplified and misleading in the equal Muslim and Christian areas.

Kadi, in addition to it being 'a large settlement which has a system of main through streets, Suqs, and nearly related agricultural units'.

82. Hakim, *Arabic-Islamic Cities*.

83. The preferred location of the *Kasbah* against the external city wall was also a characteristic of ancient Sumerian cities of Mesopotamia, as described in Chapter 1 (see A L Oppenheim *Ancient Mesopotamia*, 1977); for Islamic citadels see N Elisseff, 'Physical Layout,' in R B Serjeant (ed.) *The Islamic City*, 1980.

84. The two models as illustrated are from (1) *Settlements in Dry Countries: A Design Approach*, by A G Sheppard Fidler and Associates, Derek Lovejoy and Partners/Mauder Raikes, and Marshall, 1977; and (2) J M Wagstaffe, 'The Origin and Evolution of Towns 4000 BC – AD 1900', in G H Blake and R I Lawless (eds) *The Changing Middle Eastern City*, 1980.

85. There is as yet no comprehensively illustrated single volume English language account of the Islamic City, along the lines, for example, of a volume in the series by E A Gutkind dealing with European countries (referred to in earlier chapters). Within its necessarily limited compass, this chapter goes some way towards filling this gap.

Ghardaia and Dubai

The isolated oasis township of Ghardaia, a largely surviving historic city, differs greatly from old waterside Dubai, a counterpart of whose historic nucleus sadly, if inevitably, less and less remains. How could this have been otherwise; given that the nature of their locally dominant original determinants were so very different?

The town of Ghardaia is situated in the isolated northern Saharan oasis of the name, some 370 miles south of Algiers, as the largest of five higher ground settlements in the midst of precious, finite-extent irri-

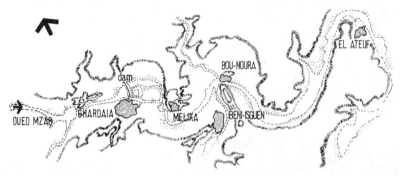

gated palm groves, where there was need for defence and with stone as a readily available construction material.[86] Old Dubai, on the other hand, was a comparatively accessible pearl fishing harbour and entrepot on the Arabian Gulf, which grew up alongside a sheltered creek on featureless, more or less flat sand without value for other than immediate waterside commerce, and with not much more than lumps of coral and palm trunks for its permanent buildings. With the exception of the ruler's fort – in reality more of a symbol than a secure fastness – Dubai had little reason to be fortified.[87]

As further distinction, Ghardaia belongs to the effectively closed, breakaway Islamic society of the Mozabites, which admitted of minimal 'foreign' access, whereas Dubai was a comparatively open trading community which welcomed, among others, the immigrant Persian merchants from Bastak, across the Gulf, whose 'imported' windtower house designs gave their Bastakia district its unique historic character.[88] Further description of these two cities is given in the captions to Figures 11.25 to 11.28.

Figure 11.25 – Ghardaia, Algeria: the locations of the five historic towns of the Ghardaia oasis which occupies the dry bed of the Oued Mzab, in the desert about 610 km by road south of Algiers. The town of Ghardaia is the youngest, having been founded in 1053; the oldest is El Ateuf of 1011, followed by Bou Noura, Melika and Beni Isguen. Until recently access for foreigners into the cities was strictly controlled and there are still constraints on visitors entering Beni Isguen – the 'holy city'. The region is known as the M'Zab, the inhabitants Mozabites and their religion Mozabite. They have been described by George Gerster as 'Islamic Protestants, Puritans of the Desert' (Sahara. 1960). They originated when the persecuted breakaway Ibadite sect were forced to flee first Kairouan, then Tiaret and Sedrata before finding sanctuary in the Ghardaia oasis.

Rainfall occurs on average once in seven years and the oasis originally depended on irrigation from deep wells hewn out of the limestone down as deep as 60 metres; latterly these have been largely replaced by deep-bore pumped and artesian wells. The lush oasis gardens, where most Mozabite families have a summer-house, are densely cultivated at the lower levels beneath the shade of the date palms. The five towns are situated against the steep sides of the oasis ravine above flood level when it rains. Their walls, which are difficult to distinguish from the natural rock, contain few openings as defence against human and climatic attack.

The plan and section of a typical Mozabite house is shown in Figure 11.14. Massively constructed with small courtyard openings, these houses can be likened to underground dwellings excavated in order to minimize heat gain from the Saharan sun. The winding streets of the Mozabite city, as they climb uphill to the mosque at the highest point, are often roofed over giving a welcome impression of walking through shaded underground tunnels. Architecturally the Mozabite city owes most to southern traditions, in particular those of the Sudan.

Figure 11.26 – Ghardaia, Algeria: the skyline massing of this personally most fascinating of all Islamic cities. Approached after dark by the autobus from Algiers, and then with a walk along the straight streets of an evidently European quarter, there was no preparation for this incomparable view at dawn across from the tourist hotel room balcony. Neither, for some unaccountable reason, had I previously known of these cities. (For a superbly illustrated account see Ravereau, Le M'Zab: une leçon d'architecture, 1981).

86. For Ghardaia see A Ravereau, Le M'Zab: une leçon d'architecture, Sindbad, Paris, 1981 (French, with excellent illustrations).
87. For historic Dubai see P H T Unwin, 'The Contemporary City in the United Arab Emirates', in I Serageldin and S El-Sadek (eds) The Arab City, 1982. I am also indebted to Mark Harris for 'The Bastakia Experience', Harvard Graduate School of Design, thesis paper 1980, and 'The Bastakia: Dubai', Oxford Polytechnic, 1978 (both unpublished).
88. See E Beazley and M Harverson, Living with the Desert: Working Buildings of the Iranian Plateau, 1982; Harris, 'The Bastakia Experience' and 'The Bastakia: Dubai'.

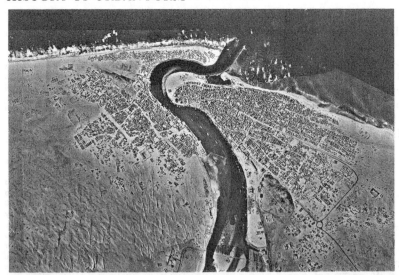

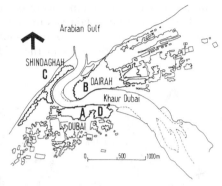

Figure 11.27 – Dubai, UAE: a vertical aerial photography of 1963, with a diagrammatic key map locating the original twin towns of Dubai (A) and Diera (B) on either side of the Khor Dubai – a splendid serpentine salt-water creek; the Shindaghah quarter (C) on the spit of land between the Creek and the Arabian Gulf, and the Bastakia quarter (D).

Figure 11.28 – Dubai: an oblique aerial photograph of 1952 looking across Dubai towards Deira in the Creek bend, and Shindaghah. The Bastakia district with its distinctive wind-tower houses is to the right in Dubai. The early history of Dubai is that of an obscure fishing village (probably the original Shindaghah) which was a dependency of the sheaikhdom of Abu Dhabia from where, in 1833, two prominent members of the Beni Yas 'Ubaid bin Sa'id and Maktum bin Buti, emigrated with 800 followers. The latter ruled from 1836 until his death in 1852, establishing the al Maktum dynasty in Dubai. Pearling, increased trade, and attractions as a centre for immigration laid the foundations of economic prosperity; by 1900 the population in the three quarters had exceeded 10,000 and the Suq of Deira with about 350 shops was the biggest on the Trucial Coast. In 1939, with about 20,000 inhabitants, Dubai was the most populous of the Trucial Shayhkdoms.

Frauke Heard-Bey notes that 'a large number of inhabitants of Dubai lived in palm-frond (*barasti*) houses until well into the 1960s [foreground of photograph] ... in the quarters where coral and mud-brick houses predominated, the alleys were nowhere near wide enough to let a car pass through ... traffic between Dubai and Deira was largely by rowing-boats ('*abrah*). There were, and still are, a number of fixed landing points on either side of the Creek. On Fridays the passage was free for people from Deira who went to attend the midday prayer in the big mosque on the Dubai side' (*From Trucial States to United Arab Emirates*, 1982).

Dubai had strong trading links across the Gulf with Persian (Iranian) ports, notably Khamir and Lingeh, and it welcomed merchants from there when they opted to emigrate in face of discriminatory taxation. They brought with them their traditional wind-tower house design, and they named their new district in Dubai after Bastak, one of their home towns. The Bastakia, alongside the Creek, which grew to contain about sixty wind-tower houses is one of the most attractive examples of traditional Islamic domestic architecture. (See Figure 11.14 for the plan and section of a wind-tower house, and A Coles and P Jackson, *A Windtower House in Dubai*, AARP, 1975 for a detailed study.)

(Barasti houses consist of mangrove or other poles, palm fronds and cord, and have the twin advantages of minimal heat storage and open texture allowing breeze penetration.)

ISLAMIC URBAN FORM: BUILDING TYPES

The main building types in historic Islamic cities are described in the sequence of first, the mosque (and related religious and teaching buildings), second residential quarters, third the street system (with reference to the urban guidelines), fourth the *suq* (and related trading buildings), fifth the *Kasbah*; sixth, the defensive wall (with its gateways), and finally, the water supply systems.

The mosque

'The mosque,' writes Welch, 'is a unique Islamic institution that is essentially different from the Jewish and Christian counterparts, the synagogue and the church.'[89] It is the pre-eminent building in the Islamic city to an arguably greater extent than that of the cathedral in Christian medieval counterparts, although that is not a comparative consideration with which we need concern ourselves. In history, the mosque – or more likely, the several mosques of a sizeable urban community – had a significance extending beyond that of a religious focus into most if not all aspects of the life of the community. 'The mosque is not exclusively a place of prayer,' explains Michon, 'but also a meeting place or forum where the city's news is exchanged. It is a centre of religious education where children and adults of all conditions sit in a circle, frequently after nightfall to chant the Qur'an or to listen to the teaching of a faqih; often it is a refuge where beggars, vagabonds and the oppressed can find shelter and asylum and receive the alms or food generously dispensed by the community at places of worship.'[90] Architecturally the mosque also has the distinction of being the most important uniquely Islamic building type.

On arrival in Medina, the Muslim community built what was to become its first mosque as the house where Muhammad and his wives lived. As shown by Figure 11.29, this took the form of a spacious courtyard entered on three sides, with a covered prayer area providing shelter from the sun; a second smaller covered area for the Prophet's friends; and individual private rooms for each of his wives.[91] Intended or otherwise, the house shortly became used for communal prayer and religious teaching; subsequently to provide the model for the first mosques that were built in early Islamic cities such as Basra (635), Jerusalem (637), Kufa (638) and al-Fustat (641). It was always exceptional to adapt existing buildings for the purpose. The Prophet's Mosque at Medina was rebuilt in 706–7 in a permanent architectural form subsequently to be followed throughout Islam. The innovation incorporated into the new design was the *mihrab*, a niche in the wall giving the *qibla* (direction of prayer towards Makkah).[92]

The general arrangement of a mosque (the word translates as 'place of prostration') has a covered prayer hall along one side of a colonnaded courtyard.[93] A mosque sometimes has an open courtyard with one or more fountains for the obligatory ablutions before prayer. The architectural ensemble is completed by the minaret, one or more of which towers provide the elevated platform from which the muezzin summons the faithful to prayer. The Jami in major cities can have as many as four minarets – one at each corner; exceptionally the mosque of Sultan Ahmed in Istanbul has six.[94] As with Christian cathedrals, in particular those in England, the architectural development of major mosques was that of progressive enlargement, as shown by Figure 11.31.[94]

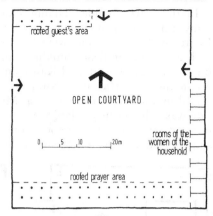

Figure 11.29 – Medina: the Prophet's House, the usual reconstruction arrangement (after Hakim, *Arabic-Islamic Cities*).

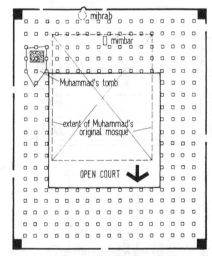

Figure 11.30 – Medina: the Prophet's Mosque.

89. Welch, 'Islam'.
90. J-L Michon, 'Religious Institutions', in R B Serjeant (ed.) *The Islamic City*, 1980.
91. See Armstrong, *Muhammad*, for a description of Muhammad's arrival at Medina and the construction of his house; see also Guillaume, *Islam*, who writes: 'from the outset Muhammad displayed the tact and diplomacy which marked his dealings with others. Beset by appeals from all sides to take up his abode with individuals who perforce belonged to one party and thus were disliked by the other, he instantly evolved a plan which would hurt the feelings of none and at the same time absolve him from the responsibility of an invidious choice. He left the choice to his camel. It came to rest in the quarter of the Najjar clan of the Khazraj'.
92. At first the Qibla direction of prayer was towards Jerusalem, until in late January 624, about eighteen months after the hijra, Muhammad was leading the Friday prayers ... when, suddenly, inspired by a special revelation, Muhammad made the whole congregation turn round and pray facing Mecca instead of Jerusalem' (Armstrong, *Muhammad*).
93, 94. See J D Hoag, *Islamic Architecture*, 1977.

HISTORY OF URBAN FORM

In the newly founded Islamic cities of conquered territories, the army commander-in-chief required that his residence, the Dar al-Imara, with the mosque, were located in the centre, sometimes also with a Diwan (audience building) and prison. This was the case at Basra, Kufa, al-Fustat, and Kairouan. At Tunis, the Jami al-Zaytuna was founded by Hassan b. Numan al-Ghassani in 703, to a plan 'inspired by the Prophet's mosque at Medina'.[95] The *Riwak* at the *Qibla* side was subsequently enlarged to create the roofed prayer area of the mosque; even so there were occasions when an excess number of worshippers had to remain in the unshaded open *Sahn* (courtyard). Hakim,

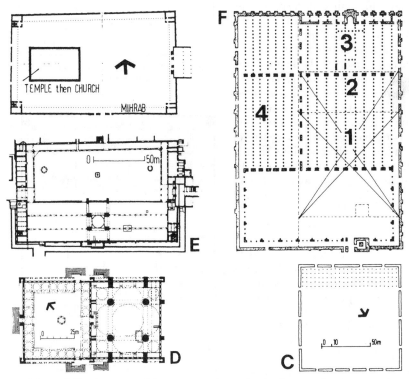

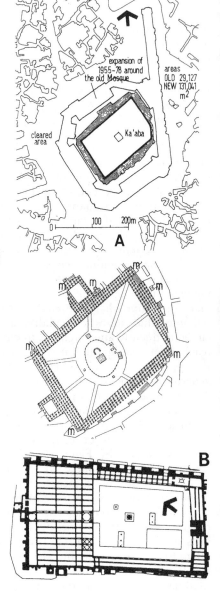

writing of Tunis, notes 'since approximately the late 15th century, the Zaytuna mosque Sahn ... provides an impressive and overwhelming sense of space and tranquility in contrast to the adjoining narrow and busy streets'. In contrast to European Christian counterparts, Muslim cities of the Middle Ages had no place of public assembly corresponding to the church square or the space in front of the town hall.[96] It was also usual for the external mosque walls to be used as supports for the suq areas and related buildings.

Related to the Jami in the mature Islamic city, there were a number of other building types, including the *Hamman* and the *Madrasa*. The *Hamman* are public bath-houses used separately by men and women during different hours of the day, or as allocated on specific days of the week.[97] Although translating as 'school', the *Madrasa* is essentially a college for advanced study of Islamic law and sciences. Hakim writes that 'the first Madrasa built as a separate facility ... was in Naisapur during the first half of the fifth century of Islam'; he gives totals of thirty in twelfth-century Baghdad, and as many as seventy-three in fifteenth-century Cairo.[98]

Figure 11.31 – Islamic urban form: the mosque, six examples: A, Mosque of the Haram (The Holy Mosque), Makkah; before and after reconstruction. (M, minarets.) See also Figures 11.11 and 11.50. B, Qairawan (Kairouan), Tunisia; C, Kufa; D, The Sultan Ahmet Mosque, Istanbul; E, Damascus; and F, Cordoba.

95. Hakim, *Arabic Islamic Cities*.
96. See the aerial photograph of Erbil (figure 1.10) where the mosque is the multi-domed flat-roofed building to the right of centre, with only a comparatively small courtyard; N Elisseff, 'Physical Layout' in R B Serjeant (ed.) *The Islamic City*, 1980, notes 'political gatherings in Damascus were held north of the citadel on the esplanade where the military parades took place ... there could well be gatherings outside the city in such

Residential quarters

The houses and their access system comprise the greater part of the city, the form of which was determined by the urban guideline requirements of domestic-domain privacy, and minimum width of the cul-de-sac lanes and through streets. The term *Dar* for the house derives from

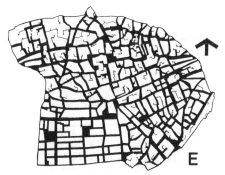

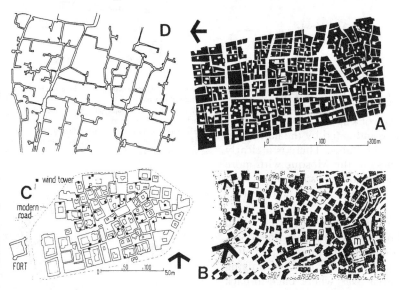

Figure 11.32 – Islamic urban form: characteristic street patterns exemplified by: A, Riyadh; B, Jiblah, street frontage (after Hirshi, *L'Architecture au Yemen du Nord*); C, The Bastakia Quarter, Dubai; D, culs-de-sac at Kerman, Iran; E, Seville, Spain, the Islamic pattern contrasted the Renaissance gridiron in the south-west.

Dara (to surround), and is a 'space surrounded by walls, buildings, or nomadic tents placed approximately in a circle'.[99] The two most commonly used terms for an individual dwelling are *Dar* and *Bayt*, which has the proper meaning of 'covered shelter where one may spend the night'. 'Symbolically,' writes Hakim, 'the first Islamic house is that built by the Prophet Mohammed on his arrival in Medina, a dwelling place for himself and his family, and as a meeting place for the Believers. The courtyard surrounded by walls is its essential feature.'[100] As described above, the function of this house was transposed into that of the first mosque.

The street system

Individual houses were grouped together around cul-de-sac lanes (variously *Derb*, *Ghair*, *Nafidh*), entered off the city's main public right-of-way thoroughfares (*Tariq Nafidh*) and which served to provide access only to that cluster of houses. The cul-de-sac is owned in common by the houses it serves and comprises a basic unit in the self-policing nature of the Islamic city. As above, the urban guidelines required a minimum cul-de-sac width that could pass one fully laden camel, taken as 4 cubits (1.84–2.00 m or 6 feet to 6 feet 6 inches). The ordinary through streets were of minimum width 7 cubits (3.23–3.50 m or 10 feet 6 inches to 11 feet approximately) which allowed for two fully laden camels to pass. Minimum width, it would appear, was often taken in effect to mean maximum, hence the comparatively narrow width of the access routes experienced by Western-world visitors, accustomed to their vehicle transport streets, both historic and modern. 'The road system in the Arab city,' Fathy has observed, 'was the result of the patterning of buildings, not a determinant as in most modern planning.'[101]

vast areas as the maydan, hippodrome or field for equestrian exercises which existed in each city; or the musalla where prayers would be said communally on the occasion of Islam's two great religious festivals'.

97. Ablution is a requirement before prayer and the *Hammam* is regarded by some Arab writers as an essential feature of the Islamic city. Public bath-houses, used separately by men and women, provided one of the communal meeting places in the city. Hakim informs us that in the Tunis Medina Central the ratio of all mosques to *Hammams* is 4.75:1 (that of Friday Khutba mosques is 0.69:1); at Damascus the ratio is 6:1 and in Baghdad 5:1 (*Arabic-Islamic Cities*). See also Elisseff, 'Physical Layout', for the need to conserve water; see also Hoag, *Islamic Architecture*.

98. Hakim, *Arabic-Islamic Cities*. For the Madrasa see Hoag, *Islamic Architecture*; H Nashabi, 'Educational Institutions' in R B Serjeant (ed.) *The Islamic City*, 1980.

99. Hakim, *Arabic-Islamic Cities*.

100. Hakim, *Arabic-Islamic Cities*. It is stressed that the enclosed courtyard form of the Prophet's house, albeit of a comparatively large extent, followed the usual practice at Medina. Although the rooms within were of only the one storey, without use of the roof as a terrace; it can be presumed that the older established Medinan houses were on one and two floors, with screened roof terraces. (It is of academically diverting interest to conjecture what might have happened had Muhammad made his journey in reverse, leaving Medina for Makkah and its (presumed) three- and four-storey street frontage houses?)

At first, it seems, the Prophet had no intention of erecting temples for daily prayer (Lewis, 'The Faith and the Faithful') preferring the alternative of domestic worship in private led by the head of the family. Mansfield writes that 'Islam had no clergy, every man is a priest and a patriarch in his own household' (*The Arabs*); to which Burton adds: 'the Moslem family, however humble, was to be the model in miniature of the State ... every father in al-Islam was able unaided to marry himself, to baptize his children, to instruct them in the law and canonically to bury himself' (R Burton, 'Terminal Essay', to his translation of *The Thousand and One Nights*, n.d.)

The comparatively large size of the Prophet's courtyard would, however, indicate awareness of need from the outset at Medina for a large enough space for gatherings of the Faithful.

101. Fathy, *Constancy, Transposition and Change in the Arab City*; see also François-Auguste de Montequin, who observes: 'the concept of the street in its combined functions as (1) artery of traffic, (2) zone of major

389

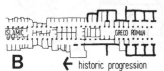

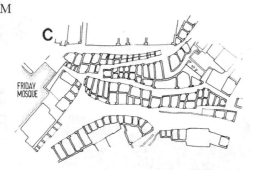

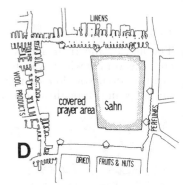

The Suq

It can be misleading to refer to the *Suq* in the singular, because the mature Islamic city contains a considerable number of *Suqs* (the Arabic plural is *Aswaq*), each providing for the sale of a specific commodity, or several more specialist ones in common, as indicated by the name, e.g. *Suq el Chechias*, the head dress market in old Tunis. There is a clearly established hierarchy in respect of nearness to the Mosque with more respectable trades such as booksellers and perfumiers so favoured, and then moving outwards with the noisy or noxious metal and leather working ones furthest distant.

Individual shops are of small size ranging upwards from 1.5 m square and they are arranged in several basic ways. First, as linear *Suqs* on either side of a through route from a city gate to the mosque, often roofed over, for shade, or beneath a Sabat giving the effect of walking through a narrow tunnel. Second, as area *Suqs* where back-to-back rows face each other and where gates can be provided for overnight security. Third, where the shops are against the perimeter wall of a large building – the mosque – or special buildings known as *Wekalas* or *Khans* (*caravanseria* is another term) which were constructed to provide secure storage for the goods of merchants, or *Funduks* with hostel accommodation.

The *Suq* is usually provided with a roof for daytime shade, the complexity of which ranges from simple canvas awnings to grand architectural designs exemplified by the Spice Market in Istanbul and parts of the Khan el Kalili in Cairo. In Isfahan the mostly dome-roofed *Suq* extends its winding route for a distance of 2.5 km from the Maidan through to the Jami (Figure 11.34).[102]

The Kasbah

With few major exceptions, the Kasbah, as the citadel of the ruling elite in an Islamic city, is positioned against or astride the city wall, a characteristic seemingly inherited from ancient Mesopotamian practice as described in Chapter 1. Even where the city originated on a mound offering early defensive potential, as at San'a (Figure 11.39), subsequent growth on the slopes and lower ground was usually on two or three sides, leaving the citadel against the perimeter. This in direct contrast to Western European form where the citadel came to be in or near the centre, as exemplified by ancient Athens. Positioned against the perimeter, the Islamic Kasbah contributed its defensive strength to the city wall and retained direct access out into the surrounding countryside. Notable examples include Cairo, Damascus, Algiers and Tunis. Aleppo is a major exception with its Kasbah more or less central, and it is cautioned that many important Islamic cities did not contain a Kasbah.

Figure 11.33 – Islamic urban form: the *Suq*, three basic configurations as

A, either side of a through route; B, evolution of Islamic form from Greco-Roman original, as at Damascus; C, an area *Suq* with back-to-back shops – exemplified by Jibla, Yemen; and D, around a core, exemplified by the Zaytuna Mosque at Tunis (after Hakim, *Arabic-Islamic Cities*).

Figure 11.34 – Isfahan, Iran: distinctively marked from above by the line of domes snaking their way across the city, the main bazaar (*Suq*) route linking the Maidan to the Masjid-i-Jami is more than 2.5 km in length (see also Figure 11.41).

architectural display, and (3) area of social intercourse is basically non-existent in the traditional Islamic urban centre. Because Muslim society contrives to avoid the exteriorization of the dwelling, which is the *raison d'être* of the street, the Islamic city does not need and therefore tends not to have such type of thoroughfares' ('The Essence of Urban Existence in the World of Islam', in the published papers on *Islamic Architecture and Urbanism*, a Symposium of 1983, King Faisal University, Damman, Saudi Arabia).

102. For the Isfahan *Suq*, see L Lockhart, 'Shah Abbas's Isfahan', in A Toynbee (ed.) *Cities of Destiny*, 1967; Elisseff, 'Physical Layout'; A V Williams Jackson, who

The defensive wall

The two-dimensional length and height characteristic of the defensive system of a mature Islamic city differed little in principle from that of its pre-artillery, medieval European contemporaries. Need for horizontal separation between attackers and defenders (and the vulnerable civilian population) was a nineteenth-century European introduction into the Middle East, essentially on the part of the French, who refortified several of their major cities, notably Algiers and Alexandria, along lines described in Chapters 5 and 6. Left to themselves, Muslims only latterly required protection against rifles, for which a comparatively simple wall (*Sur*), strengthened by towers (*Burj*, also *Burdj*) sufficed, with defensive complexities only at the gates (*Bab*). As late as 1919, when Kuwait City had to be hastily fortified, all that was required was a simple mud-brick wall and gates (Figure 11.35).[103]

Architecturally the most impressive surviving examples are the three eleventh-century Fatimid gateways of Cairo: the Bab al-Nasr, Bab al-Futuh, and Bab Zuwayla, built by Badr al-Djamali between 480–5/1087–92. These are of the simple straight-through type, compared with the bent entrance (*Bashura*) which came into later general use in order to 'prevent troops taking it by assault during a siege, and to render impossible the entry en masse of cavalry'.[104] Islamic cities were usually provided with more gateways than their European contemporaries: the Medina at Tunis numbering seven, with twelve in the outer Rabad walls.

Water supply systems

'In Islam,' writes Elisseeff, 'one of the most meritorious of pious deeds is, according to the Hadith, the gift of water; moreover, to supply a city with drinking water is one of the concerns – or rather essentials – of laying out a city.'[105] The Prophet prescribed that people must share water, and that its owner should give to others any surplus he has for drinking or irrigation. Typically the water supply system is based on a main conduit, perhaps of Roman origins, as exemplified by the qanayah of Haytan in Aleppo, or the Qanawat in Damascus. There are also the numerous instances of underground canal systems whereby loss by evaporation is minimized. That at Medina was described by Burckhardt in 1829 as taken there 'from the village of Koba, about three-quarters of an hour distant, at the expense of Sultan Solyman, the son of Saleym 1. The water is abundant and in several parts of the town steps are made down to the canal, where the inhabitants supply themselves with water, but are not, like the people at Mecca, obliged to pay for it . . the water in the canal runs at the depth of between twenty and twenty-five feet below the surface'.[106]

In Damascus in the thirteenth century, the tali were fed with water from 129 canal ducts.[107] Tunis relied in three sources of water; individual domestic cisterns to collect rainwater, aqueducts, and wells.[108]

Islamic Urban Form: major cities illustrated

RIYADH, SAUDI ARABIA

The city is located on the Njad plateau at the confluence of three major wadis which ensured the water supply for the extensive date palm oasis – Ar Riyadh in Arabic means 'gardens' and which attracted settlement from earliest times. In 1902 King Abdul Aziz captured the walled town

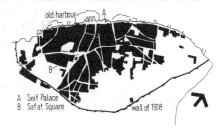

A Seif Palace
B Safat Square
old harbour
wall of 1918

Figure 11.35 – Kuwait City: originating as a fishing and pearling village in the early eighteenth century, Kuwait is one of urban history's exceptional unnatural settlements in that 'before the (modern) distillation plant came into being, fresh water was brought by ship from the Shatt el Arab and was put into goatskins and carried on the backs of camels and donkeys for sale in the town (Lt Col L W Amps, 'Kuwait Town Development', *Journal of the Royal Geographical Society*, vol XL, 1953). Nevertheless, Kuwait benefited from its strategic trading location at the head of the Arabian Gulf, and when visited in the 1760s by Neibuhr he recorded a population of about 10,000 with 800 vessels. As a pre-oil economic might-have-been, in the mid-nineteenth century Kuwait was proposed as the terminus of the overland railway section of a British inspired 'Euphrates Valley Route to India'. Later, both the Germans and the Turks had the same idea until discouraged by British strategic interests.

In 1954 the historic city of area about 8 square kilometres was still enclosed within the 14 ft high mud wall fortification of 1918, hurriedly built in eight weeks against incipient attack. The wall was about 4 miles in length, with four major gateways – Jahra, Naef, Shaab, Sabah – plus a small North gate. 'All these gates led out into the endless desert' (S G Shiber, 'Kuwait: A Case Study', L C Brown (ed.) in *From Madina to Metropolis*, 1973). The line of the wall is that of the 'green belt' surrounding the city centre, beyond which there is the first of the motor-age expediency supergrids, planned by British consultants Minoprio and Spencely in 1952.

observed: 'the bazaars in Isfahan lie behind the rows of buildings on the northern and eastern sides of the Meidan. It is possible to walk under their covered shade, or rather to push one's way through the crowded mass of camels, donkeys, packs, porters, buyers, sellers, and money-changers' (*Isfahan, Persia: Past and Present*, 1906).

103. For Kuwait City see S Gardner, *Kuwait: The Making of a City*, 1983; D Al-Yawer, 'Urban Spaces in the Islamic City', in I Serageldin and S El-Sadek (eds) *The Arab City*, 1982; L Lockhart, 'Outline of the History of Kuwait', *Journal of the Royal Geographical Society*, July/October 1947; Lt Col L W Amps, 'Kuwait Town Development', *Journal of the Royal Geographical Society*, vol XL, July/October 1953; and S G Shiba, 'Kuwait: a Case Study', in L C Brown (ed.) *From Madina to Metropolis*, 1973.

104. Al-Makrizi (d. 1441), quoted Hakim, *Arab-Islamic Cities* see also Elisseff, 'Physical Layout'.

105. Elisseff, 'Physical Layout'. Western-world technology has brought many benefits to the Islamic world, the most important of which is provision of reliable supplies of potable water.

106. Burckhardt, *Travels in Arabia*.

107. See M Ecochard, 'Damascus', in I Serageldin and S El-Sadek (eds) *The Arab City*, 1982.

108. Hakim, *Arabic-Islamic Cities*.

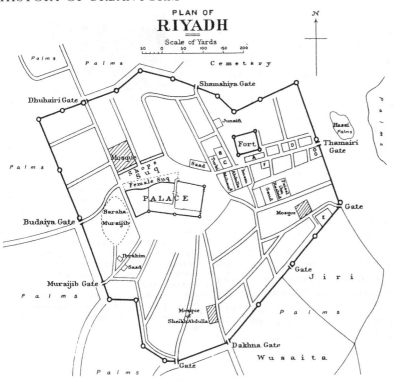

PLAN OF
RIYADH
Scale of Yards
50   0   50   100   150   200

Figure 11.36 – Riyadh; 'Philby's Plan' of Ar-Riyadh in 1918 – due allowance must be made for the compass-and-pacing basis of this plan which does, however, give a clear indication of the compact nature of the emergent Saudi capital, clustered around the Palace, the Great Mosque and the central *suq* complex; and with its encircling palm grove oasis gardens (from H StJ B Philby, *The Heart of Arabia*, 1922).

'Dickson's Plan' of Ay-Riyadh in 1937 (a properly surveyed record) confirms the general shaping within the defensive wall, which, with the exception of the King's New Palace some distance to the north, still contained the city at that date. Dickson noted that the Race Course 'would make an excellent aerodrome', (See A-M-I Daghistani, *Ar-Riyadh: Urban Development and Planning*, 1985 for this map; and Figure 11.32 for a detail plan of part of the city's historic street pattern.)

The city is completely encircled by a thick wall of coarse sun-baked mud-bricks, about twenty-five feet in height and surmounted by a fringe of plain shark's-tooth design; at frequent intervals its continuity is interrupted by imposing bastions and less pretentious guard-turrets, circular for the most part and slightly tapering toward the top but some few square or rectangular, varying from thirty to forty feet in height and generally projecting slightly outwards.

The internal arrangement of the streets is without symmetry except for the natural convergence already mentioned of all main traffic lines on the central enclave; the chief street is that which leads in a straight line from the Thumairi gate to the palace and thence through the market-place to the Budai'a outlet, with a branch going off from it at right angles from the western end of the Suq to the Dhuhairi gate.

The market-place, which occupies the whole of the open space to the north of the palace, slopes westward down a sharp incline and is divided into two sections by a partition wall, the section between this wall and the wall of the palace being reserved exclusively for the use of women – vegetable-sellers, purveyors of domestic necessaries and the like – while the other and larger section comprises about 120 unpretentious shops ranged partly along either side of a broad thoroughfare and partly back to back on a narrow island of no great length in the midst thereof; the shops along the north side of the Suq are backed in part by the outer walls of ordinary dwelling-houses and partly by the south wall of the Great Mosque itself.

The Great Mosque or Jami'a of Riyadh is a spacious rectangular enclosure about sixty yards by fifty in area, whose main entrance faces the Suq through a gap in the row of shops lining its southern wall, while the Qibla or prayer-direction, by which the whole building is oriented, is marked by a very slight south-westerly bulge in the longer western face, near which as also on the east side is a subsidiary entrance.

(H StJ Philby, The Heart of Arabia, 1922)

clustered around the palace-fort where he established the ruling Saudi dynasty.

Palgrave, in 1862, wrote on first seeing Riyadh 'before us stretched a wide open valley, and in its foreground, immediately below the pebbly slope on whose summit we stood, lay the capital, large and square, crowned by high towers and strong walls of defence, a mass of roofs and terraces, where overtopping all frowned the huge but irregular pile of Feysul's royal castle, and hard by it rose the scarce less conspicuous palace built and inhabited by his eldest son, Abdullah'.[109]

A description of the surrounds of Riyadh in 1918 is given by Philby 'all round the city, except towards the north-east, lie the dense palm-groves of the oasis, in many of which there are buildings which serve as country residences when their owners want a respite from the routine of city life. In them are the deep, and in many cases very capacious, draw-wells or Jalibs, which night and day seem from the unceasing long-drawn-out drone of their pulleys to be perpetually at work. One of these is very much like another, and that which stands in the midst of and irrigates the garden called Wusaita, a splendid grove on the south-east side of the city belonging to the Imam 'Abdulrahman, must serve as a type of the rest. The pit of this great well is about ten feet square at the top and tapers very slightly downwards to the water level, which lies between forty and fifty feet below the surface; hewn out of the solid limestone rock, its upper portion for about one-third of its depth has an additional strengthening of stone-slabs roughly cemented together. The mouth of the pit is surmounted by a ponderous triangular superstructure called 'Idda and constructed of palm-logs with Ithil-wood for the stays and subsidiary parts ...'[110]

109. W G Palgrave, *A Personal Narrative of a Year's Journey Through Central and Eastern Arabia*, 1862–63.
110. H StJ B Philby, *The Heart of Arabia*, 1922.

JEDDAH, SAUDI ARABIA

Jeddah is situated about midway along the Red Sea coast where a gap in the triple coral reefs coincides with a valley crossing of the littoral mountain range, en route to Makkah, some 75 kms inland. Although not the first choice as the port of entry for pilgrims it was early confirmed in that role, on which its economic prosperity was consolidated, with addition of the profitable Red Sea spice trade. Although well sited for trade, historic Jeddah (in common with Kuwait City) lacked a potable water supply, evidenced by Niebhur who in 1761 reported 'the city is entirely destitute of water. The inhabitants have none to drink but what is collected by Arabs in reservoirs among the hills and brought thence on camels ...'[111]

The historic wall and five gateways of the Old Town were not demolished until 1947, since when the city and port have rapidly expanded, latterly on the basis of an expediency supergrid plan drawn up by the British consultants Robert Matthew Johnson Marshall and Partners, one of whom, John Russell, has described the Al-Balad district in the eastern part of the old town as 'a dense agglomeration of two, three and four-storey town houses interspaced with mosques, merchant's palaces, carvansersai and warehouses, through which runs a system of narrow intersecting passageways, often no more than 3 metres wide and small irregularly shaped public open spaces. The buildings are constructed of coral limestone – either naturally occurring individual pieces or square cut blocks, laid to course in lime mortar and reinforced with stout timber poles (taglilat) and faced with lime plaster (tangil), and often decorated with incised floral and geometric patterns'.[112]

Philby visited Jeddah in 1918, observing 'within the city is a truly eastern jumble of wealth and poverty: great mansions of the captains of commerce and enterprise, with their solid coral walls and wide expanses of woodwork tracery, side by side with hovels broken and battered by age; mosques great and small, with pointed minarets tapering skyward amid masses of vast square buildings; and crowded bazaars, with their lines of dark shops, protected from the sun by central roofs, here of wood and canvas much the worse for wear, and there of corrugated iron. Everywhere a contrast of light and shadow, splendour and squalor, dust and dirt; and above it all flew the flags of many nations amid the countless emblems of an united Arabia'.[113]

SAN'A, YEMEN

'San'a is situated in the centre of the (Yemen) highland zone, on the edge of a basin in an elevated plateau ... between 2,350 and 2,200 m above sea level.'[114] It has a temperate climate yet day/night temperature differences of 30 degrees C are experienced. A reliable water supply was provided by wells and partly underground *ghayls* (artificial streams) down from the surrounding higher mountains. Watered, the volcanic soil is highly fertile. San'a originated at the foot of Jabal Nuqum, 2,892 m, where a rocky eminence attracted early settlement gradually to become known as the Ghumdan, located between the later citadel and the Great Mosque, as shown on the diagram accompanying Figure 11.39. The city that grew up around the Ghumdan on three sides was at first bounded on the west by the Sa'ilah, a wadi liable to flash-flooding during seasonal rains, before expanding across and up the slope. The Sa'ilah was straddled by the city wall, entering and

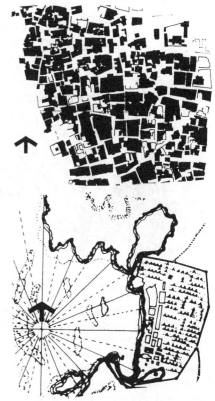

Figure 11.37 – Jeddah: (above) the historic city; (and top right) a part plan of the traditional street frontage arrangement of the multi-storey houses.

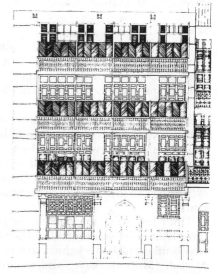

Figure 11.38 – Jeddah; the street elevations of typical multi-storey houses in Old Jeddah, showing the distinctive *rawashins* (*mashrabiya* and *shanashil* are equivalent Arabic terms).

393

leaving by way of defensive grilles. Beyond, to the west, there was the lower density 'garden suburb'. Both parts of the city were walled and within the western district there was a separately walled quarter.

'On the orders of the Prophet himself, in the year AH 6, the Great Mosque was constructed in the garden of the Ghumdan Palace.'[115] It was subsequently enlarged on three occasions and is approximately rectangular of sides 78 m × 66 m with arcaded spaces enclosing a courtyard. The central *suq* area adjoins to the north where its single-storey buildings are in marked contrast to the multi-storey *samsarah* (caravanserai/warehouses) and houses, which are an exceptionally important characteristic of the city rising to seven, eight or even nine storeys.[116] An example house is illustrated in Figure 11.14.

MUSCAT

The city of Muscat, the historic capital of Oman, is on the country's eastern, Indian Ocean coast, about 350 km south-east of the Strait of Hormuz, the entrance to the Arabian Gulf. The modern metropolis (the Capital Area) originated from the twin historic harbours of Muscat and Mutrah (al-Matrah), of which the former was renowned from earliest times as the safest harbour along the coast, favoured by seasonal monsoon winds eastwards to India and north-west to the Arabian Gulf. In 1508 they were taken by the Portuguese, who maintained a governor and a garrison in their forts until the mid-seventeenth century. A British presence dated from 1798 when, in face of French expansionism, a protection agreement was signed with the Sultan of Oman.

'Until July 1970, the gates in the town wall of Muscat were closed three hours after sunset (and) no-one was allowed to walk through the streets unless he had a lighted lantern in his hand ... the town gates of Matrah (Muttrah) were (also) closed at sunset, but as motor traffic went through a hole in the town wall it was not affected.'[117] When the first road linking Muscat and Muttran was officially opened in December 1929 there were four cars to travel on it. They were 'the chocolate

Figure 11.39 – San'a, Yemen, Von Wissmann's map of the city in the early 1930s, published in H Wissmann, *Karte des Reisegebeites in Yemen*, 1934; with an accompanying analytical diagram.

111. C Niebuhr, *Travels Through Arabia*, 1792.
112. J Palmer, 'Jeddah Renewal', *The Architectural Review*, April 1981.
113. H StJ B Philby, *The Heart of Arabia*.
114. R B Serjeant and R Lewcock (eds), *San'a: An Islamic City*, 1983.
115. P M Costa, 'San'a,' in R B Serjeant (ed.) *The Islamic City*, 1980.
116. See S Hirshi and M Hirshi, *L'Architecture au Yemen du Nord*, 1983.
117. Ministry of Information and Tourism, Oman, 1972.

brown Model A Ford Sedan which had come in November as a present to Sultan Timur ... the Delaunay Belleville belonging to Gerald Murphy, the Political Agent; a light lorry of unknown make that was the sole vehicle in the arsenal of the Muscat Levies; and the Morris of Dr Mackay, the Consulate Surgeon.[118]

Constrained by mountains immediately to the south, the Capital Area has grown to the west, along the narrow coastal strip where Seeb International Airport is 25 km from old Muscat. Oil was first discovered in western Oman in 1964 and export commenced in 1967. The 1981 population of the Capital Area is approximately 50,000.

ISFAHAN, IRAN

With the exceptions of Makkah and Medina, which have ordinarily been barred to non-Muslims for centuries, Isfahan is the Islamic city that I most regret not having visited yet. The capital of Persia (Iran since 1935) from 1598 to 1722, Isfahan embodies urban elements unique in an historic Islamic city and others that are of exceptional importance. Although Isfahan owes most to the civic improvements carried out during the reign of Shab Abbas I ('The Great' 1587–1629), Lockhart stresses that the city 'was no creation of his, for it has been a great city long before his reign; it had moreover been at times the capital of the country'.[119]

At his accession in 1587, the youthful Shah's capital was at Qazvin, about 420 km north-west of Isfahan, to where Shah Abbas moved his throne in 1598. Isfahan was the more centrally located in Persia and its well-watered fertile location was attractive for the purpose of building a grand new capital city. The Shah's 'masterplan idea' was to create for himself a new city alongside the existing ancient one which he left largely unaltered. In its way – and as a hitherto unremarked parallel – the combination at Isfahan of an ancient and a modern city is an earlier

Figure 11.40 – Muscat; the city as recorded in 1780 by C Niebuhr and published in his *Voyage en Arabie, Volume 2*.

W G Palgrave provides a pen-picture of 1868. 'Muscat Town is situated at the extremity of a small cove. The splendid little natural harbour is formed on the east by a long island which leaves only a narrow channel between it and the mainland, and on the west by precipitous rock. On the land side it is surrounded by bare and volcanic hills which greatly resemble those of Aden. There are no roads leading to the interior except narrow and rough footpaths suitable only for pedestrians and donkeys. All caravans to and from the interior use Matrah as the terminus, and goods have to be transported by sea between the two towns – a distance of two miles.

The streets of Muscat are narrow, tortuous and uneven. There are several good plain buildings, such as the consulate, the palace, and leading merchants' houses. Mosques are plain and there are no minarets. Two picturesque and strongly built Portuguese forts stand one on each side of the sandy beach. They are built on cliffs some 150 feet above the sea and were completed in 1587 and 1588.' (*Narrative of a Year's Journey through Central and Eastern Arabia*, 186B).

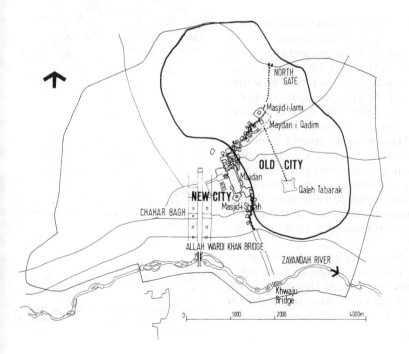

Figure 11.41 – Isfahan; the city of Shah Abbas locating its component parts, and the pre-existing city perimeter – which included the Masjid-i-Jami and the adjoining subsequently built-over Maydan-i-Qadim. (See Figure 11.34 for the detail plan of a typical intermediate part of the Grand Bazaar.)

118. W D Peyton, *Old Oman*, 1983.
119. L Lockhart, 'Shah Abbas's Isfahan; in A Toynbee (ed.), *Cities of Destiny*.

version of that which was carried out by the French at Tunis and Fez, and in recent decades with the addition of Islamabad alongside Rawalpindi as the new capital of Pakistan.

The Shah's startpoint was provided by the location of a royal pavilion situated about 100 m west of the existing centre, separated from it by an extensive open space used for temporary markets. This pavilion was reconstructed on four floors over an entranceway planned to link to the proposed royal palace to the west, with the open space – the future Maidan – to the east. When in Isfahan, the Shah resided in the Ali Qapu or Lofty Gateway.

About 600 yards to the west of the Ali Qapu and coincident in construction, the Chahar Beg – so-named because it involved purchase of four vineyards – was laid out as a grand landscaped entrance avenue to the new city. This dominant axis was extended to the south across the Zayandah by the superb extant Allah Wardi Khan Bridge, named after the Shah's commanding general, which is 968 ft in length and 45 ft wide, carried on 33 stone arches. The next project was construction of the small mosque of Masjid-i-Sadr (Shaikh Lutfullah) opposite the Ali Qapu and alongside the ancient city where its position determined the line of the eastern side of the Maidan-i-Shah, a magnificent public square of length 1,674 ft and width 540 ft.

The long north–south sides have a simple architectural uniformity, offset, respectively, by the Ali Qapu and the Lufullah Mosque.[120] The northern end has as its centrepiece the new gateway entrance – The Qaysariya – which leads into the Royal Bazaar, and at the southern end there is the new Masjid-i-Shah commenced in 1612–13, whose architect resolved, most beautifully, the different axial orientations of the Maidan and the Quibla direction of Makkah. The Royal Bazaar and its numerous caravanserai, warehouses and other related *suq* activities has been illustrated in Figure 11.34.

ALEPPO SYRIA

Halab, the old name for Aleppo, 'is an ancient Amorite word meaning copper, for which the city was famed in antiquity. It was first mentioned in the reign of Sargon Agade and has had a continuous and chequered history since'.[121] The old city is now confined within the remains of the historic wall forming a square of side about 1,500 metres. The settlement potential of the isolated rocky hill about 1,500 yards east of a small reliable river was recognized from earliest times. From fortress village it became a Greek Acropolis, a Roman sanctuary and then the Islamic citadel which dates back to AD 1029 in its present form. Between 301–281 BC the adjoining area was laid out on a gridiron plan by a Mecedonian colony, with an 11 km aqueduct bringing water into the city from the springs at Haylan. It was a Roman colony from 64 BC. In AD 636 Halab surrendered peacefully to the Muslim army. While their Great Mosque was under construction on the site of the ancient Agora, the first mosque was in a converted triumphal archway of the later Bab Antakiya.

'The *Suq*s are the most important features of the old city,' writes Bahnassi. 'There are not only streets lined with shops on both sides, but also wide footpaths covered by thick stone vaults giving protection against the summer heat and the cold of winter … these *Suq*s extend in an inter-linking manner for a distance of 15 km and concentrate at a point called al-Madinah, west of Aleppo.'[122]

Figure 11.42 – Isfahan; the detail plan of the Maidan.

Figure 11.43 – Aleppo, Syria; the early 1920s plan of Halab – prepared by the engineers of the Wilayet' (from *The Encyclopaedia of Islam*, Ist Edition, 1927). North at the top, the Great Mosque is approximately central at the western foot of the citadel mound and the main linear *suq* runs between it and the central western gate, the Bab Antakiya. The key to the map gives forty-eight Quarters of the Town; ten Gates in the wall; and twenty-two Mosques.

120. See 'Isfahan', *The Architectural Review*. May 1976.
121, 122. A Bahnassi, 'Aleppo,' in R B Serjeant (ed.), *The Islamic City*.

PLAN DE BEYROUT
d'après Julius Loÿtved.
et l'Etat-Major français

BEIRUT, LEBANON

Originally a Phoenician entrepot, succeeded by the Greeks and Romans – who raised it to colonial status under Augustus; the history of Beirut (or Berytus) 'affords an interesting proof of the power of a commercial town, favoured by position and other natural advantages, to rise superior to the worst calamities ...' Those words, which, optimistically, can be rewritten around the end of this century – after the calamitously divisive infighting of the 1970s and 1980s; were first penned by Henry Stebbing in the 1880s.[123] Beirut was taken by the Muslims in 635 and it was part of the Turkish Ottoman Empire from the sixteenth century. At the end of the eighteenth century, tightly contained within its walls and with a population of about 6,000, Beirut was known only as a small safe anchorage and for its annual silk market.

CAIRO

Surprising to some, Cairo is not an ancient city. The historic nucleus of the modern Egyptian metropolis – ranking first or second, with Mexico City, as the world's most populous cities – was not founded as Misr El Kahira until 969 AD, although there were close by predecessors to the south alongside the Nile: notably Memphis, the ancient capital of Northern Egypt, for which see Figure 1.38. El Kahira was located about 20 km south of the head of the delta, where the right bank Mukattam Hills approach nearest to the Nile, which as shown by Figure 11.46 has moved further west since then.[124] Memphis, of which there are negligible known urban remains, is about 22 km south, and in between there were four intermediate Islamic 'cities' of which there are varyingly substantial remains.

The economic and strategic importance of a site controlling the Nile and its delta had been recognized by the Ancient Egyptians, whose religion gave the west bank its sanctuary status dominated by the Pyramids of Gizeh situated about 13 km south-west of historic Cairo. (Until the first permanent Nile bridge of (1990s) Cairo remained an eastbank city; now its western suburbs extend out to the Pyramids.)

Figure 11.44 – Beirut, Lebanon; the map of the city in about 1880 (after Julius Loytved et l'Etat-Major Francais). In 1840 (when a British bombardment demolished large sections of the wall) the city's population of about 12,000 was divided equally between Greek Orthodox Christians and Sunni Muslims. 'After 1860, however, a rural exodus began which has never ceased.' (S Khalaf & P Kongstad, 'Urbanization and Urbanism in Beirut: Some Preliminary Results, in L C Brown (ed.) *From Madina to Metropolis*, 1973.) From prosperous co-existence, latterly (1920–41) under French mandated authority, differences between immigrant populations of a variety of religious and political extremes gradually deteriorated into the bitter civil wars of recent decades which have devastated large areas of the city, ruined its economy, and which, until interrupted by the 1991 Gulf War, made Beirut – for entirely the wrong reasons – the most newsworthy of Islamic cities.

By the latter part of the nineteenth century Beirut was on the brink of rapid expansion in response to greatly increasing Eastern Mediterranean trade. The port was modernized in the 1890s, filling in the old harbour, and its railway station of 1900 became a major entry for pilgrims en route to Makkah. The population was about 120,000 at that time, although the inner walls were not finally demolished until 1915 'Azmi Bey himself struck the first blow with a silver pickaxe supplied by the Town Council.' (See *Beirut Our Memory: a Guided Tour Illustrated with Postcards from the Collection of Fouad Debbas*, 1986 – the first was taken in 1839).

123. H Stebbing, *The Christians in Palestine*, 1884.
124. For Cairo see J H G Lebon, 'The Islamic City in the Near East,' in *The Islamic City*, A H Hourani and S M Stern (eds), 1970; J Abu-Lughod, *Cairo: 1001 Years of the City Victorious*; *Cairo* (Mediaeval Towns Series), 1902; Hitti, 1973; A Raymond, 'The Residential Districts of Cairo during the Ottoman Period,' I Serageldin and S El-Sadek (eds), *The Arab City*, 1982; and M Scharabi, Kairo: *Stadt und Architektur im Zeitalter des Europaischer Kolonialismus*, 1989 (German with excellent maps).

North of the largely ruined Memphis, where the Nile could be readily crossed by way of Roda Island, the Romans built a legionary fortress which, under the Byzantines, became known as Babylon. Still mainly inhabited by Coptic Christians, and with substantial Roman remains known as Kasr al-Sham, this quarter is known today as 'Old Cairo'. The Nile was then alongside the fortress and it is probable that it was bridged (possibly on boats) as two spans across the end of Roda Island. When in 638, Babylon was taken after siege by the Muslim army, its general, Amr Ibn al'As, decided to take as his garrison the army camp surrounding Babylon, rather than use Alexandria for the purpose. The new 'city' was named El Fustat, although it is doubtful if the essentially fragmented form warranted that title; the tribes and groups comprising the army settling permanently, on the basis of khitta land grants, where they had pitched camp during the siege.[125] The first mosque was built about 300 m north of the Kasr al-Sham in 642 and survives – rebuilt and much restored – as the mosque of Amr Ibn al'As.

In 74 BC another army encampment was set up at El 'Askar immediately north-east of El Fustat; to be followed to the north, in the early ninth century, by the garrison town of El Katai where the great mosque of Ibn Tulun was built in 876–79. Finally, in the northwards sequence, the Fatimids founded the new city of El Kahira, from which derives the modern name. It was laid out on the level ground between the Muquattam Hills and the Khaliq al-Masri (as Trajan's canal was then known), enclosed within an approximately rectangular defensive wall with a main north–south street flanked by two main central palaces. In contrast to El Fustat, which gradually acquired its post-conquest urban status, El Kahira was intended to be the administrative capital from the outset. It was divided into Haras (quarters), the subdivision of which was presumably left to the families concerned.

El-Fustat remained the commercial centre, and Mukaddisi, who in 985 described Fustat and its wealth in great detail, dismissed El Kahira in a few words. El Fustat was unfortified, however, and in 1168 it was systematically burnt by the defenders of El Kahira in order to deny it to the besieging Christian Crusader army. The area remained largely deserted until the later 19th and 20th century southern expansion of modern Cairo.

In 1170 Saladin repaired the defences and then in 1179 he ordered the construction of a defensive wall enclosing both El Kahira and the area of Fustat, with an imposing new citadel in the south-east corner. The wall, however, was completed only in northern parts enabling extension westwards on the defended riverplain between the Kahalidji and the Nile. When a new palace was built on the citadel in 1207 it encouraged growth in the south filling in the area west from the citadel to the Nile.

Cairo took over the trading role of El Fustat, with Bulaq gradually evolving during the fourteenth century as a separate riverport. From then until the French occupation in 1798 the extent of Cairo remained more or less constant.[126] It had three great thoroughfares parallel to the Nile, but otherwise only a maze of lanes dividing up the total of thirty-five quarters. There were seventy-one city gates and a population estimated to have fluctuated around 250,000. Modern Cairo dates from the French period. For internal security they cut the straight Shari'a al Muski street across the old city from west to east, otherwise it was left alone with new European city centre districts laid out to the west and

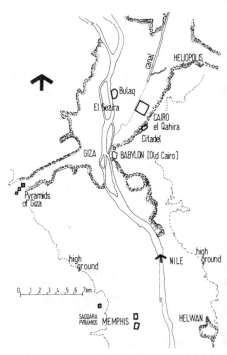

Figure 11.45 – Cairo; a map locating the sequence of cities from Memphis in the south to El-Kahira, the nucleus of the modern Islamic city, in the north. The tinted outline is that of the city in the 1960s.

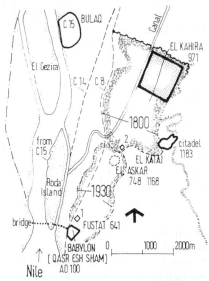

Figure 11.46 – Cairo; a detail map of El-Kahira showing its successive stages of enlargement through to the nineteenth century.

125. See Guest, 'El Fustat,' The Journal of the Royal Asiatic Society, 1907, p 63 et seq for a map of the distribution of the khitat.
126. See maps as Raymond, 'The Residential Districts of Cairo during the Ottoman Period.'

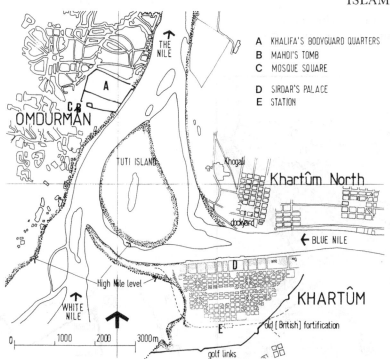

A KHALIFA'S BODYGUARD QUARTERS
B MAHDI'S TOMB
C MOSQUE SQUARE

D SIRDAR'S PALACE
E STATION

Figure 11.47 – Khartoum, Sudan; the three separate towns of 1908 at the junction of the White and Blue Niles. The present-day city comprises the three centres and their suburbs; Khartoum is the 'western-world' official district, and Omdurman that of the local Sudanese.

Khartoum, located 2165 km south of Cairo and at 381 m above sea level, was founded in 1821 as the capital of the Egyptian Sudan. The British General Gordon was killed there in 1884, after a brief siege, by the rebel Madhist army who subsequently set up their armed encampment at Omdurman, across the Nile. On 4 September 1898 a British expeditionary force under General Sir Herbert Kitchener re-entered Khartoum 'and hoisted the British and Egyptian flags over the ruined palace of Gordon ...' (A Moorhead, *The Blue Nile*, 1962).

As rebuilt according to an idea attributed to General (Field Marshal) Kitchener, the plan of Khartoum was based on a series of 'Union Jacks' (the design of the British flag) whereby a system of diagonal streets was overlaid on a gridiron. The dominant determinants of this probably unique urban recipe were those of military defence – after the mode of Haussmann – with a strong dash of imperial aggrandisement; in combination with topography and the ubiquitous gridiron.

TRIPOLI

1 : 15.000

Figure 11.48 – Tripoli (Libya); the plan of the historic city and harbour in the 1920s – with an explanatory diagram. As the census of 1914, taken three years after the Italian annexation, the population comprised 19,907 Muslim, 14,180 European, and 10,471 Jewish. In 1928 the total was about 60,000 including 25,000 Europeans. Within the early nineteenth-century European style artillery fortifications, the typical Islamic morphology is contrasted with the Italian suburban routes radiating to the south-east from the city gate adjoining the castle.

Tripoli originated as the ancient city of Oea, a Phoenician and then a Carthaginian colony which came under Roman rule in 149 BC. The original harbour was at the northern end of the bay, protected by a rocky reef, with the city located as shown, around the extant Arch of Marcus Aurelius erected in AD 163. (Nothing is known of the original layout.) For all its future importance Oea, however, was over-shadowed by Sabrata and Leptis. The name Tripoli was acquired from that of the territory of the three cities (see Chapter 3) and it was confirmed by the Muslim conquerors after 643–644, with the suffix al-Gharb to distinguish it from Tripoli in Syria. In common with other Islamic Mediterranean ports, the subsequent history of Tripoli was turbulent; periods of affluence attracting the plunderers.

For its part, during the seventeenth century the corsair navy of Tripoli was at a peak of piratical strength, out of which revenue new mosques, private baths and other urban improvements were funded until, by the end of the century, English and French policing curtailed that activity, prior to its ending in 1830. Historically Tripoli was never very big and in 1784–85 it is known that one-quarter of the city's population of about 10,000 died during an exceptionally serious outbreak of the plague. The extent of the historic city and its internal arrangement was completed during the reign of Yusuf Pasha Karamanli (1795–1832) who repaired and strengthened the fortification.

south-west. Economic activity moved away leaving medieval Cairo comparatively unscathed today but posing an enormous conservation problem.[127]

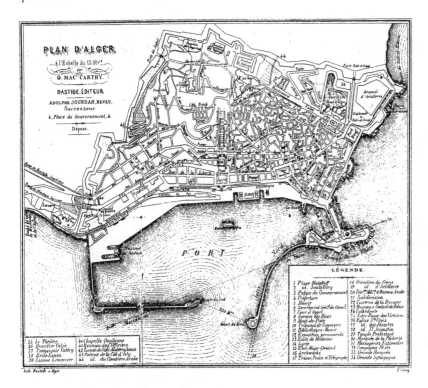

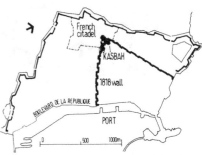

ALGIERS, ALGERIA

It is known that the location of the historic Islamic nucleus of Algiers had been a Roman colony under Vespasian, but the site subsequently became deserted and it was probably as a 'new town' that a Muslim prince of the Berber family of the Sanhadja founded Algiers during the middle of the tenth century. Its original name, Djaza'ir bani Maz-ghanna came from the rocky islets off the coast which formed a natural protective mole for a safe anchorage which had probably never com-pletely gone out of use.[128] Above the waterfront the land rises steeply. Algiers had a characteristically turbulent history requiring successive re-fortification of the close-grouped historic nucleus on the hillside above the port. Today it is notable for the impressive scale of the European redevelopment of the waterfront above a two-level corniche roadway.

Betts provides an impression of the French colonial city: in 1839 some 15,000 Europeans inhabited Algiers; in 1847 the number was nearly 43,000. The numerical growth explains what happened. Amidst the rubble and the rabble that this demographic change generated, colonial Algiers took form. By the middle of the century four news-papers greeted those who could read French; a theatre, constructed by municipal funds, played four times a week to a house of 500 – if all the seats were sold; and variety of cafes catered to the different layers of emerging algeroise society. A development plan had been considered by the government as early as 1846, but its implementation was de-

Figure 11.49 – Algiers, Algeria; the city within the French artillery fortifications constructed after annexation in 1830, with an explanatory diagram showing the extent of the Islamic city in 1818 (from Pananti, *Narrative of Residence in Algiers*.)

127. See P Bergne, 'Cairo,' *The Architectural Review*.
128. For Algiers see R Betts, *Tricoleur: The French Overseas Empire*, 1978.
129. Betts, *Tricoleur*.

layed, save for one major feature. This, a work that since its completion has received universal praises and has stood as the new city's particular architectural signature, is the Boulevard Front de Mer, which faces the sea and was constructed on a rhythmic series of arches, behind which were located warehouses for the port. Begun in 1860, with the first portion opened in 1866, the whole provides a magnificent facade before the old city and stands as an example of stylistic tastefulness in a colonial period of municipal destructiveness.' A visitor in 1860, Ernest Feydeau, offered the following description in his Alger 'Alas, when looked at close by, Algiers today seems more designed to dull the eye than to brighten it. A lot has been spoiled, made ugly, or destroyed. And the sad part of it is that the Europeans alone are guilty of these acts of vandalism.'[129]

Figure 11.50 – Makkah, Saudi Arabia: the enlarged Great Mosque (El Haram) in its urban setting. The view is from the north-east and relates to the old and new mosque plans, Figure 11.31, and Burckhardt's 1829 plan of the city, Figure 11.10. In addition to illustrating the mountainous topography and the city's multi-storey street frontage houses, the view also, and most dramatically, shows the effect, in combination, of the religious motivating force and religious aggrandizement, on the old city. The latter by way of a twentieth century Islamic equivalent of the renaissance Popes who made Rome a suitable centre for the Catholic faith.

Appendix A – China

'Chinese civilisation, and with it Chinese architecture,' writes Andrew Boyd, 'are less remarkable for their antiquity than for their continuity.'[1] Archaeological knowledge of ancient China is very recent; the first Neolithic village settlement at Yang Shao in Honan, was not identified until 1921. Daniel notes that the earliest date for these villages seems to be somewhere in the middle or at the end of the third millennium BC, and emphasizes that, 'this is of course very much later than the date of incipient agriculture and the date of the first settled peasant farmers in the ancient Near East'.[2] Considered opinion now believes that agriculture had an independent origin in northern China and was not introduced from outside, as previously postulated. Figure A.1 gives the location of the first, Shang, Chinese civilization on the river plains of the Yellow River. Its chronology, as presently accepted, is dated as: Proto-Shang (Neolithic), 2500 to 2100 BC; Early Shang, 2100 to 1750 BC; Middle Shang (Early Dynastic Period), 1750 to 1400 BC; Late Shang, 1400 to 1100 BC.

Little is known of Shang urban centres. Anyang was founded as a capital by King P'an Keng and is dated to 1384 BC but it arose late in the period[3] and there would have been earlier cities. Shang society was rigidly stratified: an urban warrior élite, based on walled cities, held absolute control over the peasant masses, a situation which only the present regime has been concerned to change. Such conditions are not favourable to widespread urban settlement and it is possible that Shang centres were few and far between. The Shang dynasty, traditionally dated 1776–1122 BC, was succeeded by the Chou dynasty (1122–255 BC) – one of the great periods, notable for the writings of Confucius and other philosophers. The second half of the third century BC, however, was a period of political turmoil; internal power struggles encouraged barbarian incursions until, under Shih Huang-ti – the first emperor of the Ch'in dynasty (221–206 BC) – reunification of northern China was possible within the Great Wall, which he commenced building around 220 BC. Eventually extending for more than 1,500 miles in length, and between 20 and 30 feet high and 12 feet wide at the top, it still defines much of the northern boundary of China.

The foundation of Ch'ang-an as a capital of China dates from the early years of the Han dynasty (206 BC–AD 220). At first, as Boyd

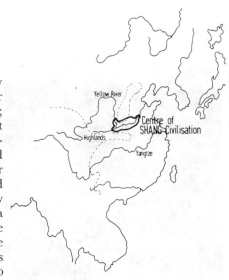

Figure A.1 – Map showing the origins of the Shang civilization in the valley of the Yellow River in northern China.

1. A Boyd, *Chinese Architecture and Town Planning 1500BC–AD1911*, 1962; see also W Eichhorn, *Chinese Civilization*, 1969.
2, 3. G Daniel, *The First Civilisations*, 1968.

explains, 'it was a makeshift capital; enclosing two pleasure palaces from the conquered Ch'in house, the city was not a rectangular one, it lacked both the north-west and south-east corners because of irregular terrain ... this was in the early years of the new dynasty while military campaigns in many areas of the country were still going on'.[4] The Han dynasty eventually disintegrated and was followed by nearly five centuries of renewed political instability, before the Sui dynasty (AD 581–618) could effect Chinese reunification. This was based on a new city of Ch'ang-an, located some distance south-east of the old Han city. The new Ch'ang-an, known at first as Tahsing, was planned before the end of the sixth century and became during the T'ang dynasty (618–907) 'the largest, richest and grandest city in the world of that time'.[5]

The seventh-century plan of Ch'ang-an is given in Figure A.2. Nelson Wu notes that, 'if it deviated from the ancient canons in many ways, it set new standards for both its size (5.7 by 5.28 miles) and its rigidity – eleven north–south streets and fourteen east–west streets – the former an odd number to provide a north–south axis'.[6] The royal palace was built at the northern end of the axis, traditionally facing south, with the administrative city adjoining it to the south. The design provided for two market places, one to the east and the other west of the central axis. 'As the great capital of Tahsing was taking shape, villages were levelled, avenues laid out, and rows of trees were planted. According to legend there was one old locust tree that was not in line. It had been held over from the old landscape because, underneath it, the architect-general had often sat to watch the progress of construction, and a special order from the emperor in honour of his meritorious official spared it from being felled. Thus, except for this tree, the total superimposition of man's order on natural terrain was complete.'[7]

Masuda notes that 'the allotment of the living quarters was quite rational. The *bo* grid blocks in Ch'ang-an, with the exception of the Imperial court and the Administrative City which were square, were all rectangular, with the long axis pointing east–west. Even the square *bo* had a single smaller road running through them from east-to-west making oblong building lots. The rectangular *bo* were subdivided by smaller crossing roads into four similar oblong blocks, and as a result it was possible to arrange every house facing the sun, and with a courtyard garden'.[8]

Peking and Rome, the world's oldest major capital cities, have both experienced extremes of urban fortune. Peking is known to have been the site of a Neolithic village around 2400 BC. Later as Boyd notes, 'it was the capital of Yen (one of the "Warring States") in the third and fourth centuries BC. A provincial town in Han times, it was lost to the northern invaders during the fourth to sixth centuries AD, and recovered by T'ang. In the tenth to twelfth centuries it was held by various nomadic peoples and from 1153–1215 became Chung-tu, the capital of the kingdom of the Golden Tartars to whom the Sung emperors paid subsidies after their removal to Hangchow'.[9] Figure A.3 shows the extent of Peking as the capital of the Ch'in (AD 265–420), and Yuan (1279–1368) dynasties; and that of the Ming (1368–1644) and Chi'ing (1644–1911) dynasties. This latter area forms the historic nucleus of what is now modern Peking.

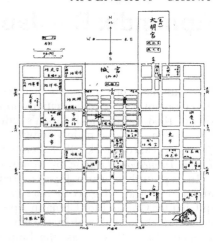

Figure A.2 – Ch'ang-an, general plan. The rectangular area added north of the original city was the Ta-ming Kung, the Pleasure Palace of the Great Luminosity. In the centre of the northern part of the city, the Imperial Palace of the Sui and T'ang dynasties occupied an area some 1½ miles long by ¾ mile wide, immediately north of the administrative city which was planned on the basis of twenty-four smaller grid blocks.

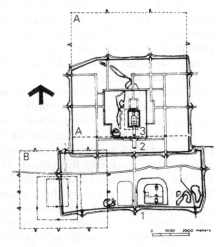

Figure A.3 – Peking, showing the extent of the Yuan dynasty capital (A) and that of the Ch'in dynasty (B). Also keyed are the Outer City (1), the Inner City (2) and the Imperial City (3). The area of the city during the Ming and Ch'ing dynasties, which forms the historic nucleus of modern Peking, is in heavy outline.

4, 5. Boyd, *Chinese Architecture and Town Planning 1500BC–AD1911.*
6, 7. N I Wu, *Chinese and Indian Architecture*, 1963.
8. T Masuda *Living Architecture: Japanese*, 1970.
9. Boyd, *Chinese Architecture and Town Planning 1500BC–AD1911.*

Appendix B – Japan

This summary of the origins of urban settlement in Japan and the development of several key city-examples down to the Mejii Restoration of 1868 – after which Japan's industrial revolution can be said to have begun – is based mainly on two sources: *Social Change and the City in Japan*, by Takeo Yazaki (Professor of Sociology, Keio University) and *Living Architecture: Japanese*, by Tomoya Masuda.

In its pre-urban history Japan passed through similar food-gathering and agricultural-village economies to those which have been described in Chapter One and for China in Appendix A. The comparison, however, ends there. Although by the latter half of the third century AD the Emperor Sujin (known as 'the Emperor who opened and ruled the country') had effected unification of almost all of present-day Japan, excluding Hokkaido, 'the Japanese, unlike the Chinese, had no experience with urban life before the institution of the capital was imported from the continent (China)'.[10] Masuda continues that, 'because of the small scale of the topography, and the agricultural livelihood of rice-cultivation in primarily swampy land, large settlements engaged in collective production were not necessary, and settlements rarely numbered more than one family (clan). This lack of urban experience made the capital merely a symbol of political authority and it scarcely functioned as a collective urban society. Most of the inhabitants were government officers, often single men who had left their families in the provinces'.[11] (Even when the capital later acquired unmistakable urban status this demographic instability persisted.)[12]

There are fascinating parallels with ancient Egypt; in Japan it was also customary for each emperor to build for himself a new imperial residence and government accordingly moved from place to place, latterly in the central Yamato region. Since Japan was relatively isolated from the Asian mainland, as well as being racially homogeneous and politically stable, resource investment in urban defensive systems was also unnecessary during this period. *Miyako* the modern Japanese word for capital city, originally meant the location of the imperial residence. Before the seventh century this was more a symbolic centre of association of clans, rather than the nucleus of the central government's administrative- machinery required to supervise the political, military, economic and religious life of a unified country. As in ancient Egypt, virtually all the population lived in self-sufficient agricultural villages with only the one quasi-urban centre.[13]

Inevitably, bureaucratic concentration did develop; the *miyako* assumed modern administrative characteristics and, 'though still shaped by the clan system, life in the capital no longer had its former flavour of an agricultural society with its small-scale system of intricate face-to-face relationships. While the ordinary people's lives advanced little beyond the primitive condition of earlier ways of the pit-dwellings, the nobility now occupied new continental homes with elevated floors, dressed stylishly in the new continental modes, and surrounded themselves with the splendid products of the fine arts and crafts of Chinese origin. The distinction between rural and urban ways of life was greatly intensified'.[14] It is hardly surprising that Chinese influence extended to

Figure B.1 – Japan, outline map locating major cities including the historic ones mentioned in this brief summary of Japanese urbanism. The country is mountainous and the proportion of flat and moderately sloping land is twenty-four per cent.

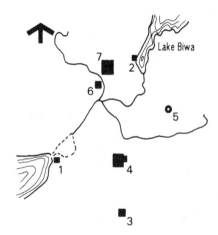

Figure B.2 – Detail map of the Yamato region, locating the following keyed ancient capitals of Japan: 1, Naniwa (Osaka); 2, Otsu; 3, Fujiwara; 4, Heijokyo (Nara); 5, Shigaraki; 6, Nagaoka; 7, Heinakyo (Kyoto).

10, 11. Masuda, *Living Architecture*: Japanese see also D Kornhauser, *Urban Japan: Its Foundations and Growth*, 1976.

12. Estimates of population in Edo in 1721 (excluding warriors and priests) give figures for men, 323,000 and women, 178,000; and in Kurume in 1699 – men, 5,143 and women 3,621.

13. See page 27 for a discussion of a comparable situation in ancient Egypt.

14. Yazaki, *Social Change and the City of Japan*, 1968.

supplying, in 645, the plan of Ch'ang-an as the model for the first Japanese capital city.

From about 590 the *miyako* had been located in the Asuka region where a complex of palaces, temples and shrines and government offices comprised a loosely formed city. Consolidation of central authority throughout western Japan required a much larger, formally structured capital and in 645 the Emperor Kotoku instigated a move to Naniwa, now part of modern Osaka. This site controlled the mouths of the Yamatogawa and Yodogawa rivers, and major regional roads. It was also the port for trading and diplomatic contacts with the T'ang court on the Chinese continent. A planner named Arataino-Hirabu was responsible for laying out the new city of Naniwa as a replica of the gridiron structure of the Chinese capital of Chang'an (see Figure A.2). Yazaki notes that, 'in addition to the buildings for the top administrative organ, the *Dajokan*, there were also constructed offices for the eight major ministries, *sho*, and hundreds of subsidiary offices. Residential lots were allocated according to the stratification of nobility and commoners, with accompanying differences in size and shapes of houses. Reform measures included an imperial edict appointing a responsible head for each city block, *bo*, and supervisors, *rei*, for every four *bo*, who were responsible for peace and order throughout the city'.[15] Naniwa was shortlived: unable to rely on adequate support from religious leaders and nobility from the Asuka district, the emperor was forced to return there, to a new palace south of Nara (Figure B.2).

Similarly in 667 the Emperor Tenchi was thwarted in his efforts to move from Asuka to a location between Mount Osaka and Lake Biwa. Unable to break free from the powerful local clans, the imperial family recognized its limitations and in 694 the Empress Jito built Fujiwara as her capital city in the south-eastern corner of the Yamato plain. Again Chang'an served as the model. Fujiwara was accepted by religious and political leaders but, hemmed in by mountains, it was unable to expand in response to continuous growth of centralized bureaucratic activity. Furthermore its communications with Japanese provinces and China were increasingly inadequate.

The next move was to Heijokyo – the famous Nara site – where in 710 the Emperor Genmei founded a capital city that eventually served seven Emperors over a period of some seventy years before in turn being superceded by Heiankyo (modern Kyoto). Heijokyo was given a well-chosen compromise location, not far enough from the Asuka district to antagonize long-established religious and political factions, yet admirably situated for communications within the northern part of the Yamato plain. 'The conciliatory approach to the Heijokyo,' writes Yazaki, 'resulted in many Asuka temples being moved to the new city ... in re-establishing these temples in Heijokyo steps were taken to leave the main temples in Asuka standing to pacify the feelings of the people in the old city who were dissatisfied with the transfer of the capital to Nara.'[16]

The plan of Heijokyo (Figure B.3) was also based on the Chang'an model. Its gridiron structure provided a total of 72 *bo*, each of which was subdivided into 16 equal-size minor blocks called *cho*. The city was eight *bo* in width from east to west (4,817 yards) and nine *bo* in length (6,667 yards). After the capital was moved to Heiankyo in the ninth century, a further twelve *bo* were laid out on the eastern side. These blocks comprise the central nucleus of modern Nara.

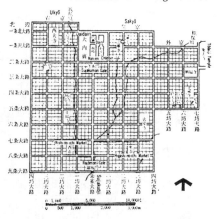

Figure B.3 – Heijokyo (Nara), general plan. The twelve grid blocks that were added to the eastern side of the plan in the early years of the city's history form the area of modern Nara. Somewhat confusingly for modern visitors, the city's modern axis runs west–east, away from the proposed area of the ancient capital which, it is believed, was only partially occupied and which lies in open country today.

Masuda notes that although 'the city blocks of Fujiwara-kyo, Heijokyo and Heiankyo were laid out on the model of Ch'ang-an ... the shape of the blocks was determined by the existing system of land division for rice cultivation, the *jori* system. One block in Ch'ang-an called a *bo*, was a rectangle with its long axis lying east–west, but in Japan the *bo* were generally square. The provincial capitals, like Dazaifu in northern Kyushu; Suho, in Hiroshima Prefecture; and Izumo, in Shimane Prefecture, were planned so that the existing *jori* land division grid coincided with the city grid ... the planners probably imagined this to be the only structural pattern for a city'.

Similarly extensions to the existing organic growth nuclei, either villages or castle towns, notably Edo (Tokyo), usually entailed building over areas of paddy fields whose modular orthogonal pre-urban cadastre patterns determined the new street structure. Gridiron planning also facilitated maintenance and development of Japan's rigidly structured, hereditary social system with its standardized hierarchy of building plot sizes.

15, 16. Yazaki, *Social Change and the City of Japan.*

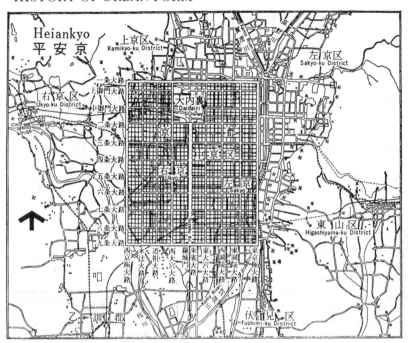

Heiankyo
平安京

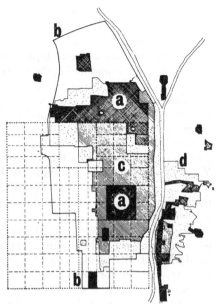

Figure B.4 – Heiankyo (Kyoto), the plan of the ancient city superimposed on the layout of modern Kyoto.

Figure B.5 – Heiankyo, showing the area occupied by the city at various stages in its history, related to the original master plan (north at the top).

Key: a, the city in the late fifteenth century; b, the 'Odoi', an earthen rampart some 15 ml in length, constructed at the end of the sixteenth century; c, the extent of the city at the beginning of the seventeenth century; d, additions of the seventeenth and eighteenth centuries (stippled).

'Residential land grants for noblemen and officials were made for life ... nobles holding the third rank or above were allocated 1 *cho* (2.45 acres), fourth and fifth rank holders ½ *cho*, and lower ranks only ¼ *cho*. Nobles of the first group were allowed frontage on the main street of the city ... commoners' lots of ¹⁄₁₅ or sometimes only ¹⁄₃₂ *cho*, allowed only about 1 *tsubo* (3.95 square yards) per head.'[17] Heijokyo soon became a major religious centre with the imperial government providing generous subsidies for the construction of temples and related buildings. Todaiji became the centre of state Buddhism and its massive wooden hall, the Daibutsuden, although reconstructed on several occasions after fire, remains one of Japan's most important historic buildings. The original area of ancient Heijokyo has been deserted for upwards of one thousand years and there are no visible signs of occupation, other than a number of temple complexes located in agricultural surroundings or modern suburbs of Nara.[18]

Short-lived early moves from Heijokyo were made in 741, 744 (to Naniwa), 745 and 749 (to Nagaoka, south-west of modern Kyoto). Finally in 794 the Emperor Kammu resolved to move his palace and capital to an entirely new site where no contending power had any claim and where the emperor's position would be neutral with respect to the rival families. The chosen location was in the Kuzuno district, in the northern corner of the Yamato plain, surrounded by mountains on three sides, open to the south and traversed by the river Kamo. In his edict announcing the move the emperor stated that 'the rivers and mountains of the imperial site in Kuzuno are beautiful to behold, may our subjects from all over the country come to see them'. It was, and remains to this day, a most beautiful location for a city, well situated for communications and imperial political strategy.

For the early establishment and construction of Heiankyo a special government agency, the *zoeishi*, was set up with a staff of some 150

17. By European standards Japanese homes have remained generally very small: e.g. in 1965 the average area of nearly 40,000 rented dwellings in the Arakawa ward in Tokyo was 22.67 sq metres at an average of 1.6 rooms per dwelling and 1.75 persons per room (see Japan issue of *Official Architecture and Planning*, August 1970).

18. See the Japan volume of the Nagel Travel Guide Series.

officers empowered to determine land and building distribution and to requisition labour and materials. The plan (Figure B.4) was essentially a larger version of Heijokyo: its eight east–west *bo* gave a width of approximately 5,100 yards, and with an extra half *bo* at the northern end it was around 6,150 yards in length. Heiankyo was divided into eastern and western halves, each with its own market, but as with Philadelphia the process of filling in the spaces in the grid was essentially a case where natural growth failed to conform to the preconceptions of planners. The western half of Heiankyo, Ukyo, was low-lying and damp compared to the higher and dryer parts of Sakyo, to the east. Figure B.5 shows how Ukyo remained undeveloped whilst the majority of the Sakyo blocks were densely built up, except for the southern area. Development east of the Kamo was partly to take advantage of the beautiful scenery and partly to escape the severe restrictions imposed on building within the city.

Heiankyo remained as the nominal *miyako* until the Mejii Restoration of 1868 but from the middle of the ninth century the 'effete court aristocracy gradually lost its political dominance and the *miyako* degenerated into a symbolic front of authority with no true power'.[19] Power from then on was firmly in the hands of the military class and in 1192 Minamoto Yoritomo, who had received the title *Seii-tai-Shogun* (generalissimo) from the thirteen-year-old puppet emperor, devised the *bakufu* form of military government, initially based on Kamakura. Heiankyo slowly acquired commercial functions but they could not make up for the loss of the bureaucrats and even the establishment of a new *bakufu* in the city in 1366 failed to promote a lasting revival. Nevertheless, as a mainly commercial city with residual imperial functions, Keiankyo survived numerous vicissitudes (fire, earthquake and military devastation) and by the seventeenth century it had a population of between 400–500,000. But this was halved when the imperial residence moved to Edo in 1869.

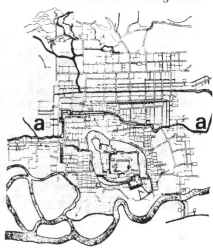

Figure B.6 – Takata, the seventeenth-century plan of this castle town in Niigata Prefecture (north at the top). The military quarter around the castle is south of the line a–a; the townsmen's quarter is north of that line.

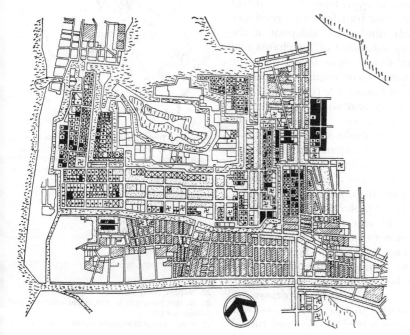

Figure B.7 – A typical castle town, after Yazaki, who observes that 'wherever the castle town was located the castle was central, and the structure of the town was schematically determined through the exercise of power in consideration of defence needs and status ... the upper military class was close to the castle, while the lower military class was located on the town's periphery. The artisans and merchants were settled in definite sections.'

卍	Temples and Shrines
▭	Townsmen (artisans and merchants)
▦	Lower class warriors
■	50 koku
▦	100 koku
◪	300 koku
◇	500 koku
▭	1000 koku
▦	streams and moats
ᵐ	hills and dirt embankments

warrior classes by ranks as indicated by yearly rice stipend (1 koku equals 4.96 bu)

19. Yazaki, *Social Change and the City of Japan.*

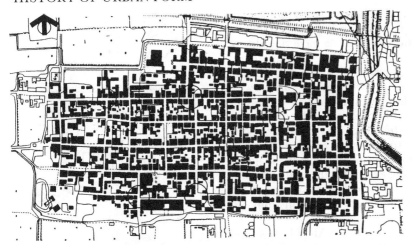

Figure B.8 – Imai, a small mid-sixteenth-century town in Nara Prefecture. The use of cranked intersections, whereby hardly any streets are continuous, is primarily to facilitate defence and is characteristic of Japanese gridiron layouts.

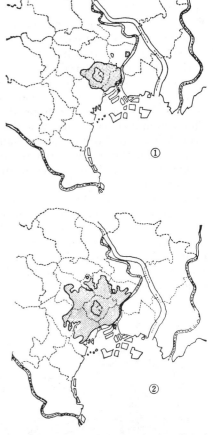

Figure B.9 – Three stages in the growth of Edo (Tokyo): 1, 1624–43; 2, 1652–55; 3, 1673–80. (A fourth map in Yazaki's *Social Change and the City of Japan* shows the area of the city in 1840–43, effectively the same as that of the third map.)

Renamed Tokyo after the Mejii Restoration of 1868, Edo was originally established as a small castle town towards the end of the twelfth century. The first castle was constructed by Edo Shigenaga, the newly appointed governor of Musashi province which included the immensely fertile plain of Kanto. Edo was centrally located at the head of a bay to control land and coastal traffic and occupied part of the area of the present-day Imperial Palace. Originally named Edojuku, it enjoyed comparative prosperity but did not attain significant size until after Ota Dokan rebuilt the castle in the years following 1456. He gave it moats of a total length of around 50 *cho* (3.4 miles) and there were three lines of defence entered through 25 stone gateways. The Hirakawa river flowed through the town and promoted a thriving coastal traffic. However Edo had not yet achieved total dominance of the Kanto plain and in 1486 its development was set back when it came under the control of its main rival, Odawara. When in 1591, following the defeat of Odawara by Tokyugawa Ieyasu, the victor transferred his residence to Edo, 'there remained only a greatly diminished Edojuku at the mouth of the Hirakawa river with only a handful of merchants and craftsmen. Only humble cottages with simple stone fences could be seen in the village called Hirakawa-mara surrounding the Edojuku nucleus ... the central, intermediary and tertiary fortresses had fallen into decay; embankments were poor earth and wood structures ... entrances consisted of four or five wooden gates ... at one place the floor of the castle was nothing more than planks salvaged from old boats'.[20] Ieyasu's ambitious rebuilding and expansion programme moved slowly until his appointment as *shogun* in 1603 made him the effective ruler of Japan and his city the centre of *bakufu* military government.[21] The next few years ensured the dominance of Edo.

Edo's rapid growth into a very large city – probably the largest in the world in the early seventeenth century with an estimated total population certainly exceeding one million – was a result of Ieyasu's enforcement of the *sankin-kotai* system requiring the *daimyo* to be resident for alternate years in Edo. 'This kept the families of the lords in Edo permanently, and brought great numbers of retainers and warriors regularly to the capital.'[22] In addition to the 260 or so *daimyo* and 50,000 standard-bearers, there were possibly in excess of 500,000 warriors resident in the city. If to this figure is added the assumed total of

20. Masuda, *Living Architecture: Japanese*.
21. By 1639 Edo castle, built by Ieyasu, was the largest in Japanese history.
22. Yazaki, *Social Change and the City of Japan*.

500–600,000 townsmen, shrine and temple communities and untabulated non-citizens, the grand total at that time would have been nearer 1.4 million. Yazaki gives the proportional areas as: military class 9,549 acres; temples and shrines 2,174 acres; and townsmen 2,201 acres.[23] Compared to the gross overall density of 55 to 60 persons per acre in the military class, there would have been a comparable density of approximately 250 persons per acre in the townsmen class.

In contrast to the gridiron clarity and simplicity of Heijokyo and Heiankyo, Edo at first followed the organic growth pattern of a castle town with its deliberately complex structure around the castle nucleus. The layout included both organic growth and planned districts, the former generally occupying regained marshy areas, whereas the development of the planned districts was determined by an earlier paddy-field pattern. Because of its seismic zone location, the reliance on charcoal braziers for cooking and heating, and the invariable use of lightweight timber and thatch for all but the most important buildings, Edo was particularly susceptible to fire and consequently did not lack opportunities for extensive rebuilding. As a result, the areas of gridiron development gradually extended until, by the early eighteenth century, most of the centre of the city had been regularly laid out. 1657 saw a particularly devastating fire. Over 50,000 people died, more than 50 miles of street were burnt out, and over 500 *daimyo* mansions were destroyed with those of 770 standard-bearers. The major buildings of Edo castle were burnt to the ground. Subsequent measures for fire prevention included the widening of most streets and lanes, and the formation of fire-break moats and earthen walls.

The rapid growth of Edo from the middle of the seventeenth century raised its consumer demands far beyond the productive capabilities of the Kanto district and required a very high level of imports. Most of this trade came through Osaka, an ancient centre of the commodity economy, which quickly consolidated its position as Edo's supply base and attained a population of 350,000 at the end of the seventeenth century. Since transport by road was very difficult, most of the trade was on the basis of coastal traffic.

Figure B.10 – Tokyo, aerial photograph of the centre of the modern city, around the extensive moated grounds of the castle, within which is located the Emperor's Palace. The gridiron districts are comparatively recent replacements of original densely packed organic growth form, including those of post World War II redevelopment.

Appendix C – Indian Mandalas

This brief summary of the role of the mandala in Indian town planning is based on 'City and Temple Layout: The Vision of a Cosmic Plan', in Andreas Volwahsen's *Living Architecture: Indian*.

An Indian text records: 'A long time ago something existed that was not defined by name or known in its form. It blocked the sky and the earth. When the gods saw it they seized it and pressed it upon the ground, face downwards. In throwing it to the ground, the gods held on to it. Brahma had it occupied by the gods and called it *vastu-purusha*.'

Volwahsen explains that 'the name given to form is mandala. Thus the so-called *vastu-purusha* mandala is the form assumed by existence, by the phenomenal world ... the *vastu-purusha* mandala is an image of the laws governing the cosmos, to which men are just as subject as is the earth on which they build. In their activities as builders men order their environment in the same way as once in the past Brahma forced the undefined *purusha* into a geometric form ... building is an act of bringing disordered existence into conformity with the basic laws that govern it. This can only be achieved by making each monument, from the hermit's retreat to the layout of a city, follow exactly the magic diagram of the *vastu-purusha* mandala'.

The *vastu-shastras* (general manuals on architecture) define thirty-two ways in which the *vastu-purusha* mandala can be formed. The basic mandala is a square; the remainder consist of this square divided into 4, 9, 16, 25, and up to 1,024 smaller squares, called *padas*. Scale is immaterial 'so far as its magic efficacy is concerned. In a plan for a large area it can regulate the disposition of the various buildings, and in the plan of a temple it can define the rhythm of the architectural members or the proportion between the thickness of the wall and the size of the interior'.

The manuals determined all aspects of town planning: the selection of the site, the choice of a suitable mandala, its subdivision according to a rigid caste system, the mystical procedures for setting out, and the detail design of important buildings. In choosing the mandala, the priest-astrologer was concerned to reconcile astrological auguries with the requirement that there should be as many *padas* as residential quarters. Volwahsen notes that mandalas with 64 and 81 *padas* were particularly popular (Figure C.1).

Early in the eighteenth century the ancient city of Jaipur was replanned by its ruler, Jai Singh II, according to the rules of the *vastu-shastras*. Discounting irregularities in the plan (Figure C.3) – the extension to the east and the incomplete north-western corner – Volwahsen describes the plan as follows: 'a splendid avenue running from east to west, an incomplete street parallel to it, and the two main streets which run from north to south divide the city (which was originally rectilinear in plan) into nine parts. The central area of the Brahma-sthana is reserved for the palace of the Maharaja; the other blocks are subdivided more or less regularly by side streets and unpaved footpaths ... individual professional groups were allotted certain quarters of the town, wholly in accordance with *vastu-vidya*. If one wants to go to a cloth-merchant, one must go to a particular street reserved for such persons ... even bicycle shops are grouped together'.

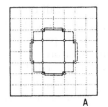

Figure C.1 – Two north Indian mandalas: A, the *manduka-mandala*, containing 8 by 8 *padas*. Around the central Brahma-sthana (4 units) are the *padas* of the inner gods (2 units) and the ring of the outer gods (1 unit). The points of intersection (*marmas*) must not be interrupted by the lines of the ground plan (shown here by thick lines). The *paramasayika-mandala*, keyed as B, contains 9 by 9 *padas*.

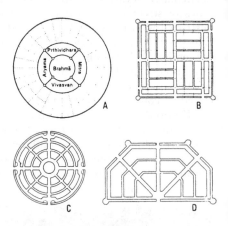

Figure C.2 – Four mandala variants: A, a circular *manduka-mandala*; B, a swastika-form city plan; C, an ideal city plan based on A; D, a *kheta* city for Shudras, low caste inhabitants, without a proper centre.

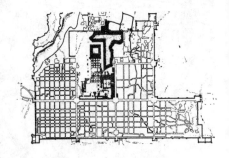

Figure C.3 – Jaipur, general plan.

Appendix D – Indonesia

This summary of historic urban development in the modern State of Indonesia relates to Chapter Eleven for the origins of Islam and the first Islamic cities of the Middle East; Chapter Four for the characteristics of historic urban development in the Netherlands; and Chapter Nine for its appended summary of the Spanish Philippines.[23]

As map Figure D.1, the modern State of Indonesia, which achieved independence in 1949, comprises about 13,700 tropical islands astride the Equator for some 3,000 miles. (The map gives the previous European and the Indonesian names for the main islands and groups.) The five major islands are Sumatra; Java and Madura; Kalimantan (two-thirds of Borneo); Sulawesi; and Irian Jaya (part of New Guinea). About 6,000 of the other islands are inhabited. Indonesia today is the world's most populous Islamic state with a 1990 population of more than 190,000,000, 90 percent of whom are Muslim, and the island of Java is one of the most densely populated areas. Djakarta, the capital, has a 1990 population variously estimated at about ten million. The language, Indonesian, is closely allied to Malay.

Islam was gradually introduced into Indonesia by peaceful missionary means and only by the end of the thirteenth century had it become substantially established in North Sumatra. From there it expanded to the harbour kingdoms of the north coast of Java and eastwards to the Moluccas.

THE DUTCH EAST INDIES

In 1594 the 'Company of Far Lands' was founded at Amsterdam with the object of sending two fleets to Indonesia for spices. This first foray proved worthwhile in bringing back valuable pepper from Bantam and encouraged other similar enterprises which were consolidated into the United Netherlands Chartered East India Company in 1602. (Similarly the West India Company of 1629 for America and West Africa). Rapid growth in the East Indian trade created the need for a rendezvous collection point where trading fleets could gather and where valuable goods could be safely collected and stored. The Straits of Malacca or Straits of Sunda (between Sumatra and Java) where trade routes and monsoon-winds converged offered the best location and the Dutch, having first failed in 1606 to wrest Malacca from Portuguese, turned to the little Javanese port of Jacatra.

The Dutch had first landed at the site at the mouth of the River Tjiliwong in November 1596, and in 1610 they signed an agreement with the Muslim Pangeran ('king') of Jacatra whereby, in return for the payment of 1,800 reals of eight, they were granted a place in the Chinese quarter on which they might build as they saw fit such stone buildings large or small as they needed.[24] As Figure D.2, their first stone building known as Nassau (later the Oude Huis) doubled as warehouse and residence and was strong enough, within a palisaded compound to serve as an emergency fort. There was a further land grant for a 'garden' in 1616 and a second stone building, Mauritz, was added in 1617. Meanwhile, the Pangeran maintained his residence near the mosque and *suq* (marketplace) of the small Islamic township of Jacatra on the west bank, some little distance upstream.

Figure D.1 – Indonesia, a map identifying the five main islands, with their modern and previous names; and the main cities.

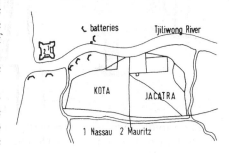

Figure D.2 – Djakarta; a sketch plan showing the embryonic Dutch city of 1619.

By 1627, there were six canals surrounding the town proper. There was also evidence of scattered building on the western side of the river on the site of the former town of Jacatra. The new fort, called Kasteel Batavia, was completed during this year and was defended by moats, so that the entire Dutch settlement, both fort and town, was surrounded by water.

During the next forty years much attention was given to the improvement of the town and by 1667 it had achieved the form it was to retain throughout the eighteenth century. Doubtless such improvements were not part of a formal town plan rigorously applied but followed the inspirations of the successive Governors-General and their Councils, the dictates of the defensive needs of a town newly established in an unfriendly land

23. The main references for this Appendix are C R Boxer, *The Dutch Seaborne Empire 1600–1800*, 1977; F Robinson, *Atlas of the Islamic World since 1500*, 1982; J L Cobban, 'Geographic Notes on the First Two Centuries of Djakarta,' *Journal of the Malaysian Branch Royal Asiatic Society*, vol 44, 1971; and P J M Nas (ed.), *The Indonesian City*, 1986.
24. J L Cobban, 'Geographic Notes on the First Two Centuries of Djakarta,' *Journal of the Malaysian Branch Royal Asiatic Society*, vol 44, 1971; see also A Henken, *The Historical Sites of Jakarta*, 1983; and S Abeyasekere, *Jakarta: A History*, 1989.

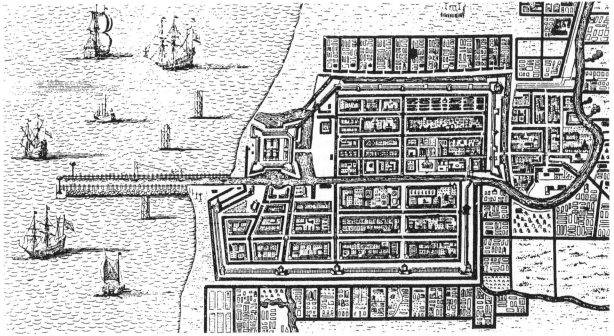

Figure D.3 – Batavia (Djakarta): the city in 1650, as mapped by Clement de Jonge (from F. de Haan, Oud Batavia, 1922.)

The rapidly increasing prosperity of Jacatra aroused the rivalry of the Sultan of Bantam and in December 1619 his troops besieged the 400 strong Dutch community in the fort, which – as an act of defiance – was renamed Batavia in the following March. The siege was raised in May and the Dutch confirmed their victory by burning Islamic Jacatra and claiming extensive new territory. There was now no obstacle to the growth of the new city of Batavia which was planned in the Dutch European image, using their tried and proven technology for dealing with marshy low-lying land (see Chapter 4).

As shown by Figure D.3, the rectangular building blocks were laid out within a system of *grachten* (access and drainage canals) entered off the main stream of the river. The first three from the north were Markgracht (later the Amesterdamschegracht); Oude Kerkgracht, and Leeuwengracht; with a fourth, the Tijgersgracht running north–south linking their ends. By 1627 the plan had been extended further south by a fourth east–west *gracht*.

Outside the Tijgersgracht an additional canal 300ft wide × 10ft deep was excavated as a defensive moat fortified on the inside. There was a bridge at the south-east corner of the extended plan providing the landside entrance through the Landpoort and along the Heerestraat, which was raised above seasonal flood level. The imposing new fort, the Kasteel Batavia, was completed in 1627.

Scattered building on the west bank was then cleared away and by 1650 Batavia had become established as a completely fortified city on both banks of the Tjiliwong. In the process the river was straightened through the town and the eastern perimeter was made parallel. Figure D.3 shows the form which Batavia was to retain until the early nineteenth century when there was major expansion to the south with the establishment of the new colonial central area about 5 km from the historic nucleus.[25] Batavia was renamed Djakarta following independence in 1949.

as well as the recollections of the towns in Holland. At least one author has suggested that the urban vistas of early Djakarta were designed to recall the images of Amsterdam carried in the minds of the European inhabitants.

A comparison of the map of Djakarta in 1628 with that of 1650 by Clemendt de Jonge shows the dramatic change in shape which the town had early assumed from one of irregular outline only partially walled on the east side of the river to one of essentially rectangular form with streets and canals intersecting at right angles surrounded by walls and bastions. The southern extension of the town, clearly observable in 1627, is no longer evident having been razed during the siege by Mataram of 1628 and 1629 and for the rest of the period under consideration, remained outside the walled ramparts of the town.

The original intent had been to include the Zuidervoorstad in the city but there was not sufficient manpower to build the walls. The sloping canal east of the Tijgersgracht has been replaced by one running in a north–south direction and the old fortifications of Nassau and Mauritz have been completely removed. The shoreline has prograded and extends farther seawards, especially noticeble west of the river mouth.
(J L Cobban, Geographical Notes on the First Two Centuries of Djakarta, Journal of the Malaysian Branch Royal Asiatic Society, Vol. 44, 1971)

25. T G McGee, 'The South-East Asian City,' *Journal of South East Asian Studies*, vol 11 no 2, 1967.

Appendix E – Comparative Plans of Cities

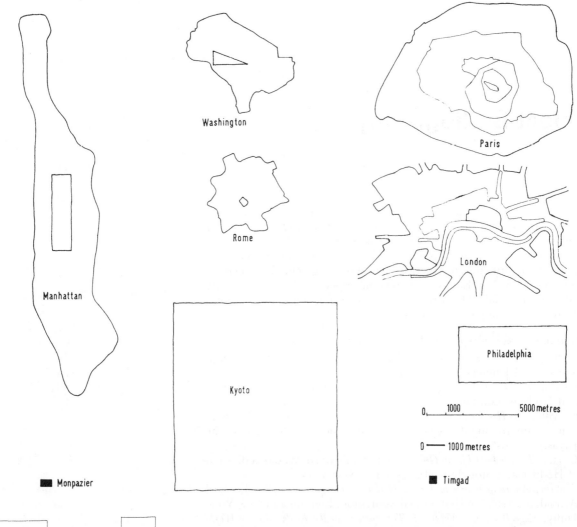

Washington

Paris

Rome

London

Manhattan

Philadelphia

Kyoto

0   1000   5000 metres

0 —— 1000 metres

■ Monpazier

■ Timgad

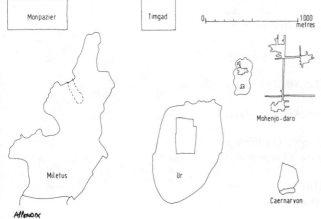

Monpazier

Timgad

0   1000
metres

Mohenjo-daro

Miletus

Ur

Caernarvon

Altenoix

Outline plans of important urban areas drawn to the same scales. Above are seven major cities; on the left are six much smaller ones at one-fifth of the scale of the large ones. For direct size comparison, Monpazier and Timgad have been drawn to both scales and shown in the upper group as solid black, for meaningful illustration. An immediate value of this comparative presentation of areas is to show how small ancient cities were.

For further comparison see Manhattan Island of New York City, including Central Park in outline (Figure 10.32); Washington DC, including the Federal Triangle, and showing the extent of L'Enfant's plan (Figure 10.42); Rome, with three growth stages down to 280 AD (Figure 3.6); Paris, with its successive rings of fortifications down to 1841 (Figure 6.13); London growth down to 1830 (Figure 8.54); Philadelphia as originally planned (Figure 10.22); Kyoto-Heiankyo as originally planned (Figure B.5); Monpazier (Figure 4.43); Miletus (Figure 2.9); Ur of Chaldees (Figure 1.7); Mohenjo-daro (Figure 1.41); and Caernarvon as in the Middle Ages (Figure 4.55).

Select Bibliography

Of the very considerable number of reference sources consulted over the years in the preparation of three editions of *History of Urban Form*, I have listed those ordinarily available books which have been of greatest assistance and which are recommended for further reading. As far as possible both British and American editions are given. References in languages other than English are given in respect of their outstanding illustrative material. (Other references have been given in the footnotes immediately accompanying the main text.)

GENERAL REFERENCE (specific chapters appended)

P Abercrombie, *Town and Country Planning*, 1933, Butterworth, London; and Oxford University Press, 1959

E Bacon, *Design of Cities*, 1978, Viking Press, New York; 1992, Thames and Hudson, London

L Benvolo, *The History of the City*, 1980, Scolar Press, London (A single-volume international history from urban origins through to the 1970s.)

H Carter, *The Study of Urban Geography*, 1975, Edward Arnold, a division of Hodder and Stoughton, London and New York

H Carter, *An Introduction to Urban Historical Geography*, 1989, Edward Arnold, a division of Hodder and Stoughton, London and New York

C Duffy, *Siege Warfare – Volume I, The Fortress in the Early Modern World 1494–1660*, 1979; and *Volume II, The Fortress in The Age of Vauban and Frederick the Great*, 1985, Routledge and Kegan Paul, London (5, 6, 7, 8)

S Giedion, *Space, Time and Architecture*, 1961, Harvard University Press, Cambridge Mass. (5)

E Jones, Metropolis: the World's Great Cities, 1990, Oxford University Press, Oxford and New York

S Kostoff, *The City Shaped: urban patterns and meanings through history*, 1991, Thames and Hudson, London

M Mosser and G Teysot (eds.) *The History of Garden Design: The Western Tradition from the Renaissance to the Present Day*, 1991, Thames and Hudson, London (5, 6, 7, 8)

L Mumford, *The City in History*, 1961, Harcourt Brace Javanovich, New York; 1968, Secker and Warburg, London; 1992, republished Penguin Books, Harmondsworth, Mddx, England

S E Rasmussen, *Town and Buildings*, 1969, MIT Press, Cambridge, Mass. and London (6, 7, 8)

J E Vance Jr, *The Continuing City*, 1990, The Johns Hopkins University Press, Baltimore and London

M Webb, *The City Square*, 1990, Thames and Hudson, London

P Wheatley, *The Pivot of the Four Quarters: a Preliminary Enquiry into the Origins and Character of the Ancient Chinese City*, 1971, The University Press, Edinburgh (1 and Appendix A)

P J Ucko, R Tringham and G W Dimbleby (eds.) *Man, Settlement and Urbanism*, 1972, Gerald Duckworth & Company, London (1, 2, 3, 4)

P Zucker, *Town and Square*, 1959, Columbia University Press, New York City; 1970 paperback, MIT Press, Cambridge, Mass. and London

As a separate mention, there are the eight volumes of E A Gutkind, *International History of City Development*, The Free Press, New York; Collier-Macmillan Limited, London. The series is limited to European countries as: Volume 1 – Central Europe; Volume 2 – Alpine and Scandinavian Countries; Volume 3 – Southern Europe: Spain and Portugal; Volume 4 – Southern Europe: Italy and Greece; Volume 5 – Western Europe: France and Belgium; Volume 6 – Western Europe: The Netherlands and Great Britain; Volume 7 – East-Central Europe: Poland, Czechoslovakia and Hungary; Volume 8 – Eastern Europe: Bulgaria, Romania and the USSR.

URBAN ATLASES (as reprints)

M C Branch, *Comparative Urban Design: Rare Engravings 1830–1843*, 1978, Arno Press Inc, New York, and University of Southern California Press, Los Angeles (Accurate reproduction cartography; the forty urban maps from the partwork Atlas published by The Society for the Diffusion of Useful Knowledge – SDUK.)

J Goss, *Braun & Hogenberg's The City Maps of Europe: a Selection of 16th Century Town Plans and Views*, 1991, Studio Editions, London (Imaginative interpretations of the contemporary reality as aerial views and 'maps'; but useful nonetheless.)

THE EARLY CITIES

B Allchin and R Allchin, *The Birth of Indian Civilization: India and Pakistan before 500 BC*, 1968, Penguin Books, Harmondsworth, Mddx, England

R W Brunskill, *Illustrated Handbook of Vernacular Architecture*, 3d Exp. ed. 1987, Faber, London

V Gordon Childe, *What Happened in History*, 1964, Penguin Books, Harmondsworth, Mddx, England

G Clark, *World Prehistory: A New Outline*, 2nd edition 1969, Cambridge University Press, Cambridge, England

H Crawford, *Sumer and The Sumerians*, 1990, Cambridge University Press, Cambridge, England

G Daniel, *The First Civilizations: the Archaeology of their Origins*, 1968, Thames and Hudson; 1971 paperback, Penguin Books; 1970, Apollo Editions, distributed Harper and Row, Scranton

R David, *The Pyramid Builders of Ancient Egypt: a Modern Investigation of Pharaoh's Workforce*, 1986, Routledge and Kegan Paul, London

O H W Dilke, *Mathematics and Measurements*, 1987, British Museum Publications, London

SELECT BIBLIOGRAPHY

H Fathy, *Natural Energy and Vernacular Architecture: Principles and Examples with Reference to Hot Arid Climates*, 1986, University of Chicago Press, Chicago, USA (And 11.)

J E Gordon, *Structures: or Why Things Don't Fall Down*, 1981, Penguin Books, Harmondsworth, Mddx, England

J Jacobs, *The Economy of Cities*, 1969, Random House, New York

A B Knapp, *The History and Culture of Ancient Western Asia and Egypt*, 1988, The Dorsey Press, New York

S A Kubba, *Mesopotamian Architecture and Town Planning 10,000–3,500 BC*, 1987, Oxford BAR, England

A L Oppenheim, *Ancient Mesopotamia: Portrait of a Dead Civilization*, revised edition (completed by Erica Reiner), 1977, University of Chicago Press, Chicago and London (And 11.)

J Mellaart, *Catal Huyuk*, 1967, Thames and Hudson, London

M Roaf, *Cultural Atlas of Mesopotamia and the Ancient Near East*, 1990, Equinox

W Alexander and A Street, *Metals in the Service of Man*, rev 9th ed 1990, Penguin Books, Harmondsworth, Mddx, England

G Urban and M Jansen, *The Architecture of Mohenjo Daro*, 1984, Aachen University

Sir L Woolley, *Ur of the Chaldees*, 1982, Cornell University Press, Ithaca, NY (revised and updated edition by P R S Moorey) (And 11.)

T M L Wigley, M J Ingram and G Farmer, *Climate and History: Studies in Past Climates and Their Impact on Man*, 1981, Cambridge University Press, Cambridge, England (And 11.)

D Zohary and M Hopf, *Domestication of Plants in the Old World*, 1988, Clarendon, Oxford, England

GREEK CITY STATES

W B Dinsmoor, *The Architecture of Ancient Greece*, 3rd edition 1950, Batsford, London; 1973, Biblo and Tannen, New York

H D F Kitto, *The Greeks*, 1951, Penguin Books, Harmondsworth, England, and New York

P Levi, *Atlas of the Greek World*, 1984, Phaidon Press, Oxford

R Martin, *L'Urbanisme dans la Grece Antique*, Editions A & J Picard, Paris (French)

J Travlos, *Pictorial Dictionary of Ancient Athens*, 1971, Praeger, New York; 1980, Hacker, New York

J Travos, *Athens au fil du temps*, 1972, J Cuenot, Boulogne (French)

R E Wycherley, *How the Greeks Built Cities*, 2nd edition 1962, Macmillan, London; W W Norton and Company, New York

ROME AND THE EMPIRE

D Breeze and B Dobson, *Hadrian's Wall*, 1987, Penguin Books, Harmondsworth, England

J Carcopino, *Daily Life in Ancient Rome*, 1956, Penguin Books, Harmondsworth, England; 1960, Yale University Press, New Haven, Conn.

Z Celik, *The Remaking of Istanbul: A Portrait of an Ottoman City in the Nineteenth Century*, 1986, University of Washington Press, Seattle and London

T Cornell and J Matthews, *Atlas of the Roman World*, 1982, Phaidon Press, Oxford

G Duby (ed.) *La ville antique: des origenes au IX siecle*, Volume 1, Histoire de la France Urbaine, 1980, Editions du Seuil, Paris

B Jones and D Mattingly, *An Atlas of Roman Britain*, 1990, Blackwell, Oxford, England

S J Keay, *Roman Spain*, 1988, British Museum Publications, London

Juvenal, trans P Green, *The Sixteen Satires*, 1967, Penguin Books, Harmondsworth, England

A King, *Roman Gaul and Germany*, 1990, British Museum Publications, London

R Lanciani, *The Ruins and Excavations of Ancient Rome*, 1897, Macmillan, London; reprinted 1968, Arno Press, New York

J Maloney and B Hobley (eds.) *Roman Urban Defences in the West*, 1983, Cncl. for Brit. Archaeol., London

A Maiuri (trans W F McCormick), *Pompeii*, 1954, Instituto Poligrafico dello Stato, Rome

G Milne, *The Port of Roman London*, 1982, Batsford, London

I A Richmond, *Roman Britain*, 1955, Penguin Books, Harmondsworth, Mddx, England

A L F Rivet, *Town and Country in Roman Britain*, 1964, Hutchinson, London

D F Robinson, *Ancient Rome: City Planning and Administration*, 1992, Routledge, London and New York

E T Salmon, *A History of the Roman World 30 BC to AD 138*, Routledge, London

E T Salmon, *Roman Colonization under the Republic*, 1969, Thames and Hudson, London

J E Stambaugh, *The Ancient Roman City*, 1988, Johns Hopkins University Press, Baltimore and London

C H V Sutherland, *The Romans in Spain*, 1939, Methuen, London

S Tlatli, *Antique Cities in Tunisia*, 1971, Les Guides Ceres

J Wacher, *The Towns of Roman Britain*, 1974, Batsford, London; 1975, University of California Press, Berkeley

G Webster (ed.) *Fortress into City: the Consolidation of Roman Britain*, 1988, Batsford, London

MEDIEVAL TOWNS

M W Beresford, *New Towns of the Middle Ages*, 1967, Lutterworth Press, Guildford, England

R H C Davis, *A History of Mediaeval Europe: from Constantine to Saint Louis*, 1987, Longman, Harlow, Essex, and London

F Divorne, B Gendre, B Lavergne and P Panerai, *Les Bastides d'Aquitaine, du Bas Languedoc and du Bearn*, 1985

G Duby (directeur) and J Le Goff (editeur), *La ville medievale: de Carolingens a la Renaissance*, Volume 2, Histoire de la France Urbaine, 1980, Editions du Seuil, Paris

D Friedman, *Florentine New Towns: Urban Design in the Late Middle Ages*, 1988, MIT Press, Cambridge, USA, and London

W G Hoslins, *The Making of the English Landscape*, 1978, Penguin Books, Harmondsworth, England

M D Lobel (ed.) *The British Atlas of Historic Towns, Volume One*, 1969, Lovell Johns, Cook Hammond and Kell, Oxford; *Volume Two*, 1975; *Volume 3*, The City of London, 1989, Oxford University Press

H Pirenne, *Mediaeval Cities: Their Origins and the Revival of Trade*, 1969, Princeton University Press, Princeton

SELECT BIBLIOGRAPHY

H Saalman, *Mediaeval Cities*, 1968, Studio Vista, London; and George
Braziller Inc, New York

C T Smith, *An Historical Geography of Western Europe before 1800*, 1967,
Longman, Harlow, London and New York

A J Taylor, *The King's Works in Wales*, 1974, HMSO, London

J E Vance Jr., *The Continuing City: Urban Morphology in Western Civilization*, 1990, Johns Hopkins University Press, Baltimore and London

D M Wilson (ed.), *The Archaeology of Anglo-Saxon England*, 1976,
Methuen & Co, London and New York

THE RENAISSANCE: ITALY

J Boswell, *Boswell on the Grand Tour: Italy, Corsica and France*; F Brady
and F A Pottle (eds.) 1955, Heinemann, London

S Giedion, *Space, Time and Architecture*, 1961, Harvard University Press,
Cambridge, Mass. (5)

H Hibbard, *Michelangelo*, 1975, Allan Lane, London

C Hibbert, *Venice: the Biography of a City*, 1989, Grafton Bks, London

C Hibbert, *Rome: the Biography of a City*, 1985, Penguin Books, Harmondsworth, England

N J Johnston, *Cities in the Round*, 1983, University of Washington Press,
Seattle, USA

J Gadol, *Leon Battista Alberti: Universal Man of the Early Renaissance*, 1969,
University of Chicago Press, Chicago, USA

M Marqusee (ed.) *Venice: an Illustrated Anthology*, 1988, Conran Octopus,
London

N Pevsner, *An Outline of European Architecture*, 1973, Allan Lane, London

J H Plumb, *The Pelican Book of the Renaissance*, 1982, Penguin Books,
Harmondsworth, England

L Di Sopra, *Palmanova: Analisi di una citta-fortezza*, 1983, Electa, Milan
(Italian)

R Wittkower, *Art and Architecture in Italy 1600–1750*, 1973, Penguin
Books, Harmondsworth, England

H Wolfflin, *Renaissance and Baroque*, 1964, Collins, London

THE RENAISSANCE: FRANCE

H Ballon, *The Paris of Louis IV: Architecture and Urbanism*, 1991, The
Architectural History Foundation and MIT Press, Cambridge,
USA, and London

L Bergeron (ed.) *Paris: Genese d'un Paysage*, 1989, Picard, Paris (French)

R Blomfield, *Sebastien le Prestre de Vauban*, 1971, Methuen, London

P Boudon, *Richelieu: Ville Nouvelle*, Collection Aspects de l'Urbanisme,
Paris (French)

J de Cars and P Pinon, *Haussmann*, 1991, Edition de l'Arsenal, Picard
Editeur, Paris (French)

J Castex, P Celeste and P Panerai, *Lecture d'une Ville: Versailles*, 1980,
Editions du Moniteur, Paris (French)

P Couperie, *Paris through the Ages*, 1970, Barrie and Jenkins, London;
1971, George Brazilier, New York

G Duby (directeur) and Le Roy Landurie (editeur) *La Ville Classique de
la Renaisance aux Revolutions*, Volume 3, Histoire de la France Urbaine,
1981, Editions du Seuil, Paris (French)

G Duby (directeur) and M Agulhon (editeur) *La ville de l'age industriel:
le cycle haussmannien*, Volume 4, Histoire de la France Urbaine, 1983,
Editions du Seuil, Paris (French)

N Evenson, *Paris: a Century of Change 1878–1978*, 1979, Yale University Press, New Haven, USA

P Francastel (ed.) *L'Urbanisme de Paris et de L'Europe 1600–1680*, 1969, Editions Klinckseick, Paris (French)

H M Fox, *Andre Le Notre: Garden Architect to Kings*, 1962, Batsford, London

P Lavedan, *French Architecture*, 1979, Scolar Press, London; Southwest Book Services, Dallas

P Lavedan, *Nouvelle Histoire de Paris*, 1975, Diffusion Hachette, Paris (French)

D Pinkney, *Napoleon III and the Rebuilding of Paris*, 1958, Princeton University Press, Princeton

B Rouleau, *Villages et Faubourges de l'Ancien Paris: histoire d'un espace urbain*, 1985, Editions du Seuil, Paris (French)

THE RENAISSANCE: EUROPE GENERALLY

K Astrom, *City Planning in Sweden*, 1967, Swedish Institute for Cultural Relations with Foreign Countries, Stockholm

J H Bater, *St Petersburg*, 1976, Edward Arnold, London

B Beirut, *The Six-Year Plan for the Reconstruction of Warsaw*, 1949, Ksiazka i Wiedza, Warsaw

K Berton, *Moscow: An Architectural History*, 1977, Studio Vista, London; 1978, St Martin's Press, New York

Bruxelles Construire et Reconstruire: Architecture et Amenagement Urbain 1780–1914, 1979, Credit Communal de Belgique, Bruxelles

G L Burke, *The Making of Dutch Towns*, 1956, Cleaver-Hulme Press, London

A Ciborowski, *Warsaw: A City Destroyed and Rebuilt*, 1964, Polonia, Warsaw

G Cotterell, *Amsterdam: The Life of a City*, 1973, D C Heath, Farnborough, England

P Gabor, *Budapest Varosepitesenek Tortenets*, Volumes 1 and 2, 1964, Kossuth Nyomda, Budapest (Hungarian)

T Hall, *Planning in European Capitals: on the evolution of town planning during the 19th century*, 1986, Graphic Systems AB, Gothenburg, Sweden (German)

E Lichtenberger, *Die Wiener Altstadt*, 1977, F Deutike, Vienna

H Lilius, *Suomalainen Punkanpunki (The Finnish Wooden Town)*, 1985, Anders Nyborg A/S, Denmark (Finnish with English summary.)

S E Rasmussen, *Kobenhavn*, 1969, GEC Gads Forlag, Kobenhavn

P Sica, *Storia dell' urbanistica l'Ottocentro*, 1991, Editori Laterza, Roma-Bari (Italian)

C Sitte, *City Planning According to Artistic Principles*, 1889, English translation 1965, Phaidon, Oxford, England; Columbia University Studies in Art, History and Archaeology, No 2

(See also the general list for the eight volumes of Gutkin, *International History of City Development*; and Branch, *Comparative Urban Design*, for the SDUK map reprints.)

THE RENAISSANCE: GREAT BRITAIN

D Defoe, *A Tour through the Whole Island of Great Britain* (ed. P Rogers), 1971, Penguin Books, Harmondsworth, Mddx, England

SELECT BIBLIOGRAPHY

J Elliot, *The City in Maps*: Urban Mapping to 1900, 1987, British Library, London

J Evelyn, *The Diary* (ed. W Bray) Vols 1 and 2, 1907, J M Dent, London; E P Dutton, New York

J Gifford, C McWilliam and D Walker, *Edinburgh*, 1984, Penguin Books, Harmondsworth, England

W Ison, *The Georgian Buildings of Bath from 1730 to 1830*, 1969, Kingsmead Reprints, Bath; Alfred Saifer, West Orange, NJ

C McWilliam, *Scottish Townscape*, 1975, Collins, London

D J Olsen, *Town Planning in London: the 18th and 19th Centuries*, 1982, Yale University Press, New Haven, Conn. and London

N Pevsner (rev. B Cherry), *London: the Cities of London and Westminster*, 1973, Penguin Books, Harmondsworth, Mddx, England

H Phillips, *Mid-Georgian London*, 1964, Collins, London and Glasgow

S E Rasmussen, *London*, revised edition 1991, MIT Press, Cambridge Mass. and London

T F Reddaway, *The Rebuilding of London after The Great Fire*, 1951, Edward Arnold, London

D Stroud, *Capability Brown*, 1975, Country Life, London

J Summerson, *Georgian London*, 1945, Pleiades Books; 1991, Penguin Books, Harmondsworth, Mddx, England

J Summerson, *Inigo Jones*, 1966, Penguin Books, Harmondsworth, Mddx, England

J Summerson, *John Nash*, Architect to King George IV, 1949, Allen and Unwin, London

J Summerson, *Architecture in Britain 1530–1830*, 1983, Pelican History of Art, Penguin Books, Harmondsworth, England

A J Youngson, *The Making of Classical Edinburgh*, 1966, Edinburgh University Press, Edinburgh; 1975, Biblio Distribution Centre, Totowa, NJ

SPAIN AND HER EMPIRE

G F Andrews, *Maya Cities: Placemaking and Urbanization*, 1977, University of Oklahoma Press, Norman, USA

L T Balbas, *Ciudades Hispano-Musulmanas*, Volumes 1 and 2, 1987, Ministero de Asuntos Exteriores, Madrid

C R Boxer, *The Portuguese Seaborne Empire 1415–1825*, 1969, Hutchinson of London

M D Coe, *The Maya*, 1966, Penguin Books, Harmondsworth, Mddx, England

R Collins, *The Arab Conquest of Spain*, 1989, Basil Blackwell, Oxford, England

M Defourneaux, *Daily Life in Spain of the Golden Age*, 1970, Allen and Unwin, London

R Delson, *New Towns for Colonial Brazil*, 1979, University Microfilms International, Ann Arbor, Michigan, USA

J H Elliott, *Imperial Spain 1469–1716*, 1963, Penguin Books, Harmondsworth, Mddx, England

C Gibson, *Spain in America*, 1966, Harper Torchbooks, New York and London

G Goodwin, *Islamic Spain*, 1990, Viking, London

J E Hardoy, *Pre-Columbian Cities*, 1973, Walker, New York

J E Hardon, *Urban Planning in Pre-Columbian America*, 1967, Studio Vista, London; and George Braziller Inc, New York

J E Hardoy and C Tabor (English translation by F M Trueblood), *La Urbanizacion en America Latina*, 1969, Editorialdel Instituto Torcuato di Tella, Buenos Aires

R L Kagan (ed.) *Spanish Cities of the Golden Age: The Views of Anton van den Wyngaerde*, 1989, University of California Press, Berkeley, USA

J Lockhart and S B Schwartz, *Early Latin America: A History of Colonial Spanish America and Brazil*, 1983, Cambridge University Press, England

J A Mason, *The Ancient Civilizations of Peru*, 1969, Penguin Books, Harmondsworth, Mddx, England

J A Michener, *Iberia*, 1968, Secker and Warburg, London and New York

J H Parry, *The Spanish Seaborne Empire*, 1977, Hutchinson of London

G Pendle, *A History of Latin America*, 1976, Penguin Books, Harmondsworth, Mddx, England

J R Scobie, *Buenos Aires*, 1974, Oxford University Press, New York and Oxford, England

G C Vaillant, *Aztecs of Mexico*, 1962, Penguin Books, Harmondsworth, Mddx, England

URBAN USA

D J Boorstein, *The Americans: the Colonial Experience*, 1958, Penguin Books, Harmondsworth, Mddx, England

H Brogan, *The Penguin History of the United States of America*, 1990, Penguin Books, Harmondsworth, Mddx, England

G Clay, *Close-Up: How to Read the American City*, 1973, Praeger Publishers, New York and Pall Mall Press, London

C W Condit, *American Building: Materials and Techniques from the Beginning of the Colonial Settlement to the Present*, 1968, University of Chicago Press, Chicago and London

W Cronon, *Nature's Metropolis: Chicago and the Great West*, 1991, W W Norton, New York

S Dallas, *Cherry Creek Gothic*, 1971, University of Oklahoma Press, Norman, Oklahoma, USA

C Gibson, *Spain in America*, Harper and Row, New York and London

C N Glaab and A T Brown, *A History of Urban America*, 1976, Macmillan Publishing Company, New York

C P Hill, *A History of the United States* (Third Edition), Hodder & Stoughton, London and Toronto

J A Kouwenhoven, *The Columbia Historical Portrait of New York*, 1972, Harper and Row, New York and London

R B Nye and J E Morpurgo, *A History of the United States*, Volume 1 – The Birth of the USA; Volume 2 – The Growth of the USA, 1955, Penguin Books, Harmondsworth, Mddx, England

J H Parry, *The Spanish Seaborne Empire*, 1977, Hutchinson of London

J W Reps, *The Making of Urban America*, 1965, Princeton University Press

J W Reps, *Cities of the American West*, 1979, Princeton University Press

J M Russell, *Atlanta 1847–1890: City Building in the Old South and New*, 1988

V Scully, *American Architecture and Urbanism*, 1969, Holt, Rinehart and Winston, New York and London

R C Simmons, *The American Colonies: From Settlement to Independence*, 1976, W W Norton & Company, New York and London

SELECT BIBLIOGRAPHY

S B Warner Jr., *The Urban Wilderness: A History of the American City*, 1972, Harper & Row, New York and London

S B Warner Jr., *The Private City: Philadelphia in Three Periods of its Growth*, 1968, University of Pennsylvania Press, Philadelphia, USA

W M Whitehill, *Boston: A Topographical History*, 1968, Belknap Press, Harvard University, Cambridge, USA and London

ISLAMIC CITIES

J Akbar, *Crisis in the Built Environment: the Case of the Muslim City*, 1988, Concept Media, Singapore

N AlSayyad, *Cities and Caliphs: on the Genesis of Arab-Muslim Urbanism*, 1991, Greenwood Press, Westport, Conn.

K Armstrong, *Muhammad: A Western Attempt to Understand Islam*, 1991, Victor Gollancz, London

E Beazley and M Harverson, *Living with The Desert: Working Buildings of the Iranian Plateau*, 1982, Aris and Phillips, Warminster, England

R Betts, *Tricouleur: The French Overseas Empire*, 1978, Gordon and Cremonesi, London and New York

L C Brown (ed.) *From Madina to Metropolis: Heritage and Change in the Near Eastern City*, 1973, The Darwin Press, Princeton, USA

G Duncan, J Russell et al (eds.) *Jeddah: Old and New*, 1982, Stacey International, London

W Facey, *Riyadh: the Old City*, 1992, Inmel Publishing, London

Y Fadan, 'Traditional Houses of Makka: the Influences of Social-Cultural Themes on Arab-Muslim Dwellings', *Islamic Architecture and Urbanism*, 1983, King Faisal University, Damman, Saudi Arabia

H Fathy, *Natural Energy and Vernacular Architecture: Principles and Examples with Reference to Hot Arid Climates*, 1986, University of Chicago Press, Chicago, USA

A Guillaume, *Islam*, 1956, Penguin Books, Harmondsworth, England

B S Hakim, *Arab-Islamic Cities: Building and Planning Principles*, 1986, KPI Limited London, and Routledge & Kegan Paul plc, London and New York

F Herd-Bey, *From Trucial States to United Arab Emirates*, 1982, Longman, London and New York

S Hirschi and M Hirschi, *L'Architecture au Yemen du Nord*, 1983

J D Hoag, *Islamic Architecture*, 1977, Harry N Abrams Inc, New York

A Hourani, *A History of the Arab Peoples*, 1991, Faber and Faber, London

A Hourani and S M Stern (eds.) *The Islamic City*, Bruno Cassirer, Oxford, 1970

G R D King, *The Historic Mosques of Saudi Arabia*, 1986, Longman, Harlow, London and New York

B Lewis (ed.) *The World of Islam*, 1976, Thames and Hudson, London

P Mansfield, *The Arabs*, 1980, Penguin Books, Harmondsworth, Mddx, England

G Michell, *Architecture of the Islamic World: Its History and Social Meaning*, 1978, Thames and Hudson, London, and William Morrow, New York

A Ravereau, *Le M'Zab, une lecon d'architecture*, 1981, Sindbad, Paris (French)

A Raymond, *The Great Arab Cities in the 16th–18th centuries*, 1984, New York University Press, New York and London

J M Richards et al, *Hassan Fathy*, 1985, Concept Media, Singapore, and the Architectural Press, London

F Robinson, *An Atlas of the Islamic World since 1500*, 1982, Phaidon, Oxford, England

I Scrageldin and S El-Sadek (eds.) *The Arab City: Its Character and Islamic Cultural Heritage*, 1982, The Arab Urban Development Institute, Riyadh, Saudi Arabia

R B Serjeant (ed.) *The Islamic City*, 1980, UNESCO, Paris

R B Serjeant and R Lewcock, *San'a*, 1983, World of Islam Festival Trust

K Talib, *Shelter in Saudi Arabia*, 1984, Academy Editions, London; and St Martin's Press, New York

APPENDICES (in sequence)

Leonardo Benevolo, *Storia della Citta Orientale*, 1988, Editori Laterza, Roma-Bari

Nancy Shatzman Steinhardt, *Chinese Imperial City Planning*, 1990, University of Hawaii Press

G W Skinner (ed.) *The City in Late Imperial China*, 1977, Stanford. (And Wheatley, as general list.)

David Kornhauser, *Urban Japan*, 1976, Longman, London and New York

Richard Storry, *A History of Modern Japan*, 1968, Penguin Books, Harmondsworth, England

Takeo Yazaki, *Social Change and the City in Japan*, 1968, Japan Publications Inc., Tokyo and New York

Andreas Volwahsen, *Living Architecture Indian*, 1969, Macdonald, London

Susan Abeyasekere, *Jakarta: a History*, 1990, Oxford University Press, Oxford

C R Boxer, *The Dutch Seaborne Empire 1600–1800*, 1965, Hutchinson of London

P J M Nas (ed.) *The Indonesian City*, 1986

General Index

Key illustrations are given in bold numerals; caption references as suffix 'c', footnote references as 'n', quotation references as 'q', and narrow column text as 'nc'. Key examples only of topics etc are given; similarly key illustrations are given in bold face in the Place Index.

GENERAL INDEX

economic motivating force – *cont.*
    Greek colonial movement 40–1; Roman imperial ambitions 56; Spanish Latin American Empire 78, 304; silver mines 312, 313n; oil 365–6nc; USA, transport and industry 335–7
Edgar, King 134
educational motivating force: Oxford UK 113c and Cambridge UK 114c
    Harvard University, Cambridge USA 331
Edward I: 'the town builder' 119–33
Edward the Elder 106, 113
egalitarian societies 17, 163q
    Greek cities 42
    ancient Rome 64n
    in burgs 106
    Paris 201q
    in Islam 17, 379, 387n, 389n
    See also social grouping
Egyptian civilization 26–30, 27n
Elizabeth I 97: and London's first 'Green Belt' 249–50, 330
Ellicot Andrew: with L'Enfant at Washington 350, 352
emigration: British, to the USA 330–1, and essential agricultural basis 333–4
    Dutch East Indies 411
    French to North America 329, 329n
    Portuguese to Brazil 319–20
    Spanish, to Florida, USA 323, 326; compared English 328, and to Latin America assessed 316
enclosure movement, England 111
encomienda 305n
Engels, Friedrich 291
English landscape. *See* landscape
Erie Canal: and economic stimulus for New York City 336
Eugenius IV 169
Euryphon 43
Evelyn, John: pollution, and the Fire of London 255, 255q; rebuilding plans for 256c, 257, 264
expediency supergrid: defined 15, 15n
    ancient precedent, El Amarna 28–9, and Mohenjo-daro 32c
    modern reliance, Abu Dhabi 368c, Islamabad 15n, Jeddah 393, Kuwait City 391c, Milton Keynes 15n

Ferdinand III of Castile 145
Ferdinand V 305, 323
Fertile Crescent 4
feudalism 93, 95
Feydeau, Ernest 401
financial provisions, urban development: Vienna 228
fire, domestic: northerly climate response 12
fire and urban devastation:
    Amsterdam 222; Dresden 17th century, and WWII 234; Warsaw 15th century and WWII 236; Helsinki 239c; Oslo 240c; The Great Fire of London 255–7; New York City 342; Atlanta, 'Gone with the Wind' 360c; Tokyo 409
    ancient Rome, Nero's fire 61, an ever present hazard, and fire-fighting watchmen 64; Londinium burnt by Boadicea 83
'first cities', claimants
    Jericho and Catal Huyuk 19–20
    Kahun 15, 29
    Mohenjo-daro 15, 29, 31, 40, 43
Fleet River, London 260–1, 260n, 261q
Florentine new towns. *See* planted towns
fortification systems – urban form determinant: introduced, defence 14–15

KEY EXAMPLES: at Ur 6n, 7; compared lack of in Egypt 27
    in Greek city planning 41; at Athens 46, 50–1
    at ancient Rome 62–3; at Pompeii 71; lack of at Colchester 82, and Londinium 85; expedient perimeter at Calleva Atrebatum 84; Constantinople, the failed Triple Walls 91
    mediaeval town walls 98; European need contrasted English absence 98, 100; Florence exemplifies 99; Oxford 113, York 115
    in mediaeval bastide planning 120, 122; Aigues Mortes 122; Carcassonne 123; Monpazier exemplifies 123–4; Kingston-upon-Hull and constraint on growth 127–8
    the Dutch water system 140; at Nijmegan 141
    Avila, Spain 148c 4.88; Cesky Budjovice 152
    renaissance artillery systems 165, exemplified Turin 165c; Cologne 166c; Naarden, pre-eminently 167c; in renaissance urban theory 168–72, and realized at Palma Nova 173
    Paris concentric rings 194c and withstands Prussian siege 202c; Nancy 207c; and Vauban, engineer pre-eminent; Toulon 218c and Le Havre 219c
    Amsterdam 222; Antwerp 225; Vienna, withstands siege by Islamic Turks 227–8; Berlin, as Dutch water-system 231
    Alfred re-fortifies London 248; English Channel replaces, but Berwick-on-Tweed 281c, and Portsmouth 284
    absence at Madrid 297; Barcelona 299; Havanna 304c; Lima 312c; Manila 318c
    St Augustine 323c; Mobile 330c; Atlanta 360c
    the French at Alexandria 381c, and Algiers 400c; the Italians at Tripoli, Libya 399c; Kuwait City, the 1918 mud wall 391c; Djakarta 412c
    expediency construction walls at Athens 63n, Rome 62, Berlin 'The Wall' 231c
Fontana, Carlo 180–1, 185–6
Fordyce, John: Regent's Park and Regent's Street, London 267, 271
Forster, von Ludwig: competition winning design for the Ringstrasse, Vienna 228
forum: in Roman imperial planning 57c
    Forum Romanum at Rome 65–7, and the imperial fora 67–8
    Pompeii 71; Dougga 75c; Londinium 86c; Constantinople 90c
Francis I 191–2, 194

Gabriel, J-A: Place de la Concorde, Paris 202–4, 208
Galvez, Jose de 326
Gallius, Allius 369
Garden City Movement, 20th century 40, 43, 70, 85, 172
geestgrondstad: Alkmaar exemplifies 140c
George II 346
Gothic architecture, contrasted renaissance 160
Gouffier, Guillaume: at Le Havre, France 219
Gordon, General: and siege of Khartoum 399c
grachten: defined 140, and *grachtenstaden*; Delft exemplifies 142, 224c
    See also Djakarta and *sloten*
Greece: climate and way of life 35
    Periclean Age 39
    regional topography 35
Greek cities 35–54
    basic planning components 41; and the Hippodamian system 42, contrasted with Islamic urban form 42
Greek civilization: and earlier 38–9

Index of Place Names

INDEX OF PLACE NAMES

INDEX OF PLACE NAMES